Artificial
DNA

Methods
and Applications

Artificial DNA

Methods and Applications

Edited by

Yury E. Khudyakov
Howard A. Fields

CRC Press
Taylor & Francis Group
Boca Raton London New York

CRC Press is an imprint of the
Taylor & Francis Group, an **informa** business

CRC Press
Taylor & Francis Group
6000 Broken Sound Parkway NW, Suite 300
Boca Raton, FL 33487-2742

First issued in paperback 2019

© 2003 by Taylor & Francis Group, LLC
CRC Press is an imprint of Taylor & Francis Group, an Informa business

No claim to original U.S. Government works

ISBN-13: 978-0-8493-1426-1 (hbk)
ISBN-13: 978-0-367-39581-0 (pbk)

Library of Congress Cataloging-in-Publication Data

Artificial DNA : methods and applications / edited by Yury E. Khudyakov and Howard A. Fields.
    p. cm.
  Includes bibliographical references and index.
  ISBN 0-8493-1426-7 (alk. paper)
   1. DNA—Synthesis. 2. DNA—Derivatives. 3. DNA probes. 4. Genetic engineering. I. Khudyakov, Yury E. II. Fields, Howard A.

QP624 .A785 2002
660.6′5—dc21

2002067416

Library of Congress Card Number 2002067416

Visit the Taylor & Francis Web site at
http://www.taylorandfrancis.com

and the CRC Press Web site at
http://www.crcpress.com

Preface

The only approach to art is art itself.

— Klaus Eidam
The True Life of Johann Sebastian Bach

Research laboratories involved in studies on molecular genetic processes or on the development of diagnostic, prophylactic, and therapeutic tools for human diseases are all exceedingly familiar with the use of synthetic oligonucleotides or synthetic DNA. Today, the application of synthetic DNA is as common as the polymerase chain reaction (PCR), DNA sequencing, or nucleic acid hybridization. What used to be the exclusive domain for a limited number of research scientists only 20 to 25 years ago has now penetrated into many different applications in research and in health products in numerous laboratories and biotechnological and pharmaceutical companies. Originally conceived as a source of artificial genetic material, synthetic DNA has found major applications in many diverse nongenetic subject areas such as molecular diagnostics, microarray analysis, computer chip research, and antisense technology, to name only a few. Several factors have contributed to the current widespread use of synthetic DNA. Some of these factors may be directly related to the *convenience* of obtaining synthetic oligo- and polynucleotides, which have become readily available in large amounts because of the development of revolutionary techniques for DNA synthesis over the last three decades; however, the major factor that has contributed to the exponential growth and use of synthetic DNA is more related to the unlimited human *control* over the structure of synthetic DNA.

This notion should not be a surprise to readers. If we look closely at the history of humankind, we will understand that gaining control over the environment is the major adaptive mechanism used by humans. The evolution of humankind may be envisioned as breaking from the grasp of natural selection to embrace the implementation of rational development directed by humans. Science is used as a steppingstone to control nature for the benefit of humankind. Being archetypal to humans, this trend can be readily noted in many other fields of human activity. It is our craving to exercise control over our lives that has led to the development of an artificial world. For example, we have constructed artificial means for transportation such as automobiles and an artificial strictly controlled environment such as highways for the optimal use of automobiles. Consider modern homes, which are actually entirely artificial entities. They provide the same purpose as caves offered to our ancestors. However, a dramatic improvement in the quality of living found in modern buildings can hardly be disputed by anyone who lives in today's world and who may have had the opportunity to visit such caves, or who has a vivid imagination to picture such primitive residences. These simple examples show that the active process of gaining control over our life is, to a large extent, built on the development of artificial means that mimic nature's function without directly mimicking nature's ways.

Since life itself is predicated on nucleic acids as carriers of genetic information, nucleic acids as the subject of biological research best exemplify these trends of human activity in the biological sciences. We began to learn how to manipulate existing genetic material using genetic engineering techniques. However, only when we learned how to synthesize DNA of any sequence did real opportunities arise for generating an artificial, strictly controlled DNA world. It has opened new avenues for designing artificial genetic systems that will help avert harm's way and promote health and prosperity for humanity.

The very first attempt to chemically synthesize short oligonucleotides from nucleotides was performed in the laboratory of Dr. H. G. Khorana more than 35 years ago. The significance of this pioneering research in the biological sciences and its enormous potential benefit to the progress of humankind cannot be overstated or underestimated. These studies provided for the first time access to unlimited modifications of the genetic material opening a new era for the unrestrained manipulation of the DNA structure, which will ultimately lead to the creation of new artificial genetic systems with strictly predetermined properties.

History shows that artificial systems may result in producing similar problems as their natural counterparts. For example, cyberspace has created an opportunity for the generation of parasitic units of information such as computer viruses and worms that have found very efficient ways to replicate themselves at the expense of the entire system. We are just now at the doorstep of exploring the potential use of artificial genetic systems, and, as should have been anticipated, there are already the first signs that these systems may create their own successfully replicating parasitic genetic units (see Chapter 4). By creating artificial systems we create a new environment and a new opportunity for self-generation of parasitic units potentially harmful to this new artificial environment. This is not the first time, however, that we must balance progress in some areas of human activity against potential harm this progress may bring. "Understand the harm and benefit in everything" teaches Miamoto Musashi, a philosopher, artist, writer, and peerless undefeated sword fighter from 17th century Japan, in his *The Book of Five Rings*. After putting his life on the line on numerous occasions in his quest for self-perfection, he came to understand the value of this notion and tried to pass it on to posterity as a piece of his wisdom.

During more than 30 years of research, synthetic DNA has been developed for use in many applications, some of which are described in this book. While the harm from using artificial DNA is still conceptual, we have already witnessed the invaluable fruits this new technology has brought to reality. Today, the list of applications for synthetic DNA is so lengthy that a cursory inspection of this list would take more time than the reading of this preface. The myriad uses for synthetic material are only limited by our own imaginations. As always, the best is still to come.

The history of research studies and applications of synthetic DNA is rather short. However, even a cursory analysis of the published literature reveals some intriguing trends. A literature search using the key words "oligonucleotide," "synthetic gene," "PCR," and "microarray" reveals very limited growth in the number of publications that used the word "oligonucleotide" at the beginning of 1970s. This growth leveled off throughout the 1970s and 1980s. However, in the late 1980s, the number of publications started expanding again, but this time growth occurred at an unprecedented rate until the early 1990s to a level where it was 20 to 30 times greater than in any other previous year. This dramatic increase was paralleled by an equally steep growth in the number of publications using the term "PCR." While the number of publications using the abbreviation "PCR" continued to increase until the turn of millennium, the number of publications using the word "oligonucleotide" declined in the mid-1990s, probably as a result of the less frequent use of this word in the context of PCR primers. This downward trend has been reversed in the last 4 to 5 years coinciding with a no less impressive growth in the number of articles using the word "microarray." What this simple analysis reveals is that the use of oligonucleotides over the last decade has been driven by the application of oligonucleotides to new technologies such as PCR, sequencing, and microarray.

The use of synthetic genes, however, has followed a very different trend. Unlike many other technologies, the application of synthetic genes did not rapidly expand after the construction of the very first one in the mid-1970s. The use of the term "synthetic gene" experienced a very mild increase in popularity around 1980 to 1981, but this blip on an otherwise flat line may be explained by the introduction of new efficient chemical methods for oligonucleotide synthesis, which made research studies on the synthesis and utility of long polynucleotides more affordable. Nonetheless, the number of publications quickly fell to zero within a couple of years. After the introduction of automatic synthesizers, which made synthetic oligonucleotides more available, the number of publications reporting assembly of synthetic genes started increasing again. Later, the number of publications using the term "synthetic gene" experienced another mild boost by the application of PCR technologies, which significantly simplified obtaining long polynucleotides from short synthetic oligonucleotides. However, by the mid-1990s the number of publications began to decline once again.

Thus, while the use of synthetic oligonucleotides matured into various applications, all examples demonstrating an increased interest in synthetic genes were associated with oligonucleotide synthesis technology improvement and the development of new approaches to the synthesis of long DNA molecules, rather than on potential applications. The expansion on the use of synthetic genes has been hampered for years by the inefficacy of methods to assemble long polynucleotides. Despite the fact that the application of PCR has significantly improved the efficiency for the synthesis of long polynucleotides, the use of synthetic genes is still declining in popularity. Therefore, we must conclude that the use of synthetic genes is not yet application driven. Designing synthetic genetic systems requires novel integrated approaches based upon a "procreative" rather than "anatomical" course and a thorough understanding of the study subject. Do we have sufficient knowledge to rationally design and construct an artificial genetic world? Even if we do not yet possess such knowledge for many sophisticated applications, are we ready to explore a new world of artificial genetics? As this book proves, so much already has been done. But, how much will be done in the future! Breakthroughs may happen at any time. This is a very exciting area of molecular science that is on the very precipice of a scientific explosion waiting for new explorers. Join in!

Yury E. Khudyakov, Ph.D.
Howard A. Fields, Ph.D.

The Editors

Yury E. Khudyakov, Ph.D., is Chief, Computational Molecular Biology Activity and Deputy Chief, Developmental Diagnostic Laboratory, Laboratory Branch, Division of Viral Hepatitis, National Center for Infectious Diseases (NCID), Centers for Disease Control and Prevention (CDC), Atlanta, Georgia. He received his M.Sc. in Genetics from the Novosibirsk State University, Novosibirsk, Russia in 1977 and his Ph.D. in Molecular Biology from the D.I. Ivanovsky Institute of Virology, Academy of Medical Sciences, Moscow in 1985. He started his research career (1977–1980) in the Laboratory of Gene Chemistry at the M.M. Schemyakin Institute of Bioorganic Chemistry, Academy of Sciences, Moscow, one of the first laboratories in Russia to develop new technologies for chemical synthesis of DNA. He was a Research Fellow (1980–1985) in the laboratory of Viral Biochemistry and Chief, Genetic Engineering Section (1985–1991) in the laboratory of Chemistry of Viral Nucleic Acids and Proteins at the D.I. Ivanovsky Institute of Virology, Moscow. In 1991, he joined the Hepatitis Branch (HB), Division of Viral and Rickettsial Diseases (DVRD)/NCID/ CDC as a National Research Council Research Associate, National Academy of Sciences, U.S.A. Since 1996 he has served as Chief, Developmental Diagnostic Unit, Molecular and Immunodiagnostic Section/HB/DVRD/NCID/CDC.

Dr. Khudyakov's main research interests are applications of synthetic DNA to the development of new diagnostics and vaccines, molecular biology and evolution of hepatitis viruses, and bioinformatics.

Dr. Khudyakov has published more than 80 research papers and book chapters. He is the author of several issued and pending patents. He is a member of the editorial board for the *Journal of Clinical Microbiology*.

Howard A. Fields, Ph.D., is Chief of the Developmental Diagnostic Laboratory, Hepatitis Laboratory Branch, Division of Viral Hepatitis, National Center for Infectious Diseases (NCID), Centers for Disease Control and Prevention (CDC), Atlanta, Georgia. Since 2002, he has been a Distinguished Consultant for CDC. He received his M.S. in Microbiology from the University of Maine, Orono, in 1971 and his Ph.D. in Virology from the University of New Hampshire, Durham, in 1974. From 1974 to 1976, after completing his Ph.D, he did postdoctoral work in immunochemistry at Baylor College of Medicine, Houston, Texas, after which he joined the Hepatitis Program at the CDC field station located in Phoenix, Arizona as an immunovirologist. In 1980, Dr. Fields became Chief of the Immunochemistry Laboratory, Hepatitis and Viral Enteritis Division, CDC. In 1983, the hepatitis program was relocated to Atlanta, Georgia, where he became Chief of the Molecular and Immunodiagnostic Section within the Division of Viral and Rickettsial Diseases.

Dr. Fields' main research interests are biochemical and biophysical characterization of antigens; epitope mapping of antigens using monoclonal antibodies, Western blots, and synthetic peptides; development of sensitive and specific immunoassays; development of molecular diagnostic assays; development of methods to assemble synthetic genes; development of highly efficient procaryotic and eucaryotic expression systems; development of animal models for viral hepatitis; molecular epidemiology using nucleotide sequence comparisons and phylogenetic analysis; and development of artificial genes encoding mosaic recombinant proteins as immunodiagnostic reagents.

Dr. Fields has published 141 research papers and book chapters. He is the author of several issued and pending patents. He is a past member of the editorial boards for the *Journal of Clinical Microbiology* and the *Journal of Applied and Environmental Microbiology*. At present, Dr. Fields is an associate editor for the *Journal of Clinical Microbiology*.

Contributors

Frederik Beck
Pantheco A/S
Copenhagen, Denmark

Roger Chammas
Department of Radiology
  and Center for Cell Therapy
Universidade de São Paulo
São Paulo, Brazil

Joy Chih-Wei Chang
Division of Laboratory Science
National Center for Environmental Health
Centers for Disease Control and Prevention
Atlanta, Georgia, U.S.A.

Howard A. Fields
Division of Viral Hepatitis
National Center for Infectious Diseases
Centers for Disease Control and Prevention
Atlanta, Georgia, U.S.A.

Lilia M. Ganova-Raeva
Division of Viral Hepatitis
National Center for Infectious Diseases
Centers for Disease Control and Prevention
Atlanta, Georgia, U.S.A.

John I. Glass
Eli Lilly and Company
Indianapolis, Indiana, U.S.A.

Elmars Grens
Biomedical Research and Study Centre
University of Latvia
Riga, Latvia

Sang Won Han
Department of Biophysics
  and Gene Therapy Center
Universidade Federal de São Paulo
São Paulo, Brazil

Beverly A. Heinz
Eli Lilly and Company
Indianapolis, Indiana, U.S.A.

Lisa S. Kelly
Department of Biochemistry
  and Molecular Biology
University of Georgia
Athens, Georgia, U.S.A.

Yury E. Khudyakov
Division of Viral Hepatitis
National Center for Infectious Diseases
Centers for Disease Control and Prevention
Atlanta, Georgia, U.S.A.

Célio Lopes Silva
Department of Biochemistry
  and Immunology
Universidade de São Paulo
São Paulo, Brazil

Salvatore A. E. Marras
Department of Molecular Genetics
International Center for Public Health
Public Health Research Institute
Newark, New Jersey, U.S.A.

Maurício Martins Rodrígues
Department of Microbiology, Immunology
  and Parasitology
Universidade Federal de São Paulo
São Paulo, Brazil

Ramón Montaño
Cellular Pathology and Molecular Biology
Instituto Venezolano
  de Investigaciones Científicas
Caracas, Venezuela

Peter E. Nielsen
Department of Medical Biochemistry
  and Genetics
The Panum Institute
University of Copenhagen
  and Pantheco A/S
Copenhagen, Denmark

Richard T. Pon
Department of Biochemistry
  and Molecular Biology
University of Calgary
Calgary, Alberta, Canada

Flor H. Pujol
Laboratory of Viral Biology
Instituto Venezolano
  de Investigaciones Científicas
Caracas, Venezuela

Paul Pumpens
Biomedical Research and Study Centre
University of Latvia
Riga, Latvia

Renu Tuteja
Malaria Group
International Centre for Genetic Engineering
  and Biotechnology
New Delhi, India

Jane Zveiter de Moraes
Department of Biophysics
Universidade Federal de São Paulo
São Paulo, Brazil

Contents

.

1 Chemical Synthesis of Oligonucleotides: From Dream to Automation

Richard T. Pon

CONTENTS

1.1 INTRODUCTION

I wish to conclude by hazarding the following rather long-range predictions. In the years ahead, genes are going to be synthesized. The next steps would be to learn to manipulate the information content of genes and to learn to insert them into and delete them from the genetic systems. When, in the distant future, all this comes to pass, the temptation to change our biology will be very strong.

— Har Ghobind Khorana, 1968[1]

Humans' ability to modify and shape their environment has been a defining characteristic of the race. In recent years, advances in life sciences have allowed us to examine and manipulate living organisms on an unprecedented scale. Indeed, there are few areas of research that have stimulated so much fascination. All of these genetic advances have resulted from our understanding of the genome and the properties of its constituent nucleic acids.

The role of chemical synthesis in determining the structure and properties of nucleic acids has always been crucial. This is because our ability to chemically synthesize molecules of DNA, RNA, and related analogues with defined structure and biological properties has provided the definitive proof for theories on nucleic acid structure, function, and properties. This was the case in the past, when the correct structures of bases and nucleotides were confirmed and the genetic code was first verified using synthetic molecules, and it continues today with the construction of artificial genes and new biological constructs. Nucleic acids are now known to have a relatively simple primary structure — five major bases linked together in a linear order via 3'-5' phosphodiester linkages, and containing either ribose (RNA) or 2'-deoxyribose (DNA) sugar rings (Figure 1.1). However, the chemistry of the three major components — bases, sugar rings, and phosphate linkages — is quite complex. This complexity and the many possible ways in which these components can be assembled make these molecules very difficult to study. Indeed, it has taken many decades and the work of a very large number of skilled chemists to determine the correct structures of these molecules and develop effective methods for synthesizing and manipulating them.

However, advances in molecular biology and synthetic chemistry have had a synergistic effect; synthetic oligonucleotides have now become very powerful tools that are used in almost all areas of the life sciences. Synthetic oligonucleotides are essential as primers in a technique that has been touted as one of the most significant and powerful discoveries of the 20th century, the polymerase chain reaction (PCR), which can amplify minute amounts of DNA.[2–4] Synthetic oligonucleotides are also essential as DNA sequencing primers, and the knowledge gained from the various genome sequencing projects is expected to radically change how future research in the life sciences is performed. Synthetic oligonucleotides can be used to create site-specific mutations or even entire synthetic genes. The resulting genetically modified organisms are not only valuable research tools, but also have the ability to dramatically affect our quality of life. Gene therapy (modification of our own genetic material) is even possible using either viral vectors, a recently introduced technique of chimeraplasty,[5,6] which uses synthetic RNA-DNA chimeric molecules to correct point mutations, or even short single-stranded oligonucleotides.[7] The unique base pairing properties of nucleic acids have been exploited to develop many specific diagnostic tools. Recently, high-density DNA arrays (gene chips) and other DNA-based biosensors have resulted in the development of many highly miniaturized devices that can produce unprecedented amounts of data in record time.[8–13] The ability of single-stranded oligonucleotides to self-assemble into complex structures may have potential in future nanoscale engineering,[14] and the information content of these molecules may also find future application in DNA-based computing.[15,16] Synthetic oligonucleotides and oligonucleotide-like molecules have also created a new class of nucleic acid medicines, which can be used to treat specific targets through either antisense and antigene effects[17–22] or immunostimulatory effects.[23–25] The first synthetic antisense oligonucleotide (Vitravene) has already been approved for pharmaceutical use, and many others are in clinical trials.

The present and future demands for synthetic oligonucleotides are so great that two new industries have arisen to supply these materials. The first industry, which caters to small-scale research demands, can now produce millions of different oligonucleotides quickly and inexpensively, while the second industry, devoted to pharmaceutical needs, is developing methods for the synthesis of multi-ton quantities of oligonucleotides.

Therefore, it is important for a book titled *Artificial DNA* to include a chapter on chemical synthesis. However, this topic is too broad for a detailed and comprehensive review. Such a treatment would only duplicate existing reviews[26–43] and interest a small number of specialists. Instead, this chapter will present a historical overview of some of the most relevant chemical contributions relating to artificial DNA and try to place them into perspective. This will be followed by an overview of protecting groups, phosphoramidite-coupling chemistry, and solid-phase synthesis, which are relevant to the most common methods used today. Special consideration will be given to how solid-phase synthesis has been automated, since this technique is the basis for today's widespread availability of custom-synthesized oligonucleotides.

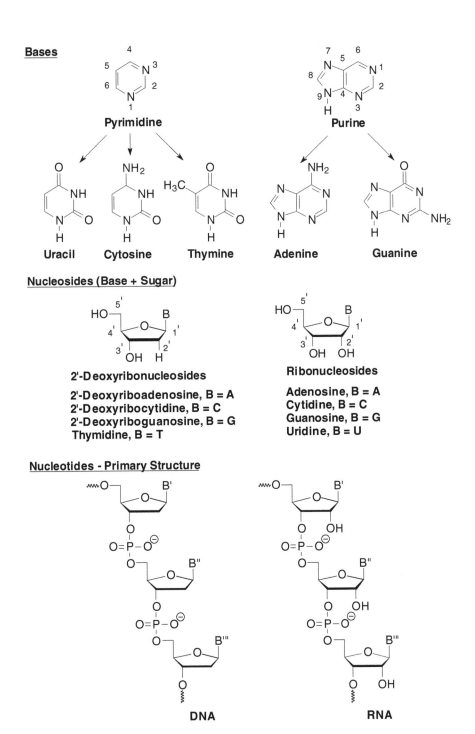

FIGURE 1.1 Nucleic acid structures. Heterocyclic bases are derived from either pyrimidine or purine. Nucleosides are sugar rings (2′-deoxyribose or ribose) attached to the N1-position of a pyrimidine base or to the N9-position of a purine base through an *N*-glycoside bond to the C1′-sugar position. Nucleotides consist of a nucleoside plus a phosphate. The primary structures of both a DNA and an RNA sequence with 3′-5′ internucleotide phosphate diester linkages are shown.

At this point, it is also appropriate to clarify the term *oligonucleotide*. Unlike *DNA*, which has entered the public vocabulary, an oligonucleotide is not as well known. A dictionary definition states that an oligonucleotide contains only a few (i.e., 2 to 10) nucleotide residues. This is clearly a case of our language not keeping up with technology, since it is now possible to synthesize oligonucleotides containing 100 to 200 nucleotides. Chemically, it is also more accurate to specify a molecule as either an *oligodeoxyribonucleotide* or an *oligoribonucleotide*. However, for simplicity's sake, the term oligonucleotide will be used whenever it is not strictly necessary to distinguish between these two classes of nucleic acid. Indeed, the large number of oligonucleotide modifications, involving different carbohydrate moieties, makes it difficult to always be so specific. Finally, readers should be aware that oligonucleotides are single-stranded molecules with properties that are quite different from those of double-stranded molecules. However, double-stranded nucleic acids are easily assembled by hybridization of two independently prepared and complementary oligonucleotides.

1.2 HISTORICAL PERSPECTIVE AND MAJOR ACCOMPLISHMENTS OF SYNTHETIC OLIGONUCLEOTIDES

The study of DNA originated with the work of Friedrich Miescher at the University of Tübingen in Germany. Miescher was the first, in 1869, to isolate a new nonprotein substance, named *nuclein*, which was later to be identified as DNA.[44] Although there was no indication back then that nuclein was a carrier of genetic information, Miescher was the first to realize that the contents of the nucleus had a unique chemical composition. His original studies were made using human pus cells, but he later switched to salmon sperm as a source for nuclear material. It is interesting to note that 130 years later salmon sperm still remains the commercial source for deoxyribonucleosides. The nuclein material Miescher isolated contained both protein and nucleic acid, and it was not until 1889 that a protein-free nuclear extract was available. This was prepared by Richard Altmann, who named the protein-free material *nucleic acid*.

The chemical composition of nucleic acids was studied in detail by Albrecht Kossel, beginning in 1879 and continuing for the next 45 years. By 1900, uracil was the last of the five major bases — after adenine, cytosine, guanine, and thymine — to be identified. However, while the elemental compositions were known, the correct structures were not established until chemical synthesis was performed. In 1897, Emil Fischer was the first to synthesize the purine bases and characterize their structures, and he coined the term *purine*. This was followed in 1901 by the synthesis of uracil and confirmation of its structure. Subsequently, in recognition of his work on sugar and purine synthesis, Emil Fischer received the 1902 Nobel Prize in Chemistry. A Nobel Prize in Physiology or Medicine was also awarded to Albrecht Kossel in 1910 for his contributions to cell chemistry. In the Nobel Prize presentation speech, it was noted that nucleic acids "certainly possess a great biological significance," even though it would be another 34 years before Oswald Avery proved that DNA was responsible for transforming activity and 43 years before Francis Crick and James Watson identified the structure of DNA. Although Avery's work was slow to be accepted and he was never recognized with a Nobel Prize, Watson and Crick became famous and, along with Maurice Wilkins, were awarded the 1962 Nobel Prize in Physiology or Medicine "for their discoveries concerning the molecular structure of nuclear acids and its significance for information transfer in living material."

During the first half of the 20th century, numerous researchers studied the structure of the nucleic acids. However, Phoebus Aaron Theodor Levene and his colleagues laid the foundation by the careful study and characterization of hydrolysis products. This group was the first to correctly identify D-ribose as the sugar in RNA and 2-deoxy-D-ribose as the sugar in DNA. However, there was a 20-year span between the first identification of D-ribose from yeast nucleic acid (RNA) in 1909 and the identification of 2-deoxy-D-ribose from thymus nucleic acid (DNA) in 1929. The time lag was due to the chemical instability of the latter sugar. Levene was also responsible for coining the term *nucleoside,* to describe a purine attached to a carbohydrate, as well as the term *nucleotide,* to describe a phosphoric ester of a nucleoside. During this time there was great confusion caused by the differences between RNA and DNA, and many incorrect proposals for their structures were put forth. In 1935, a structure with the correct 3'-5' internucleotide linkage was proposed by Levene, although he incorrectly believed that DNA was a low-molecular-weight tetranucleotide. It was not until a few years later that careful biochemical preparations and flow birefringence studies showed that DNA was actually a high-molecular-weight macromolecule. In addition, the correct 3'-5' internucleotide linkage was not able to be confirmed by chemical synthesis for another 20 years.

The history of chemical synthesis dates back to 1914, when Emil Fischer attempted nucleoside phosphorylation using phosphoryl chloride **1**. However, these early reactions were unsuccessful because of the nonspecificity of the reagents. Successful nucleotide or oligonucleotide synthesis required development of appropriate protecting groups on both the nucleoside and phosphorylating reagents to ensure that the desired products were obtained in high yield. The first time a protecting group was used on a nucleoside was in 1934, when Levene used monoacetone uridine (2',3'-*O*-isopropylideneuridine) **2** in the synthesis of uridine-5'-phosphate.[45] Improved phosphorylating reagents were obtained by first introducing phenyl and then benzyl protecting groups onto phosphoryl chloride. By 1949, A. M. Michelson and Sir Alexander R. Todd were able to unambiguously prepare 3'-ribonucleotide-monophosphates **3** by chemical synthesis and conclusively proved the structures of the natural nucleotides derived from RNA.[46] In a landmark paper in 1955, A. M. Michelson and Sir Alexander R. Todd synthesized the first dinucleotides (TpT **4** and pTpT **5**) with unambiguous 3'-5' internucleotide linkages.[47] Although the synthesis occurred 2 years after Watson and Crick proposed a double helical model for DNA,[48] the chemical synthesis and the phosphodiesterase hydrolysis studies performed on the synthetic products confirmed for the first time the correct structure of the 3'-5' phosphodiester internucleotide linkage. In 1957, in recognition of his work on nucleotides and nucleotide co-enzymes, Lord Alexander Todd was awarded the Nobel Prize in Chemistry.

Unfortunately, the reactive phosphorochloridate chemistry used for the first dinucleotide synthesis was not suitable for making longer oligonucleotides. However, milder and more suitable coupling conditions were possible when phosphodiester linkages were prepared using dicyclohexylcarbodiimide (DCC) **6** as the coupling reagent. Although this reagent was first used in nucleic acid synthesis by Todd and Har Ghobind Khorana in 1953,[49] it was the subsequent work of Khorana and his research groups, first at the British Columbia Research Council and then at the University of Wisconsin, which demonstrated the utility of this coupling reagent.[50]

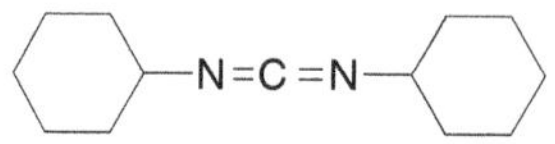

6

The vision of H. G. Khorana and the significance of his work cannot be underestimated. Khorana was the first to establish the solid framework of protecting groups, coupling strategies, purification, and analytical methods for the chemical synthesis of oligonucleotides. More importantly, he realized that many central problems of molecular biology and genetics could be tackled using synthetic oligonucleotides.

In the first phase of his career, beginning in the mid-1950s, Khorana developed the first workable scheme for the synthesis of oligodeoxyribonucleotides of any defined sequence.[1] The coupling strategy he used, which became known as the *phosphodiester approach*, was the dominant synthetic method for almost 20 years until improvements in phosphotriester and phosphite-triester chemistry led to better results (Figure 1.2). In addition, Khorana had to develop protecting groups and strategies for introducing them onto specific locations. His use of simple protecting groups, which could be removed through either mild acid or mild base treatment, created essential nucleoside and nucleotide building blocks that became widely used. Interestingly, Khorana realized that his protecting groups were less elegant than other procedures that had been developed, and in his words they were "by modern (1968) standards of organic chemistry, rather conventional in concept....

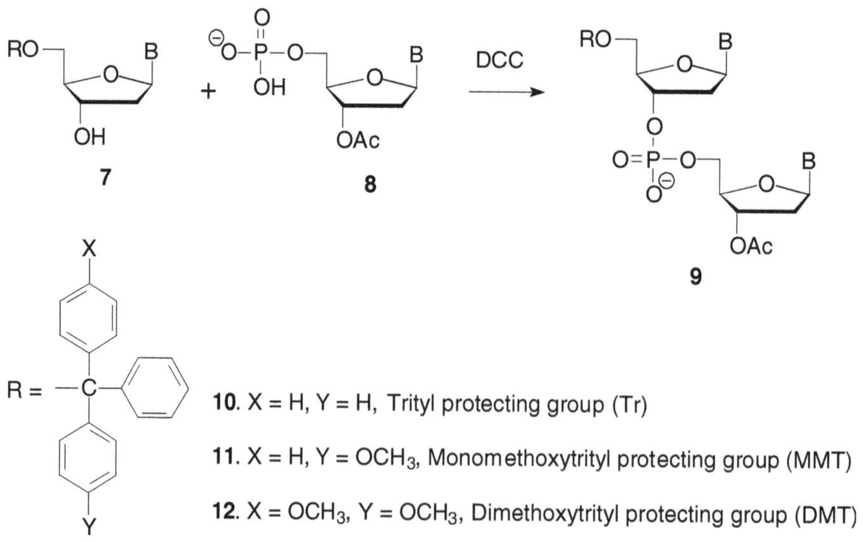

FIGURE 1.2 Phosphodiester coupling approach used by Khorana showing condensation of a deoxyribonucleoside-5′-monophosphate to the 3′-OH position of a second nucleoside. This approach can also be used to attach 3′-phosphates to 5′-hydroxyl positions. The product contains a negatively charged phosphodiester linkage.

On the other hand, it is also true that the present protecting groups, which we have had for some 8 years now, have stood the test of time: as the syntheses have become more and more demanding, especially in the duration of the total operations, their use has been completely satisfactory."[1] More than 30 years later, many of the same protecting groups, such as the monomethoxytrityl **11**, dimethoxytrityl **12**, benzoyl **13**, isobutyryl **14**, and 2-cyanoethyl **15** groups (commonly abbreviated, respectively, as MMT, DMT, Bz, iBu, and CE), still remain the preferred protecting groups and this clearly demonstrates that a simple, but well-thought-out approach can have long-lasting value.

| 13 | 14 | 15 |

The original DCC coupling reactions were very slow and inefficient (~60% yield after 2 days) and in the early days only mono-, di-, and trideoxyribonucleotides could be prepared. However, the short fragments could be chemically polymerized to produce oligodeoxyribonucleotides containing a defined, repeating sequence.[51] Later, improved conditions and faster arylsulfonyl chloride **16–17** coupling reagents[52] were developed that would allow the synthesis of longer defined sequences through either stepwise additions (one mononucleotide at a time) or through larger block additions (two oligonucleotides joined together).[1] Khorana's research group also tackled the much more difficult problem of oligoribonucleotide synthesis and eventually they were able to chemically synthesize all 64 possible ribotrinucleotides.[53]

16a X = Cl	**17a** X = Cl
16b X = imidazole	**17b** X = imidazole
16c X = triazole	**17c** X = triazole
16d X = tetrazole	**17d** X = tetrazole

After the fundamentals of chemical synthesis had been established, Khorana began a second phase of his work by asking whether chemical synthesis could make a contribution to the study of the fundamental process of biological information flow, DNA → RNA → protein. This concept was the central dogma of molecular genetics, which arose from the one gene–one enzyme hypothesis of George Beadle and Edward Tatum in 1941. Prior to this hypothesis, the study of genetic expression appeared to be one of "bewildering multiplicity."[44] Beadle and Tatum demonstrated that substances in the cell were synthesized, step by step, in long chains of chemical reactions controlled by genes, which individually regulated certain steps. For their discovery that genes act by regulating definite chemical events, Beadle, Tatum, and Joshua Lederberg received the 1958 Nobel Prize in Physiology or Medicine.

By the early 1960s, DNA-dependent RNA polymerase had been discovered and a cell-free system for amino acid synthesis was available. Khorana used his synthetic oligodeoxyribonucleotides to help decipher the genetic code.[51] Short oligonucleotides, which were the only ones available synthetically, were extended into longer DNA sequences by DNA polymerase and then converted into RNA transcripts by RNA polymerase for use in *in vitro* protein synthesis. A wide variety of mRNAs could thus be prepared, and these were used to study many aspects of protein synthesis. For their efforts interpreting and proving the genetic code, Khorana, R. W. Holley, and M. W. Nirenberg were awarded the Nobel Prize in Physiology or Medicine in 1968.

Many other important biochemical processes, such as determining the control of gene expression, still remained to be studied after solving the genetic code. Khorana realized that the total synthesis of long double-stranded DNA (dsDNA) sequences would be the key to manipulating the information content of nucleic acids, even though such molecules were much longer than could be produced by any chemical synthesis known at that time. In addition, he also knew that these long molecules needed to be produced as "bihelical" (double-stranded) DNA, and the preparation of such duplex DNA by chemical synthesis would double the amount of work, relative to synthesizing only a single-stranded molecule. Despite these difficulties, Khorana began an extremely ambitious project to perform the first total gene synthesis in 1965.[54] The target gene (77 bp) selected was the first transfer RNA gene sequenced, the major yeast alanine tRNA.

The gene synthesis strategy Khorana employed[55] was a clever combination of both chemical and enzymatic methods, which exploited the ability of short oligonucleotides to form hydrogen-bonded double-stranded structures (Figure 1.3). First, a set of short (8- to 20-base-long) oligonucleotides of defined sequence with free 3' and 5'-OH ends were chemically synthesized using his phosphodiester synthetic approach. The 5'-hydroxyl ends of each oligonucleotide were phosphorylated using T4 polynucleotide kinase and ATP. Then, the 5'-phosphorylated synthetic sequences were hybridized together to line them up in a head-to-tail duplex orientation. The synthetic fragments were designed to overlap with protruding single-stranded ends ("sticky ends") and the recently discovered T4 polynucleotide ligase was used to join these segments together. Although such an approach is straightforward by today's standards, the difficulty of chemical synthesis in the 1960s made this task extremely labor-intensive, and the project was not completed until 5 years later.

Khorana's group followed up their first gene synthesis with a second total synthesis of an even longer gene, the *Escherichia coli* tyrosine suppressor tRNA gene.[56] The 126-nucleotide-long dsDNA sequence, which was completed in 1976, was prepared from 26 individual oligonucleotides, required 8 years and the efforts of 19 coauthors to prepare. The effort made by Khorana and his research group in these two total gene syntheses was enormous, but his pioneering efforts were instrumental in demonstrating the application of chemical synthesis in solving even the most difficult biological problems. The subsequent development of cloning and other genetic engineering techniques has vindicated Khorana's vision, and now synthetic oligonucleotides and synthetic genes are very widely used.

However, a great deal of further work was still required to develop the synthesis chemistry from Khorana's tedious phosphodiester approach to today's convenient methods. This was because the phosphodiester approach suffered from two serious deficiencies. Although an ionic phosphodiester linkage is the naturally occurring backbone structure, its presence during a total chemical synthesis is troublesome. This is because all of the intermediate products contained negative charges. The charged molecules were insoluble in most organic solvents and could only be isolated or separated from one another by ion-exchange chromatography. Such separations were limited by the low capacity and resolution of the available media (DEAE-cellulose) and were very slow and tedious to perform. In addition, each internucleotide phosphodiester linkage, as well as the terminal phosphomonoester, could also be activated by the coupling reagents in subsequent chain-extension reactions, and this led to chain fragmentation. Therefore, coupling yields decreased markedly as the oligonucleotide length increased and complicated mixtures of by-products resulted. Large excesses of the incoming nucleotide block were required for reasonable yields, and this made the diester method very inefficient as well as time-consuming. Thus, each oligonucleotide synthesized using Khorana's method required several months to a year or more to produce. The amount of synthetic product produced was also very small.

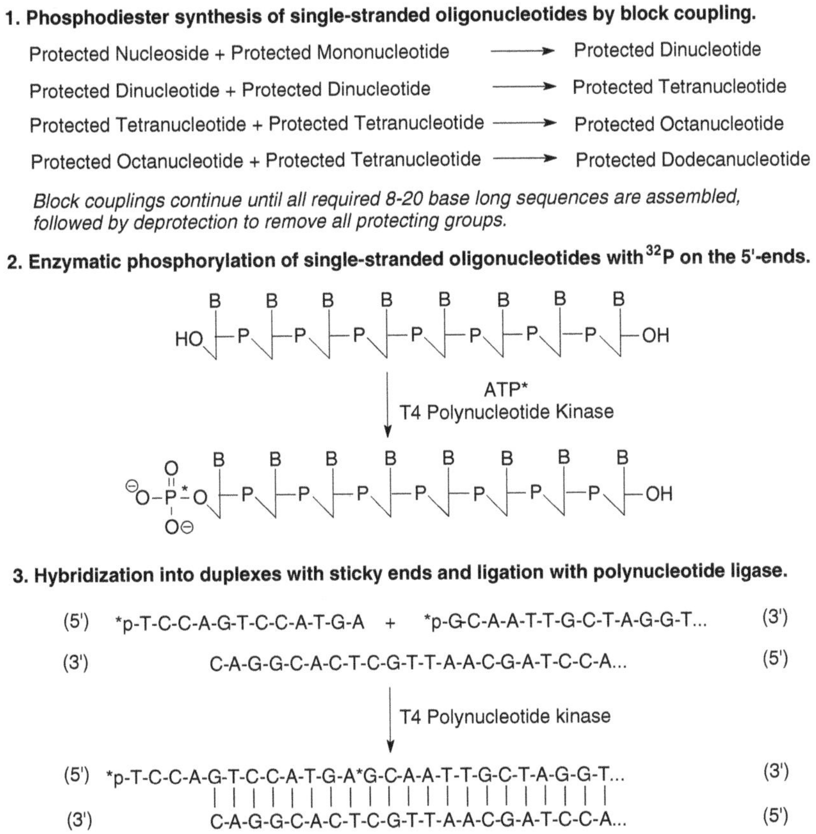

1. Phosphodiester synthesis of single-stranded oligonucleotides by block coupling.

Protected Nucleoside + Protected Mononucleotide ⟶ Protected Dinucleotide

Protected Dinucleotide + Protected Dinucleotide ⟶ Protected Tetranucleotide

Protected Tetranucleotide + Protected Tetranucleotide ⟶ Protected Octanucleotide

Protected Octanucleotide + Protected Tetranucleotide ⟶ Protected Dodecanucleotide

Block couplings continue until all required 8-20 base long sequences are assembled, followed by deprotection to remove all protecting groups.

2. Enzymatic phosphorylation of single-stranded oligonucleotides with ^{32}P on the 5'-ends.

3. Hybridization into duplexes with sticky ends and ligation with polynucleotide ligase.

FIGURE 1.3 Gene synthesis strategy employed by Khorana for total synthesis of yeast alanine tRNA (77 bp) and *E. coli* tyrosine suppressor tRNA (126 bp). (1) Oligonucleotides were assembled by phosphodiester coupling chemistry. Longer oligonucleotides were prepared by coupling previously prepared 2- to 4-base-long "blocks" together. After completion of the chemical coupling steps, all protecting groups were removed to yield single-stranded oligonucleotides with free 5'- and 3'-hydroxyl ends. (2) T4 Polynucleotide kinase and adenosine triphosphate (ATP) were used to selectively phosphorylate the 5'-ends. A ^{32}P radioactive label was incorporated during this step for detection purposes. (3) Complementary strands were annealed together (with overlapping "sticky" ends) and T4 polynucleotide kinase was used to ligate the fragments together into a continuous double-stranded gene sequence.

These oligonucleotides were very valuable materials and were kept under lock and key in the Khorana laboratory. Any use of these materials had to be fully justified before they would be given out. This limited availability was one of the reasons the first DNA amplification experiments performed by Kjell Kleppe et al. in 1970–1971 were not fully exploited.[2] In their studies on repair replication of DNA, a method for the multiplication of DNA using synthetic oligonucleotide primers and DNA polymerase was described.[57] Fourteen years later this concept, independently rediscovered by Kary Mullis,[3,4,58] was the basis of the famous and incredibly useful *polymerase chain reaction* (PCR). However, in the early 1970s, it was so difficult to synthesize large amounts of primers that specific amplification of genomic DNA was thought impractical. It remained for Kary Mullis to reduce this idea to practice, and the resulting Perkin-Elmer Cetus PCR patent has become one of the most valuable patents in biotechnology.

Fortunately, an alternative synthetic strategy to the phosphodiester method was also being investigated. This approach used an additional protecting group on the internucleotide phosphate

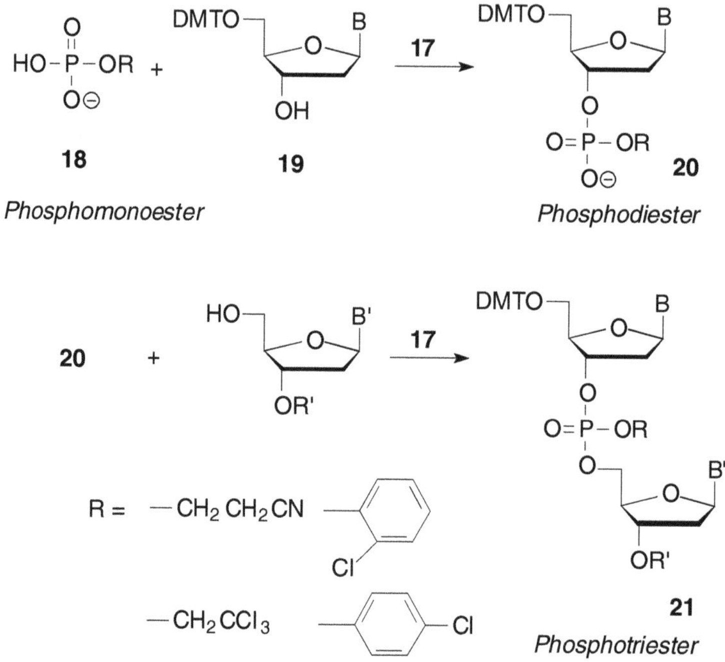

FIGURE 1.4 General strategy for the phosphotriester coupling approach. Intermediate phosphodiester reagents **20** are prepared from a protected nucleoside and a phosphomonoester. After coupling to the 5′-end of another nucleoside or nucleotide, a neutral phosphotriester linkage **21** is obtained. Removal of the phosphate-protecting group (R) after synthesis yields the natural phosphodiester linkage (see Figure 1.1).

linkages to eliminate the phosphodiester charges and was "aesthetically appealing to organic chemists in that the products of the coupling reactions contained fully substituted phosphates and thus cannot be phosphorylated further."[31] Interestingly, the first dinucleotide synthesis performed by Michelson and Todd[47] used this approach. However, the reactive phosphorochloridate reagents were unsuitable, and this approach was abandoned until 1967 when two new reports were published almost simultaneously. The work of Robert L. Letsinger and Kelvin K. Ogilvie used a 2-cyanoethyl protecting group[59,60] while the work of Fritz Eckstein and I. Rizk used the 2,2,2-trichloroethyl protecting group[61] on the phosphate linkage. The resulting linkages were neutral phosphotriesters **21** and this method was known as the *phosphotriester approach*. The general strategy in this approach (Figure 1.4) was to create building blocks by adding phosphate monoesters **18** to the 3′-position of nucleosides **19** or oligonucleotide units and then couple them to the 5′-position of another nucleoside or oligonucleotide. In this method, the oligonucleotide chain length had little pronounced effect on coupling yield since no interaction occurred between the phosphate groups and the coupling reagents. Additionally, the intermediate products were neutral uncharged molecules that could be readily purified in large quantities by chromatography on silica gel. However, the fundamental problem with this approach is that phosphodiester groups with an alkyl ester are approximately tenfold less reactive than phosphomonoester groups. In addition, a phosphate-protecting group, which could be readily and selectively cleaved from the internucleotide linkages after the synthesis was complete, was also required. Despite these difficulties, Letsinger, Ogilvie, and Miller demonstrated the synthesis of a hexathymidylate oligonucleotide in 1969.[62] This approach, along with other different phosphate-protecting groups, was subsequently studied by a variety of other research groups. In particular, aryl phosphate-protecting groups were found to be more reactive than alkyl-protecting groups.

However, this approach did not result in the synthesis of any biologically important oligonucleotides until 1973, when important modifications were introduced. In the first improvement, a modified

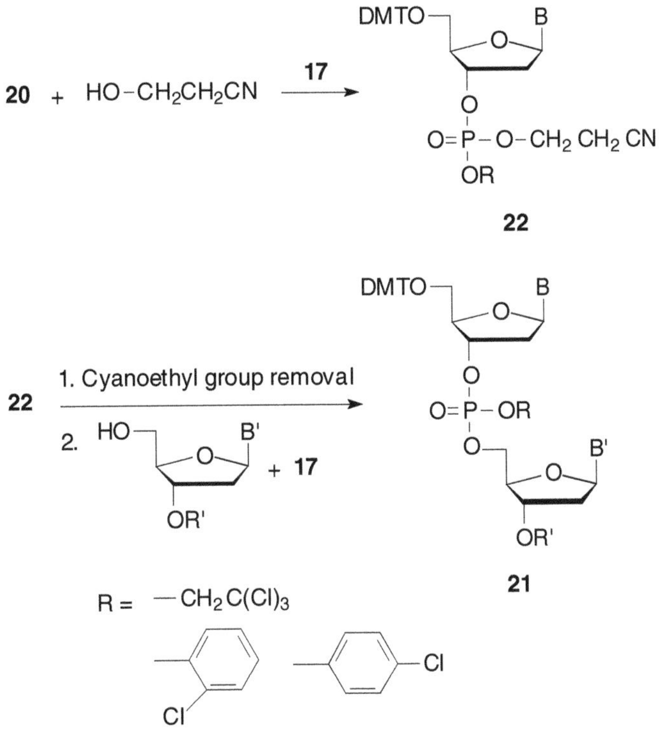

FIGURE 1.5 Modified phosphotriester coupling approach. In this strategy the charged phosphodiester building block **20** is not isolated but converted immediately into a neutral phosphotriester building block **22** by reaction with 2-cyanoethanol. Compound **22** can be easily purified by chromatography and stored. The base labile cyanoethyl protecting group on **22** is easily removed by an *in situ* treatment with mild base just prior to the next coupling step.

phosphorylation strategy was developed by Saran A. Narang, which was called the *modified phosphotriester approach*.[63] A similar modification was also reported by Joseph. C. Catlin and Friedrich Cramer.[64] In this approach (Figure 1.5), building blocks with neutral 3′-terminal phosphotriester groups containing two different phosphate-protecting groups **22** were produced. One of these was an easily removable cyanoethyl protecting group, which could be quickly removed to produce a phosphodiester group suitable for coupling to another nucleoside or oligonucleotide unit, while the other phosphate-protecting group remained on the molecule until the end of the synthesis. This strategy produced uncharged building blocks **22** that could be obtained in high purity after silica gel chromatography. It also allowed a terminal 3′-phosphotriester group to be used in place of the 3′-protecting group on oligonucleotide building blocks. Thus, a single oligonucleotide intermediate could be extended in either the 5′ direction (by removal of the 5′ dimethoxytrityl protecting group) or in the 3′-direction (by removal of a 2-cyanoethyl protecting group from the 3′-terminal phosphotriester group). The second major improvement was the discovery of arylsulfonyl triazole, **16c** and **16d**,[65] and arylsulfonyl tetrazole, **17c** and **17d**,[66] coupling reagents that were significantly more effective than the arylsulfonyl chloride, **16a** and **17a**, and arylsulfonyl imidazole, **16b** and **17b**, reagents previously used. These developments allowed faster, cleaner, and more efficient coupling reactions to be performed.

The modified phosphotriester approach with its faster and more efficient coupling and easier purification was a significant improvement from the previous phosphodiester approach. Using this approach, oligodeoxyribonucleotides of up to approximately 20 bases in length could be prepared using solution-phase techniques in only a few months (instead of a year or more) by individuals skilled in this approach. Numerous oligodeoxyribonucleotide sequences of biological significance were soon being prepared by this improved approach. Protein–DNA binding interactions were studied using a synthetic lactose operator-repressor system[67] and an *E. coli* tyrosine tRNA gene promoter system.[68] Synthetic oligonucleotides were used in various strategies for DNA sequencing and the indirect sequencing of important proteins, such as human fibroblast interferon.[69] New applications for synthetic oligonucleotides as probes for mRNAs and restriction fragments,[70] as linkers and adapters for cloning,[71] and as probes immobilized to cellulose for the isolation and purification of mRNAs were also developed.[72]

An important highlight in genetic engineering was the expression for the first time of a synthetic gene (for somatostatin) in *E. coli*, which was chemically synthesized by Keiichi Itakura et al.[73] Synthetic oligonucleotides for various other gene fragments as well as the complete genes for insulin[74] and human growth hormone[75] soon followed.

However, oligonucleotide synthesis remained in the realm of specialized and highly skilled synthetic chemists, and oligonucleotide availability was still a major problem for molecular biologists and biochemists. This was because each oligonucleotide synthesis required considerable skill and effort to prepare the protected building blocks, perform the sensitive coupling reactions, and isolate and characterize the synthetic intermediates and final products. This was a situation similar to that faced by peptide chemists almost 20 years earlier when they were faced with the difficulties involved in synthesizing polypeptide chains.

The synthesis of both oligonucleotides and polypeptides is analogous because both products are macromolecules composed of repeating monomeric units linked together in a linear fashion. In both cases, large molecules are difficult to prepare because the number of operations required is too great and purification of the intermediates becomes too difficult and time-consuming. These problems were elegantly solved by Robert B. Merrifield at the Rockefeller Institute in New York in 1962, when he devised a scheme for the solid-phase synthesis of peptides.[76] In this *solid-phase approach*, the tedious and time-consuming isolation of the product from the coupling reaction was reduced to a simple filtration step by covalently attaching one end of the product to an insoluble support (Figure 1.6). This method was amenable to mechanization since the only steps involved required delivery of reagents to a reservoir containing the insoluble support and the subsequent elution of the unreacted reagents from the support using solvent washes. With appropriate plumbing and computer control, many dozens or even hundreds of individual, repetitive steps can be programmed to run automatically, and such automation held the promise of making synthetic peptides or oligonucleotides readily available. For his development of this simple and ingenious method, Merrifield was awarded the 1984 Nobel Prize in Chemistry.

However, a successful solid-phase synthesis requires extremely high yields from every coupling reaction. Otherwise, the overall yield of the final product will be very low and its isolation from shorter failure sequences will be quite difficult. For example, Table 1.1 shows the relationship between the maximum number of coupling steps and stepwise coupling yield, if a minimum overall yield of 10% is required.

Unfortunately, obtaining very high coupling yields for nucleotide couplings was much more difficult than for peptide couplings, and this limitation severely restricted progress in solid-phase oligonucleotide synthesis for almost 20 years. For example, when the solid-phase method was initially attempted using the phosphodiester coupling chemistry in the late 1960s, the low coupling yields limited the method to the synthesis of only di- and trinucleotides.[77–79]

Part of the problem was also the lack of a good polymer support. The early syntheses used the same highly swellable, low cross-linked polystyrene resins as Merrifield used. In these supports, the coupling reactions occurred in a gel-like phase that was well suited for peptide synthesis but was unsatisfactory for more polar nucleotide coupling reactions. Eventually, better results were obtained using more rigid and polar polymers.[80,81] In 1977, Michael J. Gait and Robert C. Sheppard mechanized this method for the first time using a modified Beckman 990 peptide synthesizer and a polydimethylacrylamide support.[82] This machine was a significant improvement because it greatly

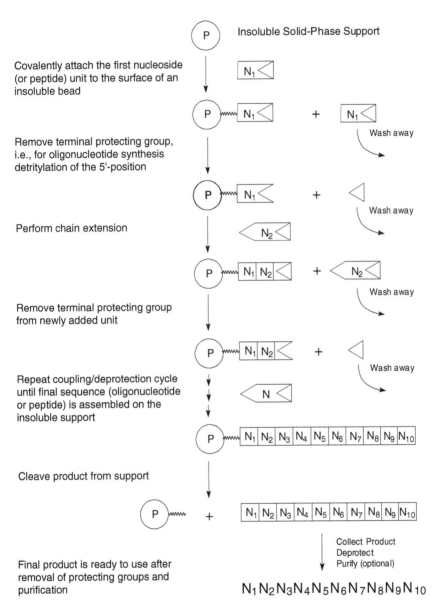

Covalently attach the first nucleoside (or peptide) unit to the surface of an insoluble bead

Remove terminal protecting group, i.e., for oligonucleotide synthesis detritylation of the 5'-position

Perform chain extension

Remove terminal protecting group from newly added unit

Repeat coupling/deprotection cycle until final sequence (oligonucleotide or peptide) is assembled on the insoluble support

Cleave product from support

Final product is ready to use after removal of protecting groups and purification

FIGURE 1.6 General strategy for solid-phase synthesis. This technique is applicable to the synthesis of any repeating linear polymer, such as a polypeptide or oligonucleotide. The first residue (amino acid or nucleoside) is covalently attached to the surface of an insoluble support so subsequent chain extension products are permanently anchored to the support. The terminal protecting group (indicated by triangles) is removed and washed away from the support to enable chain extension. The next residue is then coupled to the support through either an amide (for peptides) or phosphate linkage (for oligonucleotides). At each step, unreacted reagents or by-products remain in solution and can be easily removed by washing them from the support after each step. The deprotection/coupling cycle is easily automated and the number of residues that can be added is limited only by the efficiency of the coupling reaction. After completion of the synthesis, the covalent linkage attaching the product to the support is hydrolyzed to release the protected product. After removal of protecting groups and purification the product is ready for use.

TABLE 1.1
The Effect of Coupling Yield on the Maximum Number of Couplings Possible in a Solid-Phase Synthesis, If a Minimum Yield of Approximately 10% Is Required

Coupling Yield per Step (%)	Number of Couplings	Overall Yield of Product (%)
50	3	12.5
60	5	7.8
70	6	11.7
80	10	10.7
90	22	9.8
95	45	9.9
99	230	9.9

reduced the labor required and allowed syntheses to be performed twice as fast as nonmechanized syntheses. With this machine each cycle of nucleotide addition could be performed in approximately 24 h. However, the low coupling yields obtained using the phosphodiester coupling chemistry limited this approach to the synthesis of only hepta- and octadeoxyribonucleotides.

Solid-phase oligonucleotide synthesis did not become a viable technique until the improved coupling chemistry of the modified phosphotriester approach was applied. This strategy was principally developed by the research groups of Keiichi Itakura at the City of Hope Research Institute in Duarte, California, Michael Gait at the MRC Laboratory of Molecular Biology in Cambridge, U.K., and Alexander F. Markham at the ICI Pharmaceuticals Division in Cheshire, U.K. Each of these groups developed methods for the phosphotriester synthesis of dinucleotide and trinucleotide building blocks,[83,84] which could then be assembled into long oligodeoxyribonucleotides on polyamide[84–90] or polystyrene supports.[91] Solid-phase synthesis using dimer and trimer units (block additions) instead of mononucleotides were required to produce oligonucleotide sequences of useful length because low coupling yields were still a problem. However, by the early 1980s it was possible to assemble oligonucleotides up to 31 bases long.[87] Although, the overall yields were poor (i.e., 5 to 9% yield for 17-mers), they were still sufficient to produce the microgram quantities required for biological studies. Also, significantly, the time required to perform a dimer addition was reduced to only 2 to 3 h and a semiautomated system could produce an oligonucleotide in only 2 to 3 days.

Many different oligonucleotides were rapidly produced using the solid-phase strategy and phosphotriester solid-phase synthesis, and this marked the beginning of the modern era of oligonucleotide synthesis. Two accomplishments were particularly noteworthy. The first was the total synthesis of the human leukocyte interferon gene by the group at ICI Pharmaceuticals.[92] This gene was a 514-base-pair sequence four times longer than the tyrosine tRNA gene made by Khorana. It was assembled from 67 different oligonucleotides of approximately 15 bases in length. This group optimized the method so that two 15-mers could be prepared using only 1 man-day of labor. The rapid assembly of such a long gene convincingly demonstrated the tremendous progress chemical oligonucleotide synthesis had made. This was also the first time a synthetic gene was derived from a published nucleotide sequence (cDNA) instead of from a peptide sequence.

The second notable achievement was the extension of this method by Itakura to the synthesis of oligonucleotides of ~30 to 40 bases in length and his development of a new strategy for gene fragment synthesis using DNA polymerase.[93] In this method, long oligonucleotides were assembled and hybridized together through short overlapping regions, and DNA polymerase was used to fill in the gaps. This method was demonstrated by synthesizing a 132-base-pair region of an interferon gene using only four synthetic oligonucleotides. This approach was significant because it greatly reduced the number of synthetic oligonucleotides needed; the "fill-in-the-gap" technique has been widely used ever since.

By the early 1980s, molecular biology had evolved to the point where not only was the need for synthetic oligonucleotides clearly recognized, but there was a great outburst of demand. These synthetic oligonucleotides were used for probes, primers, linkers, site-specific mutagenesis, and other applications.[33,35,36] It was clear that a significant economic market for synthetic oligonucleotides would develop, and the first attempts to commercialize this area were begun. However, it was also apparent that further improvements in the solid-phase synthesis chemistry were still required. This was because the phosphotriester coupling chemistry was still too difficult to be used by nonchemists, and the dimer or trimer building blocks required were hard to prepare and not readily available. More importantly, the coupling yields, which were not better than 90 to 95%, were still inadequate. Therefore, an intense effort was begun to optimize every aspect of oligonucleotide synthesis chemistry.

The origins of our present oligonucleotide synthesis chemistry began with the pioneering work of Robert Letsinger. Letsinger not only reinitiated interest in phosphotriester chemistry (as discussed above), but in 1975 he began work with trivalent phosphite compounds instead of the pentavalent phosphate compounds used by Khorana and others. This synthetic approach revolutionized the chemical synthesis of oligonucleotides and has come to be known as the *chlorophosphite method* or, more generally, the *phosphite-triester approach*.[94,95] It was known that the reactivity of P^{III} derivatives was much greater than that of the corresponding P^V compounds, and Letsinger showed that coupling reactions could be performed faster and in higher yield using trivalent phosphorodichloridite reagents. In this method (Figure 1.7) a series of three steps was performed in a "one-pot" reaction. First, a phosphorodichloridite coupling reagent with a trichloroethyl ester **23** (or other protecting group) was used to phosphitylate the 3'-OH position of a 5'-protected nucleoside **19**. The nucleoside 3'-chlorophosphite intermediate **24** was not isolated but immediately reacted with the 5'-OH position of second nucleoside **26** to form a dimer with a 3'-5' *phosphite* triester linkage **27**. Finally, the phosphite triester linkage was oxidized into a P^V phosphate triester linkage **29** by the addition of an iodine/water solution. The chlorophosphite coupling reagents were so reactive that high coupling yields (95% for dinucleotide synthesis) could be obtained in only a few minutes, even though the coupling reactions were performed at a very low temperature (−78°C). Although the phosphorodichloridite reagent also produced symmetrical 3'-3' linked dimers **25** and **28**, use of an excess amount of **19** reduced the extent of these impurities. The oxidation step was extremely fast (only a few seconds) and quantitative. Therefore, even though more steps were required, the entire reaction process was fast and efficient. Indeed, in those days the yield and kinetics of the coupling reaction were unprecedented.

Then, in 1980, the chlorophosphite coupling method was independently improved and applied to the solid-phase synthesis of oligonucleotides by two different research groups (Figure 1.8). Both of these research groups developed solid-phase synthetic strategies using surface-modified silica gel supports. Silica based materials were first used as solid-phase supports for oligonucleotide synthesis by Hubert Koster in 1972,[96] but they produced unremarkable results due to the limitations of the phosphodiester coupling chemistry. However, silica gel and controlled-pore glass (CPG) later turned out to be very satisfactory solid-phase supports (i.e., **30**) for oligonucleotide synthesis using improved coupling chemistries because they were rigid, nonswelling materials that could be readily derivatized and used in continuous-flow automated synthesizers. Mark D. Matteucci and Marvin H. Caruthers synthesized nonadeoxyribonucleotides on silica gel using nucleoside-3'-O-chloromethoxyphosphite reagents **31**, derived from dichloromethoxyphosphite, under conditions that required 4 h for each nucleotide addition.[97] In a subsequent paper, this method was further refined by converting the nucleoside chlorophosphite reagents into nucleoside tetrazoyl-phosphites and reducing the cycle time for each addition to <2.5 h.[98] Coupling yields in excess of 95% per condensation were obtained.

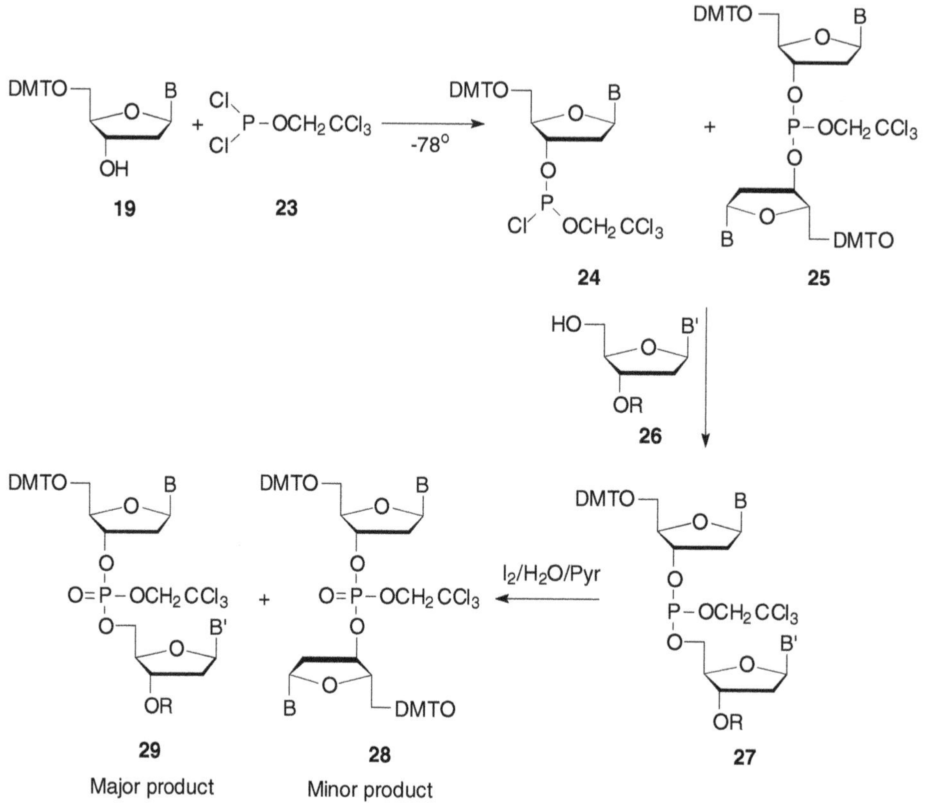

FIGURE 1.7 The chlorophosphite approach introduced by Letsinger. Phosphitylation of a nucleoside **19** with a trivalent phosphorodichloridite **23** reagent produces a very reactive nucleoside chlorophosphite **24** and a 3′-3′ dimer **25**. Intermediate **24** is not isolated but immediately reacted with the 5′-OH position of a second nucleoside or nucleotide **26** to produce a phosphite-triester internucleotide linkage **27**. *In situ* oxidation with iodine and water solution produces a pentavalent phosphotriester linkage **29**. The amount of 3′-5′ linked product **29** is much greater than the 3′-3′ linkage **28**.

The other research group using the chlorophosphite method was led by Kelvin Ogilvie at McGill University. This group first applied this method to the more difficult problem of oligoribonucleotide synthesis. Mona Nemer and Kelvin Ogilvie reported the solid-phase synthesis of a six-base-long oligoribonucleotide sequence on a silica gel support in 1980.[99] This approach was commercialized by BioLogicals (Ottawa, Canada) and applied to the automated synthesis of oligodeoxyribonucleotides.[100] Subsequently, the first commercially available, fully automated DNA/RNA synthesizers were released in 1981. These synthesizers used prepackaged nucleoside-3′-*O*-chloromethoxyphosphite reagents and a 30-min coupling cycle that could produce a 12-mer in only 5.5 h. Average coupling yields for each step of between 97.3 and 99.3% were reported. These high coupling yields and rapid coupling cycles were a dramatic improvement over all previous coupling strategies. The chlorophosphite coupling method was used to prepare the 66 oligonucleotides required to assemble a 453-bp human interferon-γ gene.[101] However, the chlorophosphite reagents were unstable with respect to moisture and too difficult to handle outside an experienced chemistry laboratory, and so this method was never widely used.

Instead, a significant breakthrough came from Serge Beaucage and Marvin Caruthers in 1981,[102] when they discovered how to use nucleoside-3′-*O*-phosphoramidite reagents **34**. These reagents contained the same reactive P^{III} oxidation state as the chlorophosphite reagents. However, a secondary amino group replaced the chlorine group to form a phosphite amide or phosphoramidite

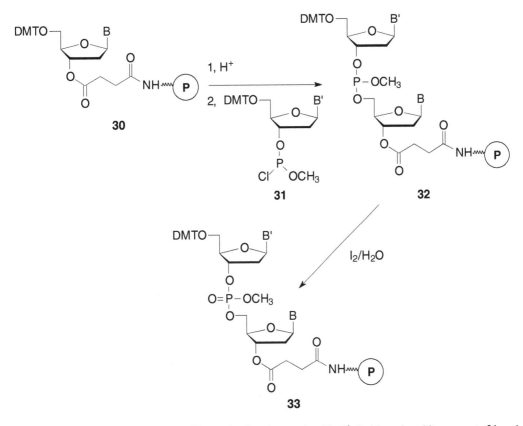

FIGURE 1.8 Solid-phase oligonucleotide synthesis using nucleoside-3′-*O*-chlorophosphite reagents **31** and derivatized silica gel supports **30**.

reagent. Thus, this method has come to be known as the *phosphoramidite approach*.[36,43,103] This simple substitution revolutionized oligonucleotide synthesis because the phosphoramidite reagents were stable, nonhygroscopic powders that could be easily prepared and handled in the open atmosphere. Therefore, it was now feasible to produce these reagents commercially because they could be easily stored for long periods and shipped long distances without decomposition. More importantly, the inherent high reactivity of the P^{III} oxidation state could be exploited by activating the stable phosphoramidite group with a weak acid such as tetrazole **35**. This activation step occurs almost instantaneously and is therefore performed *in situ* in the presence of the nucleoside or oligonucleotide residue being extended (Figure 1.9). Almost quantitative coupling yields can be obtained. The new internucleotide linkage is oxidized from the trivalent phosphite state to the pentavalent phosphate state using iodine and water, as in the chlorophosphite method.

The first phosphoramidite reagents were *N,N*-dimethylaminophosphoramidites **34a** prepared by reacting a 5′-protected nucleoside with chloro-(*N,N*-dimethylamino)methoxyphosphine.[102] These reagents were used to synthesize oligodeoxyribonucleotides up to 45 bases long in unprecedented yield.[103,104] However, the stability of the *N,N*-dimethylaminophosphoramidites in acetonitrile solution varied from hours to weeks, depending on the purity. Therefore, a series of more hindered and more stable reagents **34b–d** containing diethylamino, diisopropylamino, or morpholino substituted phosphoramidite groups were investigated.[105,106] The increased stability of these reagents allowed them to be purified by silica gel chromatography to consistent purity, and this greatly improved the reliability and reproducibility of this method. The solution stability of the phosphoramidites was also greatly improved. The *N,N*-diisopropylaminophosphoramidite compounds **34c** had the best combination of stability and reactivity. This was clearly demonstrated by Stephen P. Adams

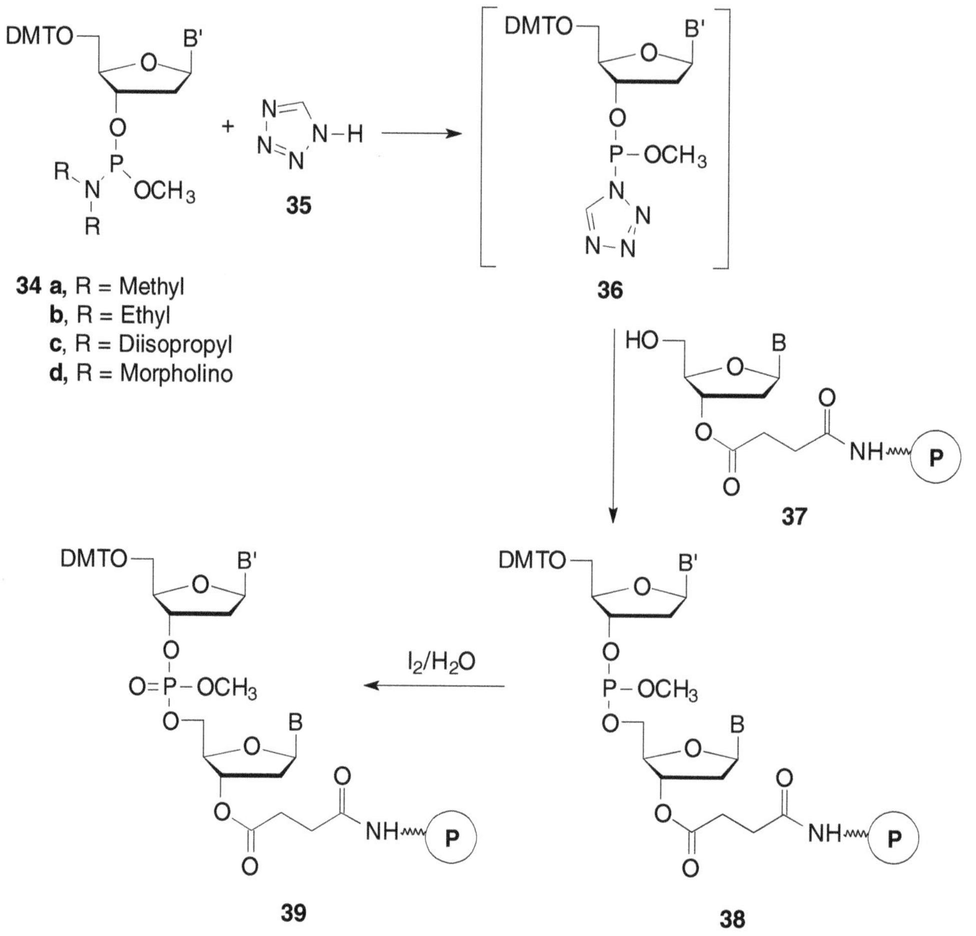

FIGURE 1.9 Activation of a nucleoside-3′-O-phosphoramidite reagent **34** with tetrazole **35** produces a very reactive tetrazolide intermediate **36**, which immediately reacts with the 5′-OH group of another nucleoside or nucleotide to produce a phosphite linkage **38**. Oxidation with iodine and water produces the desired pentavalent phosphate linkage **39**.

et al. in 1983, when two oligonucleotides of record length (51-mers) were prepared through monomer, and not dimer, additions.[106] This report was also significant because it used a long-chain alkylamine controlled-pore glass (LCAA-CPG) support for the first time. Both the N,N-diisopropylaminophosphoramidites and the new LCAA-CPG support quickly became the preferred reagents for oligonucleotide synthesis.

These two improvements were included in a new fully automated solid-phase DNA synthesizer developed by Leroy Hood et al. at the California Institute of Technology in Pasadena, California and commercialized by a new instrument company called Applied Biosystems, Inc. (Foster City, CA).[107,108] This instrument, the model 380A DNA synthesizer (introduced in 1982), was an immediate success because of its advanced features (use of argon pressure to deliver reagents, miniaturized zero-dead-volume valves, support for the automated synthesis of "mixed probes" containing degenerate base positions,[109] and its ability to synthesize up to three different oligonucleotides simultaneously) and its use of stable nucleoside-3′-phosphoramidite reagents. By using this instrument, coupling cycles could be performed in only 10 min and coupling yields of >96 to 98% were possible. This instrument was the first commercially available DNA synthesizer that could be easily

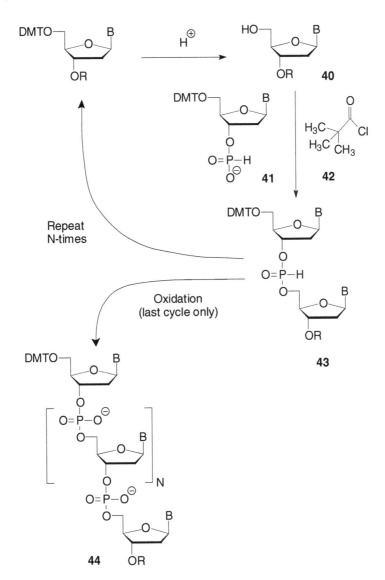

FIGURE 1.10 Nucleoside 3'-OH-phosphonate monomers **41** are the key materials in synthesis using the *H*-phosphonate method. Activation of **41** by pivaloyl chloride **42** produces intermediates with *H*-phosphonate internucleotide linkages **43** that are stable enough for use in subsequent coupling cycles. Oxidation of the linkages into the desired phosphodiester linkages **44** is performed by a single iodine/water treatment after completion of the coupling cycles.

and reliably used by operators in any laboratory, and Applied Biosystems has remained the dominant manufacturer of research-scale automated DNA/RNA instrumentation.

   Another strategy for oligonucleotide synthesis was successfully reintroduced in the mid-1980s. This was the *H-phosphonate approach*. This method was first introduced by Todd in 1957 for a diribonucleotide synthesis,[110] but it remained unused until applied to oligodeoxynucleotide synthesis by Garegg et al. and Froehler and Matteucci in 1985–1986.[111–113] In this strategy (Figure 1.10), nucleoside 3'-*H*-phosphonate monomers **41**, which are easy to prepare and stable to both hydrolysis and oxidation, are activated with an acyl chloride such as pivaloyl chloride **42**. This produces a very reactive mixed anhydride intermediate that can form an internucleotide *H*-phosphonate linkage with the 5′-hydroxyl position of another nucleoside or nucleotide **43**. Phosphonate linkages are stable enough to survive subsequent coupling cycles. Therefore, oxidation of the phosphonate linkages into phosphate linkages can be left to a single oxidation step at the end of synthesis. This makes *H*-phosphonate coupling cycles faster and simpler than phosphoramidite coupling cycles. However, *H*-phosphonate coupling reactions must be kept short (<1 min), otherwise a number of undesirable by-products begin to form.[114] Therefore, this technique has not been as widely used for solid-phase oligonucleotide synthesis as phosphoramidite chemistry. Recently, however, an *H*-phosphonate strategy for solution-phase oligonucleotide synthesis has been successful for producing large quantities of short oligonucleotide phosphorothioate sequences.[115]

Since the successful introduction of phosphoramidite reagents and LCAA-CPG supports, the basic methods for solid-phase oligonucleotide synthesis have undergone only a few incremental improvements (see following sections). Although these modifications improved the reliability, efficiency, and convenience of automated solid-phase oligonucleotide synthesis, our modern synthetic methods are still based on the solid-phase phosphoramidite coupling methods introduced in the early 1980s. However, what has changed significantly is the way chemically synthesized oligonucleotides are utilized. Automated DNA/RNA synthesizers have brought oligonucleotide synthesis out of the nucleic acid chemist's laboratory and made oligonucleotides easily and cheaply available to all researchers. More importantly, a large number of different applications for synthetic oligonucleotides have been developed, and the enormous numbers and substantial value of the oligonucleotides required have stimulated a great deal of research and economic activity.

The new applications that have been found for synthetic oligonucleotides are too numerous to describe. However, a few of the most significant can be discussed to illustrate the diversity and importance of these key materials. The first application, which has been used as an example throughout this section, has been total gene synthesis.[116,117] By the early 1980s, this application was no longer challenging to synthetic chemists because the basic chemistry was well established. However, synthetic genes were still important for the biotechnology industry, and more than 150 genes, including a 1610-base-pair-long human tissue-type plasminogen activator gene, had been synthesized by 1989.[118] The speed and ease with which synthetic genes could be prepared was further improved with the introduction of PCR-based strategies in the early 1990s and total gene synthesis from long oligonucleotides (70 to 200 nucleotides long) in a single step became possible.[119–123]

The development of the PCR was perhaps the most important application that synthetic oligonucleotides have made possible. The history of PCR is an interesting one[2] that began with the 1955 discovery of DNA polymerase and early PCR experiments by Kjell Kleppe and Ian J. Molineux in Khorana's laboratory between 1969 and 1972.[57] However, for a variety of reasons, not the least of which was the labor and expense of making synthetic oligonucleotide primers, the full potential of PCR was not realized by Khorana's group. Instead, modern PCR technology was independently developed at the Perkin-Elmer Cetus Corporation by Kary B. Mullis beginning in 1983.[3,4,58] Unlike the early work of Kleppe, Mullis was aided by the ready availability of cloned DNA polymerases and synthetic oligonucleotide primers.

In the PCR method, *in vitro* enzymatic synthesis is used to amplify specific DNA sequences located between two priming sites. Specific selection of these two priming sites requires the availability of custom-synthesized oligonucleotide PCR primers. Using this method approximately 100 billion copies of one molecule of DNA can be made in only a few hours. The PCR technique was first disclosed in 1985, and more than 1000 research reports including PCR data occurred over the next 5 years. Since then, PCR has revolutionized research in the biological sciences and medicine, provided extremely important new diagnostic techniques, greatly influenced criminology and law, and stimulated development of other amplification procedures. These methods also created a demand for millions of different oligonucleotide primers, which has further stimulated progress in the production of low-cost oligonucleotides. In 1993, Kary Mullis was awarded the 1993 Nobel Prize in Chemistry for his work in developing the PCR method.

The increased availability of synthetic oligonucleotides in the late 1970s also led to the development of a new field of *site-specific mutagenesis*.[124–127] It was realized that an entire synthetic gene was not always necessary and that specific modifications (mutations) to genes could be readily made using much shorter and more easily synthesized oligonucleotides containing single-base mismatches.[128] Once again, the flow of genetic information from DNA to RNA to proteins could be manipulated using chemically synthesized oligonucleotides. In this way, mutations could be induced in a specific manner and not randomly (as occurs in nature), so the properties of mutated proteins could be studied in a systematic fashion. This was another technique that revolutionized basic research and changed how researchers performed their experiments. In 1993, Michael Smith, of the University of British Columbia, also received the 1993 Nobel Prize in Chemistry for his fundamental contributions to the establishment of oligonucleotide-based, site-directed mutagenesis and its development for protein studies.

Determining the DNA sequence in biological samples has also been a key technology that has limited progress in the past. For example, the DNA sequence of the first gene synthesized by Khorana's group had to be deduced from the RNA sequence of the tRNA instead of by direct DNA sequencing. However, strategies for DNA sequencing using synthetic oligonucleotide primers and DNA polymerase started to be developed in the early 1970s.[129–132] These strategies used the unique base pairing properties of oligonucleotide primers to create a single specific starting point. This method, which was further developed by Frederick Sanger, and an alternate technique using chemical modifications, developed by Walter Gilbert,[133] resulted in the 1980 Nobel Prize in Chemistry (which was also shared by Paul Berg for studies on recombinant DNA). The Sanger technique has since become the most common way of sequencing DNA and has resulted in the production of many custom oligonucleotide sequencing primers.

The need for rapid and inexpensive DNA sequencing has precluded the use of radioactively labeled materials, as used in the original procedures. Instead, synthetic chemistry has been used to prepare fluorescently labeled custom oligonucleotide primers and fluorescently labeled dideoxynucleotide terminators. Since fluorescent signals can be detected quickly and extremely sensitively, this chemistry has led to the development of automated DNA sequencing instrumentation, which has revolutionized our ability to obtain DNA sequence information. Such instrumentation has made the various genome-sequencing projects practical and resulted in massive, large-scale DNA-sequencing centers. In addition, the ease of automated DNA sequence analysis has resulted in many smaller core DNA sequencing facilities that perform this task on a service basis. Again, the need for large numbers of inexpensive and reliable oligonucleotide sequencing primers has been an important factor in developing today's oligonucleotide synthesis technologies.

Nonisotopically labeled oligonucleotides have also stimulated a great deal of effort in diagnostic areas. Methods have been developed for chemically synthesizing oligonucleotides and oligonucleotide conjugates with a variety of different labels, such as biotin or fluorescent dyes, or other modifications.[37,39–42] These modified oligonucleotides have been used to develop new tests and biosensors, which promise faster and more reliable medical and environmental diagnostics. Fluorescently labeled oligonucleotides have also been used to develop new PCR applications that can be quantitatively monitored as the PCR amplification progresses ("real-time PCR").[135] Furthermore, these technologies have led to high-density DNA arrays (gene chips), which are in the process of revolutionizing our ability to simultaneously monitor the expression of multiple genes or rapidly screen samples against large numbers of known mutations or polymorphisms.[136–139] The potential rewards from such devices are very great, and this is also stimulating a great deal of work into the miniaturization and fabrication of oligonucleotide- or nucleic-acid-based devices.

Finally, an entire new industry devoted to nucleic acid medicine has emerged over the last 10 years. This industry is primarily based upon *antisense* or *antigene* oligonucleotide activity.[17,20,21,140–142] These activities target single-stranded mRNAs or double-stranded genes, respectively, to block gene expression. Already, Vitravene, the first antisense oligonucleotide pharmaceutical (Isis Pharmaceuticals, Carlsbad, CA) has been approved and many other oligonucleotides are being tested. The ability of oligonucleotides containing CpG motifs to act as powerful immuno-stimulants is also the focus of pharmaceutical[24,143–146] and new vaccine[147] development. Unlike the above applications, pharmaceutical oligonucleotides must be manufactured on much larger scales (1-million- to 10-million-fold higher), under strict purity and quality controls, and still be produced economically.[148–151] Industrial production of such complicated molecules had never been attempted before, and so the large-scale market demand for nucleic acid medicines has stimulated a great deal of research and innovation in both synthetic chemistry and manufacturing methods. This has led to many improvements in protecting groups, coupling efficiency, solid-phase supports, and very large-scale automated oligonucleotide synthesizers. Entire factories are being designed and constructed just for large-scale oligonucleotide synthesis, and the products they will produce hold the promise of new therapeutic agents of unprecedented specificity.

It is impossible to predict the number of new applications that will be developed in the coming years. Such applications may include exotic applications in areas not imagined at this time as well as novel applications in more traditional fields. In any event, it is certain that advances in oligonucleotide chemistry will lead to the development of new and improved technologies. Demand for faster, better, or different oligonucleotides from newly emerging applications will also stimulate new approaches to oligonucleotide synthesis. This type of synergy promises an interesting future for both molecular biologists and chemists.

1.3 PROTECTING GROUPS FOR SOLID-PHASE OLIGONUCLEOTIDE SYNTHESIS

As mentioned in the introduction, successful chemical synthesis of oligonucleotides depends on satisfactory schemes for protecting the various functional groups on the heterocyclic bases, sugar rings, and phosphate groups. These protecting groups must be selectively introduced at specific locations, be resistant to all chain extension conditions, be compatible with all the other moieties present, and be readily removable when required. Over the last 40 to 50 years, an enormous number of different protecting groups have been proposed, and the search for the optimum protecting group strategy still continues. This brief review can discuss only a few of these strategies, and only those protecting groups most relevant to modern solid-phase oligonucleotide synthesis are included. Information on the many other protecting groups can be found in the various reviews of oligonucleotide synthesis.

1.3.1 PROTECTING GROUPS FOR THE HETEROCYCLIC BASES

The exocyclic amino groups on adenine, cytosine, and guanine are most commonly acylated to prevent these groups from interfering in phosphorylation or coupling reactions and to improve solubility.[1] Thymine and uracil do not require acylation because they lack an exocyclic amino group. Usually, benzoyl protecting groups **13** are used on the N-6 position of adenine and the N-4 position of cytosine, while the N-2 position of guanine is protected with an isobutyryl group **14**.

Introduction of these acyl protecting groups is usually performed using the acid anhydride, although acid chloride derivatives can also be used. Specific acylation of the N-4 amino group on cytosine is possible without affecting the sugar hydroxyl groups.[152–154] However, this is not the case with either adenine or guanine, and either "per-acylation" or "transient protection" is required.[155] In the per-acylation method, all of the amino and hydroxyl groups are acylated, and then the hydroxyl groups are selectively deacylated using a strong base. In the transient protection method, the hydroxyl groups are selectively silylated with trimethylsilyl chloride before acylation of the amino group.[156] The trimethylsilyl groups are then easily removed with a mild aqueous base.

The benzoyl and isobutyryl protecting groups were originally introduced by Khorana more than 30 years ago and are still the most popular protecting groups for the exocyclic amines. However, removal of these protecting groups after synthesis requires an overnight treatment in 55°C ammonium hydroxide ($t_{1/2}$ = 3 to 11 h, depending on the nucleoside). Although these deprotection conditions are compatible with all of the standard deoxyribonucleosides, the demand for rapid oligonucleotide synthesis (≤ 24 h) has prompted development of more labile protecting groups that can be rapidly removed.

45   **46**

The phenoxyacetyl **45** and the similar t-butylphenoxyacetyl **46** protecting group were introduced as more labile base protecting groups. These groups allowed the half-time for deprotection, using room temperature (r.t.) ammonium hydroxide, to be reduced to only 2 to 8 min. This fast deprotection was particularly advantageous in the synthesis of oligoribonucleotides, which can undergo chain cleavage when treated with aqueous ammonium hydroxide for long periods of time.[157] One minor problem with such labile protecting groups is their accidental replacement with more stable acetyl groups (from the acetic anhydride capping reagent) during solid-phase synthesis. However, use of a base labile capping reagent instead of the usual acetic anhydride reagent solves this problem. Phosphoramidite reagents containing base labile phenoxyacetyl and *t*-butylphenoxyacetyl protecting groups are now commercially available from a variety of sources for the synthesis of both oligodeoxyribonucleotides and oligoribonucleotides. However, most oligodeoxyribonucleotides, especially long oligonucleotides, are still prepared using the more stable benzoyl and isobutyryl protecting groups.

Phthaloyl protecting groups **47**, which form cyclic diacyl groups, have also been proposed as amino protecting groups for the bases.[158,159] These groups have the advantage of being very rapidly released (≤1 min) using either ammonium hydroxide or the non-nucleophilic base DBU (1,8-diazabicyclo[5.4.0]undec-7-ene). Cleavage with DBU also has the advantage of allowing the oligonucleotide to remain attached to the solid-phase support during the entire deprotection process. However, a special set of protecting groups and a modified linker arm are required. Since these reagents are not readily available, the phthaloyl strategy has not been widely used.

47   **48**

Amino protecting groups that can be removed under mild or neutral conditions are also desirable for the synthesis of oligonucleotide analogues, which cannot withstand strongly basic cleavage conditions. The N-pent-4-enoyl (NPT) protecting group **48**, which can be easily removed by treatment with either an iodine reagent (r.t., 30 min), methanolic potassium carbonate (r.t., 4 to 8 h), or ammonium hydroxide (r.t., 15 min), is particularly noteworthy.[160] This protecting group has been used to prepare methylphosphonate analogues as well as regular oligonucleotides.[161]

An alternative route toward faster deprotection is the use of a more potent deprotection reagent. M. P. Reddy et al. developed a very rapid deprotection methods that uses a solution of 1:1 aqueous 40% methylamine and ammonium hydroxide (AMA reagent) as the deprotection reagent.[162–164]

This reagent is compatible with benzoyl and isobutyryl protected deoxyadenosine and deoxyguanosine nucleosides but cannot be used with N4-benzoyl protected deoxycytidine since unwanted
alkylation occurs. This problem can be easily solved by substituting a simple N-acetyl protecting
group for the N-benzoyl group; in addition; N4-acetyl-protected deoxycytidine nucleosides are now
commercially available. Using this system, the base deprotection can be completed in as little as
5 min. However, the AMA reagent has an unpleasant odor and, despite its increased speed, this
reagent is still not as widely used as ammonium hydroxide.

One consequence of acylating the exocyclic amino groups of adenine and guanine is an
increased sensitivity to depurination under acidic conditions. Cleavage of the glycoside bond
connecting the purine group to the sugar ring creates an abasic site, which leads to chain fragmentation under mildly basic conditions. In the case of oligodeoxynucleotides, which are approximately
100 times more sensitive to depurination than oligoribonucleotides, this is one of the most troublesome side reactions and it can significantly reduce yields of full-length products. Therefore, a
number of alternate amino protecting groups, such as various amidine protecting groups,[165–167]
which can stabilize the glycoside bond, have been introduced. However, amidine protecting groups
were not widely accepted and they are no longer commercially available.

Regardless of the type of amino protecting group, their use requires at least two additional
steps (their introduction and removal), and ultimately it would be preferable to avoid protection in
the first place. Selective O-phosphitylation is possible with unprotected guanine bases but not with
adenine and cytosine. However, schemes for solid-phase oligonucleotide synthesis using nucleoside
phosphoramidites without any N-protecting groups have been proposed. In 1991, Sergei Gryaznov
and Robert Letsinger described a method for the synthesis of oligonucleotides of moderate length
by following the condensation reaction with a pyridine hydrochloride/aniline wash to reverse
unwanted N-phosphitylation.[168] Later, they described how more selective O-phosphitylation could
be obtained using pyridine hydrochloride and imidazole as the phosphoramidite activator.[169]
Recently, another method for the solid-phase synthesis of oligodeoxyribonucleotides without nucleoside base protection has been developed by Yoshihiro Hayakawa and Masanori Kataoka.[170] In this
method, coupling is highly O-selective and any unwanted N-phosphitylation is reversed by a brief
treatment with benzimidazolium triflate. This is a promising approach and has the potential for
simplifying oligonucleotide synthesis (no N-deprotection steps are required), reducing the cost of
synthesis, and improving product quality (decreased sensitivity to depurination). It will be interesting to see how this method is accepted in the future.

The imide and lactam functions of thymine, uracil, and guanine are also sensitive to chemical
modification under certain conditions. In phosphotriester and H-phosphonate coupling methods,
protection of the O-4 and O-6 positions of these bases is sometimes recommended to prevent
substitution at these sites. Also, it has been shown that the O-6 position of guanine is readily
phosphitylated and causes chain cleavage during phosphoramidite coupling reactions.[171] This base
modification was the reason early solid-phase oligonucleotide synthesis attempts using phosphoramidite chemistry were unable to produce sequences containing many guanine bases. Fortunately,
however, this modification can be easily remedied by simply cleaving the O-6 adduct with water
or acetate ion to regenerate the guanine base. In practice, this is done by performing the acetic
anhydride capping step immediately after the phosphoramidite coupling reaction, instead of following the coupling reaction with the oxidation step.[172] Therefore, the phosphoramidite coupling
method does not require any additional protecting groups on the amide or lactam functions. This
is a significant advantage because such starting materials are less expensive and no additional
deprotection steps are required.

1.3.2 PROTECTING GROUPS FOR THE 5′-HYDROXYL POSITION

In the original phosphodiester coupling method, 2′-deoxyribonucleoside-5′-phosphates were used
because these nucleotides were available as starting materials. However, because the primary

5′-hydroxyl group is a much better nucleophile than a secondary 3′-hydroxyl group, almost all subsequent synthetic approaches formed the internucleotide phosphate linkage by coupling 5′-hydroxyl groups to 3′-phosphorylated (or phosphitylated) nucleosides or nucleotides. Therefore, it was important to develop protecting groups for the 5′-hydroxyl position that could be selectively removed without affecting protecting groups on other sites.

During early carbohydrate syntheses, the bulky triphenylmethyl or *trityl* protecting group **10** was widely used because it produced crystalline products that were easy to handle. However, the acidic conditions required to remove the trityl protecting groups were too harsh for oligonucleotide synthesis, and more labile trityl derivatives were introduced by Khorana in the early 1960s.[173] Khorana took advantage of the fact that a tenfold increase in acidic sensitivity could be obtained for each methoxy substituent added to the phenyl rings and created the monomethoxytrityl **11** (MMT) and dimethoxytrityl **12** (DMT) protecting groups. Although the MMT group has seen some use as an acid labile 5′-hydroxyl protecting group, the most preferred protecting group for the 5′-hydroxyl position is the DMT group. Today, this type of 5′-protection is used on virtually every commercially available phosphoramidite reagent. Although other 5′-protecting groups have been proposed, including groups that can be removed under neutral or basic conditions, none of them has turned out to be as satisfactory or as widely used as the DMT group.

The success of the DMT protecting group is due to several factors. First, the bulky nature of the group enables selective tritylation of primary 5′-hydroxyl groups over more hindered secondary 3′-hydroxyl groups. Second, quantitative and rapid detritylation is possible using a mild acid, such as dichloroacetic acid, without significant depurination.[174,175] Finally, under acidic conditions, intensely orange-colored trityl cations are produced, which can be easily measured and quantitated by spectrophotometry.[176] The intensity of the color released from the solid-phase support provides an easy method for fast and quantitative monitoring of solid-phase coupling reactions, and this feature is built into many automated DNA/RNA synthesizers.[177] The color of the trityl cation depends on the nature of the phenyl substituents, with dimethoxytrityl solutions orange and monomethoxytrityl solutions yellow. A variety of other differently trityl protecting groups have also been proposed, including a set producing different colors,[178] but these are not widely used. One small disadvantage of the dimethoxytrityl protecting group is the fact that the dimethoxytrityl cation is highly stabilized and detritylation is therefore reversible. This reversibility is not a problem in solid-phase synthesis because the cations are washed out of the synthesis column. However, when solution-phase-synthesis conditions are used, an excess of triethylsilane is recommended to scavenge dimethoxytrityl cations and increase the overall yield.[179]

The success of the acid labile dimethoxytrityl protecting group is somewhat surprising given the sensitivity of deoxyadenosine to depurination under acidic conditions. This problem is made worse when the N-6 amino position is protected (as described above). Consequently, a great deal of work has also gone into determining the optimum conditions for detritylation. In particular, detritylation using either protic acids or Lewis acids in various solvents has been examined.[174,175,180] Fortunately, detritylation during solid-phase synthesis is more efficient than during solution-phase synthesis because the released trityl cations are swept out of the synthesis column. Also, deoxyadenosine residues attached to a solid-phase support seem to be less susceptible to depurination. Therefore, good results, even for long oligonucleotides, can be obtained using dichloroacetic acid solutions in either dichloromethane or 1,2-dichloroethane as the solvent. However, in very large-scale synthesis, halogenated solvents are an environmental problem, and recently toluene has been used as a more economical and environmentally friendly solvent.[149,181]

1.3.3 PROTECTING GROUPS FOR THE 3′-HYDROXYL POSITION

The secondary hydroxyl group at the 3′-position does not require a protecting group when oligo-nucleotides are synthesized on a solid-phase support in a stepwise manner using single-base additions. This is because the growing oligonucleotide is covalently linked to the solid-phase support through the 3′-terminal hydroxyl group and the incoming monomers have 3′-phosphoramidite groups. However, it is also possible to perform solid-phase oligonucleotide synthesis in a stepwise manner using di- or trinucleotide phosphoramidite "blocks" instead of single nucleoside-3′-phosphoramidites. The solution-phase synthesis of the necessary phosphoramidite building blocks can be performed by coupling a 3′-phosphodiester unit to a nucleoside with unprotected 5′- and 3′-hydroxyl groups using the phosphotriester method.[182,183] However, such coupling reactions produce a mixture of both 5′-3′ and 3′-3′ linkages. Also, the aryl phosphate-protecting group required for the phosphotriester coupling method is unsatisfactory because it can lead to significant (>4% per linkage) chain cleavage during the final product deprotection.

Therefore, strategies for the synthesis of phosphoramidite building blocks with specific 5′-3′ linkages from phosphoramidite monomers were developed. In the first strategy, the 3′-hydroxyl position of a nucleoside is protected with a levulinyl group **49**. This approach was first used by G. Kumar and M. S. Poonian in 1984 with methyl phosphoramidites to produce dimer building blocks.[184] Later this synthetic strategy was adopted to perform codon-based mutagenesis using trinucleotide phosphoramidites with 2-chlorophenyl phosphate protection.[185] However, synthesis of triplet building blocks is long and complicated and 20 different triplets are required to encode all 20 natural amino acids. Fortunately, the amount of work required for codon-based mutagenesis could be significantly reduced by using a combinatorial (resin-splitting) strategy that required only seven dinucleotide phosphoramidite building blocks to encode all 20 amino acids.[186] The levulinyl group has also been used to prepare both dinucleotide and trinucleotide building blocks with the more labile 2-cyanoethyl phosphoramidite group.[186–189] In the synthesis of pharmaceutically important antisense oligonucleotide phosphorothioate sequences, stepwise synthesis with dimer or trimer blocks improved the purity of the final product.[188,189]

The base labile phenoxyacetyl group **45** has been used as a 3′-hydroxyl protecting group for the synthesis of 20 trinucleotide phosphoramidites.[190] However, the basic conditions (ammonia/methanol) required to remove the phenoxyacetyl group from the 3′-hydroxyl position were incompatible with 2-cyanoethyl phosphate protection, and these reagents could only be synthesized as methyl protected phosphoramidites.

The *t*-butyldimethylsilyl group has also been used as a 3′-hydroxyl protecting group **50**. In early work by Kumar and Poonian,[184] this group was advantageous because it conferred greater solubility than a 3′-levulinyl protecting group. In later work, by Matthew Lyttle et al. several trinucleotide phosphoramidites were also prepared using the *t*-butyldimethylsilyl group.[191] Although traditional silyl group cleavage using tetrabutylammonium fluoride was incompatible with base labile 2-cyanoethyl phosphate protection, this group found that acidic deprotection (using HCl) could be used. However, acidic depurination resulted in very low yields (20%) for sequences containing a 3′-terminal deoxyadenosine nucleoside.

1.3.4 PROTECTING GROUPS FOR THE INTERNUCLEOTIDE PHOSPHATE LINKAGE

A protecting group for the phosphate linkage is an essential part of both the phosphotriester and phosphite-triester coupling methods. However, unlike the phosphotriester method, which requires an aryl protecting group for sufficient reactivity, protecting groups for the phosphite-triester (phosphoramidite) method are not constrained. In the first chlorophosphite syntheses by Letsinger and Lunsford,[95] the 2,2,2-trichloroethyl group was the preferred phosphate-protecting group, i.e., **23**. However, removal of this group using Zn or Zn/Cu was difficult and not compatible with solid-phase synthesis. Therefore, a variety of other protecting groups were evaluated.

A simple methyl ester protecting group was first used in oligonucleotide synthesis by G. W. Daube and E. E. Van Tamelen in 1977.[192] This phosphorus-protecting group was adopted for the early solid-phase chlorophosphite syntheses on silica gel supports. Methyl ester protection was also widely used in the solid-phase phosphoramidite coupling methods that soon followed. The low molecular weight of the methyl substituent made the required chlorophosphite and chlorophosphoramidite starting materials easy to prepare, and the small size of the methyl group did not inhibit the coupling reactions. After synthesis, the methyl protecting group could be selectively and completely removed by treatment with thiophenol without any chain cleavage. Furthermore, these deprotection conditions did not affect the succinyl linkage to the solid-phase support, so this deprotection step could be automatically performed by the DNA synthesizer. This was important because the smell of the thiophenol reagent was very strong and unpleasant.

51a, R = CH$_3$
51b, R = CH$_2$CH$_2$CN

Once synthesis using nucleoside-3′-*N,N*-diisopropylaminomethoxyphosphoramidites was established and the first reliable, automated DNA synthesizers were commercially available, methyl protected phosphoramidite reagents **51a** became a significant commercial market. These reagents were used in all automated solid-phase oligonucleotide syntheses for most of the 1980s. However, as more applications for oligonucleotides developed, a greater emphasis was placed on the speed of synthesis and the elimination of even minor side reactions. The methyl protecting group was eventually found to cause small amounts of thymine-base methylation at the N-3 position.[193] This modification, along with the unpleasant and lengthy (1 h) conditions required to remove the methyl protecting group, eventually caused the oligonucleotide synthesis industry to abandon the methyl group as a phosphate-protecting group.

The 2-cyanoethyl protecting group, which can be easily and quickly removed by mild base through a beta-elimination mechanism, was first used in oligonucleotide synthesis by Letsinger and Ogilvie[59,60] and Catlin and Cramer[64] in early phosphotriester coupling reactions. However, it was not until this protecting group was reintroduced by Nanda D. Sinha et al.[194] for phosphoramidite reagents **51b** that this protecting group became widely used. The 2-cyanoethyl protecting group is rapidly and quantitatively removed at the same time the oligonucleotide is cleaved from the solid-phase support with ammonium hydroxide. This eliminates the need for the unpleasant thiophenol reagent and cuts 1 h from the deprotection time. These benefits enabled 2-cyanoethyl phosphoramidite reagents to completely replace methyl phosphoramidites in the oligonucleotide synthesis market within only a short time. Since then, 2-cyanoethyl protected phosphoramidite reagents have been the preferred reagents for virtually all oligonucleotide synthesis.

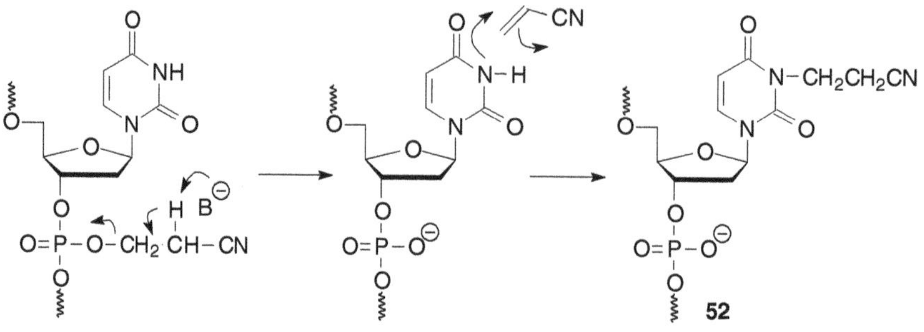

FIGURE 1.11 Mechanism for alkylation of thymine residues with acrylonitrile during removal of 2-cyanoethyl phosphate-protecting group.

It was also widely believed that thymine-base modification no longer occurred with 2-cyanoethyl phosphoramidite reagents. However, very recently, significant alkylation of thymine bases at the N-3 position has been observed. This occurs when the thymine bases react, under strongly basic conditions, with the acrylonitrile, which is produced upon beta-elimination of the 2-cyanoethyl protecting group (Figure 1.11). This type of alkylation was first studied in 1988.[195] Although this early study did not find any detectable modifications under typical oligonucleotide synthesis conditions, recent work on a T-rich oligonucleotide therapeutic sequence (using more sensitive CE and MS analysis) has discovered thymine alkylation. Such base modification has serious consequences for products intended for pharmaceutical use. Fortunately, the 2-cyanoethyl protecting group is labile enough to be selectively removed using a mild treatment with a secondary amine such as diisopropylamine. This allows the 2-cyanoethyl groups to be released from the phosphate linkages and the resulting acrylonitrile washed away from the synthesis column. Cleavage from the support and deprotection can then be performed without thymine modification.

A large number of other phosphate-protecting groups have also been developed for special applications. One interesting example is the *p*-nitrophenylethyl phosphate-protecting group introduced by Wolfgang Pfleiderer.[196] This protecting group is removed by treatment with DBU. When used in conjunction with *p*-nitrophenylethyloxycarbonyl and *p*-nitrophenylethyl base protecting groups and a support derivatized with a secondary *N*-methyl amine group, oligonucleotides can be completely deprotected with DBU while still attached to the solid-phase support. Such a scheme has the potential to completely automate all of the synthesis and deprotection steps in solid-phase oligonucleotide synthesis.

1.3.5 PROTECTING GROUPS FOR THE 2′-HYDROXYL POSITION: RNA SYNTHESIS

The structural difference between RNA and DNA is due to the presence of an additional secondary hydroxyl group on the 2′-position of the sugar ring. This modification may seem to be relatively minor, but it severely complicates the chemical synthesis of oligoribonucleotides. The 2′-hydroxyl group, through neighboring group assistance, greatly increases the susceptibility of RNA to both chemical and enzymatic cleavage. Thus, RNA sequences cannot be exposed to basic conditions and great care must be taken to avoid accidental exposure to the many RNAases that are easily spread by skin contact. Also, the 2′-hydroxyl group can cause isomerization of the 3′-5′ phosphate linkage to an unnatural 2′-5′ linkage. Successful chemical synthesis of oligoribonucleotides with exclusive 3′-5′ phosphate linkages requires the use of a protecting group that can be selectively added to the 2′-hydroxyl position. Furthermore, this protecting group cannot interfere with the coupling reactions and must be stable to the chain extension conditions, and be easily and quantitatively removed without causing chain cleavage or isomerization.

The search for satisfactory 2′-hydroxyl protecting groups and oligoribonucleotide synthesis conditions has challenged the ingenuity of many chemists. Furthermore, the difficulties imposed by the above conditions have always caused progress in oligoribonucleotide synthesis to lag behind oligodeoxyribonucleotide synthesis. Originally, acyl protecting groups, such as acetyl or benzoyl groups, were used. However, these groups are unsatisfactory, and only very short oligoribonucleotides could be prepared. Nevertheless, in a feat reminiscent of Khorana's early gene synthesis, an entire yeast alanine tRNA molecule was assembled from 2- to 6-base-long oligoribonucleotides using a combination of chemical phosphodiester and enzymatic coupling methods. This massive effort was achieved by Yu Wang and six research groups in China between 1968 and 1981.[197]

53   **54**   **55**   **56**   **57**

Acid-labile acetal and ketal protecting groups for the 2′-hydroxyl position have been used more extensively. The tetrahydropyranyl (Thp) group **53**, which is cleaved by strong acid (0.01 N HCl, 3 to 4 h), was widely used because it is reasonably compatible with acid labile 5′-dimethoxytrityl protection.[173] However, a small amount of Thp loss occurred during each detritylation step, and this restricted syntheses to oligoribonucleotides of less than ~20 bases. More acid labile protecting groups, such as the tetrahydrofuranyl (Thf) **54** and 4-methoxytetrahydropyran-4-yl (Mthp) **55**, have also been developed for use in strategies that avoid acid labile 5′-hydroxyl group protection.[198,199] Two acid labile protecting groups that can be used with 5′-dimethoxytrityl protection are the 1-[(2-chloro-4-methyl)phenyl]-4-methoxypiperidin-4-yl (Ctmp) **56** and 1-(2-fluorophenyl)-4-methoxypiperidin-4-yl (Fpmp) **57** groups introduced by Colin Reese.[200–205] Ribonucleoside phosphoramidites with 2′-O-Fpmp group protection have become commercially available, and their 2′ deprotection using an aqueous acidic buffer has been claimed to be easier and more efficient than 2′ deprotection of silylated oligoribonucleotides (see below). However, despite these claims 2′-O-Fpmp protected reagents do not seem to be widely used.

58   **59**

Protecting groups for the 2'-hydroxyl position that can be removed under neutral conditions have also been developed. Photolabile o-nitrobenzyl **58** and o-nitrobenzyloxymethyl **59** protecting groups, which can be removed by ultraviolet (UV) irradiation, have been developed.[206,207] However, quantitative removal of these protecting groups is difficult, and this has limited their use to short oligoribonucleotides. A number of other strategies for protecting groups that can be removed under neutral conditions have also been reported, but most of these were never widely used.

The major exception has been the 2'-alkylsilyl protecting groups developed by Kelvin Ogilvie and his research group.[280,209] Bulky alkylsilyl groups, such as the t-butyldimethylsilyl group **60**, are resistant to hydrolysis under either acidic or basic conditions. However, they can be rapidly removed under neutral conditions by treatment with fluoride ions. Originally, tetrabutylammonium fluoride (TBAF) in tetrahydrofuran (THF) was the preferred deprotection reagent, but this has recently changed to triethylammonium trihydrofluoride (which can be easily removed by evaporation).[210,211] Nucleosides with 2'-O-t-butyldimethylsilyl group protection are also easy to synthesize and very soluble in organic solvents. Although silylation of ribonucleosides usually produces a mixture of both 2'-O-silyl and 3'-O-silyl isomers, these compounds can be readily separated by silica gel chromatography. Greater regioselectivity for the 2'-O-silyl isomer can also be obtained using silver nitrate as a silylation additive[212] or a different purine protection route.[213] Once purified, the silyl isomers are stable unless exposed to prolonged methanolic or basic conditions, which can produce silyl group migration.[214] Fortunately, these conditions are easy to avoid, and methods have been developed for the preparation of 2'-O-t-butyldimethylsilyl ribonucleoside-3'-O-phosphoramidites with ≥99.5% isomeric purity.[215] The bulkiness of 2'-O-t-butyldimethylsilyl protecting groups affects the neighboring 3'-hydroxyl position and results in coupling reactions that are slower and slightly less effective than those in oligodeoxyribonucleotide synthesis. However, slightly longer coupling times (10 min) and more effective phosphoramidite activators, such as 5-(4-nitrophenyl)-tetrazole, 5-methylthiotetrazole, 5-ethylthiotetrazole, or 5-trifluoromethyltetrazole[210,211,216–218] can be used to boost coupling efficiency to >98%. The 2'-O-t-butyldimethylsilyl groups are also slowly removed during the prolonged ammonium hydroxide hydrolysis conditions (16 h, 55°C) required for the removal of N-benzoyl and N-isobutyryl amino protecting groups and this leads to chain cleavage and reduced product yields. Again, however, this problem can be greatly reduced by changing the deprotection conditions from aqueous ammonium hydroxide to either 3:1 ammonium hydroxide/ethanol or anhydrous ammonia/methanol.[157] An improvement can also be obtained by using more labile base protecting groups such as the phenoxyacetyl groups **45** and **46**, which are removed more rapidly.[157]

Despite these apparent difficulties, oligoribonucleotides synthesis using 2'-O-silyl protection and solid-phase phosphoramidite coupling chemistry has been very successful. The power of this approach was clearly demonstrated in 1988 when a 77-base-long tRNA[fmet] analogue (without the minor bases) was synthesized in a single stepwise solid-phase synthesis.[219] Since then, other tRNA molecules have also been prepared[220–222] along with synthetic ribozymes[211] and many other shorter oligoribonucleotide sequences. At the present time, ribonucleoside phosphoramidites with 2'-O-silyl protecting groups are commercially available from a number of sources and these materials are the most widely accepted and used reagents for solid-phase oligoribonucleotide synthesis.

However, improvements in 2′-protecting groups and oligoribonucleotide synthesis continue. Recently, the triisopropylsilyloxymethyl (TOM) group **61** was introduced. This group is claimed to improve coupling efficiency to >99.5% by reducing steric hindrance around the 3′-hydroxyl position. Although no independently published data on oligoribonucleotides synthesis using the TOM protecting group have appeared,* commercial oligoribonucleotide synthesis using 2′-*O*-triisopropylsilyloxymethyl protected ribonucleoside phosphoramidites is now available (Xeragon AG, Zurich).

A more radical, and potentially much more useful, improvement on 2′-hydroxyl protecting groups was recently described by Stephen Scaringe et al. at the University of Colorado.[223,224] This group decided to make a break from the conventional approach to solid-phase synthesis, which used 5′-dimethoxytrityl nucleoside phosphoramidites, since synthetic schemes compatible with the acidic deprotection of the dimethoxytrityl group seemed to have progressed as far as possible. Instead, they developed a *de novo* strategy that would be optimal for RNA synthesis. The key to this approach was a new 2′-*O*-bis(2-acetoxyethoxy)methyl (ACE) orthoester protecting group **62**. The 2′-ACE orthoester is stable to both nucleoside and oligonucleotide synthesis conditions. However, this stability is changed when the ester groups are hydrolyzed from the ACE protecting group and the resulting 2′-*O*-bis(2-hydroxyethoxy)methyl orthoester becomes ten times more acid labile. The hydrolysis of the ACE protecting group occurs under the same conditions as hydrolysis of the amino protecting groups on the bases. After this hydrolysis, the resulting oligoribonucleotides with the 2′-*O*-bis(2-hydroxyethoxy)methyl orthoester protecting groups can be easily handled and purified because they are still resistant to enzymatic degradation. Once purified, a simple incubation in a pH 3 sodium acetate buffer (10 min, 55°C) removes the remaining 2′-protecting groups and produces RNA in high yield.

The acid sensitivity of the ACE protecting group requires replacement of the conventional 5′-dimethoxytrityl protecting group with novel 5′-silyl protecting groups,[224] which can be rapidly removed with HF in triethylamine/DMF (35 sec). The fluoride ion deprotection conditions are not compatible with base labile 2-cyanoethyl phosphate protection and so this scheme also has to use the older methyl group protection for the phosphoramidite reagents. Despite these changes, this method is amenable to rapid solid-phase synthesis and a preliminary report has shown that 36-base-long oligoribonucleotides can be prepared in unprecedented yield and purity. Although, the necessary reagents for this strategy are not yet commercially available, custom oligoribonucleotide synthesis services using this method are available from Dharmacon Research, Inc. (Boulder, CO). This method appears to be very promising, and it will be interesting to see if the ACE protecting groups become more popular than the *t*-butyldimethylsilyl protecting groups, which have been so widely used over the last 15 to 20 years.

1.4 PHOSPHORAMIDITE COUPLING CHEMISTRY

1.4.1 PREPARATION OF NUCLEOSIDE-3′-PHOSPHORAMIDITE REAGENTS

In any phosphite–triester coupling strategy the starting material is phosphorus trichloride (PCl_3), which is a readily available, yet highly reactive, liquid. Reaction of this material with one equivalent of an alcohol, such as methanol or 2-cyanoethyl, yields a dichloroalkoxyphosphine **63** that can be distilled off and produced in large quantities (Figure 1.12). In the chlorophosphite coupling method, this reagent is reacted with the 3′-hydroxyl group of a suitably 5′- and N-protected nucleoside to yield a nucleoside-3′-chlorophosphite reagent (Figure 1.7).[225] However, even at −78°C the high reactivity of the dichlorophosphite reagent produces significant amounts of unwanted 3-3′ dimers **25**. Also, the nucleoside-3′-chlorophosphites **24** are too reactive to be easily handled or isolated.

---

* Note added in proof: Pitsch, S. et al., Reliable chemical synthesis of oligoribonucleotides (RNA) with 2′-*O*-[(tri-isopropylsilyl) oxyl]methyl-(2′-*O*-tom)-protected phosphoramidities, *Helv. Chim. Acta,* 84, 3773, 2001.

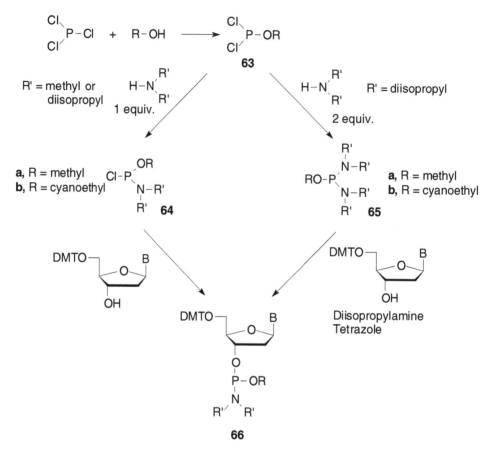

FIGURE 1.12 Preparation of nucleoside-3′-*O*-phosphoramidite reagents.

These problems were eventually overcome by reacting the dichloroalkoxyphosphine reagent **63** with one equivalent of a secondary amine (initially *N,N*-dimethylamine but later *N,N*-diisopropyl-amine) to produce a chloro-(*N,N*-dialkylamino)-phosphine **64**. This new type of reagent with only one reactive Cl group can then be used to produce nucleoside-3′-phosphoramidite reagents **66** without producing any 3′-3′ dimers.[102] Furthermore, the resulting phosphoramidites (especially the *N,N*-diisopropylamino reagents) are stable and can be easily handled without decomposition. How-ever, the chlorophosphite **64** reagents are reactive liquids that are difficult to handle, and so the preparation, isolation, and purification of nucleoside-3′-phosphoramidites are still somewhat difficult.

A variety of improved bis-triazolyl[226,227] or bis-(dialkylamino)alkoxyphosphine reagents, which were very stable to oxidation and hydrolysis and thus much easier to handle, were subsequently introduced. The first bis-dialkylamino reagent was bis-(pyrrolidino)methoxyphosphine, introduced by Beaucage in 1984,[228] and it could be monoactivated with 4,5-dichloroimidazole to reduce the amount of 3′-3′ dimer to only 7%. Shortly afterward, bis-(diisopropylamino)methoxyphosphine **65a** or bis-(diisopropylamino)-2-cyanoethoxyphosphine **65b** reagents that could be selectively activated by *N,N*-diisopropylamine hydrotetrazolide were introduced[229,230] (Figure 1.12). With these reagents, the salt produced by *N,N*-diisopropylamine and 1-*H*-tetrazole was sufficient to selectively activate only one group on the reagent, and the desired nucleoside-3′-phosphoramidite **66** could be prepared in high yield with only very small amounts (~1%) of 3′-3′ dimer formation. This method has since become the preferred synthetic group for nucleoside-3′-phosphoramidite synthesis and is used for the commercial production of up to 100 kg at a time.

67

It is interesting to note that the earliest accounts of this method for making phosphoramidites envisaged it as an *in situ* method,[229,231] which would allow nucleosides to be converted into their phosphoramidite derivatives without having to isolate them. However, nonchemists wanting to make their own oligonucleotides were only interested in purchasing reagents that could be used immediately and as simply as possible. Therefore, *in situ* phosphoramidite synthesis was never popular and a large industry devoted to phosphoramidite synthesis soon developed. However, after 12 years, the *in situ* method was eventually adopted for the large-scale synthesis of pharmaceutical oligonucleotides at Hybridon, Inc. (Worcester, MA). In this work a novel phosphitylating reagent, 2-cyanoethoxy-(N,N-diisopropylamino)-3-nitro-1,2,4-triazolylphosphine **67** was reported.[232] This reagent could be selectively reacted with the 3′-hydroxyl group of a suitably protected nucleoside without the need for any additional activating reagent and without the formation of any 3′-3′ dimer. The resulting solution of nucleoside-3′-phosphoramidites could then be used directly in solid-phase oligonucleotide synthesis. This method is commercially attractive because it uses much less expensive nucleosides instead of nucleoside-3′-phosphoramidite reagents.

1.4.2 THE PHOSPHORAMIDITE COUPLING CYCLE

Extension of an oligonucleotide on the solid-phase support requires several steps for each coupling reaction. These steps and the necessary washes between each of them are collectively referred to as a coupling or synthesis cycle (Figure 1.13).

The first step is removal of the 5′-dimethoxytrityl protecting group from the immobilized nucleoside or oligonucleotide. This exposes the 5′-hydroxyl group to which the next internucleotide linkage will be added. Detritylation is best accomplished using a solution of 2 to 5% dichloroacetic acid (pKa = 1.5) in either dichloromethane or dichloroethane. In our laboratory, dichloroethane is the preferred solvent because it is less volatile and produces less depurination. Nevertheless, use of halogenated solvents is becoming difficult, and more environmentally friendly aromatic solvents, such as toluene, are likely substitutes for future use.[181]

The kinetics of detritylation have been carefully studied because it is important to minimize exposure of purine bases to acidic conditions (otherwise depurination leading to chain cleavage will occur). These studies have shown that the most susceptible position toward depurination is a 5′-terminal deoxyadenosine, which is approximately five times more sensitive than a 5′-terminal deoxyguanosine and ten times more sensitive than an internal deoxyadenosine position. The rate of detritylation for the four most common protected nucleosides is 5′-dimethoxytrityl-N2-isobutyryl-deoxyguanosine > 5′-dimethoxytrityl-N6-benzoyl-deoxyadenosine > 5′-dimethoxytrityl-N4-benzoyl-deoxycytidine > 5′-dimethoxytritylthymidine. As a result, detritylation steps should be shorter for purine bases than for pyrimidine bases. The exact detritylation time varies with the synthesizer model, but is usually about 45 to 60 sec. Recently, a somewhat surprising recommendation for reducing depurination was proposed that entailed tripling the dichloroacetic acid concentration from 5 to 15% (v/v).[174] This is because the acid strength is depleted by complex formation with residual acetonitrile and by binding to the growing oligonucleotide. Therefore, an increased dichloroacetic acid concentration is recommended to achieve quick saturation. This results in faster and more complete detritylation with minimum exposure of the oligonucleotide to acid. Detritylation solutions containing 3% trichloroacetic acid in dichloromethane (v/v) are also commercially

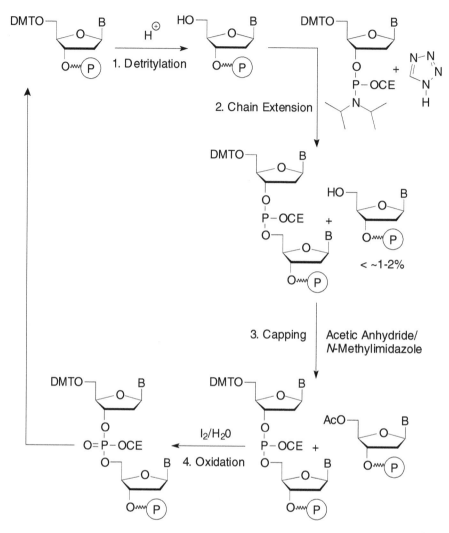

FIGURE 1.13 The phosphoramidite coupling cycle. (1) Detritylation with acid. (2) Chain extension using a nucleoside-3′-*O*-phosphoramidite activated with tetrazole. (3) Capping (acetylation) of unreacted residues. (4) Oxidation of phosphite linkages into phosphate linkages with iodine and water.

available. However, these cannot be recommended, since it has been well proven that more depurination occurs with this stronger acid (pKa = 0.7).

The trityl cation that is released can be measured as it flows out of the synthesis column by either colorimetric or conductivity detectors,[177] which are built in to many automated DNA synthesizers, so coupling failures can be immediately detected. These online detectors provide only an estimate of the coupling efficiency, and more accurate measurement requires collection of the trityl effluent and subsequent measurement in a spectrophotometer.[176] Most automated DNA synthesizers provide connections for external trityl collection in a fraction collector. Although quantitative trityl measurements are very useful for testing reagent quality and monitoring or troubleshooting synthesizer performance, they are not essential for day-to-day operation.

After the acidic detritylation reagent has been washed from the support, the next step is the *in situ* activation of the nucleoside-3′-phosphoramidite and coupling of the activated species to the growing oligonucleotide. The rapid conversion of the stable nucleoside phosphoramidite derivatives into highly reactive activated compounds (Figure 1.9) is one of the most remarkable aspects of phosphoramidite synthesis chemistry.[223,234] This activation is performed by mixing a solution of nucleoside phosphoramidite with a solution of a very mild acid (activator). Initially, the activators were tertiary amine hydrochloride salts such as *N,N*-dimethylaniline hydrochloride, but since these reagents were hygroscopic and difficult to work with 1*H*-tetrazole soon became the preferred activator.[102] 1*H*-tetrazole is a stable, nonhygroscopic solid that can be readily obtained in high purity by either crystalization or sublimation and 0.45 to 0.5 *M* solutions in acetonitrile are widely used. Although tetrazole and its substituted derivatives are potentially explosive (especially when attempting to sublime 1*H*-tetrazole), the room-temperature solutions required for DNA synthesis are not considered unduly hazardous.

The first step in the mechanism of phosphoramidite activation is protonation of the trivalent phosphorus, followed by slow displacement of the secondary amine (usually *N,N*-diisopropylamine) by tetrazolide. The tetrazolide then immediately reacts with the 5′-hydroxyl group of the oligonucleotide attached to the solid-phase support. The activation and coupling reaction is very fast, and for deoxyribonucleoside phosphoramidites coupling times of between 30 to 60 sec are generally used. Oligoribonucleotide synthesis requires longer coupling times (up to 600 sec) because the phosphoramidites are more hindered by the adjacent 2′-hydroxyl protecting group (especially with the more bulky 2′-*O*-*t*-butyldimethylsilyl protecting group). However, the 1*H*-tetrazole solutions are acidic enough to cause small amounts of detritylation during prolonged coupling reactions.[235] This detritylation can result in the formation of unwanted oligonucleotide impurities that are longer than the expected product (i.e., N + 1, N + 2, etc. long-mers). Fortunately, however, normal small-scale synthesis conditions produce undetectable, or barely detectable, amounts of these impurities.

As with every other step in solid-phase oligonucleotide synthesis, the activation step has been extensively studied, and a variety of other activating reagents have been proposed. More acidic 5-(*p*-nitrophenyl)-1*H*-tetrazole[205,216,218] and 5-ethylthio-1*H*-tetrazole[210,211,221,236] have been used for RNA synthesis; benzimidazolium triflate,[237] imidazolium triflate,[170] and pyridine hydrochloride/imidazole[169] have been used for O-selective phosphitylation; and more nucleophilic *N*-methylanilinium trifluoroacetic,[227] 1*H*-tetrazole/DMAP,[238] 1*H*-tetrazole/*N*-methylimidazole[239] solutions, as well as 4,5-dicyanoimidazole[239] and other azole activators[240] have been developed for difficult couplings. Other activators include pyridinium salts,[241] pyridinium trifluoroacetate/*N*-methylimidazole,[242] trimethylchlorosilane,[243] and 2,4-dinitrophenol.[244] These activators are generally faster and more soluble than 1*H*-tetrazole but are also usually more expensive or more hygroscopic, and so none of them has replaced 1*H*-tetrazole for general oligodeoxyribonucleotide synthesis.

The high reactivity of the activated nucleoside phosphoramidites also leads to phosphitylation of guanine bases at the O-6 position.[171] These guanine base modifications are unstable as long as the phosphorus atom remains in the trivalent oxidation state. Therefore, the newly formed internucleotide phosphite linkages cannot be oxidized into the desired pentavalent phosphate oxidation state until the guanine base modifications are removed. Fortunately, these modifications are simultaneously removed at the same time unreacted 5′-hydroxyl groups (typically <0.5 to 1% after each coupling reaction) are acetylated.

Therefore, the next step in the synthesis cycle is treatment with an acetic anhydride reagent to cap (block) off residual 5′-hydroxyl sites and reverse the guanine base modifications.[172] This capping step is very important because it prevents unreacted molecules from participating in further chain extension reactions. If this step were omitted, then a complex mixture of oligonucleotide impurities containing random deletions would result and isolation of the desired full-length product would be more difficult. Therefore, a fast and efficient capping step is required. Acetic anhydride is an inexpensive and effective acylating reagent as long as it is used in the presence of a nucleophilic catalyst such as 4-dimethylaminopyridine (DMAP) or *N*-methylimidazole (NMI). DMAP catalysis was used originally,[245] but subsequent analysis found that it produced trace amounts of mutagenic 2,6-diaminopurine base modifications.[246] Since catalysis with NMI does not produce these modifications, DMAP is no longer used. The acetylation reagent used for the capping step is packaged into two solutions: Cap A, acetic anhydride/2,6-lutidine/THF; and Cap B, NMI/THF; these are individually installed on the DNA synthesizer. During the capping step, the synthesizer simultaneously delivers both reagents to the synthesis column, where they mix and rapidly (10 to 60 sec) acetylate unreacted sites.

A variety of other capping reagents have also been proposed, but the only alternate reagent of significance is the *t*-butylphenoxyacetic anhydride reagent required for use with more labile *t*-butylphenoxyacetyl or phenoxyacetyl base protecting groups.[247] In this case, small amounts of the base labile protecting groups may be lost during synthesis, and if they were replaced by the much more stable acetyl groups, then incomplete base deprotection would result.

The last step in the coupling cycle is oxidation of the trivalent phosphite linkage into a more stable pentavalent phosphate linkage. The phosphite compounds are easily oxidized and in Letsinger's original phosphite–triester coupling reactions[95] the oxidation step was performed using a solution of iodine and water in THF. This oxidation is very fast (<10 sec), quantitative, and convenient. The progress of this step can be observed by the almost instantaneous disappearance of the dark iodine color on contact with the phosphite compounds. In early solid-phase oligonucleotide synthesizers, an iodine/water oxidation reagent containing 2,6-lutidine as an acid scavenger was employed.[36] Unfortunately, these solutions had a limited shelf life because a black iodine containing precipitate eventually formed. However, iodine/water solutions containing pyridine as the acid scavenger are stable indefinitely[248] and all automated oligonucleotide synthesizers now use these solutions for the oxidation step. This is somewhat surprising considering the high sensitivity of coupling reactions to moisture. However, in practice any residual moisture from the oxidation reagent is easily washed away from synthesis column using anhydrous acetonitrile. Therefore, nonaqueous oxidation reagents based on organic peroxides, such as *m*-chloroperbenzoic acid[249] or *t*-butylhydroperoxide,[250] have never been widely used.

The need for a separate oxidation step, which was originally considered an inconvenience, has turned out to be extremely beneficial because the oxidation step allows the specific introduction of a variety of internucleotide backbone modifications to be easily accomplished.[40,41] In the simplest case, isotopically labeled phosphate linkages can be obtained by using water labeled with an oxygen isotope. Iodine and a substituted amine can be used instead of iodine/water to produce oligonucleotides with modified phosphoramidate linkages useful for the synthesis of oligonucleotide conjugates. Elemental sulfur or selenium can also be used to produce phosphorothioate or phosphoroselenoate linkages, which have one of the nonbridging oxygen atoms replaced with either a sulfur or selenium atom.

The phosphorothioate modification (Figure 1.14) is especially significant,[251] since this is a rather conservative change. The native charge on the phosphate group is retained and the size of a sulfur atom is only slightly larger than an oxygen atom. Even though the modified phosphorus center is now chiral and each phosphorothioate linkage produces a set of two diastereoisomers, these compounds are excellent analogues for studying both the stereochemical course of reactions occurring at phosphorus and antisense inhibition of gene expression. The latter application is possible because phosphorothioate linkages are considerably more stable toward nucleases than are phosphodiester linkages.

Elemental sulfur was first used in 1978 by Burgers and Eckstein to synthesize dinucleoside monophosphorothioates from phosphite triesters.[252] Longer oligonucleotide phosphorothioates were later prepared by solid-phase techniques using similar sulfurization conditions.[253] However, the sulfurization step with elemental sulfur was relatively slow (7.5 min) and required the use of unpleasant smelling carbon disulfide as the solvent. Since that time a great deal of effort has gone into the development of improved sulfurization reagents to increase the speed and efficiency of this

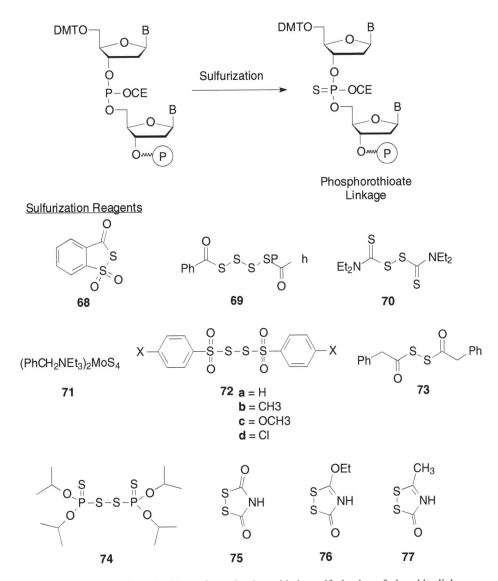

FIGURE 1.14 Phosphorothioate backbone formation by oxidative sulfurization of phosphite linkages.

step. One early sulfurization reagent that has become commercially available and quite widely used is 3*H*-1,2-benzodithiol-3-one-1,1-dioxide **68** (the *Beaucage* reagent).[254] Other sulfurization reagents that have been developed include dibenzoyl tetrasulfide **69**,[255] *N,N,N′,N′*-tetraethylthiuram **70** (TEDT),[256] benzyltriethylammonium tetrathiomolybdate **71** (BTTM),[257] bis(arylsulfonyl) disulfide **72**,[258] phenylacetyl disulfide **73** (PADS),[259,260] bis(*O,O*-diisopropoxyphosphinothioyl) disulfide **74** (S-Tetra),[261] 1,2-dithiazolidine-3,5-dione **75** (DtsNH),[262] 3-ethoxy-1,2,4-dithiazoline-5-one **76** (EDITH),[262,263] and 3-methyl-1,2,4-dithiazoline-5-one **77** (MEDITH).[264]

Regardless of the type of oxidation, the resulting pentavalent internucleotide linkage completes the phosphoramidite coupling cycle. At this point, if the oligonucleotide sequence is incomplete, then the cycle can be repeated as many times as necessary until the full-length sequence has been assembled. Oligonucleotides of up to approximately 150 bases in length can be prepared in a single automated synthesis, although most applications only require shorter oligonucleotides (20 to 30 bases). When the last coupling cycle has been completed, either the oligonucleotide can be cleaved from the support with the 5′-terminal dimethoxytrityl protecting group still attached or a final detritylation step can be automatically performed to remove the last dimethoxytrityl protecting group. This choice will depend upon the type of purification that will eventually be performed. Oligonucleotides that will be used directly or purified by polyacrylamide gel electrophoresis or anion-exchange HPLC should be detritylated on the synthesizer, while oligonucleotides that will be purified by reverse-phase HPLC or solid-phase extraction require the lipophilic dimethoxytrityl group to remain present.

1.4.3 REMOVAL OF PROTECTING GROUPS

After completion of the solid-phase synthesis, the oligonucleotide product must be cleaved from the surface of the solid-phase support and any residual protecting groups removed. Certain DNA synthesizers can automatically follow the coupling steps with an ammonium hydroxide treatment, which cleaves the product from the support and delivers it, in ammonium hydroxide solution, into a collection vial. Other instrumentation requires manual transfer of the support from the synthesis column into a separate vial for the cleavage and deprotection steps. Removal of the cyanoethyl phosphate-protecting groups and the acyl protecting groups on the heterocyclic bases occurs simultaneously in aqueous ammonium hydroxide, although prolonged treatment (overnight) and elevated temperature (50 to 55°C) may be required, depending on the type of base protection. Significantly faster processing can also be performed using a solution of methylamine and aqueous ammonium hydroxide (AMA reagent) as long as deoxycytidine and cytidine nucleosides with N4-acetyl protecting groups are used.[162–164,265]

A clever alternative to deprotection in aqueous ammonium hydroxide is the use of anhydrous ammonia or methylamine in the gas phase.[266,267] This method is especially suited for the batchwise deprotection of large numbers of solid-phase syntheses. A stainless steel pressure chamber, which can hold multiple oligonucleotides in either individual synthesis columns or 96-well microtiter plates, is simply pressurized with either anhydrous ammonia (~10 bar) or methylamine (~2.5 bar) and left to sit at room temperature. The gaseous reagent permeates the supports and simultaneously cleaves the oligonucleotide from the support and removes phosphate- and base-protecting groups. The deprotected products are left sitting on the surface of the support, and after the appropriate time (2 min to ~7 h, depending on the protecting groups and reagent used) the supports are removed from the pressure vessel and washed with water to collect the oligonucleotide product. This method is much faster than using aqueous ammonium hydroxide because the cleavage reactions are faster in the gas phase and the lengthy evaporation step to remove ammonium hydroxide is no longer required. When first introduced, this method seemed very attractive because of its speed and ability to simultaneously handle large numbers of oligonucleotides. Very recently, however, this method has been shown to produce oligonucleotides with modified thymine bases. The modifications result from the alkylation of the thymine bases with the acrylonitrile produced as a by-product of the cyanoethyl group removal.[195] This is a serious modification, and thus gas-phase deprotection cannot be recommended without prior removal of the cyanoethyl groups from the support. Unfortunately, however, this extra step negates some of the speed and handling advantages of the gas-phase method.

In practice, the length of overnight deprotection is not a serious restriction. A more important consideration is the amount time that synthesis instrumentation is tied up performing the automated cleavage of the oligonucleotide from the support. In addition, certain oligonucleotide modifications cannot withstand the strongly basic conditions required for cleavage and deprotection. In these cases, replacement of the traditional succinic acid linker arm, i.e., **37** with the more labile hydroquinone-*O,O′*-diacetic acid (*Q-Linker*) linker arm, i.e., **78** can be helpful.[268] This allows automated cleavage to be significantly shortened (2 min vs. 60 min cleavage time) and, in conjunction with labile base protecting groups, allows base labile oligonucleotide modifications.

Finally, in the case of oligonucleotides purified by reverse-phase HPLC or solid-phase extraction cartridges, the 5′-terminal dimethoxytrityl group must be removed by manually performing an acidic detritylation step to complete the deprotection process. However, because the base protecting groups have been removed by this stage, the oligonucleotide is much less susceptible to depurination, and this step can be safely performed with either 80% acetic acid or trifluoroacetic acid.

1.5 SYNTHESIS ON SOLID-PHASE SUPPORTS

1.5.1 THE SOLID-PHASE APPROACH

Although *solution-phase* synthetic methods for coupling small units together were developed many years ago, the large number of couplings needed to assemble useful sequences presented a very daunting task. This was because each step required some type of workup, extraction, or purification, and the labor and cumulative loss of material from these manipulations rapidly became significant problems. The problems involved in performing so many repetitive steps were addressed by R. B. Merrifield with the introduction of *solid-phase* synthesis.[76,269,270]

In this strategy (see Figure 1.6), a large insoluble support is covalently linked to the end of the sequence being assembled. The product on the surface of the support is available to react with reagents in the surrounding solution phase. The extended products remain covalently linked to the insoluble support, while unreacted reagents remain free in solution. Therefore, at the completion of each step the products can be rapidly and conveniently isolated by simply washing the unbound reagents away from the support. This can be performed as easily as filtering off the support and washing it with solvent. The support with its attached product is then ready for immediate use in the next step, as long as unwanted moisture contamination has not been introduced (in which case the support must be dried before use). In practice, it is convenient to handle the supports inside sealed reactors or columns so exposure to the atmosphere is minimized. This is also ideal for automation and the necessary reagent additions and solvent washes are readily mechanized. The process of adding each unit is repeated over and over until the desired sequence has been assembled on the surface of the support. The product can then be released from the support by cleavage of the covalent attachment linking the product to the surface. Removal of any remaining protecting groups completes the synthesis.

The big advantage of solid-phase synthesis is the ease with which immobilized products can be separated from other reactants and by-products. Solid-phase supports also permit relatively small quantities of material to be synthesized. This is because the additional physical bulk of the support, which is ≈10 to 100 times the mass of the attached nucleoside, can be handled more easily than the nucleoside alone. Also, confinement of the support inside a synthesis column eliminates handling losses. A small synthesis scale is important because of the high cost of reagents. Very little material is required for many biochemical applications and, as instrumentation has improved, the synthesis scale has decreased. The simplicity and similarity of the steps required for each chain extension reaction also greatly facilitate the synthesis of modified oligonucleotides, and a large number of modified substituents are available as phosphoramidite derivatives. Chimeric oligonucleotides containing peptide or PNA (peptide nucleic acids) sequences can also be prepared.[271,272] Finally, multiple bases ("mixed bases") can be incorporated at defined positions by using a mixture of different monomers, instead of a single monomer, in the chain-extension reaction. This is also the procedure used in "base doping," when only one base, at random, within a particular section needs to be mutated.[273]

Although a powerful technique, solid-phase synthesis has its drawbacks. The main limitation is the need for very high coupling yields in every chain-extension step. This is because the overall yield of product decreases rapidly as the number of consecutive chain extension steps increases. For example, if each base addition step had a yield of 90%, then the amount of dinucleotide produced (one base addition) is 90%, the yield of trinucleotide (two base additions) is $0.90 \times 0.90 \times 100\% = 81\%$, the yield of tetranucleotide (three base additions) is $0.90 \times 0.90 \times 0.90 \times 100\% = 73\%$, and so on. Note that the first nucleoside is attached to the insoluble support prior to the start of oligonucleotide synthesis and the efficiency of that step is not included in the calculation. The mathematical relationship between the overall yield (OY) and the average coupling efficiency (AY) is either

$$OY = \frac{-AY}{100}^{n} \times 100\% \text{ or } OY = \frac{-AY}{100}^{N-1} \times 100$$

where n = the number of coupling steps and N is the length of the oligonucleotide. The second equation assumes that the synthesis was performed by extending the product by one base at a time, as is usual with phosphoramidite reagents. Therefore, very high yields are required in every step and coupling yields that would be acceptable for most solution-phase reactions (such as the 90% yield assumed in the above example) are not adequate. Phosphoramidite reagents, which can produce average coupling efficiencies of 98 to 99%, are the most preferred, and oligonucleotides of up to 150 to 200 bases in length have been prepared.

Another consequence of having less than 100% coupling efficiencies is the accumulation of failure sequences containing deletions. The number of these failure products can be greatly reduced by the addition of a capping step after each chain-extension reaction. This step, which typically uses acetic anhydride to acetylate nonextended molecules, prevents these failure sequences from participating in any further reactions. However, a series of failure sequences, each one base shorter than the desired full-length product, will be present at the end of the synthesis.

Separating the full-length product (of length N) from the shorter failure sequences, and especially the $N-1$ failure sequence, is another significant problem. This purification problem becomes more difficult as oligonucleotide length increases, and for oligonucleotides greater than ~30 bases long, only PAGE (polyacrylamide gel electrophoresis) has sufficient resolving power to separate the full-length product from the $N-1$ component. Fortunately, however, many biochemical applications do not have stringent purity requirements and, if the coupling efficiency is high enough, the mixture of products produced can often be used with either minimal (desalting) or no purification.[274] However, the use of di- or trinucleotide blocks, which was a common approach in the older phosphotriester coupling method, has recently been proposed for phosphoramidite synthesis as well. In this strategy, the closest impurity to the full-length product (N-mer) is only $N-b$ (b = block size) bases long instead of $N-1$, and this makes large-scale purification much easier.[188,189]

Analysis of the synthetic products, while still attached to the surface of the insoluble support, also presents a major difficulty for those researchers developing new techniques or new solid-phase supports. This is an especially significant problem for applications using immobilized arrays because removal of the products for characterization is often difficult, if not impossible. Only a few studies have been performed, using techniques such as magic angle spinning NMR, ellipsometry, interferometry, or optical wave guides on immobilized oligonucleotides,[275–277] but these techniques have not been able to provide specific information about the quality of the oligonucleotide syntheses.

Therefore, analysis usually takes place after the products are cleaved from the support. Identification of the desired product (*N*-mer) is usually quite easy because, if coupling efficiencies are high enough, the full-length product is the most dominant product. Separation of the desired product from the other shorter impurities can be readily accomplished using either polyacrylamide gel electrophoresis, reverse-phase[278,279] or ion-exchange HPLC, or solid-phase extraction.[280] However, many biochemical applications are not compromised by the presence of small amounts of impurities and the crude reaction mixture is often used.[274] Characterization of the full-length product, other than a qualitative assessment of its purity, is not usually performed because it is difficult to sequence short oligonucleotides and because it is usually assumed that the automated synthesizer does not make mistakes in assembling the correct sequence. However, recent advances in MALDI-TOF mass spectrometry have provided fast and reliable methods for verifying both the molecular weight and the sequence of oligonucleotides.[281–284] Also, for nucleic acid medicines, there has been increased emphasis on characterizing the impurities present.[285] This has resulted in methods for determining the identity of oligonucleotide and $N - 1$ deletion sequences by either sequencing[286,287] or hybridization to an immobilized probe array.[288]

Finally, before a solid-phase synthesis can begin, a support with the first nucleoside attached to it must be prepared. In some instances the cost and time required to prepare and install supports can become a significant factor. For example, when large numbers of different oligonucleotides are produced in parallel on high-throughput DNA synthesizers, installation of supports with the correct 3′-nucleoside in the correct location becomes a concern. Therefore, new *universal* supports have been developed that allow the first 3′-nucleoside to be incorporated using the same phosphoramidite coupling chemistry as used for the rest of the sequence.[289–300] Alternatively, new linker phosphoramidite reagents[301] can be used to add the first nucleoside to less expensive, underivatized amino supports. When very large quantities of pharmaceutical oligonucleotide are required, the cost of the support also becomes a major factor. Typically, ~3 g of support are required for each gram of oligonucleotide product. These supports are expensive (~$20 to $30/g) and are usually used only once before being discarded. Since the latest large-scale DNA synthesizers (200- to 500-mmol scale) consume several kilograms of support per run, there is a strong economic incentive to lower the cost of the support. This can be done by developing less expensive resins with higher loadings or by developing methods for regenerating and reusing the supports.[302–304]

1.5.2 SOLID-PHASE SUPPORTS

Selection of the optimum solid-phase support is very important in oligonucleotide synthesis.[305,306] The ideal support should act as an inert carrier without hindering coupling reactions and provide sufficient surface functionality to yield synthesis capacities of between 10 and 50 µmol/g (for small-scale requirements) or 100 to 200 µmol/g (for very large-scale requirements). Generally, rigid macroporous supports work best under the polar coupling conditions required for oligonucleotide synthesis. These supports are based on inorganic materials such as silica gel and controlled pore glass or highly cross-linked polymers such as polystyrene or polymethacrylate. These supports do not become swollen with solvent and have permanent porosity. Their rigidity allows them to be used in packed, continuous-flow columns, and they have properties similar to the packing materials used in HPLC separations. Well-defined pores are created on the surfaces of these supports to increase the surface area and loading capacity. Pore sizes of 500 to 1000 Å are required to satisfactorily accommodate oligonucleotides up to ~50 to 60 bases in length, and pore sizes of 1000 Å or more are required to prepare longer oligonucleotides.

The amount of surface derivatization (loading) on the support determines the maximum amount of product that can be prepared. Optimum results are obtained on supports with less than ~40 µmol/g of nucleoside. This is because the efficiency of coupling decreases as the number of molecules on the surface increases. Since the amount of support can usually be increased to accommodate the scale required, supports with loadings greater than 50 µmol/g are not commonly used. The lower coupling efficiencies obtained with higher loaded supports can actually make it counterproductive to use these materials in most automated DNA synthesizers.

Although a large number of different materials have been used in the past as supports,[306] only three materials are widely used. Small-scale oligonucleotide syntheses are almost always performed on supports of highly cross-linked polystyrene or controlled pore glass (CPG). The polystyrene supports are advantageous because the surface is hydrophobic and very inert.[307,308] However, the most popular polystyrene supports (from Applied Biosystems) have a low nucleoside loading (8 µmol/g) and are not available in columns >200 nmol. Long-chain alkylamine CPG[106,176] is available with higher nucleoside loadings (~40 µmol/g) and is quite widely used for both small- and large-scale synthesis (0.05 to 15 µmol). Very large-scale (1 to 500 mmol) pharmaceutical oligonucleotides are generally only prepared on a proprietary, high-loading polystyrene resin (Amersham Pharmacia Biotech's HL-30) since very large-scale pilot and process-scale synthesis instrumentation has been designed around this material. Obtaining high coupling yields on higher loading supports is difficult, and a great deal of process optimization is required. At present, the 80 to 100 µmol/g loading of HL-30 is the most widely used industrial scale support. However, research continues on developing rigid supports with higher loading capacities, and a new support with a loading of ~200 µmol/g is expected to replace the HL-30 support in the near future.

1.5.3 LINKER ARMS

79   **80**

The structure of the compound(s) used to join the surface of the support to the first nucleoside is also of critical importance.[306] This attachment is composed of two distinct portions. The first portion is the spacer, which connects the active functional group (usually NH$_2$ or OH groups) to the matrix of the insoluble support. This spacer can be as simple as a single methylene group (i.e., aminomethyl polystyrene **79**) or it can be a lengthy alkyl or alkyloxy chain (i.e., long-chain alkylamine CPG **80**). Generally a long chain is preferred to distance the terminal functional group from the support's surface. Usually, the supports are sold with a satisfactory spacer, but sometimes additional spacers are added to change the terminal functional group or increase the overall length.[309,310]

A second difunctional molecule is then used as a linker arm to connect the amino group on the support to the first nucleoside unit. In oligonucleotide synthesis, the linker arms are usually dicarboxylic acids such as succinic acid[311–313] **81** or hydroquinone-O,O-diacetic acid[268,303] **82** which connect the nucleoside to the support via ester and amide bonds. The length, rigidity, and hydrophobicity of the linker arm can affect coupling efficiency, and the chemical stability restricts which conditions can be used during synthesis.[306] This affects the choice of protecting groups. Most linkers for oligonucleotide synthesis are resistant to acidic conditions and cleavable by basic conditions. This allows the most popular combination of protecting groups — acid labile 5′-dimethoxytrityl groups and base labile N-acyl and 2-cyanoethyl phosphate-protecting groups — to be used. Strategies that require removal of oligonucleotide protecting groups without cleavage from the support require linker arms that are either very stable or removable using conditions orthogonal to the deprotection conditions. Finally, different linker arms can be used to prepare oligonucleotides with terminal end modifications, such as 3′-phosphate, amino, carboxyl, thiol, or other substituents.

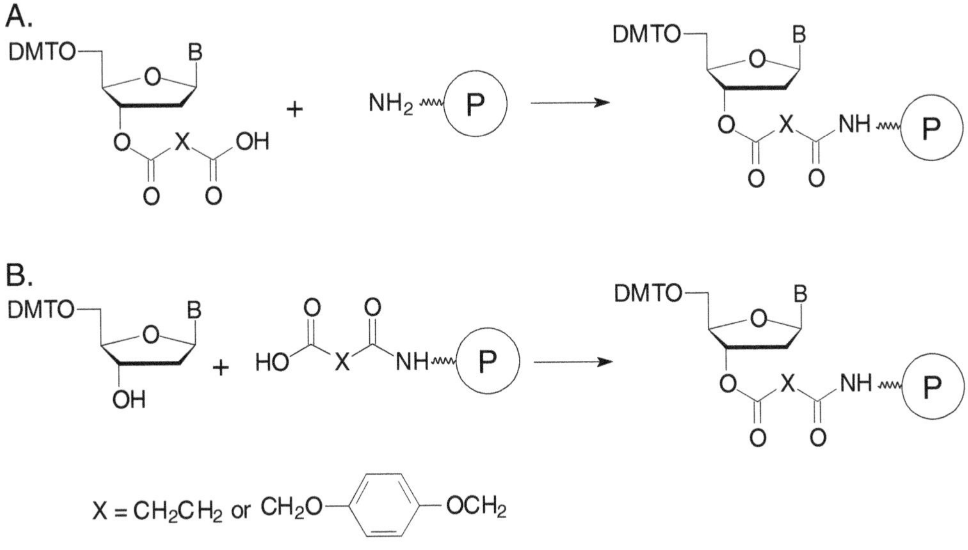

FIGURE 1.15 Two strategies for attaching the first nucleoside to the surface of a solid-phase support. (A) Through a primary amide linkage. (B) Through a secondary ester linkage.

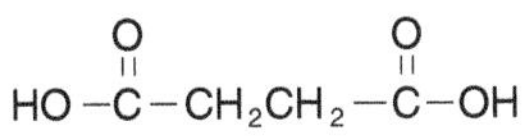

The most commonly used linker arm in oligonucleotide synthesis is succinic acid **81**. This linker was used as early as 1971[311] and has remained very popular because of the low cost and ease of incorporation. Nucleosides can be succinylated at the 3′-hydroxyl position and the resulting 3′-*O*-hemisuccinate coupled to an amino or hydroxyl derivatized support[176,313–318] (Figure 1.15). Alternatively, the support can be succinylated first and then coupled to a nucleoside.[312,317,318] This method has the advantage of not requiring the synthesis of an inventory of succinylated nucleosides. However, coupling of a nucleoside to a succinylated support is more difficult and usually gives lower nucleoside loadings than attaching a presynthesized nucleoside 3′-*O*-hemiester.

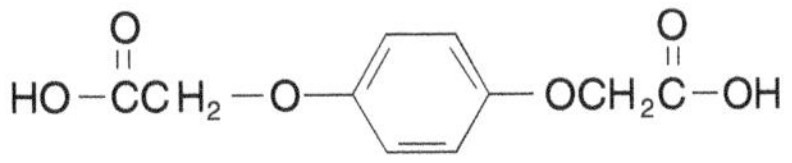

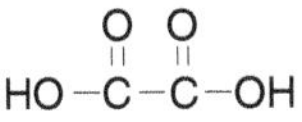

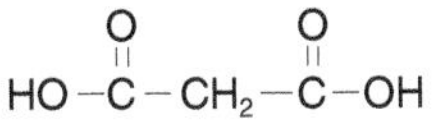

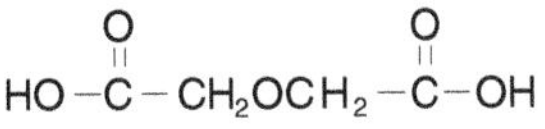

After completion of the oligonucleotide assembly, the protected products can be cleaved from succinylated supports by hydrolysis of the succinyl linker arm with either concentrated aqueous ammonium hydroxide (1 to 2 h) or gaseous ammonia[266,267] at 10 bar pressure (15 min). Faster hydrolysis can be performed by including stronger reagents, such as methylamine[162–164,265] or sodium hydroxide,[292,293,319,320] with the ammonium hydroxide. These reagents can reduce the cleavage time to 5 min and speed the removal of base protecting groups. However, there are potential problems with the modification of cytosine bases through either aminoalkylation or deamination with these reagents.

Although the succinic acid linker has been widely used for a long time, the succinyl linker is unnecessarily stable for oligonucleotide synthesis. The relatively harsh conditions required to hydrolyze the succinyl linker are incompatible with a number of base-sensitive minor bases, backbone modifications, and dye labels, and the time required to cleave the succinyl linker with NH_4OH is unnecessarily long.[268,321] Therefore, a number of more easily cleavable linker arms have been investigated. Photolysis offers a very mild method for cleavage, and photolabile linker arms, based on o-nitrobenzyl groups, have been used to synthesize oligonucleotides with 3'-hydroxyl, 3'-phosphate and other 3'-end modifications.[322–326] However, the photolysis can cause small amounts (<3%) of thymine-thymine photodimers and alkaline or other conditions still need to be employed to remove base protecting groups. In addition, N-benzoyl protected dA and dC nucleosides must be avoided.

The most labile dicarboxylic acid linker reported has been oxalic acid[327] **83**. This was completely cleaved by concentrated NH_4OH in only a few seconds and cleavage with a number of other milder reagents was also possible. However, the oxalyl linker was too labile for routine use and significant spontaneous nucleoside loss occurred during storage. More stable linkages have been created using either malonic **84**[328] or diglycolic acid **85**[329,330] as the linker arm. Treatment of diglycolic and malonic acid linkers with room-temperature concentrated NH_4OH for 10 min was sufficient to hydrolyze, respectively, 68 and 90% of these linker arms, conditions that caused only 15% cleavage of the succinyl linker.[268] However, a more satisfactory replacement for succinic acid is hydro-quinone-O,O'-diacetic acid **82**, which is used to create a *Q-Linker* arm. This linker is sufficiently stable so that decomposition during room-temperature storage is not a problem. However, the *Q-Linker* can be cleaved much faster than either the succinyl or diglycolic acid linkers. For example, cleavage using NH_4OH requires only 2 min.[268] Moreover, for routine use, supports derivatized with the *Q-Linker* can be used without any modifications to either protecting groups, reagents, or synthesis procedures (other than a reduction in cleavage time). Thus, the *Q-Linker* can serve as a general replacement for the succinyl linker in the synthesis of either unmodified or base-sensitive oligonucleotides. However, the main advantage of the *Q-Linker* is the improved productivity that results from the decreased cleavage time. Unlike postsynthesis deprotection, which is performed off the automated synthesizer, the cleavage step is usually performed by the instrument, and subsequent runs cannot be started until the cleavage is completed. Since typical oligonucleotide syntheses are usually completed within 2 h, waiting an additional hour or two for cleavage of a succinyl linker represents a significant bottleneck.[331]

In the above supports, attachment of the first nucleoside is always done separately from the actual oligonucleotide synthesis, and synthesis laboratories maintain an inventory of pre-derivatized supports. Recently, however, high-throughput DNA synthesizers[332–335] have created a need for *universal* supports, which have the terminal nucleoside added as part of the automated synthesis. This is required, not so much for inventory purposes, but because manual setup of prederivatized supports is time-consuming and error-prone. One approach to this problem has been the development of supports with universal linker arms. In these supports the first nucleoside is added as a conventional 3'-phosphoramidite derivative to form a phosphate linkage to the support. After synthesis the product is released from the support and the 3'-terminal phosphate group is removed via an intramolecular attack by a neighboring hydroxyl group. This strategy was first used in the early 1980s on cellulose supports,[319,336] and the universal support concept was first fully examined when ribonucleosides were attached to CPG through 5'-succinate linkages.[289,290,292,295] Oligonucle-otide synthesis, using nucleoside-3'-phosphoramidites, was then performed from the 2'- (or 3'-) hydroxyl position of the ribonucleoside linker in the normal manner. Since then, a number of different strategies for universal linker arms with adjacent hydroxyl groups have been reported[291,293,294,296–300] and "universal" supports are now commercially available from vendors such

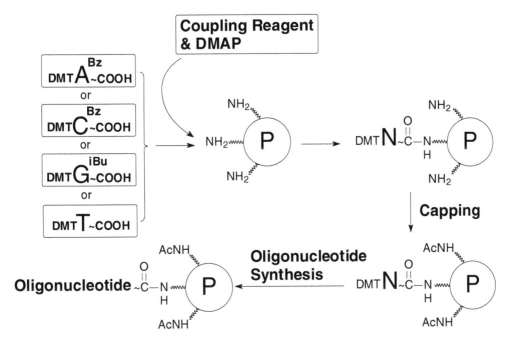

FIGURE 1.16 Strategy for automated online coupling of nucleosides to underivatized (universal) amino solid-phase supports using uronium or phosphonium salt coupling reagents and DMAP for rapid (~4 sec) amide bond formation.

as Glen Research, Clontech, and Beckman Coulter. However, cleavage from the linker involves two steps: (1) hydrolysis of the succinyl linker to release the material from the support and (2) elimination of the terminal residue as the 2′,3′-cyclic phosphate. The latter step often requires the addition of metal salts to the ammonium hydroxide cleavage reagent (which must be removed afterward). Although both steps, along with removal of base protecting groups, can be performed simultaneously, elimination of the terminal cyclic phosphate is slow, and the total cleavage and dephosphorylation take much longer than conventional supports.

A second approach is to simply speed up and automate the nucleoside coupling reaction so that amide or ester linkages can be used instead of phosphate linkages. However, coupling reactions with carbodiimide coupling reagents, such as dicyclohexylcarbodiimide,[315,316] 1-(3-dimethylamino-propyl)ethylcarbodiimide,[176,312] or diisopropylcarbodiimide[313] are too slow (1 to 24 h) for automation. A faster coupling reaction — involving reaction of a nucleoside-3′-O-hemisuccinate with 2,2′-dithiobis(5-nitropyridine) (DTNP) and dimethylaminopyridine (DMAP), followed by the addition of triphenylphosphine (TPP) and LCAA-CPG — can reduce the coupling time to between 2 and 30 min.[337] However, much faster coupling reactions (~4 sec) can be obtained using a variety of phosphonium or uronium coupling agents and DMAP.[317,338] Such fast coupling reactions allow this step to be readily included as part of the automated synthesis without significantly increasing the time required (Figure 1.16). Although implementation of this strategy requires an additional five reagents (four nucleosides and one coupling reagent) on the synthesizer, this method allows inexpensive amino- or hydroxyl-derivatized supports to be used and eliminates the lengthy cleavage and dephosphorylation steps required with universal linker supports. When the above coupling chemistry is used to attach a nucleoside to the 5′-OH end of an existing oligonucleotide (still attached to the support), then it is also possible to synthesize multiple oligonucleotides, linked end-to-end in tandem, in a single synthesis.[339,340] Such tandem syntheses are potentially very useful for preparing sets of oligonucleotides such as PCR primer pairs or dsDNA and RNA fragments, which are used together.

86

87

Finally, recent work has combined the advantages of using phosphoramidite reagents with easily cleavable 3′-ester linkages by preparing new linker phosphoramidite reagents.[301] These reagents differ from other phosphoramidite reagents by inserting a cleavable linker arm between the nucleoside residue and the phosphoramidite group. In the simplest form, a readily available nucleoside-3′-O-succinate is attached to a phosphoramidite group through an ethylene glycol spacer **86**. After synthesis, the succinate linker is hydrolyzed to yield products with 3′-OH groups, and the phosphate residue left attached to the support is discarded. In another form, a sulfonyl diethanol spacer is used to connect the phosphoramidite group to the succinate linker **87**. The sulfonyl diethanol spacer allows rapid release from the support, using a 5-min ammonium hydroxide treatment, because it undergoes a rapid β-elimination. Hydrolysis of any residual succinate linker occurs during the subsequent base deprotection step. Linker phosphoramidite reagent **87** can also be used to produce 5′-phosphorylated oligonucleotides and multiple oligonucleotides linked in tandem.

The linker phosphoramidites allow inexpensive underivatized amino supports to replace previously used universal supports. The first coupling cycle adds the linker phosphoramidite to the support through a phosphoramidate linkage. After oligonucleotide synthesis is completed with conventional 3′-phosphoramidite reagents, treatment with ammonium hydroxide releases the product with only the desired 3′-OH end. Since no 3′-dephosphorylation is required, all previous difficulties involved with dephosphorylation from universal supports are eliminated. These reagents are available from Transgenomic under the trade name First Base™ Phosphoramidites.

1.6 AUTOMATED AND SEMI-AUTOMATED DNA SYNTHESIS

The primary advantage of solid-phase synthesis is the ease with which the solid-phase supports can be manipulated. Successful chemical strategies for oligonucleotide synthesis on solid-phase supports were a vast improvement over the tedious methods previously performed in solution. In the earliest work, solid-phase synthesis was manually performed using simple funnels or test tubes to hold the supports.[104,341] However, researchers sought easier and faster ways to perform the syntheses. One such method used a syringe to hold the insoluble support.[342,343] This reduced exposure of the support to atmospheric moisture and made it easier to work on small synthesis scales. Alternately, manually operated reagent manifolds were used to deliver the necessary reagents and solutions.[344,345] Although, these improvements were more convenient when compared to previous methods, they were not automated and not suitable for use outside of an organic chemistry laboratory.

The first automated solid-phase synthesizers were developed to support peptide synthesis, since this chemistry matured more quickly than oligonucleotide synthesis. One of the first attempts at

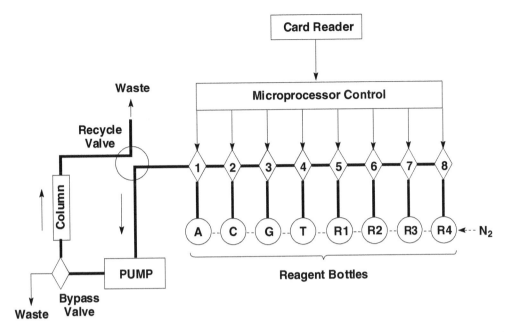

FIGURE 1.17 Schematic diagram of an early prototype DNA synthesizer used at McGill University in 1980.

automated solid-phase oligonucleotide synthesis (using phosphodiester coupling chemistry) was performed on a modified Beckman 990 peptide synthesizer.[82] This instrument, which was very large and programmed using a punched paper tape, was considered a remarkable innovation and state of the art for peptide synthesis in the 1970s. However, this technology did not provide much advantage to nucleic acid chemists because of the significant differences between peptide and oligonucleotide synthesis in reagents, supports, and operating scales. Instead, successful automated oligonucleotide synthesis required the independent development of instrumentation specific to oligonucleotide chemistry.

Early automated DNA synthesizers with chlorophosphite and phosphoramidite coupling chemistry used solid-phase supports packed into high-pressure columns similar to HPLC columns.[98] This was, not surprisingly, because the early silica gel supports were based on HPLC media. It was a relatively simple matter to assemble a linear manifold of three-way valves to deliver the appropriate synthesis reagents and solvents to a high-pressure HPLC pump under some type of programmable control. The pump then forced the reagents through the synthesis column and into a waste receptacle. Addition of another valve on the outlet end of the column allowed reagents to be recycled through the synthesis column. A schematic diagram of an early semi-automated DNA/RNA prototype synthesizer used by the author during his graduate studies (early 1980s) is shown in Figure 1.17. This instrument was programmed by shading appropriate lines on a paper card with a pencil.

More sophisticated instruments were soon offered commercially from vendors such as Vega, Bio Logicals, KabiGen AB, and Biosearch.[100,103,346] Although these instruments were significant because they were the first generation of fully automated DNA synthesizers available to the general research population, they all shared certain shortcomings. The valves, fittings, and pumps were based on readily available HPLC components and were not particularly reliable when used with the organic reagents required for DNA synthesis. This problem was compounded by the relatively high operating pressures of the HPLC-like column configuration. These synthesizers were also limited to the synthesis of one oligonucleotide at a time. Consequently, none could be considered a market success.

General acceptance of automated DNA/RNA synthesizers as reliable, easy-to-use laboratory instruments traces its roots to the laboratory of Leroy Hood at the California Institute of Technology. In this laboratory a microchemical facility for the analysis and synthesis of genes and proteins was established.[107] These researchers developed a collection of improved instrumentation that began with a protein microsequenator and eventually included an automated DNA synthesizer. The original DNA synthesizer used a conical Pyrex flask as a prereactor to mix the nucleoside phosphoramidite reagents with tetrazole before delivering the activated mixture to a second Pyrex reactor containing the support.[108] However, mixing in a prereactor was unnecessary because of the speed of the phosphoramidite activation, and this feature was later eliminated. However, two key innovations from these early instruments were destined to greatly simplify and improve the reliability of future DNA synthesizers. First, an innovative use of positive argon gas pressure, instead of a mechanical pump, was used to move reagents and solvents. This eliminated the problems with leaky check valves, piston seals, and air bubbles that plagued earlier pump-based synthesizers. Second, new zero-dead volume valves with chemically inert Kalrez diaphragms, which were mounted side by side on a single Kel-F block, were developed. These valves were highly reliable and the availability of multiple valves (up to 12) in a single valve block significantly reduced line volumes and eliminated the number of connections required.

This technology was commercialized with the formation of Applied Biosystems, Inc. (Foster City, CA) in the early 1980s. The model 380A DNA synthesizer, introduced by Applied Biosystems in 1982, was an immediate success because of its low reagent consumption, reliable design, built-in computer control, and ability to simultaneously prepare up to three different oligonucleotides at the same time (on scales of either 0.2, 1, or 10 μmol). Eventually, the model 380A synthesizer was succeeded by the 380B synthesizer, which added a touch-screen CRT for programming and a floppy disk drive for data storage. A more economical single column 381A synthesizer was also introduced. These synthesizers quickly dominated the market for automated DNA/RNA synthesizers and became very widely used. Later, the synthesizer design was further refined with the introduction of the model 391 (single-column), 392 (two-column), and 394 (four-column) synthesizers, which introduced optional online trityl conductivity measurement[177] and external control from a Macintosh computer. Operating costs were also significantly reduced with the implementation of low-volume (LV) columns[347] and smaller (0.04 μmol) synthesis scales. The 390 series synthesizers have proven to be highly reliable, and many core DNA synthesis laboratories, including the author's, have relied upon them. For example, at Amgen, Inc., a high-throughput, highly automated oligonucleotide production facility was developed with over 20 394 DNA synthesizers and a variety of robotic workstations.[348] With each synthesizer capable of synthesizing four different oligonucleotides at a rate of one synthesis cycle every 7 min, this facility could synthesize more than 300 20-base-long oligonucleotides each day.

Other automated DNA synthesizers, using the same phosphoramidite chemistry, were also introduced by vendors such as Pharmacia, Beckman, PerSeptive Biosystems (now part of Applied Biosystems), and Eppendorf. However, these instruments never gained the market acceptance of the Applied Biosystems synthesizers.

The rapidly increasing demand for oligonucleotides has been a constant incentive for the development of techniques and instrumentation that can produce more oligonucleotides faster and cheaper. One approach to the synthesis of more oligonucleotides per instrument is to simply increase the number of column positions. In addition to the above Applied Biosystems instruments, which could perform simultaneous synthesis on up to four synthesis columns,[334] other DNA synthesizers have had up to 10 positions available in a row, or up to 24 positions available in a cylindrical cartridge. However, the 10-position instrument was never commercialized, and the 24-position instrument was a proprietary design for the exclusive use of Genset Oligos, a commercial DNA synthesis company.

Another high-capacity DNA synthesizer was the Applied Biosystems 3948 instrument, introduced in 1996,[334] which was described as a high-throughput automated oligonucleotide production system. In this instrument, three synthesis columns were used in parallel for the actual solid-phase synthesis. However, a revolving carousel containing 16 sets of three columns allowed continuous production of up to 48 different primer-length oligonucleotides in each 24-h period. This instrument completely automated the ammonium hydroxide cleavage and deprotection steps. Each synthesis column contained both a derivatized polystyrene solid-phase support for synthesis and an underivatized polystyrene support. Solid-phase synthesis occurred on the former material, while the underivatized polystyrene support acted as a solid-phase extraction medium to automate the purification of the final 5′-tritylated product. Thus, one instrument produced deprotected and purified oligonucleotides suitable for immediate use. However, this instrumentation was expensive and best suited for the synthesis of small amounts of oligonucleotide primers. A less expensive alternative was the Expedite 8909 DNA synthesizer (which replaced the 391/392/394 models in 1998) with the multiple oligonucleotide synthesis (MOSS) option. This instrument allowed up to 12 oligonucleotide columns to be configured at one time. However, synthesis was only performed on two columns at a time and automatic cleavage and deprotection were not performed.

The above instruments can be considered as employing a *closed reagent* delivery system, in which all reagents flow through Teflon tubing, common manifolds, and sealed reaction columns. Although reliable, this type of delivery system has two problems. First, increasing the number of synthesis columns significantly increases the number of valves required and the complexity of the tubing network. Second, such a design leads to excess reagent consumption because of the necessity of rinsing common pathways and expelling dead volume during synthesis. A more efficient *open reagent* delivery system was developed at the Lawrence Berkeley Laboratory at the University of California, Berkeley in 1995.[333] In this design, a stationary dispensing manifold drops reagents into an open, multichannel reaction chamber mounted on a movable linear table inside an argon atmosphere. Each reaction chamber contained 12 individual wells holding insoluble support. Up to eight reaction chambers could be used for a total synthesis capacity of 96 oligonucleotides. This design significantly reduced the plumbing complexity, increased throughput by a factor of 24, and reduced reagent consumption and hazardous waste generation by approximately 70%.

A similar but improved high-throughput DNA synthesizer was also independently developed at the Stanford DNA Sequencing and Technology Center at Stanford University.[332] This instrument used readily available 96-well microtiter plates as the reaction chambers, 11 banks of valves (each mounted in a set of eight), and a linear movable stage to position each well under the appropriate reagent dispenser. This automated multiplex oligonucleotide synthesizer (AMOS) instrument had an 11-min cycle time (96 couplings) and could produce a plate of 96 different 20 base-long oligonucleotides in <4.5 h (not including deprotection and lyophilization times). Installation of the correct prederivatized support (i.e., A, C, G, or T) in the correct well of multiwell synthesis plates was a new difficulty arising from this type of technology. However, this problem can be solved by either robotic support dispensing or the use of universal solid-phase supports or the alternative *in situ* nucleoside derivatization methods described in the previous section.

Later, 96-well plate synthesis technology was further improved by including two-dimensional movement. In this modification, the reagent dispensers moved in the y-axis, and up to 64 different reagents could be rapidly delivered into any of the 96 wells for automated ribozyme or combinatorial solid-phase organic synthesis.[349] Although these new designs represented a dramatic improvement, the inventors had initial difficulties finding commercial backing, and eventually exclusive rights to this design were awarded to a single company, ProtoGene Laboratories (purchased later by Life Technologies, Inc.). However, once available the cost savings and advantages of this instrumentation were quickly realized. Indeed within 1 year (1994–1995) of producing the first oligonucleotide Life Technologies became the largest custom oligonucleotide-synthesis company in the world.

This new competition also started a major price war among commercial oligonucleotide synthesis services, and oligonucleotide users benefited from a several-fold decrease in cost. Although this further increased the availability of synthetic oligonucleotides, it reversed the trend toward oligonucleotide synthesis on local, laboratory-scale instrumentation in favor of synthesis in highly specialized, high-throughput commercial facilities. For example, by 1997 Life Technologies had a daily oligonucleotide synthesis capacity in excess of 7500 oligonucleotides *per day* on approximately 25 96-well DNA synthesizers. Other large commercial DNA synthesis companies also expanded rapidly (using different and often proprietary instrumentation), and a synthesis capacity of thousands or tens of thousands of oligonucleotides per day soon became common. For example, at Illumina (San Diego, CA), a proprietary DNA synthesizer called the Oligator 768™ uses a centrifugal rotor to produce eight 96-well plates of oligonucleotides per run. Twelve of these instruments, operating within an in-house synthesis facility known as the Oligator Farm™, provide Illumina with an annual oligonucleotide synthesis capacity of greater than 5 million oligonucleotides per year. Such incredible progress in speed and capacity must surely be astonishing to the researchers who toiled so hard in Khorana's laboratory making the first synthetic genes.

Other 48- or 96-oligonucleotide-capacity and commercially available DNA synthesizers have also been described. In one, a pipetting robot — originally designed for peptide synthesis[350] — was used to deliver reagents to a rack of 48 disposable pipet tips containing a derivatized membrane support. In others, either a single 96-well synthesis plate (PolyPlex DNA Synthesizer, GeneMachines) or a dual 96-well plate oligonucleotide synthesizer (the "MerMade," BioAutomation) were developed[335] and made commercially available. A totally different approach based on the concept of "tea-bag" synthesis (originally developed for peptide synthesis) on movable pins has appeared. In the PrimerStation 960 DNA synthesizer from Intelligent Automation Systems, Inc. (Cambridge, MA), the solid-phase support is enclosed in a porous pouch and attached to the end of a piston-like rod. The 96 independent rods move up and down to dip the pouch in and out of the reagent-filled tray. However, this tray must be rinsed and refilled with new reagent for every step, and this leads to high reagent consumption. Also, the method is only suitable for small-scale synthesis due to the small size of the pouches (~2 to 3 mg), and automated cleavage or deprotection is not performed. Finally, in 2000 Applied Biosystems introduced its model 3900 DNA synthesizer. This instrument was a hybrid synthesizer using individual prepacked columns instead of multiwell synthesis plates. However, the columns were mounted in an open-reagent system using a circular 48-column rotor. Thus, this instrument combined the speed and economy of the above 96- and 384-well synthesizers with the flexibility and convenience of individual prepacked synthesis columns. Up to 288 primer-length oligonucleotides (six runs of 48) could be prepared in a 10-h period using any combination of 40-, 200-, and 1000-nmol scales.

Another method for speeding simultaneous synthesis of large numbers of different oligonucleotides is known as *segmental solid-phase* synthesis. In this approach, the synthesis is interrupted after each coupling cycle. Each solid-phase support is sorted into one of four groups (A, C, G, or T), depending on which base is required for the next addition. The supports in each group are then stacked together, and the next coupling cycle is performed on the entire stack. Although each coupling must be performed on four different stacks, up to 100 or more oligonucleotides can be processed at a time. When first introduced, this method used paper disks as the insoluble support[345,351–354] (cellulose has many surface hydroxyl groups that can anchor growing oligonucleotide chains). These paper disks were inexpensive, easy to handle, and easy to label with a pencil. However, they were not suitable for automation. Subsequently, simultaneous synthesis of multiple oligonucleotides using segmental units composed of various stackable cartridges or wafers has been reported.[344,355–357] This has led to the development of the Abacus commercial high-throughput oligonucleotide synthesis employed by Sigma-Genosys (Woodlands, TX). However, fully automated sorting is still very difficult. Consequently, the systems developed have only been semimechanized, and the actual sorting steps have been left to human operators. However, even with this manual intervention, one instrument and one operator can produce 144 25-base-long oligonucleotides in only 4 h. After solid-phase synthesis, the wafers are moved to a separate automated cleavage, and deprotection instrument and a laboratory with only ten instruments can produce >2500 oligonucleotides per day. Segmental synthesis is also not limited to macroscale syntheses, as in the above examples. Use of fluorescent dye encoding can allow the size of a single support to be reduced to a single 8.8-μm microsphere. Sorting of these individual microspheres can be performed at sort rates of approximately 25,000 events/sec by flow cytometry. This concept has recently been demonstrated,[358] but it remains to be fully developed.

The above strategy represents a simple combinatorial approach to solid-phase synthesis. However, more elaborate combinatorial strategies can also be employed. The most powerful of these is the approach developed by Stephen Fodor and others at Affymetrix to prepare high-density oligonucleotide arrays on glass wafers (Genechips).[359–364] In this approach, photolithography, light-sensitive protecting groups, and a series of combinatorial masking steps are employed to produce two-dimensional arrays of oligonucleotides permanently immobilized on a flat glass surface. Very large numbers of different oligonucleotides with defined sequence and defined location can be rapidly produced because the combinatorial process allows 4^N different sequences to be prepared using only $4N$ cycles. The precision of modern photolithography also allows extremely small cell dimensions (10 μm × 10 μm) in the arrays. This technology has made it possible to mass-produce oligonucleotide arrays containing hundreds of thousands to more than a million different oligonucleotide sequences on small (1.28 cm^2) glass chips. The high cost of fabricating individual photolithographic masks makes this technology somewhat inflexible and quite expensive. However, new micro-mirror technology may make maskless photolithography possible. These micro-mirrors were developed for video projection systems and employ one electronically controlled mirror for each pixel in the image. A maskless array synthesizer using this technology has recently been used to prepare oligonucleotide arrays containing more than 76,000 16 μm^2 features.[365] The diagnostic potential for these arrays holds enormous promise for applications in DNA sequencing, gene expression studies, and health care, and a great deal of investment in and development of DNA array technology is under way.[10,11,136,366–378]

An alternative *in situ* method to the synthesis of oligonucleotide arrays on glass chips uses ink-jet printing technology. In one approach piezoelectric capillary jets delivered phosphoramidite monomers in propylene carbonate microdroplets to exact positions on a hydrophobic glass wafer.[379–381] Conventional phosphoramidite synthesis conditions, i.e., no photolithography, were used to produce arrays of ~25,000 different oligonucleotides. In another approach photolithography is used to create spatially addressable, circular features containing amino-terminated organosilanes on glass.[382] Piezoelectric ink-jet reagent delivery of conventional, dimethoxytrityl protected phosphoramidite reagents is used to synthesize the oligonucleotide features. This method, uses the differential surface tension between the modified and unmodified glass sites to define each synthesis site and has been termed a surface tension array.

The amount of synthetic oligonucleotide produced in a single cell of a high-density DNA array is very small (at an average loading of 20 pmol/cm,^{2} a 10 μm^2 cell contains only 20 attomoles of DNA) and represents the smallest end of the scale for automated oligonucleotide synthesis. However, the various applications for synthetic oligonucleotides span a very large range. On the high end of the scale are the synthetic requirements for nucleic acid medicines, which are expected to exceed 1000 kg/year in the near future. In order to produce such large quantities, very high capacity automated DNA synthesizers operating at up to a 2-mol scale are being developed.[383] Thus, chemical oligonucleotide synthesis spans a scale of 17 orders of magnitude.

However, despite the therapeutic promise of antisense oligonucleotides, development of synthetic methods and instrumentation capable of a 1-million- to 10-million-fold increase in synthesis scale has been difficult. These large-scale syntheses must be performed under the strictly controlled and validated conditions required for pharmaceutical products, and, most importantly, the final products have to be produced as cheaply as possible in order to be affordable to the patient. In particular, there are three factors that make very large scale oligonucleotide synthesis very challenging.

First of all, synthetic oligonucleotides are much larger and more complex than any other type of pharmaceutical prepared by organic synthesis. More than 80 individual steps and over 24 raw materials are required, and any single failure can result in the loss of an entire synthetic run. Second, the cost for materials and reagents is very high, and certain materials (such as the deoxyribonucleosides, which are commercially obtained from salmon sperm) have limited availability. Consequently, oligonucleotide manufacturing methods that have high yields, few and simple steps, high atom efficiency, and use materials that are readily accessible, inexpensive, and environmentally acceptable have had to be developed.[149] Finally, scale-up of reactor (i.e., column) and instrument design over several orders of magnitude is not straightforward, and significant problems must be addressed to maintain synthesis efficiency, reliability, and economy.

The earliest commercial DNA synthesizers had a maximum synthesis capacity of about 10 to 15 μmol or 0.01 to 0.015 mmol (~400 to 500 mg of support). However, by the early 1990s commercial synthesizers for larger scales were introduced. The Biosearch (later Millipore) model 8800 DNA synthesizer was the most widely used of these early instruments for production of oligonucleotides in gram quantities.[384] This instrument used a "mixed-bed" (fluidized) glass reactor that agitated the CPG support with argon gas bubbles. With this type of reactor only a fivefold excess of phosphoramidite reagent (instead of the tenfold excess commonly used for smaller synthesis scales) was necessary for satisfactory coupling efficiency. Initially, this instrument had a synthesis capacity of 0.03 to 0.4 mmol (1 to 10 g of support) using CPG with a loading of 30 to 40 μmol/g. Later, this capacity was increased to 0.6 mmol, and the amount of excess phosphoramidite was reduced to only 4 to 5 equivalents. Researchers at Hybridon, Inc. further increased the 8800's synthesis capacity by modifying it with larger reaction vessels (up to 250 ml, capable of holding 65 g of CPG) and using higher loading (~80 μmol/g) CPG.[150] With these modifications, oligonucleotides were synthesized on either 1-, 2-, or 5-mmol scales with respective yields of 2.4, 4.8, or 12 g of purified product.

The Applied Biosystems 390Z synthesizer,[385] introduced in 1991, also used a mixed-bed plastic reactor, but this was attached to a vortex mixer to keep the support agitated. This instrument had an initial synthesis capacity of only 0.025 to 0.2 mmol. Using this instrument and multiple 0.2 mmol scale syntheses, 3 g of an anti-*rev* (HIV III) phosphorothioate oligonucleotide was reported.[148] The synthesis capacity was eventually increased to 1 mmol through the use of high-loaded (150 to 190 μmol/g), polyethylene glycol-polystyrene (TentaGel) supports.[236,386]

Other proprietary synthesizers with stirred-bed reactors have also been recently developed. Lynx Therapeutics has reported an instrument for phosphorothioate oligonucleotides synthesis on 1 to 10 mmol scales using high-loaded TentaGel (150 to 170 μmol/g) supports.[151] Synthesis was performed using only 2.5 to 3.5 equiv. of phosphoramidite. Another large-scale stirred-bed reactor, for use with the CPG supports, was also developed at NexStar Pharmaceuticals (now part of Gilead Pharmaceuticals), but details are not available.

A different approach to reactor design was taken by the Large Scale Biology Corporation, which developed a production centrifugal oligonucleotide synthesizer (PCOS) with a zonal centrifuge rotor as the reactor.[387] In this design a combination of centrifugal force and density differences between solutions is used to displace one solution by another quantitatively. The solid-phase support is packed in an annular bed (68 to 400 ml). This type of design was advantageous because the CPG solid-phase support could be exposed to exactly the same conditions for the same time intervals, regardless of the synthesis scale. In collaboration with Isis Pharmaceuticals a PCOS-2 large-scale synthesizer was satisfactorily evaluated.[149] These tests showed that the solid-phase support was rugged enough for packed-bed use and that a packed-bed reactor was more efficient in both solvent and phosphoramidite consumption. However, the centrifugal reactor design was too complex (and the idea of large, industrial-scale reactors spinning at high speeds was unsettling from a safety perspective), and their development was halted. Instead, packed-bed reactor design using a flow-through, fixed-bed, low-aspect chromatography column was investigated instead.

A series of fixed-bed, flow-through DNA synthesizers was developed by Pharmacia (later Amersham Pharmacia Biotech) beginning in 1993. These synthesizers used packed-bed, stainless-steel columns to hold a proprietary high-loading (~80 μmol/g), polystyrene-based, solid-phase support known as Primer Support 30, HL. The synthesis capacity of these instruments was successively increased from 0.4 mmol (OligoPilot I, 1993), 5 mmol (OligoPilot II, 1996), 100 to 200 mmol (OligoProcess I, 1996), to 500 mmol (OligoProcess II, 1999), as both pilot- and process-scale instrumentation was developed. An even larger production-scale synthesizer, the OligoMax, which can produce over 6 kg of pure oligonucleotide per day on a 2000-mmol (2 mol) synthesis scale, has also been designed but not yet produced.[383] Significantly, this series of synthesizers was designed so that method development could be linearly scaled up between instruments. Proprietary DNA synthesizers, such as the Hybridon 601, with fixed-bed reactors have also been developed by Hybridon for large-scale synthesis of oligonucleotides on a contract basis. The synthesizers have a capacity of up to ~100 mmol and use either CPG or proprietary high-loading solid-phase supports.

Between 1990 and 1995, Isis Pharmaceuticals (among others) was able to obtain a 20,000-fold scale-up using OligoPilot and OligoProcess instrumentation. Furthermore, the excess phosphoramidite required was reduced to only 1.5 equiv. without any decrease in coupling efficiency. This has allowed Isis Pharmaceuticals to produce kilogram quantities of phosphorothioate oligonucleotides for its various clinical trials. A large-scale contract oligonucleotide manufacturing facility was established in 1999 using OligoProcess II instrumentation in a new £3.5 million oligonucleotide manufacturing suite at Avecia (formerly Zeneca Specialties, Grangemouth, Scotland). Operating on a 500-mmol scale, this suite can produce 1.5 kg of oligonucleotide every 10 h for annual capacity of 750 kg. This capacity is expected to exceed 3000 kg per year when the 2-mol scale OligoMax is available.

The success of the above fixed-bed reactor synthesizers has shown that even complicated, high-molecular-weight oligonucleotide compounds can be satisfactorily manufactured on pharmaceutical scales. However, the cost of producing these materials is still quite high. Prior to 1990, synthesis of gram quantities of oligonucleotides was unthinkable because of the extremely high cost (>$100,000/g). However, with the introduction of the Applied Biosystems 390Z and Millipore 8800 DNA synthesizers it became possible to prepare the first gram-scale quantities. However, the cost was still very high (~$10,000 to 40,000/g). Since then, refinements in virtually every aspect of large-scale synthesis have led to substantial cost reductions. For example, in the years 1993, 1994, and 1995 the cost of raw materials required to synthesize 1 g of oligonucleotide decreased to only $2000, $1600, and $500, respectively. Actual costs for present synthesis are not available, but when synthesis reaches tons per year, an ultimate raw materials cost of ~$50/g or less is hoped for.

A major obstacle to reducing the cost of synthesis is the relatively high cost of the insoluble support. This is the most expensive single material required, and in 1998 it accounted for ~40% of the total raw materials cost. The most commonly used solid-phase support (Pharmacia HL-30) costs around $25,000/kg and can be used only once before being discarded. On a 500-mmol scale, the cost of support is more than $150,000 per run. Since manufacturing operations will perform two runs per day, the annual cost for support will be very high. Therefore, there is great incentive to substantially reduce the cost of this material.

One way to do this would be to abandon solid-phase synthesis altogether and return to solution-phase methods for industrial production. Recently, impressive improvements in coupling efficiency, sulfurization efficiency, and workup conditions using an *H*-phosphonate synthetic strategy have been achieved.[115] This new solution-phase approach, combined with a block-coupling strategy, has been successfully used on industrial scales to produce short oligonucleotides (≤10 bases long). However, yields drop off rapidly as chain length increases, and so far this method has not been competitive with solid-phase synthesis for the longer (~20-base-long) oligonucleotides required for antisense therapeutics.

An alternative method to reduce the relative cost of solid-phase supports is development of supports that can be used more than once. This objective has been a goal of the author's laboratory for the last several years. During this time a new linker arm, the *Q-Linker*, was developed that can be rapidly removed under mild conditions.[268] In addition, very fast methods for attaching the first nucleoside to a solid-phase support were developed that could be performed automatically by a DNA synthesizer.[317,338] When combined with novel hydroxyl-derivatized solid-phase supports (instead of the amino-derivatized supports commonly used) these developments allow a strategy for multiple oligonucleotide syntheses on reusable solid-phase supports[302–304] (Figure 1.18). The

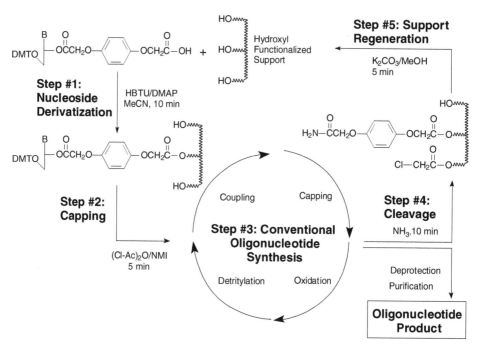

FIGURE 1.18 Process for multiple oligonucleotide syntheses on reusable hydroxyl-derivatized solid-phase supports.

key to this method is the formation of a more easily cleavable ester linkage between the first nucleoside and the surface of the solid-phase support. The milder cleavage conditions required to cleave the ester linkage and the *Q-Linker* reduce both hydrolytic damage to the support and the time required to recycle the synthesis column. The support regeneration and rederivatization steps require less than 1 h to perform, and the support never needs to be removed from the synthesis column. Multiple syntheses on more than a dozen different hydroxyl-derivatized solid-phase supports have been performed during small-scale (1 µmol) testing. The best supports have been satisfactorily recycled up to 12 times. The process is also compatible with tandem oligonucleotide synthesis[339,340] so that multiple copies of the same sequence can be made to further increase the yield from a synthesis column. Scale-up of this work to larger pilot and process scales on OligoPilot and OligoProcess instrumentation in collaboration with Isis Pharmaceuticals is also in progress. Hopefully, reusable supports will allow significant materials and cost reductions so future oligonucleotide pharmaceuticals will be more affordable.

1.7 FUTURE TRENDS IN OLIGONUCLEOTIDE SYNTHESIS

The remarkable progress that has been made in our understanding of nucleic acid chemistry and biology over the last two or three decades has substantially changed our perspective on the world. The work described in this chapter represents only a small part of the achievements of numerous researchers in many different fields. Therefore, it is difficult to foresee future developments. However, it is certain that synthetic oligonucleotides and their analogues will continue to be widely used.

Also, the recent trend toward instrument miniaturization, which has resulted in high density DNA chips and various attempts at "lab-on-a-chip" designs, will continue to be very important. This will result in new and very much smaller diagnostic tests and instrumentation. Consequently, much smaller oligonucleotide quantities will be required, and substantial miniaturization of oligonucleotide synthesis instrumentation may occur. This might be in the form of a "single-chip" synthesizer. Alternatively, synthesis scales may be reduced to a single particle of an insoluble support (~10 ng).[358] Peptide[338] and oligonucleotide applications on single beads have already been reported.[389,390] The fascination with miniaturization and nanotechnology will also lead to the use of oligonucleotides in both self-assembling and nanoscale devices.

Many new types of oligonucleotide analogues will also be developed. The variety of these will be limited only by the chemist's imagination. As our ability to select desired activities *in vitro* from combinatorial libraries or to design molecules with specific activities *de novo* improves, many new hybrid molecules with either enzymatic or therapeutic activity will be produced. Many RNA (ribozyme) and DNA molecules with catalytic activity are already known. Eventually, useful catalysts and therapeutics containing the desirable properties of both nucleic acids and proteins will be produced, in much the same way that peptide nucleic acids (PNA) analogues have found use as novel diagnostic tools.[271,391,392] These molecules should have greatly increased resistance to both enzymatic and thermal degradation and may find use in both medical and industrial applications.

Of course, interest in unmodified synthetic DNA will also continue. The completion of the various genome sequencing projects, especially the Human Genome Project, will provide us with an unprecedented explosion of genetic information. This will trigger an enormous desire to analyze and manipulate the genetic content of various organisms, including ourselves (see the opening quotation from Ghobind Khorana). As long as our synthetic ability falls far below that of the simplest biological organisms, synthetic chemists and molecular biologists will continue to push the limits of artificial DNA. Eventually, improvements in our understanding of synthesis, folding, molecular self-assembly, and biology should enable us to manipulate or modify much larger and more complex cellular structures such as ribosomes, whole genes, or even entire chromosomes. There is already a great deal of interest in gene therapy, and improved methods for this type of intervention have the potential for massive improvements in our lifestyles. However, many social and ethical issues outside the domain of the laboratory will also need to be resolved, and progress in these powerful new technologies will ultimately be limited more by social acceptance than by science.

REFERENCES

1. Khorana, H. G., Nucleic acid synthesis, *Pure Appl. Chem.,* 17, 349, 1968.
2. Templeton, N. S., The polymerase chain reaction: history, methods, and applications, *Diagn. Mol. Pathol.,* 1, 58, 1992.
3. Mullis, K. B., The polymerase chain reaction (Nobel Lecture), *Angew. Chem. Int. Ed. Engl.,* 33, 1209, 1994.
4. Mullis, K. B., The unusual origin of the polymerase chain reaction, *Sci. Am.,* 56, 1990.
5. Gura, T., Repairing the genome's spelling mistakes, *Science,* 285, 316, 1999.
6. Igoucheva, O. and Yoon, K., Targeted single-base correction by RNA-DNA oligonucleotides, *Hum. Gene Ther.,* 11, 2307, 2000.
7. Igoucheva, O., Alexeev, V., and Yoon, K., Targeted gene correction by small single-stranded oligonucleotides in mammalian cells, *Gene Ther.,* 8, 391, 2001.
8. Burke, D. T., Burns, M. A., and Mastrangelo, C., Microfabrication technologies for integrated nucleic acid analysis, *Genome Res.,* 7, 189, 1997.
9. Konings, D. A. et al., Strategies for rapid deconvolution of combinatorial libraries: comparative evaluation using a model system, *J. Med. Chem.,* 40, 4386, 1997.
10. Cho, R. J. et al., Parallel analysis of genetic selections using whole genome oligonucleotide arrays, *Proc. Natl. Acad. Sci. U.S.A.,* 95, 3752, 1998.
11. Case-Green, S. C. et al., Analysing genetic information with DNA arrays, *Curr. Opin. Chem. Biol.,* 2, 404, 1998.

12. Cheng, J. et al., Degenerate oligonucleotide primed polymerase chain reaction and capillary electrophoretic analysis of human DNA on microchip-based devices, *Anal. Biochem.,* 257, 101, 1998.
13. Waters, L. C. et al., Multiple sample PCR amplification and electrophoretic analysis on a microchip, *Anal. Chem.,* 70, 5172, 1998.
14. Seeman, N. C., DNA nanotechnology: novel DNA constructions, *Annu. Rev. Biophys. Biomol. Struc.,* 27, 225, 1998.
15. Liu, Q. et al., DNA computing on surfaces: encoding information at the single base level, *J. Comput. Biol.,* 5, 269, 1998.
16. Frutos, A. G., Smith, L. M., and Corn, R. M., Enzymatic ligation reactions of DNA "words" on surfaces for DNA computing, *J. Am. Chem. Soc.,* 120, 10277, 1998.
17. Crooke, S. T., Progress in antisense therapeutics, *Med. Res. Rev.,* 16, 319, 1996.
18. Agrawal, S., Antisense oligonucleotides: towards clinical trials, *Trends Biotech.,* 14, 376, 1996.
19. Akhtar, S. and Agrawal, S., *In vivo* studies with antisense oligonucleotides, *Trends Pharm. Sci.,* 18, 12, 1997.
20. Gewirtz, A. M., Sokol, D. L., and Ratajczak, M. Z., Nucleic acid therapeutics: state of the art and future prospects, *Blood,* 92, 712, 1998.
21. Agrawal, S. and Zhao, Q., Antisense therapeutics, *Curr. Opin. Chem. Biol.,* 2, 519, 1998.
22. Hogrefe, R. I., An antisense oligonucleotide primer, *Antisense Nucleic Acid Drug Dev.,* 9, 351, 1999.
23. Krieg, A. M., Immune stimulation by oligonucleotides, in *Antisense Research and Application,* Crooke, S. T., Ed., Springer-Verlag, Berlin, 1998, 243.
24. Krieg, A. M., Immune stimulation by CpG DNA, *Antisense Nucleic Acid Drug Dev.,* 9, 429, 1999.
25. Weeratna, R. D. et al., Optimization strategies for DNA vaccines, *Intervirology,* 43, 218, 2000.
26. Zhdanov, R. I. and Zhenodarova, S. M., Chemical methods of oligonucleotide synthesis, *Synthesis,* 222, 1975.
27. Amarnath, A. and Broom, A. D., Chemical synthesis of oligonucleotides, *Chem. Rev.,* 77, 183, 1977.
28. Wu, R., Bahl, C. P., and Narang, S. A., Synthetic oligodeoxynucleotides for analyses of DNA structure and function, *Prog. Nucleic Acid Res. Mol. Biol.,* 21, 101, 1978.
29. Reese, C. B., The chemical synthesis of oligo- and poly-nucleotides by the phosphotriester approach, *Tetrahedron,* 34, 3143, 1978.
30. Itakura, K., Synthesis of genes, *Trends Biochem. Sci.,* 114, 1980.
31. Gait, M. J., Polymer-supported synthesis of oligonucleotides, in *Polymer-Supported Reactions in Organic Synthesis,* Hodge, P. and Sherrington, D. C., Eds., John Wiley & Sons, New York, 1980, 435.
32. Ohtsuka, E., Ikehara, M., and Söll, D., Recent developments in the chemical synthesis of polynucleotides, *Nucleic Acids Res.,* 10, 6553, 1982.
33. Davies, J. E. and Gassen, H. G., Synthetic gene fragments in genetic engineering — the renaissance of chemistry in molecular biology, *Angew. Chem. Int. Ed. Engl.,* 22, 13, 1983.
34. Narang, S. A., DNA synthesis, *Tetrahedron,* 39, 3, 1983.
35. Itakura, K., Rossi, J. J., and Wallace, R. B., Synthesis and use of synthetic oligonucleotides, *Annu. Rev. Biochem.,* 53, 323, 1984.
36. Caruthers, M. H., Gene synthesis machines: DNA chemistry and its uses, *Science,* 230, 281, 1985.
37. Goodchild, J., Conjugates of oligonucleotides and modified oligonucleotides: a review of their synthesis and properties, *Bioconjugate Chem.,* 1, 165, 1990.
38. Beaucage, S. L. and Iyer, R. P., Advances in the synthesis of oligonucleotides by the phosphoramidite approach, *Tetrahedron,* 48, 2223, 1992.
39. Beaucage, S. L. and Iyer, R. P., The functionalization of oligonucleotides via phosphoramidite derivatives, *Tetrahedron,* 49, 1925, 1993.
40. Beaucage, S. L. and Iyer, R. P., The synthesis of modified oligonucleotides by the phosphoramidite approach and their applications, *Tetrahedron,* 49, 6123, 1993.
41. Verma, S. and Eckstein, F., Modified oligonucleotides: synthesis and strategy for users, *Annu. Rev. Biochem.,* 67, 99, 1998.
42. Iyer, R. P. et al., Modified oligonucleotides — synthesis, properties and applications, *Curr. Opin. Mol. Ther.,* 1, 344, 1999.
43. Caruthers, M. H., Chemical synthesis of DNA and DNA analogues, *Acct. Chem. Res.,* 24, 278, 1991.
44. Portugal, F. H. and Cohen, J. S., *A Century of DNA,* MIT Press, Cambridge, MA, 1977.

45. Levene, P. A. and Tipson, R. S., The partial synthesis of ribose nucleotides I. Uridine 5-phosphoric acid, *J. Biol. Chem.,* 106, 113, 1934.
46. Michelson, A. M. and Todd, A. R., Nucleotides. Part III. Mononucleotides derived from adenosine, guanosine, cytidine, and uridine, *J. Chem. Soc.,* 2476, 1949.
47. Michelson, A. M. and Todd, A. R., Nucleotides. Part XXXII. Synthesis of a dithymidine dinucleotide containing a 3′:5′-internucleotide linkage, *J. Chem. Soc.,* 2632, 1955.
48. Watson, J. D. and Crick, F. H. C., A structure of deoxyribose nucleic acid, *Nature,* 171, 737, 1953.
49. Khorana, H. G. and Todd, A. R., Studies on phosphorylation. Part XI. The reaction between carbodiimides and acid esters of phosphoric acid. A new method for the preparation of pyrophosphates, *J. Chem. Soc.,* 2257, 1953.
50. Gilham, P. T. and Khorana, H. G., Studies on polynucleotides. I. A new and general method for the chemical synthesis of the C5′-C3′ internucleotidic linkage. Synthesis of deoxyribo-dinucleotides, *J. Am. Chem. Soc.,* 80, 6212, 1958.
51. Khorana, H. G. et al., Polynucleotide synthesis and the genetic code, *Cold Spring Harbor Symp. Quant. Biol.,* 31, 39, 1966.
52. Lohrmann, R. and Khorana, H. G., Studies on polynucleotides. LII. The use of 2,4,6-triisopropylbenzenesulfonyl chloride for the synthesis of internucleotide bonds, *J. Am. Chem. Soc.,* 88, 829, 1966.
53. Lohrmann, R. et al., Studies on polynucleotides. LI. Syntheses of the 64 possible ribotrinucleotides derived from the four major ribomononucleotides, *J. Am. Chem. Soc.,* 88, 819, 1966.
54. Agarwal, K. L. et al., Total synthesis of the gene for an alanine transfer ribonucleic acid from yeast, *Nature,* 227, 27, 1970.
55. Khorana, H. G., Total synthesis of a gene, *Science,* 203, 614, 1979.
56. Khorana, H. G. et al., Total synthesis of the structural gene for the precursor of a tyrosine suppressor transfer RNA from *Escherichia coli, J. Biol. Chem.,* 251, 565, 1976.
57. Kleppe, K. et al., Studies on polynucleotides XCVI. Repair replication of short synthetic DNAs as catalyzed by DNA polymerases, *J. Mol. Biol.,* 56, 341, 1971.
58. Mullis, K. et al., Specific enzymatic amplification of DNA *in vitro*: the polymerase chain reaction, *Cold Spring Harbor Symp. Quant. Biol.,* 51, 263, 1986.
59. Letsinger, R. L. and Ogilvie, K. K., A convenient method for stepwise synthesis of oligothymidylate derivatives in large-scale quantities, *J. Am. Chem. Soc.,* 89, 4801, 1967.
60. Letsinger, R. L. and Ogilvie, K. K., Synthesis of oligothymidylates via phosphotriester intermediates, *J. Am. Chem. Soc.,* 91, 3350, 1969.
61. Eckstein, F. and Rizk, I., Synthesis of oligonucleotides by use of phosphoric triesters, *Angew. Chem. Int. Ed.,* 6, 695, 1967.
62. Letsinger, R. L., Ogilvie, K. K., and Miller, P. S., Developments in syntheses of oligodeoxyribonucleotides and their organic derivatives, *J. Am. Chem. Soc.,* 91, 3360, 1969.
63. Itakura, K. et al., A modified triester method for the synthesis of deoxyribopolynucleotides, *Can. J. Chem.,* 51, 3649, 1973.
64. Catlin, J. C. and Cramer, F., Deoxyoligonucleotide synthesis *via* the triester method, *J. Org. Chem.,* 38, 245, 1973.
65. Katagiri, N., Itakura, K., and Narang, S. A., Novel condensing reagents for polynucleotide synthesis, *J. Chem. Soc. Chem. Commun.,* 325, 1974.
66. Stawinski, J., Hozumi, T., and Narang, S. A., Arylsulphonyltetrazoles as highly efficient condensing reagents for polynucleotide synthesis, *Can. J. Chem.,* 54, 670, 1976.
67. Marians, K. J. et al., Cloned synthetic *lac* operator DNA is biologically active, *Nature,* 263, 744, 1976.
68. Bahl, C. P. et al., Chemical and enzymatic synthesis of lactose operator of *Escherichia coli* and is binding to lactose operator, *Proc. Natl. Acad. Sci. U.S.A.,* 73, 91, 1976.
69. Houghton, M. et al., The amino-terminal sequence of human fibroblast interferon as deduced from reverse transcripts obtained using synthetic oligonucleotide primers, *Nucleic Acids Res.,* 8, 1913, 1980.
70. Szostak, J. W. et al., Specific binding of a synthetic oligodeoxyribonucleotide to yeast cytochrome *c* mRNA, *Nature,* 265, 61, 1977.
71. Heyneker, H. L. et al., Synthetic *lac* operator DNA is functional *in vivo, Nature,* 263, 748, 1976.
72. Blanks, R. and Mclaughlin, L. W., Oligodeoxynucleotides for affinity chromatography, in *Oligonucleotides and Analogues. A Practical Approach,* Eckstein, F., Ed., Oxford University Press, New York, 1991, 241.

73. Itakura, K. et al., Expression in *Escherichia coli* of a chemically synthesized gene for the hormone somatostatin, *Science,* 198, 1056, 1977.

74. Crea, R. et al., Chemical synthesis of genes for human insulin, *Proc. Natl. Acad. Sci. U.S.A.,* 75, 5765, 1978.

75. Goeddel, D. V. et al., Direct expression in *Escherichia coli* of a DNA sequence coding for human growth hormone, *Nature,* 281, 544, 1979.

76. Merrifield, R. B., Automated synthesis of peptides, *Science,* 150, 178, 1965.

77. Letsinger, R. L. and Mahadevan, V., Oligonucleotide synthesis on a polymer support, *J. Am. Chem. Soc.,* 87, 3526, 1965.

78. Hayatsu, H. and Khorana, H. G., Studies on polynucleotides. LXXII. Deoxyribooligonucleotide synthesis on a polymer support, *J. Am. Chem. Soc.,* 89, 3880, 1967.

79. Melby, L. R. and Strobach, D. R., Oligonucleotide syntheses on insoluble polymer supports. I. Stepwise synthesis of trithymidine diphosphate, *J. Am. Chem. Soc.,* 89, 450, 1967.

80. Cramer, F. and Koster, H., Synthesis of oligonucleotides on a polymeric carrier, *Angew. Chem. Int. Ed.,* 7, 473, 1968.

81. Gait, M. J. and Sheppard, R. C., Rapid synthesis of oligodeoxyribonucleotides: a new solid-phase method, *Nucleic Acids Res.,* 4, 1135, 1977.

82. Gait, M. J. and Sheppard, R. C., Rapid synthesis of oligodeoxyribonucleotides. II. Machine-aided solid-phase syntheses of two nonanucleotides and an octanucleotide, *Nucl. Acids Res.,* 4, 4391, 1977.

83. Broka, C. et al., Simplifications in the synthesis of short oligonucleotide blocks, *Nucl. Acids Res.,* 8, 5461, 1980.

84. Gait, M. J. et al., Rapid synthesis of oligodeoxyribonucleotides. V. Further studies in solid phase synthesis of oligodeoxyribonucleotides through phosphotriester intermediates, *Nucl. Acids Res. Symp. Ser.,* 7, 243, 1980.

85. Miyoshi, K. et al., Solid-phase synthesis of polynucleotides. II. Synthesis of polythymidylic acids by block coupling phosphotriester method, *Nucleic Acids Res.,* 8, 5473, 1980.

86. Miyoshi, K., Huang, T., and Itakura, K., Solid-phase synthesis of polynucleotides. III. Synthesis of polynucleotides with defined sequences by the block coupling phosphotriester method, *Nucleic Acids Res.,* 8, 5491, 1980.

87. Dembek, P., Miyoshi, K., and Itakura, K., Solid-phase synthesis of hentriacontanucleotide, *J. Am. Chem. Soc.,* 103, 706, 1981.

88. Gait, M.J., Singh, M., and Sheppard, R. C., Rapid synthesis of oligodeoxyribonucleotides. IV. Improved solid phase synthesis of oligodeoxyribonucleotides through phosphotriester intermediates, *Nucleic Acids Res.,* 8, 1081, 1980.

89. Markham, A. F. et al., Solid phase phosphotriester synthesis of large oligodeoxyribonucleotides on a polyamide support, *Nucleic Acids Res.,* 8, 5193, 1980.

90. Duckworth, M. L. et al., Rapid synthesis of oligodeoxyribonucleotides. VI. Efficient, mechanized synthesis of heptadecadeoxyribonucleotides by an improved solid phase phosphotriester route, *Nucleic Acids Res.,* 9, 1691, 1981.

91. Miyoshi, K. et al., Solid-phase synthesis of polynucleotides. IV. Usage of polystyrene resins for the synthesis of polydeoxyribonucleotides by the phosphotriester method, *Nucleic Acids Res.,* 8, 5507, 1980.

92. Edge, M. D. et al., Total synthesis of a human leukocyte interferon gene, *Nature,* 292, 756, 1981.

93. Rossi, J. J. et al., An alternate method for the synthesis of double-stranded DNA segments, *J. Biol. Chem.,* 257, 9226, 1982.

94. Letsinger, R. L. et al., Phosphite coupling procedure for generating internucleotide links, *J. Am. Chem. Soc.,* 97, 3278, 1975.

95. Letsinger, R. L. and Lunsford, W. B., Synthesis of thymidine oligonucleotides by phosphite triester intermediates, *J. Am. Chem. Soc.,* 98, 3655, 1976.

96. Koster, H., Polymer support oligonucleotide synthesis. VI. Use of inorganic carriers, *Tetrahedron Lett.,* 1527, 1999.

97. Matteucci, M. D. and Caruthers, M. H., The synthesis of oligodeoxypyrimidines on a polymer support, *Tetrahedron Lett.,* 21, 719, 1980.

98. Matteucci, M. D. and Caruthers, M. H., Synthesis of deoxyoligonucleotides on a polymer support, *J. Am. Chem. Soc.,* 103, 3185, 1981.

99. Ogilvie, K. K. and Nemer, M. J., Silica gel as solid support in the synthesis of oligoribonucleotides, *Tetrahedron Lett.,* 21, 4159, 1980.

100. Alvarado-Urbina, G. et al., Automated synthesis of gene fragments, *Science,* 214, 270, 1981.

101. Jay, E. et al., Chemical synthesis of a biologically active gene for human immune interferon-γ, *J. Biol. Chem.,* 259, 6311, 1984.

102. Beaucage, S. L. and Caruthers, M. H., Deoxynucleoside phosphoramidites — a new class of key intermediates for deoxypolynucleotide synthesis, *Tetrahedron Lett.,* 22, 1859, 1981.

103. Caruthers, M. H. et al., Chemical synthesis of deoxyoligonucleotides by the phosphoramidite method, in *Recombinant DNA Part E,* Wu, R. and Grossman, L. Eds., Academic Press, San Diego, 1987, 287.

104. Caruthers, M. H. et al., New methods for synthesizing deoxyoligonucleotides, in *Genetic Engineering Principles and Methods,* Setlow, J. and Hollaender, A., Eds., Plenum Press, New York, 1982, 1.

105. McBride, L. J. and Caruthers, M. H., An investigation of several deoxynucleoside phosphoramidites useful for synthesizing deoxyoligonucleotides, *Tetrahedron Lett.,* 24, 245, 1983.

106. Adams, S. P. et al., Hindered dialkylamino nucleoside phosphite reagents in the synthesis of two DNA 51-mers, *J. Am. Chem. Soc.,* 105, 661, 1983.

107. Hunkapiller, M. et al., A microchemical facility for the analysis and synthesis of genes and proteins, *Nature,* 310, 105, 1984.

108. Horvath, S. J. et al., An automated DNA synthesizer employing deoxynucleoside 3'-phosphoramidites, in *Recombinant DNA Part E,* Wu, R. and Grossman, L., Eds., Academic Press, San Diego, 1987, 314.

109. Zon, G. et al., Analytical studies of "mixed sequence" oligodeoxyribonucleotides synthesized by competitive coupling of either methyl- or β-cyanoethyl-*N,N'*-diisopropylamino phosphoramidite reagents, including 2'-deoxyinosine, *Nucleic Acids Res.,* 13, 8181, 1985.

110. Hall, A. R., Todd, A., and Webb, R. F., Nucleotides. Part XLI. Mixed anhydrides as intermediates in the synthesis of dinucleoside phosphates, *J. Chem. Soc.,* 3291, 1957.

111. Garegg, P. J. et al., Formation of internucleotidic bonds via phosphonate intermediates, *Chem. Scr.,* 25, 280, 1985.

112. Froehler, B. C. and Matteucci, M. D., Nucleoside H-phosphonates: valuable intermediates in the synthesis of deoxyoligonucleotides, *Tetrahedron Lett.,* 27, 469, 1986.

113. Froehler, B. C., Ng, P. G., and Matteucci, M. D., Synthesis of DNA via deoxynucleoside H-phosphonate intermediates, *Nucleic Acids Res.,* 14, 5399, 1986.

114. Froehler, B. C., Oligodeoxynucleotide synthesis, in *Methods in Molecular Biology, Vol. 20: Protocols for Oligonucleotides and Analogs,* Agrawal, S., Ed., Humana Press, Totowa, NJ, 1993, 63.

115. Reese, C. B. and Song, Q. L., A new approach to the synthesis of oligonucleotides and their phosphorothioate analogues in solution, *Bioorg. Med. Chem. Lett.,* 7, 2787, 1997.

116. Engels, J. W. and Uhlmann, E., Gene synthesis, *Angew. Chem. Int. Ed. Engl.,* 28, 716, 1989.

117. Edwards, M., Total gene synthesis, *Am. Biotech. Lab.,* 5, 38, 1987.

118. Bell, L. D. et al., Chemical synthesis, cloning and expression in mammalian cells of a gene coding for human tissue-type plasminogen activator, *Gene,* 63, 155, 1988.

119. Michaels, M. L., Hsiao, H. M., and Miller, J. H., Using PCR to extend the limit of oligonucleotide synthesis, *Biotechniques,* 12, 44, 1992.

120. Ivanov, I., Christov, C., and Alexciev, K., Two methods for rapid assembly and oligomerization of synthetic genes — construction of human calcitonin-encoding sequences, *Gene,* 95, 295, 1990.

121. Dillon, P. J. and Rosen, C. A., A rapid method for the construction of synthetic genes using the polymerase chain reaction, *Biotechniques,* 9, 299, 1990.

122. Horton, R. M. et al., Gene splicing by overlap extension: tailor-made genes using the polymerase chain reaction, *Biotechniques,* 8, 528, 1990.

123. Ciccarelli, R. B. et al., Construction of synthetic genes using PCR after automated DNA synthesis of their entire top and bottom strands, *Nucleic Acids Res.,* 19, 6007, 1991.

124. Hutchison, C. A. et al., Mutagenesis at a specific position in a DNA sequence, *J. Biol. Chem.,* 253, 6551, 1978.

125. Razin, A. et al., Efficient correction of a mutation by use of chemically synthesized DNA, *Proc. Natl. Acad. Sci. U.S.A.,* 75, 4268, 1978.

126. Gillam, S. and Smith, M., Site-specific mutagenesis using synthetic oligodeoxyribonucleotide primers. I. Optimum conditions and minimum oligodeoxyribonucleotide length, *Gene,* 8, 81, 1979.

127. Bhanot, O. S., Khan, S. A., and Chambers, R. W., A new system for studying molecular mechanisms of mutation by carcinogens, *J. Biol. Chem.*, 254, 12684, 1979.

128. Smith, M., *In vitro* mutagenesis, *Annu. Rev. Genet.*, 19, 423, 1985.

129. Loewen, P. C. and Khorana, H. G., Studies on polynucleotides. CXXII. The dodecanucleotide sequence adjoining the C-C-A end of the tyrosine transfer ribonucleic acid gene, *J. Biol. Chem.*, 248, 3489, 1973.

130. Schott, H., Fischer, D., and Kossel, H., Synthesis of four undecanucleotides complementary to a region of the coat protein cistron of phage fd, *Biochemistry*, 12, 3447, 1973.

131. Sanger, F. et al., Use of DNA polymerase I primed by a synthetic oligonucleotide to determine a nucleotide sequence in phage f1 DNA, *Proc. Natl. Acad. Sci. U.S.A.*, 70, 1209, 1973.

132. Wu, R., Tu, C. P. D., and Padmanabhan, R., Nucleotide sequence analysis of DNA XII. The chemical synthesis and sequence analysis of a dodecadeoxynucleotide which binds to the endolysin gene of bacteriophage lambda, *Biochem. Biophys. Res. Commun.*, 55, 1092, 1973.

133. Maxam, A. M. and Gilbert, W., A new method for sequencing DNA, *Proc. Natl. Acad. Sci. U.S.A.*, 74, 560, 1977.

134. Englisch, U. and Gauss, D. H., Chemically modified oligonucleotides as probes and inhibitors, *Angew. Chem. Int. Ed. Engl.*, 30, 613, 1991.

135. Heid, C. A., Livak, S. J., and Williams, P. M., Real-time quantitative PCR, *Genome Res.*, 6, 986, 1996.

136. Hoheisel, J. D., Oligomer-chip technology, *Trends Biotech.*, 15, 465, 1997.

137. McKenzie, S. E. et al., Parallel molecular genetic analysis, *Eur. J. Hum. Genet.*, 6, 417, 1998.

138. Graves, D. J., Powerful tools for genetic analysis come of age, *Trends Biotech.*, 17, 127, 1999.

139. Gerhold, D., Rushmore, T., and Caskey, C. T., DNA chips: promising toys have become powerful tools, *Trends Biochem. Sci.*, 24, 168, 1999.

140. Uhlmann, E. and Peyman, A., Antisense oligonucleotides, *Chem. Rev.*, 90, 544, 1990.

141. De mesmaeker, A. et al., Antisense oligonucleotides, *Acc. Chem. Res.*, 28, 366, 1995.

142. Luyten, I. and Herdewijn, P., Hybridization properties of base-modified oligonucleotides within the double and triple helix motif, *Eur. J. Med. Chem.*, 33, 515, 1998.

143. Krieg, A. M., From bugs to drugs: therapeutic immunomodulation with oligodeoxynucleotides containing CpG sequences from bacterial DNA, *Antisense Nucleic Acid Drug Dev.*, 11, 181, 2001.

144. Agrawal, S. and Kandimalla, E. R., Antisense and/or immunostimulatory oligonucleotide therapeutics, *Curr. Cancer Drug Tar.*, 1, 197, 2001.

145. Kandimalla, E. R. and Agrawal, S., Therapeutic potential of synthetic CpG DNA — current status and future directions, *IDrugs*, 4, 963, 2001.

146. Hartmann, G. et al., Delineation of a CpG phosphorothioate oligodeoxynucleotide for activating primate immune responses *in vitro* and *in vivo*, *J. Immunol.*, 164, 1617, 2000.

147. Liu, M. A. and Ulmer, J. B., Gene-based vaccines, *Mol. Ther.*, 1, 497, 2000.

148. Geiser, T., Large scale economic synthesis of antisense phosphorothioate analogues of DNA for preclinical investigations, *Ann. N.Y. Acad. Sci.*, 616, 173, 1990.

149. Sanghvi, Y. S. et al., Chemical synthesis and purification of phosphorothioate antisense oligonucleotides, in *Manual of Antisense Methodology*, Hartmann, G. and Endres, S., Eds., Kluwer Academic Publishers, Boston, 1999, 3.

150. Padmapriya, A. A., Tang, J., and Agrawal, S., Large scale synthesis, purification and analysis of oligodeoxynucleotide phosphorothioates, *Antisense Res. Dev.*, 4, 185, 1994.

151. Fearon, K. L. et al., Phosphorothioate oligodeoxynucleotides: large-scale synthesis and analysis, impurity characterization, and the effects of phosphorus stereochemistry, *CIBA Found. Symp.*, 209, 19, 1997.

152. Watanabe, K. A. and Fox, J. J., A simple method for selective acylation of cytidine on the 4-amino group, *Angew. Chem. Int. Ed.*, 5, 579, 1966.

153. Igolen, J. and Morin, C., Rapid syntheses of protected 2'-deoxycytidine derivatives, *J. Org. Chem.*, 45, 4802, 1980.

154. Bhat, V. et al., A simple and convenient method for the selective N-acylations of cytosine nucleosides, *Nucleosides Nucleotides*, 8, 179, 1989.

155. Jones, R. A., Preparation of protected deoxyribonucleosides, in *Oligonucleotide Synthesis: A Practical Approach*, Gait, M. J., Ed., IRL Press, Oxford, 1984, 23.

156. Ti, G. S., Gaffney, B. L., and Jones, R. A., Transient protection: efficient one-flask syntheses of protected deoxynucleosides, *J. Am. Chem. Soc.*, 104, 1316, 1982.

157. Wu, T., Ogilvie, K. K., and Pon, R. T., Prevention of chain cleavage in the chemical synthesis of 2'-silylated oligoribonucleotides, *Nucleic Acids Res.,* 17, 3501, 1989.

158. Kume, A. et al., Cyclic diacyl groups for protection of the N6-amino group of deoxyadenosine in oligodeoxynucleotide synthesis, *Nucleic Acids Res.,* 12, 8525, 1984.

159. Beier, M. and Pfleiderer, W., Phthaloyl strategy: a new concept of oligonucleotide synthesis, *Helv. Chim. Acta,* 82, 633, 1999.

160. Iyer, R. P. et al., Methyl phosphotriester oligonucleotides: facile synthesis using *N*-pent-4-enoyl nucleoside phosphoramidites, *J. Org. Chem.,* 60, 8132, 1995.

161. Habus, I. et al., Improved synthesis of oligonucleoside methylphosphonate analogs, *Bioorg. Med. Chem. Lett.,* 6, 1393, 1996.

162. Reddy, M. P., Hanna, N. B., and Farooqui, F., Fast cleavage and deprotection of oligonucleotides, *Tetrahedron Lett.,* 35, 4311, 1994.

163. Reddy, M. P., Hanna, N. B., and Farooqui, F., Ultrafast cleavage and deprotection of oligonucleotides synthesis and use of C-Ac derivatives, *Nucleosides Nucleotides,* 16, 1589, 1997.

164. Reddy, M. P., Farooqui, F., and Hanna, N. B., Methylamine deprotection provides increased yield of oligoribonucleotides, *Tetrahedron Lett.,* 36, 8929, 1995.

165. Froehler, B. C. and Matteucci, M. D., Dialkylformamidines: depurination resistant N6-protecting group for deoxyadenosine, *Nucleic Acids Res.,* 11, 8031, 1983.

166. McBride, L. J. et al., Amidine protecting groups for oligonucleotide synthesis, *J. Am. Chem. Soc.,* 108, 2040, 1986.

167. Vu, H. et al., Fast oligonucleotide deprotection phosphoramidite chemistry for DNA synthesis, *Tetrahedron Lett.,* 31, 7269, 1990.

168. Gryaznov, S. M. and Letsinger, R. L., Synthesis of oligonucleotides via monomers with unprotected bases, *J. Am. Chem. Soc.,* 113, 5876, 1991.

169. Gryaznov, S. M. and Letsinger, R. L., Selective O-phosphitylation with nucleoside phosphoramidite reagents, *Nucleic Acids Res.,* 20, 1879, 1992.

170. Hayakawa, Y. and Kataoka, M., Facile synthesis of oligodeoxyribonucleotides via the phosphoramidite method without nucleoside base protection, *J. Am. Chem. Soc.,* 120, 12395, 1998.

171. Pon, R. T., Damha, M. J., and Ogilvie, K. K., Modification of guanine bases by nucleoside phosphoramidite reagents during the solid phase synthesis of oligonucleotides, *Nucleic Acids Res.,* 13, 6447, 1985.

172. Pon, R. T. et al., Prevention of guanine modification and chain cleavage during the solid phase synthesis of oligonucleotides using phosphoramidite derivatives, *Nucleic Acids Res.,* 14, 6453, 1986.

173. Smith, M. et al., Studies on polynucleotides. XIV. Specific synthesis of the C3'-C5' inter-ribonucleotide linkage. Syntheses of uridylyl-(3'-5')-uridine and uridylyl-(3'-5')adenosine, *J. Am. Chem. Soc.,* 84, 430, 1962.

174. Paul, C. H. and Royappa, A. T., Acid binding and detritylation during oligonucleotide synthesis, *Nucleic Acids Res.,* 24, 3048, 1996.

175. Septak, M., Kinetic studies on depurination and detritylation of CPG-bound intermediates during oligonucleotide synthesis, *Nucleic Acids Res.,* 24, 3053, 1996.

176. Pon, R. T., Usman, N., and Ogilvie, K. K., Derivatization of controlled pore glass beads for solid phase oligonucleotide synthesis, *Biotechniques,* 6, 768, 1988.

177. Kaufman, J. et al., Trityl monitoring of automated DNA synthesizer operation by conductivity: a new method of real-time analysis, *Biotechniques,* 14, 834, 1993.

178. Fisher, E. F. and Caruthers, M. H., Color coded triarylmethyl protecting groups useful for deoxypolynucleotide synthesis, *Nucleic Acids Res.,* 11, 1589, 1983.

179. Ravikumar, V. T., Krotz, A. H., and Cole, D. L., Efficient synthesis of deoxyribonucleotide phosphorothioates by the use of DMT cation scavenger, *Tetrahedron Lett.,* 36, 6587, 1995.

180. Habus, I. and Agrawal, S., Improvement in the synthesis of oligonucleotides of extended length by modification of detritylation step, *Nucleic Acids Res.,* 22, 4350, 1994.

181. Krotz, A. H., Cole, D. L., and Ravikumar, V. T., Synthesis of an antisense oligonucleotide targeted against C-raf kinase: efficient oligonucleotide synthesis without chlorinated solvents, *Bioorg. Med. Chem.,* 7, 435, 1999.

182. Miura, K. et al., Blockwise mechanical synthesis of oligonucleotides by the phosphoramidite method, *Chem. Pharm. Bull.,* 35, 833, 1987.

183. Kayushin, A. L. et al., A convenient approach to the synthesis of trinucleotide phosphoramidites — synthons for the generation of oligonucleotide/peptide libraries, *Nucleic Acids Res.*, 24, 3748, 1996.

184. Kumar, G. and Poonian, M. S., Improvements in oligodeoxyribonucleotide synthesis: methyl *N,N*-dialkylphosphoramidite dimer units for solid support phosphite methodology, *J. Org. Chem.*, 49, 4905, 1984.

185. Zehl, A. et al., Efficient and flexible access to fully protected trinucleotides suitable for DNA synthesis by automated phosphoramidite chemistry, *J. Chem. Soc. Chem. Commun.*, 2677, 1996.

186. Neuner, P., Cortese, R., and Monaci, P., Codon-based mutagenesis using dimer-phosphoramidites, *Nucleic Acids Res.*, 26, 1223, 1998.

187. Wolter, A., Biernat, J., and Koster, H., Polymer support oligonucleotide synthesis XX: synthesis of a henhectacosadeoxynucleotide by use of a dimeric phosphoramidite synthon, *Nucleosides Nucleotides*, 5, 65, 1986.

188. Krotz, A. H. et al., Phosphorothioate oligonucleotides: largely reduced (N − 1)-mer and phosphodiester content through the use of dimeric phosphoramidite synthons, *Bioorgan. Med. Chem. Lett.*, 7, 73, 1997.

189. Eleuteri, A. et al., Oligodeoxyribonucleotide phosphorothioates: Substantial reduction of (N − 1)-mer content through the use of trimeric phosphoramidite synthons, *Nucleosides Nucleotides*, 18, 475, 1999.

190. Virnekäs, B. et al., Trinucleotide phosphoramidites: ideal reagents for the synthesis of mixed oligo-nucleotides for random mutagenesis, *Nucleic Acids Res.*, 22, 5600, 1994.

191. Lyttle, M. H. et al., Mutagenesis using trinucleotide β-cyanoethyl phosphoramidites, *Biotechniques*, 19, 274, 1995.

192. Daub, G. W. and van Tamelen, E. E., Synthesis of oligoribonucleotides based on the facile cleavage of methyl phosphotriester intermediates, *J. Am. Chem. Soc.*, 99, 3526, 1977.

193. Gao, X. et al., Methylation of thymine residues during oligonucleotide synthesis, *Nucleic Acids Res.*, 13, 573, 1985.

194. Sinha, N. D. et al., Polymer support oligonucleotide synthesis. XVIII. Use of β-cyanoethyl-*N,N*-dialkylamino-/*N*-morpholino phosphoramidite of deoxynucleosides for the synthesis of DNA frag-ments simplifying deprotection and isolation of the final product, *Nucleic Acids Res.*, 12, 4539, 1984.

195. Mag, M. and Engels, J. W., Synthesis and structure assignments of amide protected nucleosides and their use as phosphoramidites in deoxyoligonucleotide synthesis, *Nucleic Acids Res.*, 16, 3525, 1988.

196. Weiler, J. and Pfleiderer, W., An improved method for the large scale synthesis of oligonucleotides applying the NPE/NPEOC strategy, *Nucleosides Nucleotides*, 14, 917, 1995.

197. Wang, Y., A total synthesis of yeast alanine transfer RNA, *Acc. Chem. Res.*, 17, 393, 1984.

198. Iwai, S. and Ohtsuka, E., 5′-Levulinyl and 2′-tetrahydrofuranyl protection for the synthesis of oligor-ibonucleotides by the phosphoramidite approach, *Nucleic Acids Res.*, 16, 9443, 1988.

199. Reese, C. B., Saffhill, R., and Sulston, J. E., 4-Methoxytetrahydropyran-4-yl: a symmetrical alternative to the tetrahydropyranyl protecting group, *Tetrahedron*, 26, 1023, 1970.

200. Reese, C. B., The problem of 2′-protection in rapid oligoribonucleotide synthesis, *Nucleosides Nucle-otides*, 6, 121, 1987.

201. Reese, C. B., Serafinowska, H. T., and Zappia, G., An acetal group suitable for the protection of 2′-hydroxy functions in rapid oligoribonucleotide synthesis, *Tetrahedron Lett.*, 27, 2291, 1986.

202. Reese, C. B. and Thompson, E. A., A new synthesis of 1-arylpiperidin-4-ols, *J. Chem. Soc. Perkin Trans. 1*, 2881, 1988.

203. Sakatsume, O. et al., Solid phase synthesis of oligoribonucleotides using the 1-[(2-chloro-4-methyl)phenyl]-4-methoxypiperidin-4-yl (Ctmp) group for the protection of the 2′-hydroxy functions and the H-phosphonate approach, *Nucleic Acids Res.*, 17, 3689, 1989.

204. Rao, M. V. et al., Use of the 1-(2-fluorophenyl)-4-methoxypiperidin-4-yl (Fpmp) protecting group in the solid-phase synthesis of oligoribonucleotides and polyribonucleotides, *J. Chem. Soc. Perkin Trans. 1*, 43, 1993.

205. Beijer, B. et al., Synthesis and applications of oligoribonucleotides with selected 2′-*O*-methylation using the 2′-*O*-[1-(2-fluorophenyl)-4-methoxypiperidin-4-yl] protecting group, *Nucleic Acids Res.*, 18, 5143, 1990.

206. Tanaka, T., Tamatsukuri, S., and Ikehara, M., Solid phase synthesis of oligoribonucleotides using o-nitrobenzyl protection of 2′-hydroxyl via a phosphite triester approach, *Nucleic Acids Res.*, 14, 6265, 1986.

207. Schwartz, M. E. et al., Rapid synthesis of oligoribonucleotides using 2′-*O*-(ortho-nitrobenzyloxyme-thyl)-protected monomers, *Bioorg. Med. Chem. Lett.*, 2, 1019, 1992.

208. Ogilvie, K. K. et al., The synthesis of oligoribonucleotides. II. The use of silyl protecting groups in nucleoside and nucleotide chemistry. VII, *Can. J. Chem.,* 56, 2768, 1978.

209. Ogilvie, K. K., Schifman, A. L., and Penney, C. L., The synthesis of oligoribonucleotides. III. The use of silyl protecting groups in nucleoside and nucleotide chemistry. VIII, *Can. J. Chem.,* 57, 2230, 1979.

210. Sproat, B. et al., An efficient method for the isolation and purification of oligoribonucleotides, *Nucleosides Nucleotides,* 14, 255, 1995.

211. Wincott, F. et al., Synthesis, deprotection, analysis and purification of RNA and ribozymes, *Nucleic Acids Res.,* 23, 2677, 1995.

212. Hakimelahi, G. H., Proba, Z. A., and Ogilvie, K. K., Nitrate ion as a catalyst for selective silylations of nucleosides, *Tetrahedron Lett.,* 22, 4775, 1981.

213. Song, Q. L. et al., High yield protection of purine ribonucleosides for phosphoramidite RNA synthesis, *Tetrahedron Lett.,* 40, 4153, 1999.

214. Ogilvie, K. K. and Entwistle, D. W., Isomerization of *tert*-butyldimethylsilyl protecting groups in ribonucleosides, *Carbohydrate Res.,* 89, 203, 1981.

215. Wu, T. and Ogilvie, K. K., A study on the alkylsilyl groups in oligoribonucleotide synthesis, *J. Org. Chem.,* 55, 4717, 1990.

216. Froehler, B. C. and Matteucci, M. D., Substituted 5-phenyltetrazoles: improved activators of deoxynucleoside phosphoramidites in deoxyoligonucleotide synthesis, *Tetrahedron Lett.,* 24, 3171, 1983.

217. Hering, G. et al., Preparation and properties of chloro-*N,N*-dialkylamino-2,2,2-trichloroethoxy- and chloro-*N,N*,-dialkylamino-2,2,2-trichloro-1,1-dimethylethoxyphosphines and their deoxynucleoside phosphiteamidates, *Nucleosides Nucleotides.,* 4, 169, 1985.

218. Sproat, B. S. et al., Highly efficient chemical synthesis of 2′-*O*-methyloligoribonucleotides and tetrabiotinylated derivatives — novel probes that are resistant to degradation by RNA or DNA specific nucleases, *Nucleic Acids Res.,* 17, 3373, 1989.

219. Ogilvie, K. K. et al., Total chemical synthesis of a 77-nucleotide-long RNA sequence having methionine acceptance activity, *Proc. Natl. Acad. Sci. U.S.A.,* 85, 5764, 1988.

220. Goodwin, J. T., Stanick, W. A., and Glick, G. D., Improved solid-phase synthesis of long oligoribonucleotides: application to tRNA(phe) and tRNA(gly), *J. Org. Chem.,* 59, 7941, 1994.

221. Ohtsuki, T. et al., Automated chemical synthesis of biologically active tRNA having a sequence corresponding to *Ascaris suum* mitochondrial tRNA(Met) toward NMR measurements, *J. Biochem. Tokyo,* 120, 1070, 1996.

222. Persson, T. et al., Chemical synthesis and biological investigation of a 77-mer oligoribonucleotide with a sequence corresponding to *E. coli* tRNA(Asp), *Bioorg. Med. Chem.,* 9, 51, 2001.

223. Scaringe, S. A., Wincott, F. E., and Caruthers, M. H., Novel RNA synthesis method using 5′-*O*-silyl-2′-*O*-orthoester protecting groups, *J. Am. Chem. Soc.,* 120, 11820, 1998.

224. Scaringe, S. A., RNA oligonucleotide synthesis via 5′-silyl-2′-orthoester chemistry, *Methods,* 23, 206, 2001.

225. Letsinger, R. L., Chemical synthesis of oligodeoxyribonucleotides: a simplified procedure, in *Genetic Engineering Principles and Methods,* Setlow, J. K. and Hollaender, A., Eds., Plenum Press, New York, 1983, 191.

226. Fourrey, J. L. and Varenne, J., A new and general procedure for the preparation of deoxynucleoside phosphoramidites, *Tetrahedron Lett.,* 24, 1963, 1983.

227. Fourrey, J. L. and Varenne, J., Improved procedure for the preparation of deoxynucleoside phosphoramidites: arylphosphoramidites as new convenient intermediates for oligodeoxynucleotide synthesis, *Tetrahedron Lett.,* 25, 4511, 1984.

228. Beaucage, S. L., A simple and efficient preparation of deoxynucleoside phosphoramidites *in situ*, *Tetrahedron Lett.,* 25, 375, 1984.

229. Barone, A. D., Tang, J. Y., and Caruthers, M. H., *In situ* activation of bis-dialkylaminophosphines — a new method for synthesizing deoxyoligonucleotides on polymer supports, *Nucleic Acids Res.,* 12, 4051, 1984.

230. Lee, H. J. and Moon, S. H., Bis-(*N,N*-dialkylamino)-alkoxyphosphines as a new class of phosphate coupling agent for the synthesis of oligonucleotides, *Chem. Lett.,* 1229, 1984.

231. Nielsen, J. et al., Application of 2-cyanoethyl N,N,N′,N′-tetraisopropylphosphorodiamidite for *in situ* preparation of deoxyribonucleoside phosphoramidites and their use in polymer-supported synthesis of oligodeoxyribonucleotides, *Nucleic Acids Res.,* 14, 7391, 1986.

232. Zhang, Z. and Tang, J. Y., A novel phosphitylating reagent for *in situ* generation of deoxyribonucleoside phosphoramidites, *Tetrahedron Lett.*, 37, 331, 1996.

233. Dahl, B. H., Nielsen, J., and Dahl, O., Mechanistic studies on the phosphoramidite coupling reaction in oligonucleotide synthesis. I. Evidence for nucleophilic catalysis by tetrazole and rate variations with the phosphorus substituents, *Nucleic Acids Res.*, 15, 1729, 1987.

234. Berner, S., Muhlegger, K., and Seliger, H., Studies on the role of tetrazole in the activation of phosphoramidites, *Nucleic Acids Res.*, 17, 853, 1989.

235. Krotz, A. H. et al., On the formation of longmers in phosphorothioate oligodeoxyribonucleotide synthesis, *Tetrahedron Lett.*, 38, 3875, 1997.

236. Wright, P. et al., Large scale synthesis of oligonucleotides via phosphoramidite nucleosides and a high-loaded polystyrene support, *Tetrahedron Lett.*, 34, 3373, 1993.

237. Hayakawa, Y., Kataoka, M., and Noyori, R., Benzimidazolium triflate as an efficient promoter for nucleotide synthesis via the phosphoramidite method, *J. Org. Chem.*, 61, 7996, 1996.

238. Pon, R. T., Enhanced coupling efficiency using *N,N,*-dimethylaminopyridine (DMAP) and either tetrazole, 5-(*o*-nitrophenyl)tetrazole, or 5-(*p*-nitrophenyl)tetrazole in the solid phase synthesis of oligoribonucleotides by the phosphoramidite procedure, *Tetrahedron Lett.*, 28, 3643, 1987.

239. Vargeese, C. et al., Efficient activation of nucleoside phosphoramidites with 4,5-dicyanoimidazole during oligonucleotide synthesis, *Nucleic Acids Res.*, 26, 1046, 1998.

240. Hayakawa, Y. et al., Acid/azole complexes as highly effective promoters in the synthesis of DNA and RNA oligomers via the phosphoramidite method, *J. Am. Chem. Soc.*, 123, 8165, 2001.

241. Beier, M. and Pfleiderer, W., Nucleotides. Part LXII. Pyridinium salts — An effective class of catalysts for oligonucleotide synthesis, *Helv. Chim. Acta*, 82, 879, 1999.

242. Eleuteri, A. et al., Pyridinium trifluoroacetate/*N*-methylimidazole as an efficient activator for oligo-nucleotide synthesis via the phosphoramidite method, *Org. Proc. Res. Dev.*, 4, 182, 2000.

243. Dabkowski, W. et al., Trimethylchlorosilane: a novel activating reagent in nucleotide synthesis via the phosphoramidite route, *Chem. Commun.*, 877, 1997.

244. Dabkowski, W. et al., 2,4-Dinitrophenol: a novel activating reagent in nucleotide synthesis via the phosphoramidite route. Design of new effective phosphitylating reagents, *Tetrahedron Lett.*, 41, 7535, 2000.

245. Chow, F., Kempe, T., and Palm, G., Synthesis of oligodeoxyribonucleotides on silica gel support, *Nucleic Acids Res.*, 12, 2807, 1981.

246. Eadie, J. S. and Davidson, D.S., Guanine modification during chemical DNA synthesis, *Nucleic Acids Res.*, 15, 8333, 1987.

247. Sinha, N. D. et al., Labile exocyclic amine protection of nucleosides in DNA, RNA and oligonucleotide analaog synthesis facilitating N-deacylation, minimizing depurination and chain degradation, *Bio-chimie*, 75, 13, 1993.

248. Pon, R. T., An improved iodine/water oxidation reagent for automated oligonucleotide synthesis, *Nucleic Acids Res.*, 15, 7203, 1987.

249. Ogilvie, K. K. and Nemer, M., Nonaqueous oxidation of phosphites to phosphates in nucleotide synthesis, *Tetrahedron Lett.*, 22, 2531, 1999.

250. Hayakawa, Y., Uchiyama, M., and Noyori, R., Nonaqueous oxidation of nucleoside phosphites to the phosphates, *Tetrahedron Lett.*, 27, 4191, 1986.

251. Eckstein, F. and Gish, G., Phosphorothioates in molecular biology, *Trends Biochem. Sci.*, 14, 97, 1989.

252. Burgers, P. M. J. and Eckstein, F., Synthesis of dinucleoside phosphorothioates via addition of sulfur to phosphite triesters, *Tetrahedron Lett.*, 3835, 1978.

253. Stein, C. A. et al., Physicochemical properties of phosphorothioate oligodeoxynucleotides, *Nucleic Acids Res.*, 16, 3209, 1988.

254. Iyer, R. P. et al., The automated synthesis of sulfur-containing oligodeoxyribonucleotides using 3*H*-1,2-benzodithiol-3-one-1,1-dioxide as a sulfur transfer reagent, *J. Org. Chem.*, 55, 4693, 1990.

255. Rao, M. V., Reese, C. B., and Zhao, Z. Y., Dibenzoyl tetrasulphide — a rapid sulphur transfer agent in the synthesis of phosphorothioate analogues of oligonucleotides, *Tetrahedron Lett.*, 33, 4839, 1992.

256. Vu, H. and Hirschbein, B. L., Internucleotide phosphite sulfurization with tetraethylthiuram disulfide — phosphorothioate oligonucleotide synthesis via phosphoramidite chemistry, *Tetrahedron Lett.*, 32, 3005, 1991.

257. Rao, M. V. and Macfarlane, K., Solid phase synthesis of phosphorothioate oligonucleotides using benzyltriethylammonium tetrathiomolybdate as a rapid sulfur transfer reagent, *Tetrahedron Lett.*, 35, 6741, 1994.

258. Efimov, V. A. et al., New efficient sulfurizing reagents for the preparation of oligodeoxyribonucleotide phosphorothioate analogues, *Nucleic Acids Res.*, 23, 4029, 1995.

259. Kamer, P. C. J. et al., An efficient approach toward the synthesis of phosphorothioate diesters via the Schonberg reaction, *Tetrahedron Lett.*, 30, 6757, 1989.

260. Cheruvallath, Z. S. et al., Use of phenylacetyl disulfide (PADS) in the synthesis of oligodeoxyribonucleotide phosphorothioates, *Nucleosides Nucleotides*, 18, 485, 1999.

261. Stec, W. J. et al., Bis-(*O,O*-diisopropoxyphosphinothioyl) disulfide — a highly efficient sulfurizing reagent for cost-effective synthesis of oligo(nucleoside phosphorothioates), *Tetrahedron Lett.*, 34, 5317, 1993.

262. Xu, Q. et al., Use of 1,2,4-dithiazolidine-3,5-dione (DtsNH) and 3-ethoxy-1,2,4-dithiazoline-5-one (EDITH) for synthesis of phosphorothioate-containing oligodeoxyribonucleotides, *Nucleic Acids Res.*, 24, 1602, 1996.

263. Ma, M. Y. et al., Evaluation of 3-ethoxy-1,2,4-dithiazoline-5-one (EDITH) as a new sulfurizing reagent in combination with labile exocyclic amino protecting groups for solid-phase oligonucleotide synthesis, *Nucleic Acids Res.*, 25, 3590, 1997.

264. Zhang, Z. et al., Solid-phase synthesis of oligonucleotide phosphorothioate analogues using 3-methyl-1,2,4-dithiazolin-5-one (MEDITH) as a new sulfur transfer reagent, *Tetrahedron Lett.*, 40, 2095, 1999.

265. Strobel, S. A. et al., The 2,6-diaminopurine riboside 5-methylisocytidine wobble base pair: an isoenergetic substitution for the study of G:U pairs in RNA, *Biochemistry*, 33, 13824, 1994.

266. Boal, J. H. et al., Cleavage of oligodeoxyribonucleotides from controlled-pore glass supports and their rapid deprotection by gaseous amines, *Nucleic Acids Res.*, 24, 3115, 1996.

267. Iyer, R. P. et al., The use of gaseous ammonia for the deprotection and cleavage steps during the solid-phase synthesis of oligonucleotides, and analogs, *Bioorg. Med. Chem. Lett.*, 7, 1443, 1997.

268. Pon, R. T. and Yu, S., Hydroquinone-*O,O'*-diacetic acid ('*Q-linker*') as a replacement for succinyl and oxalyl linker arms in solid phase oligonucleotide synthesis, *Nucleic Acids Res.*, 25, 3629, 1997.

269. Merrifield, B., Solid phase synthesis, *Biosci. Rep.*, 5, 353, 1985.

270. Merrifield, B., Concept and early development of solid-phase peptide synthesis, in *Methods for Solid-Phase Assembly of Peptides*, Fields, G. B., Ed., Academic Press, New York, 1997, 3.

271. Hyrup, B. and Nielsen, P. E., Peptide nucleic acids (PNA): synthesis, properties and potential applications, *Bioorg. Med. Chem.*, 4, 5, 1996.

272. Van der Laan, A. C. et al., A convenient automated solid-phase synthesis of PNA-(5')-DNA-(3')-PNA chimera, *Tetrahedron Lett.*, 38, 2249, 1997.

273. Hermes, J. D. et al., A reliable method for random mutagenesis: the generation of mutant libraries using spiked oligodeoxyribonucleotide primers, *Gene*, 84, 143, 1989.

274. Pon, R. T. et al., Multi-facility survey of oligonucleotide synthesis and an examination of the performance of unpurified primers in automated DNA sequencing, *Biotechniques*, 21, 680, 1996.

275. Macdonald, P. M. et al., Phosphorus 31 solid state NMR characterization of oligonucleotides covalently bound to a solid support, *Nucleic Acids Res.*, 24, 2868, 1996.

276. Gray, D. E. et al., Ellipsometric and interferometric characterization of DNA probes immobilized on a combinatorial array, *Langmuir*, 13, 2833, 1997.

277. Stimpson, D. I. et al., Real-time detection of DNA hybridization and melting on oligonucleotide arrays by using optical wave guides, *Proc. Natl. Acad. Sci. U.S.A.*, 92, 6379, 1995.

278. Huang, G. J. and Krugh, T. R., Large-scale purification of synthetic oligonucleotides and carcinogen-modified oligodeoxynucleotides on a reverse-phase polystyrene (PRP-1) column, *Anal. Biochem.*, 190, 21, 1990.

279. Germann, M. W., Pon, R. T., and van de Sande, J. H., A general method for the purification of synthetic oligodeoxyribonucleotides containing strong secondary structure by reversed-phase high-performance liquid chromatography on PRP-1 resin, *Anal. Biochem.*, 165, 399, 1987.

280. Johnson, B. A. et al., Rapid purification of synthetic oligonucleotides: a convenient alternative to high-performance liquid chromatography and polyacrylamide gel electrophoresis, *Biotechniques*, 8, 424, 1990.

281. Dai, Y. et al., Accurate mass measurement of oligonucleotides using a time-lag focusing matrix-assisted laser desorption/ionization time-of-flight mass spectrometer, *Rapid Commun. Mass Spectrom.*, 10, 1792, 1996.

282. Keough, T. et al., Antisense DNA oligonucleotides. II: The use of matrix-assisted laser desorption/ionization mass spectrometry for the sequence verification of methylphosphonate oligodeoxyribonucleotides, *Rapid Commun. Mass Spectrom.*, 7, 195, 1993.

283. Van Ausdall, D. A. and Marshall, W. S., Automated high-throughput mass spectrometric analysis of synthetic oligonucleotides, *Anal. Biochem.*, 256, 220, 1998.

284. Ball, R. W. and Packman, L. C., Matrix-assisted laser desorption ionization time-of-flight mass spectrometry as a rapid quality control method in oligonucleotide synthesis, *Anal. Biochem.*, 246, 185, 1997.

285. Keough, T. et al., Detailed characterization of antisense DNA oligonucleotides, *Anal. Biochem.*, 68, 3405, 1996.

286. Temsamani, J., Kubert, M., and Agrawal, S., Sequencing of synthetic oligonucleotides and analogs by homopolymeric tailing, *Nucleic Acids Res.*, 23, 1271, 1995.

287. Temsamani, J., Kubert, M., and Agrawal, S., Sequence identity of the n − 1 product of a synthetic oligonucleotide, *Nucleic Acids Res.*, 23, 1841, 1995.

288. Chen, D. H. et al., Analysis of internal (N − 1)mer deletion sequences in synthetic oligodeoxyribonucleotides by hybridization to an immobilized probe array, *Nucleic Acids Res.*, 27, 389, 1999.

289. Gough, G. R., Brunden, M. J., and Gilham, P. T., Phosphocytidines as versatile 3′ protecting groups in triester synthesis of oligodeoxyribonucleotides, *Tetrahedron Lett.*, 24, 5317, 1983.

290. Gough, G. R., Brunden, M. J., and Gilham, P. T., 2′(3′)-*O*-Benzoyluridine 5′ linked to glass: an all-purpose support for solid phase synthesis of oligodeoxyribonucleotides, *Tetrahedron Lett.*, 24, 5321, 1983.

291. Cosstick, R. and Eckstein, F., Synthesis of d(GC) and d(CG) octamers containing alternating phosphorothioate linkages: effect of the phosphorothioate group on the B-Z transition, *Biochemistry*, 24, 3630, 1985.

292. Debear, J. S. et al., A universal glass support for oligonucleotide synthesis, *Nucleosides Nucleotides*, 6, 821, 1987.

293. Scott, S. et al., A universal support for oligonucleotide synthesis, in *Innovation and Perspectives in Solid-Phase Synthesis. Peptides, Proteins, and Nucleic Acids, Biological and Biomedical Applications*, Epton, R. Ed., Mayflower Worldwide, Birmingham, 1994, 115.

294. Hardy, P. M. et al., Reagents for the preparation of two oligonucleotides per synthesis (TOPS™), *Nucleic Acids Res.*, 22, 2998, 1994.

295. Schwartz, M. E. et al., A universal adapter for chemical synthesis of DNA or RNA on any single type of solid support, *Tetrahedron Lett.*, 36, 27, 1995.

296. Lyttle, M. H., Hudson, D., and Cook, R. M., A new universal linker for solid phase DNA synthesis, *Nucleic Acids Res.*, 24, 2793, 1996.

297. Scheuerlarsen, C. et al., Introduction of a universal solid support for oligonucleotide synthesis, *Nucleosides Nucleotides*, 16, 67, 1997.

298. Nelson, P. S. et al., Rainbow™ universal CPG: a versatile solid support for oligonucleotide synthesis, *Biotechniques*, 22, 752, 1997.

299. Azhayev, A. V., A new universal support for oligonucleotide synthesis, *Tetrahedron*, 55, 787, 1999.

300. Azhayev, A. V. and Antopolsky, M. L., Amide group assisted 3′-dephosphorylation of oligonucleotides synthesized on universal A-supports, *Tetrahedron*, 57, 4977, 2001.

301. Pon, R. T. and Yu, S., Linker phosphoramidite reagents for oligonucleotide synthesis on underivatized solid-phase supports, *Tetrahedron Lett.*, 43, 8943, 2001.

302. Pon, R. T. et al., Reusable solid-phase supports for oligonucleotide synthesis using hydroquinone-*O,O′*-diacetic acid (Q-Linker), *Nucleosides Nucleotides*, 18, 1237, 1999.

303. Pon, R. T. et al., Multiple oligodeoxyribonucleotide syntheses on a reusable solid-phase CPG support via the hydroquinone-*O,O′*-diacetic acid *(Q-Linker)* linker arm, *Nucleic Acids Res.*, 27, 1531, 1999.

304. Pon, R. T. et al., Reusable solid-phase supports for oligonucleotides and antisense therapeutics, *J. Chem. Soc. Perkin Trans. 1*, 2638, 2001.

305. Pon, R. T., Solid-phase supports for oligonucleotide synthesis, in *Protocols for Oligonucleotides and Analogs*, Agrawal, S., Ed., Humana Press, Totowa, NJ, 1993, 465.

306. Pon, R. T., Solid-phase supports for oligonucleotide synthesis, in *Current Protocols in Nucleic Acid Chemistry*, Beaucage, S. L., Glick, G. D., Bergstrom, D. E., and Jones, R. A., Eds., John Wiley & Sons, New York, 2000, 3.1.1.

307. McCollum, C. and Andrus, A., An optimized polystyrene support for rapid, efficient oligonucleotide synthesis, *Tetrahedron Lett.,* 32, 4069, 1991.
308. McCollum, C. and Andrus, A., A new support for automated oligonucleotide synthesis, *Nucleosides Nucleotides,* 10, 573, 1991.
309. Katzhendler, J. et al., Spacer effect on the synthesis of oligonucleotides by the phosphite method, *Reactive Polym.,* 6, 175, 1987.
310. Van Aerschot, A., Herdewijn, P., and Vanderhaeghe, H., Silica gel functionalised with different spacers as solid support for oligonucleotide synthesis, *Nucleosides Nucleotides,* 7, 75, 1988.
311. Yip, K. F. and Tsou, K. C., A new polymer support method for the synthesis of ribooligonucleotide, *J. Am. Chem. Soc.,* 93, 3272, 1971.
312. Damha, M. J., Giannaris, P. A., and Zabarylo, S. V., An improved procedure for derivatization of controlled pore glass beads for solid-phase oligonucleotide synthesis, *Nucleic Acids Res.,* 18, 3813, 1990.
313. Bhongle, N. N. and Tang, J. Y., A convenient and practical method for derivatization of solid supports for nucleic acid synthesis, *Synth. Commun.,* 25, 3671, 1995.
314. Kumar, P. et al., Improved methods for 3′-O-succinylation of 2′-deoxyribo- and ribonucleosides and their covalent anchoring on polymer supports for oligonucleotide synthesis, *Nucleosides Nucleotides,* 12, 565, 1993.
315. Montserrat, F. X., Grandas, A., and Pedroso, E., Predictable and reproducible yields in the anchoring of DMT-nucleoside-succinates to highly loaded aminoalkyl-polystyrene, *Nucleosides Nucleotides,* 12, 967, 1993.
316. Walsh, A. J., Clark, G. C., and Fraser, W., A direct and efficient method for derivatisation of solid supports for oligonucleotide synthesis, *Tetrahedron Lett.,* 38, 1651, 1997.
317. Pon, R. T. and Yu, S., Efficient and rapid coupling of nucleosides to amino derivatized solid-phase supports, *Synth. Lett.,* 1778, 1999.
318. Pon, R. T., Attachment of nucleosides to solid-phase supports, in *Current Protocols in Nucleic Acids Chemistry,* Beaucage, S. L., Glick, G. D., Bergstrom, D. E., and Jones, R. A., Eds., John Wiley & Sons, New York, 2000, 3.2.1.
319. Van der Marel, G. A. et al., Phosphotriester synthesis of DNA fragments on cellulose and polystyrene solid supports, *Recl. Trav. Chim. Pays-Bas Belg.,* 101, 234, 1982.
320. Pochet, S., Huyn-Dinh, T., and Igolen, J., Synthesis of DNA fragments linked to a solid phase support, *Tetrahedron,* 43, 3481, 1987.
321. Avino, A. et al., A comparative study of supports for the synthesis of oligonucleotides without using ammonia, *Nucleosides Nucleotides,* 15, 1871, 1996.
322. Greenberg, M. M. and Gilmore, J. L., Cleavage of oligonucleotides from solid-phase supports using O-nitrobenzyl photochemistry, *J. Org. Chem.,* 59, 746, 1994.
323. McMinn, D. L. and Greenberg, M. M., Novel solid phase synthesis supports for the preparation of oligonucleotides containing 3′-alkyl amines, *Tetrahedron,* 52, 3827, 1996.
324. Venkatesan, H. and Greenberg, M. M., Improved utility of photolabile solid phase synthesis supports for the synthesis of oligonucleotides containing 3′-hydroxyl termini, *J. Org. Chem.,* 61, 525, 1996.
325. Dell'Aquila, C., Imbach, J. L., and Rayner, B., Photolabile linker for the solid-phase synthesis of base-sensitive oligonucleotides, *Tetrahedron Lett.,* 38, 5289, 1997.
326. McMinn, D. L., Hirsch, R., and Greenberg, M. M., An orthogonal solid phase support for the synthesis of oligonucleotides containing 3′-phosphates and its application in the preparation of photolabile hybridization probes, *Tetrahedron Lett.,* 39, 4155, 1998.
327. Alul, R. H. et al., Oxalyl-CPG — a labile support for synthesis of sensitive oligonucleotide derivatives, *Nucleic Acids Res.,* 19, 1527, 1991.
328. Guzaev, A. and Lonnberg, H., A novel solid support for synthesis of 3′-phosphorylated chimeric oligonucleotides containing internucleosidic methyl phosphotriester and methylphosphonate linkages, *Tetrahedron Lett.,* 38, 3989, 1997.
329. Mullah, B. and Andrus, A., Automated synthesis of double dye-labeled oligonucleotides using tetramethylrhodamine (TAMRA) solid supports, *Tetrahedron Lett.,* 38, 5751, 1997.
330. Mullah, B. et al., Efficient synthesis of double dye-labeled oligodeoxyribonucleotide probes and their application in a real time PCR assay, *Nucleic Acids Res.,* 26, 1026, 1998.

331. Bellenson, J. and Smith, A. J., Increasing DNA synthesizer throughput via off-instrument cleavage and deprotection, *Biotechniques,* 12, 219, 1992.

332. Lashkari, D. A. et al., An automated multiplex oligonucleotide synthesizer: development of high-throughput, low-cost DNA synthesis, *Proc. Natl. Acad. Sci. U.S.A.,* 92, 7912, 1995.

333. Sindelar, L. E. and Jaklevic, J. M., High-throughput DNA synthesis in a multichannel format, *Nucleic Acids Res.,* 23, 982, 1995.

334. Baier, J. et al., Synthesis and purification in a single column on a high-throughput automated oligonucleotide production system, *Biotechniques,* 20, 298, 1996.

335. Rayner, S. et al., MerMade: an oligodeoxyribonucleotide synthesizer for high-throughput oligonucleotide production in dual 96-well plates, *Genome Res.,* 8, 741, 1998.

336. Crea, R. and Horn, T., Synthesis of oligonucleotides on cellulose by a phosphotriester method, *Nucleic Acids Res.,* 8, 2331, 1980.

337. Kumar, P. et al., Express protocol for functionalization of polymer supports for oligonucleotide synthesis, *Nucleosides Nucleotides,* 15, 879, 1996.

338. Pon, R. T., Yu, S. Y., and Sanghvi, Y. S., Rapid esterification of nucleosides to solid-phase supports for oligonucleotide synthesis using uronium and phosphonium coupling reagents, *Bioconjugate Chem.,* 10, 1051, 1999.

339. Pon, R. T., Yu, S., and Sanghvi, Y. S., Multiple oligonucleotide synthesis in tandem on solid-phase supports for small and large scale synthesis, *Nucleosides Nucleotides Nucleic Acids,* 20, 985, 2001.

340. Pon, R. T., Yu, S., and Sanghvi, Y. S., Tandem oligonucleotide synthesis on solid-phase supports for the production of multiple oligonucleotides, *J. Org. Chem.,* 67, 856, 2002.

341. DeHaseth, P. L. et al., Chemical synthesis and biochemical reactivity of bacteriophage lambda PR promoter, *Nucleic Acids Res.,* 11, 773, 1983.

342. Tanaka, T. and Letsinger, R. L., Syringe method for stepwise chemical synthesis of oligonucleotides, *Nucleic Acids Res.,* 10, 3249, 1982.

343. Seliger, H., Scalfi, C., and Eisenbeiss, F., An improved syringe method for the preparation of oligonucleotides of defined sequence, *Tetrahedron Lett.,* 24, 4963, 1983.

344. Seliger, H. et al., New preparative methods in oligonucleotide chemistry and their application to gene synthesis. II. Recent developments of apparatus, *Chem. Scr.,* 26, 569, 1986.

345. Frank, R. et al., Simultaneous synthesis and biological applications of DNA fragments: an efficient and complete methodology, *Methods Enzymol.,* 154, 221, 1987.

346. Kaplan, B. E., The automated synthesis of oligodeoxyribonucleotides, *Trends Biotechnol.,* 3, 253, 1985.

347. McCollum, C. et al., Rapid and efficient oligonucleotide synthesis with low reagent consumption via a new synthesis column design: preparation of fluorescent dye labelled primers for application in PCR, *Biomed. Pept. Prot. Nucleic Acids,* 1, 25, 1994.

348. Marshall, W. S. and Boymel, J. L., Oligonucleotide synthesis as a tool in drug discovery research, *Drug Discov. Today,* 3, 34, 1998.

349. Brennan, T. et al., Two-dimensional parallel array technology as a new approach to automated combinatorial solid-phase organic synthesis, *Biotechnol. Bioeng.,* 61, 33, 1998.

350. Gausepohl, H. et al., Automated multiple peptide synthesis, *Pept. Res.,* 5, 315, 1992.

351. Frank, R. et al., A new general approach for the simultaneous chemical synthesis of large numbers of oligonucleotides: segmental solid supports, *Nucleic Acids Res.,* 13, 4365, 1983.

352. Matthes, H. W. D. et al., Simultaneous rapid chemical synthesis of over one hundred oligonucleotides on a microscale, *EMBO J.,* 3, 801, 1984.

353. Frank, R., Strategies and techniques in simultaneous solid phase synthesis based on the segmentation of membrane type supports, *Bioorg. Med. Chem. Lett.,* 3, 425, 1993.

354. Matthes, H. W., Staub, A., and Chambon, P., The segmented paper method: DNA synthesis and mutagenesis by rapid microscale "shotgun gene synthesis," *Methods Enzymol.,* 154, 250, 1987.

355. Seliger, H. and Rosch, R., Simultaneous synthesis of multiple oligonucleotides using nucleoside H-phosphonate intermediates, *DNA Cell Biol.,* 9, 691, 1990.

356. Beattie, K. L. et al., Gene synthesis technology: recent developments and future prospects, *Biotechnol. Appl. Biochem.,* 10, 510, 1988.

357. Beattie, K. L. and Fowler, R.F., Solid-phase gene assembly, *Nature,* 352, 548, 1991.

358. Nanthakumar, A. et al., Solid-phase oligonucleotide synthesis and flow cytometric analysis with microspheres encoded with covalently attached fluorophores, *Bioconjugate Chem.*, 11, 282, 2000.

359. McGall, G. et al., Light-directed synthesis of high-density oligonucleotide arrays using semiconductor photoresists, *Proc. Natl. Acad. Sci. U.S.A.*, 93, 13555, 1996.

360. Pease, A. C. et al., Light-generated oligonucleotide arrays for rapid DNA sequence analysis, *Proc. Natl. Acad. Sci. U.S.A.*, 91, 5022, 1994.

361. Lipshutz, R. J. et al., Using oligonucleotide probe arrays to access genetic diversity, *Biotechniques*, 19, 442, 1995.

362. McGall, G. H. et al., The efficiency of light-directed synthesis of DNA arrays on glass substrates, *J. Am. Chem. Soc.*, 119, 5081, 1997.

363. Lipshutz, R. J. et al., High density synthetic oligonucleotide arrays, *Nat. Genet.*, 21, 20, 1999.

364. Barone, A. D. et al., Photolithographic synthesis of high-density oligonucleotide probe arrays, *Nucleosides Nucleotides Nucleic Acids*, 20, 525, 2001.

365. Singh-Gasson et al., Maskless fabrication of light-directed oligonucleotide microarrays using a digital micromirror array, *Nat. Biotechnol.*, 17, 974, 1999.

366. Hacia, J. G. et al., Evolutionary sequence comparisons using high-density oligonucleotide arrays, *Nat. Genet.*, 18, 155, 1998.

367. Hacia, J. G. et al., Strategies for mutational analysis of the large multiexon ATM gene using high-density oligonucleotide arrays, *Genome Res.*, 8, 1245, 1998.

368. Hacia, J. G. et al., Two color hybridization analysis using high density oligonucleotide arrays and energy transfer dyes, *Nucleic Acids Res.*, 26, 3865, 1998.

369. Hacia, J. G., Resequencing and mutational analysis using oligonucleotide microarrays, *Nat. Genet.*, 21, 42, 1999.

370. Vahey, M. et al., Performance of the Affymetrix GeneChip HIV PRT 440 platform for antiretroviral drug resistance genotyping of human immunodeficiency virus type 1 clades and viral isolates with length polymorphisms, *J. Clin. Microbiol.*, 37, 2533, 1999.

371. Ishii, M. et al., Direct comparison of GeneChip and SAGE on the quantitative accuracy in transcript profiling analysis, *Genomics*, 68, 136, 2000.

372. Salamon, H. et al., Detection of deleted genomic DNA using a semiautomated computational analysis of GeneChip data, *Genome Res.*, 10, 2044, 2000.

373. Eisen, M. B. and Brown, P. O., DNA arrays for analysis of gene expression, in *cDNA Preparation and Characterization*, Weissman, S.M., Ed., Academic Press, San Diego, 1999, 179.

374. Hauser, N. C. et al., DNA arrays for transcriptional profiling, in *Methods in Microbiology*, Vol. 28, Craign, A. G. and Hoheisel, J. D., Eds., Academic Press, San Diego, 1999, 193.

375. Lockhart, D. J. and Winzeler, E. A., Genomics, gene expression and DNA arrays, *Nature*, 405, 827, 2000.

376. Mei, R. et al., Genome-wide detection of allelic imbalance using human SNPs and high-density DNA arrays, *Genome Res.*, 10, 1126, 2000.

377. Thibault, C. et al., DNA arrays and functional genomics in neurobiology, in *International Review Neurobiology*, Vol. 48, Bradley, R. J., Harris, R. A., and Jenner, P., Eds., Academic Press, San Diego, 2001, 219.

378. Zarrinkar, P. P. et al., Arrays of arrays for high-throughput gene expression profiling, *Genome Res.*, 11, 1256, 2001.

379. Blanchard, A. P., Kaiser, R. J., and Hood, L. E., High-density oligonucleotide arrays, *Biosens. Bioelectr.*, 11, 687, 1996.

380. Theriault, T. P., Winder, S. C., and Gamble, R. C., Application of ink-jet printing technology to the manufacture of molecular arrays, in *DNA Microarrays. A Practical Approach*, Schena, M., Ed., Oxford University Press, Oxford, 1999, 101.

381. Hughes, T. R. et al., Expression profiling using microarrays fabricated by an ink-jet oligonucleotide synthesizer, *Nat. Biotechnol.*, 19, 342, 2001.

382. Butler, J. H. et al., *In situ* synthesis of oligonucleotide arrays by using surface tension, *J. Am. Chem. Soc.*, 123, 8887, 2001.

383. Brown, K. D., Process specialists see promise in oligonucleotide therapeutics, *Genet. Eng. News*, 18, 1, 1998.

384. Lyttle, M. H., Cook, R. M., and Wright, P. B., Large scale automated DNA synthesis, *Biopharm. Manufact.*, 1988.

385. Andrus, A. et al., Large scale automated synthesis of oligonucleotides, *Nucleic Acids Res. Symp. Ser.,* 41, 1991.

386. Tsou, D. et al., Large scale synthesis of oligonucleotides via phosphoramidite nucleosides and a high loaded polystyrene support, in *Solid Phase Synthesis of Peptides, Proteins, Nucleic Acids, Biological and Biomedical Applications, Collected Papers, Third International Symposium,* 1993, Oxford, U.K., Epton, R. Ed., Mayflower Worldwide, Birmingham, 1994, 125.

387. Anderson, N. G. et al., Large scale oligonucleotides synthesizers. I. Basic principles and system design, *Appl. Biochem. Biotechnol.,* 54, 19, 1995.

388. Needels, M. C. et al., Generation and screening of an oligonucleotide-encoded synthetic peptide library, *Proc. Natl. Acad. Sci. U.S.A.,* 90, 10700, 1993.

389. Seliger, H. et al., Synthetic oligonucleotide combinatorial libraries — Tools for studying nucleic acid interactions, *Nucleosides Nucleotides,* 16, 703, 1997.

390. Dorre, K. et al., Techniques for single molecule sequencing, *Bioimaging,* 5, 139, 1997.

391. Uhlmann, E., Peptide nucleic acids (PNA) and PNA-DNA chimeras: from high binding affinity towards biological function, *Biol. Chem.,* 379, 1045, 1998.

392. Uhlmann, E. et al., PNA: synthetic polyamide nucleic acids with unusual binding properties, *Angew. Chem. Int. Ed.,* 37, 2797, 1998.

2 Artificial Hybridization Probes

Salvatore A. E. Marras

CONTENTS

2.1 OVERVIEW

Nucleic acid hybridization, in which a DNA or RNA strand binds its complement to form a duplex structure, is a fundamental process in molecular biology. The hybrids that are formed are the strongest and most specific macromolecular complexes known. A critical aspect of this process is the specificity of molecular recognition of one strand by the other. Detection of nucleic acids using nucleic acid probes, synthesized *in vitro* or via automated synthesis, is a powerful technique used in molecular biology diagnostics and research. Two hybridization types can be distinguished — heterogeneous hybridization, as is used in *in situ* hybridization, and homogeneous hybridization, as is used for real-time monitoring of amplification reactions. In order to improve specificity and sensitivity, new detection methods have been developed within the last 10 years. This overview discusses the development and application of artificially synthesized nucleic acid probes, with a focus on novel probes such as padlock probes, peptide nucleic acid probes, molecular beacon probes, adjacent probes, and 5′-nuclease probes.

2.2 *IN SITU* HYBRIDIZATION

The ability to identify DNA and RNA sequences in cells and tissue is becoming more important as links between phenotypes and genotypes are established. Nucleic acid hybridization forms the basis for the diagnosis of genetic and infectious diseases and reveals information on the storage, transfer, and expression of genetic information in living cells. When two single strands of nucleic acids collide, they can align in an antiparallel fashion and hybridize (anneal) to each other. Two forces control this process — the formation of hydrogen bonds between base pairs and the mutual repulsion of negative charges on the triphosphate groups of the nucleic acid backbones. Hybridization is dependent on environmental conditions such as temperature and salt concentration, which counteract the repulsive forces. If enough bases match, a stable hybrid will form. By selecting nucleic acid sequences that are complementary to a target sequence, nucleic acid probes can be designed for the detection of any gene. The principles of nucleic acid hybridization were first described by Sol Spiegelman and his colleagues in 1964.[1-3] Later, in 1969, *in situ* hybridization (ISH) was developed.[4,5]

ISH is a technique for the localization of specific nucleic acid sequences and the detection of gene expression within individual cells, either in tissue sections or in whole cell preparations. It is based on the complementary binding of a nucleic acid probe to a specific target sequence of DNA or RNA in the cell. The added nucleic acid probe is labeled with a reporter molecule and the sites of binding are visualized by microscopy. The initial use of double-stranded DNA (dsDNA) probes and RNA probes with radioactive labels limited their application to research. The earliest phase of ISH relied on autoradiographic detection of abundant sequences such as the localization of DNA in amplified polytene chromosomes or the identification of highly repeating sequences in metaphase chromosomes.[4,5]

In 1981, it was shown that it was possible to localize single sequences on metaphase chromosomes by autoradiography.[6] This technique was limited by poor resolution and by the requirement that localization necessitated the statistical analysis of many metaphase cells. Autoradiography also lacks the ability to distinguish more than one nucleic acid target simultaneously. In addition, autoradiography is often time-consuming, requiring several weeks before results can be obtained, and safety measures must be taken into account. To overcome these limitations, nonautoradioactive detection methods were developed. The earliest nonautoradioactive techniques used antibodies to recognize RNA-DNA hybrids or used avidin to detect bound biotin-labeled nucleic acid probes.[7,8] Later, methods for the chemical modification of nucleotides and for the recognition of these modified nucleotides by fluorescent molecules, gold particles, or enzymatic reporter molecules were developed to increase the assay sensitivity.[9]

Enzymatic detection methods, such as those that utilize horseradish peroxidase or alkaline phosphatase, require extra steps to produce a visible product. However, they have advantages over fluorescence in that the reaction can be prolonged to amplify the signals and the signals do not fade. For example, tyramide conjugates have been introduced as substrates for horseradish peroxidase.[10] The production of free oxygen radicals by horseradish peroxidase produces an activated tyramide conjugate that binds in close proximity to the horseradish peroxidase at an electron-rich moiety, such as tyrosine, phenylalanine, or tryptophan. The conjugate on the tyramide, a hapten or fluorophore, can be used for direct or indirect fluorescence signal development. This amplification technique results in a 500- to 1000-fold increase in sensitivity. Figure 2.1 shows an example of a tyramide signal amplification hybridization.[11]

Fluorescent dyes, such as derivatives of rhodamine, fluorescein, or Texas red, provide high resolution with the light microscope and can be adapted for multicolor labeling. Multicolor fluorescence *in situ* hybridization (FISH) is used to detect multiple nucleotide sequences simultaneously.[12,13] In the last 5 years, FISH has been improved by the development of interphase ISH, with the advantage that no cell culturing is necessary.[14]

2.2.1 CHOICE OF HYBRIDIZATION PROBES

The choice of probes for ISH depends on the desired specificity and sensitivity, stability of the hybrids, ease of tissue penetration, and reproducibility of the technique. Another important factor is the ease with which the probes can be synthesized. Different types of probes can be used, each with its own advantages and disadvantages.

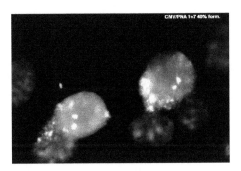

FIGURE 2.1 (Color figure follows p. 82.) mRNA detection using tyramide signal amplification hybridization. Human cytomegalo virus (HCMV-IE) mRNA in a transformed rat fibroblast cell line is detected using fluorescence microscopy and fluorescein-labeled PNA probes. The fluorescein signals are amplified using antifluorescein-biotin-tyramide-based detection. HCMV-IE mRNAs appear as green fluorescent spots. Only some cells exhibit expression of this mRNA. (Courtesy of Dr. Hans Tanke, Department of Cytochemistry and Cytometry, University of Leiden, the Netherlands.)

2.2.1.1 RNA Probes

RNA probes, or riboprobes, are very specific and sensitive. In addition, RNA:RNA hybrids are more stable than DNA:RNA hybrids. They are generated by cloning DNA in a vector, followed by a transcription. When used in an assay, posthybridization ribonuclease A treatment removes much of the nonhybridized and nonspecifically hybridized probe. The disadvantages are that, in general, RNA probes produce higher, relatively stable background signals. In addition, RNA probes used in *in situ* hybridization are easier to degrade than DNA probes. For example, Ainger and his co-workers used RNA probes to study the transport and localization of exogenous myelin basic protein mRNA in oligodendrocytes in culture by microinjecting labeled mRNA into living cells and analyzing the intracellular distribution of the injected RNA by confocal microscopy.[15]

2.2.1.2 Double-Stranded DNA Probes

Double-stranded DNA (dsDNA) probes are generated by cloning and amplification of specific sequences of DNA or cDNA, derived by reverse transcription of mRNA. The amplified sequences are extracted and labeled using nick translation or by the direct incorporation of reporter molecules during amplification. The greater the number of incorporated reporter molecules, the higher the sensitivity of the assay. dsDNA probes produce a lower background than RNA probes and are more resistant to degradation than RNA probes. Disadvantages are that dsDNA probes require denaturation to produce single-stranded DNA (ssDNA) before hybridization and can reanneal to each other, resulting in a weaker hybridization signal. Because of their double-stranded nature, these probes have more difficulty penetrating into cells. Furthermore, the hybrids that DNA probes form are less stable than the hybrids that RNA probes form, and assays that utilize DNA probes are not as sensitive as assays that utilize RNA probes.

2.2.1.3 Single-Stranded DNA Probes

Single-stranded DNA (ssDNA) probes are in general shorter in length than RNA probes and dsDNA probes. Because of their shorter length, they offer a high level of specificity and can differentiate single base pair differences. In addition to their short length, they find easier access to complementary target sequences and they have good cell penetration properties. This eliminates pretreatment steps that might damage cell morphology. They can be synthesized by automated synthesis, producing large quantities. The probes are stable for a relatively long time. Since a relatively high molar concentration can be used, the rate of hybridization is higher. Compared to larger probes, the level of sensitivity is lower because only a small number of reporter groups can be linked to each probe.

In 1978, Montgomery and her colleagues[16] demonstrated the usefulness of synthetic oligonucleotide probes in the isolation of the cloned yeast iso-L-cytochrome *c* gene. Wallace and his colleagues published reports in 1979 and 1981 that showed that synthetic oligonucleotides hybridized specifically to complementary DNA sequences.[17,18] Under optimized hybridization conditions, only perfectly complementary oligonucleotide:DNA duplexes formed; duplexes containing a single mismatched base pair were not stable and did not form. This high degree of specificity has led to the development of a general method for using oligonucleotides as specific probes to identify cloned DNA coding for proteins of interest. In its early stage of development, this technique was used to isolate human β2-microglobulin cDNA.[19] A mixture of oligonucleotides was synthesized that represented all possible codon combinations for a small portion of the amino acid sequence of human β2-microglobulin. Within the mixture, one sequence formed a perfect base pair duplex with the DNA sequence of human β2-microglobulin, whereas the other sequences formed mismatched duplexes. Under the stringent hybridization conditions employed, only perfectly matched duplexes formed, allowing the mixture of oligonucleotides to serve as a specific hybridization probe. In 1983, Conner and his colleagues used the β-globin gene as a model system to test the ability of synthetic oligonucleotides to be used for the detection of point mutations within a single copy gene that causes sickle cell anemia.[20]

2.2 *IN SITU* TARGET AMPLIFICATION

Different methods have been applied to increase the sensitivity of *in situ* hybridization. Although selection of the type of probe can increase the sensitivity of a particular application, in general the sensitivity of an assay is limited to between 20 to 200 copies of a nucleic acid's target per cell. To increase the detection limit, either the nucleic acid target sequence must be amplified prior to hybridization or the signals produced by the hybridization must be amplified. *In situ* polymerase chain reactions (IS-PCR) have been used to increase the number of target DNA sequences. Alternatively, an initial reverse transcription reaction can be used to generate cDNA from a target mRNA, which is then amplified by a polymerase chain reaction (IS-RT-PCR).[21] Reporter molecules can be incorporated into the amplification products, or the amplified products can contain a target sequence for another probe. A disadvantage of IS-PCR and IS-RT-PCR is that, because the amplification steps involve thermal cycling, the morphology of the cell can be lost, making localization of specific targets difficult. Isothermal amplification techniques have been developed, such as the *in situ* self-sustained sequence reaction (IS-3SR) or the similar *in situ* nucleic acid sequence based amplification reaction (IS-NASBA).[22] These reactions take place at a constant temperature of 37 to 42°C and use RNA as target. The basis of this technique, first described by Guatelli and his colleagues in 1990, is the incorporation of a T7-promotor sequence into the target with the help of a reverse transcription reaction.[23] This is followed by transcription of the cDNA product with T7 RNA polymerase, generating 10 to 1000 copies of the target sequence. Product accumulating is exponential, since the newly synthesized cDNAs and RNAs function as templates for a continuous series of reverse transcription and transcription reactions (see Figure 2.2). This method is able to generate as many as 10^9 copies of each RNA target molecules in 90 min.[24] The transcripts can be detected by the incorporation of reporter molecules, or they can contain a target sequence for another probe.

The next section describes the development and applications of new probes, such as padlock probes, peptide nucleic acid (PNA) probes, and fluorescent probes used for real-time monitoring of the PCR, that markedly improve the specificity and sensitivity of hybridization assays.

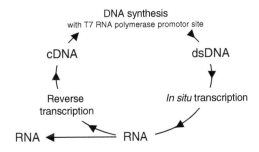

FIGURE 2.2 Schematic overview of *in situ* nucleic acid amplification.

2.3 PADLOCK PROBES

Enhanced high specificity in *in situ* hybridization is obtained by applying several washing steps, most under stringent conditions, in order to remove nonspecific hybridized probes and access of nonhybridized probes. Although most of the nonspecific signals will be removed, specific signals might fade, because some specific hybridized hybridization probes will not resist the stringent washing conditions.

Nillson and his colleagues developed an assay where a linear probe could be converted to a circular molecule, clamping itself around the target sequence and thereby making it more resistant to stringent washing conditions.[25] The method, first described by Landegren and his co-workers, is based on the ability of oligonucleotides to hybridize to a specific sequence and on the ability of T4 DNA ligase to distinguish mismatched nucleotides in a DNA double helix.[26] In order to identify single nucleotide differences, they developed an assay where two oligonucleotides are hybridized to a denatured DNA target, whereby the 3′ end of one oligonucleotide hybridizes immediately adjacent to the 5′ end of the other oligonucleotide. T4 DNA ligase will only join the two adjacent oligonucleotides if both of them are correctly base paired to the target strand. Thus, the generation of ligation products indicates the presence of a target sequence that is perfectly complementary to the two oligonucleotides (see Figure 2.3).

This method was further developed to make possible its use for localized DNA detection. Nillson and his colleagues used an oligonucleotide probe that consisted of two target-complementary segments (one at the 3′ end and the other at the 5′ end of the oligonucleotide) and a linker segment between them.[25] When both complementary segments hybridize to adjacent positions on the target sequence, T4 DNA ligase joins the 3′ and 5′ ends, forming a circular DNA. Because of the requirement that there must be precise base pairing at the ligation junction, these probes are highly specific. The probes are called "padlock probes," since they become catenated to the target upon ligation. Padlock probes are insensitive to exonucleases and resist extreme washing conditions, thus lowering the background of nonspecific hybridizations and making them excellent tools for ISH.

In order to detect the formation of a padlock probe, the oligonucleotide can be tagged with a radioactive or fluorescent label, or the linker sequence can be the target sequence for a detection probe. In addition, the circular oligonucleotide can be used as a template for enhanced detection by rolling circle amplification.

2.3.1 ROLLING CIRCLE AMPLIFICATION

Rolling circle amplification (RCA), which is driven by DNA polymerase, can replicate circularized oligonucleotide probes under isothermal conditions (see Figure 2.4). Linear oligonucleotides, which do not form a padlock, will not yield amplification products. Using a single primer, RCA generates

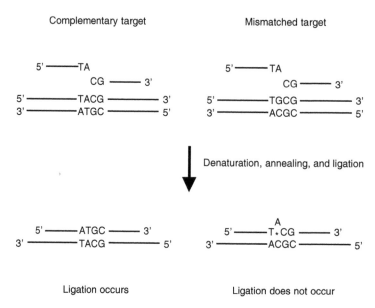

FIGURE 2.3 Overview of ligation of two matching oligonucleotides binding to a complementary target.

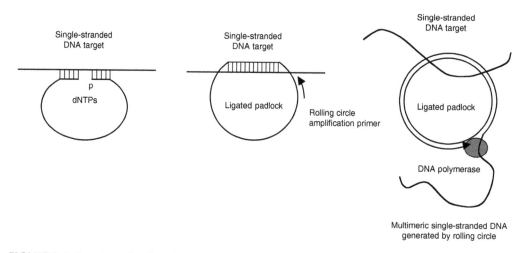

FIGURE 2.4 Overview of rolling circle amplification.

hundreds of tandemly linked copies of a covalently closed circle in a few minutes. The use of phage Ø29 DNA polymerase, a highly processive enzyme, ensures good strand displacement and the enzyme maintains polymerization for up to 12 h. Banér and his co-workers showed that the RCA could improve by introducing a free end into the DNA target sequence, which allows the padlock probe to slip off the target.[27] Free padlock probes are more efficient templates for rolling circle amplification, although precise localization of the target sequence becomes more difficult for ISH. Introducing a second primer, specific for a sequence in the tandemly repeated DNA generated by RCA, initiates a second primer extension reaction. As each of these extending upstream primers runs into the product of a downstream primer, strand displacement occurs, generating single-stranded tandem repeats containing the same sequence as the original padlock probe. This displaced strand contains multiple binding sites for the downstream primer. This process creates a continuously expanding pattern of DNA branches connected to the original padlock probe and is termed "hyperbranched rolling circle amplification." The amplification product can be tagged with a fluorescence probe or a radioisotope. Since the extended products are localized near the target sequence, they can be imaged as a single light point.[28]

2.4 PEPTIDE NUCLEIC ACID PROBES

In the last 10 years, attempts to enhance the properties of oligonucleotide probes (for example, DNA and RNA can rapidly be degraded by nucleases) have resulted in the synthesis of a variety of new oligonucleotide derivatives. Chemical modifications introduced into the phosphate groups, in the ribose sugar, or in the nucleic acid bases generally improve the biological stability of nucleotides. However, in some cases chemical modification lowers the binding affinity of the probe. In 1991, Nielsen and co-workers studied the effects of the replacement of the entire sugar-phosphate backbone by an *N*-(2-aminoethyl)glycine-based polyamide structure (see Figure 2.5).[29] The resulting polyamide or peptide nucleic acid (PNA) contained all four natural bases, enabling it to hybridize to complementary oligonucleotide targets. PNA obeys the Watson–Crick base-pairing rules and mimics DNA in terms of base pair recognition.[30] In contrast to DNA, PNA can bind in both parallel and antiparallel orientation, whereby the PNA C-terminus corresponds to the 3′ end of an oligonucleotide, and the PNA N-terminus corresponds to the 5′ end of an oligonucleotide. Antiparallel binding, as is the case for nucleic acid duplexes, is favored. In addition, PNA shows a higher binding affinity for complementary nucleic acids than does DNA or RNA. The increased thermal stability of PNA-DNA and PNA-RNA duplexes relative to the corresponding DNA-DNA and DNA-RNA duplexes is due primarily to the lack of electrostatic repulsion between the two strands.[30] The introduction of a single base pair mismatch reduces the T_m of the probe-target hybrid by 8 to 20°C, sometimes doubling the T_m difference that is observed for DNA-DNA and DNA-RNA mismatch duplexes.[31]

It was also shown that PNA forms triplexes with DNA by binding to the major groove of double-stranded DNA. In the case of C-rich PNAs and CG-rich DNA duplexes, (DNA)$_2$-PNA triplexes are observed. Homopyrimidine PNA oligomers bind to double-stranded DNA, resulting in the formation of (PNA)$_2$-DNA triplexes. These very stable complexes ($T_m > 70$°C for decamer hybrids) are formed by both Watson–Crick and Hoogsteen base pairing.[32] The displaced DNA strand forms a single-stranded loop and is so far only observed at salt concentrations below 50 mM and in AT-rich regions of DNA.[33] With the exception of the bases, PNA and DNA have no functional groups in common. The chemical stability is therefore completely different. Unlike DNA, PNA does not depurinate after treatment with strong acids and is completely acid stable. PNA is also

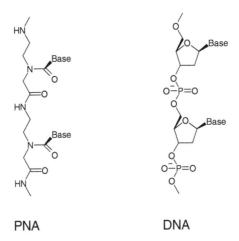

FIGURE 2.5 Comparison of the chemical structure of PNA and DNA.

reasonably stable in weak bases. Only the free amino group at the N-terminus is subject to chemical degradation. PNA has a lower solubility in water because it is neutrally charged. This solubility decreases as the length of the oligomer or the purine:pyrimidine ratio increases. Replacement of glycine in the backbone by lysine (which is positively charged) increases the solubility of PNA.[34] An important feature of PNA is its high biostability. PNA oligomers are not degraded by either nucleases or proteases.[35] Furthermore, PNA oligomers can be assembled by solid-phase peptide synthesis and reporter groups and other modifiers can be attached through either the N-terminal amino group or the C-terminal carboxylic acid.

The DNA and RNA binding properties of PNA, as well as its chemical and biological stability, has made PNA an attractive tool for use as antisense and antigene therapeutic agents, as well as a useful tool in diagnostics and molecular biology. PNA is able to efficiently block translation (protein synthesis) when it binds to mRNA. Transcription can be blocked when PNA binds to the promoter region or coding region of a gene.[36] As a molecular biology tool, Ørum and his co-workers developed a method for the detection of single nucleotide polymorphisms in a PCR assay, called PCR-clamping.[37] This method combines the ability of PNA to form more stable hybrids with a DNA target than DNA does and the inability of PNA to function as a primer for DNA polymerases. A PNA-DNA duplex effectively blocks the formation of a PCR product when a PNA probe binds to one of the PCR primer sites, or a PNA probe binds to a sequence between the two primers, resulting in an elongation arrest by the DNA polymerase. The primer exclusion principle is favored over the elongation arrest, since a smaller percentage of the DNA strands escapes PNA clamping. In their report, they show that PNA can discriminate at the level of a single base pair polymorphism. Perry-O'Keefe and her co-workers used labeled PNA oligomers as probes for pre-gel hybridization, as a simple alternative to Southern blot hybridization.[38] In this technique, the PNA probe is hybridized to a denatured DNA sample at low ionic strength and the mixture is loaded onto a gel for direct electrophoretic analysis. Gel electrophoresis separates the single-stranded DNA fragments by length, and PNA-DNA duplexes have a lower mobility than the excess of unbound PNA. Detection of the bound PNA is possible by direct fluorescence detection with capillary electrophoresis, or the PNA:DNA hybrids can be blotted onto a membrane and detected with standard chemilumines-cent techniques.

PNA is an attractive tool for *in situ* hybridization, because of its single-stranded nature, relative small size (good penetration properties), and ability to introduce fluorescent labels into the probe directly in a controlled synthesis process. For example, the standard procedure for measurement of telomere length was Southern blot hybridization with a telomere-specific probe. The number of telomeric repeats in human somatic cells decreases with age and with the number of cell divisions that occur. In tumor cells, however, higher levels of the enzyme telomerase can be found, resulting in the elongation of the telomeres. For Southern blot analysis, at least a million cells are digested with restriction enzymes that leave the telomeres intact. The fragments are then electrophoretically separated and hybridized with a sequence-specific telomeric probe. The smear offers only a crude estimation of the average telomeric length and does not provide information on variation in telomeric fragments due to interchromosomal, intrachromosomal, and intracellular differences. Lansdorp and his colleagues developed a method by which they used digital fluorescence microscopy with fluorescently labeled PNA probes to visualize nucleotide repeats in telomeres in metaphase chromosomes[39] (see Figure 2.6). De Pauw and his colleagues developed the method further to assess the length of telomeres in interphase chromosomes.[40]

2.5   REAL-TIME POLYMERASE CHAIN REACTION

The polymerase chain reaction (PCR), first described by Mullis and Saiki in 1985,[41,42] has made it possible to detect and quantitate rare target sequences isolated from cell, tissue, or blood samples. The basis of this technique is the ability of DNA polymerase to extend an oligonucleotide primer that is specifically hybridized to a single-stranded DNA sequence. Repeating thermal cycles allows

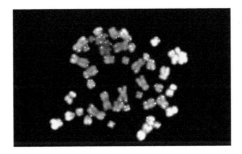

FIGURE 2.6 (Color figure follows p. 82.) Hybridization of metaphase chromosomes with a telomere-specific PNA probe. Digital fluorescence microcopy is used to visualize telomeric repeats in metaphase chromosomes using Cy3-labeled PNA probes (red fluorescence signals). Cy3 was excited with green light (515 to 560 nm) and the red emission was selected with a long-pass 580 nm filter. The chromosomes were counterstained with DAPI, resulting in a blue fluorescence signal. DAPI was excited with UV light (340 to 380 nm), and blue emission was selected with a long-pass 430 nm filter. (Courtesy of Dr. Hans Tanke, Department of Cytochemistry and Cytometry, University of Leiden, the Netherlands.)

newly formed DNA molecules to serve as a template for a new round of oligonucleotide primer hybridization and extension, resulting in an exponential accumulation of PCR products. There is a quantitative relationship between the amount of starting target sequence and the amount of PCR product at any given thermal cycle. However, in practice, it is common for different reactions to yield different amounts of PCR product. The development of real-time quantitative PCR has eliminated the variability associated with quantitative PCR. It was first described by Higuchi and his co-workers.[43] Homogeneous detection of PCR products can be performed using double-stranded DNA binding dyes, such as ethidium bromide, or fluorescent hybridization probes, such as molecular beacons, adjacent probes, and 5′-nuclease probes. During the amplification, the accumulating products generate an increase in fluorescence signal. By plotting the increase in fluorescence vs. cycle number, the *threshold cycle* can be determined. The threshold cycle is the number of thermal cycles required to generate a significant fluorescent signal above the fluorescent background. The threshold cycle is plotted as a function of the logarithm of the number of DNA template molecules added to each assay tube; unknown numbers of initial DNA template molecules can be calculated by interpolation. Compared to endpoint quantitation methods, real-time PCR offers reproducible results and has a larger dynamic range.

The use of novel developed fluorescent hybridization probes enhances overall assay sensitivity by eliminating background signals due to the synthesis of false amplicons and the formation of primer-dimer. The synthesis of false amplicons can be significant when the targets are rare or absent and the sample contains an abundant and diverse nucleic acid population. In these situations, the primers anneal to unintended sequences, and the false amplicons are the primary PCR products. Moreover, the primers can occasionally bind to each other, generating primer-dimers that can also be mistaken for the expected amplicon.[44] However, the hybridization probe sequence is chosen to be complementary to a target sequence within the expected amplicon, so they do not bind to the false amplicons or primer-dimers, thus enhancing the specificity of the assay. In addition, the use of the novel developed fluorescent probes enables real-time PCR assays to be carried out in sealed tubes, thereby eliminating the source of carryover contamination.[45]

2.5.1 MOLECULAR BEACON PROBES

Sanjay Tyagi and Fred Russell Kramer have developed novel nucleic acid hybridization probes that undergo a conformational change when they bind to their target that causes them to fluoresce brightly, thereby eliminating the need to isolate the probe-target hybrids. These probes are called molecular beacons.[46]

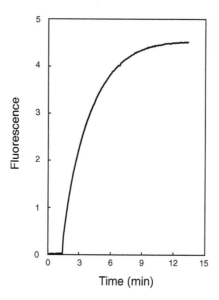

FIGURE 2.8 Fluorescence increase upon hybridization of a molecular beacon.

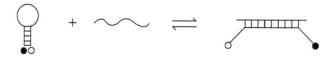

FIGURE 2.7 Interaction of a molecular beacon with its complementary target.

Molecular beacons form a stem-and-loop structure (see Figure 2.7). The loop portion of the molecule is a probe sequence (15 to 25 nucleotides long) that is complementary to a target sequence in a target nucleic acid. The probe sequence is embedded between two "arm" sequences. There is no relationship between the arm sequences and the probe sequence. The arm sequences (5 to 8 nucleotides long) are complementary to each other. Under assay conditions, the arms can bind to each other to form a double-helical stem hybrid that encloses the single-stranded probe sequence, forming a hairpin structure. A fluorophore is covalently linked to one end of the oligonucleotide and a nonfluorescent "quencher" moiety is covalently linked to the other end of the oligonucleotide. When the fluorophore is illuminated with light of a particular wavelength, it absorbs the energy of the light. After storing the energy for approximately 10 ns, the fluorophore releases the energy as fluorescent light of a longer wavelength. When the quencher is sufficiently close to the fluorophore, it interacts with the fluorophore in such a manner that the energy of the fluorophore is transferred to the quencher and is released as heat. When a molecular beacon encounters a target molecule, it forms a relatively rigid probe-target hybrid that is longer and more stable than the stem hybrid. The rigidity of the probe-target hybrid precludes the simultaneous existence of the stem hybrid. The molecular beacon therefore undergoes a spontaneous conformational reorganization that forces the stem hybrid to dissociate and the fluorophore and the quencher to move away from each other, restoring fluorescence. Since molecular beacons are dark when not hybridized and brightly fluorescent when hybridized, the course of their hybridization can be followed in real time with a spectrofluorometer. Figure 2.8 shows the results of an experiment in which the addition of an excess of complementary oligonucleotide targets to a dilute solution of molecular beacons causes a 900-fold increase in fluorescence intensity. The binding of a molecular beacon to its target follows second-order kinetics, and the rate of the reaction depends on the concentration of the probe, the concentration of the target, the temperature, and the salt concentration. Under assay conditions, hybridization is

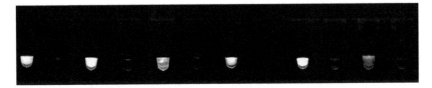

FIGURE 2.9 (Color figure follows p. 82.) Fluorescence of differently colored molecular beacons. (Courtesy of Dr. Sanjay Tyagi, Department of Molecular Genetics, Public Health Research Institute, Newark, New Jersey.)

spontaneous and rapid, reaching completion in only a few seconds, and the intensity of the resulting fluorescence is linearly proportional to the amount of target present.

In 1998, Tyagi and his colleagues showed that almost any fluorophore serves as a good label in a molecular beacon.[47] This means that a fluorophore can be chosen that is well suited for use with the available light source and the available emission detector. Moreover, the same quencher moiety, dabcyl, can be used, irrespective of the choice of fluorophore. Tyagi and his colleagues demonstrated by absorption spectroscopy the mechanism for quenching in molecular beacons; it is achieved when the formation of the stem hybrid brings the fluorophore and the quencher so close to one another that they share electrons, thereby forming a nonfluorescent compound. This interaction does not require the overlap of the emission spectrum of the fluorophore with the absorption spectrum of the quencher, as is the case with fluorescence resonance energy transfer (FRET).[48] Dabcyl thus serves as a universal quencher. Figure 2.9 shows an experiment carried out with molecular beacons that differ only in the identity of their fluorophore. Their emission spectra span the visible range from blue to red. A solution of each of the six different molecular beacons was added to a pair of test tubes, and an excess of complementary oligonucleotide targets was added to one tube of each pair. The tubes were illuminated with an ultraviolet light source. The results demonstrate that in the absence of targets each molecular beacon is well quenched, while in the presence of targets each molecular beacon fluoresces brightly in its own characteristic color. When used for *in vitro* and *in vivo* assays, the color of the resulting signal indicates which target sequences are present, and the intensity of each color indicates how many target molecules are present. When used in cells, the differently colored molecular beacons enables the multiplex detection of localization patterns of selected target sequences.

To design a molecular beacon for a given assay condition the probe sequence is selected to be sufficiently long to form a stable probe-target hybrid, and the complementary arm sequences are selected on the basis of their being sufficiently strong so that in the absence of target the stem hybrid remains closed. However, the length of the stem hybrid is also chosen so that it is sufficiently weak, that when the loop sequence hybridizes to a target, the stem hybrid is able to dissociate. When it is desirable for the molecular beacon to be highly specific, so that it will only form a hybrid with a perfect complementary target sequence, then the length of the probe sequence and the length of the arm sequences can be selected so that the stability of the perfectly complementary probe-target hybrid is just sufficient to overcome the stability of the stem hybrid. If such a molecular beacon encounters a target that contains a single nucleotide polymorphism (which would create a mismatched base pair within the hybrid), then the probe-target hybrid will not form and the stem hybrid will not dissociate. On the other hand, when it is desirable to design a molecular beacon that will form a probe-target hybrid even when there are one or more polymorphisms in the target sequence, then the length of the probe sequence can be increased to enhance the stability of the probe-target hybrid. To estimate the melting temperature of the probe-target hybrids that would form with different putative probe sequences a computer program that utilizes the "percent-G:C rule" can be employed.[49] A DNA folding program can be used to estimate the melting temperature of putative stem hybrids.[50]

Molecular beacons enable real-time PCR assays to be carried out. Figure 2.10 shows the quantitative relationship between the amount of target molecules added prior to amplification and

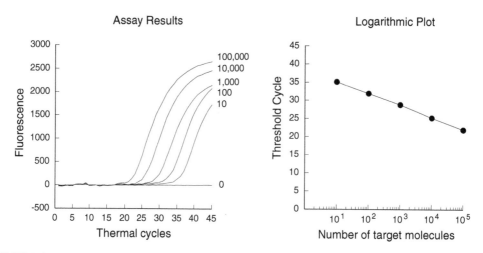

FIGURE 2.10 Inverse linear relationship between the logarithm of the number of targets and the number of thermal cycles that are required to generate a visible fluorescent signal appears.

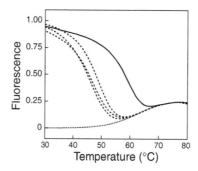

FIGURE 2.11 Fluorescence of an allele-discriminating molecular beacon as a function of temperature, in the absence of targets (dotted line), in the presence of perfectly complementary targets (solid line), and in the presence of each of three targets containing a mismatched nucleotide at the same position (dashed lines).

the threshold cycle. Five reactions were initiated with either 10^5, 10^4, 10^3, 10^2, or 10^1 DNA template molecules and one reaction was initiated without any DNA template added. The threshold cycle is plotted as a function of the logarithm of the number of DNA template molecules added to each assay tube. The results demonstrate that the assay provides quantitative results over an extremely wide range of target concentrations.

To test the ability of molecular beacons to specifically identify polymorphism in a nucleic acid population we developed a multiplex molecular beacon-PCR assay.[51] Four different target DNA templates were prepared that were identical, except that the nucleotide at one position was either adenosine, cytidine, guanosine, or thymidine. One pair of PCR primers was used that generates amplicons from any of the four templates. Four different molecular beacons were designed, each of which possessed a probe sequence that was perfectly complementary to a target sequence within one of the four different templates. All of the molecular beacons had the same arm sequences and all utilized dabcyl as the quencher. Four different fluorophores — fluorescein, tetrachlorofluorescein, rhodamine, and tetramethylrhodamine — were used as labels, one for each of the four different molecular beacons. The molecular beacons designed for this assay formed perfectly complementary probe-target hybrids whose melting temperature was about 13°C higher than the melting temperature of probe-target hybrids that contained one of the three possible mismatched base pairs (see Figure 2.11). Four different PCR assays were carried out, each initiated with one of the four different

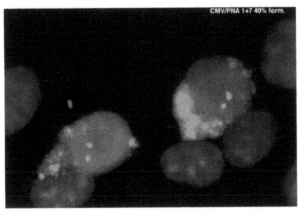

COLOR FIGURE 2.1. mRNA detection using tyramide signal amplification hybridization. Human cytomegalo virus (HCMV-IE) mRNA in a transformed rat fibroblast cell line is detected using fluorescence microscopy and fluorescein-labeled PNA probes. The fluorescein signals are amplified using antifluorescein-biotin-tyramide-based detection. HCMV-IE mRNAs appear as green fluorescent spots. Only some cells exhibit expression of this mRNA. (Courtesy of Dr. Hans Tanke, Department of Cytochemistry and Cytometry, University of Leiden, the Netherlands.)

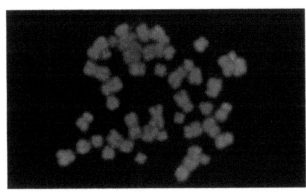

COLOR FIGURE 2.6 Hybridization of metaphase chromosomes with a telomere-specific PNA probe. Digital fluorescence microcopy is used to visualize telomeric repeats in metaphase chromosomes using Cy3-labeled PNA probes (red fluorescence signals). Cy3 was excited with green light (515 to 560 nm) and the red emission was selected with a long-pass 580 nm filter. The chromosomes were counterstained with DAPI, resulting in a blue fluorescence signal. DAPI was excited with UV light (340 to 380 nm), and blue emission was selected with a long-pass 430 nm filter. (Courtesy of Dr. Hans Tanke, Department of Cytochemistry and Cytometry, University of Leiden, the Netherlands.)

COLOR FIGURE 2.9 Fluorescence of differently colored molecular beacons. (Courtesy of Dr. Sanjay Tyagi, Department of Molecular Genetics, Public Health Research Institute, Newark, New Jersey.)

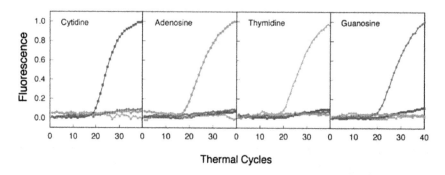

COLOR FIGURE 2.12 Multiplex detection of single-nucleotide polymorphisms utilizing different colored molecular beacons in a real-time PCR assay. (From Marras, S. A., Kramer, F. R., and Tyagi, S., *Genet. Anal.*, 14, 151, 1999. With permission.)

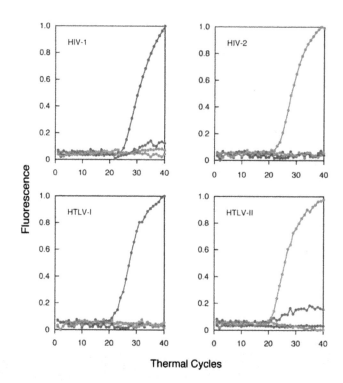

COLOR FIGURE 2.13 Real-time detection of four different retroviral DNAs in a multiplex PCR assay. (From Vet, J.A. et al., *Proc. Natl. Acad. Sci. U.S.A.*, 96, 6394, 1999. With permission.)

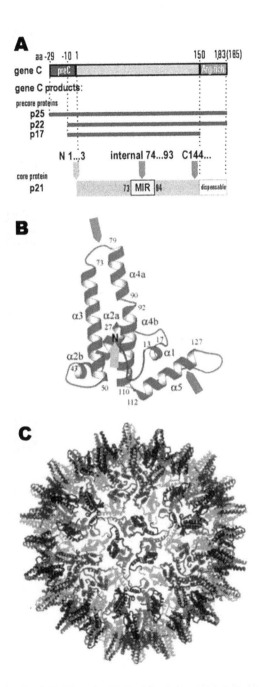

COLOR FIGURE 8.4 HBcAg as a VLP carrier. (A) Products encoded by the C gene with localization of the insertion sites for foreign epitopes. (B) A schematic representation of the fold of the HBc monomer derived from the crystal structure. The dimer interface is nearest to viewer. Insertion sites for foreign epitopes are marked by arrows. (C) The T = 4 HBc capsid viewed down an icosahedral threefold axis. (The maps are the generous gift of R. Anthony

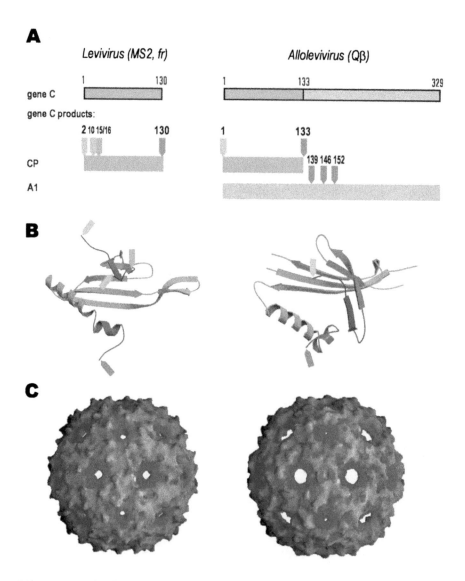

COLOR FIGURE 8.6 RNA phages as VLP carriers. (A) Products encoded by the C gene of the Leviviruses (left) and Alloleviviruses (right) with localization of the insertion sites for foreign epitopes. (B) A schematic representation of the fold of the CP monomers derived from the crystal structure. Insertion sites for foreign epitopes are marked by arrows. N terminus (blue)

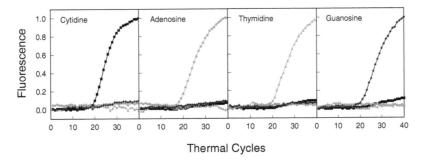

FIGURE 2.12 (Color figure follows p. 82.) Multiplex detection of single-nucleotide polymorphisms utilizing different colored molecular beacons in a real-time PCR assay. (From Marras, S.A., Kramer, F.R., and Tyagi, S., *Genet. Anal.*, 14, 151, 1999. With permission.)

target DNA templates. Each assay contained the same set of primers and a mixture of the four differently colored molecular beacons. The results (see Figure 2.12) show that only one of the four differently colored molecular beacons in each reaction formed probe-target hybrids during the course of the amplification. Only the molecular beacon possessing the perfectly complementary probe sequence formed a stable hybrid. Thus, the color of the fluorescence that developed in each reaction identified the variant nucleotide that was present in the target. The results of this assay indicate that molecular beacons can be designed for use in PCR reactions that are sufficiently specific to distinguish sequence differences as small as a single nucleotide substitution.

To show the use of molecular beacons in extremely sensitive, high-throughput, clinical tests we developed an assay for the detection of retroviral nucleic acids in blood samples and in tissues for transplantation.[52] A multiplex PCR assay was developed that uses four differently colored, mismatch-tolerant, molecular beacons for the simultaneous detection of amplicons generated from unique sequences found in four different pathogenic retroviruses. The assay contained four compatible sets of PCR primers[53] that are specific to the *gag* gene of HIV-1, the *env* gene of HIV-2, the *tax* gene of HTLV-I, and the *pol* gene of HTLV-II. Each of the four molecular beacons was designed to hybridize to a more or less conserved sequence within one of the four amplicons. To allow the molecular beacons to form stable hybrids, even if there are polymorphisms in the target sequences, the probe sequence in each molecular beacon was increased in length. The HIV-1, fluorescein-labeled, molecular beacon probe was designed to detect HIV-1 subtypes A, B, C, D, F, and G. The HIV-2, tetrachlorofluorescein-labeled, molecular beacon probe was designed to detect HIV-2 subtypes A, D, and SD. The HTLV-I, tetramethylrhodamine-labeled, molecular beacon probe was designed to detect all HTLV-I subtypes; and the HTLV-II, rhodamine-labeled, molecular probe was designed to detect HTLV-II subtypes A and B. Figure 2.13 shows that the individual retroviruses could be distinguished from one another in a multiplex format. Four reactions carried out in parallel were initiated with 100,000 molecules of one of the four retroviral DNAs. Each reaction contained all four molecular beacons and all four primer pairs (one pair for each retrovirus). The only significant fluorescence that appeared in the course of the amplification reactions carried out in each assay tube was fluorescence from the molecular beacon that was complementary to the sequence of the retroviral DNA that was originally added to the assay mixture. No significant fluorescence developed in a control assay that did not contain any template DNA. These results demonstrate that each molecular beacon is specific for its intended target amplicon.

To evaluate the ability of the assay to detect a rare retroviral DNA in the presence of an abundant retroviral DNA five reactions were initiated with 10^5 molecules of HTLV-I DNA and either 10^5, 10^4, 10^3, 10^2, or 10^1 molecules of HIV-2 DNA, and a sixth reaction did not contain any template DNA. The results (see Figure 2.14) show that both a fluorescent signal from tetramethylrhodamine (indicative of the presence of HTLV-I amplicons) and a fluorescent signal from tetrachlorofluorescein (indicative of the presence of HIV-2 amplicons) developed in every assay, except in the control

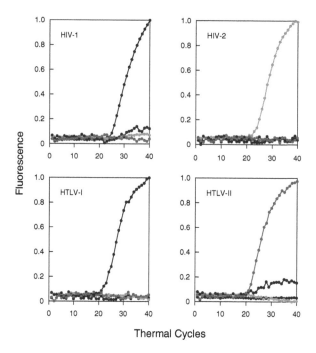

FIGURE 2.13 Real-time detection of four different retroviral DNAs in a multiplex PCR assay. (From Vet, J.A. et al., *Proc. Natl. Acad. Sci. U.S.A.,* 96, 6394, 1999. With permission.)

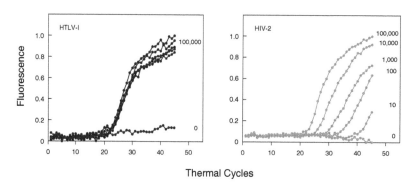

FIGURE 2.14 Detection of a rare retroviral target in the presence of an abundant retroviral target.

reaction, which did not contain any template DNA. The results show that the number of thermal cycles required for a significant tetramethylrhodamine signal to develop from the 100,000 HTLV-I target molecules was unaffected by the number of HIV-2 target molecules, and the number of thermal cycles required for a significant tetrachlorofluorescein signal to develop was indicative of the number of HIV-2 target molecules, irrespective of the presence of a relatively large number of HTLV-I target molecules.

In other studies, molecular beacons have been utilized for the detection of point mutations in human populations,[54–59] to identify fungal pathogens,[60] to detect drug resistance in *Mycobacterium tuberculosis*,[61–63] and to compare the expression of different *M. tuberculosis* mRNAs when the bacteria are grown in culture to their expression when they infect human macrophages.[64] They have been used to follow nucleic acid sequence-based amplification reactions.[65,66] They have also proved useful as a tool to study the control of ribosome release from mRNA,[67] to study the microflora that reside within intestines,[68] and to detect mRNA in cultured human cells.[69,70]

2.5.2 ADJACENT PROBES

Adjacent probe assays utilize two oligonucleotides that bind to neighboring sites on a target.[71,72] One probe is labeled with a donor fluorescent moiety at its 3′ end, and the other probe is labeled with an acceptor fluorescent moiety at its 5′ end. The donor and acceptor fluorophores are chosen so that efficient fluorescence resonance energy transfer (FRET) can take place from the donor to the acceptor, which occurs when the distance between the two probes is sufficiently small (about 70 Å).[48] No energy transfer should occur when the two probes are apart from each other and are free in solution. Upon hybridization of the oligonucleotides, energy transfer is measured by the decrease in donor fluorescence or the increase in acceptor fluorescence. Wittwer and his co-workers showed that the use of adjacent probes can be combined with rapid cycle DNA amplification.[73] For rapid cycling, they used a thermal cycler that employs capillaries and forced air heating. A total of 30 thermal cycles of DNA amplification and simultaneous monitoring of energy transfer take place in less than 15 min. The use of adjacent probes has also been demonstrated for the detection of specific mRNAs in living cells.[74] A benefit of this approach is higher specificity, as two probes, rather than one, must bind to the target in order to generate a signal.

2.5.3 5′-NUCLEASE PROBES

The 5′-nuclease activity of Taq polymerase cleaves linear hybridization probes that are hybridized specifically for accumulating PCR products.[75] The probe is labeled with a donor-acceptor fluorophore pair that interact via FRET. Probes that are free in solution transfer energy from the donor fluorophore to the acceptor fluorophore, resulting in a low fluorescence signal from the donor fluorophore. Cleavage of the probe results in the separation of the donor-acceptor pair and an increase in the fluorescence signal from the donor fluorophore. With each cycle, additional donor fluorophores are cleaved from their respective probes and the fluorescence intensity is monitored during the annealing/extension step of PCR. 5′-Nuclease assays are designed to amplify relatively short amplicons (75 to 150 base pairs), the primer T_m is chosen to be between 58 and 60°C, the probe T_m is chosen to be between 65 and 67°C, and the probe does not contain a guanosine on its 5′ end.[76] Using this guideline, Livak and his co-workers showed the detection of two allelic variants using the 5′-nuclease assay.[76] It has been widely employed for high-throughput assays, utilizing a 96-well thermal cycler that monitors fluorescence in real time. For example, it has been used to study mutations in the chemokine receptor 2 and chemokine receptor 5 genes that are associated with pathogenesis in AIDS.[77] One study describes the use of a fluorescent *in situ* 5′-nuclease assay for detecting HIV-1 DNA.[78]

The development of multiplex assays using more than one hybridization probe based on fluorescence resonance energy transfer is limited, since efficient FRET-dye pairs must have overlapping emission and absorption spectra. In addition, these methods use linear probes, which are less discriminatory than hairpin-shaped probes. Bonnet and his colleagues, utilizing a thermodynamic analysis, showed that hairpin-forming probes have an enhanced specificity compared to corresponding linear probes, because of their ability to form a stem-and-loop structure.[79] In a direct comparison, Täpp and co-workers observed better allelic discrimination with molecular beacons than with linear 5′-nuclease probes in real-time PCR amplifications.[80]

Other groups have used the enhanced specificity of hairpin-forming oligonucleotides to develop detection probes. Nazarenko and his co-workers described a method for the direct incorporation of a labeled hairpin primer into the reaction product.[81] At the 5′ end of an oligonucleotide primer a hairpin structure is introduced. This hairpin structure contains a fluorophore and quencher moiety in the stem of the hairpin. Hairpin primers not yet incorporated into an amplification product do not give a fluorescence signal, because the hairpin structure keeps the reporter and quencher in close proximity. During the amplification, the hairpin primer is linearized and becomes incorporated into the double-stranded amplification product. A fluorescence signal is generated, because the fluorophore and quencher are no longer in close proximity. A disadvantage of this method is that hairpin primers can generate "false" amplicons or primer-dimers, resulting in false positive fluorescence signals.

Whitcombe and his co-workers developed Scorpion primers, a method based on an oligonucleotide primer with a hairpin structure attached to its 5′ end by a linker that prevents copying of the 5′ extension.[82] The hairpin structure contains a fluorophore and quencher moiety in the stem of the hairpin. The hairpin structure contains a probe sequence that hybridizes to a complementary sequence incorporated into the extended product of the Scorpion primer. Scorpion primers that are free in solution do not give a fluorescence signal, because the hairpin structure keeps the fluorophore and quencher moiety in close proximity. Only a fluorescence signal is generated when the loop sequence hybridizes to a complementary sequence in the amplification product, opening the hairpin structure and forcing the fluorophore and quencher moieties apart. Although this method does not require a separate detection probe, as is the case with molecular beacon and FRET probe assays, the design and synthesis are more laborious and expensive because Scorpion primers contain a linker, fluorophore, and quencher.

Today's development of molecular technologies includes the production of DNA chips. Thousands of small DNA fragments can be placed on a chip surface and can be hybridized to fluorescently labeled targets or probes. Hacia and his colleagues describe the development of a high-density oligonucleotide array for scanning polymorphisms that occur in the *BRCA1* gene by differential hybridization.[83] High-throughput, multiplexed analysis of single-mismatch polymorphisms (SNP) on an array was demonstrated by Armstrong.[84] The attachment of molecular beacons to a solid surface will enable even more complex multiplex SNP analysis.[85] The advantages are that the array is self-reporting and the hairpin structure in the probe ensures high specificity. In addition, the incorporation of several copies of each probe decreases the chance of both false negatives and false positives.

With the further development of nucleic acid probes, labels, and detection devices, DNA chip-based assays will provide a valuable technology for high-throughput, cost-efficient detection of genetic alterations and gene expression.

ACKNOWLEDGMENTS

The studies on molecular beacons described in this chapter are the result of a collaboration with Dr. Fred Russell Kramer and Dr. Sanjay Tyagi and were supported by National Institutes of Health Grants HL-43521 and ES-10536.

REFERENCES

1. Gillespie, D. and Spiegelman, S., A quantitative assay for DNA-RNA hybrids with DNA immobilized on a membrane, *J. Mol. Biol.*, 12, 829, 1965.
2. Hall, B. D. and Spiegelman, S., Sequence complementarity of T2-DNA and T2-specific RNA, *Proc. Natl. Acad. Sci. U.S.A.*, 47, 137, 1961.
3. Spiegelman, S., Hybrid nucleic acids, *Sci. Am.*, 183, 1964.
4. Gall, J. G. and Pardue, M. L., Formation and detection of RNA-DNA hybrid molecules in cytological preparations, *Proc. Natl. Acad. Sci. U.S.A.*, 63, 378, 1969.
5. John, H. A. et al., RNA-DNA hybrids at the cytological level, *Nature*, 223, 582, 1969.
6. Harper, M. E. et al., Localization of the human insulin gene to the distal end of the short arm of chromosome 11, *Proc. Natl. Acad. Sci. U.S.A.*, 78, 4458, 1981.

7. Broker, T. R. et al., Electron microscopic visualization of tRNA genes with ferritin-avidin: biotin labels, *Nucleic Acids Res.*, 5, 363, 1978.
8. Rudkin, G. T. and Stollar, B. D., High resolution detection of DNA-RNA hybrids *in situ* by indirect immunofluorescence, *Nature*, 265, 472, 1977.
9. Langer, P. R. et al., Enzymatic synthesis of biotin-labeled polynucleotides: novel nucleic acid affinity probes, *Proc. Natl. Acad. Sci. U.S.A.*, 78, 6633, 1981.
10. Bobrow, M. N. et al., Catalyzed reporter deposition, a novel method of signal amplification. Application to immunoassays, *J. Immunol. Methods*, 125, 279, 1989.
11. Van de Corput, M. P. et al., Fluorescence *in situ* hybridization using horseradish peroxidase-labeled oligodeoxynucleotides and tyramide signal amplification for sensitive DNA and mRNA detection, *Histochem. Cell Biol.*, 110, 431, 1998.
12. Dirks, R. W. et al., Simultaneous detection of different mRNA sequences coding for neuropeptide hormones by double *in situ* hybridization using FITC- and biotin-labeled oligonucleotides, *J. Histochem. Cytochem.*, 38, 467, 1990.
13. Nederlof, P. M. et al., Multiple fluorescence *in situ* hybridization, *Cytometry*, 11, 126, 1990.
14. Raap, A. K., Advances in fluorescence *in situ* hybridization, *Mutat. Res.*, 400, 287, 1998.
15. Ainger, K. et al., Transport and localization of exogenous myelin basic protein mRNA microinjected into oligodendrocytes, *J. Cell. Biol.*, 123, 431, 1993.
16. Montgomery, D. L. et al., Identification and isolation of the yeast cytochrome *c* gene, *Cell*, 14, 673, 1978.
17. Wallace, R. B. et al., The use of synthetic oligonucleotides as hybridization probes. II. Hybridization of oligonucleotides of mixed sequence to rabbit beta-globin DNA, *Nucleic Acids Res.*, 9, 879, 1981.
18. Wallace, R. B. et al., Hybridization of synthetic oligodeoxyribonucleotides to phi chi 174 DNA: the effect of single base pair mismatch, *Nucleic Acids Res.*, 6, 3543, 1979.
19. Suggs, S. V. et al., Use of synthetic oligonucleotides as hybridization probes: isolation of cloned cDNA sequences for human beta 2-microglobulin, *Proc. Natl. Acad. Sci. U.S.A.*, 78, 6613, 1981.
20. Conner, B. J. et al., Detection of sickle cell beta S-globin allele by hybridization with synthetic oligonucleotides, *Proc. Natl. Acad. Sci. U.S.A.*, 80, 278, 1983.
21. Haase, A. T. et al., Amplification and detection of lentiviral DNA inside cells, *Proc. Natl. Acad. Sci. U.S.A.*, 87, 4971, 1990.
22. Mueller, J. D. et al., Self-sustained sequence replication (3SR): an alternative to PCR, *Histochem. Cell Biol.*, 108, 431, 1997.
23. Guatelli, J. C. et al., Isothermal, *in vitro* amplification of nucleic acids by a multienzyme reaction modeled after retroviral replication, *Proc. Natl. Acad. Sci. U.S.A.*, 87, 7797, 1990.
24. Bush, C. E. et al., Detection of human immunodeficiency virus type 1 RNA in plasma samples from high-risk pediatric patients by using the self-sustained sequence replication reaction, *J. Clin. Microbiol.*, 30, 281, 1992.
25. Nilsson, M. et al., Padlock probes: circularizing oligonucleotides for localized DNA detection, *Science*, 265, 2085, 1994.
26. Landegren, U. et al., A ligase-mediated gene detection technique, *Science*, 241, 1077, 1988.
27. Baner, J. et al., Signal amplification of padlock probes by rolling circle replication, *Nucleic Acids Res.*, 26, 5073, 1998.
28. Lizardi, P. M. et al., Mutation detection and single-molecule counting using isothermal rolling-circle amplification, *Nat. Genet.*, 19, 225, 1998.
29. Nielsen, P. E. et al., Sequence-selective recognition of DNA by strand displacement with a thymine-substituted polyamide, *Science*, 254, 1497, 1991.
30. Egholm, M. et al., PNA hybridizes to complementary oligonucleotides obeying the Watson–Crick hydrogen-bonding rules, *Nature*, 365, 566, 1993.
31. Jensen, K. K. et al., Kinetics for hybridization of peptide nucleic acids (PNA) with DNA and RNA studied with the BIAcore technique, *Biochemistry*, 36, 5072, 1997.
32. Eriksson, M. and Nielsen, P. E., PNA-nucleic acid complexes. Structure, stability, and dynamics, *Q. Rev. Biophys.*, 29, 369, 1996.
33. Peffer, N. J. et al., Strand-invasion of duplex DNA by peptide nucleic acid oligomers, *Proc. Natl. Acad. Sci. U.S.A.*, 90, 10648, 1993.

34. Haaima, G. et al., Peptide nucleic acids (PNAs) containing thymine monomers derived from chiral amino acids: hybridization and solubility properties of D-Lysine PNA, *Angew. Chem. Int. Ed. Engl.*, 35, 1939, 1996.

35. Demidov, V. et al., Sequence selective double strand DNA cleavage by peptide nucleic acid (PNA) targeting using nuclease S1, *Nucleic Acids Res.*, 21, 2103, 1993.

36. Hanvey, J. C. et al., Antisense and antigene properties of peptide nucleic acids, *Science*, 258, 1481, 1992.

37. Orum, H. et al., Single base pair mutation analysis by PNA directed PCR clamping, *Nucleic Acids Res.*, 21, 5332, 1993.

38. Perry-O'Keefe, H. et al., Peptide nucleic acid pre-gel hybridization: an alternative to Southern hybridization, *Proc. Natl. Acad. Sci. U.S.A.*, 93, 14670, 1996.

39. Lansdorp, P. M. et al., Heterogeneity in telomere length of human chromosomes, *Hum. Mol. Genet.*, 5, 685, 1996.

40. De Pauw, E. S. et al., Assessment of telomere length in hematopoietic interphase cells using *in situ* hybridization and digital fluorescence microscopy, *Cytometry*, 32, 163, 1998.

41. Mullis, K. et al., Specific enzymatic amplification of DNA *in vitro*: the polymerase chain reaction, *Cold Spring Harb. Symp. Quant. Biol.*, 51, 263, 1986.

42. Saiki, R. K. et al., Enzymatic amplification of beta-globin genomic sequences and restriction site analysis for diagnosis of sickle cell anemia, *Science*, 230, 1350, 1985.

43. Higuchi, R. et al., Kinetic PCR analysis: real-time monitoring of DNA amplification reactions, *Biotechnology (N.Y.)*, 11, 1026, 1993.

44. Erlich, H. A. et al., Recent advances in the polymerase chain reaction, *Science*, 252, 1643, 1991.

45. Kwok, S. and Higuchi, R., Avoiding false positives with PCR, *Nature*, 339, 237, 1989.

46. Tyagi, S. and Kramer, F. R., Molecular beacons: probes that fluoresce upon hybridization, *Nat. Biotechnol.*, 14, 303, 1996.

47. Tyagi, S. et al., Multicolor molecular beacons for allele discrimination, *Nat. Biotechnol.*, 16, 49, 1998.

48. Haugland, R. P. et al., Dependence of the kinetics of singlet-singlet energy transfer on spectral overlap, *Proc. Natl. Acad. Sci. U.S.A.*, 63, 23, 1969.

49. Lathe, R., Synthetic oligonucleotide probes deduced from amino acid sequence data. Theoretical and practical considerations, *J. Mol. Biol.*, 183, 1, 1985.

50. SantaLucia, J., Jr., A unified view of polymer, dumbbell, and oligonucleotide DNA nearest-neighbor thermodynamics, *Proc. Natl. Acad. Sci. U.S.A.*, 95, 1460, 1998.

51. Marras, S. A. et al., Multiplex detection of single-nucleotide variations using molecular beacons, *Genet. Anal.*, 14, 151, 1999.

52. Vet, J. A. et al., Multiplex detection of four pathogenic retroviruses using molecular beacons, *Proc. Natl. Acad. Sci. U.S.A.*, 96, 6394, 1999.

53. Heredia, A. et al., Development of a multiplex PCR assay for the simultaneous detection and discrimination of HIV-1, HIV-2, HTLV-I and HTLV-II, *Clin. Diagn. Virol.*, 7, 85, 1996.

54. Giesendorf, B. A. et al., Molecular beacons: a new approach for semiautomated mutation analysis, *Clin. Chem.*, 44, 482, 1998.

55. Gonzalez, E. et al., Race-specific HIV-1 disease-modifying effects associated with CCR5 haplotypes, *Proc. Natl. Acad. Sci. U.S.A.*, 96, 12004, 1999.

56. Kostrikis, L. G. et al., A chemokine receptor CCR2 allele delays HIV-1 disease progression and is associated with a CCR5 promoter mutation, *Nat. Med.*, 4, 350, 1998.

57. Kostrikis, L. G. et al., Spectral genotyping of human alleles, *Science*, 279, 1228, 1998.

58. Lewin, S. R. et al., Use of real-time PCR and molecular beacons to detect virus replication in human immunodeficiency virus type 1-infected individuals on prolonged effective antiretroviral therapy, *J. Virol.*, 73, 6099, 1999.

59. Vogelstein, B. and Kinzler, K.W., Digital PCR, *Proc. Natl. Acad. Sci. U.S.A.*, 96, 9236, 1999.

60. Park, S. et al., Rapid identification of *Candida dubliniensis* using a species-specific molecular beacon, *J. Clin. Microbiol.*, 38, 2829, 2000.

61. Piatek, A. S. et al., Genotypic analysis of *Mycobacterium tuberculosis* in two distinct populations using molecular beacons: implications for rapid susceptibility testing, *Antimicrob. Agents Chemother.*, 44, 103, 2000.

62. Piatek, A. S. et al., Molecular beacon sequence analysis for detecting drug resistance in *Mycobacterium tuberculosis*, *Nat. Biotechnol.*, 16, 359, 1998.

63. Rhee, J. T. et al., Molecular epidemiologic evaluation of transmissibility and virulence of *Mycobacterium tuberculosis*, *J. Clin. Microbiol.*, 37, 1764, 1999.

64. Manganelli, R. et al., Differential expression of 10 sigma factor genes in *Mycobacterium tuberculosis*, *Mol. Microbiol.*, 31, 715, 1999.

65. Ehricht, R. et al., Monitoring the amplification of CATCH, a 3SR based cooperatively coupled isothermal amplification system, by fluorimetric methods, *Nucleic Acids Res.*, 25, 4697, 1997.

66. Leone, G. et al., Molecular beacon probes combined with amplification by NASBA enable homogeneous, real-time detection of RNA, *Nucleic Acids Res.*, 26, 2150, 1998.

67. Gao, W. et al., Messenger RNA release from ribosomes during 5′-translational blockage by consecutive low-usage arginine but not leucine codons in *Escherichia coli*, *Mol. Microbiol.*, 25, 707, 1997.

68. Schofield, P. et al., Molecular beacons: trial of a fluorescence-based solution hybridization technique for ecological studies with ruminal bacteria, *Appl. Environ. Microbiol.*, 63, 1143, 1997.

69. Matsuo, T., *In situ* visualization of messenger RNA for basic fibroblast growth factor in living cells, *Biochim. Biophys. Acta*, 1379, 178, 1998.

70. Sokol, D. L. et al., Real time detection of DNA:RNA hybridization in living cells, *Proc. Natl. Acad. Sci. U.S.A.*, 95, 11538, 1998.

71. Cardullo, R. A. et al., Detection of nucleic acid hybridization by nonradiative fluorescence resonance energy transfer, *Proc. Natl. Acad. Sci. U.S.A.*, 85, 8790, 1988.

72. Heller, M. J. and Morrison, L. E., *Rapid Detection and Identification of Infectious Agents*, Academic Press, New York, 1985.

73. Wittwer, C. T. et al., Continuous fluorescence monitoring of rapid cycle DNA amplification, *Biotechniques*, 22, 130, 1997.

74. Tsuji, A. et al., Direct observation of specific messenger RNA in a single living cell under a fluorescence microscope, *Biophys. J.*, 78, 3260, 2000.

75. Holland, P. M. et al., Detection of specific polymerase chain reaction product by utilizing the 5′-3′ exonuclease activity of *Thermus aquaticus* DNA polymerase, *Proc. Natl. Acad. Sci. U.S.A.*, 88, 7276, 1991.

76. Livak, K. J. et al., Oligonucleotides with fluorescent dyes at opposite ends provide a quenched probe system useful for detecting PCR product and nucleic acid hybridization, *PCR Methods Appl.*, 4, 357, 1995.

77. Yuan, C. C. et al., 5′-Nuclease assays for the loci CCR5-+Delta32, CCR2-V64I, and SDF1-G801A related to pathogenesis of AIDS, *Clin. Chem.*, 46, 24, 2000.

78. Patterson, B. K. et al., Detection of HIV-1 DNA in cells and tissue by fluorescent *in situ* 5′-nuclease assay (FISNA), *Nucleic Acids Res.*, 24, 3656, 1996.

79. Bonnet, G. et al., Thermodynamic basis of the enhanced specificity of structured DNA probes, *Proc. Natl. Acad. Sci. U.S.A.*, 96, 6171, 1999.

80. Täpp, I. et al., Homogeneous scoring of single-nucleotide polymorphisms: comparison of the 5′-nuclease TaqMan assay and molecular beacon probes, *Biotechniques*, 28, 732, 2000.

81. Nazarenko, I. A. et al., A closed tube format for amplification and detection of DNA based on energy transfer, *Nucleic Acids Res.*, 25, 2516, 1997.

82. Whitcombe, D. et al., Detection of PCR products using self-probing amplicons and fluorescence, *Nat. Biotechnol.*, 17, 804, 1999.

83. Hacia, J. G. et al., Detection of heterozygous mutations in BRCA1 using high density oligonucleotide arrays and two-colour fluorescence analysis, *Nat. Genet.*, 14, 441, 1996.

84. Armstrong, B. et al., Suspension arrays for high throughput, multiplexed single nucleotide polymorphism genotyping, *Cytometry*, 40, 102, 2000.

85. Steemers, F. J. et al., Screening unlabeled DNA targets with randomly ordered fiber-optic gene arrays, *Nat. Biotechnol.*, 18, 91, 2000.

3 Peptide Nucleic Acid (PNA): A DNA Mimic with a Pseudopeptide Backbone

Frederik Beck and Peter E. Nielsen

CONTENTS

3.1 INTRODUCTION

The peptide nucleic acid (PNA) was originally conceived and designed as a ligand for major groove recognition of double-stranded DNA (dsDNA) via major groove interactions analogous to those employed by triple-helix-forming oligonucleotides.[1] It was immediately clear, however, that PNA is a very effective structural mimic of DNA (or RNA)[2,3] despite the fact that it is a pseudopeptide (polyamide) and, therefore, from a chemical point of view, is much more closely related to peptides and proteins than to nucleic acids[4] (Figure 3.1). Thus, PNA is truly an artificial DNA in terms of structure but not in terms of biological function. Not surprisingly, these properties of PNA have attracted widespread interest for the development of gene therapeutic antisense and antigene drugs, genetic diagnostics, and molecular tools as well as in bioorganic chemistry for studying biomolecular recognition and DNA structure and function (for recent reviews, see References 3 through 10).

3.2 STRUCTURAL ASPECTS OF PNA

At present, four high-resolution structures of PNA complexes are available. The three-dimensional structure of a PNA-RNA[11] and a PNA-DNA duplex[12] were solved by NMR techniques, while the structures of a PNA$_2$DNA triplex[13] and a PNA-PNA duplex[14] were solved by X-ray crystallography (Figure 3.2). The heteroduplex structures of PNA make it clear that it is largely able to adapt to its

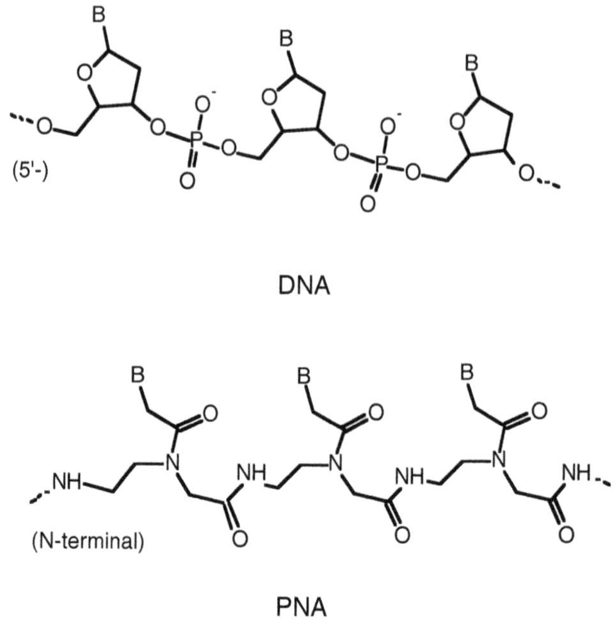

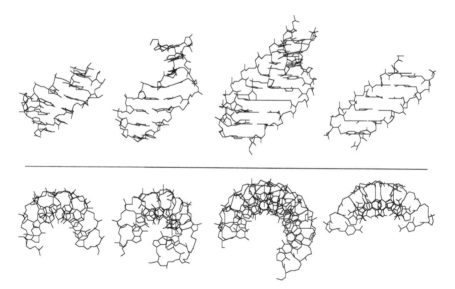

FIGURE 3.1 Chemical structures of DNA and PNA.

FIGURE 3.2 Structures of PNA complexes shown in side view (upper panel) or top view (lower panel). From left to right, the complexes are PNA-RNA, PNA-DNA duplex, PNA-DNA-PNA triplex, and PNA-PNA duplex.

nucleotide partner. The nucleotide part of these duplexes adapts close to their preferred conformation. For instance, the DNA deoxyribose adapts the 2′-endo conformation,[12] whereas the RNA ribose is 3′-endo.[11] On the other hand the triplex and particularly the pure PNA duplex structures clearly show that the PNA oligomers prefer a helical conformation, the P-form, which is distinctly different from the natural A- or B-form helices in that it has very large helical pitch (18 bp per turn) and a wider diameter (28 Å) as compared to, e.g., B-DNA (10 bp per turn and 20 Å diameter).

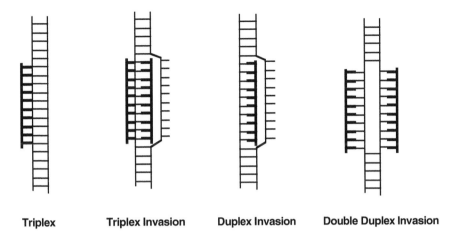

| Triplex | Triplex Invasion | Duplex Invasion | Double Duplex Invasion |

FIGURE 3.3 Schematic drawing of the modes by which PNA has been shown to bind to duplex DNA targets. PNA oligomers are shown in bold.

3.3 PNA HYBRIDIZATION

The initial results indicated that at physiological ionic strength (140 mM K$^+$/Na$^+$), the thermal stability of PNA-DNA and PNA-RNA duplexes was approximately 1°/bp higher than that of the corresponding DNA-DNA or DNA-RNA duplexes.[15,16] However, a more detailed analysis revealed that the sequence-dependent stability of a PNA-DNA duplex is more complex than that of a DNA-DNA duplex, which initial estimates indicate is directly related to the GC/AT ratio. The PNA-DNA duplex is, of course, sequence asymmetrical, and duplexes of identical GC/AT ratio, in which the purines are present in the PNA strand, turn out to be far more stable than their opposite. This is quite readily seen from the empirical equation that was derived from over 300 T_m measurements and that relates the T_m of a PNA-DNA duplex to that of the corresponding DNA-DNA duplex:[17]

$$T_{m_{\text{pred}}} = 20.79 + 0.83 * T_{m_{\text{nnDNA}}} - 26.13 * f_{\text{pyr}} + 0.44 * \text{length}$$

in which $T_{m_{\text{nnDNA}}}$ is the T_m as calculated using a nearest-neighbor model for the corresponding DNA-DNA duplex, applying ΔH^0 and ΔS^0 values as described by SantaLucia et al.[18] without taking end effects into account; f_{pyr} denotes the fractional pyrimidine content; and length is the PNA sequence length in bases. The above-mentioned asymmetry is reflected by the f_{pyr} factor in the equation.

3.4 BINDING TO DOUBLE-STRANDED DNA

Most surprisingly, targeting of dsDNA with a homopyrimidine PNA results in a triplex invasion complex (Figure 3.3) rather than the expected triplex (Figure 3.3).[1,19] Conventional PNA-DNA$_2$ triplexes have been detected with cytosine-rich PNAs,[20,21] but these are much less stable than the corresponding triplex invasion complexes. The latter, however, show remarkable stability, with a half-life of many days for a 10-mer PNA.[22] On the other hand, such invasion complexes show very slow on-rates because a transient opening (denaturation) of the DNA double helix is required to initiate binding, and the binding is also dramatically inhibited by elevated ionic strength, which stabilizes the DNA duplex.[23,24] In fact, simple homopyrimidine PNA oligomers show virtually no binding to dsDNA under physiological ionic conditions. The binding rate can, however, be dramatically enhanced by using bis-PNAs, in which the Watson–Crick and the Hoogsteen binding-PNA strands are chemically joined, and in particular if three to four positively charged residues

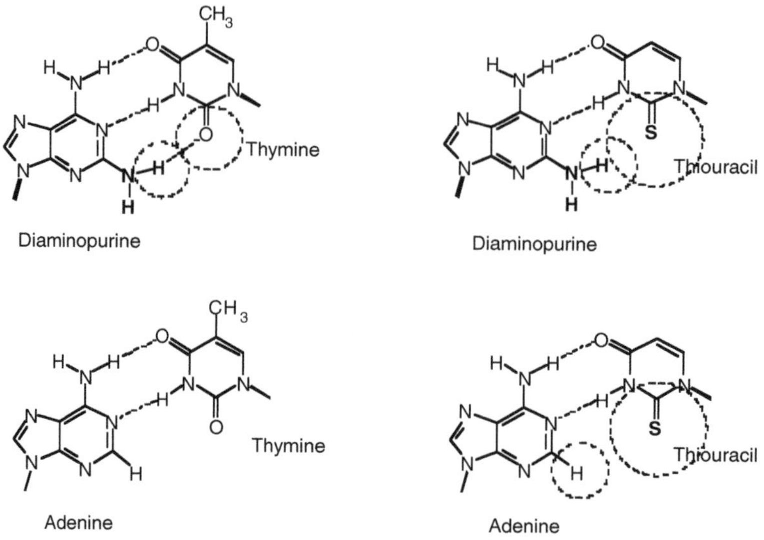

FIGURE 3.4 Chemical structures of diaminopurine-thymine, diaminopurine-thiouracil, adenine-thymine, and adenine-thiouracil base pairs illustrating the steric clash in the diaminopurine-thiouracil pair which is not present in the adenine-thiouracil pair.

such as lysines are also included.[22,25,26] Such cationic bis-PNAs do indeed bind their targets in dsDNA at 140 mM K+.[22,26] The substitution of cytosines with pseudoisocytosine (a nucleobase that mimics N3-protonated cytosine) in the Hoogsteen PNA strand of a bis-PNA will further improve binding at neutral pH.[25]

The slow binding kinetics, combined with the extremely slow dissociation kinetics of the triplex invasion complexes (at least for 8-mer or longer targets), usually prevents equilibrium from being reached under most binding conditions. Therefore, the binding and, most importantly, the sequence discrimination of the dsDNA binding of triplex invasive PNAs is kinetically controlled.[26,27]

As indicated in the previous section, duplexes between homopurine PNAs (especially those with high G-content) and complementary DNA have unusually high thermal stability; such PNAs can form relatively stable duplex invasion complexes.[28] Sterically compromised nucleobase pairs such as diaminopurine-thiouracil (Figure 3.4) are pseudo-complementary PNA oligomers that are unable to form stable duplexes with each other, but they are still able to hybridize strongly to complementary DNA. Such pseudo-complementary PNA oligomers form very stable double-duplex invasion complexes (Figure 3.3) when targeting dsDNA, thereby breaking the restriction to homopurine targets. When diaminopurine-thiouracil bases are used, all sequences containing at least 50% AT base pairs may be targeted[29] and, provided an analogous sterically compromised guanine-cytosine base pair can be designed, no sequence restrictions should be imposed on PNA targeting of duplex DNA.

3.5 ANTISENSE APPLICATIONS

It is now well established that PNA oligomers are very potent sequence-specific inhibitors of translation in cell-free systems.[30–34] This *in vitro* antisense effect is RNase H-independent because PNA-RNA duplexes are not substrates for this enzyme.[30–32] The inhibition of translation is therefore believed to be a steric blocking of or interference with the translational machinery (ribosomes, etc.). This fully accords with the general observation that targets around or upstream from the translation initiation (AUG) site are most sensitive to inhibition by PNA.[32,33] However, efficient antisense inhibition has been demonstrated in some systems with PNAs recognizing targets within the coding region of a mRNA.[34] Furthermore, triplex-forming homopyrimidine PNAs are able to arrest elongating ribosomes.[32]

PNA oligomers are generally taken up very poorly by eukaryotic cells, and as a result experiments on live cells or animals have progressed very slowly. However, within the past couple of years a variety of methods for cellular delivery of PNAs have been developed. These include the use of PNAs conjugated to uptake peptides, such as penetratin;[35,36] transportan,[35] or nuclear localization signal (NLS) peptide;[37] electroporation;[38,39] or delivery via cationic liposomes of weak PNA-DNA complexes[40] or PNA-fatty acid conjugates.[41] An increasing number of laboratories that utilize these techniques report biological effects of antisense PNA, both *ex vivo* on cells in culture[35–39] as well as *in vivo* on rats. The latter experiments were done by direct injection of the PNAs into the brain.[35] Finally, antisense inhibition of genes in the bacterium *Escherichia coli* has been demonstrated,[42] thereby presenting the possible development of PNA-based novel antibiotics.

3.6   ANTIGENE APPLICATIONS

The extremely stable complexes formed upon triplex invasion binding of homopyrimidine PNAs to dsDNA efficiently arrest elongating RNA polymerases.[30,43] They also hinder the access of proteins such as transcription factors, DNA-restriction enzymes, and methylases.[44–46] Therefore, PNAs may be developed into antigene agents that modulate gene expression at the transcriptional level.[30,43,44] Most interestingly, it has been found that a P-loop from a triplex invasion complex is recognized by RNA polymerase as a transcription initiation site, i.e., a promoter for which the PNA functions as an artificial transcription factor.[47] This principle could lead to the development of gene-activating drugs.[38,47]

3.7   GENETIC DIAGNOSTICS

Genetic diagnostic application has taken advantage of two particular PNA techniques. PNA probes for *in situ* hybridization have been used with great success for the quantitative analysis of telomere length,[48] which is of interest in cancer-related studies. Furthermore, infections with a number of viruses and bacteria can now be diagnosed by PNA *in situ* hybridization.[49,50] PCR amplification of the target DNA is a crucial technique in most genetic diagnostic protocols. However, in cases where the diagnosis or analysis is dependent on the discrimination between two genes that differ by only a point mutation — a single base change — this cannot be done by simple PCR analysis. In such cases "PCR-clamping" by PNA that is targeted, for example, to the wild-type gene can fully suppress the PCR amplification of this gene, while allowing full amplification of the homologous gene with the one base mutation.[51] In fact, even in cases in which, for example, tumor tissue infiltrates normal tissue where the oncogene may constitute only a few percent of the normal gene, it is possible to suppress the signal from the normal gene vs. that of a mutated oncogene using PCR-clamping.[52,53] Finally, array technologies[54] and various gene sensors[55] may eventually also benefit from PNA chemistry.

3.8   BIOMOLECULAR TOOLS

Besides the techniques described above, the unique properties of the PNA-dsDNA invasion complexes have been exploited in a variety of biomolecular tools. The high sequence specificity of the binding combined with the very high stability of the complexes has allowed the development of an Achilles' heel [DNA restriction methylase (nuclease treatment)] for single cutting of (yeast) chromosomes.[45,56] The single-stranded DNA (ssDNA) in a P-loop is very sensitive to cleavage by nuclease S1 (or mung bean nuclease), and if two closely positioned PNA sites are targeted, S1 will efficiently cleave both DNA strands at this position.[57] The single-stranded piece of DNA between two such "PNA openers" (forming a so-called PD-loop) can also be targeted by hybridization of an oligonucleotide[58] that can either function as a capture probe for isolation/purification of specific DNA fragments (for further analysis) or be used as a primer, e.g., for sequencing.[59]

Finally, (homopyrimidine bis-) PNAs can be employed as efficient tags for plasmid DNA used in transfections or gene therapy. The tag may be simply a fluorescent probe for tracing and quantifying the DNA[60,61] on a cellular marker for targeting a specific cell type carrying a receptor for the marker. The marker could, for instance, be transferrin[62] or a hormone.[63]

3.9 PREBIOTIC "LIFE"

Our knowledge of the present-day biological world is based on the central tenet that genetic information is stored in the DNA of cells and that this information is transcribed to an RNA messenger, which is in turn translated to the functional molecules of the cell, the proteins. It is now widely recognized that this present-day biological world was preceded by an RNA world.[64] RNA has the capability of carrying genetic information as well as expressing a biological function in terms of catalytic activity. However, whereas it is possible to form both amino acids and DNA-RNA nucleobases under conditions thought to be present on the young Earth ("prebiotic soup" experiments),[65,66] it is very difficult to envisage how ribose and nucleosides could have formed. On the other hand, PNA-like monomers and oligomers, which are peptide-like molecules, could have formed much more easily. Thus, PNA has been suggested as a model for the prebiotic genetic material that preceded RNA.[67] A number of model experiments have indeed demonstrated that chemical information transfer between PNA oligomers and from PNA to RNA is actually possible,[68] and PNA precursors can be formed under prebiotic conditions.[69] Therefore, one cannot exclude the possibility that life on Earth originated from a PNA-like molecule.

3.10 PNA ANALOGUES

A large variety of PNA analogues and derivatives have been made since the first publication on PNAs in 1991,[1] including backbone and nucleobase modifications. Studies of these PNA analogues are yielding a better understanding of the structural and biological properties of PNA as well as providing PNA oligomers with specifically improved properties for various of the above applications.

3.10.1 BACKBONE MODIFICATIONS

The backbone modifications published to date, along with the changes in T_m, per incorporated analogue when hybridized to a complementary DNA or RNA (if available), are presented in Table 3.1. The data are not quantitatively comparable because the T_m measurements are made in different systems (sequence context, oligomer length, etc.).

As Table 3.1 shows, much effort has been put into elucidating the effects of changes made in the original aminoethylglycine PNA backbone.

The effect of changing the number of bonds in the PNA backbone or in the linker from the backbone to the nucleobase is depicted in entries 2 through 4. The change in backbone structure from aminoethylglycine to aminopropyl glycine[70] caused a significant loss in hybridization efficiency. However, some sequence selectivity was retained when incorporating a single modified unit in a PNA decamer as a mismatch incorporated in the DNA strand opposite to the modified monomer caused a further decrease in T_m of the duplex. In the case of a propionyl linker from the backbone to the nucleobase (entry 4), a similar pattern was observed: a significant decrease in T_m while retaining sequence selectivity. In the case of an oligomer comprised solely of propionyl-linker-modified units (entry 4) with the nucleobase T, a T_m of 22°C was observed. For an oligomer

TABLE 3.1
PNA Backbone Modifications

Entry	Structure	Backbone	ΔT_m DNA °C	ΔT_m RNA °C	Ref.
1		Ethylglycine	0	0	1
2		Propylglycine	−8.0	−6.5	70
3		Ethyl-β-alanine	−10	−7.5	70
4		Propionyl linker	−20	−16	70
5		Ethyl linker	−22	−18	71
6		Retro inverso	−6.5	n.d.	72–74
7		(S,S)-cyclohexyl	−1.3	0.5	75
8		(R,R)-cyclohexyl	−7	−7.5	75
9		L-Ornithine	n.d.	−8	76,77
10		2-Me-ethylglycine	n.d.	n.d.	78

TABLE 3.1 (CONTINUED)
PNA Backbone Modifications

Entry	Structure	Backbone	ΔT_m DNA °C	ΔT_m RNA °C	Ref.
11		Ethyl-lysine	L-Lys: −1 D-Lys: +1	L-Lys: −1.3 D-Lys: 0	79
12		L-Proline	n.d.	n.d.	80
13		D-Proline	n.d.	n.d.	80
14		Glycine backbone/ethyl linker	n.d.	n.d.	80
15		L-4-*trans*-Amino proline	+2 (toward C-term), +2 (toward N-term)	n.d.	81
16		L-4-*cis*-Amino proline	−14	n.d.	81
17		D-4-*trans*-Amino proline	−7	n.d.	81
18		β-Alanine/proline	+10	n.d.	81

TABLE 3.1 (CONTINUED)
PNA Backbone Modifications

Entry	Structure	Backbone	ΔT_m DNA °C	ΔT_m RNA °C	Ref.
19		Glycylglycine/ethyl linker	−12.5	n.d.	81–83
20		Glycine/ethyl linker	n.d.	n.d.	84
21		Proline-glycine	No hybridization	No hybridization	85
22		β-Amino-alanine	−3.6	n.d.	86
23		E-OPA	−6.5	n.d.	87, 88
24		Z-OPA	−14.2	n.d.	87, 88
25		APNA	No hybridization	n.d.	89
26		Serinol-ethyl-methyl linker	n.d.	−2.5	90
27		Serinol-ethyl-ethyl linker	n.d.	−3	90, 91
28		α-Methyl-serinol-ethyl-ethyl linker	n.d.	−1.6	90

TABLE 3.1 (CONTINUED)
PNA Backbone Modifications

Entry	Structure	Backbone	ΔT_m DNA °C	ΔT_m RNA °C	Ref.
29		Aminopentan	n.d.	n.d.	92
30		Hydroxyethyl phophono glycine	See text	See text	93–96
31		Aminoethyl phosphono glycine	See text	See text	95–96
32		Lysine	n.d.	n.d.	97
33		Aminoethyl prolyl	See text	n.d.	98
34		Serinyl methylene	See text	n.d.	99–101

n.d. = data not available.

comprised solely of the units shown in entry 2 or 3,[70] i.e., aminopropyl glycine and ethyl β-alanine, no hyperchromicity was detected when the oligomer was mixed with a complementary DNA, that is, no duplex was formed. When hybridizing a mixed sequence oligomer H-(proT)$_4$-(proC)-(proT)$_5$-Lys NH$_2$ to a complementary DNA target, a T_m of 25°C was observed, and upon introducing a mismatch in the complementary DNA strand, no complex was detected (T_m < 10°C). As all three of the above-mentioned analogues differed from the "6 + 3" rule — six bonds in the backbone, three bonds in the linker between the backbone and the nucleobase — the lowered hybridization efficacy of an oligomer containing one of these modified units may very well be the result of changed spatial geometry as compared to the unmodified PNA. However, the modifications also result in a significant increase in molecular flexibility.

When changing the linker between the backbone from acyl to alkyl[71] (entry 5), a tertiary amine is introduced into the PNA backbone. This amine is protonated at physiological pH and could thus assist in hybridization to a complementary DNA due to the electrostatic attraction between the positive charge of the amine and the negative charges of the DNA phosphate backbone. However, a duplex between a PNA incorporating one of the modified units hybridized to a complementary DNA showed a dramatic decrease in T_m of 22°C (18°C when hybridized to complementary RNA). But the sequence selectivity was retained, as a mismatch in the DNA strand opposite to the modified unit caused a further decrease in T_m. The significant decrease in T_m of a duplex is most likely caused by the introduction of the very flexible ethylene linker in the modified unit. Any electrostatic attraction between the DNA and the PNA containing the modified unit was too weak to compensate for the loss of stability caused by the ethylene linker.

Krotz et al.[73,74] synthesized the "retro-inverso" PNA (entry 6). This monomer was based on a backbone of N-(aminomethyl)β-alanine with a carbonyl-methylene linker from the backbone to the nucleobase. Even though this analogue does obey the 6 + 3 rule, to keep the correct interbase distance the retro-inverso PNA does not seem to constitute a potent DNA mimic.

Lagriffoul et al.[75] have synthesized an analogue with increased rigidity in the backbone (entries 7 and 8) by incorporating, respectively, (S, S)-cyclohexane and (R, R)-cyclohexane into the ethylene part of the backbone. The (S, S)-cyclohexyl analogue had almost the same hybridization properties toward a complementary DNA as an unmodified PNA analogue, whereas the (R, R) analogue showed a significant decrease in T_m. Thermodynamic measurements showed that the (S, S) monomer hybridized with a reduced loss of entropy indicating that a better preorganization of the single-stranded PNA had indeed been attained. The more favorable entropy change was, however, compensated by a less favorable enthalpy gain upon binding.[102] Interestingly, the (S, S) analogue showed excellent selectivity when a mismatch was introduced into the complementary DNA strand, while the (R, R) analogue showed very little sequence discrimination.

Several papers have been published with the ornithine (entry 9) backbone derivative.[76,77,103] However, only one paper published melting point data[103] showing that a 10-mer comprised entirely of ornithine thymine units formed a triplex when hybridized to poly A RNA with a stoichiometry of PNA$_2$-RNA, as determined by a Job plot. The T_m of this triplex was 21°C. When mixing this T_{10} ornithine PNA with poly (dA) no melting transition could be detected. This suggests that the ornithine PNA recognizes RNA better than it recognizes DNA. This was substantiated by the fact that the T_m of an ornithine 15-mer PNA hybridized to a complementary A$_{10}$ RNA was 23°C. When hybridized to a poly (A) target, the melting temperature was found to be 31°C. The lower hybridization efficiency of this analogue can very well be attributed to the extra bond in the linker between the backbone and the nucleobase.

Haaima et al.[79] synthesized a PNA Boc-T-monomer with an aminoethyl lysine instead of an aminoethyl glycine backbone (entry 11). As mentioned earlier, this could aid the binding of the PNA strand to the target DNA by electrostatic attraction between the protonated side-chain amino group and the phosphate groups in the DNA. Also, monomers based on the D and L forms of serine, D-glutamic acid, L-aspartic acid, and L-isoleucine were synthesized. An oligomer with the sequence H-GT$_x$AGAT$_x$CACT$_x$Lys-NH$_2$ was synthesized, where T$_x$ denotes the modified monomer, and the oligomer was subsequently hybridized to its complementary DNA target. Only the oligomer containing the D-lysine-based oligomer gave a slight increase in T_m when compared to the unmodified PNA oligomer. Analogous experiments with oligomers based on the other monomers synthesized showed only a moderate negative effect on the T_m of the duplex (−0.6 to −3.3°C/monomer). As would be expected, the monomers with negatively charged side chains (D-glutamic acid, L-aspartic acid) produced the greatest effect on T_m. Also, Püschl et al.[104] synthesized Boc-PNA T-monomers based on amino acids other than glycine. The T monomers based on the L-amino acids depicted in Table 3.2 were synthesized and hybridized to both complementary RNA and DNA. The data from these experiments are outlined in the table.

Püschl et al. obtained data similar to those obtained by Haaima et al.[79] As shown in Table 3.2, the lowering of T_m by incorporating modified monomers is not additive; in addition, the sterically hindered monomers (entries f, g, h, i, j, and k) have a greater effect on T_m than do the other monomers. It

TABLE 3.2
Thermal Stabilities of Duplexes with
Backbone-Modified PNAs

Entry	Backbone Modification	T_m (RNA antiparallel)	T_m (DNA antiparallel)
a	Glycine	54.0	50.5
b	Arginine	49.0	45.5
c	Leucine	49.0	45.5
d	Glutamine	49.0	43.5
e	Lysine	48.5	46.0
f	Tyrosine	47.5	42.5
g	Histidine	47.5	42.5
h	Threonine	47.5	44.0
i	Tryptophan	47.0	42.5
j	Phenylalanine	46.5	42.0
k	Valine	46.5	42.0
l	Alanine	49.0	46.5
m	Lysine	47.5	47.0
n	Leucine	47.0	45.0
o	Histidine	46.0	42.0

Note:

Entries a–k: only one modified monomer was incorporated in the sequence H-GTAGAT$_X$CACT-NH$_2$.

Entries l–o: three modified monomers were incorporated in the sequence H-GT$_X$AGAT$_X$CACT$_X$-NH$_2$.

should be mentioned that these results were obtained with the L-isomers; also, it seems that D-isomers are in general better adapted within the PNA-DNA and PNA-RNA duplex structures.

Jordan et al.[81] (entries 15 through 18) synthesized 8-mers based on alternating units of D-4-*trans* (or the L-4-*trans*) amino proline and unmodified PNA. T_m measurements clearly showed that the oligomers containing the L-4-*trans* aminoproline units produced a significant increase in T_m (+2°C/unit), whereas the D-4-*trans*-aminoproline-containing oligomer showed a decrease of 7°C/unit. These increases in T_m were evaluated on the basis of the T_m of a PNA-T octamer, which the authors claim showed a T_m of 40 to 42°C when hybridized to its complementary DNA. Others have found this complex to have a T_m of 50 to 50.2°C,[105] in accordance with the data by Egholm et al.[106] on a PNA T$_{10}$ oligomer hybridized to a DNA d(A$_{10}$), which gave a T_m of 72°C. This could indicate that the L-4-*trans* aminoproline containing oligomer actually hybridizes to the complementary DNA target with a decrease in T_m of 1°C/monomer. The α-PNA monomer (entry 19) was synthesized by Jordan et al.[85] as the T monomer and by Howarth et al.[82] as the A, T, C, and G monomers. Howarth et al. did not report any T_m measurements for the synthesized oligomers.

One reason for the large decrease in T_m reported by Jordan et al. may be the introduction of the very flexible ethylene linker in the modified unit, which reduces the degree of preorganization in the system. Considerable increase in backbone rigidity is also introduced with this analogue.

Jordan et al.[85] also synthesized the monomeric unit depicted in entry 21. An oligomer based on this monomer showed no binding to a complementary DNA target. The authors speculate, based on molecular modeling, that this may be because oligomers containing unmodified PNA units adopt a preorientated conformation fitting helical DNA very well, whereas the fully modified PNA adopts a random three-dimensional structure.

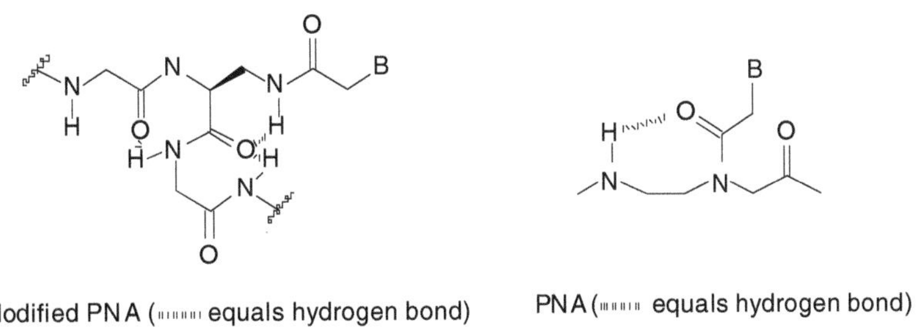

Modified PNA (⸽⸽⸽⸽ equals hydrogen bond)   PNA (⸽⸽⸽⸽ equals hydrogen bond)

FIGURE 3.5 Hydrogen bonding in modified monomer (entry 22) compared to hydrogen bonding in unmodified PNA.

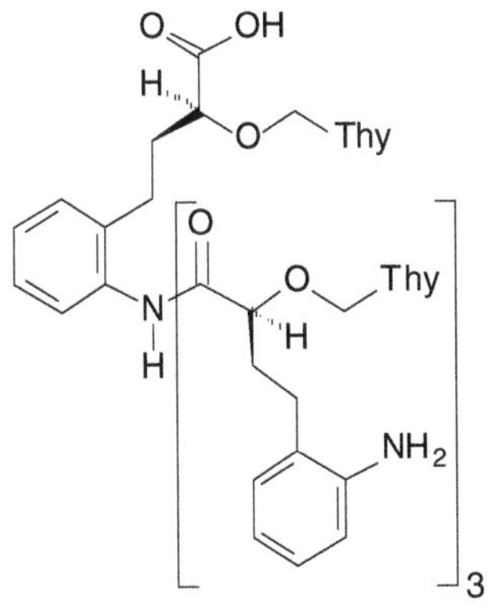

FIGURE 3.6 Aromatic peptide nucleic acid (APNA).

Fujii et al.[86] synthesized the C and T monomers based on the structure illustrated in entry 22. The linker from the backbone is longer than the original three atoms, and this may account for the lower T_m (36°C, 3.6°C/unit) when hybridized to complementary DNA (dA$_{10}$) as compared to unmodified PNA hybridized to complementary DNA (dA$_{10}$). The authors speculate that the base moieties may adopt a conformation as shown in Figure 3.5, in which an intramolecular hydrogen bond is possible; this bond stabilizes the structure, even though the T_m is lower than for the unmodified PNA.[107,108]

It has been speculated regarding unmodified PNA[70] that if such a conformational organization did occur, it should be retained in the N-(3-aminopropyl)glycine backbone (entry 2) and disrupted in the N-(2-aminoethyl)-β-alanine backbone (entry 3). However, Hyrup et al.[70] found no significant difference in the hybridization properties between PNAs containing the monomers entry 2 or 3. Likewise, the available structural data (NMR, X-ray) on PNA complexes[12,13,109-11] do not at all support the importance of intrabackbone hydrogen bonds.

Cantin et al.[87] synthesized the E- and Z-OPA monomer units based on thymine and adenine (entries 23 and 24). These structures should simulate the geometry of the amide bond in the carboxymethylene linker between the backbone and the nucleobase. Based on X-ray studies of PNA-XNA duplexes, the carboxymethylene linker exists primarily in the Z-rotameric form, but as a mixture of E and Z when the oligomer is single-stranded[11,13,14,112] T_m experiments show that both the E- and Z-monomer induce a decrease in T_m and that the decrease induced by the Z conformer is greater than that induced by the E conformer.[88]

Tsantrizos et al.[89] synthesized the aromatic peptide nucleic acid (APNA) tetramer depicted in Figure 3.6. The aromatic character of the backbone was incorporated in order to test if the possible π-stacking of the aniline moieties would have a stabilizing effect on a duplex with complementary

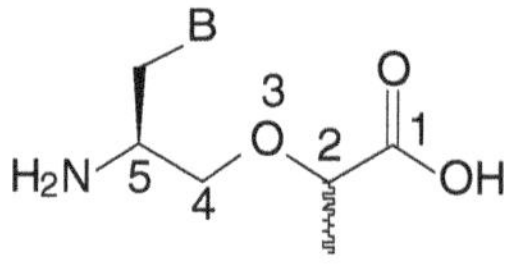

FIGURE 3.7 α-Me-serinol-ethyl backbone with an ethyl linker to the nucleobase.

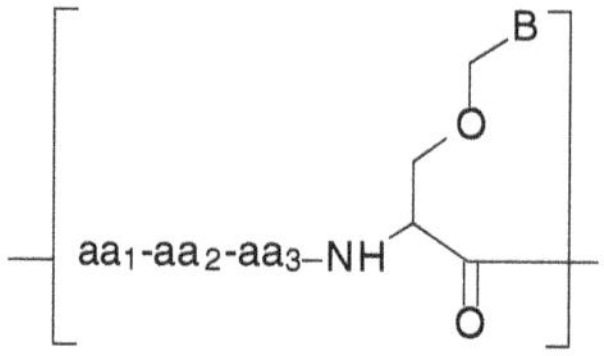

FIGURE 3.8 α-PNA by Garner et al.[99–101]

DNA. It was reported only that hybridization experiments were made with the tetramer shown in the figure, and no thermal transition above 5°C was detected.

Altmann et al.[90,113] synthesized the three serine-derived analogues depicted in entries 26 through 28. Entries 26 and 27 were synthesized as the thymine and cytosine monomers, and 28 was synthesized as the thymine monomer in two isomers: (2R, 5S) and (2S, 5S) (see Figure 3.7). A 10-mer containing three monomers hybridizes to complementary RNA, with a lower affinity than the corresponding aeg-PNA. The T_m data showed that oligomers containing the (2R, 5S) analogue produced by far the more stable duplex. T_m of the (2R, 5S)-containing duplex was 23.2°C vs. the (2S, 5S) analogue with a T_m of 9°C. This indicates a steric interference of a substituent in the S configuration.

Finally, Diederichsen et al.[114–116] synthesized an analogue with a backbone based on alternating units of alanine and homoalanine. While these oligomers do not hybridize to DNA or RNA, complementary oligomers based on these modified units formed a linear hybrid. For example, the self-complementary oligomer H-(AlaG-AlaG-AlaC-AlaG-AlaC-AlaC)-Lys-NH$_2$ displayed a T_m of 40°C.

Efimov et al.[95,96] as well as Peyman[93,94] synthesized phosphono analogues of PNA (entries 30 and 31). The aim was for these monomers to improve the solubility of the PNA oligomers, reduce any tendency of self-aggregation, and improve cellular uptake. Hybridization experiments showed that PNA oligomers containing either of these monomers form stable complexes with complementary DNA and RNA. When hybridizing a thymine 14-mer comprised of the monomeric unit depicted in entry 30 to a complementary DNA, a triplex with a T_m of 49°C was formed. When hybridizing to an RNA target, the observed T_m was 28°C. For comparison, the corresponding PNA$_2$-DNA triplex [(T$_{10}$)$_2$/dA$_{10}$] has a T_m of 73°C.[1]

A thymine 14-mer formed by alternating units of the monomeric unit outlined in entry 31 and regular PNA T monomers had a T_m, when hybridized to complementary DNA, of 52°C and a T_m of 53°C when hybridized to complementary RNA. When Peyman[93,117] synthesized a T$_9$-mer based on entry 30, T_m measurements showed a T_m of 21°C. By comparison, the T_m of d(T$_9$)-d(A$_9$) is 22°C. However, the corresponding PNA-DNA hybrid has a T_m of 70°C, i.e., a considerable decrease in T_m (–4°C/monomeric unit). Efimov found that these oligomers also showed that sequence specificity as a mismatch (T-C) in the middle of the DNA strand resulted in a decrease in T_m of 12 to 15°C. The authors[95] state that the solubility of the oligomer and chimeras that forms between the monomers in entries 30 and 31 and DNA-RNA is very high and that they are also resistant toward nucleases.

D'Costa et al.[98] synthesized the aminoethylprolyl T monomer outlined in entry 33 (aepPNA). When hybridizing a homo T oligomer to complementary DNA, UV titration experiments showed a 2/1 stoichiometry similar to the stoichiometry of a PNA_2-DNA triplex. Introducing one T-aepPNA unit at the C-terminal of a PNA T 8-mer produced only a slight effect. However, when increasing the number of aepPNA units, the stability of the complex increased by approximately 5 to 8°C/aep-PNA unit. Thus, the aepPNA vastly increases the stability of the PNA_2-DNA triplex. Surprisingly, changing the stereochemistry from R to S in the aepPNA had little or no effect on the T_m experiments. When introducing a single T-T mismatch in the middle of the homo T aep strand, no sigmoidal transition could be observed. It was thus shown that not only was the stability of the triplex increased, but the sequence selectivity was also retained.

The remarkable high increase in T_m suggests that the aepPNA backbone is correctly conformationally constrained, which in combination with the positive charge ($pK_a = 6.5$ for the monomer) contributes electrostatic attraction to the negatively charged phosphate backbone of DNA to give efficient hybridization. It would be very interesting to obtain data (including thermodynamic measurements) on aepPNA-RNA duplexes.

Garner et al.[99–101] synthesized the serinyl-methyl or α-PNA monomer outlined in entry 34. The monomer was incorporated in an oligomer as outlined in Figure 3.8.

When hybridizing (Ac-Cys-Lys-(Ser^T-Ala$_2$Lys)$_4$-Ser^T-Gly-Lys-NH$_2$)$_2$ to d(A$_{10}$), a T_m of 46°C was observed with a second transition of 54°C, i.e., a decrease of 2.4°C/monomeric unit as compared to a PNA T$_{10}$. When using only the Ac-Cys-Lys-(Ser^T-Ala$_2$Lys)$_4$-Ser^T-Gly-Lys-NH$_2$ in hybridization to d(A$_{10}$), a T_m of 17°C was observed. One could expect that only very poor hybridization would be the result of the very long interbase distances but, due to the α-helical nature of the peptide backbone of the oligomer, the nucleobases of the α-PNA seem to be able to line up with the corresponding bases of the DNA. When introducing a T-G mismatch in the cytosine containing an oligomer listed above, a decrease in T_m of 16°C was observed, thus suggesting sequence-specific hybridization.

3.10.2 NUCLEOBASE MODIFICATIONS

Table 3.3 presents most of the nucleobase modifications published in a PNA context, along with the changes in T_m, per incorporated modified nucleobase, when hybridized to a complementary DNA or RNA, if available.

Egholm et al.[25] used the pseudo-isocytosine (or J-base) (entry 1) to mimic protonated cytosine, which is necessary to obtain Hoogsteen-type base pairing in triplexes. For example, the J-base was incorporated into a bis-PNA, two PNA 7-mer strands linked by three 8-amino-3,6-dioxaoctanoic acid units. With the J-base in the Hoogsteen strand of the bis-PNA, the DNA binding was pH-independent, i.e., the J-base does *not* require protonation to form Hoogsteen-type base pairing. When hybridizing the J-base containing bis-PNA to a DNA target, the T_m of the triplex was 61.5°C at pH 5 and 59°C at pH 9. The corresponding triplex containing the C-base had T_m of 68.5°C at pH 5 and 41°C at pH 9, showing the improved binding of the J-base at neutral and high pH as compared to the C-base.

Haaima et al.[118] synthesized a diaminopurine (D) PNA monomer (entry 2) that is capable of making three hydrogen bonds to T instead of the two by A, thereby leading to increased duplex stability. When substituting an A for a D in a PNA 10-mer, an increase in T_m of a PNA-DNA duplex of 5°C was observed. When a 15-mer PNA was used, the ΔT_m was +3°C. Similar results were obtained with an RNA target. Also, a slight increase in sequence discrimination was found, depending on the type of mismatch (C-T yielding the largest change).

Eldrup et al.[119] designed and synthesized the modified PNA nucleobase depicted in entry 3 ("E-base"), aiming to expand triplex targeting beyond homopurine stretches. The idea was that this E-base could form a hydrogen bond with the T-A base pair of the duplex (Figure 3.9). The E-base was incorporated into the Hoogsteen strand of a 10-mer bis-PNA[22,25] at two positions facing A in the Watson–Crick strand. The T_m of the bis-PNA triplexed with the DNA target was found to be 57°C. When substituting the E-base with 2-acetyl-aminoethylglycine, the T_m fell to 47.5°C. A C-G

TABLE 3.3
Nucleobased-Modified PNA Analogues

Entry	Structure	Nucleobase	ΔT_m DNA °C	ΔT_m RNA °C	Ref.
1		Pseudo iso cytosine	See text	n.d.	25
2		Diamino purine	+3 to +5	+2.5 to +6	118
3		3-Oxo-2,3-dihydrophyri-dazine	See text	n.d.	119
4		Isocytosine	n.d.	n.d.	120
5		5-Bromo uracil	n.d.	n.d.	120, 121
6		5-Methyl cytosine	See text	n.d.	121
7		4-Thiothymine	n.d.	n.d.	122, 123

TABLE 3.3 (CONTINUED)
Nucleobased-Modified PNA Analogues

Entry	Structure	Nucleobase	ΔT_m DNA °C	ΔT_m RNA °C	Ref.
8		N^3-Methyl-4-thiothymin	n.d.	n.d.	122
9		5-Nitroindole	−9 (See text)	n.d.	124
10		3-Nitropyrrole	−13 (See text)	n.d.	124
11		2-Thiouracil	−2.5	n.d.	29
12		Thioguanine	−8.5 antiparallel, −7.5 parallel	n.d.	125
13		N^4-Benzoyl cytosine	−2.5	n.d.	126
14		2-Amino purine	−2 to −3	n.d.	127

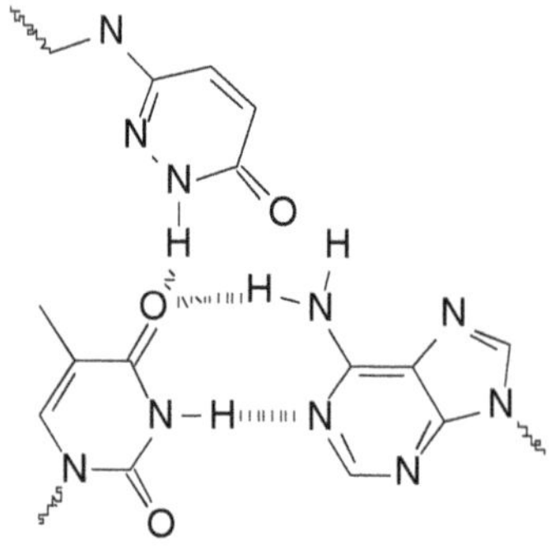

FIGURE 3.9 E-base hybridizing to a T-A base pair.

base pair in the target resulted in a T_m of 42°C, thus indicating that the E-base does in fact hybridize specifically to a T-A base pair. Sequence discrimination was in the range of ΔT_m = 12 to 16°C when the center strand of the triplex was opposite both Watson–Crick strands.

PNA 1 (X=N$_p$)
PNA 2 (X =N$_i$)
H-TGT ACG **X** CACAAC TA-NH$_2$
3'-ACA TGC **Y** CTG TTG AT-NH$_2$
Y=A, C, G, T

FIGURE 3.10 Nitroindole/nitropyrrole-containing oligomers.

TABLE 3.4
Hybridization Experiments Using the Oligonucleotides Illustrated in Figure 3.6[124]

Indole Substitution	Nucleobase	T_m (°C)
X = N$_p$	A	56.2
	G	54.7
	C	55.9
	T	55.1
X = N$_i$	A	58.4
	G	58.9
	C	59.7
	T	59.3
X = T	A	68.5

Ferrer et al.[121] synthesized the 5-bromouracil monomer as well as a 5-methylcytosine monomer. These two monomers were incorporated into PNA-DNA chimeras and not into pure PNA oligomers. A T-hydroxyethylglycine unit joined the chimeras. In the melting experiments T was substituted by 5-bromouracil and C by 5-methylcytosine. When the 5-methylcytosine was incorporated into the PNA part of a chimeric 11-mer, a ΔT_m of −1°C/monomeric unit was found when compared to the corresponding DNA-DNA duplex. In a similar experiment with the 5-bromouracil, a ΔT_m of −2.5°C/monomeric unit was found. Only when 5-methylcytosine was incorporated into both the PNA and DNA part of the chimera was a stabilizing effect observed. This could be due to increased stacking interactions.

Challa et al.[124] synthesized nitroindole and nitropyrrole monomers (entries 9 and 10) in order to introduce nucleobase analogues that would bind without base discrimination at the site where these monomers were incorporated in analogy with results obtained with deoxyribonucleotides. This would be useful as, for example, degenerate hybridization probes. When hybridizing the PNA-DNA outlined in Figure 3.10, the results in Table 3.4 were obtained.

The variation in T_m is only 1.5°C when using the nitropyrrole monomer and 1.3°C when using the nitroindole monomer opposite the four naturally occurring nucleobases. However the incorporation of a nitropyrrole monomer gives a mean ΔT_m of –13°C, while the incorporation of a nitroindole monomer is slightly less destabilizing, with a mean $\Delta T_m = -9.4$°C. Therefore, as anticipated, these monomers can be employed as single promiscuous bases.

Lohse et al.[29] synthesized the thiouracil monomer in entry 11, in order to use it in combination with diaminopurine (entry 2) for "double-duplex invasion." Double-duplex invasion is a concept according to which the DNA duplex is unwound and both DNA strands are targeted simultaneously by two PNAs that are pseudocomplementary, i.e., they recognize their DNA targets but cannot bind each other. When a thiouracil monomer was introduced opposite adenine, a decrease in T_m of 2 to 3°C was observed. The diaminopurine produced an expected increase of 5 to 6°C in T_m. When two PNAs (H-GTAGATCACT-NH$_2$ and H-AGTGATCTAC-NH$_2$) were synthesized, and all thymines were substituted with thiouracil and all adenines with diaminopurine, these two PNAs were found to be hybridized better to their DNA complements than the original PNAs, but they did not hybridize to each other. Gel-mobility assays confirmed that the principle did in fact work when a 204-base-pair dsDNA fragment containing the target sequence was targeted. Experiments showed that, in order to obtain the double-duplex invasion, the critical A/T content was about 40% A/T. The double duplex complexes were found to effectively block access of the RNA polymerase to the DNA but, unlike PNA, triplex-invasion complexes could not arrest elongating phage RNA polymerases.

Hansen et al.[125] synthesized a PNA monomer based on 6-thioguanine (entry 12). Experiments showed that incorporation of a single thioguanine unit in a PNA 10-mer resulted in a decrease in T_m $(\Delta T_m = -8.5)$°C, a greater effect than in the DNA-DNA case. Interestingly, when hybridizing to a parallel target the effect was a less pronounced decrease in T_m $(\Delta T_m = -5.5$°C$)$. When introducing a mismatch opposite the thioguanine moiety, a decrease in T_m $(\Delta T_m = -7.5$°C$)$ was observed, i.e., some sequence discrimination was retained.

Christensen et al.[126] synthesized an N^4-benzoylated cytosine PNA monomer (entry 13). When incorporating this monomer into a mixed-sequence PNA oligomer, only a slight decrease in T_m was observed $(\Delta T_m = -2.5$°C$)$. However, when incorporated into the middle of a homopyrimidine PNA oligomer, the T_m of the resulting PNA$_2$-DNA triplex was reduced by 30°C as compared to the T_m of the unmodified PNA$_2$-DNA triplex, due to a dramatic destabilization of the triplex.

Gangamani et al.[127] synthesized the fluorescent 2-aminopurine monomer depicted in entry 14. When H-TAT TAT TAT T-βala-OH (A being the 2-aminopurine monomer) was hybridized to a complementary DNA, the T_m of the complex was decreased by 2 to 3°C in both the parallel and the antiparallel orientations.

3.11 CONCLUDING REMARKS

Clearly, the introduction of PNA has inspired many chemists to pursue this type of DNA mimic, as exemplified by the vast number of derivatives and modifications synthesized thus far. Furthermore, great ingenuity has been invested in exploring the potential of PNA in a variety of applications within molecular biology, diagnostics, and drug development. The future will undoubtedly further expand the chemical scope of PNA as well as the present and novel applications. It is also hoped that safe and efficient PNA-based gene therapeutic drugs will be developed to treat genetically based diseases such as cancer as well as to combat the increasing threat from infectious diseases by microorganisms that have developed resistance to existing drugs.

REFERENCES

1. Nielsen, P. E. et al., Sequence-selective recognition of DNA by strand displacement with a thymine-substituted polyamide, *Science*, 254, 1497, 1991.
2. Egholm, M. et al., Peptide nucleic acids (PNA). Oligonucleotide analogues with an achiral peptide backbone, *J. Am. Chem. Soc.*, 114, 1895, 1992.
3. Hyrup, B. and Nielsen, P. E., Peptide nucleic acids (PNA): synthesis, properties and potential applications, *Bioorg. Med. Chem.*, 4, 5, 1996.
4. Nielsen, P. E., Peptide nucleic acid: a molecule with two identities, *Acc. Chem. Res.*, 32, 624, 1999.
5. Nielsen, P. E. and Egholm, M., *Peptide Nucleic Acid (PNA): Protocols and Applications*, Horizon Scientific Press, Norfolk, VA, 1999.
6. Ray, A. and Nordén, B., Peptide nucleic acid (PNA): its medical and biotechnical applications, *FASEB J.*, 14, 1041, 2000.
7. Soomets, U., Antisense properties of peptide nucleic acids, *Frontiers Biosci.*, 4, D782, 1999.
8. Nielsen, P. E., Peptide nucleic acids as therapeutic agents, *Curr. Opin. Struct. Biol.*, 9, 353, 1999.
9. Nielsen, P. E., Applications of peptide nucleic acids, *Curr. Opin. Biotechnol.*, 10, 71, 1999.
10. Corey, D. R., Peptide nucleic acids: expanding the scope of nucleic acid recognition, *TIBTECH*, 15, 224, 1997.
11. Brown, S. C. et al., NMR Solution structure of a peptide nucleic acid complexed with RNA, *Science*, 265, 777, 1994.
12. Eriksson, M. and Nielsen, P. E., Solution structure of a peptide nucleic acid-DNA duplex, *Nat. Struct. Biol.*, 3, 410, 1996.
13. Betts, L. et al., A nucleic acid triple helix formed by a peptide nucleic acid-DNA complex, *Science*, 270, 1838, 1995.
14. Rasmussen, H. and Sandholm, J., Crystal structure of a peptide nucleic acid (PNA) duplex at 1.7 Å resolution, *Nat. Struct. Biol.*, 4, 98, 1997.
15. Egholm, M. et al., PNA hybridizes to complementary oligonucleotides obeying the Watson–Crick hydrogen-bonding rules, *Nature*, 365, 566, 1993.
16. Kilså Jensen, K. et al., Kinetics for hybridization of peptide nucleic acids (PNA) with DNA and RNA studied with the BIAcore technique, *Biochemistry*, 36, 5072, 1997.
17. Giesen, U. et al., A formula for thermal stability T_m prediction of PNA/DNA duplexes, *Nucleic Acids Res.*, 26, 5004, 1998.
18. SantaLucia, J., Allawi, H. T., and Seneviratne, P. A., Improved nearest-neighbor parameters for predicting DNA duplex stability, *Biochemistry*, 35, 3555, 1995.
19. Nielsen, P. E., Egholm, M., and Buchardt, O., Evidence for $(PNA)_2$/DNA triplex structure upon binding of PNA to dsDNA by strand displacement, *J. Mol. Recognition*, 7, 165, 1994.
20. Praseuth, D. et al., Peptide nucleic acids directed to the promotor of the α-chain of the interleukin-2 receptor, *Biochim. Biophys. Acta*, 1309, 226, 1996.
21. Wittung, P., Nielsen, P., and Nordén, B., Extended DNA-recognintion repertoire of peptide nucleic acid (PNA): PNA-dsDNA triplex formed with cytosine-rich homopyrimidine PNA, *Biochemistry*, 36, 7973, 1997.
22. Griffith, M. C. et al., Single and bis peptide nucleic acids as triplexing agents: binding and stoichiometry, *J. Am. Chem. Soc.*, 117, 831, 1995.
23. Bentin, T. and Nielsen, P. E., Triplexes involving PNA in *Triple Helix Forming Oligonucleotides*, Malvy, C., Barel-Bellan, A., and Pritchard, L., Eds., Kluwer Academic Publishers, Dordrecht, 1999, 245.
24. Bentin, T. and Nielsen, P. E., Enhanced peptide nucleic acid binding to supercoiled DNA: possible implications for DNA "breathing" dynamics, *Biochemistry*, 35, 8863, 1996.
25. Egholm, M. et al., Efficient pH-independent sequence-specific DNA binding by pseudocytocine-containing bis-PNA, *Nucleic Acids Res.*, 23, 217, 1995.
26. Kuhn, H. et al., Kinetic sequence discrimination of cationic bis-PNAs upon targeting of double stranded DNA, *Nucleic Acids Res.*, 26, 582, 1998.
27. Demidov, V. V. et al., Kinetics and mechanism of polyamide ("peptide") nucleic acid binding to duplex DNA, *Proc. Natl. Acad. Sci. U.S.A.*, 92, 2637, 1995.
28. Nielsen, P. E. and Christensen, L., Strand displacement binding of a duplex-forming homopurine PNA to a homopyrimidine duplex DNA target, *J. Am. Chem. Soc.*, 118, 2287, 1996.

29. Lohse, J., Dahl, O., and Nielsen, P. E., Double duplex invasion by peptide nucleic acid: a general principle for sequence-specific targeting of double-stranded DNA, *Proc. Natl. Acad. Sci. U.S.A.*, 96, 11804, 1999.

30. Hanvey, J. C. et al., Antisense and antigene properties of peptide nucleic acids, *Science*, 258, 1481, 1992.

31. Bonham, M. A. et al., An assessment of the antisense properties of RNase H-competent and steric-blocking oligomers, *Nucleic Acids Res.*, 23, 1197, 1995.

32. Knudsen, H. and Nielsen, P. E., Antisense properties of duplex- and triplex-forming PNAs, *Nucleic Acids Res.*, 24, 494, 1996.

33. Mologni, L. et al., Additive antisense effects of different PNAs on the *in vitro* translation of the PML/RAR, *Nucleic Acids Res.*, 26, 1934, 1998.

34. Dias, N. et al., Antisense PNA tridecamers targeted to the coding region of ha-ras mRNA arrest polypeptide chain elongation, *J. Mol. Biol.*, 294, 403, 1999.

35. Pooga, M. et al., Cell penetrating PNA constructs regulate galanin receptor levels and modify pain transmission *in vivo*, *Nat. Biotechnol.*, 16, 857, 1998.

36. Aldrian-Herrada, G. et al., A peptide nucleic acid (PNA) is more rapidly internalized in cultured neurons when coupled to a *retro-inverso* delivery peptide. The antisense activity depresses the target mRNA and protein in magnocellular oxytocin neurons, *Nucleic Acids Res.*, 26, 4910, 1998.

37. Cutrona, G. et al., Effects in live cells of a c-*myc* antigene PNA linked to a nuclear localization signal, *Nat. Biotechnol.*, 18, 300, 2000.

38. Wang, G. et al., Peptide nucleic acid (PNA) binding-mediated induction of human γ-globin gene expression, *Nucleic Acids Res.*, 27, 2806, 1999.

39. Shammas, M. A. et al., Telomerase inhibition by peptide nucleic acids reverses "immortality" of transformed human cells, *Oncogene*, 18, 6191, 1999.

40. Hamilton, S. E. et al., Cellular delivery of peptide nucleic acids and inhibition of human telomerase, *Chem. Biol.*, 6, 343, 1999.

41. Ljungstrøm, T., Knudsen, H., and Nielsen, P. E., Cellular uptake of adamantyl conjugated peptide nucleic acids, *Bioconjugate Chem.*, 10, 965, 1999.

42. Good, L. and Nielsen, P. E., Antisense inhibition of gene expression in bacteria by PNA targeted to mRNA, *Nat. Biotechnol.*, 16, 355, 1998.

43. Nielsen, P. E., Egholm, M., and Buchardt, O., Sequence-specific transcription arrest by peptide nucleic acid bound to the DNA template strand, *Gene*, 149, 139, 1994.

44. Vickers, T. A. et al., Inhibition of NF-κB specific transcriptional activation by PNA strand invasion, *Nucleic Acids Res.*, 23, 3003, 1995.

45. Veselkov, A. G. et al., A new class of genome rare cutters, *Nucleic Acids Res.*, 24, 2483, 1996.

46. Nielsen, P. E. et al., Sequence specific inhibition of DNA restriction enzyme cleavage by PNA, *Nucleic Acids Res.*, 21, 197, 1993.

47. Møllegaard, N. E. et al., Peptide nucleic acid-DNA strand displacement loops as artificial transcription promoters, *Proc. Natl. Acad. Sci. U.S.A.*, 91, 3892, 1994.

48. Rufer, N. et al., Telomere length dynamics in human lymphocyte subpopulations measured by flow cytometry, *Nat. Biotechnol.*, 16, 743, 1998.

49. Stender, H. et al., Direct detection and identification of *Mycobacterium tuberculosis* in smear-positive sputum samples by fluorescence *in situ* hybridization (FISH) using peptide nucleic acid (PNA) probes, *Int. J. Tuberc. Lung Dis.*, 3, 830, 1999.

50. Stender, H. et al., Fluorescence *in situ* hybridization assay using peptide nucleic acid probes for differentiation between tuberculous and nontuberculous *Mycobacterium* species in smears of mycobacterium cultures, *J. Clin. Microbiol.*, 37, 2760, 1999.

51. Ørum, H. et al., Single base pair mutation analysis by PNA directed PCR clamping, *Nucleic Acids Res.*, 21, 5332, 1993.

52. Thiede, C. et al., Simple and sensitive detection of mutations in the ras proto-oncogenes using PNA-mediated PCR clamping, *Nucleic Acids Res.*, 24, 983, 1996.

53. Weiler, J. et al., Hybridisation based DNA screening on peptide nucleic acid (PNA) oligomer arrays, *Nucleic Acids Res.*, 25, 2792, 1997.

54. Behn, M. et al., Facilitated detection of oncogene mutations from exfoliated tissue material by a PNA-mediated "enriched PCR" protocol, *J. Pathol.*, 190, 69, 2000.

55. Wang, J., DNA biosensors based on peptide nucleic acid (PNA) recognition layers. A review, *Biosensors Bioelectron.*, 13, 757, 1998.

56. Izvolsky, K. I. et al., Yeast artificial chromosome segregation from host chromosomes with similar lengths, *Nucleic Acids Res.*, 26, 5011, 1998.

57. Demidov, V. et al., Sequence selective double strand DNA cleavage by peptide nucleic acid (PNA) targeting using nuclease S1, *Nucleic Acids Res.*, 21, 2103, 1993.

58. Bukanov, N. O. et al., PD-loop: a complex of duplex DNA with an oligonucleotide, *Proc. Natl. Acad. Sci. U.S.A.*, 95, 5516, 1998.

59. Demidov, V. et al., Artificial primosome: design, functioning and applications, *Chembiochem. Eur. J. Chem. Biol.*, 2, 133, 2001.

60. Zelphati, O. et al., Gene chemistry: functionally and conformationally intact fluorescent plasmid DNA, *Hum. Gene Ther.*, 10, 15, 1999.

61. Zelphati, O. et al., PNA-dependent gene chemistry: stable coupling of peptides and oligonucleotides to plasmid DNA, *Biotechniques*, 28, 304, 2000.

62. Liang, K. W., Hoffmann, E. P., and Huang, L., Targeted delivery of plasmid DNA to myogenic cells via transferrin-conjugated peptide nucleic acid, *Mol. Ther.*, 1, 236, 2000.

63. Boffa, L. C. et al., Dihydrotestosterone as a selective cellular/nuclear localization vector for anti-gene peptide nucleic acid in prostatic carcinoma cells, *Cancer Res.*, 60, 2258, 2000.

64. Gesteland, R., Cech, T. R., and Atkins, J. F., *The RNA World*, 2, Cold Spring Harbor Laboratory Press, Cold Spring Harbor, NY, 1999.

65. Oró, J., Synthesis of adenine from ammonium cyanide, *Biochem. Biophys. Res. Commun.*, 2, 407, 1960.

66. Miller, S. L., A production of amino acids under possible primitive earth conditions, *Science*, 117, 528, 1953.

67. Nielsen, P. E., Peptide nucleic acid (PNA) a model structure for the primordial genetic material, *Origins Life*, 23, 323, 1993.

68. Böhler, C., Nielsen, P. E., and Orgel, L. E., Template switching between PNA and RNA oligonucleotides, *Nature*, 376, 578, 1995.

69. Nelson, K. E., Levy, M., and Miller, S. L., Peptide nucleic acids rather than RNA may have been the first genetic molecule, *Proc. Natl. Acad. Sci. U.S.A.*, 97, 3868, 2000.

70. Hyrup, B. et al., Structure-activity studies of the binding of modified peptide nucleic acids (PNAs) to DNA, *J. Am. Chem. Soc.*, 116, 7964, 1994.

71. Hyrup, B. et al., A flexible and positively charged PNA analogue with an ethylene-linker to the nucleobase: synthesis and hybridization properties, *Bioorg. Med. Chem. Lett.*, 6, 1083, 1996.

72. Krotz, A. H. et al., A "retro-inverso" PNA: structural implications for DNA and RNA binding, *Bioorg. Med. Chem.*, 6, 1983, 1998.

73. Krotz, A. H., Buchardt, O., and Nielsen, P. E., Synthesis of "retro-inverso" peptide nucleic acids: 1. Characterization of the monomers, *Tetrahedron Lett.*, 36, 6937, 1995.

74. Krotz, A. H., Buchardt, O., and Nielsen, P. E., Synthesis of "retro-inverso" peptide nucleic acids: 2. Oligomerization and stability, *Tetrahedron Lett.*, 36, 6941, 1995.

75. Lagriffoul, P., Nielsen, P. E., and Buchardt, D., Denmark, WO 96/20212, 1, 1996.

76. Van der Laan, A. C. et al., *Innovation and Perspectives in Solid Phase Synthesis and Combinatorial Synthesis*, Mayflower Scientific, Birmingham, 1995, 1.

77. Lioy, E., and Kessler, H., Synthesis of a new chiral peptide analogue of DNA using ornithine subunits and solid phase peptide synthesis methodologies, *Liebigs Ann. Chem.*, 201, 1996.

78. Kosynkina, L., Wang, W., and Liang, T. C., A convenient synthesis of chiral peptide nucleic acid (PNA) monomers, *Tetrahedron Lett.*, 35, 5173, 1994.

79. Haaima, G. et al., Peptide nucleic acids (PNAs) containing thymine monomers derived from chiral amino acids: hybridization and solubility properties of D-lysine PNA, *Angew. Chem. Int. Ed. Eng.*, 35, 1939, 1996.

80. Lowe, G. and Vilaivan, T., Amino acids bearing nucleobases for the synthesis of novel peptide nucleic acids, *J. Chem. Soc. Perkin Trans.*, 1, 539, 1997.

81. Jordan, S. et al., New hetero-oligomeric peptide nucleic acids with improved binding properties to complementary DNA, *Bioorg. Med. Chem. Lett.*, 7, 687, 1997.

82. Howarth, N. M. and Wakelin, L. P. G., α-PNA: a novel peptide nucleic acid analogue of DNA, *J. Org. Chem.*, 62, 5441, 1997.

83. Szyrwiel, J. et al., Chiral peptide nucleic acids: a new family of ligands for metal ions, *J. Chem. Soc. Dalton Trans.*, 1263, 1998.

84. Ciapetti, P., Soccolini, F., and Taddei, M., Synthesis of N-Fmoc-a-amino acids carrying the four DNA nucleobases in the side chain, *Tetrahedron*, 53, 1167, 1997.

85. Jordan, S. et al., Synthesis of new building blocks for peptide nucleic acids containing monomers with variations in the backbone, *Bioorg. Med. Chem. Lett.*, 7, 681, 1997.

86. Fujii, M., Yoshida, K., and Hidaka, J., Nucleic acid analog peptide (NAAP) 2. Synthesis and properties of novel DNA analog peptides containing nucleobase linked β-aminoalanine, *Bioorg. Med. Chem. Lett.*, 7, 637, 1997.

87. Cantin, M., Schütz, R., and Leumann, C. J., Synthesis of the monomeric building blocks of Z-olefinic PNA (Z-OPA) containing the bases adenine and thymine, *Tetrahedron Lett.*, 38, 4211, 1997.

88. Schütz, R. et al., Olefinic peptide nucleic acids (OPAs): new aspects of the molecular regognition of DNA by PNA, *Angew. Chem. Int. Ed. Eng.*, 39, 1250, 2000.

89. Tsantrizos, Y. S. et al., Stereoselective synthesis of a thymine derivative of (*S*)-2-hydroxy-4-(2-aminophenyl)butanoic acid. A novel building block for the synthesis of aromatic peptidic nucleic acid oligomers, *J. Org. Chem.*, 62, 5451, 1997.

90. Altmann, K. H., Chiesi, C. S., and García-Echeverría, C., polyamide based nucleic acid analogs. Synthesis of δ-amino acids with nucleic acid bases bearing side chains, *Bioorg. Med. Chem. Lett.*, 7, 1119, 1997.

91. Kuwahara, M., Arimitsu, M., and Sisido, M., Novel peptide nucleic acid that shows high sequence specificity and all-or-none-type hybridization with the complementary DNA, *J. Am. Chem. Soc.*, 121, 256, 1999.

92. Bergmeier, S. C. and Fundy, S. L., Synthesis of oligo(5-aminopentanoic acid)-nucleobases (APN): potential antisense agents, *Bioorg. Med. Chem. Lett.*, 7, 3135, 1997.

93. Peyman, A. et al., Phosphonic ester nucleic acids (PHONAs): oligonucleotide analogues with an achiral phosphonic acid ester backbone, *Angew. Chem. Int. Ed. Eng.*, 35, 2636, 1996.

94. Uhlmann, E. et al., Synthesis of polyamide nucleic acids (PNAs), PNA/DNA chimeras and phosphonic ester nucleic acids (PHONAs), *Nucleosides Nucleotides*, 16, 603, 1997.

95. Efimov, V. A. et al., Synthesis and evaluation of some properties of chimeric oligomers containing PNA and phosphono-PNA residues, *Nucleic Acids Res.*, 26, 566, 1998.

96. Efimov, V. A. et al., Synthesis and binding study of phosphonate analogues of PNAs and their hybrids with PNAs, *Nucleosides Nucleotides*, 17, 1671, 1998.

97. Zhang, L., Min, J., and Zhang, L., Studies on the synthesis and properties of new PNA analogs consisting of L- and D-lysine backbones, *Bioorg. Med. Chem. Lett.*, 9, 2903, 1999.

98. D'Costa, M., Kumar, V. A., and Ganesh, K. N., Aminoethylprolyl peptide nucleic acids (aepPNA): Chiral PNA analogues that form highly stable DNA:aepPNA$_2$ triplexes, *Org. Lett.*, 1, 1513, 1999.

99. Garner, P. and Yoo, J. U., Peptide-based nucleic acid surrogates incorporating Ser[CH$_2$B]-Gly subunits, *Tetrahedron Lett.*, 34, 1275, 1993.

100. Garner, P. et al., Modular nucleic acid surrogates. Solid phase synthesis of α-helical peptide nucleic acids (αPNAs), *Org. Lett.*, 1, 403, 1999.

101. Garner, P., Dey, S., and Huang, Y., α-helical peptide nucleic acids (α-PNAs): a new paradigm for DNA binding molecules, *J. Am. Chem. Soc.*, 122, 2405, 2000.

102. Lagriffoul, P. et al., Peptide nucleic acids (PNAs) with a conformationally constrained chiral cyclohexyl derived backbone, *Chem. Eur. J.*, 3, 912, 1997.

103. Petersen, K. H., Buchardt, O., and Nielsen, P. E., Synthesis and oligomerization of N^δ-Boc-N^α-(thymin-1-ylacetyl)ornithine, *Bioorg. Med. Chem. Lett.*, 6, 793, 1996.

104. Püschl, A. et al., Peptide nucleic acids (PNAs) with a functional backbone, *Tetrahedron Lett.*, 39, 4707, 1998.

105. Eldrup, A. B., Modified Nucleobases in Peptide Nucleic Acids, Ph.D. thesis, Faculty of Science, University of Copenhagen, 1998, 91.

106. Egholm, M. et al., Recognition of guanine and adenine in DNA by cytosine and thymine containing peptide nucleic acids (PNA), *J. Am. Chem. Soc.*, 114, 9677, 1992.

107. Almarsson, Ö. and Bruice, T. C., Peptide nucleic acid (PNA) conformation and polymorphism in PNA-DNA and PNA-RNA hybrids, *Proc. Natl. Acad. Sci. U.S.A.*, 90, 9542, 1993.

108. Almarsson, Ö. et al., Molecular mechanics calculations of the structures of polyamide nucleic acid DNA duplexes and triple helical hybrids, *Proc. Natl. Acad. Sci. U.S.A.*, 90, 7518, 1993.

109. Eriksson, M., Structure of PNA-nucleic acid complexes, *Nucleosides Nucleotides*, 16, 617, 1997.

110. Torres, R. A. and Bruice, T. C., Interresidue hydrogen bonding in a peptide nucleic acid-RNA heteroduplex, *Proc. Natl. Acad. Sci. U.S.A.*, 93, 649, 1996.

111. Kastrup, J. S. et al., Crystallization and preliminary X-ray analysis of a PNA-DNA complex, *Fed. Eur. Biochem. Soc. Lett.*, 363, 115, 1995.

112. Chen, S. et al., Molecular dynamics and NMR studies of single-stranded PNAs, *Tetrahedron Lett.*, 35, 5105, 1994.

113. García-Echeverría, C. et al., Novel polyamide based nucleic acid analogs. Synthesis of oligomers and RNA-binding properties, *Bioorg. Med. Chem. Lett.*, 7, 1123, 1997.

114. Diederichsen, U. and Schmitt, H. W., Self-pairing PNA with alternating alanyl/homoalanyl backbone, *Tetrahedron Lett.*, 37, 475, 1996.

115. Diederichsen, U., Alanyl-PNA oligomers: a new system for intercalation, *Bioorg. Med. Chem. Lett.*, 7, 1743, 1997.

116. Diederichsen, U., Alanyl PNA: evidence for linear band structures based on guanine-cytosine base pairs, *Angew. Chem. Int. Ed. Eng.*, 36, 1886, 1997.

117. Uhlmann, E. and Peyman, A., Antisense oligonucleotides: a new therapeutic principle, *Chem. Rev.*, 90, 543, 1990.

118. Haaima, G. et al., Increased DNA binding and sequence discrimination of PNA oligomers containing 2,6-diaminopurine, *Nucleic Acids Res.*, 25, 4639, 1997.

119. Eldrup, A. B., Dahl, O., and Nielsen, P. E., A novel peptide nucleic acid monomer for recognition of thymine in triple-helix structures, *J. Am. Chem. Soc.*, 119, 11116, 1997.

120. Egholm, M. et al., U.S. Patent WO 96/02558, 39, 1995.

121. Ferrer, E., Shevchenko, A., and Eritja, R., Synthesis and hybridization properties of DNA-PNA chimeras carrying 5-bromouracil and 5-methylcytosine, *Bioorg. Med. Chem.*, 8, 291, 2000.

122. Clivio, P. et al., Synthesis and photochemical behaviour of peptide nucleic acid dimers and analogues containing 4-thiothymine: unprecedented (4–5) photoadduct reversion, *J. Am. Chem. Soc.*, 120, 1157, 1998.

123. Clivio, P. et al., A photochemical approach to highlight backbone effects in PNA, *J. Am. Chem. Soc.*, 119, 5255, 1997.

124. Challa, H., Styers, M. L., and Woski, S. A., Nitroazole universal bases in peptide nucleic acids, *Org. Lett.*, 1, 1639, 1999.

125. Hansen, H. F. et al., 6-Thioguanine in peptide nucleic acids. Synthesis and hybridization properties, *Nucleosides Nucleotides*, 18, 5, 1999.

126. Christensen, L. et al., Inhibition of PNA triplex formation by N^4-benzoylated cytosine, *Nucleic Acids Res.*, 26, 2735, 1998.

127. Gangamani, B. P., Kumar, V. A., and Ganesh, K. N., 2-Aminopurine peptide nucleic acids (2-*ap*PNA): intrinsic fluorescent PNA analogues for probing PNA-DNA interaction dynamics, *Chem. Commun.*, 19, 1913, 1997.

4 Synthetic DNA Used in Amplification Reactions

Lisa S. Kelly

CONTENTS

0-8493-1426-7/03/$0.00+$1.50
© 2003 by CRC Press LLC

4.1 INTRODUCTION

Because nucleic acids constitute the structure and information of genomes, reproduction of organisms can be considered a means of biological amplification of DNA or RNA. For hundreds of millions of years, amplification had been confined within organisms and limited to their evolved sequences. Within the last 50 years, the physical structure of DNA[1] and the nature of its templated, primed replication has been deciphered. Today this knowledge allows replication to occur outside of cells, easily controlled and manipulated by the researcher. This *in vitro* amplification has transformed the life sciences. It is an extremely powerful tool, allowing any sequence, natural or synthetic, to be manufactured to the necessary abundance for purposes of analysis, manipulation, or diagnostics.

The first intimation of exponential enzymatic amplification came in the early 1970s. Because artificial synthesis of DNA was so cumbersome as to render impossible the fabrication of large amounts of long DNA, Kleppe[2] endeavored to replicate the genes of transfer-RNA enzymatically in order to generate sufficient quantities to study. Using short synthetic DNA primers, he performed several iterations of primer annealing, extension, and melting — essentially a few rounds of the polymerase chain reaction.[3] So why did Kleppe's experiment not captivate the scientific community as did the polymerase chain reaction 15 years later? At that time, it was not feasible to synthesize even the primers in enough quantity for many laboratories to realistically perform the reaction. Ironically, only with the introduction of automated phosphoramidite synthesis[4] could a more ambitious form of the replication reaction, which was originally designed to circumvent inefficient synthesis, become widespread and well known. Thus, the power to create DNA synthetically on demand conferred the power to create DNA enzymatically on demand.

Once both the concept and the capability converged, numerous reactions with varying amplification mechanisms were developed. The profusion of amplification reactions assures that no single chapter, including this one, can serve as a comprehensive review (for a review of recent advances, see Reference 5). Rather, the purpose of this chapter is to briefly describe the various amplification reactions, highlight the use of nonamplified DNA in amplification reactions, and recount how *in vitro* amplification reactions can be used to mimic evolution.

4.2 AMPLIFICATION REACTIONS

4.2.1 POLYMERASE CHAIN REACTION

The polymerase chain reaction (PCR) was the first well-known exponential amplification reaction, and it remains the most prominent. Its utility has been recognized from its inception. This reaction, and its extended family of modifications, has been employed in literally hundreds of thousands of published experiments. Its success has triggered a bevy of competitors,[6] none of which has approached its popularity. In the 1980s, Mullis and co-workers[3] reinvented the repair replication reaction of Kleppe. Their essential contribution was developing the reaction not just for replicating purified DNA of known sequence, but as a tool to extract sequences within the background of genomic DNA or unknown sequence. The subsequent modifications and techniques are too numerous to describe here (see, e.g., Reference 7), but an overview of some of the possibilities will be presented. PCR enables practically any procedure requiring the amplification of DNA such as gene construction, probe labeling, sequence detection, DNA sequencing, DNA profiling, etc.

4.2.1.1 Reaction Mechanics

The basic reaction (Figure 4.1A) requires two oligonucleotides whose 3′ termini are capable of annealing to the downstream edges of the target sequence, each on a separate strand.[8] Extension of the primers by DNA polymerase generates overlapping double-stranded products. Heat denaturation regenerates the original binding sites and also exposes each product strand as a new template for primer binding. Use of a thermostable polymerase simplifies the protocol by delegating to a thermocycling machine the management of the temperature oscillations.[9] DNA corresponding to the sequence from practically any target source can be acquired. Information from RNA targets may be gathered by first performing reverse transcription to generate double-stranded cDNA.[3]

The simplicity of the reaction belies its power. Theoretically, the amount of target sequence doubles with each cycle. However, this maximum rate is not realized because of competitive inhibition through self-annealing of product strands, which precludes new primer binding. The reaction usually expands exponentially before reaching a linear phase at 10^{-8} M of product, and essentially stopping at 10^{-7} M[10] as the performance plummets.[11] Because of this inhibition, the reaction efficiency during the last rounds of PCR is unpredictable.

Quantitation of the original template is often desired in diagnostics or expression studies. In order to relate the total product to the amount of introduced template, internal competitive control targets must be included in the reaction over a range of known concentrations. They are susceptible to the same efficiencies and constraints and are identical to the true target except for random internal sequence differences that allow differentiation.[12-18] Alternatively, because the earlier exponential phase is more predictable, measuring the unknown target compared to similar standards at this phase using sensitive, real-time detection with fluorescent molecules (e.g., Reference 19 and see below) gives reproducible quantitation.

4.2.1.2 Oligonucleotides

The basic rules for oligonucleotide primers originated for use in PCR but are applicable to any DNA-polymerase-dependent amplification reaction. The primers should be short, about 16 to 24 bases long with a GC-AT ratio paralleling or superior to the template.[20] They should frame a target sequence of 200 to 400 bases for highest efficiency, although this can be easily increased with optimization. To avoid nonspecific priming, the 3′ end should not be extremely stable (i.e., GC-rich).

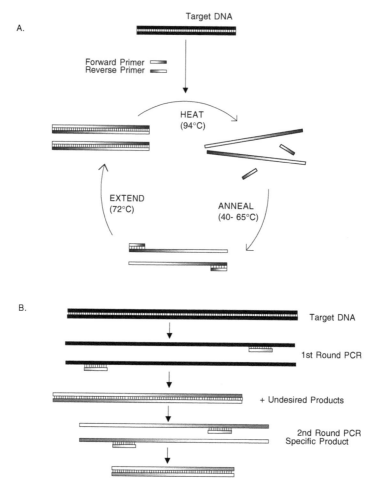

FIGURE 4.1 (A) Polymerase chain reaction (PCR). Forward and reverse primers are designed flanking a segment of target DNA on opposite strands. Heat denatures the target, providing access for primer annealing upon cooling. A heat-stable polymerase extends the primers across the target region, resulting in double-stranded target sequence that can reenter the cycle. (B) Nested PCR. One round of PCR with external primers is followed by a second round with primers complementary to more internal sequences. The dual amplification increases sensitivity and improves detection of many viral targets of low abundance.

The polymerase can extend and stabilize a mismatched primer very rapidly,[21] and any sequence between two primers will be amplified. To prevent adventitious binding of the more stable primer, primer pairs should have a similar annealing temperature.[22] The primers should enjoy limited 3′ interaction with each other or themselves. To increase specificity, nested primers may be used by which external primers are used for approximately 30 rounds (Figure 4.1B), and a portion of that product is then amplified with an internal set of primers to give a shorter, specific sequence.[23] Software packages are available to perform the hybridization analysis and assist in primer selection. Performing the two reactions in one vessel separated by an oil phase reduces cross-contamination.[24]

    Not all primers must be specific to a single target. Degenerate primers can be designed from protein sequences or conserved regions of genomes[25] to allow amplification when the exact sequence is unknown or variable. The primer may contain one of several nucleotides inserted during random synthesis at a particular location, introducing numerous sequences of related primers in the reaction. An alternative to a random base is a universal nucleotide that can pair with all four bases during hybridization.[26,27] Completely unknown sequences can be amplified by ligating adapters of known

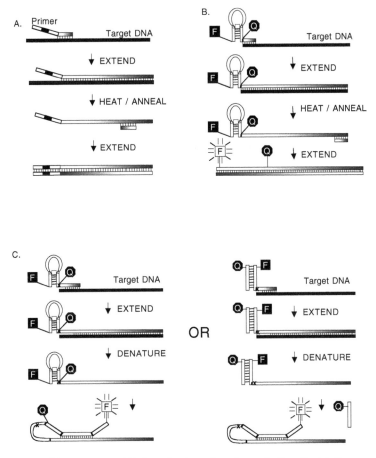

FIGURE 4.2 Primers with a purpose. (A) Introduction of engineered sites. Restriction enzyme sites, transcription promoters, purification tags, or any other desired sequence is introduced on the target sequence by one primer after opposite primer extension. (B) Hairpin primer for detection. A quiescent fluorophore is dampened by a quencher held close with a hairpin structure until forcibly separated by opposite strand extension, releasing fluorescence. (C) Scorpion primer for detection. Similar to (B), a fluorophore is held in close proximity to a quencher. However, the primer contains a blocker past which the opposite strand does not extend. Instead, the primer self-probes its own extension, releasing fluorescence.

sequence onto the end of targets truncated with restriction enzymes. Amplification then proceeds with adapter-specific primers (e.g., Reference 28).

The goal of using degenerate primers is typically the same as with wholly specific primers — to distill one region of interest from a brew of extraneous sequence. When a snapshot of the entire genome is desired over any particular sequence, arbitrary primers of 5 to 32 nucleotides corresponding to random, repetitive, or dispersed elements in genomes amplify DNA polymorphisms[29] to generate signature fingerprints of individuals or related organisms for use in forensics or epidemiology (e.g., Reference 30).

The exceptional utility of PCR derives from the fact that only the 3′ end of the primer confers specificity. Modifications at the 5′ end, while not complementary to the original target, will be assimilated in the product strands after the first round of opposite primer extension (Figure 4.2A). Large amounts of mutated template are generated by incorporating a few changes in the primers away from the 3′ end.[31] Other modifications include transcription promoters,[23] restriction enzyme sites,[3] and translation initiation sites.[32]

Detection mechanisms may also be incorporated into the primer if appended upstream of the specific binding site. These primers typically employ fluorescence resonance energy transfer (FRET). FRET is the transfer of energy from one fluorescent dye molecule (the donor) to another dye molecule (the acceptor), which may or may not be fluorescent.[33,34] Energy transfer falls off rapidly with distance and occurs only if the acceptor lies within 10 to 75 Å of the donor. If the acceptor is fluorescent, this fluorescence is measured as the acceptor comes into close proximity to the donor. If the acceptor is nonfluorescent, its propinquity quenches donor fluorescence, which only escapes as the two dyes diverge. Numerous commercially available options exist for both donor and acceptor moieties. FRET primers include Sunrise and Scorpion primers.

The fluorescence of Sunrise primers is concealed by forcing the donor and quencher moieties together with hairpin secondary structure until the primer is made double-stranded in opposite strand extension (Figure 4.2B). With this method, as few as ten molecules of original target are detected. Fluorescence directly expresses primer incorporation and is used for detection or quantitation of the amplified target.[35-37] Scorpion primers also contain a hairpin loop, but they do not unfold during polymerization as do Sunrise primers. The unraveling of Scorpion hairpin primers occurs through intramolecular self-probing during the annealing phase after its own extension (Figure 4.2C). The loop of the hairpin complements a downstream sequence on the same strand, acting as a target-specific probe. Binding separates the donor and quencher moieties, releasing fluorescence. The 3′ end of the hairpin connects to the target-specific primer with a blocking hexethylene glycol, beyond which complementary strand synthesis does not proceed, diminishing nonspecific fluorescence due to mispriming or primer-dimers. Because the probe and target are united on the same molecule, the kinetics of binding are very fast. The probe is likely displaced by an oncoming second strand during the subsequent extension phases, so real-time measurements must be taken at the beginning of this phase.[38] Interaction between the two ends of the primer will lead to high background and must be avoided through careful design. Quantitation from Scorpion primers encompasses a large dynamic range,[39] and their specificity grants allele-specific multiplexing.[40] Dissection of the stem-loop into a duplex (Figure 4.2C) increases output of fluorescence while preserving the rapid cycling possible with an intramolecular probe.[41]

Transformation from a resting to an active state for the hairpin primers results from alterations of their secondary structure. Primary structure may also camouflage an active primer until it is devoted to amplification. The predictable base pairing of DNA replication means single-stranded catalytic sequences (DNAzymes) may lie quiescent as their own complement, to be activated upon polymerization and melting.[42] The active DNAzyme cleaves the RNA portion of DNA/RNA/DNA reporter substrate, releasing fluorescence. The substrate is blocked from extension to eliminate nonspecific interaction on the template.

4.2.1.3 Conclusions

The myriad primers echo the numerous protocols and applications developed for PCR since its appearance. The importance of PCR and its impact on molecular biology cannot be overstated. It dominates the amplification field and wields a practical monopoly outside of diagnostic applications. Its temperature-dependent mechanics limit its speed and specificity, but its widespread use ensures those limitations will be addressed, just as the thermostable polymerase and temperature-cycling machine removed the tedium of manual cycling in the original protocols. The reaction vessel itself advances toward miniaturization, microfabrication, and microarrays (e.g., References 43 and 44). Temperature cycling imparts a volume-dependent pace, so that the time required can be reduced either by performing the reaction in narrow capillaries instead of tubes (e.g., Reference 45) or by moving the sample volume through temperature zones for total reaction times of several minutes.[46] Real-time detection systems permit rapid and reproducible quantification. Detection of multiple targets is limited by primer interactions but may be promoted with multiplex detection mechanisms or primers linked to a solid phase.[47,48] PCR currently reigns over the amplification reactions.

Although competitive reactions surpassing PCR in speed or specificity have been developed for some circumstances, the popularity of PCR will likely remain undiminished for years.

4.2.2 Ligase Chain Reaction

There are means and mechanisms for assaying some specific alleles directly with PCR,[49] but the reaction must be rigorously optimized, and it may not succeed. One method for discriminating single-base polymorphisms indirectly is to amplify the region of interest with PCR and then anneal allele-specific hybridization probes to the products under stringent conditions.[50] However, hybridization alone indulges some mismatches, leading to spurious amplification. The enzyme ligase displays consummate skill in differentiating a mismatch from a true complement when it joins adjacent sequences. The oligonucleotide ligation assay (OLA)[51] augments the specificity of hybridization with ligase by annealing two oligonucleotide probes adjacent to each other on the amplified target.[52] Only if the 3' end of the upstream probe is complementary will they be joined and detected. The ligase chain reaction (LCR) is an extension of the OLA that uses the probes not only for detection, but also as the instrument of amplification.[53-56]

4.2.2.1 Reaction Mechanics

LCR is an exponential amplification that, like PCR, relies on iterative thermal denaturation and annealing to achieve its effect. However, because there is no extension between the primers, each cycle contains only these two steps, not three. The reaction[57] begins with the first heat-denaturation of the double-stranded template DNA. Four primers are designed such that two will anneal adjacent to each other and opposite to the remaining pair on the target strands (Figure 4.3A) when the temperature is lowered. If the adjacent probes are exactly complementary near their junction, ligase will unite them. Once joined, the probes can bind the alternate pair of primers, serving as a template for further amplification in the subsequent 20 to 30 rounds. Like PCR, with perfect efficiency, template sequence doubles with each temperature cycle.[56,57] Described originally in 1989 as the ligation amplification reaction (LAR) with the temperature-sensitive T4 DNA ligase,[57] the reaction required the employment of a thermostable enzyme[58,59] to make it practical.[60] This obviates the need for adding fresh enzyme after each round and diminishes nonspecific reactions because the reaction occurs near the melting temperature of the probes. Since the primer sequences must be completely known, the reaction is useful only for amplifying identified sequences, not for discovering new ones.

Detection depends on discrimination of separate probes from those united. Originally, gel electrophoresis with radioactively labeled primer separated the product from the substrate probes.[57-60] More often, one upstream probe is labeled with a capture entity to secure it to a microtiter plate, and the other upstream probe is labeled with a detection entity. The capture entity can comprise any specific recognition system such as streptavidin-biotin,[61] antigen-antibody,[62] or arbitrary tails for nucleotide hybridization.[63,64] Detection labels have been radioactive,[65] fluorescent,[66,67] chemiluminescent,[68] or colorimetric.[61]

Alternate detection means more compatible with high-throughput assays include laser-induced fluorescence after capillary electrophoresis separation from a reaction on a microchip,[69] mass spectroscopy,[70] or universal microarrays.[71] Improvements in efficiency gained with new detection technologies represent the future of LCR diagnostics.

4.2.2.2 Oligonucleotides

The specifics of primer design have been described.[54] The primers are synthetic oligonucleotides about 20 to 30 nucleotides in length, describing a unique sequence or set of sequences in the genome. Often the primers are labeled to allow solid-phase detection. For optimum specificity, all of the probes should exhibit similar melting temperatures. By annealing near this temperature, the

A.

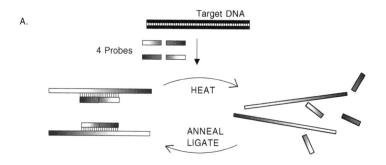

B.

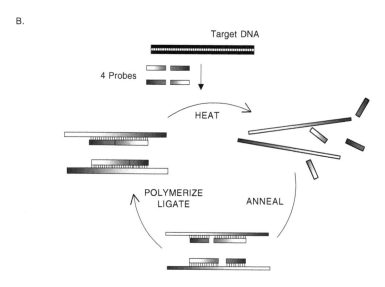

C.

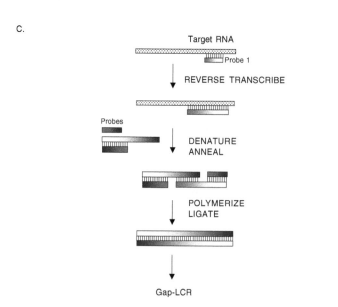

FIGURE 4.3

ligase not only discriminates mismatches using both the energetics of the mismatch and the destabilization of the helix, but it also achieves higher specificity.[60] Also, the typical thermostable *Thermus thermophilus* DNA ligase is inherently more discriminating than that of the T4 phage.[72] High background is the bane of sensitivity. Blunt-end ligation is particularly damaging because the result is completely indistinguishable from the correct product.[56] Nonspecific blunt-end ligation has been a problem for some researchers,[73] but it can be reduced with single-base-pair overhangs at the juncture site and unpaired tails at the terminal downstream ends. Alternatively, the downstream probes may contain 5′ overhangs. The extension is cleaved by a flap endonuclease (see invasive probe reaction below) once annealed to its target.[74]

The discriminating base pair at the 3′ end lends both the specificity and efficiency[54] to the reaction. The fact that the primer sequences are restricted to those immediately adjacent to the mutations means that the sensitivity will vary greatly from several hundred targets[73] to over 1 billion,[75] depending on the particular mutation.

4.2.2.3 Additional Ligation-Dependent Reactions

Problems of specificity and sensitivity have led to the development of several ligation-dependent reactions that also require a polymerase.

4.2.2.3.1 *PCR-Coupled LCR*

Coupling the sensitivity of PCR with the single-base discrimination of LCR, the protocol proceeds similarly to a nested PCR in that the sequence of interest is first expanded by PCR, and those amplicons are investigated with LCR. For example, even if products could not be detected after the PCR step, the addition of LCR brought base-pair-specific detection for *Listeria monocytogenes*[65] down to 10 colony-forming units.[76] The initial PCR allows reduction in the number of LCR cycles, which diminishes false positive results.[77] This procedure has been used to investigate sequence changes in bacteria, viruses, oncogenes, human inherited disease, and fruit-fly recombination.[66,75,77-81]

4.2.2.3.2 *Gap- and Asymmetric Gap-LCR*

These modifications increase the sensitivity of the reaction by decreasing the background noise. The primers are designed such that when they anneal they leave nonoverlapping gaps of one to several nucleotides between them on each strand[82] to be filled in by a thermostable, exonuclease-minus polymerase before ligation. The gap precludes target-independent ligation of blunt ends. Not all of the dNTPs are required to fill the gap, and only those necessary are supplied[82,83] to prevent nonspecific extension (Figure 4.3B). The diagnostic nucleotide lies ultimate or penultimate to the 3′ end near either gap.[84] Compared to the original LCR, which can detect one mutation in a background of 100 normal sequences,[61,73] Gap-LCR can detect mutants in 10,000-fold excess normal DNA.[84] It is available commercially for testing of *Chlamydia trachomatis* (e.g., Reference 85).

Asymmetric gap-LCR is a modified version of gap-LCR[86] that allows detection of RNA targets. Instead of an equal gap on both the sense and antisense strands, there is a larger gap (9 to 15) between the probes that are complementary to the RNA and a small gap (2 to 4) between the probes

---

FIGURE 4.3 (Opposite) Ligase Chain Reaction (LCR). (A) Two pairs of adjacent probes bind opposite each other on denatured target DNA. If the nucleotides at their juncture are complementary to the target, the probes are ligated and serve as a target for probe annealing in subsequent rounds. The reaction is susceptible to non-specific blunt-end ligation. (B) Gap-LCR. To reduce blunt-end ligation, the probes are not designed immediately adjacent to each other, but are separated by a short gap. The gap requires 1 to 3 of the possible 4 nucleotides and is filled by a polymerase. The now abutting probes are then ligated and reenter the cycle as targets. (C) Asymmetric Gap-LCR. Success of ligation is limited on RNA targets. RNA targets are made amenable for amplification by first annealing one probe, which is extended by 9 to 15 nucleotides. Three additional probes are added, one of which contains the complement to the recently appended 9 to 15 nucleotides. The probes bind to leave a short gap and amplification continues as in Gap-LCR.

that are of the same sequence as the RNA (Figure 4.3C). Again, only the missing nucleotides are supplied, which limits both the cDNA and the probe extension. The antisense upstream probe is added first and extended to synthesize a short cDNA. The duplex is heat-denatured, and the remaining three probes, ligase, and polymerase are added. The cDNA furnishes the hybridization site for the opposite upstream probes, admitting extension and ligation. This new DNA serves as a template for subsequent cycles.

4.2.2.3.3 Ligation-Dependent PCR

This technique essentially reverses PCR-coupled LCR. Two adjacent hemiprobes are ligated together on an RNA target[87] to form a single template. Each free end of the ligated probe functions as a primer-binding site for PCR, obviating the need for reverse transcription. This technique promotes detection in formalin-fixed tissue, which contains cross-links refractory to polymerization and detrimental to sensitivity.[88] This method rescues the sensitivity of fixed tissue to that of reverse-transcription PCR in frozen tissue.[89] LD-PCR is a progenitor of rolling-circle amplification (see below), which condenses the two linear hemiprobes into one circle[90] before polymerization through rolling-circle synthesis.

4.2.2.4 Gene Construction

The artificial synthesis of large oligonucleotides is limited to just over 100 bases, as the possibility of construction errors heightens with each base. Longer oligonucleotides must be assembled from smaller segments. This may be accomplished with PCR, overlapping the edges of the long primers in serial reactions. With LCR, large synthetic genes can be constructed by combining numerous oligonucleotides or PCR products in the same reaction.[91] Mutated probes can introduce alterations throughout the sequence of a gene, instead of merely at the two primer ends as in PCR.[92,93]

4.2.2.5 Conclusions

LCR evolved from an auxiliary reaction for PCR detection. In many cases, it remains entwined with polymerases due to its limited sensitivity. Although beneficial for detection of some mutations, the fidelity of the enzyme is about 1,000- to 10,000-fold for the correct vs. the incorrect base, which may be inadequate for distinguishing somatic mutations. The sensitivity cannot always be controlled because the discriminatory base must be situated exclusively at the 3′ end of the upstream primer.[72] If the secondary structure of the probes or alternate hybridization sites reduces the sensitivity of the reaction, the researcher has no recourse but to choose primers in a different location. In addition, the reaction performs best for DNA targets. Ligase inefficiently recognizes DNA-RNA hybrids, making it difficult to detect RNA without performing the cumbersome asymmetric gap-LCR. The enzyme itself is also susceptible to inhibition.[85,94]

On the other hand, LCR exhibits high sensitivity and specificity when optimized. Thus, diagnostic kits for perpetually investigating the same sequence make good use of its strengths. Detection currently displays simplicity but not speed. Real-time or high-throughput detection will have to be established to stay competitive with faster reactions. Otherwise, the use of LCR will likely be preempted by rivals such as rolling-circle amplification.

4.2.3 Rolling-Circle Amplification

Rolling-circle amplification (RCA) is an isothermal technique that exploits a circular probe to yield additional amplification from each primer. Amplification can be either linear or exponential. Although better described for DNA targets, the reaction may also detect RNA.[95]

Circular probes were first employed without primers, directly hybridized to *in situ* targets. The so-called padlock probes are long (>80 nt) probes, with target-specific sequence at each end flanking noncomplementary sequence. The extremities bind adjacent to each other, like a belt, to denatured

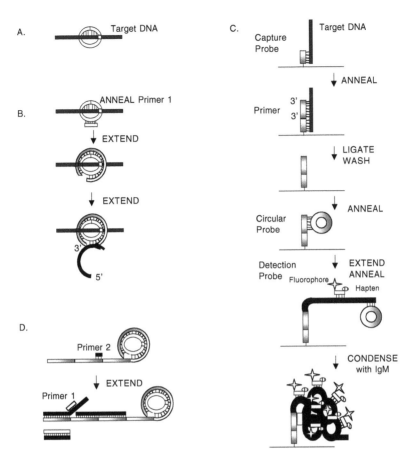

FIGURE 4.4 Rolling-circle amplification (RCA). (A) Padlock probe. A linear strand of DNA circularizes as it binds to its target. Due to the helical nature of DNA, ligation of the ends of the probe locks it onto the target. (B) Linear RCA. A primer binds the single-stranded region of the circular probe, which may be of any sequence. New DNA synthesis displaces the original sequence, and elongation may approach 12 kB. (C) RCA-CACHET (RCA-Condensation of Amplified Circles after Hybridization of Encoding Tags) Denatured target DNA is captured to a solid surface. A primer with two 3′ ends — one for ligation and one for polymerization — is annealed. Ligation of the primer is followed by a stringent washing. A premade circular probe is introduced, and the primer begins linear RCA. Detection probes with a fluorophore and a hapten anneal to each repeat of the extension. Their diffuse binding is condensed with IgM for visual detection. (D) Exponential RCA. A second primer binds to each repeat of the single-stranded product of linear amplification, with upstream primers displacing extended downstream primers. Because this second strand comprises the same sequence as the original circle, it accepts binding and extension of Primer 1. Upstream Primer 1 also extends and displaces downstream primers, resulting in a complicated tangle of branches (not represented).

or single-stranded targets.[96] Ligating the ends together locks the probe onto the target where it resists stringent washes. Labeled nucleotides in the nonhybridized region allow for detection. This method was used to discriminate single-nucleotide differences in the centromere repeat sequences of human chromosomes 13 and 21[97] (Figure 4.4A). However, without amplification, single-copy gene detection remained elusive.

In nature, circular DNA molecules such as small genomes and plasmids replicate (i.e., amplify) through rolling-circle replication,[98,99] but they require a suite of enzymes and a long template. It was uncertain that diminutive circular molecules such as padlock probes would support replication because the templates might be smaller than the polymerase itself! Interestingly, polymerases alone are capable of replicating these circles,[100] running laps around the circle like an ultramarathoner on an indoor track, producing DNA composed of concantameric repeats up to 12,000 bases or more, depending on the polymerase.[101] Because polymerization is not contingent on primary or secondary structure of the template,[100] DNA of any sequence can be circularized and amplified.

4.2.3.1 Reaction Mechanics

Polymerization itself is a linear reaction. Thus, the kinetics of this reaction depends on whether one (linear) or two (exponential) primers are used. As is true for other reactions, a thermostable polymerase improves the specificity of primer binding, although mesophilic enzymes are also used. The reactive circle may be specifically formed as a padlock probe on the target, or a preformed circle may be specifically bound by a primer. The amplification runs at a single temperature, with no need of thermocycling.

4.2.3.1.1 Linear Rolling-Circle Amplification

The amplification is usually performed on a surface such as a microwell. The probe is designed such that the two ends will bind in juxtaposition to a target molecule, just as a padlock probe (Figure 4.4B). A surface-bound primer hybridizes at its 3' end to the middle of the probe sequence. Upon target binding and ligation, the primer extends, displacing the target and making end-to-end single-stranded repeats of the probe sequence, often tens of thousands of bases long.[102] If ligation does not occur, the minimal extension ceases at the terminus of the probe.

An alternate method (RCA-CACHET) uses preformed circles and assembles the primer according to the presence of the target[103] (Figure 4.4C). A probe is linked to the solid surface by its 3' end; its 5' end is target-specific. The target- and allele-specific probes are added. This second probe continues the target-specific sequence of the first, and they are ligated together against the target sequence. A strong washing removes the target and any unligated probes. Allele-specific circles are added to the reaction. The second probe now acts as a primer, with polymerase generating single-stranded concatamers. Detector probes, complementary to every repeat of a portion of the circle, are tagged with a hapten and a fluorescent moiety. The fluorescence scattered along the elongated molecule is condensed with antibodies against the haptens.[103]

4.2.3.1.2 Exponential Rolling-Circle Amplification

Also called hyperbranched,[103] or cascade[104] rolling-circle amplification, or ramification extension amplification,[95] this method uses two primers (Figure 4.4D) to achieve million-fold amplification in about 1 h.[95] The ends of the probe are hybridized to the target and ligated together as described for padlock probes. The forward primer complements part of the spacer region, and the reverse shares sequence with the circular probe. The forward primer makes concatamers of the circle to which the reverse primers bind. Downstream sites emerge first and are bound by primers, which can then be displaced by actively polymerizing upstream primers. Each strand generated by a forward primer can be bound by a reverse primer, and vice versa, resulting in a complicated tangle of branches. A branch terminates when the furthest upstream binding site is occupied and every offshoot is double stranded. The system is amenable to homogenous detection with fluorescent primers.[104]

4.2.3.2 Oligonucleotides

The probes to be circularized typically are 60 to 100 bases long. The 5' end and 3' end are designed such that 10 to 20 bases on each side bind adjacent to each other on the target molecule, separated by a linker or spacer sequence of 30 to 50 arbitrary nucleotides. The 5' end must be phosphorylated for ligation. The 5' and 3' ends must be synthesized exactly as intended in order to get efficient ligation against a target under relatively stringent conditions, and special chemistries have been developed to fulfill this requirement.[105]

The linker sequence allows for the binding of primers and detection tags. Primers work best with a melting temperature slightly greater than the reaction temperature. The primers may be allele-specific if desired. If using two primers, one primer is complementary to the circle; the other has the same sequence of a different part of the linker region of the circle. Obviously, the primers should not complement each other.

RCA-CACHET is interesting in that the second ligation probe has two 3′ ends, one for ligation and one for polymerization.[103] The purpose is to allow the bound primer to expose a 5′ end, thus decreasing nonspecific polymerization without ligation. The circles used are preformed with a complementary helper oligonucleotide, which forces the ends together for ligation.

4.2.3.3 Conclusions

Undoubtedly, a polymerase running circuits around synthetic circles can generate tremendous amounts of DNA in a short time, especially with ramification. Although exponential amplification allows single-copy genes to be detected with a sensitivity of about 17,000 molecules,[103] the greatest amplification comes from artificial templates,[104] with sensitivities down to ten copies. However, this sensitivity is not realized if the circle lacks free rotation by being padlocked onto a template. It can move linearly down the template[96] a few hundred nucleotides, but polymerization with the usually processive Phi29 polymerase is sterically inhibited if an end is too distant.[106] However, recently conditions for exponential amplification have been described with an alternate polymerase (from *Bacillus stearothemophilus*) that escapes these topological constraints observed with the original polymerase,[107] allowing sensitive detection down to ten target molecules.[108] Blocked DNA amplification *in situ* has been overcome by digesting the nontarget DNA with restriction enzymes and endonucleases.[109]

Currently, linear rolling-circle amplification appears its most popular use. Linear amplification may be sufficiently sensitive if a detection molecule is incorporated for each repeat. The major benefits of RCA are the allele specificity granted by the ligation and the multiplexing allowed by surface amplification such as on a microarray.[110,111] RCA on microarrays may be used for direct amplification of target sequence[112] or as a signal amplification method from targets previously amplified by PCR.[113,114] Conjugating a primer to an antibody allows signal amplification for sensitive protein detection[115,116] or *in situ* localization of mRNA hybridized to hapten-labeled DNA probes.[117]

4.2.4 STRAND-DISPLACEMENT AMPLIFICATION

RCA uses strand displacement for exponential amplification. However, each new upstream strand made is eventually suffocated by the closure of a double strand. Given that the original strand can be kilobases long, the amplification is substantial. For amplifying only short linear sequences, a method is needed for reviving the template for reuse. PCR reclaims its template through heat denaturation. Strand-displacement amplification (SDA) uses a restriction enzyme to restore the double-stranded templates during the isothermal reaction.

4.2.4.1 Reaction Mechanics

SDA is continuous and exponential amplification of target sequence between two primers using a restriction enzyme, an exonuclease-deficient polymerase, one thiolated nucleotide, and three unmodified nucleotides[118] (Figure 4.5A). The restriction enzyme nicks the unprotected strand of a hemithiolated asymmetrical recognition site introduced from a primer upstream of the target sequence. The polymerase commences nucleotide addition at the nick, copying a new target strand while displacing its predecessor. The 3′ end of the displaced strand binds a complementary primer and is then extended to regenerate a double-stranded but half-shielded restriction site. This begins a new cycle of primer extension, cleavage, and displacement. The cycling occurs continuously, not

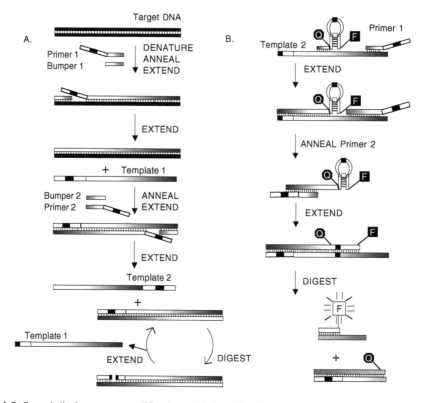

FIGURE 4.5 Strand displacement amplification. (A) Amplification from only one strand of target DNA is shown. A primer containing a Type IIs restriction enzyme (REN) site anneals and is extended with thiolated nucleotides by polymerase. A bumper primer dislodges the single-stranded Template 1. Template 1 binds Primer 2 and Bumper 2, resulting in free template 2 and a double-stranded REN site protected on the new strand from cleavage by the thiolated nucleotides. The REN digests the primer, generating an opening for polymerase extension and strand displacement. Released Template 1 can reenter the cycle by binding Primer 2, and free Template 2 undergoes similar amplification with Primer 1 (not shown). (B) Real-time detection. A hairpin probe with a quenched fluorophore and a REN site in its loop binds downstream of Primer 1. The probe is elongated and then displaced by primer extension. The elongation of the probe complements and binds Primer 2, whose extension straightens the hairpin. Due to its designed sequence, neither strand of the double-stranded REN site of the loop is protected from digestion, and fluorescence is released.

discretely, and each displaced strand from an extended primer functions as a template for the alternate primer. Thus, the reaction is self-perpetuating, exhausting only when the enzymes are no longer in excess.[119]

Only the 3′ end of the primer is specific to the target, yet amplification requires that the 5′ end of the primer, containing the restriction enzyme site, become double-stranded. Displaced strands generated during amplification have their 3′ ends extended upon primer binding to form the restriction site. However, the 5′ end of the primers that bind to the original template have no inherent means of becoming double-stranded. A mechanism had to be designed to displace the original binders and introduce a double-stranded restriction enzyme site to initiate the reaction. Originally, the target was first trimmed with another restriction enzyme to generate known extendible ends to serve as the target-specific 3′ annealing site.[118] This procedure was time-consuming and restrictive because the target had to be selected between appropriate restriction sites and only generated million-fold amplification in 4 h. A preferred solution was devised by adding bumper primers.[120] These primers bind specifically just upstream of the amplification primers. They do not necessarily contain a restriction site and are at about one tenth the concentration of the amplification primers, which must extend first. When the bumper primers are extended, they displace the amplification primers, launching the self-sustaining cycle. This allows 100-million-fold amplification in just 2 h.

The amplification was immense; however, the background was so high that it prevented the products from being directly detected on a gel. The high background was aggravated by the low stringency possible with mesophilic enzymes. Short primer-dimers made 3′ of the restriction site are amplified very efficiently, much more than target sequences. For example, mesophilic strand displacement with the polymerase Klenow exo- loses tenfold efficiency with every 50 nucleotides of target span.[120] Background is decreased by using thermophilic enzymes at about 50 to 65°C,[121,122] which allows more specific primer binding. In addition, the efficiency of thermophilic SDA increases to produce more than 1-billion-fold amplification in 15 min. The reaction can be used for detection of RNA only by first preparing cDNA.[123,124]

4.2.4.2 Oligonucleotides

4.2.4.2.1 *Primers*

There are four primers in the current SDA system. The internal primers are at tenfold higher concentration than the external primers. The external primers are only incorporated during the first round in order to bump the internal primers and give defined ends to the target.[125] The 3′ ends of the primers must not form primer-dimers because any made 3′ to the restriction site will be amplified exponentially, probably even more efficiently than the product. The 5′ ends of the primers are not sequence specific, but must be long enough to stabilize the restriction enzyme on the succeeding nonpalindromic restriction enzyme site.[124] A nonpalindromic enzyme must be used so that the thiolated nucleotide will protect only one strand. A symmetrical site would grant protection to all strands made after original target cleavage, halting the reaction. Obviously, the target sequence between the primers cannot contain the restriction enzyme recognition site.

4.2.4.2.2 *Detection*

In the beginning, the promiscuity of the mesophilic reaction necessitated probing with a labeled specific probe. Desired products were separated from the rubbish by a specific solid-phase capture oligonucleotide. Alkaline phosphatase-labeled detection probes then bound to impart a sensitive signal.[126] Capturing the amplicons on beads and incorporating DIG-dUTP in the reaction allowed the use of flow cytometry, which gave quantification over three orders of magnitude. An attempt to make the reaction homogenous with fluorescence polarization[127-129] required adapting complex machinery and time-consuming equilibration, and the results generated were not easily quantifiable. To attain mass acceptance by clinical laboratories, a real-time fluorescent assay was needed.

Nadeau et al.[119] have developed a detection probe assay that uses only the enzymes inherent in the reaction, the displacing polymerase, and the restriction enzyme (Figure 4.5B). The probe structure consists of single-stranded DNA (ssDNA) with a restriction enzyme site between the quiescent fluorophores. Either a hairpin loop holds the two labels close together, or the dynamics of a single strand in solution may maintain the quenching. The two labels should not be closer than 9 nucleotides; otherwise they hinder the restriction enzyme. Hairpin probes with a loop of 15 nucleotides maximized sensitivity.

The 3′ end (1 to 30 nucleotides) of the probe is complementary to the target and is extended during the reaction to include a primer-binding site. A primer will bind and make the probe double-stranded, thus producing a cleavage site for the restriction enzyme. The site is designed such that neither strand is protected by the thiolated nucleotide. Cleavage separates the fluorophore and the quencher and releases detectable fluorescence. Because both of the strands are severed if the probe is extended, it cannot act as an amplification primer and does not support amplification. Spurious extension may lead to some loss of efficiency, but not to higher background. Using this method, ten molecules of a spiked sample were detected in half an hour. This technology has been incorporated in an automated system capable of reliably detecting *Chlamydia trachomatis* or *Neisseria gonorrhoeae* in clinical samples within 1 h.[130]

4.2.4.3 Conclusions

SDA is a very rapid, isothermal amplification reaction. It progresses exponentially until the restriction enzyme concentration becomes limiting.[119] Background DNA competes for restriction enzyme binding and can inflate the amount of expensive enzyme required. The use of a restriction enzyme affects the choice of target because the sequence cannot contain an internal recognition site. Another amplification method has been described that relies on strand displacement and bumper primers but does not require a restriction enzyme. Loop-mediated isothermal amplification (LAMP) generates inverted repeats flanking single-stranded loop sequences available for primer binding. Amplification is extensive but has not been proven with clinical samples.[131,132]

Although target sequence is amplified, the product is not amenable for further manipulation (i.e., cloning, sequencing) because of its heterogeneity. Also, four primers must be optimized for every new reaction. Its utility probably lies in diagnostics, where its speed grants great advantage. Real-time detection allows semiquantitation if a competitor of a known quantity is coamplified.[119,124] Quantitative estimates based on a standard curve were valid over a wide range (500 to 5 million) of original target concentration.

Running this internal standard is a form of multiplexing. In solution, up to three sequences may be amplified together if they use the same primer pairs with specific detection.[133-135] Otherwise, the quadruple primers needed per sequence drastically increase the possibility of cross-reacting with the wrong target or other primers. Multiplexing might be better accomplished on a solid surface. Here, specific primers are electronically placed in discrete zones of amplification, which would allow up to ten or more sequences to be amplified simultaneously.[136,137] Discrimination among four bacterial pathogens and allele-specific detection have been accomplished with this method.[136]

4.2.5 TRANSCRIPTION-BASED AMPLIFICATION

RCA and SDA reflect some aspects of plasmid and circular viral replication. Transcription based-amplification closely mirrors the process of retroviral replication. An RNA template is reverse-transcribed into a double-stranded template of DNA, which is itself transcribed into numerous copies of an RNA template ready to reenter the cycle. The reaction proceeds through numerous continuous cycles of transcription and reverse transcription, catalyzed by RNA polymerase and reverse transcriptase. In an early incarnation of the reaction, the researcher needed to cycle the reaction vessel through a temperature denaturation step to separate the DNA-RNA hybrids,[138] but the current reaction is isothermal and self-perpetuating. The same basic reaction operates under several names: nucleic acid sequence-based amplification (NASBA), transcription-mediated amplification (TMA), and self-sustained sequence replication (3SR).

4.2.5.1 Reaction Mechanics

Although double-stranded DNA targets can be amplified with additional denaturation steps (e.g., Reference 139), the reaction works most efficiently with single-stranded RNA targets, heated moderately to destroy the secondary structure. Reverse transcription commences from a complementary downstream primer. The RNA of the newly formed RNA-DNA hybrid is destroyed by RNAse H activity, obviating the need for thermal cycling and leaving a ssDNA capable of binding the upstream primer. The primer is extended through the target sequence and to the 5′ end of the downstream primer, which encodes a promoter sequence for a bacteriophage RNA polymerase.

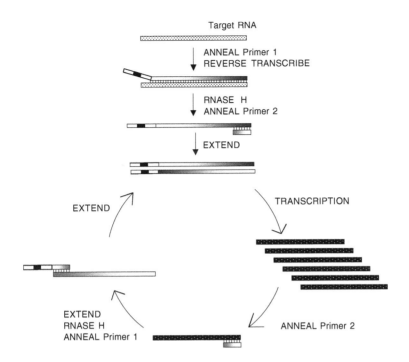

FIGURE 4.6 Transcription-based amplification. A primer containing a transcription promoter site anneals to an RNA target and initiates reverse transcription. RNase H (or an enzyme that carries this activity) degrades the target RNA. A second primer is extended on the cDNA, generating a double-stranded active promoter. Transcription of up to 100 or more copies of the target generates multiple targets for Primer 2 extension. RNase H activity removes the RNA portion for Primer 1 binding, whose extension renews the cycle.

This enzyme begins transcription of antisense RNA from the functional promoter. With this RNA, a cycle begins anew of upstream primer binding, double-stranded cDNA synthesis with the downstream primer, and transcription (Figure 4.6). Because each DNA can propagate numerous copies of RNA, 1-million- to 1-billion-fold amplification can be achieved in 1 or 2 h.[139-143]

The reaction depends on four enzymatic activities — an RNA-dependent DNA polymerase (reverse transcriptase), RNase H, DNA-dependent DNA polymerase, and promoter-dependent RNA polymerase. Under the proper conditions, the reverse transcriptase may be able to perform all of the activities save the promoter-dependent RNA polymerization.[140,144] Otherwise, reverse transcriptase has to be supplemented with RNase H. The two-enzyme reaction benefits in terms of cost, and more importantly, with purer products, it likely benefits because RNase H degrades the RNA product if present.[140] The trade-off is a twofold increase in the doubling time of amplification, which itself is diminished.[144] The cumulative infidelity of the two polymerases is sufficiently accurate (2×10^{-4} mutations) for cloning or sequencing of 90% of the products[143,145] but not as error-free as PCR with a single polymerase.[146]

The reaction is performed at a constant temperature of 41°C. An array of the necessary thermostable enzymes has not yet been established for a stringent, high-temperature reaction. Thus, the resulting products cannot be observed directly with a universal stain on a gel but need to be probed specifically to distinguish the desired sequences. Traditionally, an enzyme-linked gel assay was performed with horseradish peroxidase-labeled probes incubated to bind with the products and then run on an agarose gel (e.g., Reference 147). More recently, homogenous assays based on molecular beacons (see below),[148,149] fluorescence correlation spectroscopy,[150] or acridinium ester[151] have been developed.

4.2.5.2 Oligonucleotides

Like PCR, the reaction amplifies target sequence between two primers. The primer regions complementary to the target may be between 15 and 30 nucleotides long and follow the standard rules of hybridization for most primers — a balanced GC-AT ratio, similar melting temperatures, and not self-complementary.[142] Either both of the primers or just the downstream primer may contain the promoter sequence at its 5′ end. The transcriptional efficiency of the downstream promoter-primer is greater, and the addition of a second promoter does not contribute significantly to amplification,[140,152] so most often only one promoter is used. Transcription efficiency is enhanced with a four-base purine-rich sequence consecutive to the promoter.[142,153] At least seven nucleotides have to be appended in front of the promoter sequence to maintain full promoter activity in the face of slippage during termination of transcription.[142] Sensitivity of the reaction can be increased by using a nested set of primers.[139,141]

4.2.5.3 Conclusions

The utility of a transcription-based amplification method peaks when the species to be amplified begins as RNA or when only RNA expression is being investigated,[154-156] for then it removes the necessity of a separate reverse-transcription step. One consequence of facile RNA amplification is that DNA amplification requires cumbersome denaturation steps to initiate the reaction.

Typically, the input and output of the reaction is single-stranded RNA (ssRNA). Due to the particularities of RNA polymerase,[157] the length of the target sequence is inversely related to the amplification efficiency. The secondary structure of ssRNA can obstruct polymerization when amplifying G-C-rich templates;[158] the use of inosine in the amplification reaction counteracts this problem. Because the reaction is run at a moderate temperature, desired products are resolved from background amplification through sequence-specific detection.[159] The product RNA, already single stranded, is amenable to hybridization techniques often used in detection.[160,161] New homogeneous assays simplify detection, although careful optimization must be done to ensure that the detection probes do not compete with reactive products (e.g., Reference 148). The isothermal amplification gave tantalizing expectation of preserving structural integrity during *in situ* applications. However, the reaction exhibited merely tenfold amplification due to lack of product specificity.[157,162]

The fidelity of the reaction is satisfactory for sequencing mostly defined targets,[163] but it is not sufficient to uncover novel sequences between the primers with assurance. This uncertainty essentially relegates the reaction to diagnostic use. In fact, it has found use as a tool in detecting disease-causing viruses, slow-growing bacteria, and other common bacteria.[164-174] Transcription-based amplification is similar in sensitivity to PCR, with the exact sensitivity dependent on the assay.[175-177]

Quantitation of viremia can be crucial in the management of chronic diseases like HCV and HIV infection.[178,179] Typically, quantitation is based on competition with a dilution series of three internal standards spiked into the sample before nucleic acid extraction. The standards make use of the same primers, their sequence differing from the target only by randomization of a short interior segment to allow differential detection.[161,180] Accurate quantification is difficult with transcription-based amplification because each round of promoter-based transcription displays different efficiency of RNA production; the reaction does not march in lock-step like the production of DNA in PCR.[148]

Overall, the strength of transcription-based amplification reactions is the exclusive detection of RNA, especially from within a background of DNA. The reaction will not supplant PCR for investigating unknown sequences, but likely will continue to expand its role in RNA diagnostics and investigation of gene expression.

4.2.6 BRANCHED DNA

All of the above reactions lead to amplification of some portion of the sequence of the target. Even LCR, ostensibly a probe-amplification reaction, causes magnification of a short segment of the target sequence. Branched DNA (bDNA) entails signal amplification and is a more modern, powerful version of classic hybridization. It is founded on layered, multivalent DNA probe hybridization, with no enzymatic amplification. Nucleic acid amplification results from the condensation of numerous oligonucleotides upon the target.

4.2.6.1 Reaction Mechanics

The bDNA assay arose from earlier hybridization protocols in an attempt to construct a sensitive nonradioactive probe. A single signal per target, such as from a traditional hybridization probe, lacks sensitivity for many applications. Multiple signals must emanate from each target molecule. The bDNA assay begins with oligonucleotide probes called capture extenders, which link the target to a solid surface adorned with capture oligonucleotides. After extensive washing, the target sequence binds numerous label extenders. Each label extender attracts an amplification multimer, which in turn bind numerous labeled probes.[181]

The first assay toward this end used an almost circular amplification multimer[182] made by crosslinking oligonucleotides at their modified (*N*-alkylamino deoxycytidine) residues. This multimer gave a 50- to 100-fold signal increase over singly labeled probes with multiple hybridization sites. Larger assemblages of oligonucleotides did not increase sensitivity because steric interference obstructed additional labeled probe from binding in the interior of the spherical aggregations. (A method of organizing DNA into spheres called dendritic DNA such that the external sites can be highly labeled has recently been described.[183-186]) The asteroid shape of the amplifier oligonucleotide was changed into a comblike structure to maximize the available surface area,[187] which increased sensitivity about 50-fold.[188,189] However, the method was not as sensitive as PCR, or as sensitive as necessary for viral detection.[190]

Sensitivity is based on the signal-to-noise ratio. The second generation of the assay attempted to increase the detectable signal. In this assay, the label extenders do not bind immediately to the amplifier molecule. Rather, two label extenders must adjoin each other on the target molecule. Stacking interactions allow a new probe, the preamplifier oligonucleotide, to bind, forming a cruciate structure (Figure 4.7). Each preamplifier can bind eight or more amplifiers. The amplifiers have 15 branches that can each bind three labeled probes, giving over 10,000-fold illuminators per target molecule.[191] Consequently, this generation is at least 20-fold more sensitive than its predecessor. However, increased numbers of probes give increased probability of nonspecific binding interference.

The third generation of the assay attempted to increase sensitivity by decreasing the noise due to nonspecific binding. Synthetic nucleotides, 5'-methyl-2'deoxy-isocytidine and isoguanosine (isobases), which bind to each other but not any of the natural bases, supplement the nucleic acid code to six bases in preamplifier, amplifier, and labeled probes.[192] Although they attach to each other, they cannot bind to nonspecific background DNA. This alteration endowed the requisite sensitivity to monitor viremia[193] with detection around 100 molecules.[192]

4.2.6.2 Oligonucleotides

The assay employs a number of oligonucleotides to effect specific solid-phase detection with signal amplification. The solid surface of a microplate or microbeads displays capture probes of about 20 nucleotides.[194] There are at least two sets of probes that bind specifically to the target, each about 30 bases long. One set bridges the capture probes and target, with half of the probe binding to each. The other set has a similar structure and bridges the amplifier or preamplifier to the target. The integrity of any RNA target is preserved by enveloping the entire exposed length with protective probes called spacer oligonucleotides.[195,196] For efficient preamplifier binding, the target probes are designed to abut each other, lending stacking energy and stabilizing the preamplifiers. The target probes should bind at conserved portions of the genome for comprehensive detection. The preamplifiers are almost 250 bases long,[191] constructed by ligation of three pieces. They contain sequence

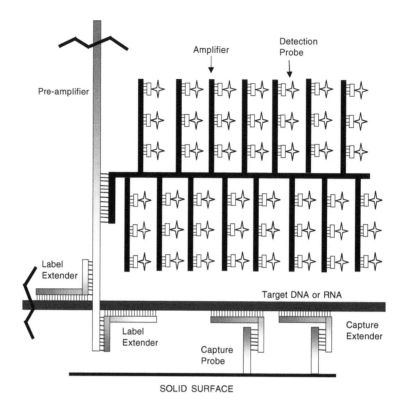

FIGURE 4.7 Branched DNA (bDNA). Target DNA or RNA is immobilized to a solid surface through hybridization of 5 to10 capture extender probes specific to both the target and the capture probes. Multiple (typically 9 to 20) sets of label extenders bind adjacent to each other and bridge the target to the preamplifier molecules. Each preamplifier molecule can bind up to 14 identical amplifiers (only one shown). The amplifier contains 15 branches, each capable of attracting three detection probes labeled with alkaline phosphatase. The preamplifier, amplifier, and detection probes may contain non-natural nucleotides to reduce nonspecific binding.

repeats to promote binding of numerous amplifier molecules, which themselves bind the enzyme-labeled short (15 to 20 bases) label probes. Sequences of the third-generation preamplifier, amplifier, and label probes contain about 30% isobases.

All probes except the amplifiers are linear oligonucleotides. The amplifier molecules embody the branching from which the assay takes its name and required developing new methods of oligonucleotide synthesis. First, a linear primary sequence is made, incorporating modified nucleotides to serve as branching monomers spaced throughout.[187] These nucleoside analogues contain two protected hydroxyl functions that can be removed to permit simultaneous short secondary sequence addition as branches.[197,198] Longer sequences are ligated to the short additions. This method allows up to 30 branches to be appended to the trunk[199] with three binding sites for label probes per branch.[200] The amplifier, preamplifier, and detection probes do not interact with the target and may be universal, provided their sequence does not substantially hybridize with the target sequence and contribute to background noise. Only the target-specific probes must be redesigned for each new assay, and commercial software is now available for designing the target-specific capture and label extenders.[201]

4.2.6.3 Conclusions

Branched DNA procures signal amplification through multiple rounds of probe hybridization in order to detect targets less than 100 molecules/ml in abundance. Relying entirely on hybridization grants

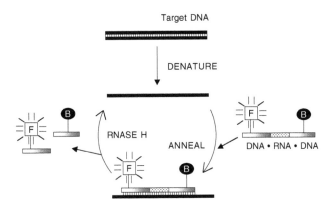

FIGURE 4.8 Cycling probe technology. Denatured target DNA binds the hybrid DNA•RNA•DNA probe labeled with fluorescein (F) and biotin (B). Disintegration with RNase H destabilizes the remaining portions of the probe, renewing the target for another cycle. For detection, cleaved probes are separated from uncleaved probes through the biotin moiety.

several advantages. Its efficiency is relatively independent of target volume. Because there is no enzymatic amplification, the assay is not as susceptible to inhibitors present in samples, allowing easier sample preparation and permitting *in situ* applications (e.g., Reference 202). Any enzymes used for detection of the label are applied only after all inhibitors are washed away. Altering the hybridization conditions allows specific detection of RNA or DNA targets.[203] The multitude of target-specific probes allows detection across viral genotypes or subtypes.[204-206] The assay is inherently quantitative[207] because the signal generated is directly proportional to the target present, permitting quantitative measurement of mRNA (e.g., Reference 208). Because the target itself is not amplified, there are not millions of amplicons present to contaminate later assays.

Although many of the required probes are universal, designing the target-specific probes requires that the entire sequence of the target be known. Because of this, bDNA will probably be used predominantly for repetitive diagnostic assays in clinical laboratories using kit-based applications. Its disadvantages for diagnostics include the requirement for an extended hybridization, prolonging the time to results. The assay is not as reproducibly sensitive as target-based amplifications, and it does not have as wide a quantitative scope,[209,210] each vital for the management of chronic infections.[204,211]

4.2.7 PROBE-DEGRADATION REACTIONS

Nucleic acid amplification is often perceived as a process of synthesis — constructing or consolidating swarms of oligonucleotides. Paradoxically, amplification may also be achieved by the destruction of probe oligonucleotides. Several approaches to enzymatic probe degradation have been designed. These reactions are isothermal, relying on the destabilization of the degraded probe to regenerate the single-stranded target.

4.2.7.1 Cycling Probe Reaction

4.2.7.1.1 Reaction Mechanics

This reaction centers on a chimeric probe composed of DNA flanking a short run of RNA;[212] the probe is entirely complementary to an ssDNA target. RNase H, which digests the RNA portion of DNA-RNA hybrids, degrades the central portion of the probe upon its binding to the target. Separately, the DNA ends are too unstable to remain bound to the target and return to solution (Figure 4.8). The turnover time of the probe determines the rate of reaction, which typically runs for 30 min. The target is now free to renew the cycle by binding another probe, while the cleaved probes accumulate awaiting detection. Similar reactions exist with DNA-only probes, in which an exonuclease digests the target-bound probe but ignores the single-stranded probe in solution.[213,214]

Originally, the cycling probe reaction was run at 50°C with the RNase H from *E. coli*.[212] At this temperature the reaction suffers from high background due to nonspecific binding and degradation of the probes. As seen previously for the other reactions, using a thermostable enzyme at a higher temperature allows for an improvement in specificity.[215] Unfortunately, this enzyme is relatively inefficient and requires a high concentration for reactions with low numbers of target DNA.[216] The RNase from *Thermus thermophilus* was tagged with streptavidin to increase its affiliation with a biotinylated probe at low concentrations of target.[215] The enzyme was capable of cleaving only the probe with which it became permanently associated, and this avenue appears to have been abandoned.

Other attempts have been made to make the reaction more sensitive and specific. Nonhomologous DNA distracts the probe and the enzyme, inhibiting the reaction. The target DNA may be isolated by temporary attachment to a solid support.[217] This technique increases the sensitivity at least 100-fold. Additives to the reaction environment such as spermine and EDTA impart additional sensitivity. Spermine is a polyamine that aids the release of RNase H from ssDNA, thereby increasing cleavage rates by pushing the enzyme off nonhomologous DNA, effectively granting a higher enzyme concentration.[218] Unfortunately, the enzyme increases its activity on both specific and nonspecific DNA. The presence of EGTA lowers the background and increases specificity. The probe is typically isotopically labeled for detection.

4.2.7.1.2 Oligonucleotides

Naturally, probes should be fashioned to have a minimum of interaction with anything but their assigned target.[216] They are typically about 30 nucleotides long, with an equal number of DNA residues flanking four purine ribonucleotides. Minimizing the number of ribonucleotides diminishes nonspecific cleavage by RNase H.[212]

Originally, cleaved probes were segregated from unreacted probes with electrophoresis. Recently, the probe has been redesigned to permit nonradioactive detection. Labeling the 5′ end with fluorescein permits capillary electrophoresis and fluorescent detection on a microfluidic chip.[219] For solid-phase assay, a biotin tag on the 3′ end of the probe enables surface capture and detection.[220,221] The solid-phase assay has been simplified further with the introduction of a lateral flow strip for the detection of methicillin-resistant *Staphylococcus aureus*.[222]

4.2.7.1.3 Conclusions

The cycling probe reaction lacks the sensitivity of exponential target-based amplification. Theoretically, its linear amplification would allow ease of quantification. However, its susceptibility to inhibition at varying levels of nonhomologous DNA makes that linearity unrealizable in practice.[218] High background from inter- or intraspecific molecular interactions makes probe design challenging. Noise also arises from RNases present in the sample.[220] The target is limited to ssDNA, and the reaction is unable to distinguish single-nucleotide polymorphisms. One advantage of the reaction is its speed. It may find utility in operations that require specificity and speed, but not sensitivity, such as testing an entire colony of bacteria for harboring a particular gene.[221,223–225] Signal amplification reduces contamination concerns compared to target amplification.

4.2.7.2 Invasive Probe Reaction

DNA is always synthesized enzymatically by sequential nucleotide addition to its 3′ end. During DNA replication, recombination, or repair, this lengthening end displaces the occupant downstream DNA for a short distance. The dislocated strand forms a single-stranded flap apart from the duplex. The anatomy of invasion and displacement is recognized by repair enzymes, which amputate the

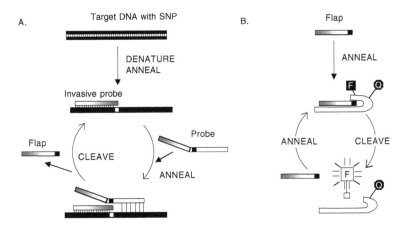

FIGURE 4.9 Invasive probe amplification. (A) Linear amplification. An invasive probe oligonucleotide anneals to target DNA so that its 3′ terminal nucleotide overlaps a single nucleotide polymorphism (SNP). A probe containing a 5′ flap of any sequence hybridizes to the target. If it contains a complementary nucleotide to the SNP, it will be cleaved, severing the flap. The downstream portion of the probe is relatively unstable and disassociates, relinquishing its position to a new probe. (B) Squared amplification. The severed flap from the first reaction acts as the invasive probe to a universal combined secondary probe and target. Cleavage of the probe-target releases fluorescence from its nearby quencher.

free arm. Enzymes possessing this activity reside in eukaryotes[226,227] and prokaryotes,[228,229] and include several thermophilic polymerases and flap endonucleases.[230-232] Severance depends exclusively on the structure of the oligonucleotides and does not require the presence of free nucleotides. Invader technology reproduces this recognized structure with an invader and signal probe hybridized to the target sequence.

4.2.7.2.1 Reaction Mechanics

The reaction begins with a single-stranded target bound with two probes. The terminal nucleotide of the upstream invader oligonucleotide, which may be any nucleotide, invades the complementary duplex formed by the trailing half of the downstream signal probe. The 5′ region of the signal probe is not complementary to the target and forms a single-strand arm or flap. The cleavage enzyme recognizes the resultant structure and migrates down the 5′ arm of the flap to cleave the signal probe one base beyond the invasion site.[228] A mismatched base one nucleotide upstream of the cleavage site on the signal probe will destroy overlap and preclude cleavage.[233,234] After cleavage, the invader oligonucleotide persists on the target, while the abbreviated signal probe dissociates, allowing a new signal probe to hybridize (Figure 4.9A). The reaction runs for 2 h within a narrow temperature window near the melting temperature of the probe. If the reaction is too cool, cycling of the signal probe is retarded, and if the reaction is too hot, endurance of the signal probe is insufficient for cleavage.[105] The turnover rate for the probe is 10 to 40/min/enzyme molecule,[105,235] or even faster on synthetic targets.[233] Thus, linear amplification is achieved.

Exponential amplification occurs by repeating the invasive cycle (at a lower temperature for the first-generation assay) with a synthetic target, with the severed 5′ flap of the primary signal probe as the invading probe in the secondary reaction (Figure 4.9B). Uncleaved primary probes may be checked through hybridization to an arrestor oligonucleotide.[235,236] The synthetic target and secondary signal probe may be compatible with many primary reactions. The amplification is approximately squared,[235] yet remains quantitative because it is proportional to the amplification achieved in the primary reaction.[105]

A variety of methods have been produced for detection of the cleaved signal probe. The labeled signal probe may spontaneously exhibit FRET fluorescence upon cleavage,[236] or different signal sequences may be analyzed with mass spectrometry.[235] The spectrometry is very fast; however, sample preparation requires an additional hour of preparation. For an additional level of discrimination against thermal or nonspecific cleavage, biotinylated signal probes may be extended after the reaction on a synthetic target with DIG-labeled nucleotides, which are probed with a labeled antibody after attachment to a streptavidin microwell plate.[234] This procedure enhances specificity because the 3′ end of the probe must be complementary for successful extension and nucleotide incorporation.

4.2.7.2.2   Oligonucleotides

The probes for the primary reaction consist of the upstream invader probe and the downstream signal probe. The invader probe needs to be only two to three bases long for enzymatic recognition;[233] practically, its length of about 30 nucleotides ensures secure binding on the target with a melting temperature 10°C higher than the reaction temperature. The target-specific portion of the signal probe is 12 to 20 nucleotides in length with a melting temperature near the optimum for the cleavage enzyme.[105] The sequence of the 5′ arm of the primary probe has little effect on enzyme activity[228] and can be designed for detection or advancement into the secondary reaction. The invader probe must overlap the target-complementary region of the downstream probe by at least one base for cleavage to occur, but the 3′ terminal nucleotide of the invader does not have to complement either the probe or the target. A lone base opposite this nucleotide in the signal probe decides invasion and thus imparts single-nucleotide specificity to this reaction.

The synthetic template of the first-generation cascade reaction comprised an arbitrary 30 bases and was either generic for several reactions or allele specific to allow multiplexing within one reaction. The brevity of the synthetic target enhanced the possibility of its 3′ end wrapping around to act as a nonspecific invader. To deter this, the 3′ terminus was blocked with an amino group, and the last five bases were substituted with 2′-O-methyl RNA.[235] This chemistry is not recognized as a substrate by the enzyme. To further reduce background, an arrestor oligonucleotide,[235,236] also composed of 2′-O-methyl ribonucleic acid, was introduced. Its obstructive binding prevented unreacted primary signal probes from entering the second phase. The second-generation assay removes much of the complexity of the first generation by combining the secondary target and signal probe into a single oligonucleotide that loops back upon itself (Figure 4.9B). Because uncleaved probes do not persist on the secondary template as in the first generation, an arrestor nucleotide is unnecessary, and a single temperature can be maintained throughout the reaction.[237] The secondary signal probe is labeled according to the desired detection system, typically FRET detection.

4.2.7.2.3   Conclusions

The structure-specific endonuclease reaction is an isothermal amplification system capable of detecting single-nucleotide polymorphisms or deletions[105] in conserved regions within background genomic DNA. All alleles may be investigated because specific initiation is based on both hybridization stability and enzymatic recognition.[234] The reaction is sensitive enough to detect polymorphisms in single-copy genes, even when the mutant is at one to one hundredth the wild-type concentration.[105] Depending on the reaction conditions and enzyme used, either DNA or RNA can be examined.

However, the reaction is relatively slow, taking 2 h for the primary amplification and another 2 h for the secondary amplification, not including additional time for detection. The pace is due to the linear kinetics of each phase, bestowing quantitative results. Measuring the generation of products during the reaction affords a wide dynamic range, which allows quantification of viral titer and gene expression.[229] Terminally cleaved products are a dead end for amplification and will not contribute to contamination. Cleaved primary probes could escape and contaminate subsequent reactions, but their linear expansion should mitigate any detrimental effect. Linear amplification has been accomplished on a solid phase, an essential achievement if the technology is to be adapted to high-throughput microarrays.[238]

4.3 NONAMPLIFIED DNA IN AMPLIFICATION REACTIONS

Not every synthetic oligonucleotide present in an amplification reaction directly contributes to the amplification. The presence of synthetic oligonucleotides may assist specific amplification or advance expeditious detection. Many of these auxiliary oligonucleotides are found as accessories to PCR. This is due more to the dominance of PCR than to its singularity. Many of the following oligonucleotides can and have been adapted to other target-amplification systems.

4.3.1 INHIBITORS

Paradoxically, one way to enhance the sensitivity of a reaction is to use synthetic DNA as inhibitors of polymerization. Polymerization arises from both specific and nonspecific extension of primers. Sometimes the latter surpasses the former, especially when the target is in low abundance or the target is similar to background sequences. Focusing the attention of the polymerase improves the ratio of true signal to unwanted noise, reducing the lower detection limit. The inhibitors may interfere with the active site of the polymerase directly or block adventitious priming sites on the background DNA.

4.3.1.1 Polymerase Inhibitors

One cause of nonspecific bands in PCR is the residual activity of *Thermus aquaticus* (Taq) DNA polymerase at room temperature. Before the high reaction temperature is realized, the opportunity exists for mispriming due to low stringency annealing and extension. For highest specificity, the polymerase must remain functionally separate or inert until the stringent conditions are attained.

The first "hot start" PCRs constituted withholding an essential component of the reaction, which was manually added later at the higher temperature.[239] The additional step increases hands-on time and exposes the tube to contamination. Improvements include placing the polymerase atop a wax layer that melts upon first cycle heating,[240] an anti-Taq DNA polymerase antibody that is irreversibly denatured by heating,[241] or a modified Taq polymerase that is activated by 15 to 20 min of heating.[242] Although all of these methods effectively reduce nonspecific priming, they require manual manipulation or time-consuming activation steps before initiating cycling.

Using *in vitro* selection techniques (SELEX,[243] see below), Dang and Jayasena[244] have isolated two families of 78-base single-stranded deoxynucleotides that show more than 1000-fold affinity and inhibition of Taq polymerase at room temperature than random oligonucleotides of this length. Inhibition vanishes at temperatures greater than 40°C but reactivates whenever the temperature is lowered, making them the only reversible inhibitors. Truncating and linking representatives of the two inhibition families into a heterologous molecule with a combined total of 81 bases led to a molecule capable of inhibiting several thermostable polymerases, which neither family could achieve separately.[245] Combining domains may lead to a single reagent compatible with any commercially available enzyme. Although the sequence of the oligonucleotides was developed with *in vitro* selection, they are easily manufactured with standard synthetic techniques.

4.3.1.2 Extension Blockers

Often, nonspecific amplification comes not from random mispriming, but by competition from closely related sequences. Especially daunting is distinguishing between mutant and wild-type sequences differing by a single nucleotide. There are at least three basic mechanisms of impeding this unwanted amplification using terminally blocked oligonucleotides incapable of extension. The blocking chemistries may be dideoxynucleotides, amine or sulfur groups, or inorganic phosphates.

The simplest includes these oligonucleotides exactly complementing the spurious priming site, thus outcompeting the primers for hybridization at that location.[246,247] This method allowed the detection of a single mutant among a normal population of 100,000.[248]

The second method recruits the blocked wild-type primer to amplify mutant sequence through the 3′ to 5′ exonucleolytic activity of the *Pyrococcus furiosis* polymerase. Primers are synthesized so that the contested nucleotide resides at the 3′ terminus, which is blocked from extension with a dideoxynucleotide. Mismatched primers trigger efficient proofreading by the polymerase, with removal of the obstructing base and subsequent lengthening of the primer. Matched primers are inefficiently removed and remain blocked. This protocol was used to detect a single relevant transition within an oncogene.[249]

The blocked oligonucleotide of the final technique hybridizes to competing interior sequences rather than primer sequences.[250,251] This allows extraction of a particular sequence from an array of closely related sequences. For example, the protocol allows the use of generic 16S rDNA primers, common to several related bacteria, to amplify only a species-specific sequence between the primers.[251] The polymerase may not contain any 5′-3′ exonuclease or strong-strand displacement activity that would negate the blocking effect of the internal oligonucleotide.

4.3.2 DETECTION

To the naked eye, a reaction tube at the end of an amplification reaction appears as at the beginning — a vial of transparent fluid. The procedure is meaningless without a detection mechanism. Traditional methods have included gel-based (e.g., see Reference 252) and solid-phase formats (e.g., see Reference 253). Although serviceable for the research laboratory, these methods are slow, insensitive, labor-intensive, and susceptible to contamination. Routine clinical diagnostics and other large-scale applications require a homogenous, real-time detection format amenable to automation. Several methods employing enzymes or inorganic compounds have been proposed for this duty,[254–257] but this section focuses on detection formats incorporating synthetic DNA. These formats include enzymatically cleaved probes, molecular beacons, and other homogenous hybridization probes that are compatible with several amplification reactions. Probes entirely specific to a particular reaction are described with that reaction.

4.3.2.1 5′-Nuclease Assay

Taq DNA polymerase is the most common enzyme used in PCR. It does not have any 3′ → 5′ proofreading ability, but it possesses a 5′ → 3′ nucleolytic domain.[258,259] The structure-specific domain functions in repair replication, cleaving the displaced downstream DNA protruding from the actively forming duplex, ignoring ssDNA. The polymerase and nuclease domains working in concert allow the concomitant extension of a primer and degradation of a hybridization probe. In the presence of dNTPs, the enzyme releases mostly mono- or dinucleotides.

Holland et al.[260] reported the first use of degradative probes to measure amplification during PCR. The assay requires the single-stranded probe to bind to the template during the annealing phase before primer extension. Comparatively, the probe must have a lower melting temperature or a higher concentration than the primers. Polymerization from its 3′ end is precluded by a terminal blockage. Lengthening of an upstream primer concurrently displaces the 5′ end of the probe, which is serially cleaved by the oncoming polymerase (Figure 4.10A). Degradation separates a donor-quencher pair of fluorophores resident on the probe sequence,[261] leading to amplification-dependent fluorescence. The donor typically sits near the 5′ end. The position of the quencher is flexible, provided it is not immediately adjacent to the donor,[262] because they may then be severed together. One strategy situates the quencher at the 3′ end of the probe, relying on the elasticity of the single-stranded backbone to keep the fluorophores within effective range. This guarantees their separation upon any cleavage. The rigidity imposed by hybridization increases fluorescence even in the absence of cleavage. The probe should not contain either a guanine residue at the 5′ end or stretches of guanines throughout as they interfere with the donor fluorescence.[263] Probes can be internally

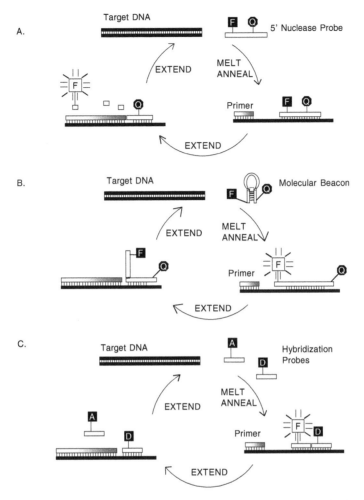

FIGURE 4.10 Fluorescent detection. (A) 5′-Nuclease probe. Degradation of the probe by the 5′ → 3′ exonuclease activity of the extending polymerase removes the fluorescent moiety from its quencher. (B) Molecular beacon. The hairpin molecular beacon stretches along the single-stranded target during the annealing phase, physically separating the fluorescent moiety from its quencher. Strand displacement activity of the polymerase displaces the beacon for subsequent cycles. (C) Hybridization probes. A pair of probes hybridizes adjacent to each other, allowing fluorescence donation [D] to an acceptor [A].

modified with pyrimidine C-5 propyne nucleotides, which grant higher melting temperatures for shorter sequences and improve specificity and sensitivity, especially in A-T rich regions.[264]

The first assays were not truly homogeneous because the products had to be moved to a fluorometer to be read at the end of the reaction. Heid et al.[19] used a sequencing machine to monitor real-time fluorescence during the PCR reaction. Measurements taken during exponential expansion allow quantification of the reaction before it enters the plateau stage of concentration inhibition. The cycle at which an arbitrary fluorescence threshold is crossed (usually 10 SD above mean baseline) is predictive of the quantity of input target and decreases linearly with increasing target. The dynamic range is 100-fold more than with a direct measurement in fluorescence.

An important aspect of the 5′-nuclease assay is that it inherently donates increased specificity to the PCR reaction. Spurious products do not hybridize the probe, rendering them invisible and allowing single-nucleotide differences in alleles to be detected.[265–267] The use of different reporter dyes among specific probes allows multiplexing of several alleles amplified with the same primers.

The assay is based on the thermodynamic contribution of a mismatch or deletion in the middle of the probe under competitive conditions, and this modification promotes disassociation before cleavage.[263] Although robust probes may be difficult to design,[265] once optimized this method performs well and has been adopted extensively for diverse applications.[190,268–272,273–276]

4.3.2.2 Molecular Beacons

The 5′-nuclease assay acts by destruction of the primary structure of the probe. Molecular beacons may also be considered degradative probes, but only their secondary structure is dissolved. First described in 1996,[277] a molecular beacon is a partially single-stranded probe that is self-complementary for 5 to 8 nucleotides at the 3′ and 5′ ends such that it spontaneously forms a hairpin (Figure 4.10B). The loop portion consists of 15 to 30 nucleotides complementary to the target 20 to 30 nucleotides downstream of a primer,[278] and the 3′ end is blocked against extension. When the probe is free in solution, its latent fluorophores at either end are held together in close proximity. The stem of the hairpin does not have to be tightly held because the average molecular motion of the flexible probe is more rapid than the fluorescent energy transfer. Target binding is much more stable than hybridization within the stem. When the probe binds to the target sequence, its two fluorophores are forcibly separated and cannot interact, liberating the fluorescence of the donor. Similar to the 5′ nuclease assay, unhybridized probes are self-quenching and do not have to be separated from hybridized probes.

Astonishingly, a single nucleotide difference or deletion in the center of the loop section can easily determine if a beacon will hybridize. This discrimination is impossible with a standard linear hybridization probe of the same length (not 5′ nuclease probes, which gain specificity through their interaction with the polymerase). The enhanced specificity is due to the multiple conformations of the probe[279] — double stranded (bound to target), closed hairpin, and open random coiled. Linear hybridization probes exist only in double-stranded or open random-coiled phases. When melting off the target, the mismatched probe converts to the intermediate hairpin structure much earlier than to the random coil, thereby granting a wider temperature range between when the matched and mismatched probes bind. For example, there was a temperature differential of 14°C between the release of mismatched and perfectly matched molecular beacons, but of only 8°C for linear probes.[279]

Because their action is not mediated by an enzyme, molecular beacons may be used for direct detection of mRNA affiliated with the purified ribosomal fractions of cells,[280] or within the living cells themselves.[281] Direct detection paves the way for rapid assays using biosensors, which translate hybridization events into electrical signals. Recently, molecular beacons have been affixed to a solid surface toward this end.[282,283] Molecular beacons are easily multiplexed by labeling probes with one of several available fluorophores and a universal quencher.[284–287] Molecular beacons have been used to detect viral[288–290] and bacterial sequences,[291–293] and in typing human alleles.[294] They may prove to be very useful in automated diagnostics.[295]

4.3.2.3 Other Hybridization Probes

The final set of hybridization probes measures the change in fluorescence upon hybridization of more than one linear probe per assay.[296] Fluorescence is measured after energy transfer (FRET, see above) from a donor to an acceptor moiety as two blocked hybridization probes bind adjacent to each other during the PCR cycles[297] or onto the PCR products.[298] One end-point assay monitors the melting curve through fluorescent changes of labeled probes over a wide range of temperature.[299,300] After PCR product accumulation and probe annealing, the fluorescent profile is monitored as the temperature is increased (Figure 4.10C). The temperature at which the probes melt, denoted by a rapid change in fluorescence, discriminates between matched and mismatched probes. This technique can be used to investigate a variety of sequences within a mixture,[301] but multiplexing is limited to the resolution of the fluorescent peaks distinguishable within the range of probe melting temperatures.[300]

4.3.2.4 Conclusions

By using hybridization probes that bind during the reaction, researchers can minimize the time to detection, quantitate the amount of target, reduce the chance of contamination, and enhance the specificity of the reaction. The 5′-nuclease probes and molecular beacons enjoy an advantage over melting-curve probes in that only one oligonucleotide has to be produced and purified. Molecular beacons and hybridization probes do not require enzymatic cooperation for activity and may be used with DNA or RNA products.

4.4 *IN VITRO* EVOLUTION

Once a diagnostic protocol is optimized, the oligonucleotides, enzymes, and reaction conditions may together be considered a single machine with the purpose of detecting specific target sequence. As with other machines, a consumer who follows instructions may successfully operate the reaction with no understanding of its functional mechanisms. The reactions may be purchased as all-complete kits which produce consistent, reliable, and predictable results.

Despite their predictability as domesticated molecular machines, many of the target amplification systems harbor characteristics reminiscent of the natural processes on which they were based, such as survival and linear inheritance of genetic sequence. When replicating, these sequences can be subject to selection, competition, and mutation, mirroring natural evolution. *In vitro* evolution enjoys the advantage of generation times on the order of seconds, with the "genome" of the species fully defined, allowing rigorous testing of evolutionary theory. Based on the disparity or synchronicity of the generations, evolution is continuous or discrete.

4.4.1 CONTINUOUS EVOLUTION

Evolution requires differential survival and reproduction of diverse replicating entities. The first *in vitro* evolution took place more than 30 years ago with the RNA-directed polymerase (replicase) and single-stranded RNA genome of the bacteriophage QB. Replicase does not initiate from a primer or promoter. It recognizes the secondary structure of both its cognate template and the complementary negative strand, performing asymmetrically alternating, end-to-end synthesis from each. The first reaction tube contained the replicase, template, and dNTPs in a suitable buffered environment. Generated sequences competed for reproduction with the limited resource of nucleotides. Subsequent tubes of replicase and nucleotides were seeded with products from the immediately preceding reaction. Incubation times between the serial transfers were progressively shortened. The experiment rewarded survival of the fastest, without requiring any biological activity of the genome beyond replication. The virus lost 83% of its sequence, retaining just enough structural information for recognition by the replicase.[302] With an additional constraint of replication in the presence of increasing ethidium bromide concentration, successful sequences developed resistance, reducing the drug binding by 35%.[303]

RNA polymerases offer ample opportunity for evolution studies because of their high error rate leading to the sequence variation on which selection acts. Sequences descended from natural RNA are limited by the structure-specific recognition of the replicase and may have a restricted evolutionary landscape. In contrast, sequences downstream of promoters do not suffer this limitation. Promoters within synthetic DNA direct the synthesis of numerous copies of RNA, as seen in the transcription-based amplification reactions. In fact, this reaction has been modified and employed extensively for *in vitro* evolution.

The complicated life cycle of 3SR is amenable to studying ecological interactions beyond simple competition, such as cooperation and predator–prey relationships. Cooperation studies are based on two overlapping cycles of 3SR, which combine to construct the double-stranded DNA template that transcribes the two species from promoters oriented in opposite directions.[304,305] In predatory situations, the ssDNA intermediate of the prey cycle is consumed as a primer in the predator cycle.[306] Because they are simple, these systems can be rigorously defined, mathematically predicted, and then empirically tested.[307,308] These biophysical results are beyond the scope of this chapter, but they illuminate models on the genesis of complex systems such as the subassembly of components necessary for the production of life.

Every molecular ecosystem develops parasites that usurp the polymerase and overwhelm the cooperative process (e.g., Reference 304). These ubiquitous parasites bypass the cooperative inheritance of a promoter by modifying their secondary structure to encode an active double-stranded promoter in the form of a hairpin.[309] In fact, seeding the reaction with this hairpin structure of random sequence leads to the development of promoters similar in sequence to the cognate promoters of the RNA polymerase.[310] These parasites are typically shorter than their predecessors, but longer genomes evolve if they replicate faster under the reaction conditions,[311] such as being less likely to fold into inactive conformations.[312] The parasites often interfere with cooperation at low concentrations of the original template,[304] and thus the system could benefit investigations into how cooperative systems originally kept such parasites at bay.[306]

The parasites of the cooperative system are usually modified but recognizable descendants of the nucleic acid sequences placed in the reaction. However, with no externally added nucleic acid, the RNA polymerase will fabricate RNA from free ribonucleotides. Every reaction tube evolves a dominant family of sequences unique to that reaction,[313] related as *quasi-species*.[314] This term is commonly used to define a population of RNA viruses within an organism. Thus, understanding the simple *in vitro* system could lead to understanding the molecular evolution of viruses such as hepatitis C virus.

Transcription-mediated amplification is not the only continuous amplification system useful for studying evolution.[313,315] For example, successful products selected for rapid reproduction during the strand-displacement amplification reaction discarded much of their sequence.[316] All continuous amplification systems are susceptible to parasites[311] that circumvent the control mechanisms requiring intermolecular interactions (cooperation) such as primer binding. In contrast, during the stepwise PCR, the discrete phases disallow more than one copy per single-strand per cycle, thus removing any selective pressure for fast replicators and limiting the damage of any mutants.[317] So while useless for studying model ecological molecular systems,[307] its immunity to parasites renders it invaluable for generating and maintaining functional sequences.

4.4.2 Discrete Evolution

Functional RNA or DNA capable of specific binding[318] or catalysis[319,320] can be evolved and screened from large pools of random sequences using PCR-based *in vitro* selection techniques.[321–324] The 30- to 200-nucleotide random sequences are generated by random synthesis between PCR primer-binding sequences.[324] The oligonucleotides are amplified with PCR and expressed in the desired form (RNA, ssDNA, dsDNA). RNA requires an additional transcription step after amplification. Selective pressure relevant to the described function[243,326] is applied (e.g., affinity chromatography for binding activity) to extract the more competitive sequences. Because not all sequences will be represented in the original random pool, variation may be introduced during error-prone polymerization.[326] The selected sequences are reamplified and subjected to increasingly stringent selection for 5 to 20 iterations. The resultant functional molecule is called an aptamer.

Many useful molecules are made with only natural nucleotides. However, circumstances exist in which modified nucleotides are essential for high efficiency. There are hundreds of nucleotide modifications available that can be incorporated during polymerization or added later during artificial synthesis.[327] For example, oligonucleotides to be used *in vitro* are structurally stabilized with disulfide cross-links and protected from exonucleases with phosphorothioates.[328] Defensive *in vivo* modifications include changing the 2' OH of ribonucleotides to $2'NH_2$, $2'F$, or $2'CH_3$, appending a $3'-3'$ dT cap for exonuclease resistance, or linking to macromolecules to avoid clearance.[329] To increase their diagnostic potential, the aptamers may include 5' bromo or 5' iodo nucleotides capable of cross-linking to bound protein targets. Cross-linking allows stringent washing to remove non-specific background, increasing the signal-to-noise ratio among closely spaced determinants in microarrays.[330,331] Proteins are differentiated by the location of their binding and may be detected with a universal protein stain.

Aptamers show great promise as analytical[332,333] and preparative[334–336] reagents, therapeutic compounds,[328,337–340] diagnostic imaging instruments,[341] and windows into the history of molecular evolution.[342–346]

4.4.3 CONCLUSIONS

All natural replicators are subject to evolution. The mechanisms of isothermal target amplification systems are often based on natural processes such as retroviral reproduction (transcription-based amplification) or plasmid replication (SDA), and thus it is not unusual that these *in vitro* shadows of replicating systems support evolution. Rigorous study of these defined systems may shed light on the development and sustainment of ecological interaction such as parasitism, predation, competition, and mutualism in the prebiotic world. The discrete target amplification of PCR, while not representative of the natural world, is highly relevant for finding and cultivating functional oligonucleotides and may elucidate the connection between structural and functional properties of nucleic acid sequences. Automated robots are being developed for applications using continuous or discrete evolution.[347–349]

4.5 SUMMARY

Although different mechanistically, the above reactions may be described as comprising three phases — initiation, progression, and resolution.[6,350] Initiation describes the triggering of the reaction, progression the mechanism of amplification, and resolution either the termination or regeneration of a template. The bDNA reaction is the only reaction that is not internally cyclical in that one initiation event progresses to a single resolution. In all others, a reaction product reenters as a substrate, initiating a new cycle. The cycles may be invisible to the operator if not externally controlled by temperature changes.

All of the reactions share a common initiation event of Watson–Crick hybridization, or a closely related facsimile (i.e., modified nucleotides). The initiation event imparts most of the reactions with their specificity. Progression and initiation are intertwined because the hybridization event (initiation) is immediately followed by extension, ligation, cleavage, etc. (progression). In these cases, initiation is inseparable from progression, limiting the reaction to detection of complementary nucleic acids. If initiation wholly involves specific recognition disassembled from progression, detection of proteins or analytes is possible. For example, immuno-PCR[351] combines the specific recognition of proteins with the high sensitivity of PCR through a chimeric molecule consisting of an antibody cross-linked to a synthetic template. Although effecting sensitive protein detection (e.g., Reference 352), the procedure requires the *in vivo* production of antibody and tedious construction of the hybrids. On the other hand, portions of wholly synthetic aptamers generated through *in vitro* evolution to bind proteins could be directly introduced into nucleic-acid-amplification reactions as an artificial template. Multiple recognition sequences, discovered separately, could reproducibly be synthesized on the same molecule in practically endless amounts. In addition, nucleotides can be designed to accomplish several purposes simultaneously; the recognition and signaling attributes can be combined on molecular beacon aptamers.[353,354]

Increasing speed is tantamount to improving progression. Currently, the kinetics of synthesis are much faster than condensation (hybridization) or degradation (melting). Improvements can be made through enzymatic engineering, such as for invasive probe assay, or physical improvements, such as reduced-volume PCR.

The mechanism of the resolution of the reaction determines whether it expands linearly or exponentially. If resolution consistently generates only signal and not new target, then the reaction will be linear (CPT, primary invasive-probe assay). If new targets or probes are generated at resolution, then the reaction is exponential. The stoichiometry of new target resolved determines the rate of the reaction. Thus, transcription based on SDA, with numerous products manufactured per template molecule, gives faster amplification than PCR.

The plethora of amplification reactions invented over the last 15 years has made possible previously unattainable explorations into all fields of the life sciences. Many reactions are primarily applicable to diagnostics, but as the sequence of the human genome is revealed, predictive diagnostics will require rapid, nonintrusive assays (e.g., Reference 355) capable of detecting single-nucleotide polymorphisms.[356] In addition to revealing information about ourselves, the reactions allow a glimpse into the prebiotic world and the development of ecology at the genesis of our evolution.

REFERENCES

1. Watson, J. D. and Crick, F. H., Molecular structure of nucleic acids: a structure for deoxyribose nucleic acid, *Nature*, 171, 737, 1953.
2. Kleppe, K. et al., Studies on polynucleotides: XCVI. Repair replication of short synthetic DNAs as catalyzed by DNA polymerases, *J. Mol. Biol.*, 56, 341, 1971.
3. Mullis, K. et al., Specific enzymatic amplification of DNA *in vitro*: the polymerase chain reaction, *Cold Spring Harb. Symp. Quant. Biol.*, 51, 263, 1986.
4. Horvath, S. J. et al., An automated DNA synthesizer employing deoxynucleoside 3'-phosphoramidites, *Methods Enzymol.*, 154, 314, 1987.
5. Schweitzer, B. and Kingsmore, S., Combining nucleic acid amplification and detection, *Curr. Opin. Biotechnol.*, 12, 21, 2001.
6. Landegren, U., The challenges to PCR: a proliferation of chain reactions, *Curr. Opin. Biotechnol.*, 7, 95, 1996.
7. Newton, C. R. and Graham, A., Polymerase chain reaction, in *Introduction to Biotechniques*, 2nd ed., Springer-Verlag, New York, 1997.
8. Saiki, R. K. et al., Enzymatic amplification of B-globin genomic sequences and restriction site analysis for diagnosis of sickle cell anemia, *Science*, 230, 1350, 1985.
9. Saiki, R. K. et al., Primer-directed enzymatic amplification of DNA with a thermostable DNA polymerase, *Science*, 239, 487, 1988.
10. Bloch, W., A biochemical perspective on the polymerase chain reaction, *Biochemistry*, 30, 2735, 1991.
11. Raeymaekers, L., Quantitative PCR: theoretical considerations with practical implications, *Anal. Biochem.*, 214, 582, 1993.
12. Raeymaekers, L., A commentary on the practical application of competitive PCR, *Genome Res.*, 5, 91, 1995.
13. Sachadyn, P. and Kur, J., The construction and use of a PCR internal control, *Mol. Cell. Probes*, 12, 259, 1998.
14. Jin, C. F., Mata, M., and Fink, D. J., Rapid construction of deleted DNA fragments for use as internal standards in competitive PCR, *PCR Methods Appl.*, 3, 252, 1994.
15. Diviacco, S. et al., A novel procedure for quantitative polymerase chain reaction by coamplification of competitive templates, *Gene*, 122, 313, 1992.
16. Zenilman, M. E. et al., Competitive reverse-transcriptase polymerase chain reaction without an artificial internal standard, *Anal. Biochem.*, 224, 339, 1995.
17. Freeman, W. M., Walker, S. J., and Vrana, K. E., Quantitative PCR: pitfalls and potentials, *Biotechniques*, 26, 112, 1999.

18. Connolly, A. R., Cleland, L. G., and Kirkham, B. W., Mathematical considerations of competitive polymerase chain reaction, *J. Immunol. Methods*, 187, 201, 1995.
19. Heid, C. A. et al., Real time quantitative PCR, *Genome Res.*, 6, 986, 1996.
20. Rychlik, W., Selection of primers for polymerase chain reaction, *Mol. Biotechnol.*, 3, 129, 1995.
21. Wu, D. Y. et al., The effect of temperature and oligonucleotide primer length on the specificity and efficiency of amplification by the polymerase chain reaction, *DNA Cell. Biol.*, 10, 233, 1991.
22. Rychlik, W., Spencer, W. J., and Rhoads, R. E., Optimization of the annealing temperature for DNA amplification *in vitro*, *Nucleic Acids Res.*, 18, 6409, 1990.
23. Mullis, K. B. and Faloona, F. A., Specific synthesis of DNA *in vitro* via a polymerase-catalyzed chain reaction, *Methods Enzymol.*, 155, 335, 1987.
24. Berg, J. et al., Single-tube two-round polymerase chain reaction using the LightCycler™ instrument, *J. Clin. Virol.*, 20, 71, 2001.
25. Compton, T., Degenerate primers for DNA amplification, in *PCR Protocols: A Guide to Methods and Applications*, Innis, M. A. et al., Eds., Academic Press, San Diego, 1990.
26. Knoth, K. et al., Highly degenerate inosine-containing primers specifically amplify rare cDNA using the polymerase chain reaction, *Nucleic Acids Res.*, 17, 10932, 1988.
27. Hill, F., Loakes, D., and Brown, D. M., Polymerase recognition of synthetic oligodeoxyribonucleotides incorporating degenerate pyrimidine and purine bases, *Proc. Natl. Acad. Sci. U.S.A.*, 95, 4258, 1998.
28. Schlayer, H. J. et al., Amplification of unknown DNA sequences by sequence-independent nested polymerase chain reaction using a standardized adaptor without specific primers, *J. Virol. Methods*, 38, 333, 1992.
29. Williams, J. G. et al., DNA polymorphisms amplified by arbitrary primers are useful as genetic markers, *Nucleic Acids Res.*, 18, 6531, 1990.
30. Welsh, J. and McClelland, M., Fingerprinting genomes using PCR with arbitrary primers, *Nucleic Acids Res.*, 18, 7213, 1990.
31. Ho, S. N. et al., Site-directed mutagenesis by overlap extension using the polymerase chain reaction, *Gene*, 77, 51, 1989.
32. Kain, K. C., Orlandi, P. A., and Lanar, D. E., Universal promoter for gene expression without cloning: expression-PCR, *Biotechniques*, 10, 366, 1991.
33. Morrison, L. E., Detection of energy transfer and fluorescence quenching, in *Nonisotopic DNA Probe Techniques*, Kricka, L. J., Ed., Academic Press, San Diego, 1992, 358.
34. Selvin, P. R., Fluorescence resonance energy transfer, in *Biochemical Spectroscopy*, Sauer, K., Ed., Academic Press, San Diego, 1995, 300.
35. Nazarenko, I. A., Bhatnagar, S. K., and Hohman, R. J., A closed tube format for amplification and detection of DNA based on energy transfer, *Nucleic Acids Res.*, 25, 2516, 1997.
36. Winn-Deen, E. S., Direct fluorescence detection of allele-specific PCR products using novel energy-transfer labeled primers, *Mol. Diagn.*, 3, 217, 1998.
37. Ju, J. et al., Fluorescence energy transfer dye-labeled primers for DNA sequencing and analysis, *Proc. Natl. Acad. Sci. U.S.A.*, 92, 4347, 1995.
38. Whitcombe, D. et al., Detection of PCR products using self-probing amplicons and fluorescence, *Nat. Biotechnol.*, 17, 804, 1999.
39. Saha, B. K., Tian, B., and Bucy, R. P., Quantitation of HIV-1 by real-time PCR with a unique fluorogenic probe, *J. Virol. Methods*, 93, 33, 2001.
40. Thelwell, N. et al., Mode of action and application of Scorpion primers to mutation detection, *Nucleic Acids Res.*, 28, 3752, 2000.
41. Solinas, A. et al., Duplex Scorpion primers in SNP analysis and FRET applications, *Nucleic Acids Res.*, 29, e96, 2001.
42. Todd, A.V. et al., DzyNA-PCR: use of DNAzymes to detect and quantify nucleic acid sequences in a real-time fluorescent format, *Clin. Chem.*, 46, 625, 2000.
43. O'Donnell-Maloney, M. J. and Little, D. P., Microfabrication and array technologies for DNA sequencing and diagnostics, *Genet. Anal.*, 13, 151, 1996.
44. Shandrick, S., Ronai, Z., and Guttman, A., Rapid microwell polymerase chain reaction with subsequent ultrathin layer gel electrophoresis of DNA, *Electrophoresis*, 23, 591, 2002.
45. Kuypers, A. W. H. M. et al., Contamination-free and automated composition of a reaction mixture for nucleic acid amplification using a capillary electrophoresis apparatus, *J. Chromatogr. A*, 806, 141, 1998.

46. Kopp, M. U., de Mello, A. J., and Manz, A., Chemical amplification: continuous-flow PCR on a chip, *Science*, 280, 1046, 1998.

47. Lee, L. G. et al., Seven-color, homogeneous detection of six PCR products, *Biotechniques*, 27, 342, 1999.

48. Shapero, M. H. et al., SNP genotyping by multiplexed solid-phase amplification and fluorescent minisequencing, *Genome Res.*, 11, 1926, 2001.

49. Wu, D. Y. et al., Allele-specific enzymatic amplification of ß-globin genomic DNA for diagnosis of sickle cell anemia, *Proc. Natl. Acad. Sci. U.S.A.*, 86, 2757, 1989.

50. Saiki, R. K. et al., Analysis of enzymatically amplified beta-globin and HLA-DQ alpha DNA with allele-specific oligonucleotide probes, *Nature*, 324, 163, 1986.

51. Landegren, U. et al., A ligase-mediated gene detection technique, *Science*, 241, 1077, 1988.

52. Nickerson, D. A. et al., Automated DNA diagnostics using an ELISA-based oligonucleotide ligation assay, *Proc. Natl. Acad. Sci. U.S.A.*, 87, 8923, 1990.

53. Barany, F., The ligase chain reaction in a PCR world, *PCR Methods Appl.*, 1, 5, 1991.

54. Wiedmann, M. et al., Ligase chain reaction (LCR): overview and applications, *PCR Methods Appl.*, 3, S51, 1994.

55. Landegren, U., Ligation-based DNA diagnostics, *Bioessays*, 15, 761, 1993.

56. Shimer, G. H., Jr. and Backman, K. C., Ligase chain reaction, in *Diagnostic Bacteriology Protocols*, Howard, J. and Whitcombe, D. M., Eds, Humana Press, Totowa, NJ, 1995, 269.

57. Wu, D. Y. and Wallace, R. B., The ligation amplification reaction (LAR): amplification of specific DNA sequences using sequential rounds of template-dependent ligation, *Genomics*, 4, 560, 1989.

58. Takahashi, M., Yamaguchi, E., and Uchida, T., Thermophilic DNA ligase, *J. Biol. Chem.*, 259, 10041, 1984.

59. Barany, F. and Gelfand, D. H., Cloning, overexpression and nucleotide sequence of a thermostable DNA ligase-encoding gene, *Gene*, 109, 1, 1991.

60. Barany, F., Genetic disease detection and DNA amplification using cloned thermostable ligase, *Proc. Natl. Acad. Sci. U.S.A.*, 88, 189, 1991.

61. Bourgeois, C. et al., Value of a ligase chain reaction assay for detection of ganciclovir resistance-related mutation 594 in UL97 gene of human cytomegalovirus, *J. Virol. Methods*, 67, 167, 1997.

62. Kratochvil, J. and Laffler, T. G., Nonradioactive oligonucleotide probes for detecting products of the ligase chain reaction, in *Protocols for Nucleic Acid Analysis by Nonradioactive Probes*, Isaac, P. G., Ed., Humana Press, Totowa, NJ, 1994, 243.

63. Reyes, A. A. et al., Ligase chain reaction assay for human mutations: the sickle cell by LCR assay, *Clin. Chem.*, 43, 40, 1997.

64. Zebala, J. A. and Barany, F., Implications for the ligase chain reaction in gastroenterology, *J. Clin. Gastroenterol.*, 17, 171, 1993.

65. Wiedmann, M. et al., Discrimination of *Listeria monocytogenes* from other *Listeria* species by ligase chain reaction, *Appl. Environ. Microbiol.*, 58, 3443, 1992.

66. Eggerding, F. A., A one-step coupled amplification and oligonucleotide ligation procedure for multiplex genetic typing, *PCR Methods Appl.*, 4, 337, 1995.

67. Iovannisci, D. M. and Winn-Deen, E. S., Ligation amplification and fluorescence detection of *Mycobacterium tuberculosis*, *Mol. Cell. Probes*, 7, 35, 1993.

68. Davies, P. O. and Ridgway, G. L., The role of polymerase chain reaction and ligase chain reaction for the detection of *Chlamydia trachomatis*, *Int. J. STD AIDS*, 8, 731, 1997.

69. Cheng, J. et al., Analysis of ligase chain reaction products amplified in a silicon-glass chip using capillary electrophoresis, *J. Chromatogr. A*, 732, 151, 1996.

70. Jurinke, C. et al., Analysis of ligase chain reaction products via matrix-assisted laser desorption/ionization time-of-flight- mass spectrometry, *Anal. Biochem.*, 237, 174, 1996.

71. Gerry, N. P. et al., Universal DNA microarray method for multiplex detection of low abundance point mutations, *J. Mol. Biol.*, 292, 251, 1999.

72. Luo, J., Bergstrom, D. E., and Barany, F., Improving the fidelity of *Thermus thermophilus* DNA ligase, *Nucleic Acids Res.*, 24, 3071, 1996.

73. Kälin, I., Shepard, S., and Candrian, U., Evaluation of the ligase chain reaction (LCR) for the detection of point mutations, *Mutat. Res.*, 283, 119, 1992.

74. Demchinskaya, A.V. et al., A new approach for point mutation detection based on a ligase chain reaction, *J. Biochem. Biophys. Methods*, 50, 79, 2001.

75. Minamitani, S. et al., Detection by ligase chain reaction of precore mutant of Hepatitis B virus, *Hepatology*, 25, 216, 1997.

76. Wiedmann, M., Barany, F., and Batt, C. A., Detection of *Listeria monocytogenes* with a nonisotopic polymerase chain reaction-coupled ligase chain reaction assay, *Appl. Environ. Microbiol.*, 59, 2743, 1993.

77. Balles, J. and Pflugfelder, G. O., Facilitated isolation of rare recombinants by ligase chain reaction: selection for intragenic crossover events in the *Drosophila optomotor-blind* gene, *Mol. Gen. Genet.*, 245, 734, 1994.

78. Wilson, W. J. et al., Identification of *Erwinia stewartii* by a ligase chain reaction assay, *Appl. Environ. Microbiol.*, 60, 278, 1994.

79. Pfeffer, M., Meyer, H., and Wiedmann, M., A ligase chain reaction targeting two adjacent nucleotides allows the differentiation of cowpox virus from other *Orthopoxvirus* species, *J. Virol. Methods*, 49, 353, 1994.

80. Feero, W. G. et al., Hyperkalemic periodic paralysis: rapid molecular diagnosis and relationship of genotype to phenotype in 12 families, *Neurology*, 43, 668, 1993.

81. Lehman, T. A. et al., Detection of K-*ras* oncogene mutations by polymerase chain reaction-based ligase chain reaction, *Anal. Biochem.*, 239, 153, 1996.

82. Birkenmeyer, L. G. and Mushahwar, I. K., DNA probe amplification methods, *J. Virol. Methods*, 35, 117, 1991.

83. Birkenmeyer, L. and Armstrong, A. S., Preliminary evaluation of the ligase chain reaction for specific detection of *Neisseria gonorrhoeae*, *J. Clin. Microbiol.*, 30, 3089, 1992.

84. Abravya, K. et al., Detection of point mutations with a modified ligase chain reaction (gap-LCR), *Nucleic Acids Res.*, 23, 675, 1995.

85. Berg, E. S. et al., False-negative results of a ligase chain reaction assay to detect *Chlamydia trachomatis* due to inhibitors in urine, *Eur. J. Clin. Microbiol. Infect. Dis.*, 16, 727, 1997.

86. Marshall, R. L. et al., Detection of HCV RNA by the asymmetric gap ligase chain reaction, *PCR Methods Appl.*, 4, 80, 1994.

87. Hsuih, T. C. H. et al., Novel, ligation-dependent PCR assay for detection of Hepatitis C virus in serum, *J. Clin. Microbiol.*, 34, 501, 1996.

88. Miyauchi, I. et al., Further study of hepatitis C virus RNA detection in formalin-fixed, paraffin-embedded liver tissues by ligation-dependent polymerase chain reaction, *Pathol. Int.*, 48, 428, 1998.

89. Park, Y. N. et al., Detection of Hepatitic C virus RNA using ligation-dependent polymerase chain reaction in formalin-fixed, paraffin-embedded liver tissues, *Am. J. Pathol.*, 149, 1485, 1996.

90. Brandwein, M., Li, H., and Zhang, D. Y., Detection of EBV early RNA (EBER-1) in parotid pleomorphic ademonas: a novel observation utilizing ligation-dependent PCR, in *Head and Neck Cancer: Advances in Basic Research*, Werner, J. A., Lippert, B. M., and Rudert, H. H., Eds., Elsevier Science, Amsterdam, 1996, 401.

91. Au, L.-C. et al., Gene synthesis by a LCR-based approach: high-level production of leptin-L54 using synthetic gene in *Escherichia coli*, *Biochem. Biophys. Res. Commun.*, 248, 200, 1998.

92. Rouwendel, G. J. et al., Simultaneous mutagenesis of multiple sites: application of the ligase chain reaction using PCR products instead of oligonucleotides, *Biotechniques*, 15, 68, 1993.

93. Deng, S. J., MacKenzie, C. R., and Narang, S. A., Simultaneous randomization of antibody CDRs by a synthetic ligase chain reaction strategy, *Nucleic Acids Res.*, 21, 4418, 1993.

94. Schachter, J., DFA, EIA, PCR, LCR and other technologies: what tests should be used for diagnosis of chylamydia infections? *Immunol. Invest.*, 26, 157, 1997.

95. Zhang, D. Y. et al., Amplification of target-specific, ligation-dependent circular probe, *Gene*, 211, 277, 1998.

96. Nilsson, M. et al. Padlock probes: circularizing oligonucleotides for localized DNA detection, *Science*, 265, 2085, 1994.

97. Nilsson, M. et al., Padlock probes reveal single-nucleotide differences, parent of origin and *in situ* distribution of centromeric sequences in human chromosomes 13 and 21, *Nat. Genet.*, 16, 252, 1997.

98. Espinosa, M. et al., Plasmid rolling circle replication and its control, *FEMS Microbiol. Lett.*, 130, 111, 1995.

99. Novick, R. P., Contrasting lifestyles of rolling-circle phages and plasmids, *Trends Biochem. Sci.*, 23, 434, 1998.

100. Fire, A. and Xu, S. Q., Rolling replication of short DNA circles, *Proc. Natl. Acad. Sci. U.S.A.*, 92, 4641, 1995.

101. Liu, D. et al., Rolling circle DNA synthesis: small circular oligonucleotides as efficient templates for DNA polymerases, *J. Am. Chem. Soc.*, 118, 1587, 1996.

102. Hatch, A. et al., Rolling circle amplification of DNA immobilized on solid surfaces and its application to multiplex mutation detection, *Genet. Anal.*, 15, 35, 1999.

103. Lizardi, P. M. et al., Mutation detection and single-molecule counting using isothermal rolling-circle amplification, *Nat. Genet.*, 19, 225, 1998.

104. Thomas, D. C., Nardone, G. A., and Randall, S. K., Amplification of padlock probes for DNA diagnostics by cascade rolling circle amplification or the polymerase chain reaction, *Arch. Pathol. Lab. Med.*, 123, 1170, 1999.

105. Kwiatkowski, N., Nilsson, M., and Landegren, U., Synthesis of full-length oligonucleotides: cleavage of apurinic molecules on a novel support, *Nucleic Acids Res.*, 24, 4632, 1996.

106. Banér, J. et al., Signal amplification of padlock probes by rolling circle replication, *Nucleic Acids Res.*, 26, 5073, 1998.

107. Zhang, D. Y. et al., Detection of rare DNA targets by isothermal ramification amplification, *Gene*, 274, 209, 2001.

108. Zhang, W. et al., Detection of *Chlamydia trachomatis* by isothermal ramification amplification method: a feasibility study, *J. Clin. Microbiol.*, 40, 128, 2002.

109. Christian, A. T. et al., Detection of DNA point mutations and mRNA expression levels by rolling circle amplification in individual cells, *Proc. Natl. Acad. Sci. U.S.A.*, 98, 14238, 2001.

110. Isaksson, A. and Landegren, U., Accessing genome information: alternatives to PCR, *Curr. Opin. Biotechnol.*, 10, 11, 1999.

111. Landegren, U. and Nilsson, M., Locked on target: strategies for future gene diagnostics, *Ann. Med.*, 29, 585, 1997.

112. Nallur, G. et al., Signal amplification by rolling circle amplification on DNA microarrays, *Nucleic Acids Res.*, 29, e118, 2001.

113. Ladner, D. P. et al., Multiplex detection of hotspot mutations by rolling circle-enabled universal microarrays, *Lab. Invest.*, 81, 1079, 2001.

114. Qi, X. et al., L-RCA (ligation-rolling circle amplification): a general method for genotyping of single nucleotide polymorphisms (SNPs), *Nucleic Acids Res.*, 29, e116, 2001.

115. Schweitzer, B. et al., Immunoassays with rolling circle DNA amplification: a versatile platform for ultrasensitive antigen detection, *Proc. Natl. Acad. Sci. U.S.A.*, 97, 10113, 2000.

116. Wiltshire, S. et al., Detection of multiple allergen-specific IgEs on microarrays by immunoassay with rolling circle amplification, *Clin. Chem.*, 46, 1990, 2000.

117. Zhou, Y. et al., *In situ* detection of messenger RNA using digoxigenin-labeled oligonucleotides and rolling circle amplification, *Exp. Mol. Pathol.*, 70, 281, 2001.

118. Walker, G. T. et al., Isothermal *in vitro* amplification of DNA by a restriction enzyme/DNA polymerase system, *Proc. Natl. Acad. Sci. U.S.A.*, 89, 392, 1992.

119. Nadeau, J. G. et al., Real-time, sequence-specific detection of nucleic acids during strand displacement amplification, *Anal. Biochem.*, 276, 177, 1999.

120. Walker, G. T. et al., Strand displacement amplification: an isothermal, *in vitro* DNA amplification technique, *Nucleic Acids Res.*, 20, 1691, 1992.

121. Milla, M. A. et al., Use of the restriction enzyme Ava I and exo- Bst polymerase in strand displacement amplification, *Biotechniques*, 24, 392, 1998.

122. Spargo, C. A. et al., Detection of *M. tuberculosis* DNA using thermophilic strand displacement amplification, *Mol. Cell. Probes*, 10, 247, 1996.

123. Hellyer, T. J. et al., Detection of viable *Mycobacterium tuberculsis* by reverse transcriptase-strand displacement amplification of mRNA, *J. Clin. Microbiol.*, 37, 518, 1999.

124. Nycz, C. M. et al., Quantitative reverse transcription strand displacement amplification: quantitation of nucleic acids using an isothermal amplification technique, *Anal. Biochem.*, 259, 226, 1998.

125. Walker, G. T., Empirical aspects of strand displacement amplification, *PCR Methods Appl.*, 3, 1, 1993.

126. Spargo, C. A. et al., Chemiluminescent detection of strand displacement amplified DNA from species comprising the *Mycobacterium tuberculosis* complex, *Mol. Cell. Probes*, 7, 395, 1993.

127. Walker, G. T. et al., Strand displacement amplification (SDA) and transient-state fluorescence polarization detection of *Mycobacterium tuberculosis* DNA, *Clin. Chem.*, 42, 9, 1996.

128. Walker, G. T. and Linn, C. P., Detection of *Mycobacterium tuberculosis* DNA with thermophilic strand displacement amplification and fluorescence polarization, *Clin. Chem.*, 42, 1604, 1996.

129. Spears, P. A. et al., Simultaneous strand displacement amplification and fluorescence polarization detection of *Chlamydia trachomatis* DNA, *Anal. Biochem.*, 247, 130, 1997.

130. Little, M. C. et al., Strand displacement amplification and homogeneous real-time detection incorporated in a second-generation DNA probe system, BDProbeTecET, *Clin. Chem.*, 45, 777, 1999.

131. Notomi, T. et al., Loop-mediated isothermal amplification of DNA, *Nucleic Acids Res.*, 28, e63, 2000.

132. Nagamine, K., Kuzuhara, Y., and Notomi, T., Isolation of single-stranded DNA from loop-mediated isothermal amplification products, *Biochem. Biophys. Res. Commun.*, 290, 1195, 2002.

133. Badak, F. Z. et al., Confirmation of the presence of *Mycobacterium tuberculosis* and other mycobacteria in mycobacterial growth indicator tubes (MGIT) by multiplex strand displacement amplification, *J. Clin. Microbiol.*, 35, 1239, 1997.

134. Little, M. C., Spears, P. A., and Shank, D. D., Nucleotide sequence and strand displacement amplification of the 70K protein gene from mycobacteria, *Mol. Cell. Probes*, 8, 375, 1994.

135. Walker, G. T. et al., Multiplex strand displacement amplification (SDA) and detection of DNA sequences form *Mycobacterium tuberculosis* and other mycobacteria, *Nucleic Acids Res.*, 22, 2670, 1994.

136. Edman, C. F. et al., Pathogen analysis and genetic predisposition testing using microelectronic arrays and isothermal amplification, *J. Invest. Med.*, 48, 93, 2000.

137. Westin, L. et al., Anchored multiplex amplification on a microelectronic chip array, *Nat. Biotechnol.*, 18, 199, 2000.

138. Kwoh, D.Y. et al., Transcription-based amplification system and detection of amplified human immunodeficiency virus type 1 with a bead-based sandwich hybridization format, *Proc. Natl. Acad. Sci. U.S.A.*, 86, 1173, 1989.

139. Malek, L., Sooknanan, R., and Compton, J., Nucleic acid sequence-based amplification (NASBA), in *Protocols for Nucleic Acid Analysis by Nonradioactive Probes*, Isaac, P. G., Ed., Humana Press, Totowa, NJ, 1994.

140. Guatelli, J. C. et al., Isothermal, *in vitro* amplification of nucleic acids by a multienzyme reaction modeled after retroviral replication, *Proc. Natl. Acad. Sci. U.S.A.*, 87, 1874, 1990.

141. Kievits, T. et al., NASBA isothermal enzymatic *in vitro* nucleic acid amplifcation optimized for the diagnosis of HIV-1 infection, *J. Virol. Methods*, 35, 273, 1991.

142. Fahy, E., Kwoh, D. Y., and Gingeras, T. R., Self-sustained sequence replication (3SR): an isothermal transcription-based amplification system alternative to PCR, *PCR Methods Appl.*, 1, 25, 1991.

143. Compton, J., Nucleic acid sequence-based amplification, *Nature*, 350, 91, 1991.

144. Gebinoga, M. and Oehlenschläger, F., Comparison of self-sustained sequence-replication systems, *Eur. J. Biochem.*, 235, 256, 1996.

145. Sooknanan, R. et al., Fidelity of nucleic acid amplification with Avian Myeloblastosis Virus reverse transcriptase and T7 RNA polymerase, *Biotechniques*, 17, 1077, 1994.

146. Chadwick, N. et al., Comparison of three RNA amplification methods as sources of DNA for sequencing, *Biotechniques*, 25, 818, 1998.

147. Uyttendaele, M. et al., Identification of *Campylobacter jejuni*, *Campylobacter coli* and *Campylobacter lari* by the nucleic acid amplification system NASBA, *J. Appl. Bacteriol.*, 77, 694, 1994.

148. Leone, G. et al., Molecular beacon probes combined with amplification by NASBA enable homogeneous, real-time detection of RNA, *Nucleic Acids Res.*, 26, 2150, 1998.

149. Greijer, A. E. et al., Multiplex real-time NASBA for monitoring expression dynamics of hyman cytomegalovirus encoded IE1 and pp67 RNA, *J. Clin. Virol.*, 24, 57, 2002.

150. Oehlenschläger, F., Schwille, P., and Eigen, M., Detection of HIV-1 RNA by nucleic acid sequence-based amplification combined with fluorescence correlation spectroscopy, *Proc. Natl. Acad. Sci. U.S.A.*, 93, 12811, 1996.

151. Nelson, N. C., Rapid detection of genetic mutations using the chemiluminescent hybridization assay (HPA): overview and comparison with other methods, *Crit. Rev. Clin. Lab. Sci.*, 35, 369, 1998.

152. Gingeras, T. R., Whitfield, K. M., and Kwoh, D. Y., Unique features of the self-sustained sequence replication (3SR) in the *in vitro* amplification of nucleic acids, *Ann. Biol. Clin. (Paris)*, 48, 498, 1990.

153. Brink, A. A. T. P. et al., Nucleic acid sequence-based amplification, a new method for analysis of spliced and unspliced Epstein–Barr virus latent transcripts, and its comparison with reverse transcriptase PCR, *J. Clin. Microbiol.*, 36, 3164, 1998.

154. Heim, A. et al., Highly sensitive detection of gene expression of an intronless gene: amplification of mRNA, but not genomic DNA by nucleic acid sequence based amplification (NASBA), *Nucleic Acids Res.*, 26, 2250, 1998.

155. Datson, N. A. et al., Specific isolation of 3′-terminal exons of human genes by exon trapping, *Nucleic Acids Res.*, 22, 4148, 1994.

156. Darke, B. M. et al., Detection of human TNF-a mRNA by NASBA, *J. Immunol. Methods*, 212, 19, 1998.

157. Mueller, J. D., Pütz, B., and Höfler, H., Self-sustained sequence replication (3SR): an alternative to PCR, *Histochem. Cell Biol.*, 108, 431, 1997.

158. Nakahara, K., Hataya, T., and Uyeda, I., Inosine 5′-triphosphate can dramatically increase the yield of NASBA products targeting GC-rich and intramolecular base-paired viroid RNA, *Nucleic Acids Res.*, 26, 1854, 1998.

159. Reitsma, P. H. et al., Use of the direct RNA amplification technique NASBA to detect factor V Lieden, a point mutation associated with APC resistance, *Blood Coagul. Fibrinolysis*, 7, 659, 1996.

160. Romano, J. W. et al., Genotyping of the CCR5 chemokine receptor by isothermal NASBA amplification and differential probe hybridization, *Clin. Diagn. Lab. Immunol.*, 6, 959, 1999.

161. Sillekens, P. T., Qualitative and quantitative NASBA for detection of human immunodeficiency virus type 1 and Hepatitis C virus infection, *Transplant. Proc.*, 29, 2941, 1996.

162. Höfler, H. et al., *In situ* amplification of measles virus RNA by the self-sustained sequence replication reaction, *Lab. Invest.*, 73, 577, 1995.

163. Gingeras, T. R. et al., Use of self-sustained sequence replication amplification reaction to analyze and detect mutations in zidovudine-resistant human immunodeficiency virus, *J. Infect. Dis.*, 164, 1066, 1991.

164. Chadwick, N. et al., A sensitive and robust method for measles RNA detection, *J. Virol. Methods*, 70, 59, 1998.

165. Blok, M.J. et al., Early detection of human cytomegalovirus infection after kidney transplantation by nucleic acid sequence-based amplification, *Transplantation*, 67, 1274, 1999.

166. Mainka, C. et al., Characterization of viremia at different stages of varicella-zoster virus infection, *J. Med. Virol.*, 56, 91, 1998.

167. Samuelson, A. et al., Development and application of a new method for amplification and detection of human rhinovirus RNA, *J. Virol. Methods*, 71, 197, 1998.

168. Leone, G. et al., Direct capture of potato leafroll virus in potato tubers by immunocapture and the isothermal nucleic acid amplification method NASBA, *J. Virol. Methods*, 66, 19, 1997.

169. Morré, S.A. et al., RNA amplificaiton by nucleic acid sequence-based amplificaiton with an internal standard enables reliable detection of *Chlamydia trachomatis* in cervical scrapings and urine samples, *J. Clin. Microbiol.*, 34, 3108, 1996.

170. Uyttendaele, M. et al., Detection of *Campylobacter jejuni* added to foods by using a combined selective enrichment and nucleic acid sequence-based amplification (NASBA), *Appl. Environ. Microbiol.*, 61, 1341, 1995.

171. Van der Vliet, G. M. E. et al., Nucleic acid sequence-based amplification (NASBA) for the identification of mycobacteria, *J. Gen. Microbiol.*, 139, 2423, 1993.

172. Lair, S. V. et al., A single temperature amplification technique applied to the detection of citrus tristeza viral RNA in plant nucleic acid extracts, *J. Virol. Methods*, 47, 141, 1994.

173. Uyttendaele, M. et al., Development of NASBA, a nucleic acid amplification system, for identification of *Listeria monocytogenes* and comparison to ELISA and a modified FDA method, *Int. J. Food Microbiol.*, 27, 77, 1995.

174. Uyttendaele, M., Bastiaansen, A., and Debevere, J., Evaluation of the NASBA nucleic acid amplification system for assessment of the viability of *Campylobacter jejuni*, *Int. J. Food Microbiol.*, 37, 13, 1997.

175. Vandamme, A. M. et al., Detection of HIV-1 RNA in plasma and serum samples using the NASBA amplification system compared to RNA-PCR, *J. Virol. Methods*, 52, 121, 1995.

176. Sarrazin, C. et al., Detection of residual hepatitis C virus RNA by transcription-mediated amplification in patients with complete virologic response according to polymerase chain reaction-based assays, *Hepatology*, 32, 818, 2000.

177. Bootman, J. et al., An international collaborative study on the detection of an HIV-1 genotype B field isolate by nucleic acid amplification techniques, *J. Virol. Methods*, 78, 21, 1999.

178. Damen, M. et al., Characterization of the quantitative HCV NASBA assay, *J. Virol. Methods*, 82, 45, 1999.

179. Romano, J. W. et al., Detection of HIV-1 infection *in vitro* using NASBA: an isothermal RNA amplification technique, *J. Virol. Methods*, 54, 109, 1995.

180. van Gemen, B. et al., The one-tube quantitative HIV-1 RNA NASBA: precision, acccuracy, and application, *PCR Methods Appl.*, 4, S177, 1995.

181. Nolte, F. S., Branched DNA signal amplification for direct quantitation of nucleic acid sequences in clinical specimens, *Adv. Clin. Chem.*, 33, 201, 1999.

182. Urdea, M. S. et al., A novel method for the rapid detection of specific nucleotide sequences in crude biological samples without blotting or radioactivity; application to the analysis of hepatits B virus in human serum, *Gene*, 61, 253, 1987.

183. Nilsen, T. W., Grayzel, J., and Prensky, W., Dendritic nucleic acid structures, *J. Theor. Biol.*, 187, 273, 1997.

184. Shchepinov, M. S. et al., Oligonucleotide dendrimers: synthesis and use as polylabelled DNA probes, *Nucleic Acids Res.*, 25, 4447, 1997.

185. Wang, J. et al., Dendritic nucleic acid probes for DNA biosensors, *J. Am. Chem. Soc.*, 120, 8281, 1998.

186. Stears, R. L., Getts, R. C., and Gullans, S. R., A novel, sensitive detection system for high-density microarrays using dendrimer technology, *Physiol. Genomics*, 3, 93, 2000.

187. Horn, T. and Urdea, M. S., Forks and combs in DNA: the synthesis of branched oligodeoxyribonucleotides, *Nucleic Acids Res.*, 17, 6959, 1989.

188. Urdea, M. S. et al., Branched DNA amplification multimers for the sensitive, direct detection of human hepatitis viruses, *Nucleic Acids Res.*, 24, 197, 1991.

189. Alter, H. J. et al., Evaluation of branched DNA signal amplification for the detection of hepatitis C virus RNA, *J. Viral Hepat.*, 2, 121, 1995.

190. Chen, C. H. et al., Quantitative detection of hepatitis B virus DNA in human sera by branched-DNA signal amplification, *J. Virol. Methods*, 53, 131, 1995.

191. Kern, D. et al., An enhanced-sensitivity branched-DNA assay for quantification of human immunodeficiency virus type 1 RNA in plasma, *J. Clin. Microbiol.*, 34, 3196, 1996.

192. Collins, M. et al., A branched DNA signal amplification assay for quantification of nucleic acid targets below 100 molecules/ml, *Nucleic Acids Res.*, 25, 2979, 1997.

193. Wilber, J. C., Branched DNA for quantification of viral load, *Immunol. Invest.*, 26, 9, 1997.

194. Van Cleve, M. et al., Direct quantification of HIV by flow cytometry using branched DNA signal amplification, *Mol. Cell. Probes*, 12, 243, 1998.

195. Burris, T. P. et al., A novel method for analysis of nuclear receptor function at natural promoters: peroxisome proliferator-activated receptor gamma agonist actions on aP2 gene expression detected using branched DNA messenger RNA quantitation, *Mol. Endocrinol.*, 13, 410, 1999.

196. Nargessi, R. D. et al., Quantitation of progesterone receptor mRNA in breast carcinoma by branched DNA assay, *Breast Cancer Res. Treat.*, 50, 57, 1998.

197. Chang, C. A. et al., Characterization of synthetic branched DNA (BDNA) using polyacrylamide gel-filled capillary electrophoresis, *Nucleic Acids Res.*, 24, 223, 1991.

198. Horn, T. et al., Branched oligodeoxynucleotides for use as amplification monomers in bioassays: chemical synthesis and characterization, *Nucleic Acids Res.*, 24, 201, 1991.

199. Horn, T., Chang, C. A., and Urdea, M. S., An improved divergent synthesis of comb-type branched oligodeoxyribonucleotides (bDNA) containing multiple secondary sequences, *Nucleic Acids Res.*, 25, 4835, 1997.

200. Horn, T., Chang, C. A., and Urdea, M. S., Chemical synthesis and characterization of branched oligodeoxyribonucleotides (bDNA) for use as signal amplifiers in nucleic acid quantification assays, *Nucleic Acids Res.*, 25, 4842, 1997.

201. Bushnell, S. et al., ProbeDesigner: for the design of probesets for branched DNA (bDNA) signal amplification assays, *Bioinformatics*, 15, 348, 1999.

202. Player, A. N. et al., Single-copy gene detection using branched DNA (bDNA) *in situ* hybridization, *J. Histochem. Cytochem.*, 49, 603, 2001.

203. Wang, J. et al., Regulation of insulin preRNA splicing by glucose, *Proc. Natl. Acad. Sci. U.S.A.*, 94, 4360, 1997.

204. Mellor, J., Hawkins, A., and Simmonds, P., Genotype dependence of hepatitis C virus load measurement in commercially available quantitative assays, *J. Clin. Microbiol.*, 37, 2525, 1999.

205. Kobayashi, M. et al., Usefulness of hepatitis C virus RNA counts by second generation HCV bDNA-probe in chronic hepatitis C based on the HCV genotype, *J. Gastroenterol.*, 33, 223, 1998.

206. Parekh, B. et al., Impact of HIV type 1 subtype variation on viral RNA quantitation, *AIDS Res. Hum. Retroviruses*, 15, 133, 1999.

207. Urdea, M. S., Branched DNA signal amplification, *Biotechnology*, 12, 926, 1994.

208. Zhou, L. et al., A branched DNA signal amplification assay to quantitate messenger RNA of human uncoupling proteins 1, 2 and 3, *Anal. Biochem.*, 282, 46, 2000.

209. Gretch, D. R. et al., Assessment of Hepatitis C viremia using molecular amplification technologies: correlations and clinical implications, *Ann. Intern. Med.*, 123, 321, 1995.

210. De Mendoza, C., Holguin, A., and Soriano, V., Performance of an ultasensitive assay to test undetectable viral load specimens using the branched DNA assay, *AIDS*, 12(1727–1729), 1998.

211. Murphy, D. G., Gonin, P., and Fauvel, M., Reproducibility and performance of the second-generation branched-DNA assay in routine quantification of human immunodeficiency virus type 1 RNA in plasma, *J. Clin. Microbiol.*, 37, 812, 1999.

212. Duck, P. et al., Probe amplifier system based on chimeric cycling oligonucleotides, *Biotechniques*, 9, 142, 1990.

213. Copley, C. G. and Boot, C., Exonuclease cycling assay: an amplified assay for the detection of specific DNA sequences, *Biotechniques*, 13, 888, 1992.

214. Okano, K. and Kambara, H., DNA probe assay based on exonuclease III digestion of probes hybridized on target DNA, *Anal. Biochem.*, 228, 101, 1995.

215. Bekkaoui, F. et al., Cycling probe technology with RNase H attached to an oligonucleotide, *Biotechniques*, 20, 240, 1996.

216. Beggs, M. L. et al., Characterization of *Mycobacterium tuberculosis* complex direct repeat sequence for use in cycling probe reaction, *J. Clin. Microbiol.*, 34, 2985, 1996.

217. Bhatt, R. et al., Detection of nucleic acids by cycling probe technology on magnetic particles: high sensitivity and ease of separation, *Nucleosides Nucleotides*, 18, 1297, 1999.

218. Modrusan, Z., Bekkaoui, F., and Duck, P., Spermine-mediated improvement of cycling probe reaction, *Mol. Cell. Probes*, 12, 1078, 1998.

219. Tang, T. et al., Integrated microfluidic electrophoresis system for analysis of genetic materials using signal amplification methods, *Anal. Chem.*, 74, 725, 2002.

220. Warnon, S. et al., Colorimetric detection of the tuberculosis complex using cycling probe technology and hybridization in microplates, *Biotechniques*, 28, 1152, 2000.

221. Modrusan, Z. et al., CPT-EIA assays for the detection of vancomycin resistant vanA and vanB genes in enterococci, *Diagn. Microbiol. Infect. Dis.*, 37, 45, 2000.

222. Fong, W. K. et al., Rapid solid-phase immunoassay for detection of methicillin-resistant *Staphylococcus aureus* using cycling probe technology, *J. Clin. Microbiol.*, 38, 2525, 2000.

223. Bekkaoui, F. et al., Rapid detection of the mecA gene in methicillin resistant staphylococci using a colorimetric cycling probe technology, *Diagn. Microbiol. Infect. Dis.*, 34, 83, 1999.

224. Modrusan, Z. et al., Detection of vancomycin resistant genes vanA and vanB by cycling probe technology, *Mol. Cell. Probes*, 13, 223, 1999.

225. Cloney, L. et al., Rapid detection of *mecA* in methicillin resistant *Staphylococcus aureus* using Cycling Probe Technology, *Mol. Cell. Probes*, 13, 191, 1999.

226. Harrington, J. J. and Lieber, M. R., The characterization of a mammalian DNA structure-specific endonuclease, *EMBO J.*, 13, 1235, 1994.

227. Bornarth, C. J. et al., Effect of flap modifications on human FEN1 cleavage, *Biochemistry*, 38, 13347, 1999.

228. Lyamichev, V., Brow, M. A. D., and Dahlberg, J. E., Structure-specific endonucleolytic cleavage of nucleic acids by eubacterial DNA polymerases, *Science*, 260, 778, 1993.

229. Kwiatkowski, R. W. et al., Clinical, genetic, and pharmacogenetic applications of the Invader assay, *Mol. Diagn.*, 4, 353, 1999.
230. Lyamichev, V. et al., Comparison of the 5′ nuclease activities of *Taq* DNA polymerase and its isolated nuclease domain, *Proc. Natl. Acad. Sci. U.S.A.*, 96, 6143, 1999.
231. Tombline, G., Bellizzi, D., and Sgaramella, V., Heterogeneity of primer extension products in asymmetric PCR due both to cleavage by a structure-specific exo/endonuclease activity of DNA polymerases and to premature stops, *Proc. Natl. Acad. Sci. U.S.A.*, 93, 2724, 1996.
232. Ceska, T. A. and Sayers, J. R., Structure-specific DNA cleavage by 5′ nucleases, *Trends Biochem. Sci.*, 23, 331, 1998.
233. Kaiser, M. et al., A comparison of eubacterial and archaeal structure-specific 5′-exonucleases, *J. Biol. Chem.*, 274, 21387, 1999.
234. Lyamichev, V. et al., Polymorphism identification and quantitative detection of genomic DNA by invasive cleavage of oligonucleotide probes, *Nat. Biotechnol.*, 17, 292, 1999.
235. Griffen, T. J. et al., Direct genetic analysis by matrix-assisted laser desorption/ionization mass spectrometry, *Proc. Natl. Acad. Sci. U.S.A.*, 96, 6301, 1999.
236. Ryan, D., Nuccie, B., and Arvan, D., Non-PCR-dependent detection of the Factor V Leiden mutation from genomic DNA using a homogeneous Invader microtiter plate assay, *Mol. Diagn.*, 4, 135, 1999.
237. Ledford, M. et al., A multi-site study for detection of the Factor V (Leiden) mutation form genomic DNA using a homogeneous invader microtiter plate fluorescence resonance energy transfer (FRET) assay, *J. Mol. Diagn.*, 2, 97, 2000.
238. Wilkins Stevens, P. et al., Analysis of single nucleotide polymorphisms with solid phase invasive cleavage reactions, *Nucleic Acids Res.*, 29, e77, 2001.
239. D'Aquila, R. T. et al., Maximizing sensitivity and specificity of PCR by pre-amplification heating, *Nucleic Acids Res.*, 19, 3749, 1991.
240. Chou, Q. et al., Prevention of pre-PCR mis-priming and primer dimerization improves low-copy-number amplifications, *Nucleic Acids Res.*, 20, 1717, 1992.
241. Kellogg, D. E. et al., TaqStart Antibody: "Hot Start" PCR facilitated by a neutralizing monoclonal antibody directed against Taq DNA polymerase, *Biotechniques*, 16, 1134, 1994.
242. Birch, D. E. et al., Simplified hot start PCR, *Nature*, 381, 445, 1996.
243. Klug, S. J. and Famulok, M., All you wanted to know about SELEX, *Mol. Biol. Rep.*, 20, 97, 1994.
244. Dang, C. and Jayasena, S. D., Oligonucleotide inhibitors of Taq DNA polymerase facilitate detection of low copy number targets by PCR, *J. Mol. Biol.*, 264, 268, 1996.
245. Lin, Y. and Jayasena, S. D., Inhibition of multiple thermostable DNA polymerases by a heterodimeric aptamer, *J. Mol. Biol.*, 271, 100, 1997.
246. Orou, A. et al., Allele-specific competitive blocker PCR: a one-step method with applicability to pool screening, *Hum. Mutat.*, 6, 163, 1995.
247. Puskás, L. G. and Bottka, S., Reduction of mispriming in amplification reactions with restricted PCR, *Genome Res.*, 5, 309, 1995.
248. Parsons, B. L. and Heflich, R. H., Detection of a mouse H-*ras* codon 61 mutation using a modified allele-specific competitive blocker PCR genotypic selection method, *Mutagenesis*, 13, 581, 1998.
249. Bi, W. and Stambrook, P. J., Detection of known mutation by proof-reading PCR, *Nucleic Acids Res.*, 26, 3073, 1998.
250. Seyama, T. et al., A novel blocker-PCR method for detection of rare mutant alleles in the presence of an excess amount of normal DNA, *Nucleic Acids Res.*, 20, 2493, 1992.
251. Yu, D. et al., Specific inhibition of PCR by non-extendable oligonucleotides using a 5′ to 3′ exonuclease-deficient DNA polymerase, *Biotechniques*, 23, 714, 1997.
252. Jenkins, F. J., Basic methods for the detection of PCR products, *PCR Methods Appl.*, 3, S77, 1994.
253. Lazar, J. G., Advanced methods in PCR detection, *PCR Methods Appl.*, 4, S1, 1994.
254. Mantero, G. et al., DNA enzyme immunoassay: general method for detecting products of polymerase chain reaction, *Clin. Chem.*, 37, 422, 1991.
255. Morrison, T. B., Weis, J. J., and Witter, C. T., Quantification of low-copy transcripts by continuous SYBR Green I monitoring during amplification, *Biotechniques*, 24, 954, 1998.
256. Durward, E. and Harris, W. J., Colorimetric method for detecting amplified nucleic acids, *Biotechniques*, 25, 608, 1998.

257. Kricka, L. J., Nucleic acid detection technologies — labels, strategies, and formats, *Clin. Chem.*, 45, 453, 1999.

258. Longley, M. J., Bennett, S. E., and Mosbaugh, D. W., Characterization of the 5′ to 3′ exonuclease associated with *Thermus aquaticus* DNA polymerase, *Nucleic Acids Res.*, 18, 7317, 1990.

259. Chien, A., Egdar, D. B., and Trela, J. M., Deoxyribonucleic acid polymerase from the extreme thermophile *Thermus aquaticus*, *J. Bacteriol.*, 127, 1550, 1976.

260. Holland, P. M. et al., Detection of specific polymerase chain reaction product by utilizing the 5′-3′ exonuclease activity of *Thermus aquaticus*, *Proc. Natl. Acad. Sci. U.S.A.*, 88, 7276, 1991.

261. Lee, L. G., Connell, C. R., and Bloch, W., Allelic discrimination by nick-translation PCR with fluorogenic probes, *Nucleic Acids Res.*, 21, 3761, 1993.

262. Livak, K. J. et al., Oligonucleotides with fluorescent dyes at opposite ends provide a quenched probe system useful for detecting PCR product and nucleic acid hybridization, *PCR Methods Appl.*, 4, 357, 1995.

263. Livak, K. J., Allelic discrimination using fluorogenic probes and the 5′ nuclease assay, *Genet. Anal.*, 14, 143, 1999.

264. Kuimelis, R. G. et al., Structural analogues of TaqMan probes for real-time quantitative PCR, *Nucleic Acids Symp. Ser.*, 37, 255, 1997.

265. Woo, T. H. S. et al., Identification of pathogenic *Leptospira* by TaqMan probe in a LightCycler, *Anal. Biochem.*, 256, 132, 1998.

266. Ibrahim, M. S. et al., The potential of 5′ nuclease PCR for detecting a single-base polymorphism in *Orthopoxvirus*, *Mol. Cell. Probes*, 11, 143, 1997.

267. Ibrahim, M. S. et al., Real-time microchip PCR for detecting single-base differences in viral and human DNA, *Anal. Chem.*, 70, 2013, 1998.

268. Bièche, I. et al., Novel approach to quantitative polymerase chain reaction using real-time detection: application to the detection of gene amplification in breast cancer, *Int. J. Cancer*, 78, 661, 1998.

269. Yajima, T. et al., Quantitative reverse transcription-PCR assay of the RNA component of human telomerase using the TaqMan fluorogenic detection system, *Clin. Chem.*, 44, 2441, 1998.

270. Kreuzer, K. A. et al., LightCycler technology for the quantitation of bcr/abl fusion transcripts, *Cancer Res.*, 59, 3171, 1999.

271. Leutenegger, C. M. et al., Rapid feline immunodeficiency virus provirus quantitation by polymerase chain reaction using the TaqMan fluorogenic real-time detection system, *J. Virol. Methods*, 78, 105, 1999.

272. Mercier, B., Burlot, L., and Férec, C., Simultaneous screening for HBV DNA and HCV RNA genomes in blood donations using a novel TaqMan PCR assay, *J. Virol. Methods*, 77, 1, 1999.

273. Martell, M. et al., High-throughput real-time reverse transcription-PCR quantitation of Hepatitis C virus RNA, *J. Clin. Microbiol.*, 37, 327, 1999.

274. Brandt, M. E. et al., Utility of random amplified polymorphic DNA PCR and TaqMan automated detection in molecular identification of *Aspergillus fumigatus*, *J. Clin. Microbiol.*, 36, 2057, 1998.

275. Bassler, H. A. et al., Use of a fluorogenic probe in a PCR-based assay for the detection of *Listeria monocytogenes*, *Appl. Environ. Microbiol.*, 61, 3724, 1995.

276. Ryncarz, A. J. et al., Development of a high-throughput quantitative assay for detecting herpes simplex virus DNA in clinical samples, *J. Clin. Microbiol.*, 37, 1941, 1999.

277. Tyagi, S. and Kramer, F. R., Molecular beacons: probes that fluoresce upon hybridization, *Nat. Biotechnol.*, 14, 303, 1996.

278. Mhlanga, M. M. and Malmberg, L., Using molecular beacons to detect single-nucleotide polymorphisms with real-time PCR, *Methods*, 25, 463, 2001.

279. Bonnet, G. et al., Thermodynamic basis of the enhanced specificity of structured DNA probes, *Proc. Natl. Acad. Sci. U.S.A.*, 96, 6171, 1999.

280. Gao, W. et al., Messenger RNA release from ribosomes during 5′-translational blockage by consecutive low-usage arginine but not leucine codons in *Escherichia coli*, *Mol. Microbiol.*, 25, 707, 1997.

281. Sokol, D. L. et al., Real time detection of DNA·RNA hybridization in living cells, *Proc. Natl. Acad. Sci. U.S.A.*, 95, 11538, 1998.

282. Fang, X. et al., Designing a novel molecular beacon for surface-immobilized DNA hybridization studies, *J. Am. Chem. Soc.*, 121, 2921, 1999.

283. Liu, X. and Tan, W., A fiber-optic evanescent wave DNA biosensor based on novel molecular beacons, *Anal. Chem.*, 71, 5054, 1999.

284. Marras, S. A. E., Kramer, F. R., and Tyagi, S., Multiplex detection of single-nucleotide variations using molecular beacons, *Genet. Anal.*, 14, 151, 1999.

285. Tyagi, S., Bratu, D. P., and Kramer, F. R., Multicolor molecular beacons for allele discrimination, *Nat. Biotechnol.*, 16, 49, 1998.

286. Vet, J. A. M. et al., Multiplex detection of four pathogenic retroviruses using molecular beacons, *Proc. Natl. Acad. Sci. U.S.A.*, 96, 6394, 1999.

287. El-Hajj, H. H. et al., Detection of rifampin resistance in *Mycobacterium tuberculosis* in a single tube with molecular beacons, *J. Clin. Microbiol.*, 39, 4131, 2001.

288. Poddar, S. K., Detection of adenovirus using PCR and molecular beacon, *J. Virol. Methods*, 82, 19, 1999.

289. Lewin, S. R. et al., Use of real-time PCR and molecular beacons to detect virus replication in human immunodeficiency virus type 1-infected individuals on prolonged antiretroviral therapy, *J. Virol.*, 73, 6099, 1999.

290. Manganelli, R. et al., Differential expression of 10 sigma factor genes in *Mycobactium tuberculosis*, *Mol. Microbiol.*, 31, 715, 1999.

291. Piatek, A. S. et al., Molecular beacon sequence analysis for detecting drug resistance in *Mycobacterium tuberculosis*, *Nat. Biotechnol.*, 16, 359, 1998.

292. Schofield, P., Pell, A. N., and Krause, D. O., Molecular beacons: trial of a fluorescence-based solution hybridization technique for ecological studies with ruminal bacteria, *Appl. Environ. Microbiol.*, 63, 1143, 1997.

293. Rhee, J. T. et al., Molecular epidemiologic evaluation of transmissability and virulence of *Mycobacterium tuberculosis*, *J. Clin. Microbiol.*, 37, 1764, 1999.

294. Kostrikis, L. G. et al., Spectral genotyping of human alleles, *Science*, 279, 1228, 1998.

295. Giesendorf, B. A. J. et al., Molecular beacons: a new approach for semiautomated mutation analysis, *Clin. Chem.*, 44, 482, 1998.

296. Wittwer, C. T. et al., Continuous fluorescence monitoring of rapid cycle DNA amplification, *Biotechniques*, 22, 130, 1997.

297. Cardullo, R. A. et al., Detection of nucleic acid hybridization by nonradiative fluorescence resonance energy tranfer, *Proc. Natl. Acad. Sci. U.S.A.*, 85, 8790, 1988.

298. Chen, X., Livak, K. J., and Kwok, P.-Y., A homogenous, ligase-mediated DNA diagnostic test, *Genome Res.*, 8, 549, 1998.

299. Bernard, P. S., Lay, M. J., and Wittwer, C. T., Integrated amplification and detection of the C677T point mutation in the methylenetetrahydrofolate reductase gene by fluorescence resonance energy transfer and probe melting curves, *Anal. Biochem.*, 255, 101, 1998.

300. Bernard, P. S. et al., Homogenous muliplex genotyping of hemochromatosis mutations with fluorescent hybrization probes, *Am. J. Pathol.*, 153, 1055, 1998.

301. Cane, P. A. et al., Use of real-time PCR and fluorimetry to detect lamivudine resistance-associated mutations in Hepatitis B virus, *Antimicrob. Agents Chemother.*, 1999, 1600, 1999.

302. Mills, D. R., Peterson, R. L., and Spiegelman, S., An extracellular Darwinian experiment with a self-duplicating nucleic acid molecule, *Proc. Natl. Acad. Sci. U.S.A.*, 58, 217, 1967.

303. Orgel, L. E., Selection *in vitro*, *Proc. R. Soc. Lond. Biol. Sci.*, 205, 435, 1979.

304. Ehricht, R., Ellinger, T., and McCaskill, J. S., Cooperative amplification of templates by cross-hybridization (CATCH), *Eur. J. Biochem.*, 243, 358, 1997.

305. Ellinger, T., Ehricht, R., and McCaskill, J. S., *In vitro* evolution of molecular cooperation in CATCH, a cooperatively coupled amplification system, *Chem. Biol.*, 5, 729, 1998.

306. Wlotzka, B. and McCaskill, J. S., A molecular predator and its prey: coupled isothermal amplification of nucleic acids, *Chem. Biol.*, 4, 25, 1997.

307. McCaskill, J. S., Spatially resolved *in vitro* molecular ecology, *Biophys. Chem.*, 66, 145, 1997.

308. Ackermann, J., Wlotzka, B., and McCaskill, J. S., *In vitro* DNA-based predator–prey system with oscillatory kinetics, *Bull. Math. Biol.*, 60, 329, 1998.

309. Breaker, R. R. and Joyce, G. F., Emergence of a replicating species from an *in vitro* RNA evolution reaction, *Proc. Natl. Acad. Sci. U.S.A.*, 91, 6093, 1994.

310. Breaker, R. R., Banerji, A., and Joyce, G. F., Continuous *in vitro* evolution of bacteriophage RNA polymerase promoters, *Biochemistry*, 33, 11980, 1994.

311. Marshall, K. A. and Ellington, A. D., Molecular parasites that evolve longer genomes, *J. Mol. Evol.*, 49, 656, 1999.

312. Schmitt, T. and Lehman, N., Non-unity molecular heritability demonstrated by continuous evolution *in vitro*, *Chem. Biol.*, 6, 857, 1999.

313. McCaskill, J. S. and Bauer, G. J., Images of evolution: origin of spontaneous RNA replication waves, *Proc. Natl. Acad. Sci. U.S.A.*, 90, 4191, 1993.

314. Eigen, M., Self-organization of matter and the evolution of biological macromolecules, *Naturwissenschaften*, 58, 465, 1971.

315. Walter, N. G., Modelling viral evolution *in vitro* using exo-Klenow polymerase: continuous selection of strand displacement amplified DNA that binds an oligodeoxynucleotide to form a triple-helix, *J. Mol. Biol.*, 254, 856, 1995.

316. Walter, N. G. and Strunk, G., Strand displacement amplification as an *in vitro* model for rolling-circle replication: deletion formation and evolution during serial transfer, *Proc. Natl. Acad. Sci. U.S.A.*, 91, 7937, 1994.

317. Bull, J. J. and Pease, C. M., Why is the polymerase chain reaction resistant to *in vitro* evolution? *J. Mol. Evol.*, 41, 1160, 1995.

318. Ellington, A. D., Aptamers achieve the desired recognition, *Curr. Biol.*, 4, 427, 1994.

319. Sen, D. and Geyer, C. R., DNA enzymes, *Curr. Opin. Chem. Biol.*, 2, 680, 1998.

320. Breaker, R. R., *In vitro* selection of catalytic polynucleotides, *Chem. Rev.*, 97, 371, 1997.

321. Teurk, C. and Gold, L., Systematic evolution of ligands by exponential enrichment: RNA ligands to bacteriophage T4 DNA polymerase, *Science*, 249, 505, 1990.

322. Joyce, G. F., Amplification, mutation and selection of catalytic RNA, *Gene*, 82, 83, 1989.

323. Ellington, A. D. and Szostak, J. W., *In vitro* selection of RNA molecules that bind specific ligands, *Nature*, 346, 818, 1990.

324. Wilson, D. S. and Szostak, J. W., *In vitro* selection of functional nucleic acids, *Annu. Rev. Biochem.*, 68, 611, 1999.

325. Fitzwater, T. and Polisky, B., A SELEX Primer, *Methods Enzymol.*, 267, 275, 1996.

326. Joyce, G. F., *In vitro* evolution of nucleic acids, *Curr. Biol.*, 4, 331, 1994.

327. Eaton, B. E., The joys of *in vitro* selection: chemically dressing oligonucleotides to satiate protein targets, *Curr. Opin. Chem. Biol.*, 1, 10, 1997.

328. Osborne, S. E., Matsumura, I., and Ellington, A. D., Aptamers as therapeutic and diagnostic reagents: problems and prospects, *Curr. Opin. Chem. Biol.*, 1, 5, 1997.

329. Brody, E. N. and Gold, L., Aptamers as therapeutic and diagnostic agents, *J. Biotechnol.*, 74, 5, 2000.

330. Brody, E. N. et al., The use of aptamers in large arrays for molecular diagnostics, *Mol. Diagn.*, 4, 381, 1999.

331. Golden, M. C. et al., Diagnostic potential of PhotoSELEX-evolved ssDNA aptamers, *J. Biotechnol.*, 81, 167, 2000.

332. Drolet, D. W., Moon-McDermott, L., and Romig, T. S., An enzyme-linked oligonucleotide assay, *Nat. Biotechnol.*, 14, 1021, 1996.

333. Bruno, J. G. and Kiel, J. L., Use of magnetic beads in selection and detection of biotoxin aptamers by electrochemiluminescence and enzymatic methods, *Biotechniques*, 32, 178, 2002.

334. Romig, T. S., Bell, C., and Drolet, D. W., Aptamer affinity chromatography: combinatorial chemistry applied to protein purification, *J. Chromatogr. Biomed. Sci. Appl.*, 731, 275, 1999.

335. Ghadessy, F. J., Ong, J. L., and Holliger, P., Directed evolution of polymerase function by compartmentalized self-replication, *Proc. Natl. Acad. Sci. U.S.A.*, 98, 4552, 2001.

336. Chelliserrykattil, J., Cai, G., and Ellington, A. W., A combined *in vitro/in vivo* selection for polymerases with novel promoter specificities, *BMC Biotechnol.*, 1, 13, 2001.

337. Hartmann, G. et al., Cytokines and therapeutic oligonucleotides, *Cytokines Cell. Mol. Ther.*, 3, 247, 1997.

338. Famulok, M. and Mayer, G., Aptamers as tools in molecular biology and immunology, *Curr. Top. Microbiol. Immunol.*, 243, 123, 1999.

339. Shi, H., Hoffman, B. E., and Lis, J. T., RNA aptamers as effective protein antagonists in a multicellular organism, *Proc. Natl. Acad. Sci. U.S.A.*, 96, 10033, 1999.

340. White, R. et al., Generation of species cross-reactive aptamers using "toggle" SELEX, *Mol. Ther.*, 4, 567, 2001.

341. Charlton, J., Sennello, J., and Smith, D., *In vivo* imaging of imflammation using an aptamer inhibitor of human neutrophil elastase, *Chem. Biol.*, 4, 809, 1997.

342. Welch, M., Majerfeld, I., and Yarus, M., 23S rRNA similarity from selection for peptidyl transferase activity, *Biochemistry*, 36, 6614, 1997.

343. Knight, R. D. and Landweber, L. F., Is the genetic code really a frozen accident? New evidence from *in vitro* selection, *Ann. N.Y. Acad. Sci.*, 870, 408, 1999.

344. Dai, X. and Joyce, G. F., *In vitro* evolution of a ribozyme that contains 5-bromouridine, *Helv. Chim. Acta*, 83, 1701, 2000.

345. Dubertret, B. et al., Dynamics of DNA-protein interaction deduced from *in vitro* DNA evolution, *Phys. Rev. Lett.*, 86, 6022, 2001.

346. Salehi-Ashtiani, K. and Szostak, J. W., *In vitro* evolution suggests multiple origins for the hammerhead ribozyme, *Nature*, 414, 82, 2001.

347. Cox, J. C., Rudolph, P., and Ellington, A. D., Automated RNA selection, *Biotechnol. Prog.*, 14, 845, 1998.

348. Schober, A. et al., Multichannel PCR and serial transfer machine as a future tool in evolutionary biotechnology, *Biotechniques*, 18, 652, 1995.

349. Strunk, G. and Ederhof, T., Machines for automated evolution experiments *in vitro* based on the serial-transfer concept, *Biophys. Chem.*, 66, 193, 1997.

350. Landegren, U., Molecular mechanics of nucleic acid sequence amplification, *Trends Genet.*, 9, 199, 1993.

351. Sano, T., Smith, C. L., and Cantor, C. R., Immuno-PCR: very sensitive antigen detection by means of specific antibody-DNA conjugates, *Science*, 258, 120, 1992.

352. Saito, K. et al., Detection of human serum tumor necrosis factor-alpha in healthy donors, using a highly sensitive immuno-PCR assay, *Clin. Chem.*, 45, 665, 1999.

353. Jhaveri, S., Rajendran, M., and Ellington, A.D., *In vitro* selection of signaling aptamers, *Nat. Biotechnol.*, 18, 1293, 2000.

354. Yamamoto, R., Baba, T., and Kumar, P. K., Molecular beacon aptamer fluoresces in the presence of Tat protein of HIV-1, *Genes Cells*, 5, 389, 2000.

355. Quinn, T. C., Recent advances in diagnosis of sexually transmitted diseases, *Sex. Transm. Dis.*, 21, S19, 1994.

356. Landegren, U., Nilsson, M., and Kwok, P. Y., Reading bits of genetic information: methods for single-nucleotide polymorphism analysis, *Genome Res.*, 8, 769, 1998.

5 Artificial Nucleic Acids and Genome Profiling

Lilia M. Ganova-Raeva

CONTENTS

5.1 GENOME PROFILING — WHAT, WHY, HOW

Genomics has developed through many different stages since G. Mendel's first insights into heredity. Early attempts at organizing the available information about the hereditary "units" were based on observations of abnormalities, deviations from the "regular," i.e., most common, phenotypes, and their distribution and behavior within a population. Even today the majority of known human genes, for example, are linked to a particular disease. Genetic maps based on linkage analysis, recombination frequencies, and complementation analysis were among the first tools in genomics. Chromosomes (from Greek, meaning "colored bodies") actually received their name because of specific characteristic staining; differential color banding was later used as a consistent marker and mapping tool. Physical maps that were initially based only on available physical markers followed. The pioneering Nobel work of G. Beadle and E. Tatum found the connection between genetic material and proteins. In Thomas Morgan's complementation studies done with *Drosophila* flies, the "finest tool" for genome mapping detailed the early maps down to several centiMorgans (recombination units later calculated to be equal to about 1000 nucleotides). These and the award-winning work of J. Lederberg in the field of bacterial genetic recombination are all cornerstones in the studies in genomics and are, in brief, a summary of the methods preceding the direct molecular level studies of the genetic material.

The 1953 discovery of the underlying chemical structure of the genetic material — deoxyribonucleic acid — and the advent of methods to determine and translate the actual coding sequences were major accelerators of the progress in turning the accumulated information into more precise science. Physical mapping became incredibly detailed with the development of cloning techniques. Sequencing provided a tool for measuring the genome down to its smallest chemical unit, the nucleotide base. Since the late 1950s, sequence data have been piling up rapidly. Significant parts of them represent studied physiological traits, or the actual genes. To everyone's awe, however, the genetic material turned out to contain more than a mere code that translates into proteins with specific functions. It became clear that the chromosomes possessed a much larger variety of sequences.

DNA was found to encode multiple "versions" of the same genes, i.e., alleles, to have regulatory functions of different character (such as promoters, enhancers, introns, terminators, etc.), to encode all RNAs, transposable elements, countless repeated structures, palindromes, etc., or simply sequences with no recognizable function whatsoever. A decade ago we were able to estimate that the human genome has about 5% of actual coding sequences that represent about 50,000 to 100,000 genes.[1] A considerable number of theories has been developed to explain this seeming redundancy. As an example of the amount of information that has accumulated so far, it is worth mentioning that at that time only about 6000 genes had been identified and sequenced, while today this number has grown to over 30,000 mapped genetic loci. The estimated completion for the human genome project (HGP) in 2005 by HUGO (Human Genome Organization, http://www.gene.ucl.ac.uk/hugo/) has recently been reduced by over 4 years, and a second private project by Celera Genomics has made the same progress (http://www.celera.com/). The completion of the entire sequence of chromosome 21 was reported in *Nature* (May 2000) shortly after the completion of chromosome 22 (December 1999), and only 9 months later the human genome was released for public access in February 2001. For more information, go to http://www.ncbi.nlm.nih.gov/genome/guide/human.

Many organisms have long been the focus of interest in the field of genomics. *Escherichia coli, Bacillus subtilis, Saccharomyces cerevisiae, Agrobacterium tumefaciens, Pseudomonas* sp., and *Drosophila melanogaster* are only the simplest examples, because they represent species with very low genome complexity. Even in these cases, however, the amount of accumulated information has grown too overwhelming to utilize directly. After the initial enthusiasm arising from the possibility of determining the code of every single genetic sequence seized, we realized that these data must be collected and organized in a way that will make them effectively usable. *Homo sapiens, Mus musculus, Gallus gallus, Rattus norvegicus, Arabidopsis thaliana, Zea mays, Schizosaccharomyces pombe, Aspergillus nidulans, Caenorhabditis elegans, Drosophila,* and, recently, rice, *Orysa sativa,* have become the prime focus of current research interests due to the increasing involvement of molecular biology in medicine, pharmaceuticals, and certainly agriculture. Today's trend in genomics is guided by the idea that not every sequence is needed. The motto for the large genome projects became "map first, sequence later." What seems to make more sense now is a thorough and systematic creation of databases of only a few organisms that can be used as cornerstones for comparison or as models.

During the last decades with the advance of molecular techniques to acquire and interpret genetic information science has come to a stage where we are well beyond the point of working with individual specific sequences. Today we can take advantage of the power of databases, which enables us to take a look at the larger picture and thus put our interpretation of the individual into a new and, it is hoped, more insightful perspective.

Here we will make an attempt to summarize, with the help of some examples, exactly *what* genome profiling is, *why* it is important, and *how* it could be used.

5.1.1 Organizing or Mapping Complex Genome Databases

When it comes to approaching high eukaryotic genomes, examples of the scale of the enterprise can be found in the work of many authors. For instance, a human genome bacterial artificial chromosome (BAC) library using high-molecular-weight DNA from a cell line with a normal diploid karyotype consists of 96,000 clones. Such a library, with an average DNA insert size of 110 kb, covers the human genome approximately three times.[2] Such a source of data can be screened and organized by three different, though complementary, methods: (1) Probe hybridization to high-density replica (HDR) filters; (2) probe hybridization to Southern blot filters loaded with digests of pooled BAC clones; and (3) two-step polymerase chain reaction (PCR) with specific primers. Probe hybridization is usually done first on superpools of thousands of template BAC DNA and then followed by PCR. PCR is applied to only several dozens of unique DNA samples prepared from the four-dimensionally assigned BAC clones of the particular superpool. Analysis of results generated in that manner facilitates mapping and linking of the individual clones.

PCR has been the method of choice used by many authors to screen for overlapping BAC clones and for assembly of contigs. For example, a rice BAC library equal in size to about three genomes could be pooled and amplified with arbitrary primers.[3] When PCR with 22 arbitrary primers (AP-PCR) is applied and the pooled BAC DNA surveyed, it turns out that each AP primer is able to identify one to ten loci. This method was able to find one to five overlapping clones for each locus. A total of 245 BAC clones were identified by the authors as overlapping by AP-PCR, and the identities were confirmed by DNA-DNA hybridization. These 245 BAC clones were then assembled into 80 contigs and 17 single-clone loci. Such results indicate that PCR analysis with arbitrary primers is a powerful tool in screening for overlapping BAC clones with tested high accuracy and efficiency. Thus, the use of AP-PCR analysis has been demonstrated to greatly speed the construction of physical maps of plant and animal genomes.

Today advances in technology clearly allow for total genome sequencing, along with an extensive cataloging of genes via comprehensive cDNA libraries. With the recent completion of the *Saccharomyces cerevisiae* and *Drosophila melanogaster* sequencing projects, the imminent completion of *Caenorhabditis elegans*, and the sequencing of the human genome well under way, for most authors the frequently asked question is: How much can sequence data alone tell us?[4] The answer may be that a DNA sequence taken in isolation from a single organism reveals very little. As noted, the vast majority of DNA in most organisms is noncoding; however, those areas are essential and specific genome environments for each organism. Protein-coding sequences or genes cannot function as isolated units without interaction with noncoding DNA and neighboring genes. Based on this notion, the fact that the genetic code is universal and the existing known phylogenetic relationships, a universal approach might be to take a closer look at similar genes in different organisms. This could help determine how function and position have changed over the course of evolution. By understanding these processes of change one can improve the understanding of what a gene is and how it relates to genetics and inheritance. Comparative genomics with model organisms are a key to this understanding.

An example of yet another approach to grasping the complexity of a total genome is a comparative model proposed by Deutsch and Long.[5] To investigate the distribution of intron-exon structures of eukaryotic genes, they constructed a general exon database comprising all available intron-containing genes and exon databases from ten eukaryotic model organisms: *Homo sapiens, Mus musculus, Gallus gallus, Rattus norvegicus, Arabidopsis thaliana, Zea mays, Schizosaccharomyces pombe, Aspergillus, Caenorhabditis elegans,* and *Drosophila melanogaster.* In constructing the data set the redundant genes were purged to avoid any possible bias they could introduce into the databases. Questionable introns that did not contain correct splice sites were also discarded. The final database contained over 17,000 introns, more than 20,000 exons, and almost 3,000 independent or quasi-independent genes. Eukaryotic genes contain on average about 3.7 introns per kilobase of protein coding region. Most introns are 40 to 125 nucleotides long, and the exon distribution peaks around 30 to 40 residues. This comparison of the variable intron-exon structures of the ten model organisms revealed two statistical phenomena. First, genome size seems to be correlated with the total intron length per gene. For example, invertebrate introns are smaller than those of human genes, while yeast introns are shorter than invertebrate introns. Second, introns smaller than 50 nucleotides are significantly less frequent than longer introns, possibly resulting from a minimum-intron-size requirement for intron splicing. Such an approach is an alternative idea on how to organize, use, and make sense of more complex databases by using multiple individual gene structures to derive generalized conclusions about the way genomes are built.

5.1.2 DETECT ABERRATIONS

The enormous information contained in any genome project or profile is not readily accessible and understandable to allow for straightforward conclusions about the genetics or phylogeny of any particular organism. However, approaching the genetic information by the many methods of profiling can give straightforward answers to many questions of genetic diversity. One greatly important aspect of diversity from a health standpoint is the ability to readily recognize a mutant or defective genotype. The methods for genome profiling do overlap with various diagnostic methods. These diagnostic methods could be general, to distinguish difference "in bulk" and reveal characteristic genome signatures, or specific, if the target is already known. Restriction fragment length polymorphism (RFLP) is a method that can create unique genome signatures, and, if subsequently coupled with PCR, it can actually target single-copy genes by linear amplification (LA) with one oligonucleotide as primer, extended by Taq DNA polymerase up to a restriction enzyme cleavage site. The products are arithmetically amplified by thermal cycling. Polymorphisms that differ in size by a small number of base pairs can be found. For example, such a method is described for the detection of RFLPs in single-copy genes in mammalian cells using one 5′-labeled oligonucleotide.[6] RFLP-LA is a reliable and fast diagnostic technique, even though its scope is limited to the detection of deletion/insertion mutants. It does not distinguish nonsense, missense, or possible frame-shift mutations.

Single-cell diagnosis is particularly valuable in preimplantation genetic diagnosis (PGD). For the success of PDG four main criteria must be achieved: sensitivity, reliability, accuracy, and identification/elimination of contamination.[7] Fluorescent PCR meets all these criteria and at the same time allows the simultaneous investigation of genes on multiple chromosomes. Fluorescent PCR has been demonstrated to have approximately $1000 \times$ higher sensitivity compared to conventional analysis systems, and 97% accuracy and reliability rates for both sex and cystic fibrosis (CF) diagnosis in single somatic cells. Fluorescence PCR has a low detection threshold that allows allelic dropout to be easily distinguished from PCR phenomena such as preferential amplification, thus reducing the probability of misdiagnosis. Approximately the same levels of reliability (90%) and accuracy (97 to 100%) have been achieved for diagnosis in human blastomeres. Fluorescent PCR has proven a suitable method for PGD since it can also be used to DNA fingerprint single cells (STR profiling), to identify the origin of the cell, and to detect contamination.

Molecular profiling to detect aberrations plays an emerging role in prophylaxis. It had been used to identify patients with genetic changes that indicate early events of carcinogenesis.[8] Premalignant lesions, for example, precede some oral cancers. At present there are no reliable markers to identify lesions that may progress to malignancy. Such potentially malignant oral lesions are analyzed for deletions at chromosomal regions that harbor tumor-suppressor genes for oral cancer. Over 50% of allele imbalance or loss of heterozygosity was demonstrated for these sites. Such genetic alterations do not appear in normal mucosa and hence can be good markers for early detection of the carcinogenic process.

Arbitrarily primed PCR (AP-PCR) fingerprinting is also an easy and useful method for analysis of genetic alterations in anonymous chromosomal regions. Such technology has been applied to analysis of DNA from human primary cancers and found amplification of a DNA fragment from a type of fibrosarcoma.[9] PCR-based analysis of radiation hybrid panels followed by cloning and sequencing of the fragment revealed that it belongs to chromosome 12q13-q15. MDM2 and IFNG genes in this region have been previously noted as potential candidates for involvement in human carcinogenesis. Genomic DNA from the fibrosarcoma was analyzed by Southern blot and revealed the amplification of the MDM2 gene together with the fragment locus, but not the IFNG gene. Such results demonstrate that detection of DNA alterations by AP-PCR fingerprinting without any previous knowledge of the genes and subsequent analysis of radiation hybrid panels could lead to easy identification of candidates for genes involved in carcinogenesis.

As the HGP generates an overwhelming abundance of genetic information, molecular researchers and clinical practitioners are setting new criteria for evaluating the links between newly discovered gene mutations and human disorders. These requirements necessitate the development of highly accurate and yet rapid automated systems for genetic testing. That has led to the introduction of a fully automated, sensitive, and rapid system for the detection of a single nucleotide polymorphism (SNP) WAVE designed for DNA fragment analysis; the system is highly amenable to both clinical and research facilities.[10]

5.1.3 Differential Expression

Revealing the changes in gene expression associated with any cellular processes (such as division, movement, or any specialized differentiation) is a central problem in biology. Advances in molecular and computational biology have led to the development of powerful, high-throughput methods for the analysis of differential gene expression. These tools have opened new opportunities in disciplines ranging from cell and developmental biology to drug development and pharmacogenomics. The ability to relate the physiological status of individual cells to the complement of genes they express has been limited until now. Today, however, there are at least five general differential gene expression methods: differential display (DD), expressed sequence tag (EST) sequencing, subtractive cloning, serial analysis of gene expression (SAGE), and especially cDNA microarray. The application of EST sequencing and microarray hybridization was illustrated, for instance, by the discovery of novel genes associated with osteoblast differentiation. The application of subtractive cloning has been used as a tool to identify genes regulated *in vivo* by the transcription factor pax-6. A robust and reproducible method has been developed for the amplification of 3′ sequences of mRNA derived from a single cell.[11] The amplified cDNA, derived from individual human lymphoblastoma cells, can be used for the expression profiling of up to 40 different genes per cell.

In addition, it has been demonstrated that the 3′ end amplification (TPEA) PCR, a version of DD, can be used to enable the detection of both high- and low-abundance mRNA species in samples harvested from live brain neurons. This procedure can facilitate the study of complex tissue functions at the cellular level. These and other examples emphasize the power of genomics for discovering novel genes that are important in any given biological process and that also represent new targets for drug development. Each of these approaches to identifying differentially expressed genes is useful in its own way, and the experimental context and subsequent evaluation of differentially expressed genes are the critical features that determine success. The most powerful of these techniques today is the DNA microarray, also known as the DNA chip, which provides the largest coverage for comparison of expressed genes. With its rapid development it will soon become cheaper and more accessible.

5.1.4 Phylogenetic Analysis/Genotyping

The application of genome profiling in differentiation research is just the other side of the coin for studies in phylogenetic relatedness. Many molecular methods are currently available for the identification and discrimination of bacterial strains within the same species, and these methods vary in efficiency and required labor.[12] One method for fingerprinting genomes, introduced in the early 1990s and already mentioned here, called arbitrarily primed PCR (AP-PCR), has been successfully used to delineate strains within the species *Porphyromonas gingivalis*. Random amplified polymorphic DNA (RAPD) is an analogous procedure that uses a single 10-base-long random oligonucleotide as a primer to amplify short stretches of the genome by PCR. The resulting product represents a species-specific banding profile that offers quick screening of different regions of the genome for genetic polymorphisms and for genotyping purposes.[13]

The analysis of the relationships between human genotype and phenotype provoked considerable interest with the discovery and characterization of single-nucleotide polymorphisms (SNPs). A clever strategy to find SNPs involves the utilization of the publicly available expressed sequence tag (EST) databases.[14] Picoult-Newberg and colleagues devised this strategy for a set of ESTs derived from 19 different cDNA libraries. They assembled 300,000 distinct sequences and found 850 mismatches as identified from contiguous EST data sets (candidate SNP sites), without *de novo* sequencing. The presence of the candidate SNP sites could be confirmed by a polymerase-mediated, single-base, primer-extension technique. These data gave an estimate of the allele frequencies in three human populations with different ethnic origins. Altogether, this approach provides a basis for rapid and efficient regional and genome-wide SNP discovery using data assembled from sequences from different libraries of cDNAs.

An interesting example from the field of population genetics is the Guadalupe fur seal (*Arctocephalus townsendi*), which underwent one or two severe bottlenecks due to commercial sealing in the late 19th century,[15] but since then the protected population has been growing steadily around their only rookery. To estimate the level of genetic variability of the present population both nuclear and mitochondrial genomes were probed using multilocus nuclear DNA profiling and mitochondrial DNA sequences. Surprisingly, unlike other populations that have experienced similar historical bottlenecks, such as Hawaiian monk seals and northern elephant seals, this study found high levels of genetic variability.

In many ways, DNA profiling technology is very similar to the conventional techniques used for forensic identification.[16] As with blood-grouping techniques, the molecular characteristics of a crime-scene specimen may be determined and compared with those of reference samples from suspects and victims. If the molecular signature of the crime sample does not match the one of the suspect, then it cannot be from the same person. Similar material, such as blood or semen stains, may be used for both biochemical and genetic tests, and the main applications of identification and relationship testing are shared by both techniques.

Typing systems for DNA profiling are predominantly selected according to criteria related to the robustness for typing of forensic specimens, the degree of genetic polymorphism (which influences the chances of excluding a wrongfully accused person), and the amenability to standardization as a basis to obtain reproducible results. DNA profiling is clearly far superior because it has the following characteristics: (1) It is more sensitive and able to generate data from trace amounts of even partially-degraded biological material. (2) It is capable of resolving mixtures of specimens from several individuals. (3) It has greater power of discrimination between individuals — up to 1-million-fold higher than conventional techniques. (4) It provides more accurate and detailed information on the nature of relationships, particularly relevant in paternity cases or cases of incest. DNA analysis, termed DNA profiling, has become the standard method in forensic strain typing.[17,18]

In contrast to conventional serological methods, any human tissue or body fluid can be analyzed by DNA profiling as long as it contains nucleated cells. DNA profiling allows determination of the alleles at multiple loci on an individual's genome.[19] The frequencies of these alleles can then be estimated from a sample drawn from the population. They may be multiplied together to give an estimate of the probability of DNA from a randomly chosen member of the population matching the DNA in question, if the occurrences of alleles at the different loci are independent events. A distinctive feature of forensic profiling is that the majority of genetic systems studied at the DNA level are derived from "non-coding" portions from the human genome and are located either in the vicinity of expressed genes or in stretches of DNA sequences interspersing the genes. The results from the typing are usually recorded as DNA fragment lengths and indicate the number of core repeat elements for short tandem repeat systems. These typing results do not contain any useful information that might reveal genetic traits or predisposition for inherited disease about the studied individual.

Multiple arbitrary amplicon profiling (MAAP) is a forensic technique that uses one or more short oligonucleotide primers of arbitrary sequence to initiate DNA amplification and generate characteristic fingerprints from anonymous genomes or DNA templates.[20] MAAP markers can be used in general fingerprinting as well as in mapping applications, either directly or as sequence-characterized amplified regions (SCARs). MAAP profiles can be tailored in the number of mono-morphic or polymorphic products they display by multiple endonuclease digestion of template DNA or by the use of mini-hairpin primers. PCR amplification of genome DNA with random 10-mer primers generates polymorphic amplification products that have proven useful in unambiguous cultivar identification.[21] RFLP phenograms together with random amplified polymorphic DNA (RAPD) analysis show that use of different kinds of molecular markers in gene banks is essential for characterization and classification of germ plasm collections. DNA amplification fingerprinting (DAF), with mini-hairpins harboring arbitrary "core" sequences at their 3′ termini, has also been used for fingerprinting. No correlation has been found between the sequence of the arbitrary core and the stability of the mini-hairpin structure and DAF efficiency. Mini-hairpin primers with short arbitrary cores and primers complementary to simple sequence repeats present in microsatellites were also used to generate arbitrary signatures from amplification profiles (ASAP). A variety of templates have been tested in order to establish genetic relationships between plant taxa at the interspecific and intraspecific level and to identify closely related fungal and plant material.[22]

To date, 6 eukaryotic, 12 archaic, 57 bacterial, and 687 viral genomes have been completely sequenced and published. More than 50 genomes were scheduled and completed by the year 2000.[23] This explosive growth of information is forcing changes in many scientific disciplines, creating new fields, and modifying the way to use and share scientific information. Global genome sequence comparisons appear slowly, but the infrastructure for these projects is being constantly updated. A good example of efficient engines that utilize all current available genome sequences and structure informa-tion are the databases managed by The Institute for Genomic Research (TIGR) at http://www.tigr.org/ and the Entrez Genomes database at http://www.ncbi.nlm.nih.gov/Database/index.html.

5.1.5 PHYSIOLOGICAL GENOMICS

A further step in understanding and utilizing the information stored in the genetic profiles is the physiological genomics approach. Physiological genomics are genome profile data generated by mapping, mutation analysis, differential expression, and phylogenetic analyses. They are based on the idea that organisms of common ancestral origin may be expected to have similar gene organi-zation. An example is the organization of sporulation genes in *Bacillus* and *Clostridium*. If this concept is universally correct, then a few model organisms could provide clues to the structure of unknowns. "Physiological genomics" is meant to define the functions of tens of thousands of newly discovered genes found every day with the success of, for example, human genome sequencing.[24] Such ambitious projects clearly require genome tools that allow higher efficiency of identification of gene function.

The idea of linking genomes to the overall function of the whole organism has developed gradually, starting with examples like silkworm genetic variation[25] with regard to the silkworm's adaptability to certain climates. Silkworm genetic resources that are being maintained in different countries have yet to be used to develop elite varieties that will be suited to the different geographic conditions and economic requirements of different countries of the world. This has been attributed largely to a lack of protocols to find usable genetic variability in this organism. Molecular markers can provide unam-biguous estimates of genetic variability of populations and assess intra- and interpopulation genetic diversity since they are independent of the confounding effects of environment. DNA fingerprinting assays based on RAPD and minor satellite DNA multilocus probes have been able to successfully characterize the diverse silkworm genotypes at their DNA level.

Another step in the same direction is the emerging field of structural genomics that compares genomes in terms of proteins. Both the number of genes in a genome and the total number of protein folds are limited, so they could be compared in numerical surveys much like a demographic census. Fold surveys share similarities with other whole-genome characterizations, like analyses of motifs or biochemical pathways. Structure is particularly suitable for comparing genomes because it allows for the precise definition of a basic protein module and has a relationship to sequence similarity that is defined much better than protein function alone.

One requirement for doing a structure survey is to build a library of folds or "fold families" comprised of groups of known structures. The library can be built up automatically using an existing program for structure comparison. After building such a library, one can use it to count the number of folds in genomes, expressing the results in the form of Venn diagrams and "top-10" statistics for shared and common folds. Depending on the counting methodology, the database statistics can reflect different aspects of the genome, such as extent of internal duplication or gene expression. Gerstein et al., after analyzing such fold families, have shown that common folds are shared between very different microorganisms, i.e., in different kingdoms, and have a remarkably similar structure, being comprised of repeated strand-helix-strand supersecondary structure units.[26] To overcome sampling bias that could emerge from the fact that only a small subset of structures is currently in the databases the authors suggested the application of structure prediction to the whole genome of interest prior to the analysis. Various investigators have already applied, with consistent results, many of the existing techniques for predicting secondary structure and transmembrane (TM) helices to the recently sequenced genomes. Microbial genomes have similar fractions of strands and helices, even though they have significantly different amino acid composition. Current information on this subject can be found at http://bioinfo.mbb.yale.edu/genome.

Genomic data could also be utilized by a system for a cross-genome comparison of open reading frames (ORFs) from multiple genomes. Such a genome-profiling system allows pairwise comparisons at different degrees of match similarity. The queries could use the ORF number and identity, their function or functional category, and their distribution in genomes or in biological domains and be grouped depending on the statistics of their respective match families. The authors of the idea for such an analysis suggest that it requires precise definition of the relevant classification terms and concepts such as genome signature, summary signature, biologic domain signature, domain class, match level, match family, and extended match family. Over 20,000 protein sequences inferred by conceptual translation from nine microbial genomes were used for such an analysis based on automated Pearson FASTA comparison.[27] This provides a fine example of the possibility of actually operating with a large number of unknown nucleic acid sequences and retrieving answers about their possible physiological functions.

To infer protein interactions from genome sequences based on the fact that some pairs of interacting proteins have homologues in another organism Marcotte et al. proposed a computational method.[28] They applied it to sequences from several genomes and revealed close to 7000 putative protein–protein interactions in *E. coli* and about seven times more in yeast. Some proteins have links to several other proteins; these coupled links appear to represent functional interactions such as complexes or pathways. Such interacting pairs that have been experimentally confirmed are documented in the Database of Interacting Proteins (DIP).

5.1.6 MOLECULAR EVOLUTION

A number of molecular biological methods for genome profiling have been described to aid in sequence discrimination and, which has not been mentioned thus far, in the analysis of molecular evolution. Smeets et al. describe a procedure based on selective hybridization with allele-specific oligonucleotides that has been developed for the purposes of typing apolipoprotein E variants from human genome DNAs.[29] It uses two sets of synthetic oligonucleotides to discriminate between epsilon 2, 3, and 4 alleles. The combination of the allele-specific oligonucleotide hybridization with PCR was shown to improve the sensitivity and the reliability of the procedure, and subsequent direct cloning and sequencing of the *in vitro* amplified DNA enabled rapid identification of any mutation within the gene of interest.

PCR primers of arbitrary nucleotide sequence have great utility for the identification of DNA polymorphisms that could be applicable for genetic mapping in a variety of organisms. Although technically very powerful, arbitrary primers have the peculiarity of amplifying DNA sequences of unknown function.[30] Furthermore, the amplified fragments will likely not be enriched for transcribed or promoter sequences that are of evolutionary significance. For these reasons Birkenmeier et al. have suggested the use of modified arbitrarily primers derived from conserved promoter elements and protein motifs. The application of such primers makes obvious biological sense. Results with primers of that design have been tested individually and in pairwise combinations for their ability to amplify genome DNA from a variety of mouse species, including inbred laboratory mice. The results had demonstrated that motif sequence-tagged PCR products could be used as reliable markers for genome mapping in mice, and that motif primers could be used with success for genome fingerprinting of divergent species.

The sequence of the human genome is virtually complete. Advances in technology have made possible both total genome sequencing and cataloging of genes via cDNA libraries. A large portion of the DNA in most organisms is noncoding; at the same time genes cannot function in isolation without interacting with noncoding DNA and neighboring genes. Investigating similar genetic structures in different organisms and determining how function and position have changed over the course of evolution can help to improve our understanding of functional genomes and steer us away from examining individual sequences out of their gene environment. Sequence data alone cannot tell nearly as much as could comparative genomics with model organisms.[4]

There are several aspects of the molecular evolution studies, and they are all based on genome profile data. One aspect, because of the implied sequencing, clearly revolves around studies of base substitution rates, i.e., evaluation of the "actual" rate of mutations and, hence, determination of the speed of evolution. Another focuses on studying a possible preference profile of the evolutionary process. A third aspect is the study of hereto-molecular relatedness of genes, functions, and species. And last, with enough background data one can experiment in predicting possible future changes in genes and functions, particularly in human pathogens, and thus be able to develop prevention strategies; a suitable starting point might be flu viruses.

5.1.7 Discovery of New Genes/Organisms

The methods developed to explore differential expression, understand molecular evolution, or deduce phylogenetic relationships have proved extremely versatile. They are clearly suitable not only for detection of novel, previously unidentified genes, but also new organisms. Powerful molecular techniques for identification and isolation of genetic material are the modern key to finding new genes and new organisms. For many years our ability to find and describe new species has been limited by a number of factors. Some of the most obvious are the ability to culture the organism, the ability to recognize distinct phenotypes, the ability to find/define a life cycle, the ability to cause disease, etc. A recently developed genome-profiling technique, representational differences analysis (RDA), which was originally introduced to search for differentially expressed genes, led to the discovery of two new flaviviruses, GB virus A (GBV-A) and GB virus B (GBV-B). They were identified in the plasma of a tamarind infected with the hepatitis GB agent. A third virus, GB virus C (GBV-C), was subsequently identified in humans.[31] RDA also led to the discovery of TTV and later a large number of related small single-stranded DNA (ssDNA) viruses. AP-PCR with degenerate primers was used to find SEN-V (apparently from the TTV family) in AIDS patients.

5.2 PROFILING METHODS THAT INVOLVE THE USE OF ARTIFICIAL DNA

As summarized above, many methods exist for generating and using genome profiles. These methods could be grouped into several categories according to the final objective and by the tools used to approach the genetic material. In fact, in many cases different organisms have required individual modifications of an existing method or, fortunately, the development of totally new experimental designs. This section will try to sketch the major streams of known approaches and will only mention in passing some of their modifications. The focal point of all the discussed methods will be the use of artificial nucleic acid in all its various forms (vectors, oligonucleotides, probes, primers, anchors, linkers, etc.). A list of the major known methods grouped according to the kinds of artificial DNA they employ and more or less by their chronological development is presented in Table 5.1.

Artificial nucleic acids are a crucial stepping-stone in the development of the technique variety used in genome analyses. They came about only after numerous advances in nucleotide chemistry and, most importantly, with the advent of *in vitro* DNA synthesis and its automation.[32,33] Mullis and Faloona devised the first method of sequence amplification *in vitro*, the polymerase chain reaction (PCR).[34] In this method synthetic oligonucleotide primers direct repeated deoxyribonucleic acid-synthetic reactions, which are target-specific and result in the exponential increase of the amount of the target sequence. Synthetic oligonucleotides flanking sequences of interest are used in repeated cycles of enzymatic primer extension in opposite and overlapping directions. The essential steps in each cycle are thermal denaturing of double-stranded target molecules, primer annealing to both strands, and enzymatic synthesis of DNA.

Today, PCR is a well-described method for the selective amplification of DNA or RNA segments ranging from 0.1 to 20 kilobase-pairs (kb) in length and has contributed significantly to the variety of existing profiling methods.[35] The use of the heat-stable DNA polymerase from the archaebacterium *Thermus aquaticus* (Taq polymerase) made the reaction amenable to automation. Since both strands of a given DNA segment are used as templates, the number of target sequences increases exponentially. The reaction is simple, fast, and extremely sensitive. The DNA or RNA content of a single cell is sufficient to detect a specific sequence. This key method greatly facilitates the diagnosis of mutations or sequence polymorphisms of various types in human genetics as well as in the detection of pathogenic components and conditions in the context of clinical research and diagnostics; it is also useful in simplifying complex analytical or synthetic protocols in basic molecular biology. Today there are very few methods for genome profiling of higher order eukaryotic organisms that do not involve the use of PCR in one form or another or of genome profiling without the direct or indirect use of some type of artificially generated or engineered nucleic acids. Basic RFPL is one of the lonely examples.

The enzymatic amplification of specific nucleic acid sequences *in vitro* has revolutionized the use of nucleic acid hybridization assays for viral detection. Even before attempting detection, one can now increase the copy number of a pathogen-specific sequence and markedly improve the sensitivity and specificity of the detection.[36] One of the first applications of PCR was the detection of human immunodeficiency virus type 1 and human T-cell lymphoma virus type I. Theoretically, PCR can be applied to the detection of any deoxyribonucleic or ribonucleic acid; one prerequisite is that sufficient nucleotide sequence data exist to allow the synthesis of target-specific oligonucleotide primers. Even this, however, as we will discuss later, is not an absolute requirement. The use of target amplification *in vitro* has allowed the studies of pathogens or genes that have not been feasible in the past because of the low copy number of the target nucleic acids in the starting material. This advance is of crucial importance for work with unique differentially expressed genes. For example, the PCR approach has proved particularly applicable to the study of infections caused by human retroviruses. These infections are usually chronic and persistent and the virus is usually present in the tissues in very low titer. Target amplification *in vitro* has created a boom in the development of rapid and sensitive detection and diagnostic kits for the clinical laboratory.

TABLE 5.1
Genetic Profiling Methods Involving the Use of Artificial DNA

Specific Primers	Explanation of the Abbreviated Method	Arbitrary Primers
PCR	Polymerase Chain Reaction	PCR
PCR-SSP	PCR with Sequence-Specific Primers	
LAR	Ligation Amplification Reaction	
RFLP	Restriction Fragment Length Polymorphism	RFLP
	Species-specific DNA fragment length variations generated by restriction endonucleases	
	If combined with PCR RFPL generates:	
	Amplified Fragment Length Polymorphism	AFLP
CFLP	Cleavage Fragment Length Polymorphism (Cleavase enzyme)	
	Sequence-Independent Single-Primer Amplification	SISPA
	Random Amplified Polymorphic DNA with single or multiple primers	RAPD
	Arbitrary Primed PCR	AP-PCR
IRE-PCR	Interspersed Repetitive Elements Microsatellites, Minisatellites	
SINE	Short Interspersed Sequence Elements	
HS Alu	Human *Alu* sequences	
SSR	Highly repetitive sequences that have unique distribution throughout the genome and create specific genome signatures upon amplification	
RT	Reverse Transcription for generation of cDNA	RT
	Differential Display: Reverse Transcription-based PCR with short arbitrary sequence primers and an anchor oligo dT primer that uses mRNA as a template to display the transcriptome	DDRT-PCR
TD	Targeted Display: differential display with specific primers	
Hybridization	Identification and mapping tool	Hybridization
HC	Homology Cloning for molecular taxonomy	
SNP	Single Nucleotide Polymorphism: variation in a genetic sequence that occurs at appreciable frequency in the population	
	Representational Differences Analysis: representative cDNA subtraction; could be arbitrarily primed in the selection steps	RDA AP-RDA
	Suppression Subtractive Hybridization: allows normalization of underrepresented mRNA, special adapter design	SSH
	Differential Subtraction Chain — destroys nondifferential targets	DSC
Microarrays	Differential Display in microchip format that allows for simultaneous monitoring of the temporal expression of multiple Open Reading Frames	Microarrays
EST	Expressed Sequence Tags: one-time sequence data from cDNA clones	EST
STS	Sequence-Tagged Sites: short DNA sequences (200–500 bp) functionally and structurally unique within the genome	STS
YAC, BAC, etc.	Yeast Artificial Chromosomes, Bacterial Artificial Chromosomes, and Baculovirus Vectors: are all engineered tools for cloning, mapping and expression of large genomes based on "natural" source vectors	YAC, BAC, etc.
Emerging techniques?	Whole genome microarrays	Emerging techniques?
	Neural networking with genome libraries	
	Genome forecasting of pathogens for pharmacogenetics	

5.2.1 THE APPLICATION OF OLIGONUCLEOTIDES

Powerful tool though it is, PCR relies almost exclusively on the design of the oligonucleotides used in the reaction. An important point to make here is that, in general, the primers can be subdivided into two major types: specific and nonspecific. As suggested by the name, the first kind are oligonucleotides designed with a specific target in mind; they are usually aimed at known genome regions. For the purposes of genome profiling, specific primers are used in numerous techniques such as RAPD, earlier versions of differential display, microarrays, multiplex mini-sequencing, etc. The use of specific primers is greatly advantageous for genotyping methods aiming to differentiate mutant genotypes or in methods devised for cloning, subcloning, and ordering of genomes. One potential disadvantage of specific oligos when used for the purposes of genetic profiling is the limited scope of the genome that they provide. Informational sources about existing available synthetic oligonucleotides have been compiled, and a good example of one such source is the Molecular Probe Data Base.[37]

The other major group, nonspecific oligonucleotides, is designed with functional purpose in mind, rather than based on actual known sequence. This functionality usually has to do with a design that allows for using the oligonucleotides as sequence tags or anchors or priming to random targets. For example, a bank of universal primers has been established for use as Comparative Anchor Tagged Sequences (CATS). This technique allows for comparative genome analysis across different species and, even though it uses universal primers, it is based on established knowledge of the linkage of a significant number of genome markers. Other designs involve usage of base substitutions, repeated or inverted repeat sequences, short random sequences, or ones found in RNA only and that allow use in a broad spectrum of different organisms. In addition, the oligo design may have special features at the 3' end that would allow priming reactions with a larger number of targets or special 5'-end tags that allow for capture or detection by a fluorimetric method.

Our ability to create molecules with a defined sequence has improved dramatically in terms of speed, length, and automation of the processes along with the progress of techniques for DNA synthesis. In the same manner, the ways to use synthetic DNA in molecular biology and recombinant DNA research have also been improved and enriched.[38] When first used for genome profiling, oligos were mostly applied in hybridization techniques in the form of a mixture of short, specific probes. More recently, however, a new approach to isolate the genes for proteins that are present in very small quantities became possible with the development of a technique to make long, specific oligonucleotides. Such a method enables screening of cloned DNA sequence banks with long unique oligonucleotides. It reduces the number of potentially positive clones on a primary screen and enables cloning with a minimum of amino acid sequence data. Most approaches in site-directed DNA mutagenesis are based on using synthetic oligonucleotides. A simplified example of the utility of just one oligonucleotide is its possible use as primer for amplification of nucleic acid in a PCR reaction. The same oligo can be used for site-directed mutagenesis and after 5'-labeling for Southern or Northern analysis and for sequencing applications. If the same primer is designed to contain a restriction endonuclease site, it could be used for directional cloning purposes. With the use of different temperatures and ionic strength and with the help of a powerful resolving technique (like WAVE or Invader, which will be discussed later in this chapter), the same oligo could be used for detection of point (single-base) mutations. Considering the immense number of permutations in the combination of the four bases (and their substitutes), it is easy to estimate that the larger the molecule, the higher its unique specificity. Even in the context of the human genome (3.2 Gb), however, 17 nt turns out to be the minimal required length x for the creation of a specific oligo, calculated by the formula $Nx/4^x < 1$, where N is the size of the target. For a large viral genome, e.g., $N = 100,000$ nt, this minimal length is reduced to 10 bp.

As mentioned before, prior knowledge of the nucleotide sequence is not an absolute requirement for exponential amplification of any target of interest. A perfect example is the sequence-independent, single-primer amplification (SISPA) reaction. SISPA is a primer-initiated technique that requires modification of the target sequence.[39] Unlike PCR, this method aims to achieve nonselective logarithmic amplification of heterogeneous DNA populations. A prerequisite to SISPA is the directional ligation of an asymmetric linker/primer oligonucleotide to the target population of blunt-ended DNA molecules. This step provides common end sequence and allows one strand of the double-stranded linker/primer to be used in repeated rounds of annealing, extension, and denaturing in the presence of thermostable Taq DNA polymerase. The linker-primers contain restriction endonuclease sites to facilitate the molecular cloning of as little as 1 pg of starting material after amplification. SISPA is especially useful when the nucleotide sequence of the desired molecule is both unknown and present in limited amounts, making its recovery by standard cloning procedures technically difficult. These conditions are present in the initial isolation and cloning of previously uncharacterized viral genomes, for instance. SISPA's utility is the cloning and recovery of low-abundance genetic sequences and was applied to the hepatitis C virus.

DNA amplification fingerprinting (DAF) is enzymatic amplification of arbitrary stretches of DNA directed by a very short oligonucleotide primer of arbitrary sequence. The resulting successfully amplified products comprised of arbitrary fragments from the total DNA produce highly characteristic and complex DNA fingerprints.[40] The products can be adequately resolved by polyacrylamide gel electrophoresis and visualized by sensitive methods like silver staining. Such a DAF procedure has been used to distinguish between clinical isolates *of Streptococcus, Klebsiella*, and *E. coli*. Although DAF is seemingly simple in concept, it is crucial that the amplification parameters be optimized for reproducibility because the method uses very short oligos that can contribute to inconsistent priming.

Extensive studies were performed on genome templates of different sources in order to determine the contribution of primer sequence and length of the fingerprint pattern and the effect of primer-template mismatches.[41] It has been experimentally established that primers produce unique patterns only when they are 8 nucleotides or fewer in length. Larger primers produce either identical or related fingerprints. Single-base changes, especially at the 3' terminus within this first 8-nt region alter significantly the primer's spectrum of amplification products. Increasing the annealing temperature does not change this 8-nt length requirement, but shorter primers have reduced the amplification ability because the duplexes become very unstable at higher temperatures.

It has been established that a defined 3'-terminal oligonucleotide domain that is at least 8 bases in length conditions amplification but is affected by sequences beyond it. The use of template DNA concentrations higher than 1 ng/µl and high $MgCl_2$ levels has also been found important for reproducibility. Test data indicate that only some of the template annealing sites are efficiently amplified during DAF, and this suggests that a single primer preferentially amplifies certain products due to competition for annealing sites between primer and terminal hairpin-loop structures of the template. However, a reliable window of reproducibility can be established for all parameters. It should take into consideration magnesium, primer, and enzyme concentration as well as cycle number and oligo length. DAF produces representative fingerprints quickly and reliably from bacteria regardless of prior genetic or biochemical knowledge. It has been proposed for general use as a diagnostic tool for bacterial identification and taxonomy.

Very interesting DAF primer design has been created for the study of genetic relationships between plant taxa at the interspecific and intraspecific level, and to identify closely related fungal isolates and plant accessions.[22] The PCR primers are mini-hairpins that have arbitrary "core" sequences at their 3' ends. The sequence of the arbitrary core did not appear to affect the stability of the mini-hairpin structure or DAF efficiency. This primer design was used for the fingerprinting of a variety of templates from whole genomes to PCR products in a two-step protocol termed arbitrary signatures from amplification profiles (ASAP). This protocol uses mini-hairpin primers or primers complementary to simple sequence repeats (like the ones present in microsatellites) in each amplification step. To produce distinct fingerprints primer sequences partially complementary to DAF product termini have to be avoided. This approach allows the combinatorial use of oligos based on a very limited number of primer sequences that are still able to reproducibly generate numerous ASAP fingerprinting reactions for nucleic acid screening. The authors demonstrated that mini-hairpin-based ASAP analysis can separate closely related cultivars, detect linked markers in bulked segregant analysis, and significantly increase the detection of polymorphic DNA.

Arbitrarily primed PCR is a variation of the DAF and SISPA methods for fingerprinting genomes. AP-PCR was originally developed and applied to delineate strains within the species *Porphyromonas gingivalis*.[12] A single primer used on nine different strains produced nine simple distinct banding patterns, indicative of genetic polymorphism. Amplicons shared by only some strains and common amplicons were also found, suggesting that AP-PCR can be used to generate polymorphic markers. An important observation in this work was that the genome fingerprinting obtained by AP-PCR is independent of the quality of DNA, and the assays can be performed directly from a colony using whole cells as a source of DNA template.

PCR primers of arbitrary nucleotide sequence have identified DNA polymorphisms useful for genetic mapping in a large variety of organisms. Although technically very powerful, the use of arbitrary primers for genome mapping still has the limitation of involving DNA sequences of unknown function. Thus, there is no reason to anticipate that DNA fragments amplified by the use of arbitrary primers will be enriched for transcribed sequences, promoters, or any sequences that may be conserved in evolution. For that reason the arbitrarily primed PCR method can be improved by designing oligonucleotide primers based on elements of evolutionary interest, i.e., conserved promoter elements, conserved protein motifs, or combinations chosen for their ability to amplify genome DNA from a variety of specimens.[30] A study of that kind has been done in various inbred strains of laboratory mice and *Mus spretus*. Using 29 pairs of primers the authors were able to determine the chromosomal location of 27 polymorphic fragments in the mouse genome. These results demonstrated that motif sequence-tagged PCR products are reliable markers for mapping a genome and that motif primers can be used for genome fingerprinting of divergent species.

A significant step forward has been the introduction of degenerate and universal nucleic acid bases that bind indiscriminately to all or virtually all true nucleotides.[42,43] Inosine, 3-nitropyrrole, 5-nitroindol, the pyrimidin derivative P, the purine derivative K, and various other modified bases revolutionized the ability to improve primer design. Consequently, a version of PCR, termed degenerate oligonucleotide-primed PCR (DOP-PCR), which employs oligonucleotides of partially degenerate sequence, was developed for genome-mapping studies.[44] The primer degeneracy combined with PCR protocol with low initial annealing temperature enables priming from multiple (e.g., approximately 10^6 in human genomes) evenly dispersed sites within a given genome. The method is believed to be species independent, since efficient amplification has been achieved from the genomes of all tested species using the same primer. DOP-PCR has advantages over alternative techniques like interspersed repetitive-sequence PCR (IRS-PCR), which relies on the appropriate positioning of species-specific repeat elements to ensure successful and unbiased general amplification of the target DNA. DOP-PCR has been applied to the characterization of abnormal chromosomes in conjunction with chromosome flow sorting and also to the cloning of new markers for specific chromosome regions. The degeneracy feature combines the advantage of short oligos to bind arbitrarily and at the same time allows for use of longer molecules that have increased binding stability that renders the method an improved reproducibility. DOP-PCR represents a rapid, efficient, and species-independent technique for general DNA amplification. A good example that the method is efficient and species independent is the recent discovery of the SEN-V family of TTV-like viruses in HIV patients.[45]

Another study in support of the "short-primer" approach involves the use of random amplification of polymorphic DNA (RAPD)-PCR.[46] This is a technique that allows, starting from genome DNA, the amplification of a large number of sequences with the use of a single primer of "random" base composition. The technique was used to analyze DNA from human and nonhuman primates with the help of only six different single primers of variable length. This method of DNA fingerprinting was used for individual identification and in cell lineage characterization and resulted in individual or specific electrophoretic patterns. The results obtained by RAPD-PCR give good resolution and demonstrated that it is a simple and reliable approach that can provide important information for linkage studies, gene mapping, or phylogenetic purposes.

In a "reverse" study the products of RAPD fingerprinting were examined to reveal the origin and the actual nucleotide sequences of RAPD bands[47] in an attempt to understand whether RAPD primers are involved in some type of preferential amplification. The results suggested that a number of sites in the genome are flanked by perfect or imperfect invert repeats, which permit multiple mismatch-annealing events between the single primer and the template DNA and the corresponding exponential amplification of the encompassing DNA segments. Standard RAPD procedure uses a single 10-base-long random oligonucleotide as a primer to amplify short stretches of the genome by PCR. A modified procedure uses two primers in each reaction, and RAPD tends to amplify more and smaller fragments than the standard RAPD technique. The two-primer approach generates very consistent band patterns that are completely different from the products generated in a standard one-primer reaction, as revealed by Southern analysis. The new markers are not linked to markers amplified with the same primers in the standard RAPD reactions, also suggesting that they were amplified from different genome regions. The two-primer RAPD is advantageous because it allows for more reactions to be carried out with a limited number of primers, and it generates more markers. This is easily calculated if we know that, when using a single primer, the number of reactions is equal to the number of primers (n), and so when using two primers in all possible combinations, the total number of reactions will be $n \times (n-1)/2$. Thus, it also increases the total number of markers. This approach was applied successfully in conjunction with bulked segregant analysis in a *Brassica napus* mapping project aimed at identifying a second marker linked to the gene governing low secondary metabolite concentration.[13] The method could be useful for developing high-density maps of certain chromosomal regions.

A nonmathematical approach that addresses the problem of screening BAC libraries is the use of PCR analysis with arbitrary primers (AP-PCR). Such a search for overlapping BAC clones and assembly of contigs[3] was undertaken with 22 arbitrary primers for a rice BAC library containing three genome equivalents. These primers were used on pooled BAC DNAs and individual BAC DNAs. Each primer identified 1 to 10 loci. A total of 245 BAC clones were found to be overlapping by AP-PCR and the identities confirmed by DNA-DNA hybridization. These results indicate that PCR analysis with arbitrary primers is a powerful tool in screening for overlapping BAC clones with high accuracy and efficiency. The use of AP-PCR analysis can clearly speed up construction of physical maps of genomes. The AP-PCR fingerprinting method has also been confirmed as easy and useful for analysis of genetic alterations in anonymous chromosomal regions.[9] This technology has been applied to the analysis of DNA from human primary cancers (i.e., mediastinum fibrosarcoma). AP-PCR revealed MDM2, a gene that has been previously suspected to be involved in carcinogenesis. Southern blot analysis of genome DNA of the tumor revealed duplication of the MDM2 gene together with the fragment locus, but not of the IFNG gene, which has also been a fibrosarcoma suspect agent.

The discriminatory power of a random 10-mer for the generation of useful RAPD profiles has been evaluated on 64-type and serotype reference strains and 114 isolates of *Campylobacter jejuni* and *C. coli* from food, seawater, and human feces.[48] The profiles were obtained by PCR and the genetic diversity among the strains was assessed based on a total of 118 different RAPD profiles. Overall, none was common to strains of the same Penner serogroup, and only occasional strains from different Penner serotypes shared identical band signatures. RAPD analysis was able to differentiate between the species, and after numerical analysis five main clusters were defined at the 40% similarity level, corresponding to three different *Campylobacter* species. This study came to show that RAPD profiling of *Campylobacter* is highly discriminatory and is a valuable alternative to traditional typing in epidemiological studies. A similar approach was taken to compare *Canabis sativa* plants grown simultaneously at the same site that could not be distinguished by high-performance liquid chromatography (HPLC) alone.[49] HPLC comparison was able to separate 16 out of 17 plants into three groups, while all but two specimens were differentiated using RAPD analysis with certain combinations of primers and cladistic analysis.

RAPD can also be used to generate genetic markers that are linked to a specific trait of interest. Silk worms, for example, have a number diapausing and nondiapausing varieties that represent a high degree of divergence with respect to geographic origin, and morphological, qualitative, quantitative, and biochemical characteristic genotypes.[50] RAPD with 40 random primers amplified 216 products from 13 divergent varieties of silkworm. The genetic similarity was analyzed by hierarchical clustering based on a pair-wise comparison of amplified products. The observed amplification was genotype specific, and the genotypes clustered into two groups that actually represent the diapausing and the nondiapausing types. Most interesting was the observation that the RAPD marker segregation in a backcross population turned out to be inherited as a dominant Mendelian trait, which indicates that the method identified the markers with very high reproducibility. Similar research has been done on rice[51] and it led to conclusions supporting a hypothesis of multicentered origin of rice cultivation in India and China. Analogous research has been done not only on chromosomal material but also on mitochondrial and chloroplast DNA with similar success.[52]

A potential source of error in genotyping studies based on the use of short tandem repeat (STR) markers is the peculiar tendency of the Taq DNA polymerize enzyme to catalyze nontemplated addition of a nucleotide (principally adenosine) to the 3′ end of PCR-amplified products. It has been demonstrated that this is a sequence-specific phenomenon, and the addition is favored when the enzyme encounters C at the 3′ terminus, but is not in the presence of A,[53] which implies that not all products get adenylated. Inserting known sequences close to the 5′ ends of reverse primers of markers that are adenylated and ones that are not was used to screen for consensus sequences that are able to promote or inhibit adenylation of the product. GTTTCTT, when placed at the 5′ end of reverse primers, was able to promote nearly 100% adenylation of the 3′ end of the forward strand. This (and similar) modification is called "PIG-tailing" and has been found to facilitate accurate genotyping and efficient T/A cloning.

Some of the applications discussed so far have broad scope at the expense of no or limited specificity. A method described by Kuklin et al.,[10] originally designed for studies of the human Y chromosome, solves the problems of detection of a single nucleotide polymorphism (SNP). Furthermore, the proposed approach for DNA fragment analysis is a fully automated system. This new technology, based on temperature-modulated liquid chromatography and a high-resolution matrix, is called WAVE™. The matrix is a column of hydrophobic, nonporous particles where the negatively-charged DNA fragments interact in direct proportion with the positively charged ion-pairing reagent thriethyl-ammonium acetate (TEAA). The alkyl chains of TEAA, on the other hand, interact with the hydrophobic column surface, so the longer the DNA molecule, the slower its release from the column. The elution times are 6 to 15 min, and the test is carried out under partial denaturation conditions that make it suitable for sensitive heteroduplex analysis. WAVE is extremely versatile and sensitive, and probably the fastest technique available. One minor drawback of the method is the sensitivity itself because it can detect one in a thousand single-base-pair changes; however, the resolving procedure uses PCR product that can be comprised of multiple aberrant molecules that accumulate due to mistakes of the polymerase. It is noteworthy to mention here that, even though the described techniques offer a nice variety of sensible approaches toward obtaining specific genome profiles, they can be very fastidious. Their reproducibility generally depends on the size of the given genome, the concentration of DNA, and the amenability of the specific technology for automation. Their performance is usually reliable in quite a narrow window of conditions.

5.2.2 RFLP/RAPD/AFLP/CFLP

Restriction fragment length polymorphism (RFLP) was described for the first time in the mid-1980s as a genome-profiling method that displays variations between species by use of restriction endonuclease digestion of total genome DNA. This method does not require usage of artificial nucleic acid but served as the basis for further development of a number of artificial oligo-mediated length polymorphism techniques such as AFLP, MAAP, CFLP, etc. The method has as a basis pulse field electrophoresis but much better resolution and sensitivity.

Along these lines, a method was described for the detection of RFLPs in single-copy genes in mammalian cells using one 5′-labeled oligonucleotide.[6] To detect the low copy number target it employs a single oligonucleotide as primer, which is extended by Taq DNA polymerase up to the restriction enzyme cleavage site of the restriction endonuclease of choice in a thermal cycling reaction of linear (arithmetical) amplification (LA). The size of the product depends on the sequence of the oligonucleotide and the position of the restriction enzyme cleavage site. Polymorphisms can be observed by measuring the size of the DNA products that have been separated on sequencing gels. (CA)n repeats that differ in size by a small number of base pairs are especially suitable for analysis by LA. Most genes examined by this procedure have produced satisfactory resolution regardless of the GC content of the templates. The LA was found suitable for large-scale genetic linkage analysis because it has the ability to multiplex signals under the same conditions and is of relatively low cost since it uses only one primer. One drawback of LA is that it can distinguish only deletion/insertion mutants and cannot find nonsense, mis-sense, or possible frame-shift mutations; frame-shift mutations, however, could be targeted with specific primers or by techniques such as WAVE or single-strand polymorphism analysis if they have been previously recognized as hot mutation spots. RFLP-LA is still a good diagnostic technique.

What happens when one applies multiple primers instead of one in an attempt to multiplex the screening protocol? An example can be found in the work of Bhat et al., where 60 random 10-mer primers were used in PCR amplification of genome DNA from 57 banana cultivars. They generated 605 polymorphic amplification products for the purposes of cultivar identification.[54] Unweighted pair-group method analysis of this data grouped the cultivars into specific clusters depending on their genome similarities. The results of RAPD were cross-examined by RFLPs using 19 random genome fragment clones and heterologous chloroplast DNA as hybridization of probes. DNA blots digested with three different restriction endonucleases were screened in this manner. The phenograms from the cluster analysis of data from 107 polymorphic alleles and the results from RAPD were consistent. The procedure also helped identify two multilocus probes useful in distinguishing all the analyzed cultivars and distinguish A and B types of cytoplasm. Such results show that use of different kinds of molecular markers in gene banks is essential for characterization and classification of germ plasm collections. A similar approach has been described for *Lactobacillus helveticus*, isolated from natural whey starter cultures for Italian hard cheeses. The authors characterized over 20 natural isolates and three reference strains by plasmid profiling, ribotyping, and RAPD fingerprinting.[55] The data showed strain-dependent heterogeneity in natural cheese starters related to the source of the isolate. In general, RAPD fingerprinting allowed the differentiation among strains that were not distinguished by the other two techniques.

At present most forensic databases for DNA profiling of individuals consist of DNA fragments, the sizes of which were measured using Southern blot RFLP analysis.[56] Statistical studies of these databases have shown that in cases of allele heterozygosity some individuals may still have single-band patterns, so one of the alleles can remain undetected. Positive fragment-size correlation is not proof of allelic dependence within individuals when nondetectable alleles are present. Proposed methods to resolve such allelic nondetectability could be RAPD or LA-RAPD and statistical analysis of DNA profile frequencies within the known population.

Amplified fragment-length polymorphism (AFLP) was designed as a PCR-assisted RFLP approach and is an elaboration on the RFLP-LA technique. It was originally designed to aid in the differentiation of closely related genomes.[57] AFLP also takes advantage of the restriction site's polymorphism and uses the exponential amplification power of PCR to detect it. The basic AFLP protocol involves isolation or generation of genome DNA or cDNA. Following digestion with one or two restriction endonucleases, the cohesive DNA ends are ligated to adapters of known sequence to generate chimeric molecules. This chimeric target is then amplified in several rounds of conventional PCR to ensure an adequate amount of template and reduce the complexity of the DNAs.

The next step of the procedure is a second round of selective RCP amplification. It is achieved by the use of AFLP primers that are homologous to the adapters but have extensions of 1 to 3 bases at the 3′ end into the restriction fragments. PCR using these primers amplifies only those fragments in which the primer extensions match the nucleotides flanking the restriction sites. Since in theory Taq polymerase does not tolerate mismatches at the 3′ end, this step allows for "controlled" amplification of a subset of the target molecules, and the result is a reproducible fingerprint that can be easily resolved by electrophoresis.[58] Using this method, sets of restriction fragments may be visualized by PCR without prior knowledge of the actual nucleotide sequence.[59] The method allows the specific coamplification of high numbers of restriction fragments (typically 50 to 100) that can be analyzed simultaneously. The amplified restriction fragments are resolved by electrophoresis in denaturing polyacrylamide gels. The used PCR primers are usually 5′-end labeled to facilitate detection. The AFLP technique provides a very powerful DNA fingerprinting technique for DNAs of any origin or complexity.

Multiple-endonuclease digestion of template DNA or amplification products can increase significantly the detection of polymorphic DNA in fingerprints. The coupling of endonuclease cleavage and amplification of arbitrary stretches of DNA, directed by short oligonucleotide primers, has been termed multiple arbitrary amplicon profiling (MAAP).[60] MAAP is yet another improvement of RFLP and RAPD and was devised to use one or more short oligonucleotide primers (5 or more nucleotides) of arbitrary sequence to initiate DNA amplification and generate characteristic fingerprints from anonymous genomes or DNA templates after restriction digestion with one or more endonucleases.[20] The generated MAAP signatures can be used in general fingerprinting as markers or in mapping applications, either directly or as sequence-characterized amplified regions (SCARs). MAAP profiles have great flexibility and can be adjusted by the number of used enzymes and the size and design of the short amplification primers. As pointed out before, preferred oligo length for such applications is 5 to 8 nucleotides, where the first 5 nt at the 3′ end require a perfect match. MAAP analysis of cleaved template DNA has been utilized to distinguish closely related fungal and bacterial isolates and plant cultivars, even in cases of near-isogenic soybean lines, altered via ethyl methane sulfonate mutagenesis. In experiments of direct comparison between digested and nondigested material, 25 primers failed to differentiate the soybean genotypes in the absence of digestion, but 42 DNA polymorphisms were found by endonuclease cleavage and using only 19 octamer primers. MAAP studies had revealed interesting sides of the theory and practice of primer–template interactions. MAAP also turned out to be a good evaluation method for the effect of potential mutagenes.

The method for selective amplification of genome DNA fragments (SAGF) is an elegant version of AFLP, with applications ranging from fingerprinting to representational differences analysis (RDA). SAGF can utilize two interesting classes of endonucleases — IIS or IIN. IIS enzymes have separate recognition and cleavage domains and recognize 5 to 6 nucleotide palindromic or non-palidromic sites but cleave at a different location at a fixed distance, usually 2 to 20 bp downstream of the recognition site. The IIN enzymes are also hepaxoterministic enzymes that can have an ambiguous recognition site. These enzymes also create protruding ends of unique sequence. DNA cut with such restriction endonucleases is separated into multiple fragments with unique protruding ends that could be organized into subsets by ligation to double-stranded adapters of defined sequence.[61] PCR-amplification of the ligation product produces unique genome signatures that could be varied depending on the kind of used adapter. The SAGF-method has several suggested applications: for obtaining individual bands in DNA fingerprinting, for reducing the kinetic complexity of DNA in the representational difference analysis (RDA method) of complex genomes, for cataloging DNA fragments, and for constructing physical genome maps.

An important consideration in genome profiling is the ability to obtain libraries of full-size transcription products. For the purpose, one needs a very efficient and foolproof amplification of the complete mRNAs from 5′ and 3′ and correspondingly requires generation of full-length cDNAs. This can be achieved by ligating double-stranded adapters to both ends of ds cDNA.[62] Such adapter-tagged ds cDNA can be readily used for the selective amplification of 5′- or 3′-cDNA fragments by PCR with a combination of gene-specific and adapter-specific primers in what the authors called rapid amplification of cDNA ends (RACE). To reduce the amplification background the suggested adapter design combines features of vectorette (a known sequence that can enable cloning in or amplification from a vector) and suppression PCR (inverted repeat sequence that tends to form panhandle structure when present in a short molecule, and so it suppresses amplification of shorter fragments giving advantage to longer ones) technologies. The application of long and accurate PCR (LA PCR) technology allows amplification of full-length cDNAs faithfully reproducing the original mRNA. In theory, transcripts from an expressed human genome vary in size from 1.3 to 9.1 kb and the RACE approach has been successful in encompassing that range.

Cleavage fragment length polymorphism (CFLP) is the most recently developed method for mutation screening and has a significantly improved sensitivity and speed over previously used methods like single-strand conformation analysis or heteroduplex analysis.[63] The Invader Assay introduced by Third Wave Technologies in 1999 is a linear signal amplification system that uses CFLP to allow direct detection and quantification of gene-specific sequences within the genome context. The assay uses a special nuclease, Cleavase, that digests a specific structure instead of a specific sequence. The marketed system measures fluorescence of released cleaved-off fluorophore. Because the procedure can be completely automated, and thus amenable to high-throughput analysis, its current applications range from single nucleotide polymorphism detection (SNPs), genotyping, viral load, and drug resistance testing to monitoring of gene expression. The method is extremely sensitive and fast but requires prior knowledge of the sequence of interest.

Let us deviate for a moment from the numerous PCR applications described so far. We have to mention an alternative method for sequence-dependent amplification, the ligation amplification reaction (LAR). This DNA-sequence-detection method utilizes ligation of oligonucleotide pairs that are complementary to adjacent sites on appropriate DNA templates.[64] The reaction could lead to either linear or exponential increase of the target copy number, depending on the number of nucleotides used. The products of the ligation reaction serve as templates for subsequent rounds of ligation. One advantage of the described technique, besides amplification of the target, is its sequence specificity and sensitivity down to 1 nucleotide mismatch since no ligation could occur in such cases. Another important advantage is the ability to introduce an artificial "anchor" sequence to flank the amplified product that could improve the application of LAR by further multiplexing. A disadvantage is the low efficiency of the ligation reaction itself that is able to perform well only under relatively high-DNA-concentration conditions. This technique has been used to successfully distinguish the normal from the sickle cell allele of the human beta-globin gene.

5.2.3 Microsatellites and Other Repeated Sequences

How can we use noncoding sequences and what can they tell us? Simple sequence repeats (SSR), or microsatellites, are ubiquitous in eukaryotic genomes.[65] Microsatelites are comprised of short 2- to 8-bp sequence motifs repeated numerous times. Minisatellites have similar structures except that the repetitive sequence can be up to 60 bp long. A genome profiling technique has been developed based on microsatellite-directed DNA fingerprinting by (PCR).[66] No sequencing is required to design the primers for amplification of the interrepeat region of the microsatellite. Primers anchored (for example, by a single arbitrary nucleotide) at 3′ or 5′ termini of the (CA)n repeats allow extension into the flanking sequence. If these primers have a label (radioactive, fluorescence, etc.), the amplicons can be easily analyzed by electrophoresis summarizing multiple genome loci in a single gel lane. Complex, species-specific patterns could be obtained from a variety of eukaryotic taxa. This approach also allows for observation of intraspecies polymorphisms and can be used to observe their segregation as Mendelian markers. This approach is applicable for taxonomic and phylogenetic comparisons and as a mapping tool in a wide range of organisms, and it can be extended to different microsatellites and other common dispersed elements. Microsatellite repeats incorporated in the oligonucleotide design have been used to amplify genome DNA samples from various sources (plants, human, yeast, and *E. coli* DNA). Most primers generate distinct amplification products, which can typically be resolved by electrophoresis. The result is a banding pattern representative for every organism, i.e., fingerprint. Such fingerprints allowed far more interspecific as well as intraspecific distinction. Interestingly, some primers whose designs are based on microsatellites found in plants produced bands with the *E. coli* template DNA as well. That could be explained by mismatch priming, similar to the way random amplified polymorphic DNAs are generated.[67] Data like these emphasize that most techniques have a narrow window of conditions that provide reproducible and specific enough results.

Short interspersed repetitive elements (SINEs) are abundant in humans. *Alu* sequences represent the largest family of SINE with 500,000 copies per genome. Several *Alu* subfamilies were found to be human-specific (HS) and could be used as good evolution markers.[68] Today *Alu* polymorphism is an established tool in population studies and hence also a tool in DNA fingerprinting and forensic analysis. PCR is used as a base technique for the detection of HS *Alu* insertion polymorphisms.[69] *Alu* PCR as a DNA profiling tool has been tested in both population genetics and paternity assessment. Studies have described a genotypic distribution of five polymorphic *Alu* insertions among three populations from the American continent (one of African origin and two Amerindians), both at the population and at the family level. These populations were also characterized by two other systems: PCR-sequence-specific oligonucleotide probe hybridization (PCR-SSO) and PCR-RFLP of human leukocyte antigen (HLA) class II molecules, both widely used to ascertain paternity. A similar approach has been used for population studies of the human coagulation factor XIII, the lipoprotein lipase, and CD4. All these genes have short tandem repeats (STR) within their intron structure that can be utilized as priming sites. Recently accumulated data show evidence that the two major families of interspersed repeated human DNA sequences, Alu and L1, are not randomly distributed, but they have distribution asymmetry. They also suggest a clear tendency for L1 repeats to cluster, with a higher proportion of full-length elements than expected in these two repeat classes, and confirm, at the long-range molecular level, previous studies indicating their partition within the human genome.[70] Such studies[71] have established the importance and strengths of the polymorphic HS *Alu* insertion methods as well as their amenability for use in phylogenetic studies.[71–73]

The implication of sequence-tagged sites (STSs) for generating microarray libraries made them valuable for genome profiling, especially in differential expression analysis. One way to identify STSs in humans or mice is by the pattern of interspersed repetitive element (IRE)-PCR.[74] Studies in this area have been done on yeast artificial chromosome (YAC) clones containing regions of mouse chromosomes 13 and 14, where IRE-PCR with primers based on the mouse B1 repetitive element generates simple-sequence-length polymorphisms (SSLPs). This novel approach to SSLP identification represents an efficient method for saturating a given genome fragment with polymorphic genetic markers that may also expedite positional cloning of genes of interest.

Repetitive extragenic palindromic (REP) elements have been identified in numerous bacteria. These genome sequences provide useful targets for DNA amplification. As is often the case with bacterial pathogens, the capability for rapid and accurate identification at the genetic level is essential for adequate action in case of outbreaks. A very robust method for amplifying inter-REP DNA sequences, REP-multiple arbitrary amplicon profiling (REP-MAAP), has been devised for the identification of *Acinetobacter* — an important nosocomial pathogen. The technique has been tested on about 30 isolates, and the ones displaying identical REP-MAAP patterns were found identical at the genetic level.[75]

5.2.4 RT-PCR

The methods reviewed so far focus mostly on the use of array of oligonucleotides to detect differences in the total DNA makeup. An important part of variety, however, is the simple example of differentiation. This phenomenon, found in all living organisms, actually does not represent directly the variety of the hereditary material itself but mere changes of its expression activity.

Extensive research has made it evident that the expressed portion of the genetic material at any given point in the cell's lifetime is a minor percentage of the actual coding sequences. This "minor" percentage, however, still represents, for instance in mammals, about 10,000 to 30,000 different transcripts of largely varying abundance. This variety suggested a different approach to genome profiling known as transcriptome profiling. In all approaches to genome research that use only expressed genes (i.e., messenger RNA) for nucleic acid profiling, the first and critical step in the process involves reverse transcription of RNA species.

Reverse transcriptases are enzymes that allow for back-translation of RNA to its original DNA matrix. Hence, through the process of reverse transcription one can create double-stranded DNA from single-stranded RNA. This type of DNA is known as complementary DNA (cDNA). All eukaryotic mRNAs as well as some bacterial and viral RNAs have poly(A)tail at the 5′ end. All RT methods utilize this basic feature to prime the first strand of cDNA synthesis by applying an artificial oligo. This artificial first-strand synthesis primer could have several different designs based on the use of a poly(T) (18 to 30 nucleotides long) sequence or mix of random hexamers. A nice touch introduced by CLONTECH (Clontech Laboratories, Inc.) is the SMART technology, the combination of a specific 5′-sequence with 3′-poly(T) and a "SMART"3′-GGG primer. The SMART primer takes advantage of the ability of some reverse transcriptases to add a deoxycytidine stretch to the 3′ end of the cDNA strand, thereby providing a second priming site of known sequence for efficient second-strand amplification. Many methods exist to create the second-strand cDNA, and they all involve DNA-dependent DNA-polymerases that perform strand displacement or PCR. The priming in the case of PCR synthesis may utilize either specific or arbitrary oligonucleotide sequence.

An example that involves the expression of only one protein/mRNA of interest is the interleukin-10 (IL-10). IL-10 is a T-helper type-2 (Th2) cytokine capable of suppressing cytokine synthesis by T-helper type-1 (Th1) cells. It plays a role in pregnancy immunotolerance through the establishment of a Th2 cytokine bias. The expression and production of IL-10 by normal and malignant human trophoblasts can be compared by amplifying IL-10 mRNA via the RT-PCR technique.[76] The results support the concept that IL-10 is expressed at the human maternal–fetal interface.

This is a straightforward example that makes use of known DNA sequences that target the screening only to the molecular species of interest. It is not difficult to imagine, however, how the use of an arbitrary/degenerate primer sequence could allow a much broader scope of the active genome profile or identify previously unknown genes. We find such an example in the work of Zhao et al., who investigated the molecular events associated with thymus dysfunction.[77] The difference in gene expression of the thymus between spontaneously hypertensive rats (SHR) and Wistar-Kyoto (WKY) rats was studied by the use of reverse transcription combined with subsequent cDNA representational difference analysis (RDA). As we will discuss further, RDA allows for the isolation of molecular species that are different between the two sources. The different products of cDNA-RDA were cloned for differential screening, and the positive clones were sequenced. The resulting differential screening profile showed that eight genes were overexpressed in the thymus of SHR. Comparison of those eight sequences against the databases identified four known and four novel genes. Two of the new genes were found to be members of the immunoglobulin superfamily genes, and the rest had no significant homology to nonredundant GenBank entries.

Gene identification by exon amplification is a very interesting example for the use of RT-PCR.[78] This approach suggests that one can take advantage of conserved sequence motifs within the exon portions of given genes and actually create a genome profile that can represent both conserved exon sequences (DNA coding sequences) and expressed RNA populations. Targeted display is yet another new technique for the analysis of differential gene expression and one of the more recent updates on the application of RT-PCR to EST identification.[79]

Reverse transcription is an essential basic tool behind many profiling applications. It is important to note that it actually provides only a first step into the process of compiling a genome profile.

The rest of the steps usually involve the use of even more artificial primers and an array of PCR techniques. We will discuss those techniques in more detail in the following paragraphs.

5.2.5 HYBRIDIZATION

Like reverse transcription and PCR, hybridization has emerged as a basic tool in genome profiling.[80] It is based on the affinity of complementary strands of nucleic acids to each other. If we were to tag a single strand of DNA or RNA with a known label, it could be used to find its counterpart in a rather complex mix of nucleic acid fragments. Hybridization in its original form does not require the use of artificial nucleic acids; however, with the advancements in biotechnology, cloning, and mapping, the utilization of man-made probes found its way and became a necessity.

An almost inevitable step along the way of defining a genome profile is the attempt to acquire, store, and explore the entire genome of a given organism in a way that allows us easy access and manageability. The established way to catalog a genome is to create a genome library. The DNA of any organism can be divided (digested) into smaller fragments that can be introduced into an appropriate vector. Each fragment could then be stored individually as a clone in a different organism that allows the propagation of the said vector and retrieved later when needed from the storage "library." The common problem, however, is that the original ordering of fragments is lost in the process of creating the library, regardless of its backbone — YAC,[81] BAC, phage, cosmid, or plasmid. The process of placing the clones in order according to their position along the chromosome(s) is referred to as "*in vitro* reconstruction" or "contig mapping" of an organismal genome. One very exciting use for artificial oligonucleotides combined with intelligent approach toward their design is the work of Fu et al.[82] The authors proposed that clones in a phage library could be assigned binary call numbers by scoring each clone for hybridization (0 or 1) with a battery of short manufactured DNA sequences called synthetic oligonucleotides that may also have incorporated restriction endonuclease sites. Those clones with similar call numbers are placed close together in the ordered library. This mathematical approach also suggests answers to the question of how many clones and probes to use to carry out *in vitro* reconstruction of an organism's chromosomes, and its validity has been confirmed by simulation studies.

Despite abundant library resources for many organisms, their physical mapping has been seriously limited due to a lack of efficient library screening techniques. Another highly successful strategy for the large-scale screening of genome libraries has been developed based on multiplex oligonucleotide hybridization on high-density genome filters. This strategy was originally applied to generate a BAC-anchored map of mouse chromosome 11.[83] There currently exists an available mouse simple sequence length polymorphism (SSLP) database, and it was used as a stepping-stone for the design of 320 pairs of oligonucleotide probes. The oligos were selected with an "overgo" computer program, which screens new primer sequences so as to avoid microsatellite repeats. Over 90% of the hybridization probes were able to identify positive clones from a large mouse BAC library with an average of seven positive clones per marker. BACs identified by these probes are automatically anchored to the chromosome. The clones are also confirmed for 204 markers by PCR.

These data demonstrate that a significant number of clones can be efficiently isolated from a large genome library using this strategy with minimal effort. This strategy certainly has wide application to scale-up mapping and sequencing of human and other big genomes. A similar approach has been utilized in plant genomics, enabling the monitoring of gene expression on a much larger scale than was previously possible[84] and the comparison of a gel-based, nylon filter and microarray techniques to detect differential RNA expression in plants demonstrated the high utility of this method.

The hybridization method was used to do an interesting allelic profiling study in the genetically isolated Finnish population.[85] The allele frequencies for three hypervariable DNA loci were determined, and they revealed high heterozygosity rates for all three probes (0.94 to 0.96). In many mother–child pairs, no mutations were found using two of the probes, whereas a mutation rate of

0.064 was observed for the third probe. One individual had an unexpected pattern detected using the third, suggesting complex DNA polymorphism based on both a variable number of tandem repeats and restriction site polymorphism. These findings suggest that allelic profiling by hybridization could be a valuable and relatively low-cost method for individual identification in genetically isolated populations.

Comparative genome hybridization (CGH) provides a molecular cytogenetic approach for genome-wide scanning of differences in DNA sequence copy number and has captured the interest of cancer researchers. CGH publications are already compiled in a database that covers more than 1500 types of tumors. The value of this particular data set has been that it is revealing genetic abnormalities that are characteristic of certain tumor types or stages of tumor progression. Six novel gene amplicons, as well as a locus for a cancer-predisposition syndrome, have been discovered based on CGH data. CGH has now been established as a first-line screening technique for cancer researchers and is intended to serve as a basis for ongoing efforts to develop large-scale genome scanning at very fine resolution, such as the microarray technology.[86] Another exciting field in the comparative genome research is comparative mapping, which allows a deeper look into phylogenetic and evolutionary relationships between different organisms; however, it is beyond the scope of this chapter to discuss this subject in greater detail.

An interesting twist of profiling that resulted in the detection of novel retroviruses in the human genome was introduced by hybridization with synthetic oligonucleotides complementary to putative retroviral primer-binding sites. Using probes designed based on known retrovirus-like sequences a large element with structural features of a retroviral pro-virus was isolated from a human genome library.[87] Nucleotide-sequence analysis of the terminal repeats of this element (600 bp long) revealed characteristic motifs known as regulatory signals for RNA polymerase II transcription. The putative *pol* gene of the element displayed apparent homologies to conserved regions of retroviral reverse transcriptase. The authors found a 5′-long terminal repeat termed human-retrovirus-related sequence-proline (HuRRS-P) that was flanked at its 3′ end by a putative primer-binding site for reverse transcription. Researchers found 20 to 40 copies of HuRRS-P homologous sequences in DNAs of human and simian origin; this demonstrates its utility in merging the hybridization profiling techniques with the use of conserved or repeated sequence motifs without the help of PCR

Multiplexing is important, and it also applies to hybridization data. A bundle of optical fibers, where each fiber carries a different oligonucleotide probe immobilized on its distal end, have been assembled for the simultaneous analysis of multiple DNA sequences.[88] This fiber-optic biosensor array measures the increase in fluorescence that accompanies binding/hybridization of fluorescently labeled complementary oligonucleotides to an array. The authors claim that the approach is fast (<10 min) and sensitive (10 nM), can detect multiple DNA sequences simultaneously, and has the potential for quantitative hybridization analysis.

Recent developments in cloning technologies have introduced homologous recombination as an efficient cloning tool that eliminates the need for restriction enzymes or DNA ligases. This method also enables quick and easy subcloning and modification of genomic DNA in BACs. This new form of chromosome engineering, termed recombinogenic engineering or recombineering, greatly reduces the time it takes to create transgenic mouse models.[89] Thus, recombineering may facilitate many kinds of genomic experiment that have otherwise been difficult to carry out and should enhance functional genomic studies by providing better mouse-genome models.

Another efficient strategy for simultaneous genome mapping and sequencing are the recently described slalom libraries. Slalom libraries are based on physically oriented, overlapping restriction fragment libraries that combine features of general genomic, jumping (not overlapping, containing only the extreme ends of genome fragments), and linking (overlapping, containing sequences adjacent to the recognition sites of infrequently cutting endonucleases such as NotI) libraries. Slalom libraries are adaptable to different applications. In a model experiment, they were used to map and sequence two human P1-derived artificial chromosome (PAC) clones. The model demonstrated that the approach is feasible and could improve the cost-effectiveness and speed of existing mapping/sequencing methods at least five- to tenfold.[90] The efficiency of contig assembly in the slalom approach is independent of the length of the sequence reads. This implies that even short sequences produced by rapid, high-throughput techniques would be enough to complete a physical map and a sequence scan of a small genome.

5.2.6 DD/SAGE/SNP/SSCP

Differential display (DD) is a genome profiling technique that relies exclusively on mRNA and reverse transcription combined with the use of arbitrary oligonucleotides. DD is rapid, requires small amounts of RNA, and has the ability to detect both over- and underrepresented transcripts. It takes advantage of the poly(A) tailing of the eukaryotic mRNA to serve as an "anchor" for RT-PCR. The reverse-transcription step is followed by a PCR step that uses the same poly(T) oligo as a "reverse" primer and additional arbitrary 10-mer with M13–3′tail as a "forward primer." This reaction usually incorporates a labeled dNTP or labeled PCR oligo to allow visualization of the product.[91] The resulting fragments are resolved by electrophoresis on polyacrylamide sequencing gel that usually covers a size range of 40 to 2000 nucleotides. DD is a robust approach to studying changes in gene expression.

As we will see, DD may be used not only to display overall differences but also to reveal just a subset of sequences that could be of particular interest. Such a method is described as targeted genome differential display. Targeting is accomplished by capturing DNA fragments containing specific sequence by hybridization with complementary single-stranded DNA. The captured fragments are then amplified by PCR. Elaboration on this technique has also been described where targeting is accomplished by PCR with primers specific to the target sequence.[92] It takes advantage of PCR suppression to eliminate fragments not containing the target sequence. Targeting focuses the analysis on areas of interest and serves to reduce the complexity of the amplified subset. Targeting also enables the display of 3′-end restriction fragments of double-stranded cDNAs by selective PCR amplification.[93] The method produces highly consistent and reproducible patterns and can resolve hidden differences such as bands that differ in their sequence but comigrate on a gel. Using specific nested primers can ensure the detection of virtually all mRNAs in a sample, regardless of copy number. This approach can correlate gel patterns with available cDNA databases because known cDNAs move to predictable positions on the gel. It has been applied to examine the differences in gene expression patterns during T-cell activation and was able to identify as many as 3 to 4% upregulated and 2% downregulated mRNAs in a background of over 700 bands. These and other results suggest that this approach is suitable for systematic and detailed study of subtle changes in the patterns of gene expression in cells with altered physiologic states. Ordered differential display is a similar approach for the display of 3′-end restriction fragments of cDNAs. It was first described and applied to the search for molecular regional markers in the freshwater planarian *Dugesia tigrina*.[94] First, representative pools of 3′-fragments are selectively amplified using the PCR suppression effect. Then, adapter-specific primers extended by two additional randomly picked bases at their 3′ ends amplify simplified subsets of these fragments. By testing all possible combinations of extended primers, the whole mRNA pool may be systematically investigated. This method is somewhat more costly and laborious but is certainly thorough.

Another elaboration and overall improvement of the original differential display method is the 3′-end restriction fragments cDNA method with "fly" adapters.[95] It uses DD to demonstrate levels of unknown RNA species as well as those already represented in the database. Introduced first for molecular profiling of gene expression in neutrofils, the method employs ligation of restriction endonuclease-digested cDNA to Y-shaped adapters ("Fly"-adapters). The adapters have a 4-nt single-stranded region (overhang) complementary to a 4-nt sticky end of a restriction endonuclease of choice; adjacent to the overhand is a portion that is complementary, so that the oligos anneal to each other and the rest of the adapter is noncomplementary, so it creates a Y shape. The method also uses a $dT_{[18]}V_1$ primer with a 3'-"heel" sequence for the first strand priming that has about 30% efficiency of cDNA synthesis. The adapter and the "heel" sequence introduce PCR priming sites so that the cDNA can be subsequently amplified with 12 different (differing by the two 3' bases of the forward primer) primer pairs. This makes the overall procedure quite laborious and one that requires large starting material for reproducibility, but it can potentially give very complete discrimination of unique or unknown gene products and also distinguish at least a twofold increase in gene expression.

DD is a powerful method for gene expression; hence, genome profiling can use both specific and arbitrary oligos, but it has several drawbacks. First, it requires very stringent and consistent RNA isolation from all experimental samples to ensure valid reference because it has been observed that small differences in handling can produce easily false positive/negative results. Second, it is quite possible that unique mRNA species that are expressed at very low copy number could escape analysis. Finally, the procedure of final isolation and confirmation of the difference product by Northern blot RT-PCR could be tedious.

A certain version of DD — Serial Analysis of Gene Expression (SAGE) — is a powerful expression profiling method, allowing the analysis of the expression of thousands of transcripts simultaneously. SAGE is an experimental technique designed to gain a direct and quantitative measure of gene expression. SAGE is based on the isolation of unique sequence tags 9 to 10 nucleotides in length from individual mRNAs. These tags are concatenated in series into long DNA molecules for sequencing in bulk. The SAGE method can be applied to explore any biological phenomena in which changes in cellular transcription occur. SAGE gives a global gene-expression profile of any particular type of cell or tissue of choice and facilitates the identification of specific genes related to any physiological status of the cell by comparing profiles constructed for a pair of cells maintained under different conditions. The original SAGE method contains several serious drawbacks such as procedural difficulty in library construction, short cDNA tags, high RNA-input requirement, etc.[96,97]

SAGE has been used largely for quantitative analysis of gene expression. Examples include cluster analysis on multiple SAGE libraries with a total of 60,000 transcript tags derived from premalignant epidermal tissue (actinic keratosis), normal human epidermis, and cultured keratinocytes.[98] SAGE technology was also used to identify genes regulated by estrogen and tamoxifen in an estrogen-dependent breast cancer cell (ZR75–1) line. Comprehensive analysis of gene expression profiles following estrogen or tamoxifen treatment can help to better understand the role estrogen plays in tumorigenesis.[99] As pointed out, one disadvantage of the SAGE method is the high RNA-input requirement. Because of this requirement, SAGE cannot be used for the generation of expression profiles in cases when the quantity of RNA is limited, i.e., in small biological samples such as tissue biopsies or microdissected material. One of the attempts to overcome this limitation was described by Datson et al. as a modification of SAGE, named microSAGE, which requires up to 5×10^3-fold less starting material.[100] In addition, microSAGE is simplified due to incorporation of a single-tube procedure for all steps, from RNA isolation to tag release. Furthermore, a limited number of additional PCR cycles are performed. Using microSAGE gene expression profiles can be obtained from minute quantities of tissue; the estimated need is only about 10^5 cells. This method opens up a multitude of new possibilities for the application of SAGE, for example, the characterization of expression profiles in tissue biopsies, tumor metastases or in other cases where tissue is scarce, and the generation of region-specific expression profiles of complex heterogeneous tissues.

It is well known that many gene diseases or drug resistances in pathogens are caused by a single-base mutation. If they occur in a known constant location, they are easily detectable by targeted profiling methods. They could be distinguished, however, even when the exact location of the mutation is unknown. Synthetic DNA oligonucleotides can serve as efficient primers for DNA synthesis even with a single base mismatch between the primers and the corresponding template. However, when the primer–template annealing is carried out with a mixture of primers, the binding of a perfectly matched primer is strongly favored relative to a primer differing by a single base.

Such primer competition has been observed over a range of oligonucleotide sizes from 12 to 16 bases and with a variety of base mismatches. When coupled with PCR, competitive oligonucleotide priming provides a simple general strategy for the detection of single DNA base differences.[101] Another reason to look for single nucleotide differences, especially in complex genomes of multicellular organisms like the human body, for instance, is that they are a very good representation of natural variability and can serve as excellent genotype markers *in toto*.

Single nucleotide polymorphism (SNP) has considerable value for the analysis of the genotype–phenotype relationship. Finding SNPs has a very different genetic significance. The ability to distinguish between identical sequence fragments that have one or only a few base pair substitutions is crucial in genotyping, forensic fingerprinting, and diagnostic identification of diseased alleles or escaped mutants of any pathogen. It is easy to see how SNP screening and differentiation require high-throughput technology. There are several approaches for the screening of SNPs. A homogeneous method has been described based on inhibition of spontaneous branch migration by any sequence difference between two molecules of PCR-amplified DNA.[102] The method employs a set of four PCR primers: forward primers, one biotinilated and the other labeled with digoxigenin, and two reverse primers that share a priming domain but have different "tail" sequences at their 5' ends. After PCR amplification is performed, denaturation and reannealing of the single DNA strands produce doubly labeled cruciform structures, which dissociate by strand exchange. The presence of two different alleles in one sample causes changes in the dissociation and the association of biotin, and digoxigenin is homogeneously detected using luminescent oxygen-channeling immunoassay.

Molecular beacons are sensitive fluorescent probes that can hybridize selectively to designated DNA and RNA targets. They have become irreplaceable tools for quantitative real-time monitoring of single-stranded nucleic acids. Stemless and stem-containing DNA and peptide nucleic acid (PNA) probes have been subject to comparative studies. It has been demonstrated that they respond differently to changes in salt concentration. Stemless PNA beacons hybridize rapidly to the complementary oligodeoxynucleotide and are less sensitive than the DNA beacons to changes in salt concentration. Hence, PNA molecular beacons can be used for the detection of a chosen target sequence directly in its double-stranded DNA form (dsDNA) without denaturation and deproteinization. Conditions have been found where the stemless PNA beacon strongly discriminates the complementary vs. mismatched dsDNA targets.[103] A cyanine dye, 3,3'-diethylthiadicarbocyanine displays dramatic color change upon binding to DNA-PNA duplexes. Captured, single-stranded DNA molecules prepared by PCR amplification are hybridized with PNA probes in an allele-specific fashion. Experimental results demonstrate that this diagnostic method may be sufficiently sensitive to discriminate between even a fully complementary and a single-mutation DNA sequence.[104] An alternative method to screening SNPs is matrix-assisted laser desorption/ionization time-of-flight mass spectrometry (MALDI-TOFMS). This technique can accurately detect and identify the hybridized PNA probes in a very short time.[105] The possibility of genotyping single nucleotide polymorphisms (SNPs) by primer extension and HPLC has also been investigated. The system has been found to produces a robust signal and has the added advantage that sample loading and analysis are essentially automated, analytic time is brief, and no further purification step after primer extension is required. All stages of the HPLC-primer extension genotyping can be multiplexed and automated, suggesting that this system may be suitable for large-scale linkage studies based on emerging SNP maps.

Efficient SNP genotyping methods are necessary for accomplishing many current gene discovery goals. A "bottleneck" element in large-scale SNP genotyping is the number of individual biochemical reactions that must be performed. Reducing genome complexity by using degenerate oligonucleotide primer (DOP)-PCR has been applied to SNP genotyping. In a comparative study with three complex eukaryotic genomes, human, mouse, and *Arabidopsis thaliana,* and using a single DOP-PCR primer, it has been demonstrated that SNP loci spread throughout a genome can be amplified and accurately genotyped directly from the DOP-PCR product mixture. DOP-PCRs have been found to be very reproducible.[106] The issue of SNP genotype determination has also been addressed by investigating variations with the help of pyrosequencing, a novel real-time pyrophosphate detection technology.[107–109] The method is based on indirect luminometric quantification of the pyrophosphate that is released as a result of nucleotide incorporation onto an amplified template. The method uses iterative addition of all dNTPs in each step and four different enzymatic reactions (polymerase, sulfurylase, luciferase, and apyrase) to incorporate the correct base and release a light signal. This technical platform requires automation and allows the analysis of 96 samples within 10 to 20 min. Analysis of each addressed SNP reveals a highly discriminating pattern, whereby homozygous samples produce clear-cut single-base peaks and heterozygous counterparts have half-height peaks representing both allelic positions. The system has multi-base reading capacity and allows for the determination of several closely located SNPs in a single run.

It is inevitable that regardless of the number of arbitrary or specific primers used to display genome differences one will encounter the problem of amplifying products of exactly the same length but with an entirely different sequence or of identical sequences differing by only one or a few bases. To resolve such a problem two different techniques have been introduced. Single-strand conformational polymorphism (SSCP) analysis was devised to separate electophoretically in native acrylamide gel fragments that are identical in size but completely different in sequence. The technique is based on the observation that the electrophoretic mobility of single nucleic acid strands relies both on the size of the fragment and on its conformation. The method has been used successfully for profiling human mitochondrial DNA.[110] The second technique is WAVE, which was described above.[10]

5.2.7 RDA/SSH/DSC

Representational differences analysis (RDA) was described for the first time in 1993 as an efficient approach to identifying and cloning differences between complex genomes.[111,112] The method has proven truly valuable for identifying differentially expressed or unique genes. RDA is an improved method of DNA subtraction that compares representations of the whole genome produced by whole-genome PCR. The analysis of the differences between two complex genomes is a promising tool for the discovery of infectious agents and probes useful for genetic studies.

RDA is a method by which subtractive (by hybridization) and kinetic (by PCR) enrichment is used to isolate fragments present in one population of DNA molecules but not in another. RDA requires the use of two nucleic acid sources — one that is the target of interest, referred to as the "tester," and one that is considered a "normal" or known reference/standard sample, referred to as "driver." Both are subject to the same nucleic acid extraction and purification procedures aiming for a final double-stranded DNA or cDNA product that is then digested with restriction endonuclease to reduce the complexity of the sample. In the next step, the tester sample is ligated to adapters that provide a known "anchor" or label sequence for further retrieval of fragments of interest. The tester sample is then denatured and hybridized to an excess of the driver, so that all common sequences between the two samples are now hybrid double-stranded molecules that can have one anchor sequence. Sequences that are unique to the tester will either remain single stranded or hybridize to the tester counterpart, thereby acquiring two anchor sites. Such fragments could then be exponentially amplified from the complex hybridization mix by using primers containing the anchor sequence.

RDA is an efficient approach to the identification of the differences between complex genomes and could be valuable as the first step for the positional cloning of genes of interest.[113] It can be used for the detection of genetic lesions in cancer, discovery of unknown pathogens, isolation of polymorphic markers linked to a trait, etc. RDA has been applied successfully to identify the chromosomal alterations in cancer DNA. Homozygous deletions have been found in cervical, renal, and gastric carcinoma, demonstrating that RDA is an efficient method by which to identify the possible location of a tumor-suppresser gene. As the examples below illustrate, the application of this method to DNA populations of reduced complexity ("representations") resulted in the isolation of probes to viral genomes present as single copies in human DNA, probes that detect polymorphisms between two individuals, etc. RDA may also be used for isolating probes linked to sites of genome rearrangements, whether they occurred spontaneously and resulted in genetic disorders or in cancer, or were programmed during differentiation and development.

With the use of HindIII-derived amplicons and an excess of male schistosome DNA as a driver, RDA was applied to the isolation of female-specific sequences from *Schistosoma mansoni*.[8] Following three rounds of RDA, the enriched sequences included two female-specific repetitive elements. One of these (designated SM alpha fem-1) exhibited 76% homology to the SM alpha family of retroposons and represents a W-chromosome-specific variant of that family, and the other, designated W2, represented a novel, female-specific repetitive sequence.

A version of SAGF can also be applied to RDA if the template is cleaved with IIS- or IIN-types of restriction endonucleases producing single-stranded termini of different sequences at the fragment ends.[61] The method is based on the ligation of short double-stranded adapters with single-stranded termini complementary to the termini of only a selected set of fragments followed by PCR amplification with the primer that represents a strand of the adapters. The SAGF method can be used for obtaining individual bands in DNA fingerprinting, for reducing the kinetic complexity of DNA in the representational difference analysis (RDA method) of complex genomes, for cataloging DNA fragments, and for constructing physical genome maps. RDA with arbitrary primers has also been described.[114]

Creutzfeldt-Jakob Disease (CJD) is a neurodegenerative and dementing disease caused by an incompletely characterized virus-like agent.[115] All purified infectious brain preparations contain nucleic acids; however, it has not been possible to visualize unique bands that may derive from a viral genome. RDA was used to uncover such sequences. A single sucrose-gradient peak (120S) from a CJD brain highly enriched for infectivity was used as the tester and parallel 120S fractions from uninfected brain as the driver. Representative fragments of 100 to 500 bp have been captured from cDNAs cleaved with MboI for adaptor ligation and amplification, which represents the first demonstration of apparently CJD-specific nucleic acid bands in purified infectious preparations.

The mapping of genetic suppressors, modifiers, and quantitative trait loci (QTLs) requires genetic markers that can be efficiently and inexpensively genotyped for a large number of individuals; RDA has been successfully used to isolate such markers.[44] Amplicons prepared by PCR with B1 repetitive sequence as primer (B1-amplicons) generated 48 polymorphic DNA fragments after five rounds of RDA, subtracting the B1 amplicons prepared from an ACI rat from those prepared from BUF rats, and vice versa. Dot-blotting amplicons onto filters at a high density and hybridization of the filters with these B1-RDA markers enabled the genotyping of large numbers of rats simultaneously for multiple loci.

RDA was applied in the recent identification of several new flaviviruses — GB virus A (GBV-A) and GB virus B (GBV- B) — in the plasma of a tamarind infected with the hepatitis GB agent. Subsequently, a third virus, GB virus C (GBV-C), was identified in humans.[31] Following inoculation with human plasma pool, a chimpanzee that developed acute resolving hepatitis was examined, and a new virus was uncovered that exhibited 62 to 80% identity with GBV-C and was tentatively labeled GBC troglodytes virus (GBV-Ctro). The virus has not been found in humans, even after extensive search by RT-PCR.

Gene rearrangement represents a very interesting issue in genome profiling, as it is related to the regulation of expression and oncogenesis. RDA has been applied toward the detection and cloning of a 246-bp retroviral sequence, homologous to a long terminal repeat of avian erythroblastosis virus (AEV), from Md5 strain of Marek's disease virus type 1 (MDV1). This retroviral fragment did not originally belong to Md5. Such data suggest that RDA could be useful for the detection of retroviral sequences within a viral genome.[116] Chronic persistent infections by mycoplasmas have been shown to induce the malignant transformation of C3H mouse embryo cells, which had never previously been reported to undergo spontaneous transformation.[117] The mycoplasma-mediated oncogenic process has a long latency and demonstrates a multistage progression characterized by reversibility and irreversibility of malignant properties upon removal of the mycoplasma from culture. After three rounds of RDA with three different restriction-enzyme DNA digests, no gene fragment of mycoplasmal origin was amplified or identified in the permanently transformed C3H cells, suggesting that mycoplasmas do not cause malignant transformation by integrating their gene(s) in the mammalian genome.

Interleukin-4 (IL-4) stimulates B cell growth and differentiation, such as inducing mature B cells to switch to IgG1 and IgE production.[118] The effects of IL-4 on B cells have been studied by RDA combination with magnetic bead depletion as means of physical separation. This cDNA RDA technique has allowed the subtraction of the relatively small number of highly purified B cells. Two rounds of subtraction resulted in greater than a 100-fold enhancement of IL-4-induced Cgamma1 cDNA. Most of the 37 genes found could be functionally identified by sequence similarity to genes encoding Cgamma1, cytoskeletal components, and ones involved in DNA replication, metabolism, signal transduction, transcription, translation, and transport. Seven genes had no similarity to known sequences in the available databases. This study is a good example of the unbiased ability of RDA to characterize the expression profile of a given state of differentiation.

RDA is clearly a powerful method for genome profiling, but it has a few drawbacks. The intrinsic design of the procedure selects against single-stranded species in the hybridization step, is biased toward large fragments, and has demonstrated occasional inconsistency with different restriction enzymes. This implies that a rare sequence or a low abundance transcript may be lost as a PCR template after the hybridization.

A new method, termed suppression subtractive hybridization (SSH), has been developed for the generation of subtracted cDNA libraries.[119] It is an improved version of RDA and is based on the recently described technique called suppression PCR. It combines normalization and subtraction in a single procedure. The normalization step generates conditions where the unique molecular species have amplification advantage over the abundant ones because of the later tend to reanneal to themselves, thereby equalizing the cDNA's variety within the target population. The subtraction step again excludes the common sequences between the target and driver populations. A creative improvement of the technique involves the insertion of inverted terminal repeats into the design of the adapters that are used to amplify the cDNAs. This forces short molecules representing sequences common for the driver and tester to generate pan-type structures at each cycle of PCR amplification and thus prevent them from being amplified. SSH uses two different adapters with built-in palindrome sequence. After each of the modified tester cDNAs has been hybridized to the driver they can be combined to create templates of unique molecular species that have different primer sites at both ends, a feature that additionally enhances the exponential amplification of the desired fragments. Only one round of subtractive hybridization is needed. The subtracted library is normalized in terms of abundance of different cDNAs. This method dramatically increases the probability of obtaining low-abundance differentially expressed cDNA and is intended to use very small amounts of cells or tissues as starting material.[120]

In a model system, the SSH technique has been able to enrich for rare sequences over 1000-fold in one round of subtractive hybridization.[121] A representative example of how SSH can be used for the differential screening of subtracted cDNA libraries was the construction a testis-specific cDNA library, where the subtracted cDNA mixture is used as a hybridization probe. Such an approach enabled the identification of homologous sequences in a human Y chromosome cosmid library. The human DNA inserts in the isolated cosmids were further confirmed to express in a testis-specific manner.[119]

The characteristics of the subtracted libraries, the nature and level of background nondifferentially expressed clones in the libraries, as well as the entire procedure for the rapid identification of truly differentially expressed cDNA clones speak well for the performance of SSH. These results suggest that the SSH technique is applicable to many molecular genetic and positional cloning studies for the identification of diseased, developmental, tissue-specific, or other differentially expressed genes. A good example is a recent study designed to understand the molecular mechanism underlying the earliest steps of the embryonic ectoderm subdivision into epidermis and neuroectoderm of *Xenopus laevis*.[122] SSH was used successfully to clone the cDNA of *X. laevis* genes expressed specifically in the presumptive epidermis starting from the midgastrula stage; this is a very convincing illustration of the method's ability to achieve results with a very small starting amount of material.

Ermolaeva et al. proposed a theory of subtractive hybridization based on the kinetic model of the process.[123] They were able to develop a computer-generated model of the process of subtraction and used it as a basis for this theory. Consequently, a novel method of subtractive hybridization was suggested that allows for routine comparison of genomes and products of genome expression. The method has been applied to studies of the genetic mechanisms of embryogenesis, regeneration, and cell differentiation and tumor transformation.

The most recent technique based on subtractive differentiation of unique nucleic acid species is the differential subtraction chain (DSH). The method resembles RDA and SSH in that it also utilizes driver material and tester ligated to adapters and also relies on hybridization. This approach, however, takes advantage of the unique ability of the mung bean nuclease enzyme to digest single-stranded nucleic acid. Tester molecules with adapters at both ends are hybridized to the driver without an adapter, followed by enzymatic removal (by the mung bean nuclease) of the single-stranded termini from the hybrids. After several rounds of hybridization followed by nuclease digestion, molecules that are common to the tester and the driver would eventually lose their adapter "tags." Only molecules unique to the tester would retain the adapter sequences, since they will form a double-stranded configuration resistant to nuclease cleavage. It was estimated that four rounds of the process are sufficient to eliminate the background, and the remaining material could be amplified by PCR using primer sequences complementary to the adapter.[124] This is a clean and very efficient technique, and it was used to demonstrate deletion events in human glioblastoma tumors; however, it is somewhat biased toward low-copy-number molecular species. In other words, a unique infectious agent, for instance, that is preset in low titer will be lost from the mixture at the same rate as the single-copy genes that are common for the host and are found in low quantities in both the tester and the driver. In parallel model experiments, the method is outperformed by SSH.

5.2.8 MICROARRAYS/STS/ESTs

As we saw already in the discussion about differential display, one very important aspect of genome profiling is the ability to monitor gene expression. Detecting and determining the relative abundance of diverse individual sequences in complex DNA samples has been a recurring experimental challenge in genome analysis.[125,126] Microscopic arrays (spots) of DNA fragments on glass substrates for differential hybridization analysis of fluorescently labeled DNA samples were introduced as a general experimental approach to this problem. The use of high-density DNA arrays to monitor gene expression at a genome-wide scale constitutes a fundamental advance in biology. The new method emerged with the technology that enables automated spotting of minute amounts of nucleic acid on solid support, thereby creating a "DNA chip."

Recently, this technology has been improved by the introduction of a method that allows "growing" of the desired oligonucleotides directly on the chip.[127–130] The choice and size of genetic material that could be applied on the array is enormous.[131] In fact, "microarraying" is a powerful combination of ingredients that could be customized depending on the objective. These ingredients could be PCR products, DD products, sequence-tagged sites (STS),[132,133] expressed sequence tags (ESTs),[14,132,133,135] multiple artificial/arbitrary oligos,[136] multiplex hybridization probes, chemiluminescent detection of DNA,[137] laser scanning, and computerized data acquisition and analysis in a format intrinsically designed for automation and high throughput.[127,138,139,140–142] STSs are now being recognized as the building blocks of modern complex genome maps.[143,144] They are the unique products of a pair of oligonucleotides after amplification of a genome of interest. The nucleotide pair design is usually based on available one-time sequence data from ESTs (<500 bp long) and 3′ untranslated regions chosen to improve the needed unique specificity. ESTs are derived from cDNA clones and are in essence markers for the expressed portion of the genome. STSs and ESTs will be reviewed here only as integral parts of the microarray technology.

The first organism that was subjected to the powerful microarray technique was *Arabidopsis thaliana*, and the experiment enabled the monitoring of 45 individual transcripts (from as little as 2 μg of starting mRNA) with two color fluorescent markers.[125] The system was further tested and expanded on the yeast[126] and then the human genome.[145] These first microarrays were then characterized by simultaneous hybridization of two different sets of isolated probes labeled with two different fluorophores and detected by laser fluorescent scanner. This system has found numerous applications in genome-wide genetic mapping, physical mapping, and gene-expression studies.

The expression pattern of all known genes in the favorite model organism, *Saccharomyces cerevisiae*, has been investigated using microarray analysis. Yeast cDNAs were hybridized to more than 6000 genes from the yeast genome. It is well known that complex gene expression patterns result from dynamic interacting networks of genes driven by regulated gene expression. The organization of the regulatory DNA elements is an important clue toward understanding the combinatorial control of the gene. These questions led to the creation of a comprehensive Yeast Promoter Database and initiated an attempt to develop new computational methods for mapping upstream regulatory elements.[146] Carried out in collaboration with experimental biologists this project uses cell-cycle-regulated gene expression as a specific example to demonstrate how one may extract promoter information computationally from a genome-wide screening. There has been steady progress in the computational analysis of transcription control regions, but current methods of predicting the gene regulatory features of noncoding sequences are still not accurate enough to be useful in automatic genome annotation.[147] Therefore, detailed information on the expression patterns of newly sequenced genes is more likely to come from microarray-based high-throughput mRNA quantitation technologies, which have made revolutionary progress over the past few years and which are now ready for genome-wide application.

Many genes and signaling pathways that control cell proliferation, death, and differentiation, as well as maintenance of genome integrity, are also involved in cancer development. Serial analysis of gene expression as provided by cDNA microarrays enables measurement of the expression of thousands of genes in a single experiment and can potentially reveal new important cancer genes. To establish the diagnostic, prognostic, or therapeutic importance of each emerging cancer gene candidate one needs to analyze hundreds of specimens from patients in different stages of a particular disease. This calls for an array-based high-throughput technique that could survey in a comparative manner the gene expression patters of multiple tissue biopsies. Kononen et al. proposed that biopsy sections from thousands of individual tumors can be combined in a single tumor tissue microarray and then directly screened *in situ* by DNA, RNA, or protein hybridization techniques. This technique was successfully used to detect six gene amplifications and a p53 estrogen receptor in breast cancer.[148]

Several forms of human sarcoma, lymphoma, and leukemia are characterized by somatically acquired chromosome translocations that result in fusion genes that encode chimeric transcription factors with oncogenic properties. A group of several alveolar rhabdomyosarcoma (ARMS) cell lines were used to investigate their gene-expression profile on a cDNA microarray.[149] ARMS cells demonstrated a consistent pattern of gene expression. It was found that 37 out of the 1238 genes were consistently expressed, and only 3 of these were previously reported as expressed in ARMS when compared to a reference cell line.

Microarrays have already found valuable application in vaccine research. Bacille Calmette-Guerin (BCG) vaccines are live attenuated strains of *Mycobacterium bovis* administered to prevent tuberculosis.[150] A detailed understanding of the genetic differences between closely related *Mycobacteria* can suggest rational approaches to the design of improved diagnostics and vaccines. To elucidate in detail the differences between *M. tuberculosis, M. bovis,* and the various BCG daughter strains, their genome compositions have been studied by comparative hybridization experiments on a DNA microarray. The array revealed multiple deletions in the more virulent strains in evidence of the pathogens' ongoing evolution.

The generation of a viral DNA chip for simultaneous expression measurements of nearly all known open reading frames (ORFs) in the largest member of the herpesvirus family, the human cytomegalovirus (HCMV), was described for the first time in 1999.[151] The HCMV chip was fabricated and used to characterize the temporal class of viral gene expression. It was composed of oligonucleotides for the viral DNA applied by robotic deposition on glass surface. Gene expression was monitored by hybridization of fluorescently labeled cDNAs from mock-infected or infected fibroblast cells to the oligonucleotide chip. The kinetic classes of the array elements were classified by using cycloheximide and ganciclovir to block *de novo* viral protein synthesis and viral DNA replication, respectively. The profiles provided a temporal map of the gene expression covering the entire genome of HCMV and indicated the presence of potential regulatory motifs for β, γ1, and γ2 genes. This experience with the HCMV viral chip demonstrates the great utility of the microarray approach. SSH combined with microarray technology, where the SSH product is used as hybridization probe, can also be successfully applied to identify differentially expressed genes.[152]

Thousands of genes are discovered for the first time by sequencing the genomes of model organisms. Much of the natural world remains to be explored at the molecular level. DNA microarrays clearly provide a fast vehicle for this exploration, but without the application of artificial nucleic acids their creation would have hardly been feasible. The model organisms were the first for which comprehensive genome-wide surveys of gene-expression patterns or function became possible. The results can be viewed as maps that reflect the order and logic of the genetic program, rather than the physical order of genes on chromosomes. Exploration of the genome using DNA microarrays and other genome-scale technologies should narrow the gap in our knowledge of gene function and molecular biology between the currently favored model organisms and other species.[153]

5.2.9 Quo Vadis?

Whole genome microarrays for all? The advances of technology are moving us very quickly toward the ability to completely miniaturize the whole genome of one organism, which, of course, is ironic since nature has already done this in every single cell. Fully automated data acquisition and record keeping, however, allow us to ask questions and get answers that in the near past would have been impossible to interpret even in a lifetime of devotion to scientific endeavor. Only 20 years ago the ability to compare the full genetic map of the human 21 chromosome or the entire genome map to its relative structure in chimpanzees and make phylogenetic conclusions would have been unthinkable.[154,155] Automation and high throughput are not just industrial scale words. In reality, even if we chose to ignore the concepts and insight that come from advancements in science and focus only on improving our research tools, we would see that today we are capable of single-handedly generating orders of magnitude more information.

When talking about automation and terabit data analysis we naturally come to computing. Neural networking with genome libraries will enable us to truly make use of the immense information that is already available and continues to accumulate at a rate that we cannot keep up with.[156–162] Today we still feel that we have to pick, choose, and prioritize what kind of information we want to acquire first. What do we want to "file away" and evaluate first — a specific pathogen, a valuable crop plant, or a choice breed of cattle, etc.? We hope that tomorrow it will be left to us to merely ask the right questions.

Building on the notion that in the future we will need to do PCR only electronically, genome forecasting could emerge as an interesting field for research. Based on our knowledge of genome profiles and tendencies we should be able to predict the pathogens of the future and to actually design the drugs to prevent any new plagues before they appear.

REFERENCES

1. McKusick, V. A., Genomic mapping and how it has progressed, *Hosp. Prac.,* 26(10), 74, 1999.
2. Asakawa, S. et al., Human BAC library: construction and rapid screening, *Gene,* 191, 69, 1997.
3. Xu, J. et al., Screening for overlapping bacterial artificial chromosome clones by PCR analysis with an arbitrary primer, *Proc. Natl. Acad. Sci. U.S.A.,* 95, 5661, 1998.
4. Clark, M. S., Comparative genomics: the key to understanding the Human Genome Project, *Bioassays,* 21, 121, 1999.
5. Deutsch, M. and Long, M., Intron-exon structures of eukaryotic model organisms, *Nucleic Acids Res.,* 27, 3219, 1999.
6. Murray, V. et al., Detection of polymorphisms using thermal cycling with a single oligonucleotide on a DNA sequencing gel, *Hum. Mutat.,* 2, 118, 1993.
7. Findlay, I. et al., Fluorescent PCR: a new technique for PGD of sex and single-gene defects, *J. Assist. Reprod. Genet.,* 13, 96, 1996.
8. Emilion, G. et al., Frequent gene deletions in potentially malignant oral lesions, *Br. J. Cancer,* 73, 809, 1996.
9. Kuchiki, H. et al., Detection of DNA abnormalities by arbitrarily-primed PCR fingerprinting: amplification of the MDM2 gene in a mediastinum fibrosarcoma, *Biochem. Biophys. Res. Commun.,* 258, 271, 1999.
10. Kuklin, A. et al., Detection of single-nucleotide polymorphisms with the WAVE DNA fragment analysis system, *Genet. Test.,* 1, 201, 1997.
11. Dixon, A. K. et al., Expression profiling of single cells using 3′ end amplification (TPEA) PCR, *Nucleic Acids Res.,* 26, 4426, 1998.
12. Menard, C., Brousseau, R., and Mouton, C., Application of polymerase chain reaction with arbitrary primer (AP-PCR) to strain identification of *Porphyromonas (Bacteroides) gingivalis, FEMS Microbiol. Lett.,* 74, 163, 1992.
13. Hu, J., van Eysden, J., and Quiros, C. F., Generation of DNA-based markers in specific genome regions by two-primer RAPD reactions, *PCR Methods Appl.,* 4, 346, 1995.
14. Picoult-Newberg, L. et al., Mining SNPs from EST databases, *Genome Res.,* 9, 167, 1999.
15. Bernardi, G. et al., Genetic variability in Guadalupe fur seals, *J. Hered.,* 89, 301, 1998.
16. Sullivan, K. M., Forensic applications of DNA fingerprinting, *Mol. Biotechnol.,* 1, 13, 1994.
17. Schneider, P. M., Basic issues in forensic DNA typing, *Forensic Sci. Int.,* 88, 17, 1997.
18. Brookfield, J. F., The effect of relatives on the likelihood ratio associated with DNA profile evidence in criminal cases, *J. Forensic Sci. Soc.,* 34, 193, 1994.
19. Sudbury, A. W., Marinopoulos, J., and Gunn, P., Assessing the evidential value of DNA profiles matching without using the assumption of independent loci (published erratum appears in *J. Forensic Sci. Soc.,* 33, 198, 1993), *J. Forensic Sci. Soc.,* 33, 73, 1993.
20. Caetano-Anolles, G., MAAP: a versatile and universal tool for genome analysis, *Plant. Mol. Biol.,* 25, 1011, 1994.
21. Bhat, K. V., Jarret, R. L., and Rana, R. S., DNA profiling of banana and plantain cultivars using random amplified polymorphic DNA (RAPD) and restriction fragment length polymorphism (RFLP) markers, *Electrophoresis,* 16, 1736, 1995.
22. Caetano-Anolles, G. and Gresshoff, P. M., Generation of sequence signatures from DNA amplification fingerprints with mini-hairpin and microsatellite primers, *Biotechniques,* 20(6), 144, 1996.
23. Clayton, R. A., White, O., and Fraser, C. M., Findings emerging from complete microbial genome sequences, *Curr. Opin. Microbiol.,* 1, 562, 1998.
24. Cowley, A. W., The emergence of physiological genomics, *J. Vasc. Res.,* 36, 83, 1999.
25. Nagaraju, J. G. and Singh, L., Assessment of genetic diversity by DNA profiling and its significance in silkworm, *Bombyx mori, Electrophoresis,* 18, 1676, 1997.

26. Gerstein, M. and Hegyi, H., Comparing genomes in terms of protein structure: survey of a finite parts list, *FEMS Microbiol. Rev.,* 22, 277, 1998.

27. Gaasterland, T. and Ragan, M. A., Constructing multigenome views of whole microbial genomes, *Microb. Comp. Genomics,* 3, 177, 1998.

28. Marcotte, E. M. et al., Detecting protein function and protein-protein interactions from genome sequences, *Science,* 285, 751, 1999.

29. Smeets, H. J. et al., Identification of apolipoprotein E polymorphism by using synthetic oligonucleotides, *J. Lipid Res.,* 29, 1231, 1988.

30. Birkenmeier, E. H., Schneider, U., and Thurston, S. J., Fingerprinting genomes by use of PCR with primers that encode protein motifs or contain sequences that regulate gene expression, *Mamm. Genome,* 3, 537, 1992.

31. Birkenmeyer, L. G. et al., Isolation of a GB virus-related genome from a chimpanzee, *J. Med. Virol.,* 56, 44, 1998.

32. Gilham, P. T. and Khorana, H. G., Studies on polynucleotides. I. A new and general method for the chemical synthesis of the C5′–C3′ internucleotidic linkage. Synthesis of deoxyribo-dinucleotides, *J. Am. Chem. Soc.,* 80, 6212, 1958.

33. Horvath, S. et al., An automated DNA synthesizer employing deoxynucleoside 3′-phosphoramidites, *Methods Enzymol.,* 154, 314, 1987.

34. Mullis, K. B. and Faloona, F. A., Specific synthesis of DNA *in vitro* via a polymerase-catalyzed chain reaction, *Methods Enzymol.,* 155, 335, 1987.

35. Vosberg, H. P., The polymerase chain reaction: an improved method for the analysis of nucleic acids, *Hum. Genet.,* 83, 1, 1989.

36. Guatelli, J. C., Gingeras, T. R., and Richman, D. D., Nucleic acid amplification *in vitro*: detection of sequences with low copy numbers and application to diagnosis of human immunodeficiency virus type 1 infection, *Clin. Microbiol. Rev.,* 2, 217, 1989.

37. Romano, P. et al., Molecular Probe Data Base: a database on synthetic oligonucleotides, *Nucleic Acids Res.,* 21, 3007, 1993.

38. Balland, A. et al., Use of synthetic oligonucleotides in gene isolation and manipulation, *Biochimie,* 67, 725, 1985.

39. Reyes, G. R. and Kim, J. P., Sequence-independent, single-primer amplification (SISPA) of complex DNA populations, *Mol. Cell. Probes,* 5, 473, 1991.

40. Bassam, B. J., Caetano-Anolles, G., and Gresshoff, P. M., DNA amplification fingerprinting of bacteria, *Appl. Microbiol. Biotechnol.,* 38, 70, 1992.

41. Caetano-Anolles, G., Bassam, B. J., and Gresshoff, P. M., Primer-template interactions during DNA amplification fingerprinting with single arbitrary oligonucleotides, *Mol. Gen. Genet.,* 235, 157, 1992.

42. Gade, R. et al., Incorporation of nonbase residues into synthetic oligonucleotides and their use in the PCR, *Genet. Anal. Tech. Appl.,* 10, 61, 1993.

43. Meyer, R. B. J., Incorporation of modified bases into oligonucleotides, *Methods Mol. Biol.,* 26, 73, 1994.

44. Telenius, H. et al., Degenerate oligonucleotide-primed PCR: general amplification of target DNA by a single degenerate primer, *Genomics,* 13, 718, 1992.

45. Tanaka, Y. et al., Genomic and molecular evolutionary analysis of a newly identified infectious agent (SEN virus) and its relationship to the TT virus family, *J. Infect. Dis.,* 183,359, 2001.

46. Sineo, L. et al., Analysis of genetic markers by random amplified polymorphic DNA polymerase chain reaction (RAPD-PCR), *Boll. Chim. Farm.,* 132: 201–202, 1993.

47. Venugopal, G., Mohapatra, S., and Salo, D., Multiple mismatch annealing: basis for random amplified polymorphic DNA fingerprinting, *Biochem. Biophys. Res. Commun.,* 197, 1382, 1993.

48. Hernandez, J. et al., Random amplified polymorphic DNA fingerprinting of *Campylobacter jejuni* and *C. coli* isolated from human faeces, seawater and poultry products, *Res. Microbiol.,* 146, 685, 1995.

49. Gillan, R. et al., Comparison of *Cannabis sativa* by random amplification of polymorphic DNA (RAPD) and HPLC of cannabinoids: a preliminary study, *Sci. Justice,* 35, 169, 1995.

50. Nagaraja, G. M. and Nagaraju, J., Genome fingerprinting of the silkworm, *Bombyx mori,* using random arbitrary primers, *Electrophoresis,* 16, 1633, 1995.

51. Yi, Q. M. et al., Polymorphism and genetic relatedness among wild and cultivated rice species determined by AP-PCR analysis, *Hereditas,* 122, 135, 1995.

52. Demesure, B., Sodzi, N., and Petit, R. J., A set of universal primers for amplification of polymorphic non-coding regions of mitochondrial and chloroplast DNA in plants, *Mol. Ecol.*, 4, 129, 1995.

53. Brownstein, M. J., Carpten, J. D., and Smith, J. R., Modulation of non-templated nucleotide addition by Taq DNA polymerase: primer modifications that facilitate genotyping, *Biotechniques*, 20, 1004, 1996.

54. Greer, W. L. et al., Linkage disequilibrium mapping of the Nova Scotia variant of Niemann-Pick disease, *Clin. Genet.*, 55, 248, 1999.

55. Giraffa, G. et al., Genotypic heterogeneity among *Lactobacillus helveticus* strains isolated from natural cheese starters, *J. Appl. Microbiol.*, 85, 411, 1998.

56. Chakraborty, R. and Li, Z., Correlation of DNA fragment sizes within loci in the presence of non-detectable alleles, *Genetica*, 96, 27, 1995.

57. Zabeau, M. and Vos, P., Selective restriction fragment amplification: a general method for DNA fingerprinting, European patent application 0534858A1, 1993.

58. Liscum, E., Amplified fragment length polymorphism: studies on plant development, in *PCR Applications/Protocols for Functional Genomics*, Innis, M. A., Gelfand, D. H., and Sninsky, J. J., Eds., Academic Press, San Diego, 1999, 505.

59. Vos, P. et al., AFLP: a new technique for DNA fingerprinting, *Nuc. Acids Res.*, 23, 4407, 1995.

60. Caetano-Anolles, G., Bassam, B. J., and Gresshoff, P. M., Enhanced detection of polymorphic DNA by multiple arbitrary amplicon profiling of endonuclease-digested DNA: identification of markers tightly linked to the supernodulation locus in soybean, *Mol. Gen. Genet.*, 241, 57, 1993.

61. Zheleznaia, L. A. et al., A method of selective PCR-amplification of genomic DNA fragments (SAGF method), *Biokhimiya*, 60, 1363, 1995.

62. Chenchik, A. et al., Full-length cDNA cloning and determination of mRNA 5' and 3' ends by amplification of adaptor-ligated cDNA, *Biotechniques*, 21, 526, 1996.

63. Rossetti, S. et al., Detection of mutations in human genes by a new rapid method: cleavage fragment length polymorphism analysis (CFLP), *Mol. Cell. Probes*, 11, 155, 1997.

64. Wu, D. Y. and Wallace, R. B., The ligation amplification reaction (LAR)-amplification of specific DNA sequences using sequential rounds of template-dependent ligation, *Genomics*, 4, 560, 1989.

65. Gu, Z. et al., Densities, length proportions, and other distributional features of repetitive sequences in the human genome estimated from 430 megabases of genomic sequence, *Gene*, 259, 81, 2000.

66. Zietkiewicz, E., Rafalski, A., and Labuda, D., Genome fingerprinting by simple sequence repeat (SSR)-anchored polymerase chain reaction amplification, *Genomics*, 20, 176, 1994.

67. Weising, K., Atkinson, R. G., and Gardner, R. C., Genomic fingerprinting by microsatellite-primed PCR: a critical evaluation, *PCR Methods Appl.*, 4, 249, 1995.

68. Schmid, C. W., Does SINE evolution preclude Alu function? *Nucleic Acids Res.*, 26, 4541, 1998.

69. Novick, G. E. et al., Polymorphic human specific Alu insertions as markers for human identification, *Electrophoresis*, 16, 1596, 1995.

70. Arveiler, B. and Porteous, D. J., Distribution of Alu and L1 repeats in human YAC recombinants, *Mamm. Genome*, 3, 661, 1992.

71. Clayton, T. M. et al., Further validation of a quadruplex STR DNA typing system: a collaborative effort to identify victims of a mass disaster, *Forensic Sci. Int.*, 76, 17, 1995.

72. Clayton, T. M. et al., Analysis and interpretation of mixed forensic stains using DNA STR profiling, *Forensic Sci. Int.*, 91, 55, 1998.

73. Wall, W. J. et al., Variation of short tandem repeats within and between populations, *Hum. Mol. Genet.*, 2, 1023, 1993.

74. Detter, J. C., Nguyen, Q. A., and Kingsmore, S. F., Identification of novel simple sequence length polymorphisms (SSLPs) in mouse by interspersed repetitive element (IRE)-PCR, *Nucleic Acids Res.*, 26, 4091, 1998.

75. Sheehan, C. et al., Genomic fingerprinting *Acinetobacter baumannii*: amplification of multiple inter-repetitive extragenic palindromic sequences, *J. Hosp. Infect.*, 31, 33, 1995.

76. Bennett, W. A. et al., Expression and production of interleukin-10 by human trophoblast: relationship to pregnancy immunotolerance, *Early Pregnancy*, 3, 190, 1997.

77. Zhao, X. S. et al., Identification of upregulated genes in the thymus of spontaneously hypertensive rats by cDNA representational difference analysis, *Blood Pressure*, 7, 316, 1998.

78. Church, D. M. and Buckler, A. J., Gene identification by exon amplification, *Methods Enzymol.*, 303, 83, 1999.

79. Brown, A. J. et al., Targeted display: a new technique for the analysis of differential gene expression, *Methods Enzymol.,* 303, 392, 1999.

80. Chow, V. T., Tham, K. M., and Bernard, H. U., Nucleic acid hybridization and the polymerase chain reaction in biology and medicine, with special reference to the detection of human papillomaviruses, *Ann. Acad. Med. Singapore,* 18, 387, 1989.

81. Agyare, F. D. et al., Mapping cloned sequences on YACs, *Methods Mol. Biol.,* 82, 199, 1998.

82. Fu, Y. X., Timberlake, W. E., and Arnold, J., On the design of genome mapping experiments using short synthetic oligonucleotides, *Biometrics*, 48, 337, 1992.

83. Cai, W. W. et al., An anchored framework BAC map of mouse chromosome 11 assembled using multiplex oligonucleotide hybridization, *Genomics,* 54, 387, 1998.

84. Baldwin, D., Crane, V., and Rice, D., A comparison of gel-based, nylon filter and microarray techniques to detect differential RNA expression in plants, *Curr. Opin. Plant Biol.,* 2, 96, 1999.

85. Sajantila, A. et al., DNA profiling in a genetically isolated population using three hypervariable DNA markers, *Hum. Hered.,* 42, 372, 1992.

86. Forozan, F. et al., Genome screening by comparative genomic hybridization, *Trends Genet.,* 13, 405, 1997.

87. Kroger, B. and Horak, I., Isolation of novel human retrovirus-related sequences by hybridization to synthetic oligonucleotides complementary to the tRNA(Pro) primer-binding site, *J. Virol.,* 61, 2071, 1987.

88. Ferguson, J. A. et al., A fiber-optic DNA biosensor microarray for the analysis of gene expression, *Nat. Biotechnol.,* 14, 1681, 1996.

89. Copeland, N. G., Jenkins, N. A., and Court, D. L., Recombineering: a powerful new tool for mouse functional genomics, *Nat. Rev. Genet.,* 2, 769, 2001.

90. Zabarovska, V. I. et al., A new approach to genome mapping and sequencing: slalom libraries, *Nucleic Acids Res.,* 30, E6, 2002.

91. Ito, T. and Sakaki, Y., Fluorescent differential display: a fast and reliable method for message display polymerase chain reaction, *Methods Enzymol.,* 303, 298, 1999.

92. Broude, N. E. et al., PCR based targeted genomic and cDNA differential display, *Genet. Anal.,* 15, 51, 1999.

93. Prashar, Y. and Weissman, S. M., Analysis of differential gene expression by display of 3′ end restriction fragments of cDNAs, *Proc. Natl. Acad. Sci. U.S.A.,* 93, 659, 1996.

94. Matz, M. et al., Ordered differential display: a simple method for systematic comparison of gene expression profiles, *Nucleic Acids Res.,* 25, 2541, 1997.

95. Subrahmanyam, Y. V. et al., A modified method for the display of 3′-end restriction fragments of cDNAs: molecular profiling of gene expression in neutrophils, *Methods Enzymol.,* 303, 272, 1999.

96. Scott, H. S. and Chrast, R., Global transcript expression profiling by Serial Analysis of Gene Expression (SAGE), *Genet. Eng. (N.Y.),* 23, 201, 2001.

97. Yamamoto, M., Gene expression profiling using improved SAGE, *Rinsh. Byori,* 50, 52, 2002.

98. Van Ruissen, F. et al., Differential gene expression in premalignant human epidermis revealed by cluster analysis of serial analysis of gene expression (SAGE) libraries, *FASEB J.,* 16, 246, 2002.

99. Seth, P. et al., Novel estrogen and tamoxifen induced genes identified by SAGE (Serial Analysis of Gene Expression), *Oncogene,* 21, 836, 2002.

100. Datson, N. A. et al., MicroSAGE: a modified procedure for serial analysis of gene expression in limited amounts of tissue, *Nucleic Acids Res.,* 27, 1300, 1999.

101. Gibbs, R. A., Nguyen, P. N., and Caskey, C. T., Detection of single DNA base differences by competitive oligonucleotide priming, *Nucleic Acids Res.,* 17, 2437, 1989.

102. Guo, Z., Liu, Q., and Smith, L. M., Enhanced discrimination of single nucleotide polymorphisms by artificial mismatch hybridization, *Nat. Biotechnol.,* 15, 331, 1997.

103. Kuhn, H. et al., Hybridization of DNA and PNA molecular beacons to single-stranded and double-stranded DNA targets, *J. Am. Chem. Soc.,* 124, 1097, 2002.

104. Wilhelmsson, L. M. et al., Genetic screening using the colour change of a PNA-DNA hybrid-binding cyanine dye, *Nucleic Acids Res.,* 30, E3, 2002.

105. Paracchini, S. et al., Hierarchical high-throughput SNP genotyping of the human Y chromosome using MALDI-TOF mass spectrometry, *Nucleic Acids Res.,* 30, E27, 2002.

106. Jordan, B. et al., Genome complexity reduction for SNP genotyping analysis, *Proc. Natl. Acad. Sci. U.S.A.,* 99, 2942, 2002.

107. Gustafsson, A. C. et al., Screening and scanning of single nucleotide polymorphisms in the pig melanocortin 1 receptor gene (MC1R) by pyrosequencing, *Anim. Biotechnol.*, 12, 145, 2001.

108. Ahmadian, A. et al., Single-nucleotide polymorphism analysis by pyrosequencing, *Anal. Biochem.*, 280, 103, 2000.

109. Nordstrom, T. et al., Direct analysis of single-nucleotide polymorphism on double-stranded DNA by pyrosequencing, *Biotechnol. Appl. Biochem.*, 31, 107, 2000.

110. Yap, E. P. and McGee, J. O., Nonisotopic discontinuous phase single strand conformation polymorphism (DP-SSCP): genetic profiling of D-loop of human mitochondrial (mt) DNA, *Nucleic Acids Res.*, 21, 4155, 1993.

111. Hubank, M. and Schatz, D. G., Identifying differences in mRNA expression by representational difference analysis of cDNA, *Nucleic Acids Res.*, 22, 5640, 1994.

112. Lisitsyn, N. and Wigler, M., Cloning the differences between two complex genomes, *Science*, 259, 946, 1993.

113. Aburatani, H., Representational difference analysis (RDA): finding the genetic lesions in cancer DNA, *Nippon Rinsho*, 54, 949, 1996.

114. Yoshida, Y. et al., Development of the arbitrarily primed-representational difference analysis method and chromosomal mapping of isolated high throughput rat genetic markers, *Proc. Natl. Acad. Sci. U.S.A.*, 96, 610, 1999.

115. Dron, M. and Manuelidis, L., Visualization of viral candidate cDNAs in infectious brain fractions from Creutzfeldt-Jakob disease by representational difference analysis, *J. Neurovirol.*, 2, 240, 1996.

116. Endoh, D. et al., Retroviral sequence located in border region of short unique region and short terminal repeat of Md5 strain of Marek's disease virus type 1, *J. Vet. Med. Sci.*, 60, 227, 1998.

117. Zhang, B. et al., Absence of mycoplasmal gene in malignant mammalian cells transformed by chronic persistent infection of mycoplasmas, *Proc. Soc. Exp. Biol. Med.*, 218, 83, 1998.

118. Chu, C. C. and Paul, W. E., Expressed genes in interleukin-4 treated B cells identified by cDNA representational difference analysis, *Mol. Immunol.*, 35, 487, 1998.

119. Diatchenko, L. et al., Suppression subtractive hybridization: a method for generating differentially regulated or tissue-specific cDNA probes and libraries, *Proc. Natl. Acad. Sci. U.S.A.*, 93, 6025, 1996.

120. Lukyanov, K. A. et al., Inverted terminal repeats permit the average length of amplified DNA fragments to be regulated during preparation of cDNA libraries by polymerase chain reaction, *Anal. Biochem.*, 229, 198, 1995.

121. Lonneborg, A., Sharma, P., and Stougaard, P., Construction of subtractive cDNA library using magnetic beads and PCR, *PCR Methods Appl.*, 4, S168, 1995.

122. Vasiliev, O. L. et al., A novel marker of early epidermal differentiation: cDNA subtractive cloning starting on a single explant of *Xenopus laevis* gastrula epidermis, *Int. J. Dev. Biol.*, 41, 877, 1997.

123. Ermolaeva, O. D., Lukyanov, S. A., and Sverdlov, E. D., The mathematical model of subtractive hybridization and its practical application, *Ismb.*, 4, 52, 1996.

124. Luo, J. H. et al., Differential Subtraction Chain, a method for identifying differences in genomic DNA and mRNA, *Nucleic Acids Res.*, 27, e24, 1999.

125. Schena, M. et al., Quantitative monitoring of gene expression patterns with a complementary DNA microarray, *Science*, 270, 467, 1995.

126. Shalon, D., Smith, S. J., and Brown, P. O., A DNA microarray system for analyzing complex DNA samples using two-color fluorescent probe hybridization, *Genome Res.*, 6, 639, 1996.

127. Kurian, K. M., Watson, C. J., and Wyllie, A. H., DNA chip technology, *J. Pathol.*, 187, 267, 1999.

128. Pon, R. T., Yu, S., and Sanghvi, Y. S., Tandem oligonucleotide synthesis on solid-phase supports for the production of multiple oligonucleotides, *J. Org. Chem.*, 67, 856, 2002.

129. Barone, A. D. et al., Photolithographic synthesis of high-density oligonucleotide probe arrays, *Nucleosides Nucleotides Nucleic Acids*, 20, 525, 2001.

130. McGall, G. H. and Fidanza, J. A., Photolithographic synthesis of high-density oligonucleotide arrays, *Methods Mol. Bio.*, 170, 71, 2001.

131. Tomiuk, S. and Hofmann, K., Microarray probe selection strategies, *Brief Bioinf.*, 2, 329, 2001.

132. Baysal, B. E. et al., A high-resolution STS, EST, and gene-based physical map of the hereditary paraganglioma region on chromosome 11q23, *Genomics*, 44, 214, 1997.

133. Jhiang, S. M., Chiu, I. M., and Mazzaferri, E. L., An STS in the human PTC oncogene located at 10q11.2, *Nucleic Acids Res.*, 19, 4303, 1991

134. Cirera, S., Wintero, A. K., and Fredholm, M., Why do we still find anonymous ESTs? *Mamm. Genome,* 11, 689, 2000.

135. Morelli, C. et al., Mapping of 22 new ESTs around a tumor suppressor gene and a senescence gene at 6q16-q21, *Cytogenet. Cell. Genet.,* 79, 97, 1997.

136. Gao, X. et al., A flexible light-directed DNA chip synthesis gated by deprotection using solution photogenerated acids, *Nucleic Acids Res.,* 29, 4744, 2001.

137. Akhavan-Tafti, H. et al., Chemiluminescent detection of DNA in low- and medium-density arrays, *Clin. Chem.,* 44, 2065, 1998.

138. Carulli, J. P. et al., High throughput analysis of differential gene expression, *J. Cell Biochem. Suppl.,* 286, 30, 1998.

139. Falus, A., Varadi, A., and Rasko, I., The DNA-chip, a new tool for medical genetics, *Orv. Hetil.,* 139, 957, 1998.

140. Pan, W., Lin, J., and Le, C. T., Model-based cluster analysis of microarray gene-expression data, *Genome Biol.,* 3, 9, 2002.

141. Sturn, A., Quackenbush, J., and Trajanoski, Z., Genesis: cluster analysis of microarray data, *Bioinformatics,* 18, 207, 2002.

142. Jain, A. N. et al., Fully automatic quantification of microarray image data, *Genome Res.,* 12, 325, 2002.

143. Armour, J. A. and Jeffreys, A. J., STS for minisatellite MS607 (D22S163), *Nucleic Acids Res.,* 19, 3158, 1991.

144. Armour, J. A., Neumann, R., and Jeffreys, A. J., STS for minisatellite 33.1 (D9S49)]: direct typing by PCR, *Nucleic Acids Res.,* 19, 4788, 1991.

145. Schena, M. et al., Parallel human genome analysis: microarray-based expression monitoring of 1000 genes, *Proc. Natl. Acad. Sci. U.S.A.,* 93, 10614, 1996.

146. Zhang, M. Q., Promoter analysis of co-regulated genes in the yeast genome, *Comput. Chem.,* 23, 233, 1999.

147. Bucher, P., Regulatory elements and expression profiles, *Curr. Opin. Struct. Biol.,* 9, 400, 1999.

148. Kononen, J. et al., Tissue microarrays for high-throughput molecular profiling of tumor specimens, *Nat. Med.,* 4, 844, 1998.

149. Khan, J. et al., Gene expression profiling of alveolar rhabdomyosarcoma with cDNA microarrays, *Cancer Res.,* 58, 5009, 1998.

150. Behr, M. A. et al., Comparative genomics of BCG vaccines by whole-genome DNA microarray, *Science,* 284, 1520, 1999.

151. Chambers, J. et al., DNA microarrays of the complex human cytomegalovirus genome: profiling kinetic class with drug sensitivity of viral gene expression, *J. Virol.,* 73, 5757, 1999.

152. Yang, G. P. et al., Combining SSH and cDNA microarrays for rapid identification of differentially expressed genes, *Nucleic Acids Res.,* 27, 1517, 1999.

153. Brown, P. O. and Botstein, D., Exploring the new world of the genome with DNA microarrays, *Nat. Genet.,* 21, 33, 1999.

154. Fujiyama, A. et al., Construction and analysis of a human-chimpanzee comparative clone map, *Science,* 295, 131, 2002.

155. Pletcher, M. T. et al., Use of comparative physical and sequence mapping to annotate mouse chromosome 16 and human chromosome 21, *Genomics,* 74, 45, 2001.

156. Fariselli, P. et al., Prediction of protein–protein interaction sites in heterocomplexes with neural networks, *Eur. J. Biochem.,* 269, 1356, 2002.

157. Hatzakis, G. and Tsoukas, C., Neural networks in the assessment of HIV immunopathology, *Proc. AMIA Symp.,* 249, 2001.

158. Pankratz, N. et al., Use of variable marker density, principal components, and neural networks in the dissection of disease etiology, *Genet. Epidemiol.,* 21(Suppl. 1), S732, 2001.

159. Pasquier, C., Promponas, V. J., and Hamodrakas, S. J., PRED-CLASS: cascading neural networks for generalized protein classification and genome-wide applications, *Proteins,* 44, 361, 2001.

160. Marinov, M. and Weeks, D. E., The complexity of linkage analysis with neural networks, *Hum. Hered.,* 51, 169, 2001.

161. Bhat, A., Lucek, P. R., and Ott, J., Analysis of complex traits using neural networks, *Genet. Epidemiol.,* 17(Suppl. 1), S503, 1999.

162. Farber, R., Lapedes, A., and Sirotkin, K., Determination of eukaryotic protein coding regions using neural networks and information theory, *J. Mol. Biol.,* 226, 471, 1992.

6 Site-Directed Mutagenesis

John I. Glass and Beverly A. Heinz

CONTENTS

0-8493-1426-7/03/$0.00+$1.50
© 2003 by CRC Press LLC

6.1 INTRODUCTION

Site-directed mutagenesis is a fundamental tool that is widely used in molecular biology and protein engineering. Because it allows the researcher to design selective changes to the DNA, site-directed mutagenesis provides a powerful method for probing gene regulation as well as the relationship between protein structure and function. Unlike natural mutations or random mutagenesis, which require selection by phenotypic changes, mutations constructed by genetic engineering do not require extensive screening. Site-directed mutagenesis is therefore much more efficient than these earlier approaches.

The most common application of site-directed mutagenesis is to change a gene *in vitro* followed by insertion of the modified DNA into either a bacterial or eukaryotic cell. The methods utilized for *in vitro* site-directed mutagenesis are generally of two types: those that use the polymerase chain reaction (PCR) with thermostable enzymes, and non-PCR approaches that use thermolabile polymerases. In all cases, *in vitro* DNA synthesis occurs via the addition of deoxynucleotide triphosphates (dNTPs) to the 3' end of an oligonucleotide primer annealed to the template DNA. This step is typically referred to as primer extension. The template may be double or single stranded, whereas the product is always double stranded.

In recent years, many modifications to the standard mutagenesis protocols have been reported, most proclaiming small advantages or unique adaptations to specialized circumstances. This explosion of information has made the field very complex. At the same time, the standard approaches have been simplified by the availability of commercial kits that are suitable to the most common research goals. Therefore, it is not our intention to encompass all of the techniques used in *in vitro* site-directed mutagenesis, but rather to highlight the general principles and methods typically employed. In addition, in the final sections of this chapter, we will describe three fundamentally different applications of site-directed mutagenesis: ribozymes to create precise termini of RNA molecules, and two methods using novel oligonucleotides for *in vivo* mutagenesis that have potential uses in gene therapy. We hope this review will direct readers toward the pertinent literature for their needs. At the same time, we will briefly describe some of the newest applications in the field.

6.2 HISTORY

The 1993 Nobel Prize in Chemistry was awarded to Michael Smith for the invention of site-directed mutagenesis. Smith's original work, published in 1978,[1] was not the first mutagenesis at a targeted site in a gene. Previously, N^4-hydroxycytidine, a nucleoside analogue, had been enzymatically introduced at a specific site in bacteriophage QB RNA to produce a specific mutation.[2,3] Site-specific DNA mutagenesis was accomplished via chemical modification of a specific DNA fragment that was inserted into the genome of bacteriophage φX174.[4] However, these methods were intractable in that they did not allow completely predictable incorporation of all four classes of point mutation — transitions (e.g., purines substituted for purines), transversions (e.g., pyrimidines substituted for purines), insertions, and deletions — at a specific site with high efficiency. Smith and his colleagues realized the most likely route to being able to precisely control mutagenic agents was through the use of *in vitro* synthesized oligonucleotides as site-specific mutagens.

Smith's lab had developed expertise in the stepwise enzymatic assembly of oligonucleotides. The approach involved the extension of short primers using *Escherichia coli* polynucleotide phosphorylase and deoxyribonucleoside-5' diphosphates as substrate. Oligonucleotides up to 13 bases in length could be made using this method, which was the only synthetic method available during the 1970s.[5] Smith's oligonucleotides had been part of a seminal scientific collaboration that involved chain-termination DNA-sequencing technology developed by Fred Sanger and Clyde Hutchison's expertise in bacteriophage φX174 genetics, resulting in the first completely sequenced genome.[6] From the genomic sequences of both the wild-type φX174 and a cold-sensitive mutant it was possible to design and synthesize two 12-base oligonucleotide primers, one with the wild-type sequence and one with the mutant sequence. These oligonucleotides were used to prime synthesis of complementary strands on genetically heterologous viral strand φX174 templates. Transfection of *E. coli* spheroplasts with those products yielded cold-sensitive phage mutants from the wild-type templates and wild-type phage from the mutant templates. Each site-directed mutagenesis reaction had a 15% efficiency.[1,7–10]

Site-directed mutagenesis did not become a routine procedure in molecular biology until the advent of automated chemical synthesis of oligonucleotides using the methoxytrityl protecting group and nucleoside-3′ phosphoramidites as the key intermediates.[11] As the synthesis of large oligonucleotides became routine, so too did site-directed mutagenesis. Over the last 16 years mutagenic oligonucleotides have been used in the development of a wide range of methods.

6.3 PCR-BASED APPROACHES TO SITE-DIRECTED MUTAGENESIS

6.3.1 GENERAL PRINCIPLES

The introduction of polymerase chain reaction (PCR) methodologies has had an enormous impact on the field of site-directed mutagenesis.[12] Specifically, PCR has vastly improved the efficiency and versatility of DNA synthesis and mutagenesis. In brief, the *in vitro* synthesis of DNA by PCR requires the use of two oligonucleotide primers (20 to 30 bases long) complementary to the 5′ and 3′ ends on opposite strands of the desired region of the template. A common procedure is to incorporate one or two mismatches into the primer to introduce mutations that facilitate cloning, such as the creation of a new restriction-enzyme cleavage site. During PCR, multiple heating and cooling cycles in the presence of a heat-stable polymerase (e.g., *Taq, Vent, Pfu*) allow for the rapid synthesis of DNA. On average, about 30 cycles are used, each cycle consisting of denaturation at high temperature (at ~95°C to separate the strands), primer annealing (at ~55°C), and extension (at ~72°C). As the cycles progress, both the original templates and the initial products are used as templates for exponential amplification. The final product consists of linear, double-stranded mutant DNA.

6.3.2 MUTAGENIC OLIGONUCLEOTIDE DESIGN

The stability of the complex of an oligonucleotide and a DNA template is determined by the base composition of the oligonucleotide and the conditions under which it is annealed. The fundamental concerns for designing mutagenic oligonucleotides are the same as the general rules for designing a good primer. Primers for DNA synthesis should not contain direct repeats, strong secondary structures, or runs of four or more GTP bases. For PCR, it is best to have the five bases at the 3′-end of the primer contain no more than three GTP or CTP bases.

Mismatches should normally be located in the middle of the primer, although there are some applications such as inverse PCR in which the mutations are usually at the 5′ end of the primer. Mutagenic mismatches at the 3′ terminus of oligonucleotides have been used,[13] but such mismatches usually attenuate DNA synthesis. In most instances, a 17- to 20-base oligonucleotide with a mismatch located in the center will be sufficient for single-base substitutions. This affords 8 to 10 perfectly matched base pairs on both sides of the mismatch. Two or more substitutions are usually introduced with oligonucleotides of 25 bases or longer. Deletions of four bases have been obtained with oligonucleotides that are 26 to 27 bases long. Larger insertions and deletions would normally require oligonucleotides with 20 to 30 matched bases on either side of the mutation site.[14,15]

Annealing temperature and magnesium concentration are critical factors in obtaining correct and sufficient hybridization of mutagenic oligonucleotides and templates. Annealing temperatures are normally selected that are a few degrees below the melting temperature of the primer (T_m).

There are several algorithms for calculating T_m. The most simplistic one estimates T_m in degrees Celsius of an oligonucleotide as follows: $T_m = 4 \times$ (number of G and C bases) $+ 2 \times$ (number of A and T bases). More rigorous analysis can be achieved with computer programs that calculate primer T_m and check for other potential problems such as hairpins and false priming sites in the template. Popular commercial software packages that are useful in the design of primers include Oligo 6.0 (MBI, Hanover, MD), Primer Premier 5.0 (Premier Biosoft, Palo Alto, CA), and Primer Designer 4.0 (Scientific & Educational Software, Durham, NC). Web-based applications for primer design are too numerous and rapidly evolving to iterate here. Recently, computer programs became available for design of mutagenic primers. Especially noteworthy is The Primer Generator.[16]

As a consequence of the mismatches mutagenic oligonucleotides have relative to template DNA, annealing of mutagenic primers requires reduction of annealing temperatures more than a few degrees. The extent of the reduction below the needed T_m varies with the size of the mismatch. Piechocki and Hines[14] describe criteria for the design of mutagenic primers and temperature conditions for annealing mutagenic primers based on mutagenesis experiments. For PCR-based mutagenesis, the alternative to reducing the oligonucleotide annealing temperature is to raise the magnesium concentration. Magnesium is typically 1.5 mM for perfectly matched primers. With a 23-base primer containing a single mismatch, the concentration should be 2.0 mM. Larger mismatches can require 4.0 mM magnesium.[17]

6.3.3 CHOICE OF POLYMERASE

Selection of the most suitable polymerase for a particular purpose requires an understanding of the balance between fidelity and processivity of the enzyme. Although *Taq* is the most commonly used polymerase for PCR, it lacks proofreading activity via the $3'$-$5'$ exonuclease present in other commonly used polymerases. Thus, *Taq*-based PCR tends to introduce more random errors than the other polymerases such as *Vent, Deep Vent,* or *Pfu.*[18–20] On the other hand, amplification of long PCR products is best accomplished by *Taq,* which has better processivity.[21] A popular solution is to use a combination of *Taq,* the primary enzyme, plus a small amount of a secondary enzyme to supply proofreading activity.[22–24] A number of high-fidelity DNA polymerases are commercially available. *Pfu* from Stratagene (La Jolla, CA) boasts the lowest average error rate published for a DNA polymerase (1.3×10^{-6} per nucleotide[19]); the improved version, *PfuTurbo,* reportedly can amplify even larger PCR products with the same high fidelity. Other popular enzymes include *Vent* and *Deep Vent* (isolated from a hyperthermophilic bacterium) from New England Biolabs (Beverly, MA), *Pwo* from Boehringer Mannheim (Penzberg, Germany), *UlTma* from Applied Biosystems (Foster City, CA), and *Tli* from Promega (Madison, WI). In addition to fidelity, it is important to consider the nature of the $3'$ end generated by different enzymes. *Taq* generates a single-A overhang at the $3'$ end of the PCR product, whereas *Pfu, Pwo,* and *UlTma* generate blunt ends. *Vent* generates 70% blunt ends, whereas *Tli* and *Deep Vent* generate >95% blunt ends.

6.3.4 SECONDARY STRUCTURE OF TEMPLATES

A final consideration concerns the nature of the template. Templates containing very stable secondary structure or GC-rich regions can pose difficult problems during PCR. Solutions include increasing the annealing temperature to 65 to 72°C; the addition of denaturants such as DMSO, glycerol, formamide, Tween-20, and NP-40; and the use of *E. coli* single-stranded DNA-binding protein. Alternatively, the template can be synthesized to contain small amounts of nucleotide analogues such as 7-deaza-$2'$-deoxyguanosine triphosphate or deoxyinosine triphosphate to weaken the strength of GC bonds. These varied approaches are summarized in Ling and Robinson.[25] At times, it has proven useful to employ several of these treatments simultaneously. An example is the development of a technique termed "subcycling PCR" (S-PCR[26]). In S-PCR, the annealing/elongation step consists of shuttling between a low and a high temperature (for example, shuttling four times between 60 and 65°C). According to the authors, S-PCR may have advantages over standard PCR methods in several contexts, including amplification of long segments in which the GC content varies.

PCR across AT-rich regions can also be problematic. Presumably the low T_m of the duplex DNA in localized AT-rich regions of the template leads to separation of the enzyme-nascent strand complex from the template. Reduction of the temperature in the polymerase extension step of the PCR by 3 to 7°C will usually result in successful amplification of AT-rich templates that cannot be amplified under standard PCR conditions (J. Glass, unpublished observation).[27]

6.4 PCR MUTAGENESIS: SPECIFIC APPLICATIONS

6.4.1 OVERLAP EXTENSION METHODS

A specialized technique developed for the introduction of mutations into the center of a PCR fragment is known as the overlap extension method.[28] In this technique, primers are designed for the synthesis of two different PCR fragments with a region of common sequence. Each reaction uses one flanking primer and one internal primer containing the desired mutation (see Figure 6.1). The two overlapping pieces of DNA are then annealed to generate a heteroduplex that can be extended by *Taq* polymerase and amplified by subsequent rounds of PCR to produce a full-length mutant segment. By varying the templates and primers, both mutant and chimeric DNA can be generated. This approach is applicable to the splicing and recombination of segments from different genes without relying on restriction sites. For example, DNA splicing by overlap extension was used to generate hybrid proteins of the major histocompatibility complex.[29] The technique is equally applicable to the introduction of site-specific insertion and deletion mutations and can be used to perform recombination and mutagenesis simultaneously.[30] It has also been modified to allow for the introduction of multiple mutations on the same template without loss of efficiency.[31,32] Finally, although the overlap extension method was originally intended for use with relatively small DNAs, it has proven possible to prepare mutated fragments as large as 3 kb.[33]

The overlap extension method can be modified to simplify the removal of excess primers and wild-type template after each PCR step. Biotinylated primers and streptavidin-coated magnetic beads are employed to purify PCR products from reactants. This approach improves the speed and efficiency of overlap extension mutagenesis.[34]

6.4.2 MEGAPRIMER PCR

Although primers are typically short, single-stranded oligonucleotides, it was observed by Kammann et al. in 1989[35] that long, double-stranded DNAs could also serve as primers. These larger primers, known as megaprimers, can be used to introduce any combination of point mutations, deletions, or insertions. The megaprimer method of mutagenesis uses three oligonucleotide primers (two flanking and one megaprimer) and two rounds of PCR performed on a DNA template (see Figure 6.2). The first PCR uses one outside primer and the middle mutagenic primer to form a double-stranded product (the megaprimer) containing the desired mutations. This product is then purified and used in the second round of PCR along with the other outside primer. The wild-type cloned gene serves as the template in both PCR reactions. The two flanking primers are generally designed to contain suitable restriction sites to permit convenient cloning.

Several modifications and improvements of the megaprimer technique have been reported. For example, Ke and Madison[36] have described an efficient procedure that does not require the intermediate purification of DNA between the two rounds of PCR. Their protocol is based on designing forward and reverse flanking primers with significantly different T_m values. The low-T_m primer is used with a low annealing temperature in the first PCR reaction producing the megaprimer; the second reaction uses the high-T_m primer and a high annealing temperature (which prevents priming

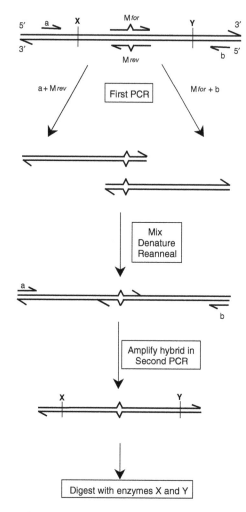

FIGURE 6.1 PCR overlap extension method of mutagenesis. Arrows on lines indicate the 5′ to 3′ orientation of the DNA. M*for* and M*rev* represent the forward and reverse mutagenic primers (the mutation is depicted by Λ); a and b represent the flanking primers. X and Y indicate restriction endonuclease sites used for cloning the mutated fragment.

by the low-T_m primer). These authors observed an average mutagenesis efficiency of 82%. A second consideration, especially with megaprimers larger then 1 kb, is that self-annealing of the megaprimer can reduce product yield.[37] This problem can be avoided by the use of microgram amounts of template in the second PCR reaction. Details of additional improvements (e.g., increasing the size of the megaprimer), simplifications accomplished by combining the cycles, and issues regarding enzyme selection were addressed recently.[25,37]

6.4.3 Construction of Large Insertions and Deletions

Although the introduction of small insertions or deletions can be readily accomplished by designing the mutations into the primer, the insertion or deletion of large segments of DNA requires special consideration. Larger deletions may be produced by looping out a portion of the DNA or by joining two PCR-amplified fragments.[38] To construct insertions greater than 200 bp, three individual fragments, including the insertion, can be produced by PCR and joined by ligation.[38] In a recently described protocol,[39] the forward and reverse PCR primers are designed to perform two functions.

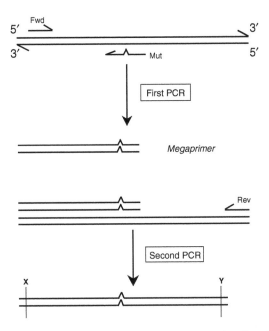

FIGURE 6.2 Megaprimer PCR mutagenesis. Arrows on lines indicate the 5′ to 3′ orientation of the DNA. The forward (Fwd), reverse (Rev), and mutagenic (Mut) primers are indicated by short lines; the mutation is depicted by Λ. X and Y indicate restriction endonuclease sites used for cloning the mutated fragment.

They carry specific sequences at their 3′ ends that are used in the initial PCR for DNA amplification, and at their 5′ ends they carry sequences complementary to the target plasmid in the region where insertion will occur. Thus, substitution of one large DNA fragment for another can be accomplished without subcloning. One shortcoming of this procedure is the requirement for single-stranded template DNA. Thus, the plasmid must contain an origin of replication of a filamentous phage (e.g., M13).

6.4.4 INVERSE PCR MUTAGENESIS

Inverse PCR is a technique used to clone unknown sequences flanking regions where the DNA sequence is known (e.g., introns, or 5′- or 3′-untranslated regions).[40] When used as a method for mutagenesis, a pair of primers is made, the 5′ termini of which hybridize end-to-end on opposite strands of the plasmid DNA to be mutagenized. One of the primers contains the desired mutation. PCR amplification of the entire double-stranded plasmid results in a linear PCR amplicon that can be circularized by ligation (Figure 6.3). The original inverse PCR mutagenesis protocol required phosphorylation of the primers and Klenow fragment treatment of the amplicon to remove any nontemplated bases added (usually an A) at the end of the newly synthesized strand.[41] Alternative approaches avoid those steps by incorporation of unique restriction sites into both primers so that endonuclease digestion yields compatible cohesive ends. The effect of the restriction sites on the final gene product is neutralized either by the use of degenerate codons for amino acids so that the peptide is not altered or through the use of a Class IIs enzyme recognition sequence.

This is a versatile mutagenesis technique that has a high mutagenic efficiency even in the absence of *in vivo* selection. Inverse PCR can be used to create deletions as large as 87 bp.[42] Its chief limitation is that it works best for plasmids smaller than 3.1 kb; however, it has been used to mutate plasmids as large as 11.5 kb.[43] A kit is available from Stratagene for application of this method (Table 6.1).

TABLE 6.1
Popular Commercially Available Kits for Site-Directed Mutagenesis

Company	Kit Name	Selection Method	Requirements	Vector	Host	Enzymes	Efficiency	Ref.
Bio-Rad (Hercules, CA)	Muta-Gene Phagemid *in vitro* Mutagenesis Kit	Resistance to uracil-*N*-glycosylase digestion	Grow in *dut*−, *ung*− *E. coli*	Any phagemid vector or M13	CJ2 *dut*−, *ung*−	T7 DNA pol	60–70%	83
Clontech (Palo Alto, CA)	Transformer Site Directed Mutagenesis Kit	Resistance to restriction enzyme digestion	Unique restriction site and antibiotic resistance	Any	Repair deficient	T4 DNA pol	70–90%	76
Clontech	Diversify™ PCR Random Mutagenesis Kit	Random mutagenesis of PCR fragment followed by cloning	Effective for PCR amplicons <4.0 kb	Any	Any	AdvanTaq Plus™ DNA pol	Controlled by varying Mn^{2+}	90
Takara (Kyoto, Japan)	Mutan™-K	Resistance to uracil-*N*-glycosylase digestion	Limited to M13 or phagemid vectors	Any ss	*dut*−*ung*−, F'	T4 DNA pol and *E. coli* DNA ligase	60–70%	83
Takara	Mutan™-Express Km	Oligonucleotide-directed Dual Amber method	Target gene must be cloned into pKF	pKF18k/19k	*sup0*, *mutS*	T4 DNA pol	70–95%	82, 91
Takara	Mutan-Super Express-Km	Oligonucleotide-directed Dual Amber method	Target gene must be cloned into pKF	pKF18k/19k	*sup0*	LA Taq pol	>80%	82
Takara	LA PCR *in vitro* Mutagenesis Kit	Resistance to restriction enzyme digestion	No need of specific host strains	pUC18/19 (118/119)	any strains	LA Taq pol	>60%	92

Promega	Altered Sites II *in vitro* Mutagenesis Systems	Conversion of antibiotic sensitive plasmid to antibiotic resistant		PAlter, pAlterEx1, pAlterEx2, pAlter-Max	ES1301 mut S JM109	T4 DNA pol	Up to 90%	93
Promega	GeneEditor	Positive selection for resistance to GeneEditor antibiotic selection mix	Vector must be ampR	Any ampR vector	BMH71-18 mut S JM109	T4 DNA pol	Up to 100%	94
Stratagene	QuickChange	Inverse PCR	PCR primer set	Any ds	XL-1	Pfu DNA pol, Dpn1	83–100%	41
Stratagene	ExSite PCR-Based Site-Directed Mutagenesis Kit	Selection against hemi-methylated parental strand	Target plasmid must be from dam+ bacteria	Any ds		Pfu DNA pol, Dpn1	>60%	95
Stratagene	Chameleon Double-Stranded Site-Directed Mutagenesis Kit	Resistance to restriction enzyme digestion	Unique restriction site in nonessential region of the plasmid	Any ds	XLmutS	Kpn1	90%	76
New England BioLabs	GPS™-LS Linker Scanning System	Variation of linker scanning using 15-bp transposon	Two frames get TAA mutations, one frame gets a 5 amino acid insert	Any ds	None	TnsABC* Transposase	99%	96

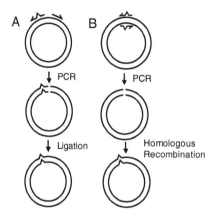

FIGURE 6.3 Inverse PCR mutagenesis. (A) A pair of head-to-tail PCR primers that contain the sought-after mutation(s) is designed so that they can PCR-amplify an entire plasmid. The linear double-stranded plasmid is then circularized by ligation and transformed into *E. coli*. (B) A variation on the technique uses complementary primers to generate a linear plasmid. The PCR amplicon is then transformed into *E. coli*, where it is circularized via homologous recombination.

6.5 NON-PCR-BASED APPROACHES TO SITE-DIRECTED MUTAGENESIS

6.5.1 GENERAL PRINCIPLES

At the heart of most of the methods described in this section is DNA synthesis by thermolabile DNA polymerases. The same factors that are germane to PCR-based methods must also be considered in the design of mutagenesis reactions in which thermolabile DNA polymerases are employed (namely, template, primer design, reaction conditions, and polymerase choice).

Some non-PCR-based mutagenesis methods use only single-stranded templates. However, as evidenced by the use of double-stranded plasmids and non-PCR-based mutagenesis in most of the commercial site-directed mutagenesis kits, there are many non-PCR-based methods in which duplex DNA is used. Although mutagenesis of double-stranded DNA is commonly performed, for all methods in which oligonucleotides are used it is necessary to denature the double helix by treatment with alkali or heat or both. Some approaches circumvent this requirement by cloning their mutagenesis targets into single-stranded vectors such as M13 bacteriophage or phagemids containing an M13 or f1 replication origin.[15,44]

The same basic criteria for PCR primer design should be followed for DNA synthesis using thermolabile DNA polymerases. In this case, it is more important to be certain that primers cannot bind to alternate sites on the template. The lower annealing and polymerization temperatures can lead to false priming. A novel method that reduces mispriming is initiation of DNA synthesis with Klenow fragment at 65°C for 5 min followed by cooling of the reaction to 37°C and the addition of more polymerase. Presumably, some primer extension occurs before the enzyme is inactivated at 65°C, thus binding the primer to the template.[45]

Since the first use of site-directed mutagenesis in 1978, the choice of suitable DNA polymerases has expanded. The *E. coli* DNA polymerase used in that experiment is now rarely used because its helicase and 5′-3′ exonuclease activities are not conducive to mutagenesis. Most methods (and most kits) include either T4 DNA polymerase or T7 DNA polymerase because those enzymes lack the extra activities. One of the parameters that influences polymerase choice is the presence of secondary structures or GC-rich regions in the template. Even after heating or alkali treatment, DNA synthesis at 37 to 42°C can be slowed or stopped by residual secondary structure or duplex DNA. Among thermolabile DNA polymerases, the T7 bacteriophage enzyme is less affected by these DNA structural conformations than the T4 bacteriophage enzyme.[46] The T7 DNA polymerase derivative Sequenase 2.0 is two to five times as efficient at generating mutations as T4 DNA polymerase.[46] The Klenow fragment of *E. coli* DNA polymerase I, which lacks a 5′-3′ exonuclease activity, may be better than T7 DNA polymerase or Sequenase 2.0 for particularly difficult templates. The helicase activity of the Klenow fragment can disrupt recalcitrant DNA conformations.[47] Another alternative for dealing with DNA secondary structures in templates is to add single-stranded DNA binding protein,[48] DMSO,[49] or one of the weak base-pairing deoxyguanosine or deoxycytosine analogues used to resolve compressions in DNA sequencing.[50]

6.5.2 MUTAGENESIS OF SINGLE-STRANDED DNA

In this method, which was originally called oligonucleotide heteroduplex mutagenesis,[51] one or more mutagenic oligonucleotides are annealed to a single-stranded circular DNA. Subsequently, the oligonucleotide is incorporated enzymatically into a double-stranded molecule using the single-stranded circle as a template. In principle, there is no DNA repair *in vivo* or asymmetry of DNA synthesis. Equivalent amounts of mutant and wild-type progeny should be produced. Although yields of 50% have been obtained,[51] more often less than 1% of the progeny will contain the mutation. Strategies for enrichment of mutants will be described later in the section on mutant selection methods.

6.5.3 MUTAGENESIS OF DOUBLE-STRANDED DNA

Most recombinant clones are propagated in host cells harboring double-stranded plasmids. A number of approaches for direct oligonucleotide mutagenesis of double-stranded DNA are available. A common feature of almost all of these methods is the creation of a single-stranded target region (gapped duplex) in the plasmid to facilitate oligonucleotide binding. After annealing, the oligonucleotide is incorporated into the duplex DNA by use of DNA polymerase and ligase. These methods are generally not as efficient as methods that use phage or phagemid vectors; however, advances in techniques for the selection of clones containing mutated plasmids have made many of these mutagenesis methods highly efficient (described later).

6.5.4 SOLID-PHASE *IN VITRO* MUTAGENESIS

Unlike other non-PCR methods, this technique can be used to efficiently mutagenize double-stranded templates in the absence of biological selection methods. A biotin-streptavidin magnetic-bead system is employed, double-stranded plasmid DNA is immobilized to the solid support, and one of the strands is selectively eluted. Thus, it becomes possible to specifically produce single-stranded products that can be used as templates to synthesize site-specific mutants. Vector and mutated inserts can be synthesized independently and then subsequently mixed and used to transform *E. coli*. By imparting a single-stranded form to the mutagenesis of a double-stranded template, the limitations on reaction efficiency imposed by the use of mismatch repair can be circumvented. Importantly, no special vectors or host strains are required for this system. Solid-phase mutagenesis can be performed by either PCR-based or non-PCR methods for *in vitro* DNA synthesis. Mutagenesis rates of more than 80% have been attained.[52]

6.5.5 LINKER SCANNING MUTAGENESIS

Linker scanning mutagenesis, or scanning mutagenesis, is a set of techniques used to map critical sequences in a defined region of a gene or genome. Systematic analysis of cloned genes and their regulatory sequences by *in vitro* mutagenesis typically progresses from crude analysis using exonuclease III to generate a series of deletion mutants up to producing mutants containing only single-base substitutions. Simple deletion methods are susceptible to the removal of critical functional regions of a gene, the loss of which could obscure the contribution of individual sequence elements.

Conversely, the effects of single-base substitution mutagenesis can be too subtle in some circumstances to yield phenotypic differences from the wild type. Linker scanning offers a powerful intermediate approach by which to survey a gene or its promoter with a set of "clustered point mutations."[53]

Linker scanning methods generate a series of mutations by replacing short sections of a gene with short common sequences of the same length, the linker. As a result, regions of the gene or transcription/replication control regions outside the linker are left intact. Linker-scan mutations generally do not alter the spatial relationships between sequence elements in the region of interest. For regulatory sequences, the exchange of wild-type sequences with a heterologous linker maintains the helical configuration of the DNA, which permits functional assessment of the contribution of specific DNA sequences in the context of native DNA topology. The mutagenic linker for coding sequences is typically 9 to 12 bp so that the reading frame is maintained. This creates a cluster of 3 to 4 point mutations, whereas the sequences on both sides of the clustered mutations remain unchanged. One example of this is alanine-scanning mutagenesis. Because alanine is a common amino acid without strong electrostatic or steric properties, systematic replacement of amino acids with alanine is an effective method by which to mildly perturb a peptide.[54] Another protein-linker-scanning mutagenesis strategy is to incorporate a protease cleavage site into the linker code so mutants can be easily identified by proteinase digestion.[55]

The original linker scanning protocol allowed its inventors, McKnight and Kingsbury, to map transcriptional control signals of the herpes simplex virus (HSV) thymidine kinase (*tk*) gene.[53] This method first used exonucleases to generate two sets of nested 5′ and 3′ deletion mutations with linkers at the ends. Each set of deletion mutants was then sequenced so that each 5′ deletion mutant could be matched with its complementary 3′ mutant. For each complementary pair, the 5′ and 3′ ends defined a 10-bp gap relative to the wild-type sequence. These matched pairs of mutants were then combined, with the junction being the 10-nt linker that replaced the 10 nucleotides that formerly separated the two deletion termini. In the 140 bp comprising the HSV *tk* transcriptional control region, McKnight and Kingsbury generated 18 linker-scan mutants.

One of the problems with this approach is that it is difficult to get perfectly matching deletions. A variation on the original method involves the construction of only one set of deletions. Oligonucleotides are synthesized to fill the gap between a nearby restriction site and the linker at the deletion endpoint.[56] In this method, the length of oligonucleotides that can be effectively synthesized determines the amount of sequence that can be scanned. Very little oligonucleotide is required for this protocol, but because yields of oligonucleotides greater than 100 bases in length are very low, the expense of synthesizing these oligos is quite high. Conversely, if an oligonucleotide is too short (~6 bases) it will not hybridize stably to its complementary oligonucleotide. As long as one strand can "bridge" the two restriction endonuclease sites on either side of the gap to be filled, the resulting single-stranded gap can be filled in after the ligation reaction with the Klenow fragment of *E. coli* DNA polymerase I. Most of the problems that occur in the making of linker scans with either protocol can be traced to incorrect oligonucleotides. It is best to have the oligonucleotides gel-purified and as clean as possible for optimal ligation. It is critical to sequence both the deletion endpoints and the final constructs in order to obtain accurate data.

Modifications to the original linker scanning method have addressed some of the difficulties. One method uses a linker-scanning mutagenesis cassette consisting of a selectable marker flanked by two oligonucleotides, each of which contains a recognition site for a different restriction endonuclease. The cleavage site for one endonuclease is within the cassette, while the other is in the target DNA beyond the end of the cassette. The linker-scanning cassette is inserted in a double-stranded break in the plasmid that is created by mild DNase I treatment in the presence of Mg^{2+} followed by S1 nuclease digestion. After antibiotic selection for clones containing the linker scanning cassette, the selectable marker is removed by digestion with the two endonucleases and the plasmid is re-ligated. In a recent report, saturation mutagenesis of a 400-bp region of the *E. coli* *rpoB* gene was performed by this method; as a result, >60% of the mutants were direct linker replacements with no additional deleted nucleotides.[57] Other approaches to linker scanning mutagenesis employ PCR-based methods such as overlap-extension or megaprimer PCR to create scanning mutants,[58,59] or they employ transposases to insert tranposons containing antibiotic resistance markers flanked by unusual restriction sites that permit deletion of all but 15 bp of the transposon.[60]

6.6 RANDOM OR EXTENSIVE MUTAGENESIS

Because the technique of site-directed mutagenesis is relatively laborious, alternate methods are preferable when the goal is to introduce multiple mutations at a specific site or simultaneously at different positions in a gene. These approaches include error-prone PCR, the incorporation of degenerate base analogues, the use of primers that contain multiple mismatched bases at a single position, and *in vitro* recombination via DNA shuffling.

6.6.1 Error-Prone PCR

Error-prone PCR involves the use of low-fidelity polymerization conditions that cause the introduction of a small number of random-point mutations in a long DNA sequence. It is well accepted that the likelihood of introducing random mutations during PCR mutagenesis is relatively high when *Taq* DNA polymerase is used. Depending on the conditions selected, the frequency can range from as high as 35%[61] to less than 1/100,000.[62] Many factors can be altered to increase the error rate of PCR for mutagenic purposes (summarized in Reference 25). These variables include high-pH buffers containing high concentrations of magnesium or 0.5 mM $MnCl_2$, excess polymerase, limited amounts of template, and a biased pool of the four dNTPs (i.e., widely differing concentrations). PCR conditions are also critical; low annealing temperatures, a large number of PCR cycles, and long extension times enhance the error rate. Typically, a random error rate of 1 per 150 bp can be achieved. Important limitations of this approach are that transitions are generally favored over transversions, and different regions of the DNA may exhibit inherently different mutation rates.

Mutagenesis via error-prone PCR has been successfully employed in many applications. For example, Vartanian and colleagues[63] used hypermutagenic PCR with Mg^{2+} and biased dNTP concentrations to create a library of mutant dihydrofolate reductase genes. These authors estimated an overall mutation frequency of 10% per amplification. The role of manganese ions in the enhancement of the proportion of transversions in this mutant library has also been explored.[64] Other recent applications of PCR random mutagenesis include the production of mutants to increase the peroxidase activity of horse heart myoglobin[65] and the structure–function analysis of the plant enzyme 1-aminocyclopropane-1-carboxylate synthetase.[66]

6.6.2 Degenerate Oligonucleotide Primers and Base Analogues

It is possible to synthesize a set of mutagenic primers that will introduce any of the three mismatches at a specified site in a single PCR run by supplying three mismatched bases during oligonucleotide synthesis. Factors that influence the synthesis of degenerate oligonucleotides for site-saturation mutagenesis were described in a recent report that analyzed the effects of different base compositions, the influence of the positions of mismatches, and the significance of non-Watson–Crick base pairs.[67]

An alternative method of introducing degeneracy utilizes deoxyinosine (dI), an ambiguous base that may pair with all four natural nucleotides during transcription. Thus, a dI-containing oligonucleotide primer is capable of introducing all three mismatched bases (plus the wild-type base) at a specified site in the template DNA. Similarly, dI may be incorporated into PCR products by replacing most of one nucleotide in the mix with dI (reviewed in Reference 68). This dI-containing template then acts as a template for subsequent rounds of PCR. An example of the successful

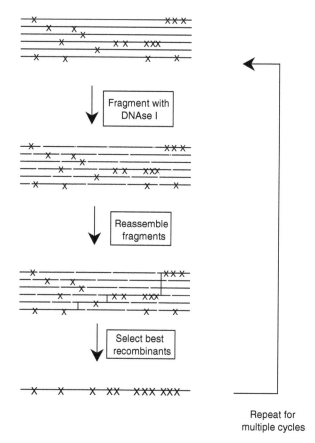

FIGURE 6.4 DNA shuffling method of mutagenesis. Lines represent homologous genes carrying different point mutations (X). Breaks in lines designate points of fragmentation caused by DNAse I treatment. A recombinant gene results from three crossover events (vertical lines). The selected pool of recombinants serves as starting material for subsequent rounds of PCR.

application to protein biochemistry of random mutagenesis using dITP is the 1998 study of transmembrane and loop domains within the P-glycoprotein of yeast.[69] In addition to dITP, other base analogues used for this purpose include 5'-triphosphates of 8-oxo-2' deoxyguanosine and 6-(2-deoxy-β-D-ribofuranosyl)-3,4-dihydro-8H-pyrimido-[4,5-C][1,2]oxazin-7-one.[70] It has been reported that the use of degenerate base analogues in PCR may be preferable to error-prone PCR methods because they enable a very high frequency of substitution (up to 1.9×10^{-1}), are less sequence dependent, and are not as biased toward transition mutations.[70]

6.6.3 DNA SHUFFLING

A technically simple approach to recombination is "DNA shuffling," a method first described by Stemmer in 1994.[70,71] In this approach (see Figure 6.4) a pool of homologous genes with different point mutations is digested with DNase I to create a pool of random DNA fragments. Because they are homologous, the fragments act as primers for each other. Oligonucleotides with ends that are homologous to the gene can also be added to the fragment mix. These fragments are reassembled by repeated cycles of thermocycling. Recombination occurs when fragments from one copy of a gene prime on another copy, causing template switch. Reassembled products are amplified by conventional PCR. A recombinant gene manifesting improved characteristics can be selected and used as the starting material for additional rounds of mutation and recombination.

In addition to the recombination events, the technique of DNA shuffling as originally described also introduces new point mutations at a relatively high rate, calculated to be about 0.7%.[72] Although in some circumstances these point mutations may provide useful diversity, in others they may be undesirable. To overcome this shortcoming, modified protocols that optimize the fidelity of shuffling have been introduced.[73] These modifications include the use of Mn^{2+} instead of Mg^{2+} during the DNase I fragmentation step, the substitution of a higher-fidelity enzyme (e.g., *Pfu* or *Pwo*) for the error-prone *Taq* polymerase, and reduction of the number of cycles during reassembly. This method has been used successfully to distinguish functional from nonfunctional or deleterious mutations in homologous genes.[74] An example of the way in which functional proteins may be improved by DNA shuffling was described recently.[75]

6.7 METHODS OF MUTANT SELECTION

When non-PCR methods of site-directed mutagenesis are used, it is necessary to reduce the background of wild-type template that will be cotransformed along with the mutant DNA. Several selection methods have been developed that result in preferential introduction or replication of the mutated plasmid.

6.7.1 RESISTANCE TO RESTRICTION ENZYME DIGESTION

Because the transformation of linear DNA into host bacteria is significantly less efficient than that of circular DNA, the loss of a unique enzyme restriction site can be used to enrich the mutant DNA. This technique is generally known as the USE (Unique Restriction Site Elimination) method and was first described by Deng and Nickoloff.[76] The USE strategy employs oligonucleotide primers to introduce two mutations simultaneously: the mutagenic primer serves to construct the desired mutation, and a selection primer results in the destruction of a unique restriction site. This approach is applicable to any double-stranded plasmid that contains a unique restriction site and an antibiotic resistance gene.

In the USE method, the mutagenic and selection primers are annealed to one strand of the denatured parental plasmid. To increase the likelihood that both primers anneal to a single DNA template, thereby ensuring that a plasmid containing a restriction site mismatch will also carry the desired mutation, both primers should be present at a 200-fold molar excess relative to the single-stranded template. The second DNA strand is synthesized with T4 DNA polymerase, and gaps are filled in with T4 DNA ligase. The subsequent step is the first round of digestion with the selection restriction endonuclease enzyme to linearize parental plasmids, followed by transformation into a *mutS*⁻ mutant, a strain of *E. coli* that is deficient in the correction of mismatches. This bacterial strain randomly corrects one strand to be complementary to the other, thereby reverting some hybrid molecules back to fully wild type. The purified DNA, now enriched for mutant plasmids, is digested a second time with the selection enzyme and used to transform a *mutS*⁺ *E. coli* strain.

The efficiency of the USE method is dependent on a number of variables, including the size of the plasmid (molecules smaller than 6 kb are reportedly optimal),[25] the efficiency of the enzyme digestion prior to each transformation (the DNA should be purified by buoyant density gradient banding, and digestion should be monitored by gel electrophoresis), and the length and GC content of the two primers (the annealing strength of the selection primer should not exceed that of the mutagenic primer). For a detailed description of the optimal conditions and troubleshooting suggestions for the USE method, see References 77 and 78.

The USE method can also be modified to generate nested deletions within plasmids.[77] In this approach, the unique restriction site must be adjacent to the gene targeted for deletion. A single oligonucleotide deletion primer serves two purposes simultaneously: it generates nested deletions and eliminates a unique restriction site. Primer design is critical to the success of this technique.

The 5′ end of the primer is complementary to the binding site on the template, whereas the 3′ end is a short tail with an arbitrary sequence of four or five nucleotides that does not anneal to the sequence adjacent to the binding site. As a result, intervening DNA sequences of varying lengths are looped out during synthesis. Because the unique restriction endonuclease site lies within the deleted segment, enzyme digestion prior to the two transformation steps as described above eliminates parental DNA.

In addition to eliminating endonuclease restriction sites, mutated DNA can also be selected by rendering it resistant to digestion via the incorporation of dNTP analogues.[45,79] During the DNA synthesis step, one or more of the dNTPs are replaced by their 5′-[a-thiol]triphosphate or methylated analogues. Parental plasmids are readily digested and removed by treatment with T7 exonuclease or exonuclease III. The resistant mutant strand then serves as a template to synthesize the second mutant strand.

6.7.2 RESTORATION OF AN ANTIBIOTIC RESISTANCE GENE

In this approach[80] a DNA plasmid containing a defective antibiotic resistance gene serves as the template for the mutagenesis. For example, the parental resistance gene may carry a small deletion. A selection primer that corrects this defect and the mutagenic primer are annealed to the template simultaneously (see Figure 6.5). To ensure that both primers anneal, they are used in large molar excess relative to the template, and annealing conditions that favor binding of the mutagenic primer are chosen. A mutagenesis efficiency of about 85% has been reported.[81]

In an analogous system, selection of the mutated DNA may result from the restoration of a defective replication origin.[82] The template carrying the defective origin is used to synthesize a hybrid plasmid in which two mutations are introduced, one to correct the defect and one to introduce the desired mutation. The isolated DNA is then used to transform bacterial cells that replicate only the restored plasmids. As before, conditions that favor binding of the mutagenic primer to the template should be used.

6.7.3 RESISTANCE TO URACIL N-GLYCOSYLASE DIGESTION

An alternative approach to selectively enrich for mutated DNA involves the incorporation of deoxyuracil into the template DNA. There have been several variations of this popular technique, often referred to as the Kunkel method.[83] In general, DNA containing a small proportion of dUMP in place of the dTMP is made by selecting for the parental plasmid in *dut⁻ ung⁻* strains of *E. coli*. The *dut⁻* cells lack the enzyme dUTPase, allowing some uracil to become incorporated into the DNA; they also lack the correcting enzyme uracil N-glycosylase (UDG; *ung⁻*), which would remove the uracil from the DNA. The mutagenic primer is annealed and the mutant strand is synthesized. Thus the nascent strand does not contain deoxyuracil. The hybrid DNA is then used to transform a *dut⁺ ung⁺* strain to eliminate the parental strand, forming double-stranded mutant plasmids.

A variation on this method is to digest the hybrid plasmids carrying deoxyuracil with UDG. The remaining (mutant) strand is then used as a template for second-strand synthesis.

6.7.4 INCORPORATION OF PHOSPHOROTHIOATE DIGESTION

The phosphorothioate technique of site-directed mutagenesis was introduced by Eckstein in 1985 and modified a few years later.[84,85] This approach uses the incorporation of nucleotide analogues into the mutant strand as a selection mechanism. Briefly, the target template is produced as a single strand using either M13 or phagemid cloning vectors (see previous sections). An oligonucleotide carrying the desired mutation is annealed, extended with DNA polymerase in the presence of dCTPαS, and ligated with T4 DNA ligase to form a closed circular heteroduplex. The dCTPαS base analogue contains a sulfur atom in place of oxygen at the alpha phosphate. After digestion of

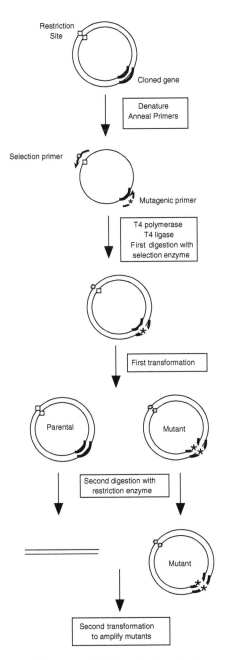

FIGURE 6.5 Unique restriction site elimination (USE) method of site-directed mutagenesis. An asterisk (★) indicates the desired mutation; the parental restriction endonuclease site is depicted by a square (□); a circle (○) depicts the selection restriction endonuclease site. Experimental details are described in the text.

the remaining single-stranded template with T5 exonuclease, the nonmutant strand is nicked by the restriction endonuclease *Nci* I. This strand is then selectively degraded by the use of exonuclease III under conditions that leave a fragment that can serve as a primer for repolymerization with DNA polymerase I and ligase.[86] The resulting product is used to transform bacterial cells. Mutation efficiencies of greater than 85% are reportedly typical.[86]

6.7.5 AMBER STOP CODON SUPPRESSION

The technique of amber (TAG) suppression offers a means by which to construct many amino acid substitutions at a specific site in a gene. *E. coli* amber suppressors have been mapped to tRNAs. These suppressors function to overcome the premature truncation of a protein by permitting the insertion of a missense amino acid at the site of an amber stop codon. The anticodons of these tRNAs recognize a sequence that is only a single base away from that of a stop codon. Some of these tRNAs have been altered by site-directed mutagenesis to allow recognition of the amber stop codon.[87,88] Overall, the collection of natural and designed amber suppressors cause the insertion of a different amino acid substitution at the site of the amber stop mutation. The mutagenic technique is briefly described as follows: A single amber stop mutation (TAG) is introduced into a plasmid by a standard method of site-directed mutagenesis (outlined elsewhere in this chapter). This DNA is then used to transform different suppressor strains of *E. coli,* and those strains are induced to express mutant genes. Selection on bacterial plates containing different antibiotics permits the use of both chromosomal- and plasmid-based suppressors. A detailed analysis of the technical hurdles involved in this approach has been published.[89]

6.8 SITE-DIRECTED MUTAGENESIS KITS

There are a number of diverse technical approaches for the creation of site-directed mutations that can be gleaned from the literature. An increasingly popular route for the generation of such mutations is through the use of commercial kits. Currently there are at least 13 different kits for site-directed mutagenesis being sold by ten different companies. Features of those kits are described in Table 6.1. Most of the kits use mutagenic oligonucleotides, which are usually user supplied. The different methods that serve as the basis of the various kits are reviewed elsewhere.[58] However, the predominant themes are either negative selection against the parental, nonmutated plasmid or positive selection for the mutated plasmid, both of which significantly raise the efficiency of mutant isolation. Most of the oligonucleotide-directed mutagenesis kits allow for all four types of mutations, namely, transitions, transversions, deletions, or insertions. Additionally, most of the kits do not require subcloning, although a second round of transformation is advised in some cases to improve efficiency.

An increasing number of companies list site-directed mutagenesis among the services they provide. These include Ana-Gen Technologies, Inc. (Richmond, VA), Genemed Synthesis, Inc., (Atlanta, GA), and Commonwealth Biotechnologies, Inc. (South San Francisco, CA).

6.9 NOVEL APPROACHES TO SITE-DIRECTED MUTAGENESIS

6.9.1 USE OF RIBOZYMES TO MODIFY RNA TERMINI

Ribozymes are catalytic RNAs with sequence-specific ribonuclease or ligation activities that have been studied extensively since their discovery in 1981 (see Reference 97; also reviewed in Reference 98). Although ribozymes have been found to exist among viruses, eukaryotes, and prokaryotes, those generally in use are designed to contain the active center structures of plant viroids or viruses. Two features characterize RNA-catalyzed reactions: magnesium ions are required at the reaction center, and target molecules are recognized via Watson–Crick base pairing in RNA regions flanking the catalytic domain. The most commonly used ribozymes belong to one of two groups: hammer-head ribozymes, as observed in the plus strand of the tobacco ringspot virus satellite RNA and in avocado sunblotch viroid RNA; and hairpin ribozymes, as derived from the negative strand of satellite tobacco ringspot virus. Both types of ribozymes can be engineered to function in *cis* or in *trans.* The secondary structure of the hammerhead ribozyme-substrate complex consists of three helical regions, a catalytic core region, and an internal loop sequence.[99] Cleavage occurs within a

35-nt catalytic core structure, with GUC'N the preferred cleavage-site sequence. The structural basis of the hammerhead self-cleavage mechanism was recently defined by X-ray crystallography.[100] The hairpin ribozyme-substrate complex, which contains a 50-nt catalytic core structure, consists of four helical regions separated by two internal-loop sequences.[101] In this case, the preferred cleavage site is 5′ of the GUC, namely, at N′GUC.

Although the primary applications of ribozymes have involved intracellular antiviral and gene therapy,[102] these molecules also provide a unique opportunity for modifying the termini of RNA molecules. For example, some tRNAs have proven difficult to study due to the sequence of the promoter recognized by the T7 polymerase. Specifically, those sequences starting with non-G residues are poor substrates for the enzyme during *in vitro* transcription. In an effort to overcome this limitation imposed by unfavorable 5′ termini, a hammerhead ribozyme gene sequence was introduced between the strong T7 polymerase promoter and the tRNA sequence.[103] Transcription of this construct produced a molecule with autocatalytic activity, liberating large amounts of a 5′-OH tRNA transcript starting with the proper nucleotide.

6.9.2 THERAPEUTIC MUTAGENESIS: STRATEGIES FOR SITE-DIRECTED MUTAGENESIS OF CHROMOSOMAL DNA

Directed mutagenesis to revert disease-causing mutations is one way to achieve long-term genome correction. Targeted homologous recombination between a defective gene and an exogenous DNA vector is generally viewed as the most promising approach for gene therapy. As a method of performing site-directed mutagenesis on whole animals, homologous recombination is the standard approach for generation of knockout mice from embryonic stem cells; however, it is too inefficient to use on differentiated mammalian cells or for gene therapy.[104,105] While homologous recombination has been a productive mutagenic method in applications that can effectively screen for successful mutagenesis, in the absence of *in vivo* selection it is always going to be a losing kinetic battle. Watson–Crick hybridization of the exogenous DNA to native duplex DNA is inherently unfavorable.

In this final section of our review, we describe two approaches to site-directed mutagenesis that are being developed as possible methods for therapeutic mutagenesis. These techniques are unique because they depart from the mutagenic paradigm of hybridizing a mutagenic DNA oligo-nucleotide to a target DNA and then utilizing either *in vitro* synthetic methods or cellular replication and repair pathways to generate the mutation. Furthermore, all the mutagenic methods described so far require some *in vitro* steps by which the gene or genome being mutagenized is manipulated as purified DNA in a test tube. The following two approaches involve the site-specific mutation of genomic DNA entirely within mammalian cells.

6.10 TRIPLE-HELIX-FORMATION-MEDIATED SITE-DIRECTED MUTAGENESIS

The capacity of oligonucleotides to bind duplex DNA and form triple helices in a sequence-specific manner has been exploited to create site-directed mutants in mammalian cells. The oligonucleotide binds in the major groove in the DNA duplex. Transcription of genes bound by these oligonucleotides is inhibited. Ordinary oligonucleotides bind only transiently, and the bonds can be disrupted by transcription at nearby sites. This problem was overcome by the use of intercalating or cross-linking agents to prolong the oligonucleotide-duplex attachment. Although originally used to block transcription, triple helix formation has also been used to mutate specific sites in selected genes that resulted in permanent, heritable changes in gene function and expression.[106–109]

Triple helix formation by oligonucleotides with duplex DNA is sequence specific. Current understanding of the phenomenon indicates the oligonucleotides can only bind to polypurine-polypyrimidine sequences. This restricted target sequence represents a significant limitation of the technique. Until either the third strand binding code is better elucidated or new nucleoside analogues are developed, targeted third-strand binding will be limited to polypurine-polypyrimidine regions.[106,108,109]

Triple-helix-mediated site-directed mutagenesis uses the third strand to juxtapose a mutagen, such as psoralen, at the mutation-targeted sequence in the DNA. In the case of psoralen, after triplex-DNA formation and site-specific intercalation of the psoralen, the psoralen is photoactivated by irradiation with long-wave ultraviolet light. This results in a covalent linkage of the psoralen to the target DNA. Subsequent cellular DNA replication and repair converts the premutagenic lesion into a permanent change. Most of the mutations are T:A to A:T transversions; however, other base pair changes and small deletions at the psoralen intercalation site have also been observed. Glazer and colleagues demonstrated targeted mutagenesis in SV40 virus-infected COS cells at a frequency of 1 to 2%[106] and more importantly have used triplex-forming oligonucleotides to induce mutations at specific genomic sites in somatic cells of adult mice.[110]

6.11 CHIMERAPLASTY

Chimeraplasty is a method of site-directed mutagenesis in which chimeric RNA/DNA oligonucleotides are used to alter genes *in vivo*. The approach is based on the observation that RNA:DNA duplexes are more stable than DNA:DNA duplexes and that single base pair mismatches in those duplexes are actively corrected by cellular DNA repair systems.[111] With regard to gene-targeting events, there is evidence to suggest that pairing is the rate-limiting step.[112] When the equilibrium constant for the formation of hybrids between the mutating oligonucleotide and the DNA target through the use of RNA:DNA duplexes is increased, there is more time for the cellular repair enzymes to make the correcting mutation.[113]

The mutagenic molecule in this method, the chimeraplast, is typically a 68-nucleotide molecule containing 48 deoxyribonucleotides and two 10-base 2′-*O*-methyl ribonucleotides. It is designed to form a 30-bp double-stranded region flanked by four thymidines that form hairpin structures at both ends. The 2′-*O*-methylation of the RNA imparts ribonuclease H, and the duplex and hairpin structures provide physical and thermal stability. The two ribonucleotides are on each side of a five-base deoxyribonucleotide that contains the mutant base (see Figure 6.6). This structure is the key to both efficient hybridization to the target sequence and for eliciting endogenous mismatch repair.[113,114]

Chimeraplast-mediated mutagenesis has worked reproducibly for a diverse set of gene targets in a broad range of cell types. In experiments with cultured cells, these molecules have been used by the initial developers of the technology to introduce single base pair mutations of episomal DNA in Chinese hamster ovary (CHO) cells and genomic DNA in lymphoblasts, human hepatoma cells, and rat hepatocytes.[114–116] In what is the most important demonstration of the therapeutic potential of this technology by Kren and colleagues, a chimeraplast designed to mutate the rat factor IX gene to a prolonged coagulation phenotype was injected into rat-tail veins. As evidenced by molecular analyses and by reduced factor IX coagulant activity, the mutagenesis was site specific and dose dependent. Under optimal conditions 40% of the rat hepatocytes were mutated and there was a 50% reduction in coagulant activity.[117] That mutagenic rate would be adequate for effective therapy on a large number of genetic diseases.

The chimeraplast is only one of two advances inherent in this *in vivo* mutagenic approach. Delivery of the mutagenic oligonucleotide is accomplished by complexing it with a polycation called polyethylenimine that is further modified by the addition of a lactose ligand. The lactose targets the oligo-polycation complex to a specific receptor, the asialoglycoprotein receptor of hepatocytes. This leads to a very high rate of delivery of the therapeutic agent to the liver. The Kren study showed that injection of a fluorescently labeled oligo-polycation complex into a liver resulted in the labeling of a majority of the cells.[117]

Although this technology is promising, it is still in its infancy, and there are many critical issues to be resolved. For instance, the mechanism of the repair has not been fully demonstrated,[113] and

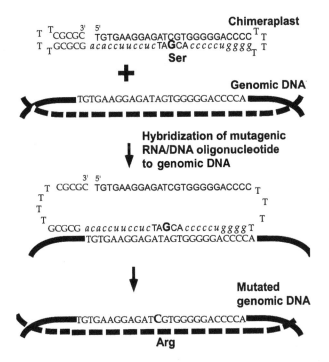

FIGURE 6.6 Site-specific mutagenesis of the rat factor IX gene using an RNA/DNA oligonucleotide called a chimeraplast. The chimeric molecule introduced a single base pair change converting a serine to an arginine in the rat gene. The uppercase letters indicate DNA. The lowercase italic letters indicate RNA. The mechanism probably involves hybridization of the RNA portion of the chimeraplast to the target sequence followed by nucleotide exchange via mismatch repair.

the optimal structure of the chimeraplast has yet to be elucidated. In 2001 two research groups other than Eric Kmiec's and Clifford Steer's demonstrated the utility of chimeraplasty in mutating eukaryotic genes of interest in CHO cells;[118,119] however, as evidenced by the difficulty other labs have reported in using chimeraplasty, the technique is not as simple as it would appear in the published literature.[120] Nevertheless, Kren et al. achieved a rate of site-directed mutagenesis without selection in a rat that is comparable to the high levels of mutagenesis observed for bacteria and cultured cells using selection.

REFERENCES

1. Hutchison, C. A., III, Phillips, S., Edgell, M. H., Gillam, S., Jahnke, P., and Smith, M., Mutagenesis at a specific position in a DNA Sequence, *J. Biol. Chem.,* 253, 6551, 1978.
2. Flavell, R. A., Sabo, D. L., Bandle, E. F., and Weissmann, C., Site-directed mutagenesis: generation of an extracistronic mutation in bacteriophage Q beta RNA, *J. Mol. Biol.,* 89, 255, 1974.
3. Sabo, D. L., Domingo, E., Bandle, E. F., Flavell, R. A., and Weissmann, C., A guanosine to adenosine transition in the 3′ terminal extracistronic region of bacteriophage Q beta RNA leading to loss of infectivity, *J. Mol. Biol.,* 112, 235, 1977.
4. Borrias, W. E., Wilschut, I. J., Vereijken, J. M., Weisbeek, P. J., and Arkel, G. A. V., Induction and isolation of mutants in a specific region of gene A of bacteriophage phi chi 174, *Virology,* 70, 195, 1976.
5. Gilliam, S., Jahnke, P., and Smith, M., Enzymatic synthesis of oligo-deoxyribonucleotides of defined sequence, *J. Biol. Chem.,* 233, 2532, 1978.
6. Sanger, F., Air, G. M., Barrell, B. G., Brown, N. L., Coulson, A. R., Fiddes, C. A., III, C. A. H., Slocombe, P. M., and Smith, M., Nucleotide sequence of bacteriophage phi X174 DNA, *Nature,* 265, 687, 1977.

7. Gillam, S. and Smith, M., Site-specific mutagenesis using synthetic oligodeoxyribonucleotide primers: I. Optimum conditions and minimum oligodeoxyribonucleotide length, *Gene,* 8, 81, 1979.

8. Gillam, S. and Smith, M., Site-specific mutagenesis using synthetic oligodeoxyribonucleotide primers: II. *In vitro* selection of mutant DNA, *Gene,* 8, 99, 1979.

9. Gillam, S., Astell, C. R., and Smith, M., Site-specific mutagenesis using oligodeoxyribonucleotides: isolation of a phenotypically silent phi X174 mutant, with a specific nucleotide deletion, at very high efficiency, *Gene,* 12, 129, 1980.

10. Gillam, S., Jahnke, P., Astell, C., Phillips, S., Hutchison, C. A., III, and Smith, M., Defined transversion mutations at a specific position in DNA using synthetic oligodeoxyribonucleotides as mutagens, *Nucleic Acids Res.,* 6, 2973, 1979.

11. Adams, S. P., Kavka, K. S., Wykes, E. J., Holder, S. B., and Galluppi., G. R., Hindered dialkylamino nucleoside phosphite reagents in the synthesis of two DNA 51-mers, *J. Am. Chem. Soc.,* 105, 661, 1983.

12. Zoller, M. J., New molecular biology methods for protein engineering, *Curr. Opin. Biotechnol.,* 2, 526, 1991.

13. Nassal, M. and Rieger, A., PCR-based site-directed mutagenesis using primers with mismatched 3′-ends, *Nucleic Acids Res.,* 18, 3077, 1990.

14. Piechocki, M. P. and Hines, R. N., Oligonucleotide design and optimized protocol for site-directed mutagenesis, *Biotechniques,* 16, 702, 1994.

15. Sambrook, J., Fritsch, E. F., and Maniatis, T., *Molecular Cloning: A Laboratory Manual,* Cold Spring Harbor Laboratory, Cold Spring Harbor, NY, 1989.

16. Turchin, A. and Lawler, J. F. J., The primer generator: a program that facilitates the selection of oligonucleotides for site-directed mutagenesis, *Biotechniques,* 26, 672, 1999.

17. Liang, Q., Chen, L., and Fulco, A. J., An efficient and optimized PCR method with high fidelity for site-directed mutagenesis, *Genome Res.,* 4, 269, 1995.

18. Kok, R. G., D'Argenio, D. A., and Ornston, L. N., Combining localized PCR mutagenesis and natural transformation in direct genetic analysis of a transcriptional regulator gene, *pobR, J. Bacteriol.,* 179, 4270, 1997.

19. Cline, J., Braman, J. C., and Hogrefe, H. H., PCR fidelity of *pfu* DNA polymerase and other thermostable DNA polymerases, *Nucleic Acids Res.,* 24, 3546, 1996.

20. Bracho, M. A., Moya, A., and Barrio, E., Contribution of Taq polymerase-induced errors to the estimation of RNA virus diversity, *J. Gen. Virol.,* 79, 1998.

21. Barnes, W. M., PCR amplification of up to 35-kb DNA with high fidelity and high yield from lambda bacteriophage templates, *Proc. Natl. Acad. Sci. U.S.A.,* 91, 2216, 1994.

22. Casas, E. and Kirkpatrick, B. W., Evaluation of different amplification protocols for use in primer-extension preamplification, *Biotechniques,* 20, 219, 1996.

23. Holterman, L., Mullins, J. I., Haaijman, J. J., and Heeney, J. L., Direct amplification and cloning of up to 5-kb lentivirus genomes from serum, *Biotechniques,* 21, 312, 1996.

24. Thongnoppakhun, W., Wilairat, P., Vareesangthip, K., and Yenchitsomanus, P. T., Long RT-PCR amplification of the entire coding sequence of the polycystic kidney disease 1 (PKD1) gene, *Biotechniques,* 26, 126, 1999.

25. Ling, M. M. and Robinson, B. H., Approaches to DNA mutagenesis: an overview, *Anal. Biochem.,* 254, 157, 1997.

26. Liu, Q. and Sommer, S. S., Subcycling-PCR for multiplex long-distance amplification of regions with high and low GC content: application to the inversion hotspot in the factor VIII gene, *Biotechniques,* 25, 1022, 1998.

27. Su, X. Z., Wu, Y., Sifri, C. D., and Wellems, T. E., Reduced extension temperatures required for PCR amplification of extremely A+T-rich DNA, *Nucleic Acids Res.,* 24, 1574, 1996.

28. Ho, S. N., Hunt, H. D., Horton, R. M., Pullen, J. K., and Pease, L. R., Site-directed mutagenesis by overlap extension using the polymerase chain reaction, *Gene,* 77, 51, 1989.

29. Warrens, A. N., Jones, M. D., and Lechler, R. I., Splicing by overlap extension by PCR using asymmetric amplification: an improved technique for the generation of hybrid proteins of immunological interest, *Gene,* 186, 29, 1997.

30. Pogulis, R. J., Vallejo, A. N., and Pease, L. R., *In vitro* recombination and mutagenesis by overlap extension PCR, in *In Vitro Mutagenesis Protocols,* Trower, M. K., Ed., Humana Press, Totowa, NJ, 1996, 167.

31. Ge, L. and Rudolph, P., Simultaneous introduction of multiple mutations using overlap extension PCR, *Biotechniques,* 22, 28, 1997.

32. Mikaelian, I. and Sergeant, A., Modification of the overlap extension method for extensive mutagenesis on the same template, in *In Vitro Mutagenesis Protocols*, Trower, M. K., Ed., Humana Press, Totowa, NJ, 1996, 193.

33. Chouljenko, V., Jayachandra, S., Rybachuk, G., and Kousoulas, K. G., Efficient long-PCR site-specific mutagenesis of a high GC template, *Biotechniques,* 21, 472, 1996.

34. Hall, L. and Emery, D. C., A rapid and efficient method for site-directed mutagenesis by PCR, using biotinylated universal primers and streptavidin-coated magnetic beads, *Protein Eng.,* 4, 601, 1991.

35. Kammann, M., Laufs, J., Schell, J., and Gronenborn, B., Rapid insertional mutagenesis of DNA by polymerase chain reaction (PCR), *Nucleic Acids Res.,* 17, 5404, 1989.

36. Ke, S. H. and Madison, E. L., Rapid and efficient site-directed mutagenesis by single-tube "megaprimer" PCR method, *Nucleic Acids Res.,* 25, 3371, 1997.

37. Barik, S., Site-directed mutagenesis *in vitro* by megaprimer PCR, in *In Vitro Mutagenesis Protocols*, Trower, M. K., Ed., Humana Press, Totowa, NJ, 1996, 203.

38. Ling, M. and Robinson, B. H., Rapid construction of three-fragment recombinant DNAs by polymerase chain reaction: application for gene-targeting in *Saccharomyces cerevisiae, Anal. Biochem.,* 242, 155, 1996.

39. Tessier, D. C. and Thomas, D. Y., PCR-assisted mutagenesis for site-directed insertion/deletion of large DNA segments, in *In Vitro Mutagenesis Protocols*, Trower, M. K., Ed., Humana Press, Totowa, NJ, 1996, 229.

40. Ochman, H., Gerber, A. S., and Hartl, D. L., Genetic applications of the inverse polymerase chain reaction, *Genetics,* 120, 621, 1988.

41. Hemsley, A., Arnheim, N., Toney, M. D., Cortopassi, G., and Galas, D. J., A simple method for site-directed mutagenesis using the polymerase chain reaction, *Nucleic Acids Res.,* 17, 6545, 1989.

42. Burke, T. F., Cocke, K. S., Lemke, S. J., Angleton, E., Becker, G. W., and Beckmann, R. P., Identification of a BRCA1-associated kinase with potential biological relevance, *Oncogene,* 16, 1031, 1998.

43. Gatlin, J., Campbell, L. H., Schmidt, M. G., and Arrigo, S. J., Direct-rapid (DR) mutagenesis of large plasmids using PCR, *Biotechniques,* 19, 559, 1995.

44. Vieira, J. and Messing, J., Production of single-stranded plasmid DNA, *Methods Enzymol.,* 153, 3, 1987.

45. Hofer, B. and Kuhlein, B., Site-specific mutagenesis in plasmids: a gapped circle method, *Methods Enzymol.,* 217, 173, 1993.

46. Ray, F. A. and Nickoloff, J. A., Site-specific mutagenesis of almost any plasmid using a PCR-based version of unique site elimination, *Biotechniques* 13, 342, 1994.

47. Redston, M. S. and Kern, S. E., Klenow co-sequencing: a method for eliminating "stops" [published erratum appears in *Biotechniques* 18(1), 68, 1995], *Biotechniques,* 17, 286, 1994.

48. Chou, Q., Minimizing deletion mutagenesis artifact during *Taq* DNA polymerase PCR by *E. coli* SSB, *Nucleic Acids Res.,* 20, 4371, 1992.

49. Seto, D., An improved method for sequencing double stranded plasmid DNA from minipreps using DMSO and modified template preparation, *Nucleic Acids Res.,* 18, 5905, 1990.

50. Schoenlein, P. V., Gallman, L. S., Winkler, M. E., and Ely, B., Nucleotide sequence of the *Caulobacter crescentus flaF* and *flbT* genes and an analysis of codon usage in organisms with G + C-rich genomes, *Gene,* 93, 17, 1990.

51. Zoller, M. J. and Smith, M., Oligonucleotide-directed mutagenesis: a simple method using two oligonucleotide primers and a single-stranded DNA template, *Methods Enzymol.,* 329, 1987.

52. Hultman, T., Murby, M., Stahl, S., Hornes, E., and Uhlen, M., Solid phase *in vitro* mutagenesis using plasmid DNA template, *Nucleic Acids Res.,* 18, 5107, 1990.

53. McKnight, S. L. and Kingsbury, R., Transcriptional control signals of a eukaryotic protein-coding gene, *Science,* 217, 316, 1982.

54. Cunningham, B. C. and Wells, J. A., High-resolution epitope mapping of hGH-receptor interactions by alanine-scanning mutagenesis, *Science,* 244, 1081, 1989.

55. Neupert, B., Menotti, E., and Kuhn, L. C., A novel method to identify nucleic acid binding sites in proteins by scanning mutagenesis: application to iron regulatory protein, *Nucleic Acids Res.,* 23, 2579, 1995.

56. Greene, J. M., Larin, Z., Taylor, I. C. A., Prentice, H., Gwinn, K. A., and Kingston, R. E., Multiple basal elements of a human hsp70 promoter function differently in human and rodent cell lines, *Mol. Cell. Biol.,* 7, 3646, 1987.

57. Dykxhoorn, D. M., Pierre, R. S., Ham, O. V., and Linn, T., An efficient protocol for linker scanning mutatgenesis: analysis of the translational regulation of an *Escherichia coli* RNA polymerase subunit gene, *Nucleic Acids Res.,* 25, 4209, 1997.

58. DeFrancesco, L., Setting your sites on genes: site-directed mutagenesis kits provide fast and efficient peeks into gene function, *The Scientist,* 11, 17, 1997.

59. Li, X. M. and Shapiro, L. J., Three-step PCR mutagenesis for "linker scanning," *Nucleic Acids Res.,* 21, 3745, 1993.

60. Biolabs, GPS™-LS Linker Scanning System Instruction Manual version 1.2, Beverly, MA, 2000.

61. Zhou, Y. H., Zhang, X. P., and Ebright, R. H., Random mutagenesis of gene-sized DNA molecules by use of PCR with *Taq* DNA polymerase, *Nucleic Acids Res.,* 19, 6052, 1991.

62. Eckert, K. A. and Kunkel, T. A., High fidelity DNA synthesis by the *Thermus aquaticus* DNA polymerase, *Nucleic Acids Res.,* 18, 3739, 1990.

63. Vartanian, J. P., Henry, M., and Wain-Hobson, S., Hypermutageneic PCR involving all four transitions and a sizable proportion of transversions, *Nucleic Acids Res.,* 24, 2627, 1996.

64. Pezo, V. and Wain-Hobson, S., Hypermutagenic *in vitro* transcription employing biased NTP pools and manganese cations, *Gene,* 186, 67, 1997.

65. Wan, L., Twitchett, M. B., Eltis, L. D., Mauk, A. G., and Smith, M., *In vitro* evolution of horse heart myoglobin to increase peroxidase activity, *Proc. Natl. Acad. Sci. U.S.A.,* 95, 12825, 1998.

66. Tarun, S. A., Tarun, J. S. L., and Theologis, A., Random mutagenesis of 1-aminocyclopropane-1-carboxylate synthase: a key enzyme in ethylene biosynthesis, *Proc. Natl. Acad. Sci. U.S.A.,* 95, 9796, 1998.

67. Airaksinen, A. and Hovi, T., Modified base compositions at degenerate positions of a mutagenic oligonucleotide enhance randomness in site-saturation mutagenesis, *Nucleic Acids Res.,* 26, 576, 1998.

68. Kuipers, O. P., Random mutagenesis by using mixtures of dNTP and dITP in PCR, in *In Vitro Mutagenesis Protocols,* Trower, M. K., Ed., Humana Press, Totowa, NJ, 1996, 351.

69. Kwan, T. and Gros., P., Mutational analysis of the P-glycoprotein first intracellular loop and flanking transmembrane domains, *Biochemistry,* 37, 3337, 1998.

70. Zaccolo, M., Williams, D. M., Brown, D. M., and Gherardi, E., An approach to random mutagenesis of DNA using mixtures of triphosphate derivatives of nucleoside analogues, *J. Mol. Biol.,* 255, 589, 1996.

71. Stemmer, W. P. C., Rapid evolution of a protein *in vitro* by DNA shuffling, *Nature,* 370, 389, 1994.

72. Stemmer, W. P. C., DNA shuffling by random fragmentation and reassembly: *in vitro* recombination for molecular evolution, *Proc. Natl. Acad. Sci. U.S.A.,* 91, 10747, 1994.

73. Zhao, H. and Arnold, F. H., Optimization of DNA shuffling for high fidelity recombination, *Nucleic Acids Res.,* 25, 1307, 1997.

74. Zhao, H. and Arnold, F. H., Functional and nonfunctional mutations distinguished by random recombination of homologous genes, *Proc. Natl. Acad. Sci. U.S.A.,* 94, 7997, 1997.

75. Moore, J. C., Jin, H. M., Kuchner, O., and Arnold, F. H., Strategies for the *in vitro* evolution of protein function: enzyme evolution by random recombination of improved sequences, *J. Mol. Biol.,* 272, 336, 1997.

76. Deng, W. P. and Nickoloff, J. A., Site-directed mutagenesis of virtually any plasmid by eliminating a unique site, *Anal. Biochem.,* 200, 81, 1992.

77. Zhu, L. and Holtz, A. E., A universal nested deletion method using an arbitrary primer and elimination of a unique restriction site, in *In Vitro Mutagenesis Protocols,* Trower, M. K., Ed., Humana Press, Totowa, NJ, 1996, 119.

78. Braman, J., Papworth, C., and Greener, A., Site-directed mutagenesis using double-stranded plasmid DNA templates, in *In Vitro Mutagenesis Protocols,* Trower, M. K., Ed., Humana Press, Totowa, NJ, 1996, 31.

79. Olsen, D. B. and Eckstein, F., High-efficiency oligonucleotide-directed plasmid mutagenesis, *Proc. Natl. Acad. Sci. U.S.A.,* 87, 1451, 1990.

80. Lewis, K. and Thompson, D. V., Efficient site directed *in vitro* mutagenesis using ampicillin selecton, *Nucleic Acids Res.,* 18, 3439, 1990.

81. Ohmori, H., A new method for strand discrimination in sequence-directed mutagenesis, *Nucleic Acids Res.*, 22, 884, 1994.

82. Hashimoto-Gotoh, T., Mizuno, T., Ogasahara, Y., and Nakagawa, M., An oligodeoxyribonucleotide-directed dual amber method for site-directed mutagenesis, *Gene*, 152, 271, 1995.

83. Kunkel, T. A., Roberts, J. D., and Zakour, R. A., Rapid and efficient site-specific mutagenesis without phenotypic selection, *Methods Enzymol.*, 154, 367, 1987.

84. Sayers, J. R., Krekel, C., and Eckstein, F., Rapid high-efficiency site-directed mutagenesis by the phosphorothioate approach, *Biotechniques*, 13, 592, 1992.

85. Taylor, J. W., Ott, J., and Eckstein, F., The rapid generation of oligonucleotide-directed mutations at high frequency using phosphorothioate-modified DNA, *Nucleic Acids Res.*, 13, 8764, 1985.

86. Dale, S. J. and Felix, I. R., Oligonucleotide-directed mutagenesis using an improved phosphorothioate approach, in *In Vitro Mutagenesis Protocols*, Trower, M. K., Ed., Humana Press, Totowa, NJ, 1996, 55.

87. Kleina, L. G., Masson, J. M., Normanly, J., Abelson, J., and Miller, J. H., Construction of *Escherichia coli* amber suppressor tRNA genes. II. Synthesis of additional tRNA genes and improvement of supressor efficiency, *J. Mol. Biol.*, 213, 705, 1990.

88. Normanly, J., Kleina, L. G., Masson, J. M., Abelson, J., and Miller, J. H., Construction of *Escherichia coli* amber suppressor tRNA genes. III. Determination of tRNA specificity, *J. Mol. Biol.*, 213, 719, 1990.

89. Lesley, S. A., Analysis of point mutations by use of amber stop codon suppression, in *In Vitro Mutagenesis Protocols*, Trower, M.K., Ed., Humana Press, Totowa, NJ, 1996, pp. 65.

90. Shafikhani, S., Siegel, R. A., Ferrari, E., and Schellenberger, V., Generation of large libraries of random mutants in *Bacillus subtilis* by PCR-based plasmid multimerization, *Biotechniques*, 23, 304, 1997.

91. Hashimoto-Gotoh, T., Yasojima, K., and Tsujimura, A., Plasmids with a kanamycin-resistance gene for site-directed mutagenesis using the oligodeoxyribonucleotide-directed dual amber method, *Gene*, 167, 333, 1995.

92. Ito, W., Ishiguro, H., and Kurosawa, Y., A general method for introducing a series of mutations into cloned DNA using the polymerase chain reaction, *Gene*, 102, 57, 1991.

93. Promega, Promega Technical Manual for Altered Sites II *in Vitro* Mutagenesis Systems (revised), Promega, Madison, WI, 1994.

94. Promega, Promega Technical Manual for GeneEditor™ *in Vitro* Site-Directed Mutagenesis System, Promega, Madison, WI, 1998.

95. Weiner, M. P., Felts, K. A., Simcox, T. G., and Braman, J. C., A method for the site-directed mono- and multi-mutagenesis of double-stranded DNA, *Gene*, 126, 35, 1993.

96. Goff, S. P. and Prasad, V. R., Linker insertion mutagenesis as probe of structure-function relationships, *Methods Enzymol.*, 208, 586, 1991.

97. Cech, T. R., Zaug, A. J., and Grabowski, P. J., *In vitro* splicing of the ribosomal RNA precursor of Tetrahymena: involvement of a guanosine nucleotide in the excision of the intervening sequence, *Cell*, 27, 487, 1981.

98. Cech, T. R., Self-splicing of group I introns, *Annu. Rev. Biochem.*, 59, 543, 1990.

99. Bratty, J., Chartrand, P., Ferbeyre, G., and Cedergren, R., The hammerhead RNA domain, a model ribozyme, *Biochim. Biophys. Acta*, 1216, 345, 1993.

100. Murray, J. B., Terwey, D. P., Maloney, L., Karpeisky, A., Usman, N., Beigelman, L., and Scott, W. G., The structural basis of hammerhead ribozyme self-cleavage, *Cell*, 92, 665, 1998.

101. Kijima, H., Ishida, H., Ohkawa, T., Kashani-Sabet, M., and Scanlon, K. J., Therapeutic applications of ribozymes, *Pharmacol. Ther.*, 68, 247, 1995.

102. Rossi, J. J., Controlled, targeted intracellular expression of ribozymes: progress and problems, *Trends Biotechnol.*, 13, 301, 1995.

103. Fecter, P., Rudinger, J., Giege, R., and Theobald-Dietrich, A., Ribozyme processed tRNA transcripts with unfriendly internal promoter for T7 RNA polymerase: production and activity, *FEBS Lett.*, 436, 99, 1998.

104. Thomas, K. R. and Capecchi, M. R., Site-directed mutagenesis by gene targeting in mouse embryo-derived stem cells, *Cell*, 51, 503, 1987.

105. Zheng, H. and Wilson, J. H., Gene targeting in normal and amplified cell lines, *Nature*, 344, 170, 1990.

106. Wang, G., Levy, D. D., Seidman, M. M., and Glazer, P. M., Targeted mutagenesis in mammalian cells mediated by intracellular triple helix formation, *Mol. Cell. Biol.,* 15, 1759, 1995.
107. Harve, P. A. and Glazer, P. M., Targeted mutagenesis of simian virus 40 DNA mediated by a triple helix-forming oligonucleotide, *J. Virol.,* 67, 7324, 1993.
108. Harve, P. A., Gunther, E. J., Gasparro, F. P., and Glazer, P. M., Targeted mutagenesis of DNA using triple helix forming oligonucleotides linked to psoralen, *Proc. Natl. Acad. Sci. U.S.A.,* 90, 7879, 1993.
109. Faria, M. and Giovannangeli, C., Triplex-forming molecules: from concepts to applications, *J. Gene Med.,* 3, 299, 2001.
110. Vasquez, K. M., Narayanan, L., and Glazer, P. M., Specific mutations induced by triplex-forming oligonucleotides in mice, *Science,* 290, 530, 2000.
111. Kmiec, E., Kmiec, B., Cole, A., and Holloman, W. K., The *REC2* gene encodes the homologous pairing protein of *Ustilago maydis, Mol. Cell. Biol.,* 14, 7163, 1994.
112. Wilson, J. H., Leung, W. Y., Bosco, G., Dieu, D., and Haber, J. E., The frequency of gene targeting in yeast depends on the number of target copies, *Proc. Natl. Acad. Sci. U.S.A.,* 91, 177, 1994.
113. Gamper, H. B. J., Cole-Strauss, A., Metz, R., Parekh, H., Kumar, R., and Kmiec, E. B., A plausible mechanism for gene correction by chimeric oligonucleotides, *Biochemistry,* 39, 5808, 2000.
114. Yoon, K., Cole-Strauss, A., and Kmiec, E. B., Targeted gene correction of episomal DNA in mammalian cells mediated by a chimeric RNA-DNA oligonucleotide, *Proc. Natl. Acad. Sci. U.S.A.,* 93, 2071, 1996.
115. Cole-Strauss, A., Yoon, K., Xiang, Y., Byrne, B. C., Rice, M. C., Gryn, J., Holloman, W. K., and Kmiec, E. B., Correction of the mutation responsible for sickle cell anemia by an RNA-DNA oligonucleotide, *Science,* 273, 1386, 1996.
116. Kren, B. T., Cole-Strauss, A., Kmiec, E. B., and Steer, C. J., Targeted nucleotide exchange in the alkaline phosphatase gene of HuH-7 cells mediated by a chimeric RNA/DNA oligonucleotide, *Hepatology,* 25, 1462, 1997.
117. Kren, B. T., Bandyopadhyay, P., and Steer, C. J., *In vivo* site-directed mutagensis of the factor IX gene by chimeric RNA/DNA oligonucleotides, *Nat. Med.,* 4, 285, 1998.
118. Graham, I. R., Manzano, A., Tagalakis, A. D., Mohri, Z., Sperber, G., Hill, V., Beattie, S., Schpelmann, S., Dickson, G., and Owen, J. S., Gene repair validation, *Nat. Biotechnol.,* 19, 507, 2001.
119. Tagalakis, A. D., Graham, I. R., Riddell, D. R., Dickson, J. G., and Owen, J. S., Gene correction of the apolipoprotein (Apo) E2 phenotype to wild-type ApoE3 by *in situ* chimeraplasty, *J. Biol. Chem.,* 276, 3226, 2001.
120. Van der Steege, G., Schuilenga-Hut, P. H., Buys, C. H., Scheffer, H. S., Pas, H.H., and Jonkman, M. F., Persistent failures in gene repair, *Nat. Biotechnol.,* 19, 305, 2001.

7 Mimotopes: An Overview

Renu Tuteja

CONTENTS

7.1 INTRODUCTION

The definition of epitopes for human B and T cells is fundamental for the understanding of the immune response mechanism and its role in the prevention and cause of human disease. This understanding can be applied to the design of diagnostics and synthetic vaccines. In the past few years, rapid advances have added new dimensions to the experimental strategies used to identify and characterize the binding of ligands (antigens) to receptors (antibodies). Mario Geyson[1] pioneered the concept that short peptides bearing critical binding residues (mimotopes) can chemically mimic the folded antigenic determinants on proteins (epitopes). And the idea that the noncovalent bonds formed between a few critical residues from an epitope and its binding molecule may make a major contribution to the total energy of binding. Before going into detail about mimotopes, their discovery, and potential uses, we must first acquire a more detailed understanding of epitopes.

An antigen can be defined as any substance that may be specifically bound by an antibody molecule or T-cell receptor. It is well known now that almost every kind of biological molecule — including sugars, lipids, hormones, as well as macromolecules such as complex carbohydrates, phospholipids, nucleic acids, and proteins — can serve as antigens. However, only macromolecules can initiate lymphocyte activation necessary for an antibody response. Molecules that generate immune responses are called immunogens. In order to generate antibodies specific to small molecules, they are attached to macromolecules before immunization. In this system, the small molecule is called a hapten and the macromolecule, usually a foreign protein, is called a carrier. This hapten-carrier complex, unlike free hapten, can act as an immunogen.

7.1.1 Epitopes or Antigenic Determinants

In general, macromolecules are much larger than the antigen-binding region of an antibody molecule. Immune cells do not interact with, or recognize, an entire immunogen molecule; instead, lymphocytes recognize discrete sites on the macromolecule called epitopes or antigenic determinants. Epitopes are immunologically active regions of an immunogen that bind to an antigen-specific receptor's surface or membrane-bound immunoglobulin on B lymphocytes and T-cell receptors on T lymphocytes or to secreted antibodies. Macromolecules typically contain multiple determinants, each of which can be bound by an antibody. The presence of multiple determinants is referred to as multivalency or polyvalency. In the case of polysaccharides and nucleic acids, identical epitopes may be regularly spaced.

The interaction between lymphocytes and a complex antigen may involve several levels of antigen structure. In the case of protein antigens, an epitope may involve elements of the primary, secondary, tertiary, and even quaternary structure of the protein. In the case of polysaccharide antigens, extensive side-chain branching via glycosidic bonds affects the overall three-dimensional conformation of individual epitopes.

In the case of either phospholipids or complex carbohydrates, the antigenic determinants are entirely a function of the covalent structure of the macromolecule. In the case of nucleic acids and proteins, the noncovalent folding of the macromolecule may also contribute to the formation of determinants. In proteins, epitopes formed by adjacent amino acid residues in the covalent sequence are called linear determinants. In a protein antigen the size of the linear determinant that forms contacts with specific antibodies is about six amino acids long. Linear determinants may be accessible to antibodies in the native folded protein if they appear on the surface or in a region of extended conformation. Usually, linear determinants may be inaccessible in the native conformation and appear only when the protein is denatured.

In contrast, conformational determinants are formed by amino acid residues from separated portions of the linear amino acid sequence that are spatially juxtaposed only upon folding. In theory, denatured proteins could transiently give rise to conformational determinants; however, such determinants are too short-lived unless they are maintained by energetically favorable interactions such as those found in native proteins.

Proteins may be subjected to covalent modification such as phosphorylation or specific proteolysis. These modifications, in altering the covalent structure, can produce new antigenic epitopes. Such epitopes, called neo-antigenic determinants, may be recognized by specific antibodies.

Studies using small antigens have revealed that B and T cells recognize different epitopes on the same antigenic molecule. For example, when mice are immunized with glucagon, a small human hormone of 29 amino acids, antibody is elicited to epitopes in the amino-terminal portion, whereas the T cells respond only to epitopes in the carboxyl-terminal portion.

B cells and T cells exhibit fundamental differences in antigen recognition. B cells recognize soluble antigen when it binds to their membrane-bound antibody. Because B cells bind antigen that is free in solution, the epitopes they recognize tend to be highly accessible sites on the exposed surface of the immunogen. T cells only recognize processed peptides that are generated by proteolytic processing of the antigen and associated with major histocompatibility (MHC) molecules on the surface of antigen-presenting cells (B cells) and altered self-cells. Because T cells exhibit MHC-restricted antigen recognition, T-cell epitopes cannot be considered apart from their associated MHC molecules.

7.1.2 PROPERTIES OF B-CELL EPITOPES

The size of a B-cell epitope is determined by the size of the antigen-binding site on the antibody molecules displayed by B cells. The evidence indicates that antibody-binding sites are generally concave and that complementary antigenic sites (epitopes) are predominantly convex.[2] In order for a strong interaction to occur, the antibody's binding site and the epitope must have a complementary conformation. Smaller ligands such as carbohydrates, nucleic acids, peptides, and haptens often bind to an antibody with a deep concave pocket. For example, angiotensin II, a small octapeptide hormone, binds within a deep and narrow groove (725 Å) of monoclonal antibody-specific to the hormone.[3]

Epitopes on globular protein antigens appear to be considerably larger and occupy a more extensive surface area on the antibody than do small antigens. B-cell epitopes in native proteins are generally hydrophilic amino acids on the protein surface that are topographically accessible to membrane-bound or free antibodies. A B-cell epitope must be accessible in order to be able to bind to an antibody. Amino acid sequences that are hidden within the interior of a protein cannot function as B-cell epitopes unless the protein is first denatured. The experimentally determined antigenic epitopes are biased toward large polar side chains, most notably arginine, lysine, and glutamine.[2]

B-cell epitopes can contain sequential or nonsequential amino acids. Epitopes may be composed of sequential contiguous residues along the polypeptide chain or nonsequential residues from segments of the chain brought together by the folded conformation of an antigen.

The surface of a protein presents a large number of potential antigenic sites. The subset of antigenic sites on a given protein that is recognized by the immune system of an individual animal is much smaller than the potential antigenic repertoire, and it varies from species to species and even among individual members of a given species. Some epitopes, which induce a more pronounced immune response than others, are referred to as immunodominant. It is thought that intrinsic topographical properties of the epitope as well as the animal's regulatory mechanisms influence the immunodominance of a particular epitope.

7.1.3 PROPERTIES OF T-CELL EPITOPES

It is well known now that T cells do not recognize soluble native antigen but rather recognize antigen that has been processed into antigenic peptides, which are presented in association with MHC molecules on a different cell.

Antigenic peptides recognized by T cells form trimolecular complexes with a T-cell receptor and an MHC molecule. Before a T cell can be activated, a trimolecular complex must form between its antigen-binding receptor on one cell and an MHC molecule associated with an antigenic peptide on the other cell. Antigens recognized by T cells, therefore, possess two distinct interaction sites: one, the epitope, interacts with the T-cell receptor, and the other, the agretope, interacts with an MHC molecule.

Epitope recognition in MHC-restricted T-cell responses involves two different binding events. First, the epitope peptides bind to the MHC molecules; second, the resulting complex is bound by the TCR. Thus, the peptides must interact simultaneously with two receptors and must contain distinct functionalities for these interactions (i.e., contact residues for the MHC molecule and the T-cell receptor, respectively).[4] The effect of peptide binding to an MHC molecule is that its conformation is fixed to correctly expose the amino acid side chains for recognition by a T-cell receptor. This conformational steering involves compensation of the terminal main-chain charges of the peptide by complementary residues of the MHC molecule, multiple hydrogen bonds to the peptide backbone, and anchoring of some of the peptide side chains into corresponding pockets in the MHC peptide-binding groove.[5-7]

Antigen processing is required to generate peptides that interact specifically with MHC molecules. Endogenous antigens are processed into peptides within the cytoplasm, while exogenous antigens are processed within the endocytic pathway. This processing yields antigenic peptides that associate with class I or class II MHC molecules; the resulting peptide-MHC complex is then presented on the cell surface where they can be recognized by T cells.

T cells tend to recognize internal peptides that are exposed during processing within antigen-presenting cells or self-cells. Although these interactions result in a strong preference for peptides, ligands with different oligomer backbone structure have been successfully designed to bind into the MHC groove and be recognized by TCRs.[8]

7.2 MAPPING CONFIRMATIONAL EPITOPES: THE MIMOTOPE APPROACH

Geysen et al.[1] synthesized peptide mixtures on plastic pins and assessed their ability to bind an antibody against a protein antigen. These workers delineated a peptide that mimics a discontinuous epitope, i.e., an antibody-binding determinant composed of residues distant in the primary sequence but adjacent in the folded structure. They called these peptide mimics "mimotopes." The mimotope approach involves synthesizing a peptide series with successively better fit to the antigen-combining site of the antibody. The end result is a peptide that, by inference, has some of the topographical equivalence of the epitope originally recognized.[9] In this way ligands can be discovered for an antibody whose specificity is not known in advance.

7.3 ANTIBODY SPECIFICITY DETERMINED BY COMBINATORIAL LIBRARY

B-cell epitopes are typically classified as either continuous or discontinuous. It is well known that the majority of the B-cell epitopes in proteins are discontinuous (i.e., they recognize only native, folded structures); however, there are exceptions in which antibodies that recognize discontinuous epitopes can also recognize linear peptide fragments of the protein. Therefore, while peptides rarely represent full epitopes of proteins, they can be used effectively to define antibody specificity.[10]

The original mimotope method postulated the need to synthesize all possible octapeptides in order to demonstrate all continuous epitopes.[11] As there are 20^8 possible octapeptides that contain only the naturally occurring L-amino acids, it is clearly not possible to test these individually. However, by synthesizing mixtures of many peptides on polypropylene pins it has been possible to test such numbers in combination. The individual peptide was deconvoluted by an iterative approach using the most active mixtures identified. The final peptide was termed a mimotope because this sequence, unrelated to the known antigen, was considered a mimic of the epitope recognized by the anti-protein antibody.

The key elements of combinatorial libraries are the randomization of the peptides, and, of course, the location of the mimotope with respect to the peptide mixture. A new type of peptide library,[12] which was synthesized using "one-bead, one-peptide" (split-synthesis method), ensures that each resin bead contains a single peptide. In the first application of this method, a library containing 2,476,099 (19^5) individual pentapeptides was synthesized and screened for recognition by a monoclonal antibody against β-endorphin. The individual peptide beads that bound antibody were identified with a secondary label, retrieved manually, and the peptide sequence was determined by microsequencing. While the native epitope was not identified, several peptide sequences (mimotopes) having good homology with the β-endorphin determinant were found to have relatively good affinities.

A more versatile approach to combinatorial libraries is to use synthetic peptide combinatorial libraries (SPCLs) composed of free peptides that can be used in a variety of bioassays.[13] This approach was validated by screening the peptide library for inhibition of monoclonal antibody binding to its peptide antigen, using a competitive enzyme-linked immunosorbent assay. Since this library contained defined and variable positions, individual peptides were identified using an iterative synthesis and screening approach. The most active sequence that was identified was found to exactly match the antigenic determinant found in earlier studies to be recognized by this monoclonal antibody.

An extension to the above technique was to synthesize six individual positional peptide libraries, each consisting of hexamers with a single position defined and five positions as mixtures.[14] This has enabled a single screening to determine the most effective amino acid residues for a position as well as the relative specificity of each position. Both SPCLs and positional scanning SPCLs can be of different lengths, ranging from three to ten residues, to accommodate the mimotope strategy.

7.4 PHAGE-DISPLAY PEPTIDE LIBRARIES

Since the early 1980s it has been suggested that peptides derived from epitopes of pathogenic agents could be fused to the coat protein of viruses to generate components of a new class of vaccines.[15] The resulting viral particles would "display" the foreign peptide on their surfaces and could be used as the principal components of cell-free vaccines. In recent years, filamentous bacteriophages have been used extensively for the display on their surface of large repertoires of peptides generated by cloning random oligonucleotides at the 5' end of the genes coding for the phage coat proteins.[16–18] Homogeneous and heterogeneous legends, such as monoclonal and poly-clonal antibodies, have been used for the affinity selection of these repertoires leading to the identification of peptides that mimic the specific ligates (mimotopes).[19–22]

Filamentous phages are the preferred vectors for the construction of genetic repertoires because of the following properties:

1. Insertion of foreign sequence at the amino terminus of either the pIII or, in certain cases, the pVIII coat protein are well tolerated; the foreign sequence is displayed on the surface of the viral particle and can bind to molecules such as antibodies and receptors.
2. Even foreign sequences that interfere with the life cycle of the virus can be displayed by using the so-called two genes or phagemid system, in which chimeric phage bearing both wild-type and recombinant coat proteins are produced.[23]
3. By using appropriate selector molecules immobilized on a solid support, it is possible to affinity-purify the phage displaying the binding protein or peptides from a mixture of >10^9 different phage particles. The selected phage can then be amplified by reinfection of *Escherichia coli* cells. The sequence of the selected peptides can be deduced from the sequence of the DNA encapsidated in the same phage particles.[23]
4. Phage display provides the linkage between phenotype (which provides the selection) and the genotype.

7.5 EPITOPE LIBRARIES: HOW THEY WORK

An epitope library comprises tens to hundreds of millions of short, variable amino acid sequences that are displayed on the surface of a bacteriophage virion. These sequences are known as epitopes to emphasize the fact that they are meant to mimic the precise determinants that ligates recognize. The term *ligates* here refers to binding molecules such as receptors, enzymes, and antibodies that are used to affinity-purify phages that have ligand epitopes.

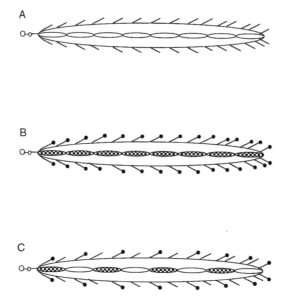

FIGURE 7.1 Schematic diagram of wild-type, recombinant, and hybrid bacteriophage viral particles. (A) Wild-type filamentous bacteriophage. (B) Recombinant virion in which each copy of the major coat protein has a small peptide insert in the N-terminal region. (C) Hybrid virion containing a mixture of the wild-type major coat protein and a modified coat protein with an insert in the N-terminal region.

Two viral proteins have been used to display epitopes — the minor coat protein pIII and the major coat protein pVIII. Both proteins are synthesized with short signal sequences that allow them to be transported to the bacterial inner membrane, where they are cleaved by signal peptidase to form mature proteins. In both coat proteins, the epitope is inserted at the amino terminus of the mature proteins, or a few residues from it. In general, the termini of proteins are more flexible and exposed than most other parts of proteins, so they are therefore likely not only to tolerate epitope insertion but also to make epitopes accessible for recognition.[24]

Libraries have been constructed by cloning random oligonucleotide sequences in frame with the genes encoding the capsid proteins pVIII or pIII. These libraries are known as random peptide libraries (RPLs), and in this way each phage displays a different peptide on its surface. There are five copies of pIII and ~2700 copies of pVIII on the coat of filamentous phage. Peptides of up to a few hundred residues in length can be displayed on pIII;[23] by contrast, the maximum length of peptide that can be displayed on pVIII is much shorter, probably no more than eight or nine residues. However if the two-gene system is used, the length of peptide that can be displayed on pVIII can be increased considerably, to 100 or more residues.[25] The two-gene system uses a mosaic of wild-type pVIII molecules from a helper phage and recombinant pVIII molecules with the foreign peptide displayed on each phage particle. The ratio of recombinant to wild-type pVIII varies from 1:1 to 1:1000, depending on the specific foreign peptide and on the experimental conditions.

Whether displayed as part of a pIII or a pVIII protein, the epitope region of the fusion protein comprises a variable region, such that each epitope in a library constitutes a different amino acid sequence[24] (see Figure 7.1). Several different codon schemes may be used to encode unspecified amino acid sequences. Each variable amino acid in the epitope region is randomly encoded by a degenerate codon, either NNK or NNS, where N stands for an equal mixture of the deoxynucleotides G, A, T, and C in the first and second positions, K stands for an equal mixture of G and T, and S for G and C in the third position, respectively. In practice the NNK codon scheme is popular because it yields only an acceptable frequency of stop codons when used to encode very large

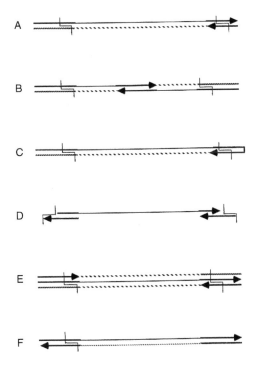

A. Sparks et al., 1996[83]
    1. Anneal oligos
    2. Extend oligos by DNA polymerase
    3. Restriction digest
    4. Ligate into vector
B. Kay et al., 1993[35]
    1. Anneal oligos
    2. Extend oligos by DNA polymerase
    3. Restriction digest
    4. Ligate into vector
C. Christian et al., 1992[84]
    1. Self-anneal oligo to form hairpin
    2. Extend oligos by DNA polymerase
    3. Restriction digest
    4. Ligate into vector
D. Cwirla et al., 1990[16]
    1. Anneal three oligos
    2. Ligate into vector
E. Scott and Smith, 1990[22]
    1. Anneal oligos
    2. Amplify by polymerase chain reaction
    3. Restriction digest
    4. Ligate into vector
F. Devlin et al., 1990[17]
    1. Anneal oligos
    2. Restriction digest
    3. Ligate into vector

FIGURE 7.2 Schematic of strategies for assembling double-stranded DNA inserts from degenerate oligonucleotides. Fixed and degenerate sequences are represented by thick and thin lines, respectively, and arrows indicate the 3′ ends of DNA fragments. Slashed lines indicate regions filled in by DNA polymerase. Restriction enzyme recognition sequences are indicated at the ends. References and steps for each process are as listed in the figure.

(>50 amino acids) peptides, and it does not require sophisticated oligonucleotide synthesis technologies. The degenerate codon motif NNK produces 32 codons, 1 for each of 12 amino acids, 2 for each of 5 amino acids, 3 for each of 3 amino acids, and 1 stop codon. Although this scheme provides the most equitable codon distribution available with standard methods of oligonucleotide synthesis, it results in a large bias against peptides containing 1-codon residues. For example, a peptide containing only 3-codon residues (e.g., hexaserine) is present 729 (3^6) times as often as a peptide containing 1-codon residues (e.g., hexaglutamine).

A variety of strategies have been used to produce clonable fragments appropriate for encoding random peptides (see Figure 7.2). Each method relies upon synthetic oligonucleotides to generate the degenerate regions within the DNA insert. To make a phage library, a degenerate oligonucleotide is synthesized using (1) single nucleotides at positions encoding the invariant amino acid sequences that surround the variable region and (2) the equimolar mixtures of nucleotides mentioned, where random amino acids are to be encoded. These degenerate oligonucleotides are then cloned as single strands into the appropriate sites in the coat protein genes of the phage or phagemid vectors.[16]

Alternatively, the complementary strand of the degenerate oligonucleotide can be synthesized chemically, with polyinosine occurring in the degenerate region[17] or enzymatically using the polymerase chain reaction to amplify the sequence. The appropriate ends for ligation with the vector can then be generated by restriction endonuclease digestion. The ligated product is introduced into bacterial cells by electroporation (with an efficiency of about 10^9 transformants per microgram of covalently closed DNA), and the phages are propagated by the cells and collected and purified.

The number of phage clones in a library is crucial to its utility. For example, a library containing all possible hexapeptides would consist of 64 million (20^6) different six-amino-acid sequences and would be encoded by about 1 billion (32^6) different hexacodons. Therefore, a hexapeptide library that includes all possible six-amino-acid sequences must contain well over a billion phage clones. Since in most cases only a few residues of an epitope predominate in binding, a 200-million-clone 15-amino-acid library will represent a larger diversity of short peptides and thus may be more effective than a shorter peptide library of the same number of clones.[24]

The next step is to purify the ligate-binding phage, which can be done by biopanning. This involves reacting biotinylated monoclonal antibody with the library, then purifying the phage that display ligand epitopes by immobilizing complexes of phage bound to biotinylated monoclonal antibody on streptavidin, washing away free phage and eluting the bound phage with acid (see Figure 7.3).[26] The eluted phage are amplified by growth in bacterial cells, then used in subsequent rounds of biopanning to further enrich for binding phage. Thus, binding phages are selectively enriched and amplified by repeated rounds of biopanning, followed by infection and propagation of enriched phage.

Individual phage clones can be isolated and their displayed epitopes tested by several criteria. The amino acid sequence of the epitope region can be deduced from sequencing single-stranded phage DNA and the presence of a unique consensus sequence among affinity-purified phage clones determined. The occurrence of similar sequences among independent phage clones indicates selective enrichment of ligate binding phage from the library. Candidate phage (either single clones or mixed, amplified phage) can be tested for binding to ligate by enzyme-linked immunosorbent assays (ELISA) or some other binding assay. Finally, free peptide bearing a consensus sequence can be synthesized and tested for binding in the absence of a phage "carrier," thereby allowing its binding kinetics, bioactivity, and free and bound structures to be precisely analyzed.

7.6 USES OF EPITOPE LIBRARIES

It should be possible to use the recently developed expression-vector and biopanning technique to identify mimotopes (phagotopes) from all possible sequences of a given length. This led to the idea of identifying peptide ligands for antibodies by biopanning epitope libraries, which could then be used in vaccine design, epitope mapping, the identification of genes, and other useful applications.[26] The use of epitope libraries could be generalized to finding mimetic peptide ligands for other receptors, enzymes, and other ligates, which would have tremendous potential in terms of drug discovery.

The epitope library can identify ligands for component antibodies in complex polyclonal antisera, whether or not the antigen is known or available. Thus, for example, autoimmune sera might be used to identify peptides that discriminate among autoimmune disease entities. Similarly, affinity purification with sera from patients infected with viruses, bacteria, or parasites may identify peptides that distinguish strains or serve as vaccines or immunogens.

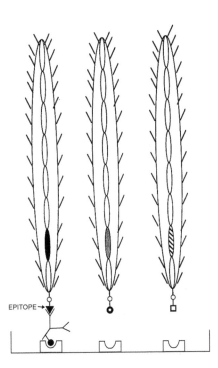

FIGURE 7.3 Biopanning, a method of affinity-purifying phages that bind biotinylated antibody. Three phage clones are shown, each bearing a different peptide sequence fused to pIII; as a result each displays a different epitope (mimotope) on the surface. Phages from the library are reacted with biotinylated ligate followed by immobilization on a streptavidin-coated plate and extensively washed to remove unbound or weakly binding phage. The bound phage is eluted from the plate in acid and base is added to neutralize the eluates. The phages are concentrated, used to infect cells, and propagated. The resulting amplified phage can be partially purified and used in further rounds of biopanning or cloned and sequenced in the epitope region and tested for binding to ligate.

By using receptors in soluble form or other binding proteins to affinity-purify phage from the epitope library, it may be possible to discover peptides that biologically mimic active determinants on hormones, cytokines, enzyme substrates, viruses, and other biomolecules. Such mimetic peptides might antagonize, agonize, or modulate the physiological action of the natural ligands and as such would be attractive candidates for drug development.

Some examples of the applications of random peptides libraries in identifying various epitopes (mimotopes) are discussed in detail.

7.6.1 PHAGOTOPES MIMICKING NATURAL EPITOPES

The use of random peptide libraries (RPLs) has enabled epitope mapping using monoclonal antibodies as selector molecules. Typically, this technique generates families of mimotopes that have easily recognizable consensus sequences, often matching the sequence found in the natural antigen.

In many cases, binding and thus selection might be impossible with a freely moving linear peptide if the loss of entropy were not compensated by the enthalpy change on binding. To overcome these problems, "constrained" peptide libraries have been constructed, in which the potential conformation of the displayed peptide is limited by the structural constraints built around it. For example, flanking the random sequence in the RPL by two cysteine residues forces, in most cases, the displayed peptide into a loop conformation.[27–29]

Specific epitopes can be selected from an RPL, only if the library contains phagotopes whose avidity-affinity for the selecting antibody is higher than the threshold determined by the stringency of the selection conditions. The most obvious thing to increase the likelihood of finding particular mimotope is to increase the complexity of the library as much as possible. The next is to optimize the stringency of the selection procedure so that phagotopes that bind specifically but with low affinity are identified. It is easier to isolate specific phagotopes from the pVIII system, in which each phagotope is displayed in hundreds of copies, than from the pIII system, in which a maximum of five copies of each epitope is displayed.

The differences between the two display systems could be exploited to implement a strategy of *in vitro* maturation of epitopes. First, low-affinity phagotopes can be selected from pVIII libraries, thus maximizing the probability of finding them. The epitopes selected in this first screen could then be mutagenized and transferred onto pIII phagemids displaying a single copy of the foreign peptide in order to identify the highest-affinity mimotopes.

A library of 3×10^8 recombinants was generated in vector fAFF1 by cloning randomly synthesized oligonucleotides.[16] The library was screened for high-avidity binding to a monoclonal antibody (3-E7) that is specific to the N-terminus of β-endorphin (Tyr-Gly-Gly-Phe). Fifty-one clones selected by three rounds of panning were sequenced and found to differ from previously known ligands for this antibody. The striking finding is that all 51 contained tyrosine as the N-terminal residue and 48 contained glycine as the second residue. The binding affinities of six chemically synthesized hexapeptides from this set ranged from 0.35 μM (Tyr-Gly-Phe-Trp-Gly-Met) to 8.3 μM (Tyr-Ala-Gly-Phe-Ala-Gln), compared with 7.1 nM for a known high-affinity ligand (Tyr-Gly-Gly-Phe-Leu). This example shows that ligands can be identified with no prior information concerning antibody specificity.

In another study by Scott and Smith,[22] the epitope library was tested with two monoclonal antibodies, A2 and M33, that are specific for the hexapeptide DFLEKI, an epitope of the protein myohemerythrin. Both monoclonal antibodies were biotinylated and used to select clones from an epitope library by successive rounds of affinity purification. Antibody-binding phages displayed peptides with striking similarity to the DFLEKI epitope, particularly in the first three residues known to be most important for antibody binding.[22] In this case also, the myohemerythrin-like sequences were isolated, even though no information about myohemerythrin was incorporated into the epitope library.

Strikingly, one clone, which bound specifically and relatively tightly to only one of the monoclonal antibodies, displayed an epitope that had no resemblance to the native antigenic sequence. This peptide appears to be a mimotope of the native peptide that can discriminate between two monoclonal antibodies that recognize the same epitope. This hexapeptide sequence, CRFVWC, as well as two others bearing the DFL consensus (CEDFLC and CRDFLC) contain Cys in the first and sixth positions; these were the only other cases in which the DFL sequence was found, besides the first three residues in the epitope region. The placement of these Cys residues is compatible with disulfide bridge formation, suggesting that binding of the mimotope as well as the DFL consensus sequence in this orientation may be enhanced by the structural constraints imposed on a looped peptide.[22]

A 37-million-clone nonapeptide library displayed by the major coat protein pVIII was generated.[20] The library was screened with an antibody that binds to residues 163 to 171 of interleukin-1β, both as free nonapeptide and in the intact protein. The monoclonal antibody binding clones revealed the consensus SND or SNE, with phage-bearing SND having a higher apparent affinity than those bearing SNE. The former tripeptide sequence is present in the interleukin-1β-related nonapeptide that was used to elicit the monoclonal antibody. All of the phages that bound monoclonal antibody contained the consensus tripeptide, thus defining this region of the interleukin-1β peptide as the critical residues for monoclonal antibody recognition.[20]

A hexapeptide phage-epitope library was used to identify epitopes for a monoclonal antibody (mAb 5.14), which is directed to a determinant within a highly immunogenic, cytoplasmic region of the alpha-subunit of acetylcholine receptor.[30] Two different peptide-presenting phages (SWDDIR phage and LWILTR-phage) that interact specifically with mAb 5.14 were selected. This interaction was specifically inhibited by acetylcholine receptor and by synthetic peptides corresponding to the hexapeptides presented by the selected phages. Although mAb 5.14 binds to acetylcholine receptor in its native as well as denatured form, the selected hexapeptides do not exist as such in the acetylcholine receptor molecule. However, three amino-acid-sequence homologies with these hexapeptides were shown to be present in the cytoplasmic region of Torpedo acetylcholine receptor. By extending the selected hexapeptide, at one or both ends, with amino acid residues flanking the hexapeptides in the phage, mimotopes with up to two order of magnitude higher affinity to the antibody were obtained. These extended peptides were able to efficiently block the binding of mAb 5.14 to both peptide-presenting phages and to acetylcholine receptor.[30]

Recently, mimotopes for monoclonal antibody CII-C1 to type II collagen were identified using a random peptide library.[31] CII-C1 is known to react with a conformational epitope on type II collagen that includes residues 359 to 363. Three rounds of selection were used to screen two nonameric phage libraries and 18 phagotopes were isolated. Of the eight most reactive phages, seven inhibited the reactivity by ELISA of CII-C1 with type II collagen. Of the 18 phages isolated, 11 encoded the motif F-G-x-Q with the sequence F-G-S-Q in 6, 2 encoded F-G-Q, and 1 the reverse motif Q-x-y-F. Most phagotopes that inhibited the reactivity of CII-C1 encoded two particular motifs consisting of two basic amino acid residues and a hydrophobic residue in the part of the insert and the F-G-x-Q or F-G-Q motif in the second part; phagotopes that contained only one basic residue in the first part of the sequence were less reactive. These motifs were not represented in the linear sequence of type II collagen and thus represent mimotopes of the epitope for CII-C1 on type II collagen. This study disclosed mimotopes for a conformational epitope of type II collagen.[31]

7.6.2 PHAGOTOPES MIMICKING DISCONTINUOUS B-CELL EPITOPES

Classical epitope mapping requires a set of overlapping synthetic peptides spanning a region of the antigen or a set of deletions obtained by genetic manipulation of the antigen-coding gene. If a peptide was found to react with the antibody, the epitope was declared to be linear or continuous. By contrast, if a reacting peptide could not be identified, the epitope was declared to be discontinuous, i.e., it was deemed to be made up of amino acids that are closely located in space but far away in the primary sequence. Thus, "complete" epitopes (i.e., the ensemble of all amino acids that make contact with antigen-binding sites) are probably all discontinuous to some degree. Cerino et al.[32] classified the epitope recognized by the monoclonal antibody B12.F8 as discontinuous because no linear synthetic peptide shorter than 20 amino acids reacted with it.

Sometimes, the consensus sequence of selected phagotopes cannot be found in the sequence of the natural antigen. In these cases, it is possible that the phagotopes are mimics of discontinuous stretches of natural epitopes. Phagotopes using monoclonal antibody 1B7 were selected.[33] This antibody is directed against the S1 subunit of the *Bacillus pertussis* toxin. Competition experiments showed that the sequence of the selected phagotopes interacted with the antigen-binding site of mAb 1B7; however, their consensus sequences did not match the sequence of the natural antigen.[33]

The consensus sequence of the selected phagotopes by using monoclonal antibody 5.5 (specific for the ligand-binding site of the nicotinic acetylcholine receptor) also did not reveal any homology with the receptor sequence.[34] However, a functional test showed that peptides derived from the phagotopes prevent interaction of the monoclonal antibody with the receptor only at very high concentrations ($IC_{50} = 100$ mM).

Another interesting study, which supports the hypothesis that phagotopes can mimic discontinuous epitopes, has been reported.[27] In this study monoclonal antibody H107, specific to the human ferritin H subunit, is used. Phagotopes selected with monoclonal antibody H107 could be classified into two groups, indicating that the two families of phagotopes are mimotopes of two different parts of the monoclonal antibody H107 epitope. Phagotopes of both groups may mimic regions that contain structural elements on either side of the discontinuity of the natural epitope.

7.6.3 Phagotopes Mimicking Non-Protein-Binding Epitopes

Phagotopes whose peptide sequences are apparently structural analogues of nonproteinaceous structures have also been isolated. Phagotopes reacting with streptavidin, which are mimics of biotin, have been isolated.[17,35] Most of these phagotopes that bind streptavidin display the consensus sequence HPQ. These studies show that peptide mimics can also be found for nonproteinaceous ligands.

In a related study, the crystal structure of the complex between streptavidin and the heptapeptide FSHPQNT (which contains the HPQ consensus sequence) was determined.[36] A major factor stabilizing the streptavidin–heptapeptide interaction is a network of hydrogen bonds that includes the following elements: the side chains of histidine, proline, and glutamine residues from the peptide; Asp 126, Thr 90, and Ser 27 of streptavidin; and a water molecule. The peptide and biotin complexes with streptavidin form an analogous hydrogen-bond network in the crystal structure, but the peptide appears to use only a subset of the biotin-interaction sites at the binding pocket.

There are some other examples of phagotopes mimicking nonpeptidic epitopes. Some studies have used the carbohydrate-binding lectin concanavalin A[37,38] and others monoclonal antibody B3, which binds to the carbohydrate antigen Lewis[21] as a selector molecule. Peptides that bind concanavalin A display a common-consensus sequence containing the amino acid YPY. The binding was Ca^{2+}-dependent and exhibited comparable affinity with methyl-alpha-D-mannopyranoside, the natural binder.[38] The phagotope binding to monoclonal antibody B3 has the sequence APWLYGPA, but mutagenesis experiments indicated that PWLY plays a crucial role.[21]

Cryptococcus neoformans is an encapsulated fungus that causes meningoencephalitis in immunocompromised individuals. It contains a capsular polysaccharide coat, and the major component of this coat is a glucuronoxylomannan. It has not been possible to generate smaller fragments from glucuronoxylomannan that are suitable for epitope studies. In an attempt to probe the fine specificity of anticryptococcal antibodies and to identify possible peptide mimotopes of glucuronoxylomannan epitope, phage display peptide libraries containing hexameric and decameric peptides were used.[39] The probe used was 2 H1, a protective IgG1K mouse monoclonal antibody against *C. neoformans*.[40]

Several 2H1-binding peptides with four different peptide motifs were identified. It was also shown that 2H1-binding peptides share the same binding site, and it overlaps partially with the binding site for glucuronoxylomannan. These peptides were classified as mimetics of the glucuronoxylomannan epitope and were potential mimotopes that would elicit antiglucuronoxylomannan antibodies.[39] In a related study,[41] the three-dimensional structure of 2H1, in both its unbound form and in complex with the 12-amino-acid-residue peptide PA1 (GLQYTPSWMLVG), has been solved. A small number of hydrogen bonds between peptide and antibody contribute to the affinity and specificity.

In a recent study the immunologic properties of P601E (SYSWMYE), a peptide from the low-affinity motif (W/YXWM/LYE) that has an extended cross-reactivity among anti-glucuronoxylomannan monoclonal antibodies, have been analyzed.[42] Further screening of a peptide library with anti-P601E monoclonal antibodies isolated peptides having a motif almost identical to the peptide motif selected by 2H1.

It has been shown that peptides can function in *in vivo* and *in vitro* models as carbohydrate surrogate antigens. To further explore the nature of the antigenic and immunogenic properties of such mimotopes, synthetic peptides with aromatic amino acids were tested to delineate reactivity patterns with several antineolactoseries monoclonal antibodies. These antibodies recognize biologically important conformations of the histoblood-group-related Lewis antigens expressed on the surface of a variety of human cancers.[43] Using ELISA, it has been demonstrated that the monoclonal antibodies can distinguish particular peptide motifs that include the sequences GGIYYPYDIYY-PYDIYYPYD, GGIYWRYDIYWRYDIYWRYD and GGIYYRYDIYYRYDIYYRYD. Substitution of arginine by proline diminished the reactivity of the anti-Lewis Y (LeY) monoclonal antibody BR55–2. The binding of LeY to BR55–2 was inhibitable by the arginine-containing peptides. Serum against all three peptides displayed reactivity with synthetic histoblood-group-related antigen probes. Immunologic presentation of the peptides as multiple antigen peptides improved a peptide's ability to induce LeY-specific immune responses. These results indicate that mapping peptide epitopes with anticarbohydrate antibodies can lead to defining antibody-fine specificities that can go undetected by screening of carbohydrate antigens alone. These results further confirm that peptides and carbohydrates can bind to the same antibody-binding site and that peptides can structurally mimic salient features of carbohydrate epitopes.[43]

In a related study it has been shown that tumor-associated carbohydrate antigens can be used as cancer vaccines. It has been suggested that peptide mimotopes of neolactoseries-related Lewis Y (LeY) and sialyl-Lewis X (sLeX) tumor-associated carbohydrate antigen and QS-21 adjuvant could be considered as an immunogenic therapeutic vaccine in carcinoma and melanoma patients.[44]

To date, the generation of anticarbohydrate Th1 immune responses, which would be useful for both tumor immunotherapy as well as in pathogen vaccine strategies, has been elusive. To augment Th1 immune responses to carbohydrate Ags, the results of DNA vaccination studies in mice using plasmid-encoding-designed peptide mimotopes (minigenes) of the neolactoseries Ag Lewis Y (LeY) have been described.[45] In contrast to LeY immunization, immunization with mimotope-encoded plasmids induced LeY cross-reactive IgG2a Abs. Minigene immunization primed for a LeY-specific response that is rapidly activated upon encounter with nominal Ag upon subsequent boost. The resulting IgG2a response mediated complement-dependent cytotoxicity of a LeY-expressing human tumor cell line in the presence of human complement. These studies establish that peptide mimotopes of carbohydrate Ags encoded as DNA plasmids are novel immunogens providing a means to manipulate carbohydrate cross-reactive Th1 responses.[45]

7.6.4 Phagotopes Mimicking T-Cell Epitopes

T lymphocytes recognize peptide fragments of antigens in association with molecules of the class I or class II MHC, which are present on the surface of antigen-presenting cells. As the peptides that bind to class-I molecules are only eight or nine amino acids long, by purifying natural complexes of peptides and class I molecules, it has been possible to identify many consensus sequences. But similar work on the binding of peptides to class II molecules cannot be easily performed because, while only eight or nine residues are involved in binding, the peptides purified from natural complexes are much longer and heterogeneous in length.

Combinatorial peptide and nonpeptide chemistry has provided new tools for the determination of T-cell epitopes. These synthetically defined epitopes are called mimotopes because they mimic the natural T-cell epitopes without being identical in structure. These mimotopes can be determined by modification of known epitopes[46,47] or defined de novo with randomized molecular libraries.[48]

The most potent approach for determining synthetic T-cell epitopes is positional scanning. This involves peptide libraries with one constant amino acid at a defined position and mixtures of all other amino acids at the remaining positions. These peptide libraries are used in assays with T-cell clones to scan every sequence position for the amino acids that can activate the T cells. These amino acids are combined with corresponding MHC-allele-specific epitope motifs to obtain defined epitopes that bind well to the MHC molecules and stimulate the T cells. This concept has proven successful for the determination of new T-cell epitopes for both experimental models and clinical applications.

In an effective way of using random peptide libraries, purified MHC class II molecules have been used. In this study, three closely related polymorphic class II molecules were used and their sequences compared with corresponding three families of epitopes.[49] All three families of phagotopes were found to have common features, as well as significant differences. The sequences can be aligned to show a conserved tyrosine residue at position 1, followed by two variable residues and a methionine residue at position 4 of the sequence. Within each family, there appears to be allele-specific conserved residue at position 6. It was proposed that tyrosine and methionine are the anchor residues, which fit into two pockets that are common to the three allelic class II molecules. In contrast, residues at position 6 in the phagotope sequence correspond to a site on the class II molecule that is responsible for allele-specific binding.[49]

The T-cell clones, obtained in a study on the specificity of three self-MHC-restricted T-cell clones that were all induced with the same natural antigen, the transplantation antigen H3, show different specificity patterns.[48] While each of them responded to a number of different peptides, only three peptides were recognized by all three clones: LIYIFNTL, LIYIYNTL, and LIYLYNTL. The most potent epitopes for each one of the clones were not recognized by either of the others.

For the allo-MHC-specific CTL clone 2C, which was raised against H-2L, two sets of epitopes were determined: an allo-MHC-restricted set, by systematic mutation of its cognate peptide LSPF-PFDL,[50] and a self-MHC-restricted set, using the peptide-library approach.[51] The two sets of epitopes were not overlapping. Molecular modeling analysis revealed that the surfaces of the two types of MHC-peptide complexes exhibit no structural similarity, and modifications in the peptides that increase the similarity of the resulting complexes reduced their capacities to stimulate T cells. The specificity of this T-cell clone is therefore defined by two distinct sets of epitopes presented by two different MHC molecules.

7.7 SELECTION OF DISEASE-SPECIFIC PHAGOTOPES

A very important and interesting area of application of epitope libraries is in the study of disease and humoral immune responses. This can be achieved by the use of polyclonal antibodies in the screening of random peptide libraries. The identification of disease-specific mimotopes using sera from patients could provide a set of novel reagents useful for the diagnosis of the disease. Furthermore, in those cases where the humoral immune response is protective, disease-related mimotopes displayed on phage (phagotopes) could represent a step in the development of acellular vaccines.

In many diseases in which the etiologic agent is unknown, disease-specific phagotopes could be used to define common features in the immune response of different individuals to the same etiological agent: its "immunofootprint." In these cases, the selected phagotopes could prove to be an invaluable tool for the identification of the etiological agent.

Sera from patients suffering from an infectious disease often contain a large number of different antibodies directed against the infectious agent. Characterizing the nature, specificity, and affinity of these antibodies is usually possible only if the natural antigen, either as a living microorganism or as the product of a cloned gene, is available and can be used as a reagent. In other diseases (e.g., autoimmune diseases), the characterization of the immunological disorder is essentially confined to those antibodies for which a self-antigen is identified.

This technology has been applied to a number of diseases: viral, bacterial, parasitic, autoimmune, and cancer. Some examples will be discussed in detail.

7.7.1 VIRAL DISEASES

It has been demonstrated that by screening phage libraries with sera from selected patients, it is possible to identify disease-related mimotopes, that is, mimotopes that react specifically with sera from patients but not from healthy control individuals. In the case of infectious diseases, this strategy leads to the recovery of some of the information left by the pathogen in the immunologic memory of the patient. Disease-related mimotopes, selected by the use of the same set of receptors, could induce the same reaction in the immune system as the original antigen; therefore, they are possible candidates for vaccine development.

7.7.1.1 Hepatitis B

The first study was carried out using hepatitis B surface antigen because it is widely known for inducing a protective immune response. Human serum was used to identify and characterize mimotopes from random peptide libraries. Phagotopes that reacted with patient sera were purified from >10^7 different sequences by affinity selection and immunological screening of plaques.[52] Disease-specific phagotopes were identified out of this pool through an antigen-independent procedure that requires only patient and normal human sera. Using this strategy, mimotopes of two different epitopes from the human hepatitis B surface antigen were selected. Immunization of mice with these mimotopes elicited a strong specific response against the hepatitis B surface antigen. These results show that for vaccine development the information about etiological agent is not required.[52]

In a related study two mouse monoclonal antibodies specific to the human hepatitis B virus surface antigen were used to screen a random peptide library of 15 amino acid residues. The phagotopes were selected by a combination of affinity selection, immunoscreening, and ELISA. The selected phage-displayed epitopes behave as antigenic mimics of hepatitis B surface antigen.[53]

The humoral immune response of animals immunized with either a recombinant hepatitis B surface antigen or mimotopes were compared to test if these mimotopes could be useful in developing a vaccine against the human hepatitis B virus.[54] The mimotopes were immunized using different carrier molecules (phage coat protein pIII and pVIII, recombinant human H ferritin, and HBV core peptides). Immunogens were also prepared by synthesizing multiple antigenic peptides carrying the mimotopes' amino acid sequences. These immunogens were injected into mice and rabbits and sera were collected and tested for the presence of hepatitis B surface-antigen-specific antibodies. The data confirmed that mimotopes can induce a humoral immune response resembling that induced by the original antigen, and HBsAg mimotopes displayed on phage prove to be the best immunogens, inducing the most reproducible and potent immunization.[54]

To characterize the primary structure of the hepatocyte-binding domain on the pre-S protein, a phage-displayed 15-mer peptide library and an 8-mer solid-phase peptide library were used to analyze the fine specificity of the pre-S-specific monoclonal antibody MA 18/7. Several mimotopes were identified with the phage-displayed peptide library, the majority of which posses a central motif with at least three identical residues present within the native pre-S1 sequence.[55] All phage mimotopes and a single mimotope from the solid-phase peptide library competed with recombinant hepatitis B surface antigen particles containing the pre-S1 region for binding to MA 18/7. These data show that the structure of the pre-S molecule around the conserved DPAF motif in the pre-S region may have a functional role in binding the hepatitis B virus to cellular receptors, and the central motif identified in mimotopes of this region may offer a novel strategy target for the improvement of existing hepatitis B vaccines.[55]

To acquire the phage-displayed mimotopes that mimic the specificity of hepatitis B virus surface antigen (HBsAg), a random peptide library expressing a linear peptide with 12 amino acids in length was used to screen with the serum from a patient infected with hepatitis B virus in the recovery phase.[56] After three rounds of biopanning, the positive phages were confirmed by competitive ELISA using HBsAg/P33. Two phagotopes were identified and one was confirmed as mimotope by competition experiment. Based on the mimotope, a multiple antigenic peptide with four branches was synthesized by solid-phase peptide synthesis. The antiginicity and specificity of the synthesized antigen was tested in BALB/c mice compared with the native epitope-based antigen. The results showed that the mimotope-based antigen could evoke higher titer of antibodies with the same specificity of the epitope-based antigen. These findings clearly indicate that mimotopes can be used in antigen and vaccine design.[56]

7.7.1.2 Hepatitis C

Hepatitis C virus (HCV) is a major cause worldwide of chronic hepatitis, liver cirrhosis, and hepato-cellular carcinoma, and the development of an effective vaccine represents a high-priority goal.

The same technology as described above for hepatitis B was also used to select antigenic and immunogenic mimics of hepatitis C virus (HCV). The sera from HCV-infected patients and non-infected subjects were used to screen random peptide libraries. Peptides specifically reacting with sera from infected patients were selected, and these were then shown to mimic distinct HCV determinants.[57] Phage-displayed HCV mimics were substitutes for the authentic HCV epitopes in inducing a strong specific response against HCV when used as immunogens in mice.

To rapidly identify mimotopes of the human HCV core protein, an anticore human monoclonal antibody (B12.F8) was used as a probe for screening phages that were affinity-selected using a serum from a HCV-infected patient.[58] Three different positive phages were isolated displaying low or no homology with the natural antigen but which still efficiently bound to the antigen-binding site of the B12.F8 antibody. Testing the reactivity of these phages with 45 sera from HCV-infected patients showed that antibodies recognizing them are present in more that 80% of this population. This study provides novel information on the potential use of the phage display technology for the characterization of antibody specificity as well as disease diagnosis and prevention.[58]

A recent study showed that mimotopes of the hypervariable region 1 of the putative envelope protein E2 of HCV can induce antibodies to cross-react with a large number of viral variants.[59] The library of more than 200 hypervariable region-1 sequences displayed on M13 bacteriophage was affinity-selected using many different sera from infected patients. Phages were identified that reacted very frequently with patient sera and bound serum antibodies that cross-reacted with a large panel of hypervariable region-1 peptides derived from natural HCV variants. When injected into experimental animals, the mimotopes with the highest cross-reactivity induced antibodies that recognized the same panel of natural hypervariable region 1 variant.[59]

The hypervariable region 1 (HVR1) of the putative envelope protein E2 of HCV contains a principal neutralization epitope, and anti-HVR1 antibodies have been shown to possess protective activity in *ex vivo* neutralization experiments. However, the high rate of variability of this antigenic fragment may play a major role in the mechanism of escape from host immune response and might represent a major obstacle to developing an HCV vaccine. Thus, even if direct experimental evidence of the neutralizing potential of anti-HVR1 antibodies by active immunization is still missing, the generation of a vaccine candidate with a cross-reactive potential would be highly desirable. To overcome the problem of HVR1 variability, cross-reactive HVR1 peptide mimics (mimotopes) have been engineered at the N terminus of the E2 ectodomain in plasmid vectors, and these are suitable for genetic immunization.[60] High levels of secreted and biologically active mimotope/E2 chimeras were obtained by transient transfection of these plasmids in cultured cells. All plasmids elicited anti-HVR1 antibodies in mice and rabbits, with some of them leading to a cross-reacting response against many HVR1 variants from natural isolates. Epitope mapping revealed a pattern of reactivity similar to that induced by HCV infection. In contrast, plasmids encoding naturally occurring HVR1 sequences displayed on either full-length E2 in the context of the whole HCV structural region or a soluble, secreted E2 ectodomain did not induce a cross-reacting anti-HVR1 response.[60]

The hypervariable region 1 (HVR1) of the second envelope protein (E2) of HCV contains a principal neutralizing determinant, but it is highly variable among different isolates and is involved in the escape from host immune response. To be effective, a vaccine should elicit a cross-reacting humoral response against the majority of viral variants. It has been shown that it is possible to achieve a broadly cross-reactive immune response in rabbits by immunization with mimotopes of the HVR1 selected from a specialized phage library using HCV patient sera.[61] Some of the cross-reacting antimimotope antibodies elicited in rabbits recognize discontinuous epitopes in a manner similar to those induced by the virus in infected patients.

7.7.1.3 AIDS

To date there is no vaccine for AIDS. Most synthetic HIV-1 gp120 V3 loop peptides that are used as immunogens in experimental HIV-1 vaccine studies are modeled from the naturally occurring viral gp120 V3 loops. In experimental animals these immunogens generally elicit type- (or variant-) specific neutralizing antibodies that are not broadly reactive among HIV-1 variants. In an attempt to find a more general structure for the V3 loop, candidates that mimic V3 loop sequences were obtained by screening random epitopes displayed in a fusion phage 15-residue epitope library. Human monoclonal antibody 447-52 D, a highly potent and broadly reactive virus-neutralizing antibody that recognizes the conserved V3 loop tip motif GPXR, was used as the probe. Over 98% of the epitopes identified contained the motif GPXR. One of these sequences was synthesized as the β-maleimidopropionyl 15-mer peptide and used to immunize rabbits after covalent conjugation to the carrier. The sera from positive rabbits neutralized HIV-1 variant *in vitro*.[62]

In a recent study HIV-specific immunogenic epitopes were isolated by screening random peptide libraries with HIV-1-positive sera.[63] After extensive counterscreening with HIV-negative sera, peptides specifically recognized by antibodies from HIV-1-infected individuals were isolated. These peptides behaved as antigenic mimics of linear or conformational HIV-1 epitopes generated *in vivo* in infected subjects. The selected peptides were immunogenic in mice, where they elicited HIV-specific antibodies that effectively neutralized HIV-1 isolates.[63]

7.7.1.4 Other Viral Diseases

Using a solid-phase 8-mer random-combinatorial peptide library, a panel of mimotopes of an epitope recognized by a monoclonal antibody to the F protein of measles virus were generated.[64] An inhibition immunoassay was used to show that these peptides were bound by the monoclonal antibody with different affinities. BALB/c mice were coimmunized with the individual mimotopes and a T-helper epitopes peptide, and the reactivity of the induced antimimotope antibodies with the corresponding peptides and with measles virus were determined. Their results indicated that appropriate immunization with mimotopes could induce protective antibody response.[64]

In a recent study synthetic peptides mimicking a conformational B-cell epitope (M2) of the measles virus fusion protein (MVF) were used for the immunization of BALB/c mice, and the anti-peptide and anti-virus antibody titres induced were compared.[65] Of the panel of tested peptides, a chimeric peptide consisting of two copies of a T-helper epitope (residues 288 to 302 of MVF), one copy of the mimotope M2 (TTM2), and a multiple-antigen peptide with eight copies of M2 (MAP-M2) induced the highest titers of anti-M2 and anti-MV antibodies. Furthermore, peptides TTM2 and MAP-M2 induced antibodies with highest affinity for the mimotope and highest avidity for measles virus. Immunization with the MAP-M2 construct induced high titers of high-affinity anti-M2 antibody despite the absence of a T-helper epitope, and lymphocyte proliferation data suggest that the addition of M2 to the MAP resulted in the generation of a structure capable of stimulating T-cell help. Sera with anti-M2 reactivity were pooled according to affinity values for binding to M2, and high- and low-affinity pools were tested for their ability to prevent MV-induced encephalitis in a mouse model. The high-affinity serum pool conferred protection in 100% of mice, whereas the lower-affinity pool conferred protection to only 50% of animals. These results indicate the potential of mimotopes for use as synthetic peptide immunogens and highlight the importance of designing vaccines to induce antibodies of high affinity.[65]

Respiratory syncytial virus (RSV) is the most important cause of bronchiolitis and pneumonia in infants and young children worldwide, and the development of a synthetic-peptide-epitope-based vaccine to induce virus-neutralizing antibodies against RSV would seem to be a valid approach to the production of an effective vaccine against infection. A combinatorial solid-phase peptide library has been screened with a virus-neutralizing, protective monoclonal antibody (MAb19) directed toward a conserved and conformationally determined epitope of the Fusion (F) protein of the virus.[66]

Two of the sequences identified from the peptide library using MAb19 reacted specifically with the antibody, and amino acid substitution experiments identified four sequences from one of the mimotopes that showed increased reactivity with MAb19. Immunization of BALB/c mice with these mimotopes, presented as MAPs, resulted in the induction of antipeptide antibodies that inhibited the binding of MAb19 to the virus and neutralized viral infection *in vitro*, with titers equivalent to those in sera from RSV-infected animals. Following RSV challenge of mimotope-immunized mice, a significant reduction in the titer of virus and a greatly reduced cell infiltration into the lungs of immunized mice compared to that in controls were observed. The induction of virus-specific cytotoxic T-lymphocyte responses as well as virus-specific antibodies is likely to be necessary in an effective vaccine. The incorporation of a peptide representing a CTL epitope from the M2 protein of the virus together with peptides inducing T-helper and antimimotope responses in a peptide cocktail vaccine resulted in a more effective clearance of the virus from immunized, challenged mice than peptide-induced humoral or cellular immunity alone.[66]

7.7.2 PARASITIC DISEASES

Malaria is the most challenging of all the parasitic diseases because no vaccine against malaria has been developed so far. Because of the potential for use in vaccine development, peptide mimotopes are of considerable interest.

The mouse monoclonal antibody 2A10, which recognizes the (NANP)n repeat of *Plasmodium falciparum* circumsporozoite surface protein, was used to screen a filamentous phage expressing random amino acid hexamers.[67] The peptide sequences obtained showed 50% homology with the native epitope (PNANPN) and therefore were considered to mimic its structure. Two of the mimotopes (TNRNPQ and NNDNPQ) inhibited the binding of monoclonal antibody 2A10 to the recombinant protein R32LR, which contains the amino acid sequence [(NANP)15NVDP]2. Immunization of mice and rabbits using the peptide (TNRNPQ)4 induced a humoral immune response[67] against *P. falciparum* sporozoites.

The phage-display technology was used to identify peptides that bind D14-3, a monoclonal antibody raised against the 42-kDa C-terminal fragment of *P. vivax* merozoite surface protein 1.[68] By screening a constrained hexapeptide library, seven independent clones binding to D14-3 were isolated. Sequence analysis of these showed that five of them shared homology with the 42-kDa C-terminal fragment and therefore appear to identify the D14-3 epitopes. These mimotopes were immunized in Biozzi mice to study their vaccine potential.[69] High titers of antibodies cross-reacting with the 42-kDa protein were generated, and two monoclonal antibodies that specifically bind the protein were isolated.[69] These data show that mimotopes selected from random repertoires do not necessarily represent structural equivalents of the original antigen but provide functional images that could replace it for vaccine development.

7.7.3 AUTOIMMUNE AND OTHER DISEASES

The application of phage-display technology to type I diabetes could provide a set of novel reagents for diabetes prediction and could also lead to the identification of novel autoantigens or even of environmental factors possibly causing the disease. Sera of prediabetic and high-risk individuals were used to select candidate peptides from phage-displayed random peptide libraries.[70] Diabetes-specific phage clones were then identified from these through screening and counterscreening, using sera from diabetic and nondiabetic individuals. Several candidate disease-related peptides were identified, one of which, clone 92, showed a significant difference in the frequency of reactivity with the sera of patients and normal controls.[71] These studies indicated that the peptide identified mimics a novel, diabetes-related self-antigen.[71]

Three monoclonal IgG2a anti-DNA polyreactive autoantibodies (derived from lupus-prone mice) — F14.6, J20.8, and F4.1 — were used for screening a constrained random peptide library displayed on M13 bacteriophages. The specific consensus motifs (mimotopes) were determined for all of these antibodies.[72] The determined consensus sequences did not match with known sequences. Peptides specific to F14.6 shared negative charges and aromatic rings that may mimic a DNA backbone, while peptides selected with J20.8 did not bear any negative charge, implying a different kind of molecular recognition, e.g., hydrogen or salt bonds. The peptides selected on J20.8 also bound serum antibodies from human patients with systemic lupus erythematosus. In addition, BALB/c mice immunized with some of the selected phages exhibited high titers of IgG3 antidouble-strand DNA antibodies, further supporting the hypothesis that peptide epitopes mimic an oligonucleotide structure.[72]

Immunohistochemical studies have shown that a unique immunoreactive molecule is present near the apical region of human biliary epithelial cells in patients with primary biliary cirrhosis. This can be visualized by confocal microscopy in primary biliary cirrhosis livers using a number of unique monoclonal antibodies to dihydrolipoamide acetyltransferase, which is the E2 component of pyruvate dehydrogenase complex, the autoantigen most commonly recognized by antimitochondrial antibodies. One such antibody, the murine monoclonal antibody C355.1, was used to identify peptide mimotopes of pyruvate dehydrogenase complex E2 by screening a random dodecapeptide phage library ON159.2.[73,74] Eight different sequences were identified in 36-phage clones. WMSYP-DRTLRTL was present in 29 clones. WESYPFRVGTSL, APKTYVSVSGMV, LTYVSLQGRQGH, LDYVPLKHRHRH, AALWGVKVRHVS, KVLNRIMAGVRH, and GNVALVSSRVNA were singly represented. Three common amino acid motifs (W-SYP, TYVS, and VRH) were shared among all peptide sequences.[73,74] These studies provide evidence that random-phage mimotopes can be recognized by monoclonal antibodies.

Mimotopes of a tumor-associated T-cell epitope were determined using randomized and combinatorial peptide libraries and a CD8[(+)] T-cell clone specific to the cutaneous T-cell lymphoma cell line MyLa.[75] Antigen recognition by this clone was found to be HLA-B8 restricted. More than 80% of HLA-matched patients with cutaneous T-cell lymphoma had mimotope-specific CD8[(+)] T cells in their peripheral blood. Mimotope-specific T cells isolated and expanded from a patient lysed MyLa cells in *in vitro* assays, thereby demonstrating their cytolytic capacity and tumor specificity. Mimotope vaccination of a patient without detectable mimotope-specific T cells induced frequencies of these cells of up to 1.82% of the peripheral blood CD8[(+)] T cells.[75]

7.8 CONCLUSIONS AND FUTURE DEVELOPMENTS

Filamentous bacteriophages have been used extensively in recent years for the "display" on their surface of large repertoires of peptides generated by cloning random oligonucleotides at the 5′ end of the genes coding for the phage coat proteins. Homogeneous and heterogeneous ligands, such as monoclonal and polyclonal antibodies, have been used for the affinity selection of these repertoires, leading to the identification of peptides that mimic the specific ligates (mimotopes). More advanced analysis of conformational B epitopes is now possible using combinatorial peptide libraries and conformationally constrained peptides. It is also possible to directly dissect out important B- and T-cell epitopes from complex antigen mixtures using bioassays with monoclonal antibodies or cytotoxic T lymphocytes, respectively, coupled with microsequencing technologies. Once a peptide mimotope is discovered, before it can be useful as a drug it must be stabilized against proteolysis so its *in vivo* activity is prolonged and also so that it cannot be presented as a T-cell epitope, which would generate an immune response.

However, even though many synthetic peptides have been used in a variety of viral, bacterial, and parasitological systems, their application as viable vaccines is just beginning to emerge. Some examples include immunization of mice with *Plasmodium falciparum* peptide construct that protected against an otherwise lethal challenge with rodent malaria,[76] protection against canine parvovirus in dogs,[77] and induction of reversible infertility in mice with synthetic peptides derived from zona pellucida glycoprotein.[78] In all these cases the synthetic peptide represented a known B-cell epitope and was coupled to another peptide or a carrier protein to provide T-cell help.

Future developments will probably focus on optimizing protective or neutralizing epitopes. The ability to synthesize large numbers of peptides by a variety of methods[79–82] will allow systemic examination of antigens for immunogens capable of eliciting protective responses that are not dominant in the native molecule. Further optimization with mutant epitopes with altered amino acid sequence may enhance the immunogenicity of the peptide or reduce the risk of unwanted T-cell responses. Perhaps the most challenging problems lie in improving peptide stability, possibly with peptidomimetics and improving vaccine efficiency using adjuvant technology. At least now some of the necessary peptide technology tools are available to select the appropriate peptide epitopes as synthetic vaccine candidates.

SUMMARY

Epitopes are immunologically active regions of an immunogen recognized by lymphocytes. It has been well established that B and T lymphocytes recognize different epitopes on the same antigenic molecule. In recent years, the understanding of the specificity of B and T cells has been advanced significantly by the development and use of combinatorial libraries made up of thousands to millions of synthetic peptides. The success of peptide libraries depends on the use of carefully selected synthetic DNA (oligodeoxynucleotides) for the synthesis of these libraries. The most important use of combinatorial libraries is for the identification of high-affinity peptides recognized by monoclonal antibodies. More recently, MHC-binding motifs and T-cell specificity have also been examined using peptide libraries. The sequences derived from the library by using various screening approaches sometimes show significant sequence divergence from the native molecules and are felt to mimic their conformation. Therefore, these epitopes are also referred to as mimotopes. The mimotope approach involves synthesizing a peptide series with successively better fit to the antigen-combining site of the antibody. This technology is a potentially powerful tool that could be applied to research in many areas of biology and medicine including drug design, the development of diagnostic markers, and vaccine production.

ACKNOWLEDGMENTS

The author is grateful to Dr. N. Tuteja (ICGEB, New Delhi, India) for his encouragement and continuous support and to Dr. O. Burrone (ICGEB, Trieste, Italy) for a critical reading of the manuscript.

REFERENCES

1. Geyson, H. M., Rodda, S. J., and Mason, T. J., *A priori* delineation of a peptide which mimics a discontinuous antigenic determinant, *Mol. Immunol.*, 23, 709, 1986.
2. Novotny, J., Handschumacher, M., and Bruccoleri, R. E., Protein antigenicity: a static surface property, *Immunol. Today*, 8, 26, 1987.
3. Garcia, K. C. et al., Three-dimensional structure of an angiotensin II-fab complex at 3 A: hormone recognition by an anti-idiotypic antibody, *Science*, 257, 502, 1992.
4. Sparbier, K. and Walden, P. T., Cell receptor specificity and mimotopes, *Curr. Opin. Immunol.*, 11, 214, 1999.
5. Sidney, J. et al., Specificity and degeneracy in peptide binding to HLA-B7-like class I molecules, *J. Immunol.*, 157, 3480, 1996.
6. Stevens, J. et al., Peptide length preferences for rat and mouse MHC class I molecules using random peptide libraries, *Eur. J. Immunol.*, 28, 1272, 1998.

7. Zarling, A. L. and Lee, D. P., Conversion of a human immunodeficiency virus cytotoxic T lymphocyte epitope into a high affinity HLA-Cw3 ligand, *Hum. Immunol.*, 59, 472, 1998.

8. Bianco, A. et al., New synthetic non-peptide ligands for classical major histocompatibility complex class I molecules, *J. Biol. Chem.*, 273, 28759, 1998.

9. Geyson, H. M., Rodda, S. J., and Mason, T. J., Synthetic peptides as antigens, in *Ciba Foundation Symposium 119*, Porter, R. and Wheelan, J., Eds., Wiley, New York, 1986, 131.

10. Pinilla, C. et al., Exploring immunological specificity using synthetic peptide combinatorial library, *Curr. Opin. Immunol.*, 11, 193, 1999.

11. Geyson, H. M. et al., Strategies for epitope analysis using peptide synthesis, *J. Immunol. Methods*, 102, 259, 1987.

12. Lam, K. S. et al., A new type of synthetic peptide library for identifying ligand binding activity, *Nature*, 354, 82, 1991.

13. Houghten, R. A. et al., Generation and use of synthetic peptide combinatorial libraries for basic research and drug discovery, *Nature*, 354, 84, 1991.

14. Pinilla, C. et al., Rapid identification of high affinity peptide ligands using positional scanning synthetic peptide combinatorial libraries, *Biotechniques,* 13, 901, 1992.

15. Dulbecco, R., U.S. patent 4,593,002, 1986.

16. Cwirla, S. E. et al., Peptides on phage: a vast library of peptides for identifying ligands, *Proc. Natl. Acad. Sci. U.S.A.*, 87, 6378, 1990.

17. Devlin, J. J., Panganiban, L. C., and Devlin, P. E., Random peptide libraries: a source of specific protein binding molecules, *Science,* 249, 404, 1990.

18. Smith, G. P., Filamentous fusion phage: novel expression vectors that display cloned antigens on the virion surface, *Science,* 228, 1315, 1985.

19. Cortese, R. et al., Identification of biologically active peptides using random libraries displayed on phage, *Curr. Opin. Biotechnol.*, 6, 73, 1995.

20. Felici, F. et al., Selection of antibody ligands from a large library of oligopeptides expressed on a multivalent exposition vector, *J. Mol. Biol.*, 222, 301, 1991.

21. Hoess, R. et al., Identification of a peptide which binds to the carbohydrate-specific monoclonal antibody-B3, *Gene*, 128, 43, 1993.

22. Scott, J. K. and Smith, G. P., Searching for peptide ligands with an epitope library, *Science*, 249, 386, 1990.

23. Smith, G. P., Surface presentation of protein epitopes using bacteriophage expression systems, *Curr. Opin. Biotechnol.*, 2, 668, 1991.

24. Scott, J. K., Discovering peptide ligands using epitope libraries, *TIBS*, 17, 241, 1992.

25. Kang, A. S. et al., Linkage of recognition and replication functions by assembling combinatorial antibody Fab Libraries along phage surfaces, *Proc. Natl. Acad. Sci. U.S.A.*, 88, 4363, 1991.

26. Parmley, S. F. and Smith, G. P., Antibody-selectable filamentous fd phage vectors: affinity purification of target genes, *Gene*, 73, 305, 1988.

27. Luzzago, A. et al., Mimicking of discontinuous epitopes by phage-displayed peptides. I. Epitope mapping of human H ferritin using a phage library of constrained peptides, *Gene*, 128, 51, 1993.

28. McLafferty, M. A. et al., M13 bacteriophage displaying disulfide-constrained microproteins, *Gene*, 128, 29, 1993.

29. Zhong, G. et al., Conformational mimicry of a chlamydial neutralization epitope on filamentous phage, *J. Biol. Chem.*, 269, 24183, 1994.

30. Barchan, D. et al., Identification of epitopes within a highly immunogenic region of acetylcholine receptor by a phage epitope library, *J. Immunol.*, 155, 4264, 1995.

31. Cook, A. D. et al., Mimotopes identified by phage display for the monoclonal antibody C II-C1 to Type II collagen, *J. Autoimmunol.,* 3, 205, 1998.

32. Cerino, A. et al., A human monoclonal antibody specific for the N terminus of the hepatitis C virus nucleocapsid protein, *J. Immunol.*, 151, 7005, 1993.

33. Felici, F. et al., Mimicking of discontinuous epitopes by phage displayed peptide. II. Selection of clones recognized by a protective monoclonal antibody against the *Bordetella pertussis* toxin from phage peptide libraries, *Gene*, 128, 21, 1993.

34. Balass, M. et al., Identification of a hexapeptide that mimics a conformation-dependent binding site of acetylcholine receptor by use of a phage-epitope library, *Proc. Natl. Acad. Sci. U.S.A.*, 90, 10638, 1994.

35. Kay, B. K. et al., An M13 phage library displaying random 38-amino-acid peptides as a source of novel sequences with affinity to selected targets, *Gene*, 128, 59, 1993.

36. Weber, P. C., Pantoliano, M. W., and Thompson, L. D., Crystal structure and ligand-binding studies of a screened peptide complexed with streptavidin, *Biochemistry*, 31, 9350, 1992.

37. Oldenburg, K. R. et al., Peptide ligands for a sugar-binding protein isolated from a random peptide library, *Proc. Natl. Acad. Sci. U.S.A.*, 89, 5393, 1992.

38. Scott, J. K. et al., A family of concanavalin A-binding peptides from a hexapeptide epitope library, *Proc. Natl. Acad. Sci. U.S.A.*, 89, 5398, 1992.

39. Udaka, K. et al., Self MHC restricted peptides recognised by an alloreactive T lymphocyte clone, *J. Immunol.*, 157, 670, 1996.

40. Mukherjee, J., Scharff, M. D., and Casadevall, A., Protective murine monoclonal antibodies to *Cryptococcus neoformans*, *Infect. Immunol.*, 60, 4534, 1992.

41. Young, A. C. M. et al., The three-dimensional structures of a polysaccharide binding antibody to *Cryptococcus neoformans* and its complex with a peptide from a phage display library: implications for the identification of peptide mimotopes, *J. Mol. Biol.*, 274, 622, 1997.

42. Valadon, P. et al., Aspects of antigen mimicry revealed by immunization with a peptide mimetic of *Cryptococcus neoformans* polysaccharide, *J. Immunol.*, 161, 1829, 1998.

43. Luo, P. et al., Antigenic and immunological mimicry of peptide mimotopes of Lewis carbohydrate antigens, *Mol. Immunol.*, 35, 865, 1998.

44. Qiu, J. et al., Towards the development of peptide mimotopes of carbohydrate antigens as cancer vaccines, *Hybridoma*, 18, 103, 1999.

45. Kieber-Emmons, T. et al., DNA immunization with minigenes of carbohydrate mimotopes induce functional anti-carbohydrate antibody response, *J. Immunol.*, 165, 623, 2000.

46. Chen, Y. Z., Matsushita, S., and Nishimura, Y., Response of a human T cell clone to a large panel of altered peptide ligands carrying single residue substitutions in an antigenic peptide, *J. Immunol.*, 157, 3783, 1996.

47. Strausbauch, M. A. et al., Identification of mimotopes for the H4 minor histocompatibility antigen, *Int. Immunol.*, 10, 421, 1998.

48. Gundlach, B. R. et al., Determination of T cell epitopes with random peptide libraries, *J. Immunol. Methods*, 192, 149, 1996.

49. Hammer, J. et al., Promiscuous and allele-specific anchors in HLA-DR-binding peptides, *Cell*, 74, 197, 1993.

50. Brock, R. et al., Molecular basis for the recognition of two structurally different major histocompatibility complexes by a single T cell receptor, *Proc. Natl. Acad. Sci. U.S.A.*, 93, 13108, 1996.

51. Valadon, P. et al., Peptide libraries define the fine specificity of anti-polysaccharide antibodies to *Cryptococcus neoformans*, *J. Mol. Biol.*, 261, 11, 1996.

52. Folgori, A. et al., A general strategy to identify mimotopes of pathological antigens using only random peptide libraries and human sera, *EMBO J.*, 13, 2236, 1994.

53. Motti, C. et al., Recognition by human sera and immunogenicity of HBsAg mimotopes selected from an M13 phage display library, *Gene*, 146, 191, 1994.

54. Meola, A. et al., Derivation of vaccines from mimotopes. Immunologic properties of human hepatitis B surface antigen mimotopes displayed on filamentous phage, *J. Immunol.*, 154, 3162, 1995.

55. D'Mello, F. et al., Definition of the primary structure of hepatitis B virus (HBV) pre-S hepatocyte binding domain using random peptide libraries, *Virology*, 237, 319, 1997.

56. Zhang, W. Y. et al., A mimotope of pre-S2 region of surface antigen of viral hepatitis B screened by phage display, *Cell Res.*, 11, 203, 2001.

57. Prezzi, C. et al., Selection of antigenic and immunogenic mimics of hepatitis C virus using sera from patients, *J. Immunol.*, 156, 4504, 1996.

58. Tafi, R. et al., Identification of HCV core mimotopes: improved method for the selection and use of disease-related phage-displayed peptides, *Biol. Chem.*, 378, 495, 1997.

59. Puntoriero, G. et al., Towards a solution for hepatitis C virus hypervariability: mimotopes of the hypervariable region 1 can induce antibodies cross-reacting with a large number of viral variants, *EMBO J.*, 17, 3521, 1998.

60. Zucchelli, S. et al., Mimotopes of the hepatitis C virus hypervariable region 1, but not the natural sequences, induce cross-reactive antibody response by genetic immunization, *Hepatology*, 33, 692, 2001.

61. Roccasecca, R. et al., Mimotopes of the hyper variable region 1 of the hepatitis C virus induce cross-reactive antibodies directed against discontinuous epitopes, *Mol. Immunol.*, 38, 485, 2001.

62. Keller, P. M. et al., Identification of HIV vaccine candidate peptides by screening random phage epitope libraries, *Virology*, 193, 709, 1993.

63. Scala, G. et al., Selection of HIV-specific immunogenic epitopes by screening random peptide libraries with HIV-1 positive sera, *J. Immunol.*, 162, 6155, 1999.

64. Steward, M. W., Stanley, C. M., and Obeid, O. E., A mimotope from a solid-phase peptide library induces a measles virus-neutralizing and protective antibody response, *J. Virol.*, 69, 7668, 1995.

65. Olszewska, W., Obeid, O. E., and Steward, M. W., Protection against measles virus-induced encephalitis by anti-mimotope antibodies: the role of antibody affinity, *Virology*, 272, 98, 2000.

66. Steward, M. W., The development of a mimotope-based synthetic peptide vaccine against respiratory syncytial virus, *Biologicals,* 29, 215, 2001.

67. Stoute, J. A. et al., Induction of humoral immune response against *Plasmodium falciparum* sporzoites by immunization with a synthetic peptide whose sequence was derived from screening a filamentous phage epitope library, *Infect. Immunol.*, 63, 934, 1995.

68. Demangel, C., Lafaye, P., and Mazie, J. C., Reproducing the immune response against the *Plasmodium vivax* merozoite surface protein 1 with mimotopes selected from a phage-displayed peptide library, *Mol. Immunol.*, 33, 909, 1996.

69. Demangel, C. et al., Phage-displayed mimotopes elicit monoclonal antibodies specific for a malaria vaccine candidate, *Biol. Chem.,* 379, 65, 1998.

70. Mennuni, C. et al., Selection of phage-displayed peptides mimicking Type 1 diabetes-specific epitopes, *J. Autoimmunol.*, 9, 431, 1996.

71. Mennuni, C. et al., Identification of a novel type1 diabetes-specific epitope by screening phage libraries with sera from pre-diabetic patients, *J. Mol. Biol.,* 268, 599, 1997.

72. Sibille, P. et al., Mimotopes of polyreactive anti-DNA antibodies identified using phage-display peptide libraries, *Eur. J. Immunol.*, 27, 1221, 1997.

73. Cha, S. et al., Random phage mimotopes recognised by monoclonal antibodies against the pyruvate dehydrogenase complex-E2 (PDC-E2), *Proc. Natl. Acad. Sci. U.S.A.*, 93, 10949, 1996.

74. Leung, P. S. et al., Inhibition of PDC-E2 human combinatorial autoantibodies by peptide mimotopes, *J. Autoimmunol.*, 9, 785, 1996.

75. Linnemann, T. et al., Mimotopes for tumor-specific T lymphocytes in human cancer determined with combinatorial peptide libraries. *Eur. J. Immunol.,* 31, 156, 2001.

76. Saul, A. et al., Protective immunization with invariant peptides of the *Plasmodium falciparum* antigen MSA2, *J. Immunol.*, 148, 208, 1992.

77. Langeveld, J. P. M. et al., First peptide vaccine providing protection against viral infection in the target animals: studies of canine parvovirus in dogs, *J. Virol.*, 68, 4506, 1994.

78. Lou, Y. et al., A zona pellucida 3 peptide vaccine induces antibodies and reversible infertility without ovarian pathology, *J. Immunol.*, 155, 2715, 1995.

79. Geysen, H. M., Meloen, R. H., and Barteling, S. J., Use of peptide synthesis to probe viral antigens for epitopes to a resolution of a single amino acid, *Proc. Natl. Acad. Sci. U.S.A.*, 81, 3998, 1984.

80. Houghten, R. A., General method for the rapid solid-phase synthesis of large numbers of peptides: specificity of antigen-antibody interactions at the level of individual amino acids, *Proc. Natl. Acad. Sci. U.S.A.*, 82, 5131, 1985.

81. Maeji, N. J., Bray, A. M., and Geysen, H. M., Multi-pin peptide synthesis strategy for T cell determinant analysis, *J. Immunol. Methods*, 134, 23, 1990.

82. Maeji, N. J. et al., Larger scale multipin peptide synthesis, *Pept. Res.*, 8, 33, 1995.

83. Sparks, A. B. et al., Distinct ligand preferences of SH3 domains from Src, Yes, Abl, Cortactin, p53b2, PLCgamma, Crk and Grb2, *Proc. Natl. Acad. Sci. U.S.A.*, 93, 1540, 1996.

84. Christian, R. B. et al., Simplified method for construction, assessment and rapid screening of peptide libraries in bacteriophage, *J. Mol. Biol.*, 227, 711, 1992.

8 Artificial Genes for Chimeric Virus-Like Particles

Paul Pumpens and Elmars Grens

CONTENTS

8.1 PROTEIN ENGINEERING OF VLPS: AN INTRODUCTION

Designed genetic reconstruction of natural proteins started in the early 1980s when Alan R. Fersht and his colleagues conceived the basic idea of protein engineering, i.e., a mutational intervention into the structure of proteins, based on spatial knowledge and oriented toward the creation of some kind of artificially improved proteins with desired functions.[1-4] Since then, protein engineering as a specific branch of gene engineering has achieved much success, constructing new forms of enzymes and their inhibitors and elucidating the basic rules of protein folding and changing their specific activities (see Reference 5).

Concurrent with the protein engineering of enzymes, i.e., monomeric or oligomeric proteins, a demand arose for the creation of artificial multimeric structures, which have been considered the primary tools for improving immunological activity of epitopes from desired proteins. It should be remembered that the early 1980s was a time when the significance of specific epitopes as antigen regions, which are necessary and sufficient for (1) the induction of antigen-specific immune response and (2) the recognition of antigens by monoclonal or polyclonal antibodies, gained general acceptance, and many epitope sequences were mapped. To become familiar with numerous epitope mapping procedures and practical mapping examples, we recommend two special issues of the *Methods in Molecular Biology* series.[6,7]

Fine mapping resulted, first, in the determination of an *epitope* sequence, which could be defined as the shortest stretch(es) within a protein molecule capable of binding to the appropriate antibodies. The next level of recognition within the epitope can be defined as an *antigenic determinant*, namely, the exact side groups of amino acid residues involved in epitope recognition by the *paratope* sequences of antibodies. The mapping techniques, based in general on the examination of libraries of short overlapping synthetic peptides or fusion proteins,[6,7] revealed the existence of *linear* and *conformational* forms of epitopes. Linear epitope means that the antigenic determinants are localized in the short stretch of amino acid residues (usually from 4 to 10) and their recognition is possible within the polypeptides deprived of specific folding, for example, short synthetic peptides and SDS-denatured fusion proteins. By contrast, conformational epitopes presuppose dispersion of antigenic determinants in the more distant polypeptide stretches or their dependence on specific protein folding.

Gene engineering offered experimental resources to construct DNA copies of both linear and conformational epitopes, to define their length and composition at a single amino acid resolution, to combine them in a different order, and to introduce them into carrier genes, i.e., genes encoding self-assembly competent polypeptides.

Constituents of capsids, envelopes, and rods of, in most cases, viral origin have been suggested as the self-assembly competent carriers (for a first review article, see Reference 8; more recent and specific reviews are indicated in the appropriate sections). In fact, supermolecular structures, frequently named VLPs (virus-like particles), built symmetrically from hundreds of proteins of one or more types, seemed to be the most efficient molecular carriers to the regular arrangement of epitope chains on the desired positions of the outer surface of the carrier. In contrast to monomeric and oligomeric protein carriers, such VLP structures were able to provide not only a high density of introduced foreign oligopeptides per particle, but also a distinctive three-dimensional conformation, which is especially important for the presentation of conformational epitopes. Therefore, the regular, repetitive pattern and correct conformation of inserted epitopes remained factors encouraging persistent work on the functional activity of epitope-carrying VLPs in inducing immunological response.

The interest in such predesigned manipulations on VLPs was reinforced by the simultaneous development of structural knowledge about viral structural units by the use of high-resolution techniques. X-ray crystallography and electron cryomicroscopy revealed high-resolution structures of such favorite VLP carriers as hepatitis B cores,[9-12] RNA phages,[13-17] tobacco mosaic virus,[18,19] poliovirus,[20] rhinovirus,[21] bluetongue virus,[22] and others.

Success in structural investigations of putative carrier moieties and insertion-intended sequences as well as the development of protein structure prediction techniques completed VLP protein

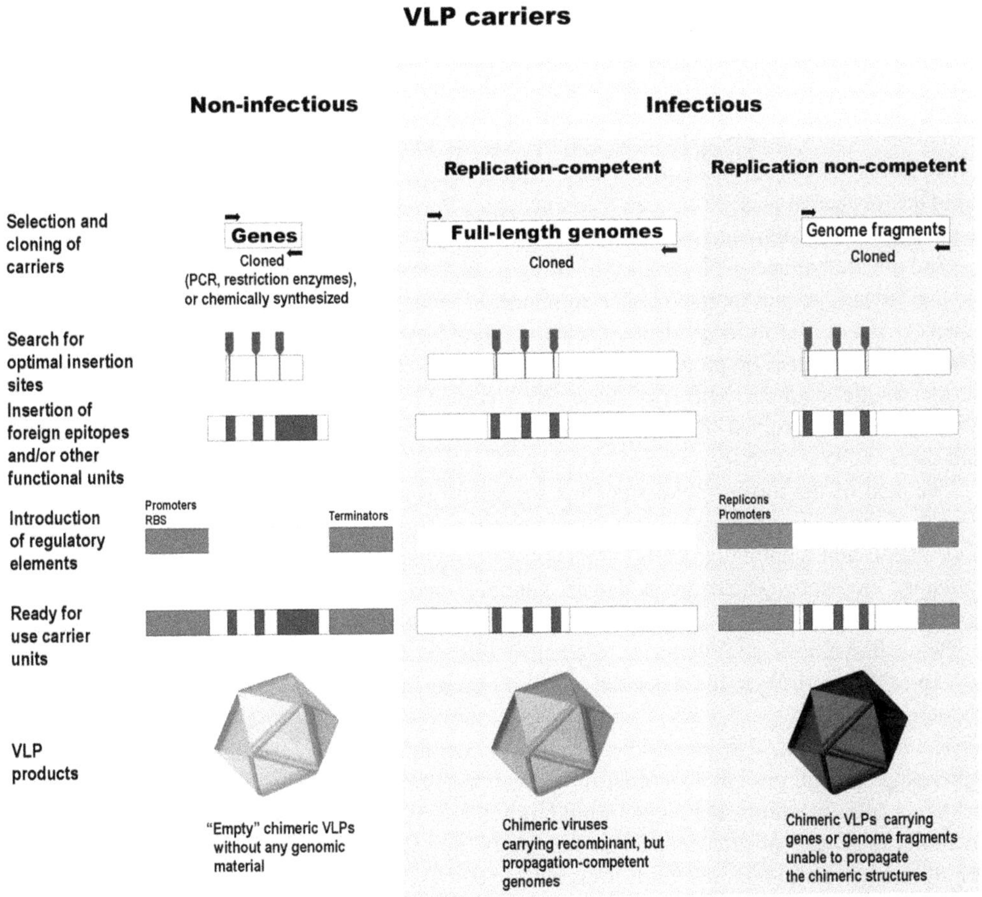

FIGURE 8.1 Major principles of VLP protein engineering, which comprise general constructing schemes of the appropriate carrier genes and genomes. VLP products are shown conditionally as icosahedrons, since the latter are prevalent in the viral world. Nevertheless, rods as well as isometric and bacilliform enveloped capsids are also possible VLP types.

engineering as a specific branch of molecular biology and biotechnology. We can define it now as a *discipline working on knowledge-based approaches to the theoretical design and experimental construction of recombinant genomes and genes that may enable the efficient synthesis and correct self-assembly of chimeric VLPs with programmed structural and functional properties.* More artistically, VLPs were defined as descendants of Chimaera, a monster in ancient Greek mythology that combined elements from different animal species in its body.[23]

Figure 8.1 presents the general outline of operations that comprise three general modes of realization of VLP protein engineering ideas. "Classical" generation of *noninfectious* VLP vectors is based on the expression of one or more isolated viral carrier genes in a suitable prokaryotic or eukaryotic expression system. After self-assembly *in vitro* or *in vivo* and subsequent purification, such classical VLPs do not contain any genomic material capable of productive infection or replication.

Another way of developing VLP techniques provides the production of *infectious* vectors, which could be divided into two main categories. First, are *replication-competent* vectors, which carry full-length infectious viral genomes with hybrid genes, allowing the production of chimeric virus progeny. A second type represents *replication-noncompetent* vectors, which are *infectious* but give rise to chimeric progeny only in special replication conditions, e.g., in the presence of special helpers. Therefore, genomes enveloped by such chimeric progeny are unable to give rise to chimeric particles.

If full-length genomes and long genome fragments are usually obtained by PCR multiplication using specific synthetic oligonucleotides as primers, individual carrier genes could also be generated by total chemical synthesis. The fully synthetic genes of hepatitis B virus (HBV) core protein, one of the most widely used VLP carriers (see Chapter 3.1.2), constructed in 1988 by Michael Nassal,[24] and of the tobacco mosaic virus coat, constructed in 1986 by Joel R. Haynes et al.,[25] furnish excellent examples of this statement.

As indicated above, DNA copies of foreign oligopeptide sequences selected for the insertion into the carrier genes comprise further components of the protein-engineering process. Since these sequences may differ in length from a few to several hundred amino acid residues, the appropriate DNA copies may also be prepared by total chemical synthesis or by PCR multiplication with the use of specific synthetic primers. These DNA copies are adjusted to specific *insertion sites* on carrier genes, which must be selected and proven theoretically and experimentally for each VLP candidate.

The next of the most crucial steps for successful realization of the chimeric VLP production consists of the choice of expression system among prokaryotic bacterial cells and eukaryotic yeast, plant, insect, and mammalian cells. The choice of the expression system determines the nature of regulatory elements, which are added to genomes or genes to be expressed. These regulatory elements (promoters, ribosome binding sites, transcription terminators, etc.) are prepared most often by chemical synthesis and comprise, after their introduction into the appropriate positions within the expression cartridge, the essential parts of sophisticated vectors responsible for high-level production of VLPs.

The latest advances in the field show that chimeric VLPs are capable of presenting not only immunological epitopes but also other functional protein motifs such as DNA or RNA binding and packaging sites, receptors and receptor binding sequences, immunoglobulins, elements recognizing low molecular mass substrates, etc. This inevitably transplants VLP ideology from the conventional area of vaccine and diagnostic tool design to genetic vaccine and therapy applications.

In the last 10 years the list of available VLP carrier candidates has been markedly extended. Tables 8.1 through 8.3, where VLP candidates are grouped in strong taxonomic order in accordance with the rules suggested in Figure 8.1, show those that have received the most attention. It is our opinion that taxonomic characteristics are very important not only to show broad coverage of viruses by the VLP ideology but also to arrange existing candidates for a better understanding of their properties. Furthermore, information on structural characteristics, including organization and symmetry of the particles, will help to assess the further potential of the VLP candidates.

8.2 FIRST GENERATION OF VLP CARRIER CANDIDATES

The early candidates for VLP protein engineering appeared in the middle 1980s and covered at once three main structural forms of carrier candidates. First, VLP candidates came from rod-shaped representatives of *Inoviridae* (filamentous bacteriophage f1)[26] and Tobamovirus (tobacco mosaic virus)[25] families. Second, the outer envelope proteins of enveloped HBV were used to expose foreign sequences on the so-called 22-nm particles formed by them.[27,28] Finally, icosahedral particles have been suggested: inner cores, or nucleocapsids, of HBV,[29–31] nonenveloped capsids of picornaviruses[32] and RNA phages,[33] and VLPs encoded by yeast retrotransposon Ty1.[34]

The first attempts were extremely fruitful not only from the viewpoint of VLP candidate symmetry, but also in laying down the course of development of the two main directions of VLP engineering based on infectious and noninfectious structures. Whereas the latter were represented by HBV envelopes and capsids, TMV, RNA phages, and Ty1 VLPs and formed the basis of a long list of replication noncompetent models, chimeric derivatives of phage f1 and poliovirus initiated

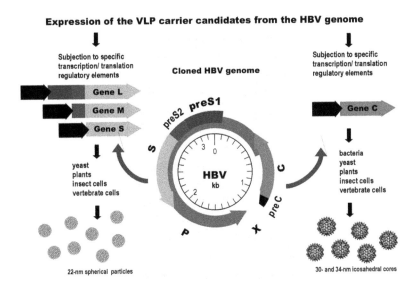

FIGURE 8.2 HBV genome as a source of genes encoding the first two promising VLP candidates: HBV envelope and cores. Size and shape of chimeric particles are depicted in accordance with the existing conceptions on the structure of HBsAg[83] and HBcAg.[12]

the line of replication-competent viruses employed as VLP models. They initiated construction of DNA phage-display vectors, not only classical, based on single-stranded DNA (ssDNA) viruses, but also novel, double-stranded DNA- (dsDNA-) based entities. Further, more poliovirus, rhinovirus, and plant (tobacco and cowpea mosaic) viruses were developed as favorite replication-competent models.

In the present chapter, we have restricted ourselves to noninfectious VLP carriers (Table 8.1).

8.3 HEPATITIS B VIRUS STRUCTURAL PROTEINS AS CARRIERS

In the late 1970s, the HBV genome was cloned, characterized, and sequenced.[35–38] In the 1980s this genome supplied the first two classical noninfectious carrier candidates. They were based on HBV surface (HBs) and core (HBc) proteins, encoded by viral genes S and C and forming two structural HBV antigens: HBsAg and HBcAg, respectively (Figure 8.2). These carriers gave onset to two basic classes of future VLPs, namely, spherical pleomorphic lipid-containing envelopes and icosahedral core capsids. From the point of view of their involvement in the construction of the actual vaccine, they have become the most advanced model species.

Since expression of HBV structural genes served as a model for further constructions, we wish to focus more attention on the development of HBV-based VLP models. Figure 8.2 presents the HBV genome and the general ways in which its structural genes were involved in VLP protein engineering. A high level of synthesis and an ability to correct self-assembly in heterologous expression systems were the decisive factors that enabled practical success of these VLP carriers. While HBV envelopes were capable of self-assembly in all the systems studied with the exception of bacteria, HBV cores formed native-like capsids in the latter as well.

8.3.1 HEPATITIS B SURFACE PROTEINS

HBsAg large (L), middle, (M) and short (S) surface proteins are encoded by a single open reading frame S of 400 (virus genotype A) or 389 (genotype D) codons (Figure 8.2). All contain the S sequence of 226 aa, whereas the M protein has a 55-aa preS2 extension at the N-terminus, and the

TABLE 8.1
Noninfectious VLP Carriers[a]

	Taxonomic and Genetic Categories of a Source of a VLP Carrier						Structural Properties of a VLP carrier		
Genome Type	Family	Genus	Species	Genome Size	Encoding Gene	Size of Protein (aa)	Morphology of Particle	Size of Particle (nm)	Resolution Level
dsDNA	Baculoviridae	Nucleopolyhedrovirus	*Autographa californica* nucleopolyhedrovirus	133,894	polyhedrin	245	Baciliform	ø 30–60; length 250–300	EM
	Papovaviridae	Papillomavirus	Human papillomavirus 16	7,931	L1	531	Icosahedral	ø 60; tubules with	cryoEM
					L2	473		ø 25–30 or 50–60	
			Bovine papillomavirus 1	7,945	L1	495	Icosahedral	ø 60	cryoEM
			Canine oral papillomavirus	8,607	L1	503	Icosahedral	ø 60	EM
		Polyomavirus	Polyomavirus Py	5,297	VP1	384	Icosahedral	ø 50	X-ray
			Hamster polyomavirus	5,366	VP1	384	Icosahedral	ø 50	EM
dsRNA	Birnaviridae	Avibirnavirus	Infectious bursal disease virus	5,294	VP2	452	Icosahedral	ø 60	cryoEM
	Reoviridae	Cypovirus	*Bombyx mori* cytoplasmic polyhedrovirus	20,500–24,000	polyhedrin	248	Icosahedral	ø 55–69	EM
		Orbivirus	Bluetongue virus	19,200	VP3	901	Icosahedral	ø 81 (single-shelled 69)	X-ray
					VP7	349			cryoEM
					NS1	552	Helical	ø 52.3; length 100	cryoEM
		Rotavirus	Bovine rotavirus	~17,000	VP6*	397	Icosahedral	ø 80	X-ray, cryoEM

Genome	Family	Genus	Virus	Genome (nt)	Protein	Length (aa)	Symmetry	Diameter	Method
Retroid	Hepadnaviridae	Orthohepadnavirus	Hepatitis B virus	3,182	surface (S)	226**	Isometric	ø 22	EM
				3,221	core (C)	183;185	Icosahedral	ø 34	X-ray
	Retroviridae	Avian type-C retroviruses	Rous sarcoma virus	9,625	Gag	701	Icosahedral?	ø 80–100	EM
		BLV-HTLV retroviruses	Bovine leukemia virus	8,714	Gag	392	Icosahedral?	ø 80–100	EM
		Lentivirus	Human immunodeficiency virus 1	9,719	Gag	512	Icosahedral?	ø 100–120	cryoEM,NMR
			Human immunodeficiency virus 2	9,671	Gag	521	Icosahedral?	ø 100–120	EM
			Simian immunodeficiency virus	9,623	Gag	512	Icosahedral?	ø 100–120	EM
			Yeast retrotransposon Ty1	~5,900	p1 protein	440	Icosahedral?	ø > 60 nm	cryo EM
ssDNA	Parvoviridae	Dependovirus	Adeno-associated virus 2	4,675	VP3	504	Icosahedral	ø 18–22; 20–26	EM
		Erythrovirus	Human parvovirus B19	5,594	VP2	554	Icosahedral	ø 18–22; 20–26	X-ray
					VP1 ***	781			
		Parvovirus	Porcine parvovirus	4,948	VP2	579	Icosahedral	ø 20–26	EM
			Canine parvovirus	5,323	VP2	584	Icosahedral	ø 20–26	X-ray
(+)ssRNA	Bromoviridae	Alfamovirus	Alfalfa mosaic virus	8,274	coat	221	Icosahedral	ø 18; 35–50	X-ray
	Caliciviridae	Calicivirus	Norwalk virus	7,598	capsid	530	Icosahedral	ø 40; 23	X-ray
	Leviviridae	Allolevivirus	RNA phage Qβ	4,219	coat****	329	Icosahedral	ø 29.4	X-ray
		Levivirus	RNA phage fr	3,569	coat	130	Icosahedral	ø 28.7	X-ray
			RNA phage MS2	3,575	coat	130	Icosahedral	ø 28.8	X-ray
	Luteovirus	Potato leaf roll virus	Potato leaf roll virus	5,883	coat	208	Icosahedral	ø 30	EM
	Nodaviridae	Nodavirus	Flock House virus	~4,500	coat	407	Icosahedral	ø 30	X-ray
	Potyviridae	Potyvirus	Johnsongrass mosaic virus	9,766	coat	303	Rod	ø 12–15; length 773–778	EM
			Plum pox potyvirus	9,741	capsid	330	Rod	ø 11; length 730	EM
	Tobamovirus	Tobamovirus	Tobacco mosaic virus	6,395	coat	158	Rod	ø 18; length 70–300	X-ray

a References are given in the text.
*non-genetic coupling; **M = S+ N-terminal preS2 (55 aa); L = M+N-terminal preS1 (108, or 119 aa); *** Parvovirus VP1 = VP2 + N-terminal 227 aa; ****with the A1 extension.

TABLE 8.2
Infectious Replication-Competent VLP Carriers

Genome Type	Taxonomic and Genetic Categories of a VLP Carrier			Structural Properties of a VLP Carrier			
	Family	Genus	Species	Genome Size	Morphology	Size of Particle (nm)	Target Protein
dsDNA	Baculoviridae	Nucleopoly hedrovirus	*Autographa californica nucleopolyhedrovirus* ("baculophages")	133,894	Bacilliform	ø 30-60; length 250–300	Glycoprotein gp64
	Tailed phages	Siphoviridae	Bacteriophage lambda ("phage display")	48,502	Icosahedral tailed	ø 60 tail 150	Capsid protein D
ssDNA	Inoviridae	Inovirus	Bacteriophages f1, fd, M13 ("phage display")	6,407	Rod	ø 6–8; length 760–1950	pVIII, pIII, pVI
(-)ssRNA	Orthomyxoviridae	Influenza virus A and B group	Influenza A virus	13,588	Spherical	ø 50–120	Hemagglutinin
(+)ssRNA	Comoviridae	Comovirus	Cowpea mosaic virus	9,870	Icosahedral	ø 28	Small coat
	Picornaviridae	Enterovirus	Sabin type 1 poliovirus	7,440	Icosahedral	ø 28–30	VP1
		Rhinovirus	Human rhinovirus 14	7,212	Icosahedral	ø 30	VP1, VP2
	Potexvirus	Potato virus X	Potato virus X	7,568	Rod	ø 13 length 515	Coat
	Potyviridae	Potyvirus	Plum pox potyvirus	9,741	Rod	ø 11 length 700	Capsid
	Tobamovirus	Tobacco mosaic virus	Tobacco mosaic virus	6,395	Rod	ø 18; length 70–300	Coat
	Togaviridae	Alphavirus	Sindbis virus	11,703	Spherical	ø 70	E2, E3
	Tombusviridae	Tombusvirus	Tomato bushy stunt virus	4,776	Icosahedral	ø 30	Coat

TABLE 8.3
Infectious Replication Noncompetent VLP Carriers[a]

Genome Type	Family	Genus	Species	Genome Size	Morphology	Size of Particle (nm)	Target Protein
						Structural Properties of a VLP Carrier	
	Taxonomic and Genomic Categories of a VLP Carrier						
dsDNA	Adenoviridae	Mastadenovirus	Human adenovirus 5, etc.	35,935	Icosahedral	ø 70	Hexon; Penton base; Fiber
	ASF-like viruses	ASFV	African swine fever virus	170,101	Spherical	ø 200–300	p54, external membrane protein
	Herpesviridae	Varicellovirus	Bovine herpesvirus 1	~125,000	Icosahedral	ø 120–200	Glycoprotein E2
	Tailed phages	Podoviridae	Bacteriophage P22 ("phage display")	~39,900	Tailed phage	ø 60 tail 17	Tailspike protein
Retroid	Retroviridae	Mammalian type C retroviruses	Murine leukemia virus	8,322	Spherical	ø 80–100	Envelope glycoproteins
			Spleen necrosis virus	~8,300	Spherical	ø 80–100	Envelope glycoproteins
ssDNA	Parvoviridae	Dependovirus	Adeno-associated virus 2	4,718	Icosahedral	ø 20–26	Capsid proteins
(−)ssRNA	Mononegavirales	Rhabdoviridae	Rabies virus	11,928	Bacilliform	ø 45–100; length 100–430	Glycoprotein G
	Rhabdoviridae		Vesicular stomatitis virus	11,161	Bacilliform	ø 45–100; length 100–430	Glycoprotein G
	Orthomyxoviridae	Influenza virus A and B group	Influenza A virus	13,588	Spherical	ø 50–120	Hemagglutinin
(+)ssRNA	Nidovirales	Coronaviridae	Murine hepatitis virus	31,357	Pleomorphic	ø 120–160	Spike glycoprotein S
	Picornaviridae	Enterovirus	Sabin type 1 poliovirus	7,440	Icosahedral	ø 28–30	VP1

[a] For different reasons, the genomes encapsidated by such chimeric virions are unable to yield the same chimeric progeny.

L protein has a preS1 N-terminal extension of 119 or 108 aa (depending on the HBV genotypes), in addition to preS2.[39] The proportion of S to L and S to M proteins in the HBV particle envelope is about 7:2:1.[40] For use as a molecular carrier HBsAg has the unique property of self-assembling with cellular lipids into noninfectious empty envelope particles, spherical or elongated, of about 22 nm diameter, which are secreted in large excess from hepatocytes during hepatitis B infection. In contrast to viral particles, the 22-nm particles consist mainly of the S protein. Eukaryotic cell lines showed the capability of producing similar 22-nm particles after expression of S and M, but not L, genes within the appropriate expression cassettes.[41–45]

Expression of the HBs proteins has been used to prove the usefulness of powerful eukaryotic vector systems based on regulatory elements taken from SV40,[46] retro,[47] vaccinia,[48] herpes simplex,[49] papilloma,[50] adeno,[51] and varicella zoster[52,53] viruses. Very early on, expression of HBsAg was achieved in transgenic mice, and animal models of the chronic HBsAg carrier state were made.[54,55] Thereafter, reasonable levels of HBsAg synthesis have been achieved in baculovirus vector-driven insect cells,[56] in silkworm larvae,[57] and in transgenic plants such as tobacco,[58] potato,[59–61] lettuce, and lupin.[62,63] The S gene was also expressed in stably transfected *Drosophila melanogaster* Schneider-2 cells under an inducible *Drosophila* metallothionein promoter.[64]

The S gene of HBV was one of the first viral genes expressed in bacteria.[35,65–67] Despite serious efforts mounted to generate "bacterial" HBsAg, self-assembly of the latter failed, possibly because specific folding and budding mechanisms were missing.[68–71] Results of the greatest practical utility were achieved by the successful expression of S, M, and L genes in yeast, where the expression products self-assembled into particles with an estimated diameter of 20 ± 4 nm.[72–74] Pioneering development of the yeast-derived recombinant hepatitis B vaccines resulted in construction of two widely and safely used human vaccine products: Engerix-B (SmithKline Beecham, Belgium) and H-B-Vax II (Merck, Sharp & Dohme), both of which are based on nonglycosylated protein S-derived particles.

Recently, plant-derived HBsAg appeared in the next pioneering step as a real edible vaccine candidate when it showed good immunogenicity in mice[60,61] and humans.[63]

Thus, the numerous expression systems and the historical record of time-tested, safe vaccination are the reasons HBsAg has attracted so much attention as a self-assembling epitope carrier candidate. Moreover, in line with its pioneering traditions, HBsAg served as one of the first DNA vaccine models,[75–81] which allowed for the easy conversion of DNA expression vectors for HBsAg-based chimeric VLPs to the corresponding DNA vaccines.

Although our knowledge of the folding and molecular architecture of the carrier particles is insufficient for the reliable prediction of preferential insertion sites, currently available data provide some hints for replacing surface-exposed protein domains by foreign sequences without destabilizing the self-assembly of the particles. Since an exact elucidation of the spatial structure of HBsAg with its lipid components is probably not available in the near future, alternative mathematical and experimental approaches have been employed. First of all, computer-aided topological models have been constructed.[82–84] These models were confirmed by the proteolytic identification of membrane-protected regions,[85] circular dichroism,[86] and small-angle neutron scattering.[87] By the latter method, yeast HBsAg was characterized as a spherical particle with a 29-nm diameter, where lipids and carbohydrates form a spherical core with a 24-nm diameter, and the S protein on the surface. Recent examination of yeast HBsAg by high-resolution negative staining electron microscopy (EM) revealed spherical to slightly ovoid character of particles with an average diameter of 27.5 nm, consisting of 4-nm subunits with a minute central pore. Ice-embedding EM gave the diameter value of 23.7 nm and a 7- to 8-nm thick cortex surrounding an electron translucent core[88] that corresponds well to the small-angle neutron scattering data. Human HBsAg particles, examined by the same EM methods, were found to be smaller in size.[88]

Briefly, according to the proposed folding of L, M, and S proteins, they consist of four membrane-spanning helices that are assembled into a highly hydrophobic complex where access to the water environment is not allowed (Figure 8.3). The proteins project their N and C termini

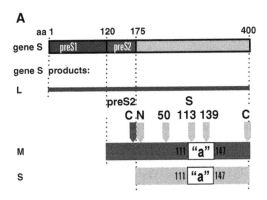

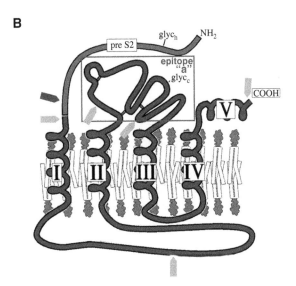

FIGURE 8.3 HBsAg as a VLP carrier. (A) Products encoded by the S gene with localization of the insertion sites for foreign epitopes. (B) Topological model of the M protein. The lipid bilayer of the ER membrane is symbolized by the lipid molecules. Five transmembrane α-helices are denoted by Roman numerals. The top of the figure corresponds to the lumen of the ER or, after multimerization and budding of the HBs protein, to the surface of the particle. The bottom of the figure corresponds to the cytosol or to the interior of the particle. Major immunodominant epitope "a" is boxed. Cysteine bridges and positions of complex biantennary glycoside (glyc$_c$) within the "a" epitope and hybrid-type glycoside (glyc$_h$) on the preS2 segment are shown. Insertion sites for foreign epitopes are marked by arrows. (The HBs model is the generous gift of Bruce Boschek and Wolfram H. Gerlich.)

(including preS1 and preS2 sequences of L and M proteins) as well as the second hydrophilic region of protein S, bearing the major immunodominant B-cell epitope "a" (aa residues 111–149) to the outer surface of particles.

Direct mutational analysis of HBsAg proteins revealed domains indispensable for self-assembly.[89,90] The role of intra- and intermolecular disulfide bonds within the hydrophilic regions of S protein, which are also responsible for its antigenic properties, was tested experimentally by site-directed *in vitro* mutagenesis substituting alanine for cysteine.[86,91,92] Moreover, mutagenesis of HBsAg led to the important conclusion that mutated, assembly noncompetent polypeptides can be rescued into mixed, or mosaic, particles in the presence of native helpers.[89]

Although human-plasma-derived and recombinant yeast HBsAg possess many similar properties, it is necessary to keep in mind that differences do exist. Besides glycosylation, which is different in yeast, such differences are clearly seen not only in the size of particles but also in both epitope presentation and anti-HBs-binding properties.[93] Despite these differences, the broad range of expression hosts stimulated numerous attempts to employ HBsAg for protein engineering studies.

Chronologically, the first example of HBsAg-based VLPs was the insertion of a long segment from herpes simplex virus 1 glycoprotein gD into the C-terminal part of the preS2 sequence and the expression of such chimera in yeast.[27]

At virtually the same time, a VP1 segment 93–103, which includes the linear part of the neutralization epitope from both poliovirus types 1 and 2 (PV-1 and PV-2), was inserted into positions 50 and 113 of the two hydrophilic domains of the S protein.[28,94–97] Recombinant genes were expressed in mouse L cells from the simian virus 40 early promoter and enabled secretion of chimeric 22-nm particles. Yields of secreted 22-nm particles were dependent on the site of insertion. Cotransfection with different plasmids carrying either modified or unmodified S genes led to the formation of mosaic particles presenting HBs-polioVP1 polypeptides on the helper HBsAg.[96] Surprisingly, mosaic particles induced much higher titers of neutralizing antibodies to poliovirus than did the homogenous ones. This study thus initiated the promising idea of designing multivalent particles carrying various peptide sequences or presenting several heterologous epitopes of interest.

Next, chimeric HBsAg derivatives contained sequences from human immunodeficiency virus, because of an overlap between populations at risk for HBV and HIV, and with them the sound hope to develop a bivalent hepatitis B/HIV vaccine.[98–100] First, an 84-aa-residue-long gp120 fragment of HIV-1 was inserted into the preS2 part of the protein M and expressed in eukaryotic cells.[98] Immunization with chimeric particles allowed not only for the generation of a humoral response in rabbits[98] but also the neutralization of antibodies and proliferative T-cell and CTL responses in rhesus monkeys to both parts of the hybrid particle, i.e., HIV and HBsAg.[101,102] The same laboratory attempted to repeat this work with the V2 region of the SIV gp140 protein.[103] Despite having consistent SIV-neutralizing antibody titers, vaccinated macaques were not protected against homologous challenge with SIV.

Further, HBsAg served as a carrier for selected epitopes of HIV-1 that were shown to elicit neutralizing antibodies. Chimeras harboring the 6-aa sequence ELDKWA from gp41 space as an internal fusion and expressed in yeast induced antibodies, which recognized HIV-1 gp160 but failed to neutralize HIV-1 *in vitro*.[100] Much attention was given to the V3 epitope of the gp120, which not only contained T-, CTL-, and B-cell epitopes including those inducing neutralizing antibodies, but also participated in coreceptor interaction and guided cell tropism of the virus. Initially, the V3 sequence of subtype MN was introduced into position 139 of the protein S and expressed in yeast.[104] Recently, it was inserted into the protein M[105] as well as at the N- and C-terminal positions of the protein S[106] for the use as DNA vaccines, which elicited rapid and sufficiently high B-cell and CTL responses in mice.

Much more successful, from the point of view of constructing vaccines, was the introduction of major preS1 and preS2 epitopes to the S protein, especially of the preS1-derived hepatocyte receptor-binding site. First attempts to enrich putative HBsAg vaccines with preS2[107] and preS1[108] peptides were realized by covalent binding of the appropriate synthetic peptides to HBsAg. Then chimeric protein, which carries preS1 region 21–47 at the C terminus of the S protein (at aa position 223), was expressed by a recombinant vaccinia virus and secreted from several mammalian cells.[109] HBV envelope was rearranged also by fusing part or all of the preS1 region to either the N or C terminus of the S protein.[110] Fusion of the first 42 residues of preS1 to either site allowed efficient secretion of the modified particles and rendered the linked sequence accessible at the surface of the particle. In opposite, fusion of preS1 sequences to the C terminus of the M protein completely blocked secretion. Although all these particles displayed preS1, preS2, and S-protein antigenicity, high titers of preS1- and preS2-, but not S-specific, antibodies were induced. Another group constructed three fusion proteins containing preS2 (120–146) and preS1 (21–47) epitopes at the N-terminus and truncated C terminus of the S protein and expressed them with the recombinant vaccinia virus system.[111] Fusion proteins were efficiently secreted in particulate form, displayed S, preS1, and preS2 antigenicity and elicited strong antibody responses against S, preS1, and preS2 in mice.

This ideology led to the production by Medeva plc of the efficient preS1- and preS2-containing vaccine Hepacare (formerly Hepagene, Hep B-3), which is the first to have a pan-European license. The Hepacare vaccine contains the preS1 sequence 20–47, the complete preS2, and the complete S region of two different subtypes, *adw* and *ayw*, and is produced in the mouse C127I clonal cell line after transfection of the cells with genes encoding the three antigens. Because these genes are expressed in mammalian cells, the Hepacare components are glycosylated and therefore represent the native viral proteins more closely. Unlike Hepacare, where desired epitopes were selected and combined within the recombinant genes, other commercial preS-containing mammalian cell-derived vaccines are produced via expression of full-length non-reconstructed M (GenHevac B Pasteur) or M and L (Bio-Hep-B, BioTechnology General Ltd, Israel) genes together with the S gene. The Hepacare vaccine was shown to stimulate stronger and more rapid cellular and humoral immune responses and to circumvent anti-HBs nonresponsiveness.[112–116]

A yeast-derived HBV vaccine containing preS1 (aa 12–52) and preS2 (aa 133–145) sequences within the rearranged L* protein, in addition to the S protein, was also constructed.[117,118] The vaccine particles consisted of S and L* at a ratio of 7:3.

Recently, modified HBsAg with a preS1 peptide 21–47 fused to its C terminus (at aa 223) was shown to induce strong humoral and cytotoxic response in transgenic mice after a single DNA injection.[119] In that way, the authors tried to model the vaccination of chronic carriers.

Development of HBsAg-based malaria vaccine candidates was also reported. Vaccine candidates have been constructed by using epitopes from three different human malaria parasite *Plasmodium falciparum* markers. First, computer-predicted B- and T-cell epitopes from the major merozoite surface antigen gp190 sequence (up to 61 aa residues in length) were introduced by replacing the proper antigenic determinants of the protein S, with further successful production of chimeric particles via vaccinia virus expression.[120] Then, chimeric particles carrying 16 repeats of tetrapeptide NANP of the circumsporozoite (CS) protein of *P. falciparum* were produced by yeast cells and tested successfully on experimentally vaccinated volunteers.[121] The next candidate, a sexual stage/sporozoite-specific antigen Pfs16, was fused to HBsAg and expressed in yeast cells in the form of mosaic particles in the presence of helper HBsAg.[122] The most promising malaria vaccine — RTS,S — contains 19 NANP repeats and the carboxy terminus (aa 210–398) of the CS antigen within the HBs-CS fusion coexpressed with the wild-type protein S as a helper.[123,124] Recently, the merozoite surface protein 1 gene of *P. vivax* was suggested as a vaccine candidate in combination with the HBsAg as a carrier.[125]

The idea of replacing the major immunodominant region of the S protein (aa 139–172) for the foreign epitopes was also used in the case of expression of the 35–98 aa segment of the human papillomavirus oncoprotein E7.[126]

In the case of expression of HCV sequences on the HBsAg carrier, both DNA-based immunization and vaccinia virus expression approaches have been tested. In the plasmid DNA vector, the 58-aa-long N-terminal fragment of the HCV nucleocapsid was fused to the M or S proteins.[127] In the vaccinia virus expression system, the five hydrophilic domains of the HCV E2 envelope as well as the hypervariable region (HVR) of E2 were presented on the HBsAg surface.[128]

Recently, yeast-expressed HBsAg fusion proteins carrying two different sized segments of hantavirus nucleocapsid protein at different positions of HBsAg or preS2 failed to form chimeric VLPs.[129] The same segments were exposed on chimeric HBV core particles, demonstrating the limitation of the HBsAg carrier in that case.

Future improvement of HBsAg as an epitope carrier may also be achieved by the addition of, for example, a Th epitope derived from tetanus toxoid.[130] Such chimeric 22-nm particles produced in a recombinant adenovirus expression system showed a several-fold enhancement of the anti-HBs response in mice, relative to native HBsAg, and were suggested for further exploitation as a new class of "highly-immunogenic HBsAg" carriers.

8.3.2 HEPATITIS B CORES

Unlike HBsAg, HBcAg as a carrier has been reviewed extensively. For detailed analyses we recommend specialized reviews dealing with the role of HBcAg as a component of HBV infection[131–133] and as a VLP carrier.[134–141]

HBcAg was first reported as a promising VLP carrier in 1986[29] and published in 1987,[30,31] essentially at the same time as HBsAg (see previous section). The HBcAg particles produced an onset of a long list of structurally well-defined icosahedral VLP carriers and to this day remain one of the most flexible and immunologically most powerful epitope carrier candidates. In contrast to HBsAg, which represents a complex and irregular lipoprotein structure, HBcAg consists of 180 or 240 copies of identical polypeptide subunits. The multifunctional character of HBcAg seems to be responsible for the unusual flexibility of the C protein. Although the HBV gene C has only two in-frame initiation AUG codons (Figure 8.2), it is responsible for the appearance of at least four different polypeptides: p25, p22, p21, and p17 (see Reference 142). The p25 precore protein, starting at the first AUG codon, becomes targeted by a signal peptide in the preC sequence to a cell secretory pathway, in which a p22 is formed by N-terminal processing. The p22 undergoes further cleavage at the C-terminal region, after position 149, to generate a p17 protein, or HBe protein, which is secreted from the cell as the HBe antigen.

The predominant p21 polypeptide is synthesized from the second AUG of the open reading frame, and constitutes a structural component of the HBcAg, or HBV nucleocapsid, and thus may be referred to as a genuine HBc polypeptide. The HBc polypeptide is able to self-assemble and was therefore selected as a target for protein engineering manipulations. Besides capsid-building, the p21 protein participates in the viral life cycle and its regulation, including synthesis of dsDNA as a cofactor of the viral reverse transcriptase-DNA polymerase (Pol), viral maturation, recognition of viral envelope proteins, and budding from the cell (for details see Reference 133). It appears that HBcAg can recognize specific sites of the envelope proteins S and L.[143,144]

Phosphorylation of serine residues by cellular protein kinase C within three repeated SPRRR motifs on the C terminus of the p21 protein[145–147] and its role in the maturation of the HBV capsid[148,149] reflect the complex function of the HBc particles. According to recent data, phosphorylation of HBc subunits induces a conformational change that exposes the C-terminal sequences, which may protrude through the holes in the capsid wall and become accessible on the surface to serve as a nuclear targeting signal.[149]

Early in the HBcAg studies, a possible self-proteolytic activity of the HBc protein was discussed.[150,151]

In many ways HBcAg holds a unique position among other VLP carriers because of its high-level expression and efficient particle formation in virtually all known homologous and heterologous expression systems, including bacteria. Correct folding of the p21 protein and formation of authentic core particles have been documented in various mammalian cell cultures[42,152–156] as well as retrovirus,[157] vaccinia virus,[158,159] and adenovirus[160] expression systems, frog *Xenopus* oocytes,[161] insect *Spodoptera* cells,[162–165] yeast *Saccharomyces cerevisiae*,[166–169] in plant *Nicotiana tabacum*,[170] and in bacteria such as *Escherichia coli*,[24,35,66,171–177] *Bacillus subtilis*,[178] *Salmonella*,[179] and *Acetobacter*.[180] Over-expressed p21 proteins showed correct self-assembly into naturally-shaped particles in the absence of any other viral component. Electron microscopy revealed the ultrastructural identity of the HBc particles derived either from HBV virions and infected hepatocytes, or from *E. coli*[181] or yeast.[182]

The fine structure of HBc particles (Figure 8.4) was revealed by electron cryomicroscopy and image reconstruction.[9–11] Finally, this three-dimensional structure was confirmed by X-ray crystallography at 3.3-Å resolution.[12] Organization of HBc particles was found largely α-helical and quite

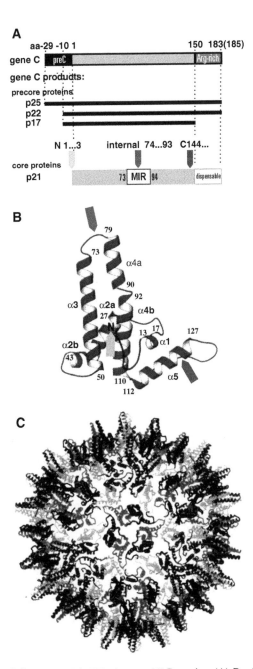

FIGURE 8.4 (Color figure follows p. 82.) HBcAg as a VLP carrier. (A) Products encoded by the C gene with localization of the insertion sites for foreign epitopes. (B) A schematic representation of the fold of the HBc monomer derived from the crystal structure. The dimer interface is nearest to viewer. Insertion sites for foreign epitopes are marked by arrows. (C) The T = 4 HBc capsid viewed down an icosahedral threefold axis.[12] (The maps are the generous gift of R. Anthony Crowther.)

different from previously known viral capsid proteins with β-sheet jelly-roll packings.[9,12] In brief, the HBc monomer fold is stabilized by a hydrophobic core that is highly conserved among human viral variants. Association of two amphipathic α-helical hairpins results in the formation of a dimer with a four-helix bundle as the major central feature. The dimers are able to assemble into two types of particles, large and small, that are 34 and 30 nm in diameter and correspond to triangulation number T = 4 and T = 3 packings, containing 240 and 180 HBc molecules, respectively. The major immunodominant region (MIR) with the central positions 76–81 is located at the tips of the α-helical hairpins that form spikes on the capsid surface.[12]

It is significant that the existence of two size classes of HBV cores, with predominant larger and less frequently occurring smaller particles, was detected earlier using particles from HBV-infected human liver, by either electron microscopy of a negatively stained specimen[181] or by gel filtration.[183]

In addition to MIR, the region 127–133 is the next exposed and accessible epitope on the particle surface. This region is located at the end of the C-terminal α-helix and forms small protrusions on the HBcAg surface.

Comparative electron cryomicroscopy and three-dimensional image reconstruction of DHBV and HBV cores of natural and bacterial origin reconfirmed at the molecular level the native core structure in bacteria, in the absence of the complete viral genome and other viral components.[184]

Of particular structural value was the clear demonstration of dispensability of the C-terminal protamine-like arginine-rich domain of the p21 protein (aa 150–183) for its self-assembly capabilities in the so-called HBcΔ particles.[185–187] The HBcΔ particles formed by C-terminally truncated polypeptides were practically indistinguishable from the HBc particles formed by full-length HBc polypeptides, as shown by electron cryomicroscopy.[9] However, unlike the full-length HBc particles, HBcΔ particles were less stable, failed to encapsidate nucleic acid, and usually accumulated as empty shells.[9,185,188–192] The unusual molecular flexibility of the C-terminal protamine-like domain has been revealed by the attempts to apply NMR spectroscopy to the structural analysis of HBc particles.[189] The C-terminal limit for self-assembly of HBcΔ particles was mapped experimentally between aa residues 139 and 144.[187,188,193] According to the most recent data,[194] it maps at position 140 and gives predominantly the T = 3 isomorph and a proportion of T = 4 isomorph of approximately 18%. The proportion of T = 4 capsids increases with the length of the HBc polypeptide, and the HBc variants truncated at positions 142, 147, and 149 aa from about 52, 79 and 94% of T = 4 capsids, respectively. The HBcΔ particles played an important role in structural modeling (see below) and seem to be very promising candidates for further vaccine and gene therapy applications.

In contrast to HBsAg, the extremely high immunogenicity of HBcAg particles has been known for a long time. Thus, HBV patients develop a strong and long-lasting humoral anti-HBc response.[195] Among the HBV polypeptides, HBc induces the strongest B-cell, T-cell, and CTL response (for review see Reference 196). HBcAg is known to function as both a T-cell-dependent and T-cell-independent antigen.[197] Following immunization, it primes preferentially Th1 cells, does not require an adjuvant,[198] and is able to mediate anti-HBs response.[199] Recently, enhanced immunogenicity of HBcAg was explained by its ability to be presented by B cells as the primary antigen to T cells in mice.[200] HBcAg elicits a strong CTL response during HBV infection,[201] and this response is maintained for decades following clinical recovery, apparently keeping the virus under control.[202]

In spite of its nature as an inner antigen, HBcAg was found in the middle 1980s to be able to provide protection against HBV infection in chimpanzees.[203–205] Attempts to include HBcAg into HBV vaccines by genetic immunization[206–208] or as a CTL epitope 18–27[209] are now in progress.

Moreover, Celltech-Medeva recently started a Hepacore project, which aims at the construction of therapeutic vaccines on the basis of chimeric HBc derivatives. A healthy volunteer study using chimeric HBc particles containing the preS1 sequence 20–47 inserted into the MIR is planned for the near future. This study will evaluate the safety and tolerability of the Hepacore product as well as provide immune response information that will be useful in designing further trials aimed at examining its potential in the immunotherapy of patients chronically infected with HBV.[210]

In general, it is widely accepted now that the HBc carrier is capable of ensuring a high level of B-cell and T-cell immunogenicity to foreign epitopes (see References 134 through 139, 141). In addition to the ability of HBc carrier moiety to provide T-cell help to inserted sequences, the

HBc capsid mediates the T-cell-independent character of humoral response against inserted epitopes, due to the high degree of epitope repetitiveness and the proper spacing between them.[211]

The major HBc B-cell epitopes c and e1 have been localized within the MIR, on the tip of the spike, around the protruding region 78–82.[212,213] The next important epitope, e2, is localized on the other surface-exposed region of the HBc particle, around position 130.[214,215] The human CTL epitope 18–27[216] deserves special notice because it was included in the HBV vaccine trial.[209,217]

Experimental search for appropriate target sites for foreign insertions pointed to the MIR region at the tip of the spike and to the N and C termini of the HBc molecule (see References 134 through 141). These findings are in general agreement with the X-ray data because these regions do not participate in the critical intra- and intermolecular interactions.[12]

The existing data on the insertions of foreign epitopes, in which the HBc carrier retained its particulate character (Figure 8.5), allow the following conclusions.

8.3.2.1 N-Terminal Insertions

A natural HBc variant with an insertion of dodecapeptide RTTLPYGLPGLD between aa residues 2 and 3 has been described for the HBV genotype G.[218] Deletions of more than 4 aa residues at the N terminus of the HBc molecule result in a protein that is not competent for self-assembly. The maximum length of N-terminal insertions is 50 aa, their accessibility on the particle surface is generally good. Exposure of a nine-N-terminal extension of the HBc molecule was shown by electron cryomicroscopy.[219] Historically, N-terminal insertions were the first ones in which chimeric HBc particles carrying the VP1 epitope 141–160 of foot and mouth disease virus (FMDV) were demonstrated in the vaccinia virus expression system,[29,30] yeast,[220] and bacteria.[221] The ability of the HBc chimera to induce FMDV-neutralizing antibodies stimulated construction of other N-terminal insertion variants with epitopes of the gp70 protein of feline leukemia virus, the VP2 protein of human rhinovirus type 2, the VP1 protein of poliovirus type 1,[221,222] the Env protein of simian immunodeficiency virus,[223] and the chorionic gonadotropin.[220] The last named was constructed as a contraceptive vaccine candidate.

The N terminus of an HBc molecule was used also as an insertion target for relatively short epitopes from HBV preS;[224–226] HIV-1 gp120 and p24,[227] gp41, p34 Pol, and p17 Gag;[228,229] and from human cytomegalovirus gp58.[230] The last named could not be purified or characterized immunologically, although it formed VLPs.

A recent remarkable breakthrough in the application of the HBcAg model for vaccine development was based on the N-terminal insertion. Chimeric particles expressed in *E. coli* and carrying 23 aa of the extracellular domain of influenza A minor protein M2 provided up to 100% protection against a lethal virus challenge in mice after intraperitoneal or intranasal administration.[231] This protection was mediated by antibodies. Due to the conserved nature of the M2 protein sequence, the HBc-M2 vaccine promises broad-spectrum, long-lasting protection against influenza A infections.

Fusion of 45 N-terminal aa of the Puumala hantavirus nucleocapsid protein to the N terminus of HBcΔ allowed for the formation of chimeric VLPs, which induced a strong antibody response and some protection in the bank vole model.[232] However, the addition of 120 N-terminal aa of the hantavirus nucleocapsid to the N terminus of HBcΔ prevented self-assembly, in contrast to their insertion into position 78[233] (see below).

8.3.2.2 C-Terminal Insertions

Regarding the C-terminal insertions, HBc positions 144, 149, and 156 were used most frequently as target sites for foreign insertions (for structural details, see Reference 141). The capacity of the constructed vectors usually exceeded 100 aa residues, depending on the structure of insertion. The C-terminal insertions involved two types of vectors, encoding either full-length or C-terminally truncated HBc. Despite the fact that capsids formed by the C-terminally truncated HBc derivatives

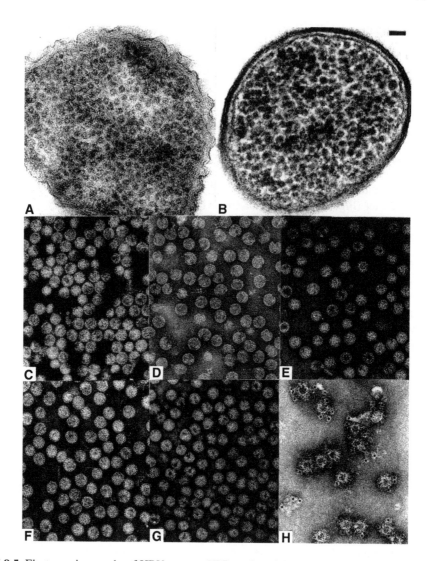

FIGURE 8.5 Electron micrographs of HBV cores as VLP carriers. Slices of *E. coli* cells filled with authentic HBc particles (A) or with HBc particles harboring C-terminal insertion of full-length preS2 segment (55 aa residues) substituting the 39 C-terminal aa residues of the HBc protein (B). Purified HBc particles formed by authentic full-length HBc protein (C), full-length HBc protein with insertion of 50 aa residues of the preS1 at position 144 (D), C-terminally truncated HBc protein with insertion of the same preS1 segment at position 144 (E), full-length HBc protein with the preS2 segment inserted at position 144 (F), C-terminally truncated HBc protein with insertion of 100 foreign aa residues (HIV-1 Gag segment) at position 144 (G), internal insertion of 11 foreign aa residues containing the preS1 DPAFR epitope, substituting the authentic HBc aa residues 79–85; preS1 epitopes exposed on the particles are labeled with monoclonal anti-preS1 antibody MA18/7 by immunogold technique (H). Scale bar, 50 nm. (Micrographs are the generous gift of Velta Ose.)

(HBcΔ) are usually less stable than the capsids formed by full-length HBc proteins, high-level synthesis in bacteria and dissociation/reassociation capabilities of the HBcΔ are advantageous. Moreover, foreign insertions at the C terminus can exert a stabilizing effect on chimeric HBcΔ derivatives, especially if internal insertions are introduced at the same construct.[234] In some cases, the inserted sequences are exposed, at least partially, at the surface of the HBc particle, but their specific B-cell immunogenicity is usually low. In general, chimeras with either the N- or C-terminal insertions retained high intrinsic HBc antigenicity/immunogenicity, which is determined primarily by intact MIR.

Full-length HBc vectors were used for inserting up to 55 aa from the HBV preS,[235-237] about 50 aa of HIV-1 gp41,[238,239] and different fragments of 20 aa of the simian immunodeficiency virus Env.[223] Furthermore, expression by vaccinia virus of chimeric HBc carrying a 9-aa-long immediate-early CTL epitope from the pp89 protein of murine cytomegalovirus (MCMV) at HBc position 179 led to the induction of T-lymphocyte-mediated protective immunity against lethal MCMV infection.[240]

C-terminally truncated vectors ensured a high level of synthesis and excellent self-assembly but only moderate specific immunogenicity of the inserted epitopes. These vectors were used for the expression of epitopes from the HBV preS[179,224,235,236,241] and S[139,241,242] regions; HIV-1 gp120,[243-246] gp41,[241] Gag,[247] and Nef;[247] bovine leukemia virus gp51;[235,239] human cytomegalovirus gp58;[230] hantavirus nucleocapsid;[248-251] and hepatitis E virus capsid.[252] Although the fusion of 45 aa of the Puumala hantavirus nucleocapsid protein allowed the formation of chimeric VLPs, they were unable to induce a protective immune response in the bank vole animal model.[232] Chimeric HBc particles carrying two C-terminally truncated virus-neutralizing epitopes from the FMDV VP1 (200–213, 131–160) showed an excellent ability to self-assemble but failed to protect animals against FMDV infection.[253] Finally, C-terminal insertions of the HCV core protein demonstrated the extraordinary capacity of HBcAg as a VLP carrier: a 559-aa-long insertion did not prevent self-assembly of chimeras, and even 741-aa-long insertions allowed for some production and self-assembly of chimeras.[254] C-terminally added HCV core[255,256] and NS3[257] sequences were used successfully for detection of specific antibodies in HCV enzyme immunoassay.

Although C-terminal additions have not met with success in terms of inducing antibodies, a new attempt was undertaken to insert a conserved sequence of 47-aa residues from several proteins of *Porphyromonas gingivalis*.[258] Although in this case the chimeric particles purified from *E. coli* were recognized by the host's immune system and induced specific antibodies, they did not protect mice against bacterial challenge. By comparison, the N-terminal insertion of a 30-aa epitope of outer membrane protein P.69 (pertactin) from other infectious bacteria, *Bordetella pertussis*, prevented the growth of *B. pertussis* in the lungs of infected mice, and protection correlated with high titers of anti-P.69 antibodies.[259]

Very recently, a 17-kDa nuclease was packaged into the interior of HBc capsids after fusion to the HBc position 155.[260] The packaged nuclease retained enzymatic activity, and the chimeric protein was able to form mosaic particles with the wild-type HBc protein.

8.3.2.3 Internal Insertions

Much more interesting are foreign insertions into the MIR, or the tip of the spike of the HBc molecule, which is now generally accepted as a target site of choice. Unlike N- and C-terminal constructions, the insertion of foreign epitopes into the MIR guarantees a high level of specific B-cell and T-cell immunogenicity. Moreover, the MIR allows for a surprisingly high number of insertions. The entire 120-aa-long immunoprotective region of the hantavirus nucleocapsid was inserted into the MIR of C-terminally truncated HBc protein, whereas N and C termini failed to accept this fragment for self-assembly.[233] It must be emphasized that the shorter, aa 1–45 segment of hantavirus nucleocapsid within the MIR also ensured the protection of bank voles against virus challenge after immunization with chimeric particles.[250,251] Furthermore, chimeric core particles were generated carrying aa 1–45 and aa 75–119 of a hantavirus nucleocapsid protein at aa position 78 and behind aa 144 of HBcΔ.[251] However, the combination of the major protective region of the nucleocapsid protein located between aa 1–45 and a second minor protective region did not improve the protective potential in the animal model when compared to the particles carrying only the first region.[232]

Recently, green fluorescent protein (GFP) of 238 aa was natively displayed on the surface of chimeric HBc particles.[261] Chimeras demonstrated not only fluorescence capabilities but also elicited a potent humoral response against native GFP. This example shows the structural importance of proper and independent folding of sequences subjected to exposure on the HBc particles and opens the way for high-resolution structural analyses of nonassembling proteins by electron microscopy.[261]

The story of MIR insertions started with the introduction of up to 27-aa-long epitopes of HBV preS,[224,262–264] 18 aa of VP2 protein from the human rhinovirus type 2,[222,265] up to 30 aa of the simian immunodeficiency virus Env,[223] and 25 aa[245,246,266] and up to 43 aa[243,267] of the V3 loop of the HIV-1 gp120. The insertion of 39 aa of the domain "a" sequence from the HBsAg (positions 111–149) was the first successful attempt to mimic a conformational epitope on the surface of chimeric particles.[268]

HBsAg and preS epitopes have been chosen for the construction of the first multivalent particles, i.e., for the simultaneous insertion of different foreign sequences from the preS1 and preS2 regions into the MIR and N terminus,[269] or into the MIR and C terminus,[224,263] or from the HBsAg and preS2 into the MIR and C terminus of the HBc protein.[267]

PreS epitopes have been chosen for the construction of the first multivalent particles, i.e., for the simultaneous insertion of different foreign sequences from the preS1 and preS2 regions into, respectively, the MIR and the N terminus of the HBc protein.[269] Later, multivalent HBcΔ particles carrying different hantavirus nucleocapsid epitopes at the MIR and C terminus were constructed.[251]

The first mosaic HBc particles carrying chimeric and wild-type HBc monomers were also constructed on the basis of full-length HBc vectors for internal insertions. In this case, an epitope of 8 aa from the Venezuelan equine encephalomyelitis virus E2 protein was inserted into position 81 of the HBc molecule.[270]

An attempt to construct a therapeutic vaccine against HPV16-associated anogenital cancer was undertaken by MIR insertions of B-cell, T-cell, and CTL epitopes from the E7 oncoprotein of the human papillomavirus type 16.[271–273] Humoral and T-proliferative responses to the chimeras were elicited successfully,[271] also in the case of *Salmonella*-driven expression (see Reference 272 and below), but the appropriate chimeric particles failed to prime E7-directed CTL responses in mice.[273]

MIR insertions into the HBc particle were thoroughly investigated for the possible construction of vaccines against infectious diseases caused by intracellular parasites. The first vaccines tested were malaria vaccines, in which chimeric HBcAg-CS particles carrying circumsporozoite (CS) protein repeat epitopes of *Plasmodium falciparum* and of two rodent malaria agents, *P. berghei* and *P. yoelli*, were expressed in *S. typhimurium*.[274] Immunization of mice with purified particles ensured not only specific B-cell and T-cell responses but also protection against a *P. berghei* challenge infection. Expression of the recombinant genes in *S. typhimurium* offered promise for the generation of oral vaccines on the basis of live, avirulent strains of *Salmonella* species.[135,136,179,263,264,272,274–279] Thus, efficacy of a single oral immunization of BALB/c mice with a recombinant *S. typhimurium* carrying HBc-preS[275] and HBc-CS[277] chimera has been shown. In this case HBc-preS chimera contained aa 27–53 of the preS1 between positions 75 and 81 of the HBc protein, and aa 133–143 of the preS2 fused C-terminally to position 156 of the HBc protein.[224,263] However, volunteers that received the oral *Salmonella* HBc-preS vaccine failed to develop humoral and cellular responses to hepatitis B antigens.[280]

Second, a C-terminal segment (SR1) of SPAG-1, a sporozoite surface antigen of *Theileria annulata*, an infectious agent of cattle theileriosis, has been expressed as an MIR-insertion.[281] The chimeric particles not only induced high titers of neutralizing antibodies and a significant T-cell response, but they also showed some evidence of protection against a sporozoite challenge,[281] which allowed for their recommendation to be included in a future multicomponent vaccine.[282]

Very recently, a promising *Salmonella* expression variant of HBc-derived chimeras was achieved with an internally inserted HBs "a" epitope.[283,284] A single rectal immunization with this HBc-HBs recombinant induced humoral and cellular immune response to HBc and HBs as well as the formation of specific mucosal immunity.[284]

Besides vaccine development, MIR insertions were used successfully for the development of immunoassays (anti-HEV)[252] and for mimicking targeting moieties, or cell-receptor-recognizing sequences.[285] For this purpose, an RGD-containing epitope from the FMDV VP1 protein was exposed within the MIR, and the HBc-RGD chimeric particles not only elicited high levels of FMDV-neutralizing antibodies in guinea pigs, but they also bound specifically to cultured eukaryotic cells and to purified integrins.[285]

Special attention is now being devoted to the construction of HBc display vectors with deletions of different length within the MIR. It must be mentioned that some of the MIR insertions that were reviewed above carried short deletions within the MIR: aa 76–80,[224,261–263] 79–81,[265] and 79–80.[245,246] Structural[9,12] and numerous experimental[243,267,286,287] data convinced us that the region between the two conserved glycines G73 and G94 can be used as a target for deletions, rearrangements, and substitutions. For optimum immunogenicity of the insert it is extremely important that deletions of proper aa residues within this region abrogate the intrinsic HBc antigenicity/immunogenicity.[286,287] Besides the ability of the HBc carrier moiety to provide T-cell help to inserted sequences, HBc carrier ensures the T-cell-independent character of humoral response against inserted epitopes in the MIR-deleted variants as well.[211]

Taking into account the unique properties of the HBc carrier, the natural HBc deletion variants (e.g., 86–93 and 77–93) occurring in patients with progressive liver disease may deserve special attention as new carrier candidates.[288] However, the insertion of 45 N-terminal aa of a hantavirus nucleocapsid protein between aa 86 and 93 of HBc abolished the formation of chimeric VLPs.[289]

An attempt to replace a more expanded fragment of the MIR, aa 72–89, within the HBcΔ by the HEV capsid epitope of 42 aa led to the production of capsomere-like 12-nm particles constituted presumably by the assembly of six dimers of the HBc protein.[252]

Two general approaches for functional studies of VLP carriers have been thoroughly developed on the HBcAg model. The first is to "scan" the gene, encoding the putative VLP subunit by a short epitope as an immunological marker in order to identify gene regions that are indifferent to foreign insertions but can ensure the desirable structural or immunological behavior of the latter.[226,248,267,287] For this purpose a short HBV preS1 epitope 31-DPAFRA-36 (or DPAFR, or DPAF), necessary and sufficient[290] to be recognized by the monoclonal antibody MA18/7,[40] has been used. The behavior of the DPAFR epitope was systematically compared after introduction into all preferred insertion sites of the HBc molecule at positions 2, 78, 144, and 183,[226] and after introduction into the MIR-carrying deletions of different length.[287]

Second, further development of the technology of mosaic particles has been achieved. A new strategy for constructing mosaic particles was based on the introduction of a linker containing translational stop codons (UGA or UAG) between sequences encoding a C-terminally truncated HBcΔ and a foreign protein sequence.[233,249,251,291] Expression of such a recombinant gene in an *E. coli* suppressor strain leads to the simultaneous synthesis of both HBcΔ as a helper moiety and a read-through fusion protein containing a foreign sequence. This technology allowed for the incorporation into and presentation onto mosaic particles of segments of hantavirus nucleocapsid that were 45,[232] 114,[251] 120,[233] and even 213[292] aa in length, although nonmosaic HBcΔ carrying the hantavirus segment at the C terminus were unable to self-assemble. However, in the animal model mosaic particles carrying 45 and 114 aa of the hantavirus nucleocapsid protein failed to induce a response, or induced only a marginally protective one.[232,251]

An important practical advantage of the HBc model consists in the fact that, because of their particulate nature, HBc-derived chimeric particles are easy to purify by gel filtration or sucrose gradient centrifugation. C-terminally truncated variants can be subjected to dissociation with subsequent reassociation, in order to remove internal impurities and produce nucleic-acid-free preparations. A special purification protocol for the preparation of HBc derivatives of vaccine quality was elaborated by the addition of 6 histidine tag to the truncated C terminus of the HBc protein at position 156.[266,293] On the other hand, the ability of full-length or special chimeric HBc derivatives to achieve controlled encapsidation of nucleic acids may be used for the further development of this carrier for gene therapy experiments.

8.4 OTHER NONINFECTIOUS VLP CARRIERS

8.4.1 DOUBLE-STRANDED DNA VIRUSES

8.4.1.1 Nucleopolyhedroviruses

Baculoviruses are widely known as excellent vectors for expressing heterologous proteins in cultured insect cells and for their production of multisubunit protein complexes (see Reference 294). Besides their employment in the VLP technologies as "baculophages," or baculovirus display tools (Table 8.2), baculoviruses are the source of polyhedrin, which may be regarded as a special class of noninfectious highly structured carriers. Polyhedrin is synthesized in massive amounts and forms a protective crystal around the viruses, which allows them to remain viable for many years outside the insect larvae.[295] Although X-ray crystallographic examination of polyhedrins was regarded as realistic as early as 1986,[295] it is not yet an accomplished fact.

Chimeric polyhedrin proteins of the *Autographa californica* nucleopolyhedrovirus (AcMNPV) carrying an influenza hemagglutinin epitope not only showed influenza-specific antigenic and immunogenic properties but also presented the foreign epitope on the surface of chimeric occlusion bodies.[296] Furthermore, chimeras of the AcMNPV polyhedrin and *Trichoplusia ni* granulosis virus (TnGV) granulin were constructed.[297] The results clearly showed that the size and structure of occlusions is affected by the primary structure of the product, although the involvement of other viral proteins in the virion occlusion body assembly and shape complicates interpretation and further application of this system.

8.4.1.2 Papillomaviruses

Interest in papillomaviruses was based not only on their favorable structural properties but also on the need for a vaccine to prevent, first, genital tract human papillomavirus (HPV) infections, especially those types associated with genital tract malignancy. There is also an additional interest in the development of a therapeutic vaccine, which, unlike a prophylactic vaccine, is used to treat individuals who are already infected.

An inability to efficiently propagate the papillomaviruses in cultured cells stimulated construction of native-like vaccine candidates. First, production of noninfectious HPV-like particles was achieved by coexpression of late capsid protein genes L1 and L2 (but not either protein alone) of HPV16 using vaccinia virus, where VLPs consisted of capsomeres similar to HPV and contained major capsid protein L1 in glycosylated form.[298] Later, the authors showed the ability of bovine papillomavirus type 1 (BPV1) L1 protein alone to self-assemble in a vaccinia virus expression system, although the presence of L2 was necessary for BPV1 DNA packaging.[299] The intrinsic self-assembly competence of the L1 protein was demonstrated for HPV16 and BPV1,[300] HPV11,[301] cottontail rabbit papillomavirus (CRPV),[302] HPV33,[303] and many other papillomaviruses in insect cells via a baculovirus vector expression, with a markedly increased outcome of VLPs during coexpression of the L1 with the L2.[302,303] Since HPV11, 16, 18, etc. VLPs were shown to be antigenically distinct, the demand for them arose for diagnostic purposes,[304] and their availability opened the way for the discovery of virus-neutralizing epitopes.[305,306] Then, HPV6 and 16 VLPs were obtained in fission yeast *Schizosaccharomyces pombe* by expression of the L1 alone and the L1/L2 genes, although L2 was not incorporated into the VLPs.[307] Meanwhile, L1 and L2 of HPV6[308] and CRPV[309] were expressed in *Saccharomyces cerevisiae* and showed clear complex formation of both proteins.

The first clear demonstration of the HPV16 L1 self-assembly in prokaryotes was achieved via its expression in *Salmonella typhimurium*.[310] The first report on the HPV L1 expression in *E. coli* did not go beyond the statement that the L1 protein was purified to near homogeneity and appeared in electron microscopy as pentameric capsomeres that self-assembled into the VLPs.[311] More recently, difficulties with bacterial expression of the HPV16 L1 were confirmed, although the authors succeeded in the solubilization and renaturation of original inclusion bodies into polymorphologic aggregations, which served further as a source of separation of correctly formed VLPs.[312]

Efficient and correct self-assembly of HPV L1 and L2 was achieved in the Semliki Forest virus expression system,[313] which allowed for the pseudotyping of papillomaviruses, i.e., BPV1 genomes coated by HPV16 L1 and L2.[314] Moreover, HPV33 VLPs carrying packed copies of a marker plasmid have been generated in COS-7 cells,[315] and the suggestion of HPV16 L1[316] and BPV1 1 L1/L2[317] VLPs as promising vehicles to package and deliver unrelated plasmid DNA X-ray crystallography of HPV particles was not undertaken, but three-dimensional reconstructions of HPV1 L1 alone or both L1 and L2 VLPs were generated by electron cryomicroscopy and image analysis techniques.[318] According to this analysis, HPV capsids have 72 pentameric capsomeres arranged on a T = 7 icosahedral lattice. Each particle (approximately 60 nm in diameter) consists of an approximately 2-nm-thick shell of protein with a radius of 22 nm with capsomeres that extend approximately 6 nm from the shell. At a resolution of 3.5 nm, recombinant capsid structures appear identical to the capsid structure of native HPV1.[319] Later, the three-dimensional structure of BPV was determined by electron cryomicroscopy at 9-Å resolution, and the location of the L2 minor capsid protein in the center of the pentavalent capsomeres was settled.[320]

An analysis of L1 sequences necessary and sufficient for self-assembly led to the observation that the BPV1 L1 protein lacking the last 24 aa was able to form VLPs in insect cells with enhanced efficiency.[321] Similar systematical analysis on canine oral papillomavirus (COPV) L1 VLPs produced in insect cells showed that 26 C-terminal aa are abundant for self-assembly, but further truncation of the L1 C terminus for 67 aa resulted in a capsid protein that formed VLPs but that failed to express conformational epitopes.[322] Deletion of the first 25 N-terminal aa of the COPV L1 also abolished expression of conformational epitopes without altering VLP formation, but the native conformation of this deletion mutant could be restored by the addition of the N terminus of the HPV11 L1.[322] Conditions for the quantitative disassembly and reassembly of purified HPV11 L1 VLPs to the level of capsomeres were elaborated after demonstration that disulfide bonds alone are essential to maintaining capsid structure, which simplifies packaging of selected exogenous compounds within the reassembled VLPs.[323]

At the very beginning, native CRPV VLPs from insect cells[324,325] and yeast[309] and BPV4 VLPs from insect cells[326] were applied for the induction of long-lasting protection against experimental papillomavirus infection in rabbits and calves, respectively. The CRPV L1 was also efficient when used in the DNA-based vaccination form.[327] Formation of HPV VLPs in *Salmonella* opened the door for the idea of oral and nasal routes for immunization[310] and led to speculation on the advantages of mucosal over parenteral immunization in mice.[328–331] Further results demonstrated the ability of the HPV16 L1/L2 VLPs to induce a CTL response in humans;[332] immunization of patients carrying genital warts with HPV6b VLPs induced immunity to L1 protein epitopes recognized during natural infection and accelerated regression of warts.[333] Phase I trials of the appropriate human vaccines are now in progress.[334]

An approach to the construction of chimeric HPV-like particles was developed by replacing the 34 C-terminal aa of the HPV16 L1 protein with foreign sequences, at first with various parts of the HPV16 E7 oncoprotein.[335] Chimeric VLPs with inserts of up to 60 aa of the E7 sequence were expressed efficiently by recombinant baculoviruses and showed their ability to induce a neutralizing antibody response.[335] These particles induced also an E7-specific CTL response in the absence of an adjuvant and prevented outgrowth of E7-expressing tumor cells in mice, even if inoculation of cells was performed 2 weeks before vaccination.[336] Therefore, the HPV16 L1-E7

VLP vaccine provided not only protection against tumor growth but also a therapeutic effect on preexisting tumors.[337]

An alternative approach to generating chimeric HPV16 VLPs was carried out by the fusion of the entire E7 or E2 proteins to the minor capsid protein L2.[338] Such chimeric L1/L2-E7/E2 VLPs were indistinguishable from the parental VLPs in their morphology and elicited high titers of neutralizing antibodies. Moreover, VLPs carrying the E7 oncoprotein protected mice from tumor challenge, and protection was mediated by MHC class I restricted cytotoxic lymphocytes.[338]

Furthermore, chimeras containing an HPV16 E7 epitope, but constructed on the basis of the L1 protein of the BPV1 as a carrier, were shown to induce mucosal as well as systemic immune responses in mice, after intranasal immunization, as well as induction of an E7-specific CTL response.[339] The ability of the BPV1 carrier to present CTL epitopes was also confirmed by the fusion of a single HIV-1 gp160 CTL epitope to the C terminus of the L1 protein.[340]

The BPV1 carrier was applied successfully to the induction of autoantibodies against self-antigens,[341] which may have remarkable clinical applications. Thus, autoantibodies were induced in mice against an extracellular loop of the mouse chemokine receptor CCR5, incorporated into an immunodominant site of the BPV1 L1 coat protein, and self-assembled into VLPs. Such antibodies not only inhibited binding of its ligand RANTES but also blocked HIV-1 infection of an indicator cell line expressing a human–mouse CCR5 chimera.[341]

Further, the antitumor potential of papillomavirus VLPs was demonstrated by the induction of a protective immune response in mice against a lethal tumor challenge with a progressor P815 tumor cell line after immunization with HPV16 L1 VLPs carrying a P1A peptide derived from the P815 tumor-associated antigen.[342]

8.4.1.3 Polyomaviruses

Although the history of studies on polyomaviruses, and in particular on murine polyomavirus Py and SV40, is much longer than that of many other viruses discussed here, their involvement as potent carriers arose only in recent years when the major capsid protein VP1 of murine and hamster polyomaviruses was used as a target for foreign insertions.

Even in 1979 X-ray diffraction resolved polyoma virion structure at 30 Å,[343] but 3.8-Å- and 3.65-Å-resolution structures of, respectively, SV40[344] and murine polyomavirus (Py)[345] were achieved by X-ray crystallography more than 10 years later. Structures of SV40 and Py were specially compared,[346] and SV40 crystal structure was refined at last to 3.1 Å.[347] The polyomavirus capsid is thus comprised of 72 pentamers of 360 copies of the VP1 protein associated with the minor virion proteins VP2 and VP3. Like the pentamers of papillomaviruses, those of polyomaviruses are arranged on a T = 7 icosahedral lattice with a right-handed symmetry (T = 7dextro),[348] with the exception of hamster polyomavirus (HaPV), which was recently reported to be left-handed (T = 7laevo).[349]

The VP1 protein of Py was expressed in *E. coli* as early as in 1985[350] and appeared as dissociated capsomeres (pentamers) capable of spontaneous *in vitro* assembly into capsid-like structures and polymorphic aggregates.[351,352] Truncation of the VP1 protein at the C terminus blocked assembly of capsomeres,[353] but its N terminus was found to be responsible for DNA binding and nuclear localization.[354] The Py internal proteins VP2 and VP3 were also expressed in *E. coli*,[355,356] and a crystal structure of a complex of the single VP2/VP3 copy with the pentameric major capsid protein VP1 has been determined at 2.2 Å resolution.[357] The VP1 proteins of other members of the polyomavirus group, namely, budgerigar fledgling disease virus[358] and human polyomavirus JC,[359] were also expressed in *E. coli*. The latter formed capsids able to deliver exogenous DNA into human cells and were suggested as a potent gene-transfer vector.[359] Of special interest is the highly efficient expression of HaPV VP1 in *E. coli,* allowing the formation of VLPs and packaging of foreign DNA,[360] since this model was one of the first in practical application as a carrier (see below).

The VP1 of Py,[356,361,362] SV40,[363,364] monkey B-lymphotropic papovavirus (LPV),[365] JC,[366,367] avian polyomavirus (APV),[368] and HaPV[369] were shown to assemble into T = 7 capsids in insect cells, independently of the minor VP2 and VP3; but if the latter were coexpressed with the VP1, they incorporated correctly into the VLPs. As in the case of the *E. coli* expression, the VLPs obtained in the baculovirus-driven expression in insect cells were found to be capable of packing DNA and were suggested as promising delivery tools for human gene therapy.[365,367,370,371] "Pseudoinfection" with polyomavirus VLPs from insect cells[372] and *E. coli*[373] was shown to be efficient enough for the transfer of small DNAs.[372]

The highly efficient expression and self-assembly of authentic HaPV VP1 and an N-terminally extended derivative (resulting from an in-frame-situated AUG codon) was shown in yeast *Saccharomyces cerevisiae*.[374] The nuclear localization of HaPV-VP1 suggests the functional activity of its nuclear localization signal in yeast cells, at with the VP1 derivative being N-terminally extended by 4 aa residues. Recently, recombinant VLPs of a set of human and nonhuman polyomaviruses were obtained in *S. cerevisiae*.[375]

Although immunological responses to polyomaviruses have been generally assumed to be T cell dependent, recent studies show the T-cell-independent character of immunoresponse against Py infection in mice.[376] However, Py VLPs, unlike the live virus, were found unable to induce the T-cell-independent IgG response.[376]

Genetic reconstruction of polyomavirus VLPs as carriers for foreign epitopes started as late as 1999. At first, all cysteine residues of the Py VP1 produced in *E. coli* were replaced by serines, and a new unique cysteine residue was introduced for the attachment of the fluorescence marker, which would be helpful for labeling virions to study virus–cell interactions preceding gene delivery.[377] Then the dihydrofolate reductase from *E. coli* was introduced as a model protein into position 293/294 of the Py VP1.[378] The formation of pentameric capsomeres and their ability to assemble into capsids were not influenced by the insertion, and the inserted dihydrofolate reductase, though less stable than the wild-type form, proved to be a fully active enzyme.[378] At last, superficially exposed regions of the HaPV VP1 were predicted on the basis of the known three-dimensional structure of the related polyomaviruses SV40 and Py to be located at aa positions 81–88, 222/223, 244–246, 289–294, and at the C terminus. Epitope mapping confirmed the C-terminal part and a more N-terminally located region of HaPV VP1 overlapping aa 81–88 as immunodominant.[349,379,380] As the C-terminal part of VP1 is responsible for the pentamer–pentamer interactions, a short model epitope DPAFR (see above) was introduced between only aa 80 and 89, 221 and 224, 243 and 247, and 288 and 295. All chimeras were capable of being self-assembled and induced a DPAFR-specific antibody response in mice.[381]

8.4.2 Double-Stranded RNA Viruses

8.4.2.1 Birnaviruses

Infectious bursal disease virus (IBDV), which since 1999 has been developed as a VLP carrier, is a pathogen of major economic importance to the world's poultry industries and causes a severe immunodepressive disease in young chickens (for a review, see Reference 382). A three-dimensional map of IBDV was determined at about a 2-nm resolution by electron cryomicroscopy.[383] The map shows that the structure of the virus is based on a T = 13 lattice and that the subunits are predominantly trimer clustered. The outer trimers seem to correspond to the protein VP2, carrying the dominant neutralizing epitope, and the inner trimers correspond to protein VP3, which has a basic carboxy-terminal tail expected to interact with the packaged RNA.[383]

Although a live attenuated IBDV vaccine, which was first introduced in 1968, remains widely used today, numerous attempts have been made to express the VP2 in yeast[384] and *E. coli*,[385] in which presence of "multimeric," but not capsid-like, particles was demonstrated. Furthermore, complete ORF A, which encodes structural proteins VP2, VP3, and VP4, was expressed in insect cells.[386] Finally, ORF A of IBDV was expressed using a vaccinia virus system, in which the A polyprotein was shown to be processed to give mature VP2, VP3, and VP4.[387] In addition, paracrystalline arrays of IBDV-like VLPs were found in the cytoplasm of cells infected with the recombinant vaccinia virus.[387] VLPs were also obtained in the insect cells after expression of the ORF A, although no VP2 was observed after polyprotein cleavage.[388]

First, chimeric IBDV VLPs were produced in insect cells coinfected with two recombinant baculoviruses, which first encoded the VP2, VP3, and VP4, and then the VP2 with five histidine residues at its C terminus.[389] Purified VLPs were mosaic and consisted of native and chimeric VP2 and VP3. Immunologically, these particles were identical to the two wild-type IBDV strains, which contributed subunits to the chimeric VLPs.[389] Furthermore, the ability of the VP2 alone to form particles of approximately 20 to 30 nm in diameter was shown in the insect cells.[390] Addition of the six histidines to the C terminus of the VP2 protein did not prevent formation of particles, which could be affinity-purified in one step, and assured a high level of protection against the challenge with the virulent IBDV.[390] It is worth noting that the VP1 protein, the putative RNA-dependent RNA polymerase, of IBDV forms a complex with the VP3 protein and efficiently incorporates into IBDV VLPs, in absence of the IBDV RNA, after coexpressing the VP1 and IBDV ORF A in vaccinia virus-driven system.[391]

8.4.2.2 Cypoviruses

The situation with cypovirus polyhedrin developed in the same direction as dsDNA nucleopolyhedrovirus polyhedrins (see previous chapter). Unlike the latter, cypovirus polyhedrin crystalline inclusion bodies were resolved by X-ray powder pattern analysis at 8.2 Å.[392] Expression of the polyhedrin gene in *E. coli* led to highly efficient synthesis of the product, which, however, did not form any crystalline structure in bacterial cells but accumulated in the form of insoluble inclusion bodies.[393] Expression of *Bombyx mori* cytoplasmic polyhedrovirus (BmCPV) polyhedrin in an improved baculovirus expression vector and its assembly into large cubic polyhedra[394] gave onset to the mutational analysis of the putative carrier. A panel of mutated polyhedrin genes was constructed, and their expression in insect cells demonstrated clearly that N- and C-terminal mutations of the polyhedrin could turn the shape and crystallization pattern of polyhedra from hexahedral to acicular, pyramidal, or amorphous,[395] and the localization of them from cytoplasmic to nuclear.[396]

8.4.2.3 Orbiviruses

These are architecturally complex nonenveloped viruses with seven structural proteins, VP1 to VP7, organized in two shells and a genome consisting of ten variously sized dsRNA segments. Significant advances in orbiviruses as well-known VLP carrier candidates have been made due to the systematic studies of Polly Roy and her group on such models as bluetongue (BTV) and African horse sickness (AHSV) viruses (for a review, see References 397 and 398). The main prototype virus, BTV, is the etiological agent of a ruminant disease that can reach epidemic proportions among sheep and cattle. The double shell of the BTV is built up by four major structural proteins, where VP2 and VP5 form the outer shell and VP3 and VP7 the inner one. Besides its unique property of being self-assembled in insect cells into the single- and double-shelled potentially chimeric VLPs, BTV offered another particulate candidate, a nonstructural NS1 protein, which generates tubules in mammalian and insect cells.[399]

Since each BTV protein is encoded by a single RNA species, there was no alternative to constructing DNA clones of all ten RNA species and expressing them in baculovirus vectors at high levels. Then, novel vectors were developed, which allowed coexpression of three, four, or five BTV genes from single recombinant vectors, and coexpressed VP3 and VP7 were shown to form BTV core-like particles (CLPs), while coexpressed VP2, VP5, VP7, and VP3 formed BTV virus-like particles.[398] Historically, this work started with the expression of the VP3 and VP7 to large amounts of BTV CLPs — similar to authentic BTV cores in terms of size, stoichiometry of VP3 to VP7, and the predominance of VP7 on the surface of the particles — in baculovirus-driven insect cells.[400] These CLPs were observed as paracrystalline arrays in infected insect cells. The next logical step consisted in assembling double-shelled, virus-like particles of BTV that had the same size and appearance as authentic BTV virions, by the simultaneous expression of four structural proteins and with the addition of the outer capsid proteins VP2 and VP5.[401] Minor structural proteins, VP1, a component of the viral RNA-directed RNA polymerase, and VP4, a guanylyl transferase, appeared as specifically incorporated into the core- and virus-like particles when expressed together with the major structural proteins.[402,403] CLPs were also formed successfully in the *in vitro* rabbit reticulocyte translation system, and each outer capsid protein synthesized *in vitro* had the capacity to bind to a preformed CLP.[404]

In order to facilitate the insertion of three or four BTV genes, multiple gene transfer vectors have been developed for the simultaneous expression of the appropriate genes.[398,405] After BTV, expression of structural genes in insect cells with the outcome of the appropriate particulate products was achieved also for AHSV[406,407] and Broadhaven virus (BRDV), a tick-borne orbivirus.[408]

The structures of several BTV and AHSV proteins were determined by X-ray crystallography. When combined with the three-dimensional image reconstruction of virions and VLPs obtained by electron cryomicroscopy, these data provided useful information not only regarding folding and assembly processes but also regarding knowledge-based engineering of their chimeric derivatives. First, the three-dimensional structure of single-shelled BTV virions has been determined to a resolution of 3 nm by using electron cryomicroscopy.[409] Such virion cores had T = 13 icosahedral symmetry in a left-handed configuration (T = 13laevo), in which 260 capsomeres in the outer layer were proposed to be made up of trimers of the VP7, whereas the inner layer seemed to be composed of the VP3.[409] The outer shell of the whole BTV particle was shown to consist of 120 globular regions of VP5, located on each of the six-membered rings of VP7 trimers, and of the sail-shaped spikes of the VP2.[410]

Electron cryomicroscopic examination of BTV VLPs, synthesized by coexpression of the VP2, VP3, VP5, and VP7 in insect cells, revealed the same T = 13laevo organization and confirmed essentially the same three-dimensional features as for the native virions.[411,412] The lack of the five VP7 trimers in the recombinant CLPs[411] was explained by the necessity of the presence of the VP2 and VP5 for adhesion of these VP7 trimers around the fivefold axes.[412] In parallel with the electron cryomicroscopy, crystallographic studies were carried on the VP7 and CLPs of BTV[413–415] and AHSV.[416] Combination of electron cryomicroscopy and X-ray crystallography data led to an atomic resolution map of the core, which consists of 780 VP7 molecules organized into the 260 trimers on a T = 13laevo icosahedral lattice.[22,417] The map showed that the β-barrel domains of VP7 are external to the core and interact with protein in the outer layer of the mature virion, whereas the lower, α-helical domains of VP7 interact with VP3 molecules, which form the inner layer of the BTV core.[417] The inner layer of the BTV and AHSV cores has T = 2 organization and contains 120 copies of large (about 100 kDa) VP3 molecules arranged as 60 approximate dimers.[418]

Electron cryomicroscopy of the BTV tubules, synthesized in insect cells and consisting of the NS1, revealed an average diameter of 52.3 nm and length of up to 100 nm. The structure of their helical surface lattice has been determined using computer image processing to a resolution of 40 Å. The NS1 protein is about 5.3 nm in diameter and forms a dimer-like structure, so that the tubules are composed of helically coiled ribbons of NS1 "dimers," with 21 or 22 dimers per turn.[419]

Although BTV CLPs and VLPs are noninfectious and lack virus (or other) DNA/RNA required for replication, unlike live virus vaccines, they are more immunogenic than killed viruses, or subunit vaccines, and are highly effective in eliciting humoral, cell-mediated, and mucosal immunities, and long-lasting protective response against challenges with infectious viruses.[397,398,420,421]

The BTV CLPs and VLPs demonstrated a remarkable ability to exchange proteins and produce mixed particles containing homologous subunits of other BTV subtypes[422] and other orbiviruses such as epizootic hemorrhagic disease virus (EHDV),[423] but not BRDV.[408]

Adaptation of BTV to the role of the molecular carrier candidate started with the deletion, point, and domain switching analyses of regions of the core proteins VP3 and VP7, which are responsible for intermolecular contacts and self-assembly.[398,424] The first chimeric BTV CLPs were constructed by the addition of the HBV preS2 fragment 1–48 to the N terminus of the VP7.[425] The chimeric VP7-preS2 protein participated in the BTV CLPs only when the insect cells were also coinfected with a recombinant baculovirus that expressed unmodified VP7 and VP3 of BTV.[425] Immunoelectron microscopy of the chimeric particles indicated that the preS2 epitope was exposed on the surface of the CLPs. Extension of the 11-aa rabies virus sequence added to the N terminus of VP7 was also found to allow for CLP formation, and several mutations critical for self-assembly of VP7 have been mapped.[426] Major VP7 epitopes mapped at the N terminus of VP7 are the most exposed on the BTV surface.[427]

Then, a series of four potent insertion vectors was constructed on the basis of the VP3 gene[428] in accordance with the deletion mapping data.[423] The insertion of 12 N-terminal aa residues of a T7 capsid protein into internal VP3 regions or the addition of 13 aa of the bovine leukemia virus (BLV) glycoprotein gp51 to the VP3 C terminus did not prevent the formation of the CLPs, if the VP7 was provided.[428] Next, artificial "recombinants" of BTV and AHSV-4 VP7 proteins were constructed on the basis of the crystal structure of the latter, but they failed to assemble into CLPs, although they were able to form trimers.[429] Recently, a 15-aa-long C-terminal CD4+ T-cell epitope of the influenza virus matrix protein M1 was added to the VP7, found to be included into CLPs in the presence of VP3 and able to induce an insertion-specific proliferative response.[430]

The tubule-forming NS1 protein was also developed as the carrier candidate for foreign epitopes. First, a series of deletion and site-directed mutants of the NS1 was generated, and the regions critical for its self-assembly, including both the N and C termini, were mapped.[431] Fine mapping of the C terminus with respect to self-assembly identified a domain between the 5th and 10th C-terminal aa as important for self-assembly.[432] The addition of foreign sequences ranging from 44 to 116 aa in length and representing a 44-aa-long sequence from *Clostridium difficile* toxin A, or a 48-aa-long sequence of the hepatitis B virus preS2 region, or the complete p15 protein of the BLV to the C terminus of the NS1 resulted in all cases in the formation of tubules, or mixed tubules, when insect cells were coinfected with three recombinant baculoviruses.[433] Chimeric tubules exposed foreign epitopes on their surface, were highly immunogenic, and therefore served as an example for the delivery of multiple epitopes on the NS1.[433]

8.4.2.4 Rotaviruses

The complex molecular architecture, morphogenesis, and history of development of rotaviruses as VLP carrier candidates are similar to those of the orbiviruses discussed above. However, being a leading cause of severe infantile gastroenteritis worldwide, rotaviruses provoke more interest for vaccine development. Therefore, rotavirus VLPs were initially brought under detailed immunological analysis as vaccine candidates, whereas construction of chimeric derivatives was set aside. As a result, rotavirus VLPs are included in this chapter only due to their role as carriers of chemically linked peptides.

Rotaviruses are icosahedral particles consisting of a double shell, or three concentric capsid layers, enclosing a genome of 11 segments of dsRNA (for a review, see Reference 434). According to electron cryomicroscopy data, the innermost layer is formed by capsid protein VP2 (analogue of VP3 in orbiviruses), which is organized in dimers of 120 quasi-equivalent molecules and acts as a scaffold for the proper assembly of the viral core.[434] The next layer of the rotavirus virion is formed by capsid protein VP6 (analogue of VP7 in orbiviruses). As its structural homolog in orbiviruses, the VP6 became the first target for studies at the atomic level by X-ray crystallography. The VP6 trimers were resolved,[435] and a model of their self-assembly was suggested.[436] The outer layer is formed by outer membrane proteins VP4 and VP7 (analogues of VP2 and VP5 in orbiviruses).

Rotavirus capsid proteins were expressed at high levels in insect cells. The bovine rotavirus VP2 protein assembled in the cytoplasm of insect cells into single-layered CLPs of 52 nm in diameter.[437,438]

The VP6 itself formed trimers,[439] whereas coexpression of the VP2 and VP6 resulted in the formation of double-layered, or single-shelled, particles. These particles were mosaic, when heterologous VP2 and VP6 were used.[437–439] Expression of the outer membrane proteins VP4 and VP7 as well as the VP2 and VP6 resulted in the formation of stable triple-layered, or double-shelled, VLPs. Formation of the mosaic particles with VP7 from different virus serotypes was also permitted.[440,441]

Recombinant rotavirus CLPs and VLPs were recognized as potent veterinary and human vaccines from the very beginning, and numerous immunological studies were based on rotavirus CLPs and VLPs. First, the protective potential of bovine rotavirus proteins was tested in a murine model, and high protective efficacy of the assembled VP6 particles harboring the outer proteins VP4 and VP7 was shown, unlike with unassembled proteins.[439] An evaluation of CLPs and VLPs from human, bovine, or simian rotavirus VP2, VP4, VP6, or VP7, as well as of chimeric VLPs harboring multiple types of the outer VP7 protein showed that they can be effective immunogens in rabbits, mice, and dairy cattle when administered parenterally.[442–444] Both CLPs and VLPs were also shown to be protective in mice and rabbits when administered orally or intranasally,[445,446] and passive immunity in newborn calves fed colostrum from VLP-immunized cows has been described.[447] The necessity of inducing a balanced Th1/Th2 response and the important role of adjuvants for full protectivity of VLPs was demonstrated recently in mice.[448]

To develop a human rotavirus vaccine, bovine VP2 was selected as a core component, and simian VP4, VP6, and VP7 were composed on it in the mosaic VLPs.[449] Then, other compositions of bovine VP6 and simian VP7, or bovine VP2 and VP6, and simian VP7 have been studied, and the combination VP2/6/7 was found to be the most promising.[450] Recent studies on the possible application of mosaic rotavirus VLPs as a human vaccine showed that immunization with VLPs containing a different human VP7 may provide sufficient priming of the immune system to induce protective heterotypic neutralizing antibody responses, and a limited number of VP7 serotypes within the VP2/6/7 VLPs may be sufficient to provide a broadly protective subunit vaccine.[451]

Bovine rotavirus VP6 particles have also been developed as a delivery system for the coupling of synthetic peptides or proteins via a "binding peptide" derived from the VP4.[452] This attachment was shown to be mediated by peptide–protein interactions and did not require additional chemicals for conjugation. The resulting macromolecular structures were highly immunogenic for both the VP6 protein and the coupled peptides, as well as in the absence of any adjuvant.[452–454]

On the other hand, secreting and anchoring derivatives of the outer VP7[455] and the inner VP6[456] proteins were synthesized in order to present them on the surface of the cell. Expression of the chimeric proteins harboring an upstream leader sequence and a downstream membrane-spanning anchor resulted in anchoring in the cell surface membrane, with the major domains of the proteins orientated externally. Such antigens, delivered by the vaccinia virus, produced efficient stimulation of both B and T lymphocytes and may be protective in mice.[455,456]

Although rotavirus VLPs have been suggested as carriers of foreign epitopes from heterologous pathogens or of drugs that need to be delivered to the gastrointestinal tract, either parenterally or orally,[457] the appropriate manipulations were limited by deletion mapping of the VP6[458] and VP2.[459] As a result, trimerization and self-assembly domains were localized on the VP6,[458] but 92 N-terminal aa residues of the VP2 were found dispensable for self-assembly. Truncated VP2 single- and double- (with the VP6) layered particles were unable to incorporate the VP1, a presumed RNA-dependent RNA polymerase, and lacked therefore a replicase activity.[459]

8.4.3 Retroid Viruses

8.4.3.1 Retroviruses

8.4.3.1.1 Oncoretroviruses

As recently as 10 years ago, the first report appeared on the successful addition of a long foreign sequence, the iso-1-cytochrome C (CYC1), from *Saccharomyces cerevisiae*, to the C-terminally truncated Pr76gag, a polyprotein precursor in Rous sarcoma virus (RSV), in which chimeric particles revealed a morphology similar to that of immature type-C retrovirions.[460]

For a long time, a representative of the BLV-HTLV retrovirus genus, bovine leukemia virus (BLV), served only as a source of epitopes for insertion into such carrier vectors as HBc (see above), until its Gag was chosen in 1999 as a potent VLP carrier candidate.[461] Expression of the BLV Gag encoding a Pr44gag precursor protein in insect cells, using the baculovirus expression system, resulted in the self-assembly and release of VLPs. Recombinant baculoviruses expressing matrix (MA) or capsid-nucleocapsid (CA-NC) proteins of BLV were generated, but neither of these domains was capable of assembling into particulate structures. Moreover, chimeras exchanging MA and CA-NC proteins of BLV and other leukemia viruses (e.g., human T-cell leukemia virus type I), or such evolutionarily divergent retrovirus group as lentiviruses (e.g., simian immunodeficiency virus) assembled efficiently and budded as VLPs.[461] Formation of VLPs in insect cells was recently achieved for human lymphotropic T-cell virus type II (HTLV-II).[462]

The Gag precursor of the feline leukemia virus (FeLV) belonging to another genus of Retroviridae, the Mammalian type-C retroviruses, was expressed in insect cells and was shown to form VLPs and release them from the cell. Coexpression of the FeLV Gag with the FeLV envelope glycoprotein gp85 resulted in the formation and budding of Gag/gp85 particles.[463] Moreover, conditions have been found under which VLPs were formed spontaneously *in vitro* from fragments of Rous sarcoma virus (RSV) Gag protein purified after expression in *E. coli*.[464]

The most advanced three-dimensional data on the oncoretroviral capsid structure have recently been achieved by NMR, which described the solution structure of the capsid protein (CA) from the HTLV-I[465] and RSV.[466]

Numerous studies have been undertaken on the construction and expression in mammalian cells of the artificial recombinants of Gag proteins of C-type retroviruses, primarily RSV and murine leukemia virus (MuLV) with other C-type oncoretroviruses and lentiviruses (for example, see References 467 through 472). However, we avoid the responsibility of referring these investigations, since they lie too far outside the general scope of the present chapter.

8.4.3.1.2 Lentiviruses

Human immunodeficiency viruses (HIV) have attracted much attention regarding the construction of promising vaccines against AIDS. Other closely related immunodeficiency viruses were developed as AIDS models. Gag proteins turned out to be necessary and sufficient for VLP self-assembly and budding, as in other retroviruses (for a review, see Reference 473). The Gag gene encodes a Pr55Gag precursor consisting, like Gag of leukemia and other retroviruses, of the domains matrix (MA), capsid (CA), nucleocapsid (NC), and p6, which are separated by the viral proteinase inside the nascent virion, leading to the morphological maturation of an infectious virus. In the mature virus, CA forms a capsid surrounding the ribonucleoprotein core consisting of NC and the genomic RNA, but a lipid bilayer containing envelope proteins surrounds the Gag capsids.

The way to structural investigations was opened by the expression of the Gag precursor of simian (SIV),[474–476] human type 1 (HIV-1)[477–480] and 2 (HIV-2),[481] bovine (BIV),[482] and feline (FIV)[483] immunodeficiency viruses in recombinant baculovirus-infected insect cells. The expression of the SIV Pr57gag[474] and the HIV-1 Pr55Gag[477] precursors were the first to demonstrate that the unprocessed Gag product can spontaneously assemble into 100- to 120-nm VLPs, indistinguishable from immature lentivirus particles, in the absence of other viral proteins, but myristylation of the N-terminal glycine guarantees its budding and extracellular release. Inclusion of the *pol* gene and expression of viral protease activity in the system resulted in efficient processing of the Gag precursor to p17, p24, p9, and p6 proteins and abolishing particle formation.[477,479,483] Moreover, the SIV matrix (MA) domain of the Gag polyprotein (p17 Gag) self-assembled into 100-nm VLPs, which were released into the culture medium. Coexpression of SIV MA and Env proteins resulted in the incorporation of gp120 and gp41 proteins into the recombinant p17-made particles.[475] Furthermore, two domains critical for VLP formation were located in the central hydrophobic α-helix of the

SIV MA.[484] The SIV MA, but not HIV-1 MA, was found later to be sufficient for VLP formation and budding in eukaryotic cells as well.[485] Coexpression of HIV Pr55 and gp160 resulted in the apparent incorporation of gp160 into Gag VLPs during the budding process.[486]

Although the baculovirus-driven expression system was the most efficient, self-assembly of the Gag precursor VLPs was observed in mammalian cells using vaccinia,[487,488] SV40,[489,490] and adenovirus[491] expression vectors. The domains of the HIV Gag precursor were expressed in and purified from *E. coli*, where *in vitro* assembly of CA yielded tubular structures with a diameter of approximately 55 nm and heterogenous length,[492] while extending CA by 5-aa residues was sufficient to convert the assembly phenotype into spherical particles.[493] The two-step self-assembly of HIV Gag protein lacking only the C-terminal p6 was modeled *in vitro* and resulted in particles equivalent to Gag VLPs obtained from Gag-expressing cells and similar to authentic immature HIV particles.[494] Three stages of Gag VLP formation have recently been described.[495] The MA domain was found to be responsible for the trimerization of the assembly-competent Gag; the p2 domain, located at the CA–NC junction and responsible for further multimerization and a change in Gag assembly morphology from tubes to spheres; and finally the N-terminal myristoylation, responsible for converting these multimers into Gag VLPs.[495]

Surprisingly, there is no clear evidence yet regarding the symmetry of Gag VLPs. The latest image processing of immature VLPs from insect cells[496] confirmed the so-called fullerene-like model for prebudding HIV-1 Gag assemblies[497] and revealed the presence of threefold symmetry and a hexagonal network of rings with a resolution of 29 Å in VLPs and greater than 25 Å in membrane-associated assemblies. These data are consistent with the hypothesis that immature HIV particles possess icosahedral symmetry with the triangulation number T = 63.[496,497] On the other hand, electron cryomicroscopy of VLPs from insect cells and immature HIV-1 particles from human lymphocytes found no evidence of icosahedral symmetry.[498]

The three-dimensional structure of the HIV-1 MA protein was determined at high resolution by the NMR method and found to be mostly α-helical and forming an icosahedral shell, which is associated with the inner membrane of the mature virus.[499] The dimers of the HIV-1 CA protein were studied by X-ray diffraction[500] and by multidimensional heteroNMR spectroscopy for the N-terminal (aa residues 1–151)[501] and C-terminal (aa residues 146–231)[502] parts of the HIV-1 CA. Unlike the structure of the previously characterized viral coats, the N terminus of HIV CA showed predominant α-helical organization, composed of seven α-helices, two β-hairpins, and an exposed partially ordered loop. The C-terminal CA domain, which contains a 20-aa major homology region (MHR) and is essential for capsid dimerization and viral assembly, consists of four α-helices and an extended N-terminal strand.[502] These data allowed for the identification of the side of the dimer that would be on the exterior of the VLPs.[500] The three-dimensional structure of HIV-1 NC bound to the genomic RNA packaging signal has been determined by heteroNMR,[503] and it showed tight binding of the NC zinc fingers to the RNA. It is worthy of note that the NC zinc fingers were the first structures determined at atomic resolution for a retroviral zinc fingerlike complex[504] (for a review, see References 505 and 506).

From the very first expression experiments in insect cells, recombinant lentivirus VLPs were highly immunogenic in experimental animals (mice, rabbits) and seemed to represent a noninfectious and attractive candidate for a basic vaccine component.[479,481,482] Further, a correlation of aa changes in the CA domain with the HIV-1 escape from CTL surveillance[507] favored the idea that CTL epitopes were capable of controlling HIV replication in long-term nonprogressors and that the HIV-specific CD8-positive CTL response could be induced by recombinant Gag VLPs (for a review, see Reference 508). Recently, the impact of therapeutic immunization with a p24 (CA) VLP vaccine candidate was evaluated in subjects with asymptomatic HIV infection as a Phase II trial. Although vaccination with the p24 VLPs plus zidovudine did not significantly alter either antibody or proliferative responses to p24, or the CD4+ cell number, immune activation, or viral load over 12 months,[509] it augmented the HIV-specific CTL activity.[510]

Deletion and mutagenesis mapping of the Gag was followed by the construction of its chimeric derivatives. Thus, the importance among lentiviruses (HIV-1, HIV-2, SIV, and FIV) of the conserved C-terminal part of the CA domain (aa 341–346 and 350–352) for VLP self-assembly was proven by the appropriate deletions within the HIV-1 protein.[511] Deletion of aa 99–154 overlapping the MA-CA cleavage site completely abolished the capacity of the HIV-1 Gag polyprotein to form VLPs, but deletions of aa 211–241 within the CA and aa 436–471 within the p6 portion had no effect on the assembly, ultrastructure, biophysical properties, or yields of mutant VLPs.[512] On the other hand, other authors found the aa 210–241 deletion to have a significant influence on Gag self-assembly, as well as on deletions 277–306 (MHR), 307–333 (within the CA), 358–374 [CA-spacer peptide 2 (sp2) junction], and 375–426 (sp2-NC junction and most of the NC).[513] Substitution of aa 341–352 within the C-terminal portion of the HIV-1 CA for alanines led to an inhibitory effect on its capacity to form VLPs, but it completely blocked the assembly and release of HIV in a replication-competent system.[514]

Mapping data on an HIV-2 Gag showed, however, that the the C-terminal p12 region of the Gag precursor protein and the zinc finger domain are dispensable for VLP assembly, but the presence of at least one of the three proline residues located between aa positions 372 and 377 of HIV-2 is required.[515] Therefore, the C-terminally truncated HIV-2 Gag protein 376 aa in length — but not the protein 372 aa in length — was found to be self-assembly competent.[515] Thorough examination of the effects of NC and p6 mutations showed that the p6 domain appeared to affect not virus release efficiency but specificity of genomic HIV-1 RNA encapsidation. The NC domain was responsible, however, not only for the specificity and efficiency of RNA encapsidation but also for particle assembly and release.[516] Replacement of both NC and p6 domains by foreign polypeptides able to form interprotein contacts permitted the efficient assembly and release of particles, although no packaged RNA was detected.[516,517] Interestingly, a new function of particle size control was attributed to the HIV-1 p6 not long ago after analysis of a large collection of Gag mutants.[518]

An attempt to construct a minimal HIV-1 Gag capable of VLP assembly and release consisted in the combination of an assembly-competent MA deletion mutant with progressive C-terminal truncations of the Gag.[519] The smallest HIV *gag* gene product still capable of VLP formation was a 28-kDa protein consisting of a few MA amino acids and the CA-sp2 domain.[519]

The potential of a Gag precursor as a carrier for the presentation of foreign epitopes was first documented by the insertion of the neutralizing domain V3 or the neutralizing V3 plus CD4 binding domains of the Env protein into the C-terminally truncated 41-kDa HIV-2 *gag* gene with further expression in baculovirus-infected insect cells.[520] It appeared that insertion of the *env* gene sequences into the C terminus, but not into the middle of the carrier, allowed for the production of VLPs, which were slightly larger than the initial Gag particles and similar to the mature HIV particles. Immunization of rabbits with these chimeric VLPs elicited high titers of neutralizing antibodies. Furthermore, Gag-V3 VLPs showed an ability to elicit a strong anti-V3 CTL response in immunized mice.[521,522] As suitable sites for insertion, replacement of two domains of a predicted high surface probability (211–241 and 436–471), but not of the aa 99–154, with foreign epitopes (V3 loop; CD4-binding-domain; Nef CTL epitope), or fusion of the above-mentioned sequences to the C terminus of the Pr55Gag were suggested in the vaccinia expression system.[512] In general, these chimeric Gag VLPs carrying T-helper and CTL epitopes from HIV-reading frames other than Gag demonstrated good capabilities to ensure a cell-mediated immunity in addition to a humoral immune response.[523] Although antisera exhibited only weak neutralizing activity, immunized mice developed a strong CTL response against a CTL epitope within the V3 domain, and the magnitude of this response was not influenced by the position of the V3 domain within the Pr55Gag carrier moiety.[524] Moreover, the Pr55Gag/V3 polypeptide retained high CTL-priming activity even after denaturation to the monomeric state.[525]

In a similar way, precise epitope-against-epitope replacement of a mapped Pr55Gag epitope 196–228, or 196–226, by the consensus V3 and/or the CD4-binding domains led to VLP production in and release from baculovirus-infected insect cells.[526,527] Although the inserted V3 was still antigenic, it was not exposed at the surface of the particles, according to immune electron micros-copy,[526] and did not elicit neutralizing antibodies as well.[527]

To regulate the specificity of the expected CTL response, two types of Gag VLP vaccines were constructed: type 1 by replacing defined Gag domains by selected HIV-1 CTL epitopes, and type 2 by stable anchoring derivatives of the HIV-1 Env protein on the surface of the VLPs.[528] In complete absence of adjuvants, type 1 and type 2 VLPs stimulated specific CTL in response, not only in mice but also in macaques. Unlike type 1 VLPs, which generate HIV-specific CTL responses in the absence of Env-specific antibodies, type 2 VLPs induced both arms of the immune system, including reasonable levels of neutralizing antibodies.[528]

Further attempts at improving V3-carrying Gag VLPs were turned to the extreme multiplication of different V3 variants and other gp160 epitopes included in particles. Thus, HIV-2 Gag chimeras with four neutralizing epitopes from HIV-1 gp160, or three tandem copies of the consensus V3 domain, which were derived from 245 different isolates of HIV-1, or four V3 domains from the most prevalent strains have been constructed and expressed in insect cells.[529] All three constructs appeared as spherical particles similar to immature HIV, but slightly larger than Gag VLPs, and induced neutralizing antibodies and strong anti-V3 CTL activity. The authors also found that the deletion of up to 143 aa at the C terminus of HIV-2 Gag, leaving 376 aa at the N terminus of the protein, did not affect VLP formation.[530] Therefore, the C-terminal p12 region of the HIV-2 Gag precursor protein and zinc finger domains are dispensable for Gag VLP assembly, but the presence of at least one of the three prolines at aa positions 373, 375, or 377 is obligatory.

In principle, another approach to constructing chimeric VLPs is possible. This involves entrap-ping chimeric Env protein derivatives into nonchimeric Gag VLPs. The first attempts in this direction were made by multiplying V3 domains by their insertion at various positions within the initial Env protein.[531] To improve such entrapping and obtain Gag VLPs containing high quantities of Env products, glycoprotein gp120 was covalently linked at different C-terminal positions to a trans-membrane (TM) domain from the Epstein–Barr virus (EBV) major Env glycoprotein gp220/350.[532] When coexpressed with Pr55Gag in insect cells, such Env-TM chimeras incorporated efficiently into the VLPs, unlike authentic Env proteins. Immunization studies carried out with such immu-nologically expanded VLPs elicited a consistent anti-Pr55Gag as well as an anti-Env antibody response, including neutralizing antibodies, and the remarkable induction of CTLs in the complete absence of additional adjuvants.[533]

The next attempts at constructing chimeric VLPs involved the combination of both genes, *env* and *gag,* in the form of in-frame, or frameshift, Gag-Env fusions.[534] Neither simple coinfection of insect cells with two Env- and Gag-expressing vectors nor the in-frame fusion strategies produced large quantities of structurally stable chimeric VLPs. The frameshift fusion method utilized the retroviral Gag-Pol ribosomal frameshift mechanism for the coexpression of Gag and Gag-frame-shift-Env fusion proteins, and reliable quantities of VLPs containing both the Gag and Env epitopes were produced in this case.[534]

This strategy was used further to translate the C-terminal 65% of the gp120 as chimeric Gag-Pol fusion protein, after insertion of the Env fragment immediately downstream of the Gag stop codon.[535] Mice inoculated with such VLPs developed CTL to both HIV-1 Gag and Env epitopes yet humoral immune responses only to Gag epitopes.

Nevertheless, despite HIV-1-specific neutralizing and CTL responses, both types of the HIV Pr55Gag VLPs, constructed either by replacing Gag sequences by the V3 loop and a linear portion of the CD4 binding domain, or by stable anchoring of modified gp120 on the surface of particles, failed to protect macaques against SHIV infection after intramuscular immunization.[536] Clearly accelerated reduction of the plasma viremia as compared to control animals was achieved after priming animals with the corresponding set of recombinant Semliki Forest viruses.[537]

Finally, a few facts of introduction of non-HIV sequences into the Gag carriers are worthy of consideration. Pseudorabies virus glycoprotein gD expressed on a separate vector was incorporated into Gag VLPs in insect cells.[538] Further studies supported the idea that the incorporation of surface glycoproteins into retroviral particles is not a specific process and that many heterologous viral and cellular glycoproteins can be incorporated as long as they do not have long cytoplasmic C-terminal regions. Thus, the incorporation of the wt human epidermal growth factor receptor (EGFR) and of its C-terminally truncated variant into Gag VLPs was achieved successfully.[519] In agreement with the above concept, a C-terminal variant of the EGFR, with only 7 C-terminal aa, was incorporated into VLPs more efficiently than the nonmodified EGFR, with 542 C-terminal cyto-plasmic aa. Gag-β-galactosidase fusions were capable of being incorporated into VLPs in the presence of intact HIV Gag protein.[539]

8.4.3.2 Yeast Retrotransposon Ty1

Yeast retrotransposon became a leading candidate for the development of VLP technologies thanks to the classic studies of Alan J. Kingsman and his group. Seventeen years ago, eukaryotic trans-posons, such as the Ty element dispersed in the *S. cerevisiae* genome, were found to show structural and functional similarities not only to prokaryotic transposons but also to retroviral proviruses. These similarities consisted in the production of a Gag-Pol analogue by a specific frameshifting event,[540] which fuses two out-of-frame open reading frames TYA and TYB,[541,542] and the expression of VLPs in yeast.[543] TYA and TYB, overlapping by either 38 or 44 nucleotides, are expressed to produce 50-kDa p1 proteins, and 190-kDa p3 proteins as a TYA:TYB fusion protein, which are processed and assembled into VLPs that contain Ty RNA and reverse-transcriptase activity.[544,545] The major structural core proteins of Ty VLPs are thus generated by proteolytic cleavage of the p1, by analogy to the Gag precursor of retroviruses, by TYB-encoded protease. Moreover, the p1 protein contains sufficient information for the assembly of pre-VLPs, which does not require the presence of either Ty protease or reverse transcriptase.[544]

Ty VLPs were found to form open structures, unable to protect the encapsulated RNA, which distinguishes them from typical viral capsids.[546] Electron cryomicroscopy examination of Ty VLPs showed that they are highly polydisperse in their radius distribution, and many of them form an icosahedral T-number series.[547] Three-dimensional reconstruction to 38-Å resolution from micro-graphs of T = 3 and T = 4 shells revealed that the single structural protein p1 encoded by the TYA gene assembles into spiky shells from trimer-clustered units.[547] Moreover, the length of the C-terminal region of the p1 was found to dictate the T number, and thus the size, of the assembled particles.[548] Particles that were assembled from the full-length immature protein p1–440, mature processed protein p1–408 named p2, and its minimal self-assembly competent variant p1–346 demonstrated the prevalence of T = 9/T = 7, T = 7/T = 4, and T = 4/T = 3 shells, respectively.[548]

Intriguingly, the p1 product assembled into spherical particles similar but not identical to Ty VLPs from yeast in *E. coli*, an organism that lacks endogenous retrotransposons.[549]

The idea of basing a new vaccine strategy on Ty VLPs appeared in 1987,[34,500] and the devel-opment of Ty VLP technologies is published in comprehensive reviews.[551–554] In order to construct the first HIV:Ty-VLPs, two HIV gp120 segments comprising aa 130–342 and 274–466 were added after aa position 381 of the p1.[34]

An attempt to map the p1 regions necessary for the formation of the Ty VLPs showed that the border of dispensability lies between aa positions 340 and 363 from the C terminus, and between aa 40 and 71 from the N terminus, but region 62–114 was identified as indispensable for self-assembly.[555] Further mapping revealed three structural determinants of the Ty VLPs in detail and explained how point mutations can lead to increasing the diameter of the particles as much as eight-fold.[556] Recently, a two-hybrid system was used to specify the p1 regions critical for Ty VLP self-assembly, where the large internal deletion 147–233 was described unexpectedly as dispensable for monomer interactions.[557]

From the first experiments, chimeric Ty VLPs were found to be capable of stimulating antibody response, T-cell proliferative responses, and CTL responses (see Reference 554). Intrinsic epitopes of the Ty VLPs were also mapped by generating a panel of monoclonal and polyclonal antibodies against the p1 particle-forming protein.[558] As a result, two N-terminal regions of the p1 protein were mapped projecting from the surface of the shell, but the C terminus was seen to be buried within the particle core and not available for antibody binding.

Immunodeficiency viruses HIV and SIV were chosen as the first targets for introduction into Ty VLP vectors. Exposition of gp120 segments on the Ty VLPs developed from the two long (approximately 200 aa gp120) stretches mentioned above[34] to the V3 loop of 40 aa, which was added to the same position 381 of the p1.[559] In this study, immunogenicity of V3:Ty-VLPs was tested for the first time in a nontoxic formulation licensed for use in humans (aluminum hydroxide), and induction of high-titer-neutralizing antibodies and an HIV-specific T-cell proliferative response was described.[559] Then, mosaic V3:Ty-VLPs were produced, which carried various V3 loops of different HIV isolates on the same particle, and such VLPs were able to induce V3-specific CTL response in BALB/c mice in the absence of any adjuvant or lipid vehicle.[560] After comparison with recombinant gp120 and gp160, and a 40mer V3 peptide, the particulate nature of chimeric Ty VLPs was found to be crucial for the efficient uptake into APC with subsequent access to the MHC class I processing pathway.[560] Interestingly, the intact particulate structure of chimeric VLPs was strictly necessary for CTL induction, since the presence of adjuvants in the V3:Ty-VLP formulations suppressed CTL response.[561]

The next object of presentation on the Ty VLPs, the HIV-1 p24 protein, was added first to position 381 of the p1 protein after the recognition sequence for the blood coagulation factor Xa.[562] Introduction of a protease cleavage site between the yeast carrier moiety and the protein of interest was thus adapted to facilitate the release of the latter after purification in particulate form.

The Ty VLPs containing all or parts of the HIV-1 proteins p24, Nef, and gp41, and HIV-2 gp36 have been constructed, purified, and used to develop a rapid immunoassay to detect and differentiate between HIV-1 and HIV-2 antibodies in a single test.[563]

The p17/p24:Ty-VLPs (for simplicity, p24:Ty-VLPs) were chosen for further immunogenicity studies as a potent vaccine candidate. Immunization of cynomolgus macaques with the p24:Ty VLPs led to the induction of a cellular response against highly conserved Th epitopes.[564] The p24:Ty-VLPs were evaluated in HIV-seronegative volunteers and found to be capable of inducing both cellular and humoral immunity to HIV-1 Gag p17 and p24 components, but no Gag-specific CTL responses were detected.[565] In the subsequent Phase I study, the p24:Ty-VLPs elicited substantial humoral and T-cell proliferative responses, although again no p24-specific CTL responses were found.[566] When p24:Ty-VLPs were tested as a potential immunotherapeutic agent with the objective of augmenting the immune response to p24 in p24 antibody-positive asymptomatic HIV-1-infected subjects, no increase in p24 antibody levels and no effect on CD4 cell counts or virus load were found.[567] Further studies of the p24:Ty-VLP vaccine documented marginal Gag-specific proliferative responses in limited numbers of HIV-1-seropositive individuals, with some showing transient elevation of HIV-1 viral load.[568] In the subsequent Phase II study, the safety, immunogenicity, and tolerability of the p24:Ty-VLPs in HIV-antibody-positive asymptomatic volunteers were studied.[569] Although the p24:Ty-VLP immunizations were well tolerated, and the CD4 changes of patients were encouraging, no significant changes in humoral or cellular responses or viral load were found. More Phase II studies on the p24:Ty-VLP vaccine are now in progress.

After exposition of HIV proteins, a large Gag segment of the related SIV was added to the same position 381 of the p1 protein in such a way that the chimera contained aa 116–363 of the SIV Gag, namely, 20 C-terminal residues of p17 contiguous with full-length p27.[570] This model vaccine was developed especially with a view to designing vaccines for the prevention of homosexual transmission of HIV infection.

Thus, the p27:Ty-VLPs were used for ororectal mucosal immunization of macaques and found to be capable of inducing not only T-cell proliferative responses but also rectal secretory IgA antibodies.[571–574] Moreover, mucosal immunization with p27:Ty-VLPs elicited SIV Gag-specific CTLs in the regional lymph nodes as well as in the spleen and PBMC of macaques.[575]

Progress in the induction of CTL responses led to the conclusion that the Ty-VLPs are a potent means of inducing CTL responses to a variety of viral CTL epitopes, including influenza virus nucleoprotein, Sendai virus, and vesicular stomatitis virus (VSV) nucleoproteins, and the V3 loop of HIV gp120, which was described above.[576] The ability of Ty VLPs carrying two different epitopes or the ability of a mixture of two different Ty VLP chimeras to prime CTL responses in two different strains of mice or against two epitopes in the same individual was shown for the first time.[576]

The Ty VLP technology was used to study Gag and Env products of an ovine/caprine lentivirus, namely, maedi-visna virus (MVV). First, the major core protein p25 of the MVV was fused, in analogy to the HIV p24, at the Factor Xa cleavage site after the p1 moiety.[577] Then, the overlapping segments of the envelope glycoprotein gp135 of the MVV were exposed on the appropriate Ty VLPs.[578] These MVV:Ty-VLPs were used to produce antibodies and map epitopes and thus to develop diagnostic assays. In the same fashion, the Ty VLPs were used to construct fusions with the rubella envelope glycoprotein E1 and to map the binding sites of the appropriate monoclonal antibodies.[579]

The Ty-VLP-driven technologies ensured an important breakthrough in herpesviruses vaccine development, namely, human herpesvirus 3, or varicella-zoster virus (VZV). First, the overlapping segments of the envelope glycoprotein G (gE) of the VZV were exposed on Ty VLPs and helped to map immunodominant and neutralizing epitopes on the gE.[580] Second, the intrinsic immunogenicity of chimeric gE:Ty-VLPs carrying aa 1–134 or 101–161, which contained the immunodominant sequences, has been proven.[581] The results demonstrated that the gE:Ty-VLPs can prime potent humoral and cellular anti-VZV responses in small animals and warrant further studies as a potential vaccine. Next, a segment of 300 aa of the VZV assembly protein (AP) was inserted into Ty VLPs, which made possible the examination of the genetic organization and proteolytic maturation of VZV assemblin.[582] Finally, both gE- and AP:Ty-VLPs were found to be capable of being recognized by lymphocytes of varicella and zoster patients, which highlighted their potential use as a candidate vaccine.[583]

Of particular value are the Ty-VLP-initiated studies on malaria vaccine candidates. The Ty VLPs carrying a well-defined CTL epitope 252-SYIPSAEKI-260 from the circumsporozoite (CS) protein of *Plasmodium berghei* were found to induce a remarkable CTL response, on a par with a lipid-tailed peptide of this same sequence, in mice.[584] Moreover, these CTLs were able to recognize a naturally processed antigen expressed by a recombinant vaccinia virus. A string of up to 15 defined CTL epitopes from *Plasmodium* species added to Ty VLP primed protective CTL responses in mice following a single administration without adjuvant, and effective processing of epitopes from the string was demonstrated *in vitro* and *in vivo* and was not affected by flanking sequences.[585] Then, the authors proposed a novel prime/boost combination of CS:Ty-VLPs and modified vaccinia virus Ankara (MVA) containing a single *P. berghei* CTL-epitope among a string of 8–15 human *P. falciparum* CTL epitopes, which provided high-level protection against malaria in mice and was suggested as a candidate for vaccination against human *P. falciparum* malaria.[586] The immunization scheme was found to be very important: Ty VLPs followed by MVA was critical for protection, but the reciprocal sequence of immunization or homologous boosting was not protective.[586] Recently, the authors confirmed their success, showing that complete protection against malaria in mice can be achieved by priming with one type of vaccine (Ty-VLPs or DNA-based) and boosting with another (MVA).[587]

Nontraditional objects for insertion into Ty VLPs are represented first by the group-I (Der p1) allergen of *Dermatophagoides pteronyssinus,* a house dust mite. Th and CTL epitopes of the Der p1 were predicted, synthesized, exposed on the Ty VLPs, and found capable not only of inducing Th and CTL responses in mice,[588] but also of suppressing allergen-specific proliferation when mice were treated with Der p1:Ty-VLPs after sensitizing them to the Der p1.[589]

Regarding the expression of noninfectious subjects on the Ty VLPs, expression and purification of human interferon-α 2 must be mentioned.[590] Then, a line of ovine cytokines — interleukins 1α and β,[591] tumor necrosis factor α,[592] and interleukin 2[593] — was purified as chimeric Ty VLPs and released from the particles by cleavage with the Factor Xa enzyme.

An interesting use of the Ty VLPs was supposed by the insertion of two nucleases into particles, which retained ability to degrade encapsidated RNA.[594] This strategy was referred to as capsid-targeted viral inactivation and was proposed as useful for interfering with the replication of retroviruses.

The Ty VLPs also assured the search, characterization, and purification of potent mitochondrial genes — reverse transcriptases from mitochondrial introns of different eukaryotic organisms[595] and mitochondrial tyrosyl-tRNA synthetase.[596]

8.4.4 Single-Stranded DNA Viruses

The Parvoviridae family has been advanced to the forefront of VLP protein engineering for two reasons. They owned the old traditions of structural investigations, and their capsids were flexible enough to allow high-level production in the form of VLPs in insect cells and to retain self-assembly properties after foreign insertions. In fact, the most promoted is a Parvovirinae subfamily, where numerous representatives of Erythrovirus and Parvovirus genera play a role as classical VLP carriers (for a review, see References 597 and 598). The adeno-associated virus of the Dependovirus genus, in turn, occupied a place of favorite gene therapy vectors (see References 599 and 600).

Erythrovirus and Parvovirus capsids are composed of 60 copies of the major capsid protein VP2 and 6 to 10 copies of the minor capsid protein VP1, where VP1 is identical to VP2 but contains an additional N-terminal "unique" region of 227 aa. A portion of that region of VP1 is external to the capsid, but the VP1 itself is not required for capsid formation.

The first described production of VP1 and VP2 of human parvovirus B19 was achieved in Chinese hamster ovary cells, where they formed VLPs.[601] After only a year, both VP1 and VP2 were expressed in insect cells by the use of baculovirus vectors,[602] where VP2 alone was able to assemble into VLPs, but stoichiometry of the capsids containing both VP1 and VP2 was similar to that previously observed in parvovirus B19-infected cells.[603,604] Immunization of rabbits with such VLPs resulted in the production of neutralizing antibodies.[604]

Recombinant B19 capsids, consisting of either VP2 alone or of both VP1 and VP2, have been shown, therefore, to be similar to native viruses in size and appearance and were thought to be an excellent source of antigen, first of all, for immunodiagnostic kits.[605]

A long list of parvoviruses followed; the capsids of these parvoviruses were reconstituted in baculovirus-driven insect-cell expression in the form of VLPs indistinguishable from regular virions: porcine parvovirus (PPV),[606] canine parvovirus (CPV) and its recombinant with feline panleuko-penia virus (FPV),[607] adeno-associated virus type 2 (AAV-2),[608] mink enteritis parvovirus (MEV),[609] muscovy duck parvovirus (DPV),[610] and the minute virus of mice (MVM).[611] The VLPs of the Aleutian mink disease parvovirus (ADV) have been obtained via vaccinia virus expression,[612] but very recently the VP2 of the CPV was expressed in insect cells using the *Bombyx mori* nucleopoly-hedrovirus (BmNPV) vector.[613] Attempts to obtain parvovirus VLPs in *E. coli* ended with the high-level synthesis of insoluble products.[614]

The first X-ray study of parvoviruses, CPV, appeared in 1988.[615] Infectious virions of CPV 25.7 nm in diameter were found to contain 60 copies of VP2 within the ordered T = 1 icosahedral capsids, and the central structural motif of VP2 had the same topology (an eight-stranded antiparallel β-barrel) as has been found in many other icosahedral viruses.[616] There was a 22-Å-long protrusion on the threefold axes, where major antigenic regions were found, a 15-Å-deep canyon circling around each of the five cylindrical structures at the fivefold axes, and a 15-Å-deep depression at the twofold axes.[616] The structure of empty CPV capsids revealed some conformational differences between the full and empty viruses in the region where some ordered DNA has been observed to bind in the CPV-full particles.[617] Finally, the structure of the CPV capsid was resolved at 2.9 Å.[618] It was documented that approximately 87% of the VP2 have N termini on the inside of the capsid, but for approximately 13% the polypeptide starts on the outside and runs through one of the pores surrounding each fivefold axis, explaining apparently conflicting antigenic data.[618]

The crystal structure of parvovirus B19 capsids from the baculovirus expression system was resolved at first to 8 Å,[619] and then the structure of the full virions and empty capsids of FPV was resolved to at least 3.3 Å.[620] Further work on the B19 structure did not improve the 8-Å resolution, but it established the significant difference of the surface structure of B19, CPV, and FPV, where B19 lacked prominent spikes on the threefold icosahedral axes.[621] Further, electron cryomicroscopy of the B19 VP2 capsids complexed with their cellular receptor, globoside, were used to confirm X-ray data.[622]

The first report of crystals from parvoviral particles that were produced in a heterologous system and diffracting X-rays to high resolution appeared recently on the MVM,[611] although the structure of MVM virions was reported much earlier.[623] The latter provided a possible mechanism to explain how the unique N-terminal region of VP1 becomes externalized in infectious parvovirions.

At last, the three-dimensional structure of baculovirus-expressed VP2 capsids of the ADV was determined to a 22-Å resolution by electron cryomicroscopy and image reconstruction techniques and compared thoroughly with the existing CPV, FPV, B19, and MVM parvovirus structures.[624] Useful information for further structural comparison of parvovirus particles may be obtained from the first three-dimensional structure of more distantly related insect parvoviruses belonging to the Densovirinae subfamily, i.e., *Galleria mellonella* densovirus (GMDNV), where the same motif of externalization of the N-terminal region of the unique VP1 sequence was found.[625] The external location and special immunological importance of the B19 VP1 was confirmed by direct immune electron microscopy investigation.[626]

The VP1 was shown to be dispensable for the encapsidation of virus ssDNA, but responsible for virus entry, subsequent to cell binding and prior to the initiation of DNA replication.[627] The B19 VLPs composed of only VP2 and VP2 containing varying amounts of VP1 protein were evaluated in mice, guinea pigs, and rabbits, and the importance of VP1 for high levels of neutralizing activity was demonstrated.[628] Epitope mapping of parvovirus capsids revealed that N-terminal portions of the unique region of the B19 VP1[629,630] and of the CPV VP2[631] were responsible for neutralizing activity. However, the CPV VP2-VLPs from insect cells were used to immunize dogs in different doses and combinations of adjuvants, and a good protective response, higher than that with a commercially available, inactivated vaccine, was obtained.[632]

Since the unique part of the VP1 appeared as an ingenuous target for foreign insertions, the experiments were undertaken to shorten it in the B19 VLPs.[633] N-terminally truncated VP1 entered capsids more efficiently than their longer versions. Progressive truncation of the unique region of the B19 VP1 with the addition of a Flag peptide DYKDDDK at the N terminus led to the formation of VLPs not only from the mosaic VP2-VP1 particles but also from the truncated Flag-VP1 proteins only.[634] Most of the VP1 unique region was confirmed to be external to the capsid and accessible to antibody binding.

N-terminal truncation of the B19 VP2 showed that aa position 25 is critical for self-assembly; VP2 truncated to aa 26 to 30 failed to self-assemble, but did participate with normal VP2 in the capsid structure, whereas truncations beyond aa 30 were incompatible with either self-assembly or coassembly.[635] The CPV VP2 allowed for N-terminal truncation of 9 and 14, but not of 24 aa, and it allowed for the deletion of loop 2, but not of the other three loops of the protein.[636]

Generation of chimeric parvovirus VLPs broke out after the construction and expression in insect cells of the B19 VP2 with defined linear epitopes from human herpes simplex virus type 1 and mouse hepatitis virus (MHV) A59 inserted at the N terminus and at a predicted surface region.[637] Immune electron microscopy indicated that the epitopes inserted in the loop were exposed on the surface of the chimeric particles. The chimeric capsids were not only immunogenic in mice, but they also induced partial protection against a lethal challenge infection with either MHV or HSV.[637] In the other approach, an external portion of the B19 VP1 was substituted with a sequence encoding the 147 aa of hen egg white lysozyme with variable amounts of retained VP1 sequence joined to the VP2 backbone.[638] The authors demonstrated the external presentation of lysozyme on the VLPs, and its enzymatic activity and immunogenicity in rabbits.

The next round of activity started with the introduction of porcine and canine parvovirus VLPs into protein-engineering studies. Thus, foreign sequences were introduced into the N and C termini of the PPV VP2, and insertion of poliovirus VP1 T- and B-cell epitopes into the N terminus did not alter the formation of the VLPs.[639] Moreover, the VLPs containing C3:T epitope of the poliovirus VP1 were able to induce a T-cell response *in vivo*, but the VLPs containing the C3:B epitope did not induce any peptide-specific antibody response. Therefore, the N terminus of the PPV VP2 was declared to be located in an internal position, useful for T-cell epitope presentation, but inadequate for the insertion of B-cell epitopes.[639]

The real breakthrough in the development of parvovirus VLPs was offered by the addition of a CTL epitope from the lymphocytic choriomeningitis virus (LCMV) nucleoprotein to the PPV VP2.[640] Immunization of mice with these VLPs carrying a single viral CTL epitope, without any adjuvant, induced a strong CTL response and complete protection of mice against a lethal LCMV infection.[640] Further detailed immunological investigations unveiled mechanisms of how the PPV VLPs can be presented by MHC class I and class II molecules.[641] Strong CTL responses and neutralizing antibodies against the LCMV were achieved after intranasal, but not oral, immunization of mice, with the PPV VLPs carrying the single viral CTL epitope.[641]

Insertion of the poliovirus C3:B epitope into the four loops and the C terminus of the CPV VP2 allowed for the recovery of capsids in all of the mutants, but only the insertion at aa 225 of loop 2 was able to elicit a significant antipeptide antibody response, but not poliovirus-neutralizing antibodies.[642] To fine-modulate this insertion site in loop 2, the authors reinserted the epitope into adjacent positions 226, 227, and 228.[643] Surprisingly, these chimeric VLPs were able to elicit a strong neutralizing antibody response against poliovirus, demonstrating that the minor displacements in the insertion place may cause dramatic changes in the accessibility of the epitope and the induction of antibody responses. The potential of combining different types of epitopes in different positions of canine and porcine parvovirus VLPs to stimulate different branches of the immune system paves the way to the elaboration of novel vaccine candidates.[644]

Finally, AAV capsids, which are composed of three proteins, VP1, VP2, and VP3, that are expressed from different initiation codons on the same open reading frame, were tested as potent VLP candidates.[645] The addition of a Flag peptide to the N terminus of the VP3 did not affect its entrapping into VLPs, although this region was found inside the chimeric VLPs.

8.4.5 SINGLE-STRANDED NEGATIVE-SENSE RNA VIRUSES

Prevalent activities in the ssRNA negative-strand viruses are directed toward the construction of replication-competent chimeric viruses (influenza A virus) and chimeric VLPs (rabies virus, vesicular stomatitis virus, influenza A virus). Although this large taxonomic unit cannot boast of impressive advances in the field of noninfectious VLPs, some of their nucleoproteins, which were expressed in various experimental systems, can be regarded as potent models for the construction of chimeric particles. As a first, the measles virus (MV), belonging to the Morbillivirus genus of the Mononegavirales family, Paramyxoviridae group, Paramyxovirinae subgroup, merits more detailed consideration. Thus, MV nucleoprotein was expressed and shown to form tubular nucleocapsid-like structures not only in mammalian systems — vaccinia virus-[646] and adenovirus-driven[647] — but also in baculovirus-directed insect cells[648,649] and even in *E. coli*[649,650] in the absence of virion RNA or other viral proteins. Very recently MV nucleoprotein was efficiently expressed in yeast.[651] Immunization of mice with recombinant adenovirus carrying the nucleoprotein gene resulted in a good humoral and CTL response and complete protection against challenge with the measles virus.[647]

Human parainfluenza virus type 1, another representative of Paramyxovirinae but belonging to the next Respirovirus genus, was studied recently for expression of its matrix and nucleoprotein genes in mammalian cells.[652] Biochemical and electron microscopic analyses of transfected cells showed that the matrix protein alone can induce the budding of virus-like vesicles from the plasma membrane and that the nucleoprotein can assemble into intracellular nucleocapsid-like structures, but the coexpression of both genes resulted in the production of vesicles enclosing NC-like structures. The nucleoprotein of the related Sendai virus of the same Respirovirus genus was synthesized in mammalian cells in the absence of other viral components and assembled into nucleocapsid-like particles, which were identical in density and morphology to authentic nucleocapsids but smaller in size.[653] Exhaustive mapping allowed for the identification of nucleoprotein domains responsible for self-assembly of the Sendai virus nucleoprotein.[653,654]

The nucleoprotein gene of Newcastle disease virus (NDV), a member of Paramyxovirinae representing the next genus Rubulavirus, was expressed to high levels in insect cells and ensured the formation of nucleocapsid-like structures in the absence of other NDV proteins, unlike the nucleoprotein of the related rubulavirus — human parainfluenza virus 2.[655] Moreover, the same structures formed by NDV nucleoprotein were also observed in *E. coli*.[655]

Very recently high-level expression of the nucleoprotein gene of mumps virus, another popular representative of Rubulavirus genus, was achieved in yeast *Saccharomyces cerevisiae*.[656] The yeast-derived mumps nucleoprotein formed VLPs and showed high immunogenicity in mice.

The nucleocapsid-like structures were formed in insect cells after expression of the nucleoprotein gene of respiratory syncytial virus (RSV), a member of Pneumovirus genus, belonging to the Pneumovirinae subgroup of the Paramyxoviridae group,[657] a rabies virus of the Lyssavirus genus belonging to Rhabdoviridae, another group of the Mononegavirales family.[658,659]

8.4.6 SINGLE-STRANDED POSITIVE-SENSE RNA VIRUSES

The ssRNA positive-strand viruses include the largest number of candidates for VLP protein engineering. Viruses from this group demonstrate an extremely broad host spectrum and can infect bacteria, plants, and animals. We will discuss them here in alphabetical order of their families, ignoring priority or significance of individual representatives of this group.

8.4.6.1 Bromoviruses

This family includes exclusively plant viruses. Recently, development of VLPs from plant viruses made substantial progress as safer alternatives to the use of bacterial and animal viruses (reviewed in Reference 660). Both noninfectious (alfalfa mosaic virus, potyvirus, tobacco mosaic virus) and replication-competent (cowpea mosaic virus, potato virus X, and the same potyvirus and tobacco mosaic virus) models are being developed.

Virions of a small spherical plant virus, cowpea chlorotic mottle virus (CCMV) of the Bromovirus genus, are formed *in vitro* from *E. coli*-expressed coat protein and *in vitro* from transcribed full-length RNA.[661] When examined with electron cryomicroscopy and image reconstruction these particles were found indistinguishable from native virions. Mapping showed that the N-terminal part of the coat protein is required for assembly of RNA-containing particles but not for the assembly of empty virions, while the C terminus is essential for coat protein dimer formation and particle assembly.

Another epitope carrier candidate, alfalfa mosaic virus (AlMV) of the Alfamovirus genus, advanced even further in vaccine development. The AlMV coat was cloned and expressed in *E. coli* as a fusion protein containing a 37-aa extension with a six-histidine region for affinity purification.[662] The "bacterial" coat protein assembled into T = 1 empty icosahedral particles. Empty particles formed hexagonal crystals that diffracted X-rays to a 5.5-Å resolution. Self-assembly of coats with the smallest AlMV RNA4 led to the formation of spherical particles, but with the largest AlMV RNA1 to bacilliform particles resembling native virions.

The T = 1 icosahedral character of empty AlMV particles with a typical β-barrel structure was determined by electron cryomicroscopy and image reconstruction methods as well as by X-ray crystallography to a 4-Å resolution[663] based on an earlier X-ray model of AlMV capsids, resolved at 4.5 Å.[664]

Furthermore, the role of specific AlMV coat aa residues for biological activity and RNA binding[665] and its structural behavior in forming of unusually long viral particles[666] was studied.

The AlMV coat protein was used as a carrier molecule to express antigenic peptides from rabies virus and HIV.[667] The *in vitro* transcripts of a recombinant virus with sequences encoding the antigenic peptides were used to generate VLPs in tobacco plants. The VLPs purified from plants were shown to elicit specific virus-neutralizing antibodies in mice.[667] Significant protection of mice against challenge infection with a lethal dose of rabies virus was found after combining oral and intraperitoneal routes of immunization with the chimeric AlMV VLPs bearing two rabies virus epitopes.[668]

8.4.6.2 Caliciviruses

These exhibit a broad host range, but the main representatives of the Caliciviridae family are Norwalk (NV) and Norwalk-like caliciviruses, which are the major etiologic agents of epidemic gastroenteritis in humans (for a review, see Reference 669). Many basic features regarding the biology and replication of the human caliciviruses are still unknown because these viruses have not yet been grown in cell culture and do not replicate in animal models other than the chimpanzee.

NV capsid protein was produced via baculovirus expression and demonstrated an ability to self-assemble into empty VLPs similar to native capsids in size and appearance.[670] These particles induced high levels of Norwalk virus-specific serum antibody in laboratory animals following parenteral inoculation. In a similar way, without requiring any other viral components, VLPs were produced in insect cells after expression of capsid genes from numerous representatives of caliciviruses: a Mexican strain of human calicivirus,[671] Lordsdale virus,[672] rabbit hemorrhagic disease virus (RHDV),[673–675] European brown hare syndrome virus (EBHSV),[676] and Toronto,[677] Hawaii,[678] Sapporo-like,[679] and Grimsby[680] viruses. Formation of the RHDV VLPs was also achieved in yeast *S. cerevisiae*.[681] Surprisingly, unlike baculovirus-driven expression, Hawaii-strain VLPs were not observed in mammalian cells,[682] whereas expression of capsid gene from feline calicivirus (FCV) in cultured feline cells resulted in VLPs.[683] However, the formation of NV VLPs in tobacco leaves and potato tubers seems to be the most promising way of delivering oral vaccines via edible plant organs.[684] The plant-expressed Norwalk VLPs were indistinguishable from insect-cell-expressed ones and were orally immunogenic in mice. When potato tubers expressing Norwalk VLPs were fed directly to mice, they developed specific serum IgG.[684] Transgenic Norwalk VLP potatoes induced immunological response in human volunteers as well.[685]

The self-assembly of VLPs of a distant relative of caliciviruses described above, i.e., hepatitis E virus (HEV), has also been achieved in insect cells.[686] The three-dimensional structure of these VLPs has been solved to a 22-Å resolution by electron cryomicroscopy and three-dimensional image reconstruction.[687] This was the first structure of a T = 1 particle with the protruding dimers at the icosahedral twofold symmetry axes, which were solved by electron cryomicroscopy.

The first fine-structure studies on caliciviruses were also carried out using electron cryomicroscopy and computer image-processing techniques. Calicivirions exhibited T = 3 icosahedral symmetry and demonstrated several architectural similarities with plant viruses such as tomato bushy stunt virus and turnip crinkle virus.[688,689] The structure of a complex between RHDV VLPs and a neutralizing monoclonal antibody has been determined at low resolution by electron cryomicroscopy.[690] The atomic coordinates of a Fab fragment were fitted to the VLPs, which had a T = 3 icosahedral lattice consisting of a hollow spherical shell with 90 protruding arches with the mAb occupation of 50%. The smaller form of Norwalk VLPs found in insect cells, 23 nm in diameter, appears to be a T = 1 symmetry variant of capsid self-assembly, consisting of 60 polypeptide molecules.[691] The X-ray structure of T = 3 NV VLPs assembled from 180 copies of coat protein with a classical eight-stranded β-sandwich motif was reported recently.[692]

Calicivirus VLPs elicited high-level immune response in experimental animals. Complete protection of RHDV VLPs against hemorrhagic disease in rabbits was reported after intramuscular[673,674,681] and oral[693] immunizations. Further studies in mice as well as in volunteers demonstrated that the NV VLPs are a promising mucosal vaccine for NV infections.[694,695] Phase I trials on healthy volunteers found the NV VLP vaccine candidate to be safe and immunogenic when administered orally without adjuvants.[696]

Calicivirus VLPs became subjects of protein engineering in 1999. First, the addition of foreign sequences has been proven on Sapporo-like virus VLPs. The addition of a short MEG tripeptide at the N terminus of the Sapporo-like virus coat did not prevent its self-assembly, whereas addition of a His-tag blocked it.[679] When the N-terminal 30-aa residues of RHDV capsid protein were substituted by well-characterized six-residue epitope from the bluetongue virus capsid protein VP7 (Btag), the fusion protein retained its ability to self-assemble into VLPs; however, the size of these particles was only 27 nm, compared to 40-nm VLPs derived from the native polypeptide.[697] However, when the Btag was fused to the C terminus of the RHDV capsid protein without deletion, the fusion proteins formed VLPs of 40 nm in size and also retained their antigenicity, but the Btag antigenicity appeared to be low in this construct.[697]

8.4.6.3   RNA Bacteriophages

These RNA viruses belong to the first classical models of molecular biology, which contributed markedly to the discovery of the genetic code, elucidation of RNA translation and replication mechanisms, virus–host interactions, and self-regulation of biological systems. The RNA phage MS2 was the first organism with the fully sequenced genome.[698] Capsids of the closely related phages R17 and f2 were among the first virions with clearly resolved symmetry.[699] Icosahedral capsids of RNA phages are composed of 180 copies of coat protein (CP) and one copy of A, or maturation, protein. Based on physical and immunological properties, RNA phages are divided into groups I to IV. Groups I and II belong to genus Levivirus, whereas groups III and IV belong to genus Allolevivirus (for a review, see Reference 700). Unlike Levivirus representatives, the Allolevivirus virions contain the protein A1 within their capsids, which is a 329-aa-long read-through variant of CP consisting of the CP body and its C-terminal, 195-aa-long extension separated from the CP body by an opal (UGA) stop codon. The A1 protein content in standard non-UGA-suppressor growth conditions averages only a few molecules for each particle, but its proportion may be raised significantly by UGA suppression. Both Qβ CP body and the A1 extension attracted attention as possible targets for introduction of foreign epitopes. The structure and principles of the use of RNA phages as VLP carriers are shown in Figure 8.6.

High-level production in *E. coli* of self-assembled capsids morphologically and immunologically indistinguishable from virions has been achieved after expression of coat protein genes of the following RNA phages: MS2[701,702] and fr[703] of group I, JP34 as an intermediate between groups I and II,[704] GA of group II,[705] Qβ of group III,[706] SP of group IV,[707] and *Pseudomonas aeruginosa* RNA phage PP7, a distant relative of the above-mentioned coliphages.[708] Figure 8.7 presents some examples of intracellular paracrystalline structures and purified samples of natural and chimeric RNA phage VLPs.

The three-dimensional structures of the most typical RNA phage representatives have been determined at high resolution by X-ray crystallography and have been found to be very similar, in spite of the marked diversity of the primary structure of their coats. First, the three-dimensional structure, showing no structural similarity to any other known RNA virus, was resolved for the MS2 virions at 3.3 Å[13,709,710] and then refined to 2.8-Å resolution.[711] According to this structure, the 180 CP subunits were arranged in dimers as initial building blocks and formed a lattice with

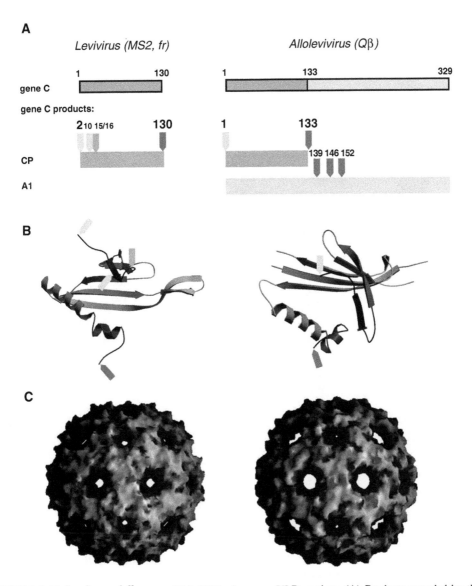

FIGURE 8.6 (Color figure follows p. 82.) RNA phages as VLP carriers. (A) Products encoded by the C gene of the Leviviruses (left) and Alloleviviruses (right) with localization of the insertion sites for foreign epitopes. (B) A schematic representation of the fold of the CP monomers derived from the crystal structure. Insertion sites for foreign epitopes are marked by arrows. (C) Three-dimensional maps of the recombinant RNA phage capsids. Peaks are bright and valleys are dark. (The maps are the generous gift of Vijay Reddi.)

the triangulation number T = 3. The CP fold differed from the fold of all other known viral coats. The CP subunit consisted of a five-stranded β-sheet facing the inside of the particle and a hairpin and two α-helices on the outside. It was suggested that the conserved P78 residue was responsible for the isomerization of a loop FG around the fivefold symmetry axis, thereby creating a channel.[711] The role of the P78 residue was confirmed further by point mutations in the recombinant MS2 capsids, self-assembled in *E. coli* cells in the absence of other viral components.[712] Determination of the MS2 structure led to an important breakthrough in the understanding of protein–RNA interactions during encapsidation of the genome, after a first crystal structure of a complex between recombinant MS2 capsids and the 19-nucleotide RNA fragment was resolved at 2.7-Å resolution.[713–715] The aa residues responsible for the protein–RNA interactions were localized by point

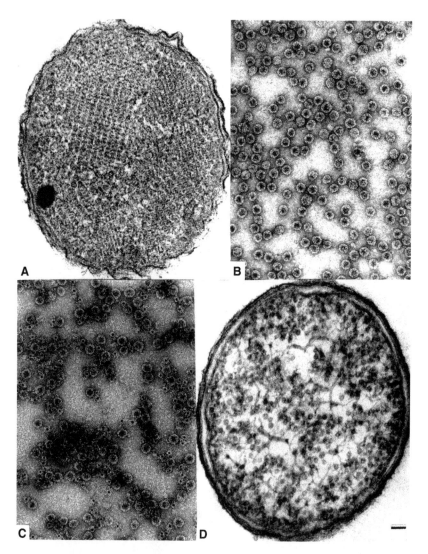

FIGURE 8.7 Electron micrographs of RNA phage coats as VLP carriers. Slices of *E. coli* cells filled with paracrystals of authentic Qβ particles (A) or with mosaic Qβ particles harboring 39 aa residues of the preS1 segment (D). Purified Qβ particles formed by authentic Qβ coat and A1 proteins (B), or by authentic Qβ coat as a helper and a read-through A1 derivative carrying the same preS1 segment of 39 aa residues; preS1 epitopes exposed on the particles are labeled with monoclonal anti-preS1 antibody MA18/7 by immunogold technique (C). Scale bar, 50 nm. (Micrographs are the generous gift of Velta Ose.)

mutations at positions 45 and 59.[716–718] Furthermore, numerous mutations allowed for the altering of the specificity[719] and efficiency[720] of translational operator complexes. The crystal structures of RNA aptamers, whose secondary structure differs from that of the wt RNA, with MS2 capsids have been reported.[721–723]

The structure of recombinant capsids of RNA phage fr has been determined by X-ray crystallography at 3.5-Å resolution and shown to be identical to the protein shell of the native virus.[14,724] This was followed by the structure of GA virions, which showed remarkable structural differences in the loop regions of the CPs, especially in the FG loops forming fivefold and quasi-sixfold symmetry contacts.[16] Then, the structure of virions and recombinant capsids of the Qβ phage was resolved at 3.5-Å resolution, which differed from the other RNA phage structures by the presence of stabilizing disulfide bonds on each side of the flexible FG loops, which link each dimer covalently to the rest of the capsid.[15] A comparison with the structure of the related phage MS2 shows that, although the fold of the Qβ coat protein is very similar, the details of the protein–protein interactions are completely different.[15] At last, the structure of the *Pseudomonas aeruginosa* phage PP7 was resolved.[17,725]

After unveiling the structure of self-assembled RNA phage capsids, the crystal structures of unassembled dimers of MS2[726] and GA[705] CP have been reported. These structures showed only minor differences in comparison to their self-assembled counterparts. Surprisingly, the phage CP sustained genetical fusion, resulting in duplicated CP, which folded normally and functioned as a translational repressor, due to the physical proximity of the N and C termini of the CP.[727]

The recombinant capsids of fr[31] and MS2[702] were first proposed as RNA-phage carriers for the presentation of foreign immunological epitopes on their surface. To search for the appropriate CP regions, the oligonucleotide linkers, coding short aa sequences and bearing convenient restriction sites, were inserted into different sites of fr CP.[33] Interestingly, this work was based on the predictions of the spatial structure of the CP and was carried out before the crystal structure of the MS2 capsid had been determined. The recombinant fr CP containing 2- to 12-aa-long inserts at the N or C termini, position 50 in the RNA-binding region, but not positions 97–111 of the αA-helix, were capable of self-assembly.[33] The majority of other fr CP mutants demonstrated reduction of capabilities to self-assemble and formed either CP dimers (mutations at residues 2, 10, 63, or 129) or both dimer and capsid structures (residue 2 or 69).[728] The loop FG of the fr CP was shortened by a 4-aa-residue deletion, and the mutant retained the ability to form capsids, although the three-dimensional structure of the mutant capsids revealed that the mutated loops were flexible and too short to interact with each other.[729] Special universal vectors were constructed for the insertion of foreign sequences into position 2 of the fr CP in all possible reading frames,[31] and the universal marker, preS1 epitope DPAFR was inserted into positions 2, 10, and 129 and exposed on the surface of particles.[730] Attempts to introduce the 40-aa-long V3 loop of the HIV-1 gp120 into the N terminus, or at positions 10, 12, and 15, or into or instead of the loop FG led to unassembled products.[731]

Other attempts to exploit the N-terminal β-hairpin exposed at the surface of the capsid, namely, aa residues 15/16, allowed for the production of MS2 capsids bearing a number of different peptide sequences, including the gp120 V3 loop (up to 24 amino acids in length).[702] Foreign epitopes expressed on these chimeras were found to be immunogenic in mice.[702] Mutational mapping revealed aa residues that are responsible for inter- and intramolecular contacts and therefore for the thermal stability of MS2 capsids.[732]

Insertion of the Flag octapeptide at the N terminus and into the N-terminal β-hairpin of the MS2 CP prevented self-assembly and proper folding, respectively. However, genetic fusion of the duplicated CP encoding sequence resulted in the synthesis of a protein considerably more tolerant to these structural perturbations and mostly corrected the defects accompanying Flag peptide insertion.[733]

A putative protective epitope T1 of 24 aa residues, from the immunodominant liver stage antigen-1 (LSA-1) of the malaria parasite *P. falciparum,* was inserted at the tip of the N-terminal β-hairpin (between positions 15 and 16) of the MS2 CP.[734] In contrast to the MS2 coat carrier, which elicited both humoral and cellular immune responses, observed as a predominance of type 2 cytokines but with a mixed profile of immunoglobulin isotypes, the LSA-1 stimulated a type 1 polarized response, with significant upregulation of interferon γ.[734]

Finally, fr CP showed unusually high capacity as a vector, when the addition of C-terminal segments of HaPV VP1 of 21, 34, and 52 aa to the N terminus of the fr CP was found to be dispensable for VLP self-assembly.[735] These recent findings restored interest in the RNA phage coats as potent vaccine candidates.

A clear limitation of these RNA phage CP models is low tolerance of their capsids to foreign insertions. To overcome these difficulties, a C-terminal extension of the Qβ CP has been proposed as a potent site for foreign insertions.[736,737] Theoretically, the 195-aa extension of the Qβ CP was considered an ideal target for insertions and exchanges. It was found to contain elements that typically protrude spikelike structures exposed on the surface of particles. First, self-assembly capabilities of capsids with mutually exchanged extensions of the Qβ and SP CPs was confirmed experimentally.[707] Second, mathematical prediction showed clear-cut colinearity of the Qβ CP extension with the superficially located HBV preS sequence.[736] The Qβ A1 proteins or CPs bearing additions at their C terminus failed to form particles in most cases. However, they were incorporated into particles in the presence of the wt "helper" CP and formed mixed, or mosaic, particles. Such mosaic particles were constructed by either enhancing the level of UGA suppression in the presence of overexpressed suppressor tRNA,[738] or changing UGA stop codon into GGA sense codon and expressing the extended and helper forms of the Qβ CP from two separate genes. These genes can be located either on the same plasmid or on two separate plasmids conveying different antibiotic resistances. The potential insertion sites were mapped by insertion of the preS1 DPAFR model epitope, and the 39-aa-long HIV-1 gp120 V3 loop.[737] By enhancing UGA suppression the mosaic particles were detected, but the proportion of A1 extended to helper CP in these particles dropped from 48 to 14%, with an increase of the length of the A1 extension.[739] A model insertion, the preS1 epitope, was located on the surface of particles and ensured specific antigenicity and immunogenicity in mice.[739] Moreover, long HBV preS insertions (full-length preS, preS1, or preS2) were able to substitute the A1 extension and appeared on the surface of mosaic particles.[740]

Regarding Qβ CP body, residues responsible for the RNA recognition were mapped.[741,742] It may be helpful in the development of gene transfer technologies based on the RNA phage model.

The ability of the phage CP to package RNA *in vivo*[743] demonstrates the principal possibility of their further application as gene delivery vectors. Another potent application of the recombinant RNA phage capsids might be drug encapsulation and targeted delivery of, for example, deglycosylated ricin A chain coupled to the RNA operator stem-loop, or antisense elements.[744–746]

Surprisingly, the icosahedral Qβ VLPs can be converted into rods after mutational intervention into the FG loop structure.[747] Appearance of alternate VLP forms of RNA phages may play an important role in the further development of this class of VLP carriers.

Very recently, the MS2 phage was found to be capable of accommodating short (pentapeptide) sequences added to the N terminus of its CP within the live virions, although not all inserts were genetically stable.[748]

8.4.6.4 Luteoviruses

These are icosahedral plant viruses, which show a definite similarity to Alloleviviruses due to a mechanism of coat protein read-through for 300 additional aa residues, with subsequent insertion of the extended coats into luteovirus capsids as minor components.[749]

This read-through coat protein of potato leafroll virus (PLRV) was used recently as a carrier for full-length green fluorescent protein, and incorporation of this chimera into PRLV particles was shown.[750] Earlier the PRLV coat was expressed in the form of VLPs in baculovirus-driven insect cells, with an N-terminal addition of a long MHHHHHHGDDDDKDAMG tag.[751]

8.4.6.5 Flock House Virus

This is an icosahedral insect virus of the family Nodaviridae. Its genome consists of two messenger-sense RNA molecules, both of which are encapsidated in the same particle. RNA1 (3.1 kb) encodes proteins required for viral RNA replication; RNA2 (1.4 kb) encodes protein α (43 kDa), the precursor of the coat protein. When *Spodoptera frugiperda* cells were infected with a recombinant baculovirus containing a cDNA copy of RNA2, coat protein α assembled into VLP precursor particles that matured normally by autocatalytic cleavage of protein α into polypeptide chains β (38 kDa) and γ (5 kDa).[752] The particles were morphologically indistinguishable from the authentic FHV and contained RNA derived from the coat protein message. Expression of mutants in which Asn-363 at the β-γ cleavage site of protein α was replaced by either Asp, Thr, or Ala residues resulted in particles that were cleavage defective.[752] Moreover, FHV was found to produce virions in plants[753] and yeast[754] after transfection with genomic RNA, which opens wide-ranging applications of the FHV model. Recently, *E. coli* was applied to express chimeric FHV capsids.[755]

Crystal structures of T = 3 FHV virion,[756] mutant cleavage-defective provirion-like FHV particles obtained in baculovirus-driven expression system,[757] and related black beetle nodavirus[758] were resolved at 2.8-Å resolution. Recently, matrix-assisted laser desorption/ionization mass spectrometry combined with time-resolved, limited proteolysis was suggested to examine dynamic FHV processes such as assembly, maturation, and cell entry.[759]

Deletion of 50 N-terminal aa residues inhibited complete self-assembly of the FHV coats, while restriction of the deletion to 31 aa resulted in a heterogenous outcome of small bacilliform-like structures and irregular structures as well as wt-like T = 3 particles, but not of T = 1 particles.[760]

As a VLP carrier candidate, the baculovirus-driven expression of FHV capsids was used first for exposure of the IGPGRAF heptapeptide, a neutralizing domain of the HIV-1 V3 loop that was inserted in two positions on the FHV coat surface.[761] Although the inserted peptide was very short, one of the chimeras was able to induce a strong neutralizing response against HIV-1.[761] Display of the V3 loop of different HIV-1 isolates at this site of the FHV capsids allowed more specific serotyping of patients' sera and was suggested as a tool for the correct evaluation of the immune response against different V3 loop core sequences.[762] Different positions on the external surface of the FHV capsids were used to insert gp41-neutralizing epitopes, IEEE and ERDRD, found at residues 735–752 of the HIV-1 gp160.[763]

Furthermore, three immunodominant regions of the HCV core protein, aa residues 1–20, 21–40, and 32–46, were displayed on the surface of the FHV capsids and used successfully as diagnostic tools.[764] Finally, the FHV display system was developed in the *E. coli* expression system, when FHV chimeras bearing epitopes of human epsilon immunoglobulin chains were used to produce specific antibody reagents.[755]

8.4.6.6 Potyviruses

Two representatives of these rod-shaped plant viruses, Johnsongrass mosaic virus (JGMV) and plum pox potyvirus (PPV), deserve mentioning as well-accepted nonreplicative VLP carrier candidates. Potyvirus coat protein (CP) is processed by proteolytic cleavage from a large polyprotein covering practically the whole genome (for a review, see Reference 765). When the full-length CP of the JGMV was expressed in *E. coli* or yeast, it assembled into potyvirus-like particles.[766] The particles were heterogeneous in length with a stacked-ring appearance and resembled JGMV particles in their flexuous morphology and width. The *E. coli* expression system allowed CP mapping and identification of key aa residues required for assembly.[767] The versatile nature of potyvirus vectors was further confirmed by high-level production of JGMV VLPs in baculovirus-driven insect cells[768] and in mammalian cells using recombinant vaccinia virus system.[769]

Concomitant with the development of the JMGV expression vectors was expression of the CP of the PPV in transgenic *Nicotiana* plants[770,771] and later in *E. coli*.[772]

The first chimeric JMGV VLPs were produced in *E. coli* after insertion of an octapeptide epitope from *P. falciparum* and a decapeptide hormone (luteinizing hormone releasing hormone) at the N or at both the N and C termini of the JMGV CP.[773] A full-length protein, Sj26-glutathione *S*-transferase of 26 kD from *Schistosoma japonicum*, was also introduced into JGMV VLPs, replacing the N-terminal 62 aa of the CP.[773] Electron microscopy of ultrathin sections of *E. coli* revealed the appearance, within the cytoplasm, of parallel strands sometimes extending the length of the cell and strung cells together, with "threads" of the JMGV VLPs appearing to connect individual bacterial cells. The JGMV VLPs bearing the 26-kD antigen Sj26 were shorter and wider and elicited an efficient Sj26-specific immune response after administration to mice. These studies led to the conclusion that the potyvirus CP can accommodate peptides or even large antigens and is able to present antigens on the surface of the VLPs.[773] Thus, the JMGV vectors allow for not only the addition of short peptides to their N or C termini, but also the fusing of large antigens to the N terminus or replacing most of the N- or C-terminal exposed regions.[774] Moreover, chimeric JGMV VLPs were highly immunogenic in mice and rabbits even in the absence of any adjuvant.[774]

Regarding the PPV model, transgenic plums transformed with the PPV CP displayed a resistance to the sharka disease.[775] The genetic modification of the PPV CP started when transgenic plants heteroencapsidated aphid nontransmissible plant viruses with the PPV CP, thereby providing them with the aphid transmissibility. The DAG triplet within the N-terminal part of the PPV CP, which was known to be involved in the aphid transmission, was removed, and self-assembly of the corresponding VLPs was achieved in *E. coli*.[772] Further, DAG-deleted and point-mutated PPV CP devoid of the aphid transmissibility were shown to be safe against the biological risks associated with the heteroencapsidation in plants.[775]

Developing chimeric PPV VLPs, the N-terminal part of the PPV CP was chosen as the site for expression of foreign antigenic peptides.[776] Modifications in this site were engineered to avoid the capability of natural transmission by aphids of this PPV vector. As a first practical attempt, different forms of an antigenic peptide (single and tandem repetition) from the VP2 capsid protein of canine parvovirus (CPV) were expressed. Both chimeras were able to infect *Nicotiana clevelandii* plants with characteristics similar to the wild-type virus. The chimeras remained genetically stable after several plant passages. Moreover, mice and rabbits immunized with chimeric virions developed CPV-specific antibodies, which showed neutralizing activity.[776]

8.4.6.7 Tobacco Mosaic Virus

Tobacco mosaic virus (TMV) has played pioneering roles in many fields (for a review, see Reference 777). TMV was the first virus for which the aa sequence of the CP was determined and the first plant virus for which structures and functions were known for all genes. It was the first virus for which activation of a resistance gene in a host plant was related to the molecular specificity of a viral gene product. In the field of plant biotechnology, TMV is one of the most promising vectors. The TMV particle was the first macromolecular structure shown to self-assemble *in vitro* (for a review, see Reference 778). Finally, TMV became one of the first candidates for the development of nonreplicative VLP carriers.[25]

The three-dimensional structure of TMV was determined by X-ray fiber diffraction methods (for a review, see Reference 779). The first TMV structure was reported more than 20 years ago,[780] but the final improvements to 2.4-Å resolution appeared very recently.[781] Parallel to this, TMV was used as a test specimen to develop techniques for high-resolution structural analysis in electron micrographs of biological assemblies with helical symmetry.[782]

The TMV CP was produced in *E. coli* and shown to assemble into virus-like rods, including RNA-free particles.[25,783] However, further studies showed that native TMV CP expressed in *E. coli* formed nonhelical, stacked aggregates *in vitro*, but coexpression of TMV CP and ssRNAs containing the TMV origin-of-assembly *in vivo* produced helical pseudovirus particles.[784,785] In 1986, Haynes and co-workers achieved the expression of the TMV VLPs bearing an epitope from poliovirus 3.[25] For this purpose they constructed a synthetic TMV CP gene containing both a convenient arrangement of restriction enzyme cleavage sites and an optimized set of codons for efficient translation in a prokaryotic system. The poliovirus-3 VP1 octapeptide epitope QQPTTRAQ was added to the C terminus of the synthetic TMV CP gene. Interestingly, self-assembly of the chimeric TMV CP-polio 3 monomers into typical rods and disks did not occur in the bacterial cells; their polymerization was initiated during dialysis of bacterial lysates at pH 5.0. Such TMV CP-polio 3 particles elicited poliovirus-neutralizing antibodies following injection into rats.[25]

Further development of the TMV model progressed in the elaboration of vectors for expression of foreign genes in plants[786,787] and engineering replication-competent, rather than noninfectious, VLPs. In plants, modified TMV were used, for example, to produce VLPs carrying a CP-angiotensin-I-converting enzyme inhibitor peptide fusion. The latter construct was synthesized via readthrough of the TMV CP stop codon and incorporated into mosaic VLPs.[787] Furthermore, TMV VLPs were used to express epitopes from murine zona pellucida ZP3 protein,[788] malaria,[789] influenza virus hemagglutinin, HIV-1 envelope protein,[790] rabies virus glycoprotein, and murine hepatitis virus S-glycoprotein.[791] The latter chimeras were shown to be protective in mice against challenge with a lethal dose of murine hepatitis virus.[792]

TMV vectors were used for the successful production in plants of such proteins as GFP,[793] heavy and light immunoglobulin chains assembled into a full-length antibody,[794] and the major immunodominant protein VP1 of the FMDV.[795] Two last examples present the first successful attempts at preparing full-size antibody and full-length mammalian antigenic protein in plants. Now, several commercial companies are planning to produce and process large quantities of plant-generated viral products in the field. The development of the TMV as a commercially important vector has been reviewed in two recent articles.[796,797]

ACKNOWLEDGMENTS

Galina Borisova, Olga Borschukova, Maija Bundule, Indulis Cielens, Andris Dishlers, Dzidra Dreilina, Edith Grene, Andris Kazaks, Tatyana Kozlovska, Velta Ose, Ivars Petrovskis, Peter Pushko, Regina Renhofa, Dace Skrastina, Irina Sominskaya, Kaspars Tars, Inta Vasiljeva. Tatyana Voronkova, and Andris Zeltins from the Biomedical Centre of the University of Latvia (Riga) and Diana Koletzki, Helga Meisel, Hassen Siray, and Rainer Ulrich from the Institute of Virology, Charité (Berlin) contributed substantially to the work on chimeric VLPs and kindly presented their unpublished data.

We wish to acknowledge Wolfram H. Gerlich (Giessen), Mark Page (London), Rainer Ulrich (Berlin), Peter Pushko (Frederick), and Kestutis Sasnauskas (Vilnius) for a long-standing collaboration and constructive reviewing and editing of the manuscript, and Edith Grene (Manassas) for constant informational support.

We thank R. Anthony Crowther (Cambridge) for the generous gift of the HBc images, Bruce Boschek (Giessen) for the HBs image, Velta Ose for electron micrographs of chimeric HBc and RNA phage derivatives, and Vijay Reddy (San Diego) for the three-dimensional images of RNA phages.

REFERENCES

1. Winter, G. et al., Redesigning enzyme structure by site-directed mutagenesis: tyrosyl tRNA synthetase and ATP binding, *Nature,* 299, 756, 1982.
2. Wilkinson, A. J. et al., Site-directed mutagenesis as a probe of enzyme structure and catalysis: tyrosyl-tRNA synthetase cysteine-35 to glycine-35 mutation, *Biochemistry,* 22, 3581, 1983.
3. Wilkinson, A. J. et al., A large increase in enzyme-substrate affinity by protein engineering, *Nature,* 307, 187, 1984.
4. Leatherbarrow, R. J. and Fersht, A. R., Protein engineering, *Protein Eng.,* 1, 7, 1986.
5. Fersht, A. and Winter, G., Protein engineering, *Trends Biochem. Sci.,* 17, 292, 1992.
6. Morris, G. E., *Epitope Mapping Protocols,* Methods in Molecular Biology, Vol. 66, Humana Press, Totowa, NJ, 1996.
7. Cabilly, S., *Combinatorial Peptide Library Protocols,* Methods in Molecular Biology, Vol. 87, Humana Press, Totowa, NJ, 1998.
8. Grens, E. J. and Pumpens, P. P., Recombinant virus capsids as a new generation of immunogenic proteins and vaccines (in Russian), *J. All-Union Mendeleyev's Chem. Soc.,* 33, 531, 1988.
9. Crowther, R. A. et al., Three-dimensional structure of hepatitis B virus core particles determined by electron cryomicroscopy, *Cell,* 77, 943, 1994.
10. Böttcher, B. et al., Determination of the fold of the core protein of hepatitis B virus by electron cryomicroscopy, *Nature,* 386, 88, 1997.

11. Conway, J. F. et al., Visualization of a 4-helix bundle in the hepatitis B virus capsid by cryo-electron microscopy, *Nature,* 386, 91, 1997.
12. Wynne, S. A. et al., The crystal structure of the human hepatitis B virus capsid, *Mol. Cell,* 3, 771, 1999.
13. Valegård, K. et al., The three-dimensional structure of the bacterial virus MS2, *Nature,* 345, 36, 1990.
14. Liljas, L. et al., Crystal structure of bacteriophage fr capsids at 3.5 Å resolution, *J. Mol. Biol.,* 244, 279, 1994.
15. Golmohammadi, R. et al., The crystal structure of bacteriophage Q beta at 3.5 Å resolution, *Structure,* 4, 543, 1996.
16. Tars, K. et al., The crystal structure of bacteriophage GA and a comparison of bacteriophages belonging to the major groups of *Escherichia coli* leviviruses, *J. Mol. Biol.,* 271, 759, 1997.
17. Tars, K. et al., The three-dimensional structure of bacteriophage PP7 from *Pseudomonas aeruginosa* at 3.7-Å resolution, *Virology,* 272, 331, 2000.
18. Namba, K. and Stubbs, G., Structure of tobacco mosaic virus at 3.6 Å resolution: implications for assembly, *Science,* 231, 1401, 1986.
19. Namba, K. et al., Visualization of protein-nucleic acid interactions in a virus. Refined structure of intact tobacco mosaic virus at 2.9 Å resolution by X-ray fiber diffraction, *J. Mol. Biol.,* 208, 307, 1989.
20. Hogle, J. M. et al., Three-dimensional structure of poliovirus at 2.9 Å resolution, *Science,* 229, 1358, 1985.
21. Rossmann, M. G. et al., Structure of a human common cold virus and functional relationship to other picornaviruses, *Nature,* 317, 145, 1985.
22. Grimes, J. M. et al., An atomic model of the outer layer of the bluetongue virus core derived from X-ray crystallography and electron cryomicroscopy, *Structure,* 5, 885, 1997.
23. Ulrich, R. et al., Chimaera and its modern virus-like descendants, *Intervirology,* 39, 126, 1996.
24. Nassal, M., Total chemical synthesis of a gene for hepatitis B virus core protein and its functional characterization, *Gene,* 66, 279, 1988.
25. Haynes, J. R. et al., Development of a genetically-engineered, candidate polio vaccine employing the self-assembling properties of tobacco mosaic virus coat protein, *Biotechnology,* 4, 637, 1986.
26. Smith, G. P., Filamentous fusion phage: novel expression vectors that display cloned antigens on the virion surface, *Science,* 228, 1315, 1985.
27. Valenzuela, P. et al., Antigen engineering in yeast: synthesis and assembly of hybrid hepatitis B surface antigen–herpes simplex 1 gD particles, *Biotechnology,* 3, 323, 1985.
28. Delpeyroux, F. et al., A poliovirus neutralization epitope expressed on hybrid hepatitis B surface antigen particles, *Science,* 233, 472, 1986.
29. Newton, S. E. et al., New approaches to FMDV antigen presentation using vaccinia virus, in *Vaccines 87. Modern Approaches to New Vaccines: Prevention of AIDS and Other Viral, Bacterial, and Parasitic Diseases,* Chanock, R. M. et al., Eds., Cold Spring Harbor Laboratory Press, Cold Spring Harbor, NY, 1987, pp. 12–21.
30. Clarke, B. E. et al., Improved immunogenicity of a peptide epitope after fusion to hepatitis B core protein, *Nature,* 330, 381, 1987.
31. Borisova, G. P. et al., Recombinant capsid structures for exposure of protein antigenic epitopes, *Mol. Gen. (Life Sci. Adv.),* 6, 169, 1987.
32. Burke, K. L. et al., Antigen chimaeras of poliovirus as potential new vaccines, *Nature,* 332, 81, 1988.
33. Kozlovskaya, T. M. et al., Genetically engineered mutants of the envelope protein of the RNA-containing bacteriophage fr (in Russian), *Mol. Biol. (Moscow),* 22, 731, 1988.
34. Adams, S. E. et al., The expression of hybrid HIV:Ty virus-like particles in yeast, *Nature,* 329, 68, 1987.
35. Burrell, C. J. et al., Expression in *Escherichia coli* of hepatitis B virus DNA sequences cloned in plasmid pBR322, *Nature,* 279, 43, 1979.
36. Charnay, P. et al., Cloning in *Escherichia coli* and physical structure of hepatitis B virion DNA, *Proc. Natl. Acad. Sci. U.S.A.,* 76, 2222, 1979.
37. Sninsky, J. J. et al., Cloning and endonuclease mapping of the hepatitis B viral genome, *Nature,* 279, 346, 1979.
38. Valenzuela, P. et al., Nucleotide sequence of the gene coding for the major protein of hepatitis B virus surface antigen, *Nature,* 280, 815, 1979.
39. Heermann, K. H. and Gerlich, W. H., Surface proteins of hepatitis B viruses, in *Molecular Biology of the Hepatitis B Virus,* McLachlan, A., Ed., CRC Press, Boca Raton, FL, 1991, pp. 109–143.

40. Heermann, K. H. et al., Large surface proteins of hepatitis B virus containing the pre-S sequence, *J. Virol.*, 52, 396, 1984.

41. Christman, J. K. et al., Amplification of expression of hepatitis B surface antigen in 3T3 cells cotransfected with a dominant-acting gene and cloned viral DNA, *Proc. Natl. Acad. Sci. U.S.A.*, 79, 1815, 1982.

42. Gough, N. M. and Murray, K., Expression of the hepatitis B virus surface, core and E antigen genes by stable rat and mouse cell lines, *J. Mol. Biol.*, 162, 43, 1982.

43. Wang, Y. et al., Expression of hepatitis B surface antigen in unselected cell culture transfected with recircularized HBV DNA, *EMBO J.*, 1, 1213, 1982.

44. Laub, O. et al., Synthesis of hepatitis B surface antigen in mammalian cells: expression of the entire gene and the coding region, *J. Virol.*, 48, 271, 1983.

45. Michel, M. L. et al., Synthesis in animal cells of hepatitis B surface antigen particles carrying a receptor for polymerized human serum albumin, *Proc. Natl. Acad. Sci. U.S.A.*, 81, 7708, 1984.

46. Liu, C. C. et al., Direct expression of hepatitis B surface antigen in monkey cells from an SV40 vector, *DNA*, 1, 213, 1982.

47. Stratowa, C. et al., Recombinant retroviral DNA yielding high expression of hepatitis B surface antigen, *EMBO J.*, 1, 1573, 1982.

48. Smith, G. L. et al., Infectious vaccinia virus recombinants that express hepatitis B virus surface antigen, *Nature*, 302, 490, 1983.

49. Shih, M. F. et al., Expression of hepatitis B virus S gene by herpes simplex virus type 1 vectors carrying alpha- and beta-regulated gene chimeras, *Proc. Natl. Acad. Sci. U.S.A.*, 81, 5867, 1984.

50. Denniston, K. J. et al., Expression of hepatitis B virus surface and e antigen genes cloned in bovine papillomavirus vectors, *Gene*, 32, 357, 1984.

51. Davis, A. R. et al., Expression of hepatitis B surface antigen with a recombinant adenovirus, *Proc. Natl. Acad. Sci. U.S.A.*, 82, 7560, 1985.

52. Shiraki, K. et al., Development of immunogenic recombinant Oka varicella vaccine expressing hepatitis B virus surface antigen, *J. Gen. Virol.*, 72, 1393, 1991.

53. Kamiyama, T. et al., Novel immunogenicity of Oka varicella vaccine vector expressing hepatitis B surface antigen, *J. Infect. Dis.*, 181, 1158, 2000.

54. Babinet, C. et al., Specific expression of hepatitis B surface antigen (HBsAg) in transgenic mice, *Science*, 230, 1160, 1985.

55. Chisari, F. V. et al., A transgenic mouse model of the chronic hepatitis B surface antigen carrier state, *Science*, 230, 1157, 1985.

56. Kang, C. Y. et al., Secretion of particles of hepatitis B surface antigen from insect cells using a baculovirus vector, *J. Gen. Virol.*, 68, 2607, 1987.

57. Higashihashi, N. et al., High-level expression and characterization of hepatitis B virus surface antigen in silkworm using a baculovirus vector, *J. Virol. Methods*, 35, 159, 1991.

58. Mason, H. S. et al., Expression of hepatitis B surface antigen in transgenic plants, *Proc. Natl. Acad. Sci. U.S.A.*, 89, 11745, 1992.

59. Ehsani, P. et al., Polypeptides of hepatitis B surface antigen produced in transgenic potato, *Gene*, 190, 107, 1997.

60. Kong, Q. et al., Oral immunization with hepatitis B surface antigen expressed in transgenic plants, *Proc. Natl. Acad. Sci. U.S.A.*, 98, 11539, 2001.

61. Richter, L. J. et al., Production of hepatitis B surface antigen in transgenic plants for oral immunization, *Nat. Biotechnol.*, 18, 1167, 2000.

62. Kapusta, J. et al., A plant-derived edible vaccine against hepatitis B virus, *FASEB J.*, 13, 1796, 1999.

63. Kapusta, J. et al., Oral immunization of human with transgenic lettuce expressing hepatitis B surface antigen, *Adv. Exp. Med. Biol.*, 495, 299, 2001.

64. Deml, L. et al., High level expression of hepatitis B virus surface antigen in stably transfected *Drosophila* Schneider-2 cells, *J. Virol. Methods*, 79, 191, 1999.

65. Charnay, P. et al., Biosynthesis of hepatitis B virus surface antigen in *Escherichia coli*, *Nature*, 286, 893, 1980.

66. Edman, J. C. et al., Synthesis of hepatitis B surface and core antigens in *E. coli*, *Nature*, 291, 503, 1981.

67. Mackay, P. et al., Production of immunologically active surface antigens of hepatitis B virus by *Escherichia coli*, *Proc. Natl. Acad. Sci. U.S.A.*, 78, 4510, 1981.

68. Fujisawa, Y. et al., Direct expression of hepatitis B surface antigen gene in *E. coli*, *Nucleic Acids Res.*, 11, 3581, 1983.

69. Pumpen, P. P. et al., Synthesis of the surface antigen of the hepatitis B virus in *Escherichia coli* (in Russian), *Dokl. Akad. Nauk S.S.S.R.*, 271, 230, 1983.

70. Smirnov, V. D. et al., Synthesis and expression of the DNA fragment coding the antigenic determinant of the surface antigen protein of hepatitis B virus (in Russian), *Bioorg. Khim.*, 9, 1388, 1983.

71. Pumpen, P. et al., Expression of hepatitis B virus surface antigen gene in *Escherichia coli*, *Gene*, 30, 201, 1984.

72. Valenzuela, P. et al., Synthesis and assembly of hepatitis B virus surface antigen particles in yeast, *Nature*, 298, 347, 1982.

73. Dehoux, P. et al., Expression of the hepatitis B virus large envelope protein in *Saccharomyces cerevisiae*, *Gene*, 48, 155, 1986.

74. Itoh, Y. et al., Expression of hepatitis B virus surface antigen P31 gene in yeast, *Biochem. Biophys. Res. Commun.*, 138, 268, 1986.

75. Davis, H. L. et al., DNA-based immunization induces continuous secretion of hepatitis B surface antigen and high levels of circulating antibody, *Hum. Mol. Genet.*, 2, 1847, 1993.

76. Davis, H. L. et al., Direct gene transfer in skeletal muscle: plasmid DNA-based immunization against the hepatitis B virus surface antigen, *Vaccine*, 12, 1503, 1994.

77. Schirmbeck, R. et al., Nucleic acid vaccination primes hepatitis B virus surface antigen-specific cytotoxic T lymphocytes in nonresponder mice, *J. Virol.*, 69, 5929, 1995.

78. Davis, H. L. et al., Use of plasmid DNA for direct gene transfer and immunization, *Ann. N.Y. Acad. Sci.*, 772, 21, 1995.

79. Michel, M. L. et al., DNA-mediated immunization to the hepatitis B surface antigen in mice: aspects of the humoral response mimic hepatitis B viral infection in humans, *Proc. Natl. Acad. Sci. U.S.A.*, 92, 5307, 1995.

80. Davis, H. L. et al., DNA-mediated immunization to hepatitis B surface antigen: longevity of primary response and effect of boost, *Vaccine*, 14, 910, 1996.

81. Davis, H. L. et al., DNA-based immunization against hepatitis B surface antigen (HBsAg) in normal and HBsAg-transgenic mice, *Vaccine*, 15, 849, 1997.

82. Stirk, H. J. et al., A topological model for hepatitis B surface antigen, *Intervirology*, 33, 148, 1992.

83. Berting, A. et al., Computer-aided studies on the spatial structure of the small hepatitis B surface protein, *Intervirology*, 38, 8, 1995.

84. Sonveaux, N. et al., Proposition of a three-dimensional representation of the constitutive protein of the hepatitis B surface antigen particles, *J. Protein Chem.*, 14, 477, 1995.

85. Sonveaux, N. et al., The topology of the S protein in the yeast-derived hepatitis B surface antigen particles, *J. Biol. Chem.*, 269, 25637, 1994.

86. Antoni, B. A. et al., Site-directed mutagenesis of cysteine residues of hepatitis B surface antigen. Analysis of two single mutants and the double mutant, *Eur. J. Biochem.*, 222, 121, 1994.

87. Sato, M. et al., Peripherally biased distribution of antigen proteins on the recombinant yeast-derived human hepatitis B virus surface antigen vaccine particle: structural characteristics revealed by small-angle neutron scattering using the contrast variation method, *J. Biochem. (Tokyo)*, 118, 1297, 1995.

88. Yamaguchi, M. et al., Fine structure of hepatitis B virus surface antigen produced by recombinant yeast: comparison with HBsAg of human origin, *FEMS Microbiol. Lett.*, 165, 363, 1998.

89. Bruss, V. and Ganem, D., Mutational analysis of hepatitis B surface antigen particle assembly and secretion, *J. Virol.*, 65, 3813, 1991.

90. Prange, R. et al., Mutational analysis of HBsAg assembly, *Intervirology*, 38, 16, 1995.

91. Mangold, C. M. et al., Secretion and antigenicity of hepatitis B virus small envelope proteins lacking cysteines in the major antigenic region, *Virology*, 211, 535, 1995.

92. Mangold, C. M. et al., Analysis of intermolecular disulfide bonds and free sulfhydryl groups in hepatitis B surface antigen particles, *Arch. Virol.*, 142, 2257, 1997.

93. Heijtink, R. A. et al., Anti-HBs characteristics after hepatitis B immunisation with plasma-derived and recombinant DNA-derived vaccines, *Vaccine*, 18, 1531, 2000.

94. Delpeyroux, F. et al., Insertions in the hepatitis B surface antigen. Effect on assembly and secretion of 22-nm particles from mammalian cells, *J. Mol. Biol.*, 195, 343, 1987.

95. Delpeyroux, F. et al., Construction and characterization of hybrid hepatitis B antigen particles carrying a poliovirus immunogen, *Biochimie,* 70, 1065, 1988.

96. Delpeyroux, F. et al., Presentation and immunogenicity of the hepatitis B surface antigen and a poliovirus neutralization antigen on mixed empty envelope particles, *J. Virol.,* 62, 1836, 1988.

97. Delpeyroux, F. et al., Structural factors modulate the activity of antigenic poliovirus sequences expressed on hybrid hepatitis B surface antigen particles, *J. Virol.,* 64, 6090, 1990.

98. Michel, M. L. et al., Induction of anti-human immunodeficiency virus (HIV) neutralizing antibodies in rabbits immunized with recombinant HIV-hepatitis B surface antigen particles, *Proc. Natl. Acad. Sci. U.S.A.,* 85, 7957, 1988.

99. Filatov, F. P. et al., Recombinant surface proteins of the hepatitis B virus, exhibiting the immunodominant membrane protein of the HIV-1 virus (in Russian), *Dokl. Akad. Nauk,* 327, 172, 1992.

100. Eckhart, L. et al., Immunogenic presentation of a conserved gp41 epitope of human immunodeficiency virus type 1 on recombinant surface antigen of hepatitis B virus, *J. Gen. Virol.,* 77, 2001, 1996.

101. Michel, M. L. et al., T- and B-lymphocyte responses to human immunodeficiency virus (HIV) type 1 in macaques immunized with hybrid HIV/hepatitis B surface antigen particles, *J. Virol.,* 64, 2452, 1990.

102. Schlienger, K. et al., Human immunodeficiency virus type 1 major neutralizing determinant exposed on hepatitis B surface antigen particles is highly immunogenic in primates, *J. Virol.,* 66, 2570, 1992.

103. Schlienger, K. et al., Vaccine-induced neutralizing antibodies directed in part to the simian immunodeficiency virus (SIV) V2 domain were unable to protect rhesus monkeys from SIV experimental challenge, *J. Virol.,* 68, 6578, 1994.

104. Sasnauskas, K., unpublished data, 1995.

105. Fomsgaard, A. et al., Improved humoral and cellular immune responses against the gp120 V3 loop of HIV-1 following genetic immunization with a chimeric DNA vaccine encoding the V3 inserted into the hepatitis B surface antigen, *Scand. J. Immunol.,* 47, 289, 1998.

106. Bryder, K. et al., Improved immunogenicity of HIV-1 epitopes in HBsAg chimeric DNA vaccine plasmids by structural mutations of HBsAg, *DNA Cell Biol.,* 18, 219, 1999.

107. Machida, A. et al., A synthetic peptide coded for by the pre-S2 region of hepatitis B virus for adding immunogenicity to small spherical particles made of the product of the S gene, *Mol. Immunol.,* 24, 523, 1987.

108. Neurath, A. R. et al., Hepatitis B virus surface antigen (HBsAg) as carrier for synthetic peptides having an attached hydrophobic tail, *Mol. Immunol.,* 26, 53, 1989.

109. Xu, X. et al., A modified hepatitis B virus surface antigen with the receptor-binding site for hepatocytes at its C terminus: expression, antigenicity and immunogenicity, *J. Gen. Virol.,* 75, 3673, 1994.

110. Prange, R. et al., Properties of modified hepatitis B virus surface antigen particles carrying preS epitopes, *J. Gen. Virol.,* 76, 2131, 1995.

111. Hui, J. et al., Expression and characterization of chimeric hepatitis B surface antigen particles carrying preS epitopes, *J. Biotechnol.,* 72, 49, 1999.

112. Jones, C. D. et al., Characterization of the T- and B-cell immune response to a new recombinant pre-S1, pre-S2 and SHBs antigen containing hepatitis B vaccine (Hepagene); evidence for superior anti-SHBs antibody induction in responder mice, *J. Viral Hepatol.,* 5 (Suppl. 2), 5, 1998.

113. McDermott, A. B. et al., Hepatitis B third-generation vaccines: improved response and conventional vaccine non-response-evidence for genetic basis in humans, *J. Viral Hepatol.,* 5, Suppl. 2, 9, 1998.

114. Pride, M. W. et al., Evaluation of B and T-cell responses in chimpanzees immunized with Hepagene, a hepatitis B vaccine containing pre-S1, pre-S2 gene products, *Vaccine,* 16, 543, 1998.

115. Waters, J. A. et al., A study of the antigenicity and immunogenicity of a new hepatitis B vaccine using a panel of monoclonal antibodies, *J. Med. Virol.,* 54, 1, 1998.

116. Jones, C. D. et al., T-cell and antibody response characterisation of a new recombinant pre-S1, pre-S2 and SHBs antigen-containing hepatitis B vaccine; demonstration of superior anti-SHBs antibody induction in responder mice, *Vaccine,* 17, 2528, 1999.

117. Leroux-Roels, G. et al., Hepatitis B vaccine containing surface antigen and selected preS1 and preS2 sequences.1. Safety and immunogenicity in young, healthy adults, *Vaccine,* 15, 1724, 1997.

118. Leroux-Roels, G. et al., Hepatitis B vaccine containing surface antigen and selected preS1 and preS2 sequences.2. Immunogenicity in poor responders to hepatitis B vaccines, *Vaccine,* 15, 1732, 1997.

119. Hui, J. et al., Immunization with a plasmid encoding a modified hepatitis B surface antigen carrying the receptor binding site for hepatocytes, *Vaccine,* 17, 1711, 1999.

120. Von Brunn, A. et al., Epitopes of the human malaria parasite *P. falciparum* carried on the surface of HBsAg particles elicit an immune response against the parasite, *Vaccine,* 9, 477, 1991.

121. Vreden, S. G. et al., Phase I clinical trial of a recombinant malaria vaccine consisting of the circumsporozoite repeat region of *Plasmodium falciparum* coupled to hepatitis B surface antigen, *Am. J. Trop. Med. Hyg.,* 45, 533, 1991.

122. Moelans, I. I. et al., Induction of *Plasmodium falciparum* sporozoite-neutralizing antibodies upon vaccination with recombinant Pfs16 vaccinia virus and/or recombinant Pfs16 protein produced in yeast, *Mol. Biochem. Parasitol.,* 72, 179, 1995.

123. Gordon, D. M. et al., Safety, immunogenicity, and efficacy of a recombinantly produced *Plasmodium falciparum* circumsporozoite protein-hepatitis B surface antigen subunit vaccine, *J. Infect. Dis.,* 171, 1576, 1995.

124. Stoute, J. A. et al., A preliminary evaluation of a recombinant circumsporozoite protein vaccine against *Plasmodium falciparum* malaria. RTS,S Malaria Vaccine Evaluation Group, *N. Engl. J. Med.,* 336, 86, 1997.

125. De Oliveira, C. I. et al., Antigenic properties of the merozoite surface protein 1 gene of *Plasmodium vivax, Vaccine,* 17, 2959, 1999.

126. Pumpens, P. et al., Evaluation of HBs, HBc, and frCP virus-like particles for expression of human papillomavirus 16 E7 oncoprotein epitopes, *Intervirology,* 45, 24, 2002.

127. Major, M. E. et al., DNA-based immunization with chimeric vectors for the induction of immune responses against the hepatitis C virus nucleocapsid, *J. Virol.,* 69, 5798, 1995.

128. Lee, I. H. et al., Presentation of the hydrophilic domains of hepatitis C viral E2 envelope glycoprotein on hepatitis B surface antigen particles, *J. Med. Virol.,* 50, 145, 1996.

129. Dargeviciute, A. et al., unpublished data, 1998.

130. Chengalvala, M. V. et al., Enhanced immunogenicity of hepatitis B surface antigen by insertion of a helper T cell epitope from tetanus toxoid, *Vaccine,* 17, 1035, 1999.

131. Gerlich, W. H. and Bruss, V., Functions of hepatitis B virus proteins and molecular targets for protective immunity, in *Hepatitis B Vaccines in Clinical Practice,* Ellis, R. W., Ed., Marcel Dekker, New York, 1993, pp. 41–82.

132. Nassal, M. and Schaller, H., Hepatitis B virus nucleocapsid assembly, in *Virus Strategies, Molecular Biology and Pathogenesis,* Doerfler, W. and Bohm, P., Eds., VCH, New York, 1993, pp. 41–75.

133. Kann, M. and Gerlich, W. H., Replication of hepatitis B virus, in *Molecular Medicine of Viral Hepatitis,* Harrison, T. J. and Zuckerman, A. J., Eds., John Wiley & Sons, New York, 1997, pp. 63–87.

134. Pumpens, P. et al., Hepatitis B virus core particles as epitope carriers, *Intervirology,* 38, 63, 1995.

135. Milich, D. R. et al., The hepatitis nucleocapsid as a vaccine carrier moiety, *Ann. N.Y. Acad. Sci.,* 754, 187, 1995.

136. Schödel, F. et al., Hybrid hepatitis B virus core antigen as a vaccine carrier moiety: I. Presentation of foreign epitopes, *J. Biotechnol.,* 44, 91, 1996.

137. Ulrich, R. et al., Core particles of hepatitis B virus as carrier for foreign epitopes, *Adv. Virus Res.,* 50, 141, 1998.

138. Pumpens, P. and Grens, E., Hepatitis B core particles as a universal display model: a structure-function basis for development, *FEBS Lett.,* 442, 1, 1999.

139. Murray, K. and Shiau, A. L., The core antigen of hepatitis B virus as a carrier for immunogenic peptides, *Biol. Chem.,* 380, 277, 1999.

140. Karpenko, L. I. et al., Analysis of foreign epitopes inserted in HBcAg. Possible routes for solving the problem of chimeric core particle self assembly (in Russian), *Mol. Biol. (Moscow),* 34, 223, 2000.

141. Pumpens, P. and Grens, E., HBV core particles as a carrier for B cell/T cell epitopes, *Intervirology,* 44, 98, 2001.

142. Scaglioni, P. P. et al., Posttranscriptional regulation of hepatitis B virus replication by the precore protein, *J. Virol.,* 71, 345, 1997.

143. Dyson, M. R. and Murray, K., Selection of peptide inhibitors of interactions involved in complex protein assemblies: association of the core and surface antigens of hepatitis B virus, *Proc. Natl. Acad. Sci. U.S.A.,* 92, 2194, 1995.

144. Poisson, F. et al., Both pre-S1 and S domains of hepatitis B virus envelope proteins interact with the core particle, *Virology,* 228, 115, 1997.

145. Kann, M. and Gerlich, W. H., Effect of core protein phosphorylation by protein kinase C on encapsidation of RNA within core particles of hepatitis B virus, *J. Virol.,* 68, 7993, 1994.

146. Liao, W. and Ou, J. H., Phosphorylation and nuclear localization of the hepatitis B virus core protein: significance of serine in the three repeated SPRRR motifs, *J. Virol.,* 69, 1025, 1995.

147. Lan, Y. T. et al., Roles of the three major phosphorylation sites of hepatitis B virus core protein in viral replication, *Virology,* 259, 342, 1999.

148. Kann, M. et al., In vitro model for the nuclear transport of the hepadnavirus genome, *J. Virol.,* 71, 1310, 1997.

149. Kann, M. et al., Phosphorylation-dependent binding of hepatitis B virus core particles to the nuclear pore complex, *J. Cell Biol.,* 145, 45, 1999.

150. Miller, R. H., Proteolytic self-cleavage of hepatitis B virus core protein may generate serum e antigen, *Science,* 236, 722, 1987.

151. Nassal, M. et al., Proteaselike sequence in hepatitis B virus core antigen is not required for e antigen generation and may not be part of an aspartic acid-type protease, *J. Virol.,* 63, 2598, 1989.

152. Hirschman, S. Z. et al., Expression of cloned hepatitis B virus DNA in human cell cultures, *Proc. Natl. Acad. Sci. U.S.A.,* 77, 5507, 1980.

153. Gough, N. M., Core and E antigen synthesis in rodent cells transformed with hepatitis B virus DNA is associated with greater than genome length viral messenger RNAs, *J. Mol. Biol.,* 165, 683, 1983.

154. Will, H. et al., Expression of hepatitis B antigens with a simian virus 40 vector, *J. Virol.,* 50, 335, 1984.

155. Roossinck, M. J. et al., Expression of hepatitis B viral core region in mammalian cells, *Mol. Cell Biol.,* 6, 1393, 1986.

156. Weimer, T. et al., Expression of the hepatitis B virus core gene *in vitro* and *in vivo, J. Virol.,* 61, 3109, 1987.

157. McLachlan, A. et al., Expression of hepatitis B virus surface and core antigens: influences of pre-S and precore sequences, *J. Virol.,* 61, 683, 1987.

158. Schlicht, H. J. and Schaller, H., The secretory core protein of human hepatitis B virus is expressed on the cell surface, *J. Virol.,* 63, 5399, 1989.

159. Kunke, D. et al., Vaccinia virus recombinants co-expressing hepatitis B virus surface and core antigens, *Virology,* 195, 132, 1993.

160. Jean-Jean, O. et al., Expression mechanism of the hepatitis B virus (HBV) C gene and biosynthesis of HBe antigen, *Virology,* 170, 99, 1989.

161. Standring, D. N. et al., A signal peptide encoded within the precore region of hepatitis B virus directs the secretion of a heterogeneous population of e antigens in *Xenopus* oocytes, *Proc. Natl. Acad. Sci. U.S.A.,* 85, 8405, 1988.

162. Takehara, K. et al., Co-expression of the hepatitis B surface and core antigens using baculovirus multiple expression vectors, *J. Gen. Virol.,* 69, 2763, 1988.

163. Hilditch, C. M. et al., Physicochemical analysis of the hepatitis B virus core antigen produced by a baculovirus expression vector, *J. Gen. Virol.,* 71, 2755, 1990.

164. Lanford, R. E. and Notvall, L., Expression of hepatitis B virus core and precore antigens in insect cells and characterization of a core-associated kinase activity, *Virology,* 176, 222, 1990.

165. Seifer, M. et al., Generation of replication-competent hepatitis B virus nucleocapsids in insect cells, *J. Virol.,* 72, 2765, 1998.

166. Kniskern, P. J. et al., Unusually high-level expression of a foreign gene (hepatitis B virus core antigen) in *Saccharomyces cerevisiae, Gene,* 46, 135, 1986.

167. Miyanohara, A. et al., Expression of hepatitis B virus core antigen gene in *Saccharomyces cerevisiae*: synthesis of two polypeptides translated from different initiation codons, *J. Virol.,* 59, 176, 1986.

168. Imamura, T. et al., Purification and characterization of the hepatitis B core antigen produced in the yeast *Saccharomyces cerevisiae, J. Biotechnol.,* 8, 149, 1988.

169. Shiosaki, K. et al., Production of hepatitis B virion-like particles in yeast, *Gene,* 106, 143, 1991.

170. Tsuda, S. et al., Application of the human hepatitis B virus core antigen from transgenic tobacco plants for serological diagnosis, *Vox Sang.,* 74, 148, 1998.

171. Pasek, M. et al., Hepatitis B virus genes and their expression in *E. coli, Nature,* 282, 575, 1979.

172. Stahl, S. et al., Hepatitis B virus core antigen: synthesis in *Escherichia coli* and application in diagnosis, *Proc. Natl. Acad. Sci. U.S.A.,* 79, 1606, 1982.

173. Borisova, G. P. et al., Structure and expression in *Escherichia coli* cells of the core antigen gene of the human hepatitis B virus (HBV) (in Russian), *Dokl. Akad. Nauk S.S.S.R.*, 279, 1245, 1984.

174. Uy, A. et al., Precore sequence of hepatitis B virus inducing e antigen and membrane association of the viral core protein, *Virology*, 155, 89, 1986.

175. Lanford, R. E. et al., Expression and characterization of hepatitis B virus precore-core antigen in *E. coli*, *Viral Immunol.*, 1, 97, 1987.

176. Khudyakov, Y. E. et al., The effect of the structure of the terminal regions of the hepatitis B virus gene C polypeptide on the formation of core antigen (HBcAg) particles, *Biomed. Sci.*, 2, 257, 1991.

177. Maassen, A. et al., Comparison of three different recombinant hepatitis B virus core particles expressed in *Escherichia coli*, *Arch. Virol.*, 135, 131, 1994.

178. Hardy, K. et al., Production in *B. subtilis* of hepatitis B core antigen and a major antigen of foot and mouth disease virus, *Nature*, 293, 481, 1981.

179. Schödel, F. et al., Hepatitis B virus nucleocapsid/pre-S2 fusion proteins expressed in attenuated *Salmonella* for oral vaccination, *J. Immunol.*, 145, 4317, 1990.

180. Schröder, R. et al., Expression of the core antigen gene of hepatitis B virus (HBV) in *Acetobacter methanolicus* using broad-host-range vectors, *Appl. Microbiol. Biotechnol.*, 35, 631, 1991.

181. Cohen, B. J. and Richmond, J. E., Electron microscopy of hepatitis B core antigen synthesized in *E. coli*, *Nature*, 296, 677, 1982.

182. Yamaguchi, M. et al., Electron microscopy of hepatitis B virus core antigen expressing yeast cells by freeze-substitution fixation, *Eur. J. Cell. Biol.*, 47, 138, 1988.

183. Gerlich, W. H. et al., Specificity and localization of the hepatitis B virus-associated protein kinase, *J. Virol.*, 42, 761, 1982.

184. Kenney, J. M. et al., Evolutionary conservation in the hepatitis B virus core structure: comparison of human and duck cores, *Structure*, 3, 1009, 1995.

185. Borisova, G. P. et al., Genetically engineered mutants of the core antigen of the human hepatitis B virus preserving the ability for native self-assembly (in Russian), *Dokl. Akad. Nauk S.S.S.R.*, 298, 1474, 1988.

186. Gallina, A. et al., A recombinant hepatitis B core antigen polypeptide with the protamine-like domain deleted self-assembles into capsid particles but fails to bind nucleic acids, *J. Virol.*, 63, 4645, 1989.

187. Inada, T. et al., Synthesis of hepatitis B virus e antigen in *E. coli*, *Virus Res.*, 14, 27, 1989.

188. Birnbaum, F. and Nassal, M., Hepatitis B virus nucleocapsid assembly: primary structure requirements in the core protein, *J. Virol.*, 64, 3319, 1990.

189. Bundule, M. A. et al., C-terminal polyarginine tract of hepatitis B core antigen is located on the outer capsid surface (in Russian), *Dokl. Akad. Nauk S.S.S.R.*, 312, 993, 1990.

190. Melegari, M. et al., The arginine-rich carboxy-terminal domain is necessary for RNA packaging by hepatitis core protein, in *Viral Hepatitis and Liver Diseases*, Hollinger, F. B. et al., Eds., Williams & Wilkins, Baltimore, 1991, pp. 164–168.

191. Hatton, T. et al., RNA- and DNA-binding activities in hepatitis B virus capsid protein: a model for their roles in viral replication, *J. Virol.*, 66, 5232, 1992.

192. Ulrich, R. et al., Immunogenicity of recombinant core particles of hepatitis B virus containing epitopes of human immunodeficiency virus 1 core antigen, *Arch. Virol.*, 126, 321, 1992.

193. Seifer, M. and Standring, D. N., Assembly and antigenicity of hepatitis B virus core particles, *Intervirology*, 38, 47, 1995.

194. Zlotnick, A. et al., Dimorphism of hepatitis B virus capsids is strongly influenced by the C-terminus of the capsid protein, *Biochemistry*, 35, 7412, 1996.

195. Hoofnagle, J. H. et al., Antibody to hepatitis-B-virus core in man, *Lancet*, 2, 869, 1973.

196. Chisari, F. V. and Ferrari, C., Hepatitis B virus immunopathogenesis, *Annu. Rev. Immunol.*, 13, 29, 1995.

197. Milich, D. R. and McLachlan, A., The nucleocapsid of hepatitis B virus is both a T-cell-independent and a T-cell-dependent antigen, *Science*, 234, 1398, 1986.

198. Milich, D. R. et al., The hepatitis B virus core and e antigens elicit different Th cell subsets: antigen structure can affect Th cell phenotype, *J. Virol.*, 71, 2192, 1997.

199. Milich, D. R. et al., Antibody production to the nucleocapsid and envelope of the hepatitis B virus primed by a single synthetic T cell site, *Nature*, 329, 547, 1987.

200. Milich, D. R. et al., Role of B cells in antigen presentation of the hepatitis B core, *Proc. Natl. Acad. Sci. U.S.A.,* 94, 14648, 1997.

201. Mondelli, M. et al., Specificity of T lymphocyte cytotoxicity to autologous hepatocytes in chronic hepatitis B virus infection: evidence that T cells are directed against HBV core antigen expressed on hepatocytes, *J. Immunol.,* 129, 2773, 1982.

202. Rehermann, B. et al., The hepatitis B virus persists for decades after patients' recovery from acute viral hepatitis despite active maintenance of a cytotoxic T-lymphocyte response, *Nat. Med.,* 2, 1104, 1996.

203. Murray, K. et al., Hepatitis B virus antigens made in microbial cells immunise against viral infection, *EMBO J.,* 3, 645, 1984.

204. Iwarson, S. et al., Protection against hepatitis B virus infection by immunization with hepatitis B core antigen, *Gastroenterology,* 88, 763, 1985.

205. Murray, K. et al., Protective immunisation against hepatitis B with an internal antigen of the virus, *J. Med. Virol.,* 23, 101, 1987.

206. Townsend, K. et al., Characterization of CD8+ cytotoxic T-lymphocyte responses after genetic immunization with retrovirus vectors expressing different forms of the hepatitis B virus core and e antigens, *J. Virol.,* 71, 3365, 1997.

207. Sällberg, M. et al., Genetic immunization of chimpanzees chronically infected with the hepatitis B virus, using a recombinant retroviral vector encoding the hepatitis B virus core antigen, *Hum. Gene Ther.,* 9, 1719, 1998.

208. Wild, J. et al., Polyvalent vaccination against hepatitis B surface and core antigen using a dicistronic expression plasmid, *Vaccine,* 16, 353, 1998.

209. Heathcote, J. et al., A pilot study of the CY-1899 T-cell vaccine in subjects chronically infected with hepatitis B virus. The CY1899 T Cell Vaccine Study Group, *Hepatology,* 30, 531, 1999.

210. Page, M., personal communication, 2000.

211. Fehr, T. et al., T cell-independent type I antibody response against B cell epitopes expressed repetitively on recombinant virus particles, *Proc. Natl. Acad. Sci. U.S.A.,* 95, 9477, 1998.

212. Salfeld, J. et al., Antigenic determinants and functional domains in core antigen and e antigen from hepatitis B virus, *J. Virol.,* 63, 798, 1989.

213. Sällberg, M. et al., Characterisation of a linear binding site for a monoclonal antibody to hepatitis B core antigen, *J. Med. Virol.,* 33, 248, 1991.

214. Sällberg, M. et al., Human and murine B-cells recognize the HBeAg/beta (or HBe2) epitope as a linear determinant, *Mol. Immunol.,* 28, 719, 1991.

215. Sällberg, M. et al., Immunochemical structure of the carboxy-terminal part of hepatitis B e antigen: identification of internal and surface-exposed sequences, *J. Gen. Virol.,* 74, 1335, 1993.

216. Bertoletti, A. et al., Definition of a minimal optimal cytotoxic T-cell epitope within the hepatitis B virus nucleocapsid protein, *J. Virol.,* 67, 2376, 1993.

217. Livingston, B. D. et al., The hepatitis B virus-specific CTL responses induced in humans by lipopeptide vaccination are comparable to those elicited by acute viral infection, *J. Immunol.,* 159, 1383, 1997.

218. Stuyver, L. et al., A new genotype of hepatitis B virus: complete genome and phylogenetic relatedness, *J. Gen. Virol.,* 81, 67, 2000.

219. Böttcher, B. et al., Finding the small difference: a new amino acid extension to the hepatitis B core protein. Electron microscopy. Paper presented at ICEM14. Symposium QQ, Vol. 1, Cancun, Mexico, 1998, pp. 737–738.

220. Beesley, K. M. et al., Expression in yeast of amino-terminal peptide fusions to hepatitis B core antigen and their immunological properties, *Biotechnology (N.Y.),* 8, 644, 1990.

221. Clarke, B. E. et al., Presentation and immunogenicity of viral epitopes on the surface of hybrid hepatitis B virus core particles produced in bacteria, *J. Gen. Virol.,* 71, 1109, 1990.

222. Brown, A. L. et al., Foreign epitopes in immunodominant regions of hepatitis B core particles are highly immunogenic and conformationally restricted, *Vaccine,* 9, 595, 1991.

223. Yon, J. et al., Stimulation of specific immune responses to simian immunodeficiency virus using chimeric hepatitis B core antigen particles, *J. Gen. Virol.,* 73, 2569, 1992.

224. Schödel, F. et al., The position of heterologous epitopes inserted in hepatitis B virus core particles determines their immunogenicity, *J. Virol.,* 66, 106, 1992.

225. Kalinina, T. I. et al., Introduction of heterologous epitopes at the N-terminal part of the hepatitis B core protein (in Russian), *Mol. Biol. (Moscow), 29*, 199, 1995.

226. Lachmann, S. et al., Characterization of potential insertion sites in the core antigen of hepatitis B virus by the use of a short-sized model epitope, *Intervirology, 42*, 51, 1999.

227. Moriarty, A. M. et al., Expression of HIV Gag and Env B-cell epitopes on the surface of HBV core particles and analysis of the immune responses generated to those epitopes, in *Vaccines 90,* Brown, F. et al., Eds., Cold Spring Harbor Laboratory, Cold Spring Harbor, NY, 1990, pp. 225–229.

228. Isaguliants, M. G. et al., HIV-1 epitopes exposed by hybrid hepatitis B core particles affect proliferation of peripheral blood mononuclear cells from HIV-1 positive donors, *Immunol. Lett., 52*, 37, 1996.

229. Isaguliants, M. G. et al., Expression of HIV-1 epitopes included in particles formed by human hepatitis B virus nucleocapsid protein (in Russian), *Biokhimiia, 61*, 532, 1996.

230. Tarar, M. R. et al., Expression of a human cytomegalovirus gp58 antigenic domain fused to the hepatitis B virus nucleocapsid protein, *FEMS Immunol. Med. Microbiol., 16*, 183, 1996.

231. Neirynck, S. et al., A universal influenza A vaccine based on the extracellular domain of the M2 protein, *Nat. Med., 5*, 1157, 1999.

232. Koletzki, D. et al., Puumala (PUU) hantavirus strain differences and insertion positions in the hepatitis B virus core antigen influence B-cell immunogenicity and protective potential of core-derived particles, *Virology, 276*, 364, 2000.

233. Koletzki, D. et al., HBV core particles allow the insertion and surface exposure of the entire potentially protective region of Puumala hantavirus nucleocapsid protein, *Biol. Chem., 380*, 325, 1999.

234. Borisova, G. P., personal communication, 1995.

235. Borisova, G. P. et al., Recombinant core particles of hepatitis B virus exposing foreign antigenic determinants on their surface, *FEBS Lett., 259*, 121, 1989.

236. Borisova, G. P. et al., Hepatitis B core antigen as a carrier of functionally active epitopes: exposure of pre-S sites on capsids (in Russian), *Dokl. Akad. Nauk S.S.S.R., 312*, 751, 1990.

237. Zakis, V. et al., Immunodominance of T-cell epitopes on foreign sequences of the hepatitis B virus nucleocapsid fusion proteins, in *Vaccines 92,* Brown, F. et al., Eds., Cold Spring Harbor Laboratory Press, Cold Spring Harbor, NY, 1992, pp. 341–347.

238. Ulrich, R. et al., Exposition of epitopes of transmembrane protein gp41 of the human immunodeficiency virus on surface capsids of the hepatitis B virus core antigen (in Russian), *Bioorg. Khim., 16*, 1283, 1990.

239. Ulrich, R. et al., Exposure of the major immunodominant epitope of the gp51 envelope protein of bovine leukemia virus on the surface of the hepatitis B core antigen capsid (in Russian), *Mol. Biol. (Moscow), 25*, 368, 1991.

240. Del Val, M. et al., Protection against lethal cytomegalovirus infection by a recombinant vaccine containing a single nonameric T-cell epitope, *J. Virol., 65*, 3641, 1991.

241. Stahl, S. J. and Murray, K., Immunogenicity of peptide fusions to hepatitis B virus core antigen, *Proc. Natl. Acad. Sci. U.S.A., 86*, 6283, 1989.

242. Shiau, A. L. and Murray, K., Mutated epitopes of hepatitis B surface antigen fused to the core antigen of the virus induce antibodies that react with the native surface antigen, *J. Med. Virol., 51*, 159, 1997.

243. Borisova, G. et al., Display vectors. I. Hepatitis B core particle as a display moiety, *Proc. Latv. Acad. Sci., 51*, 1, 1997.

244. Grene, E. et al., Relationship between antigenicity and immunogenicity of chimeric hepatitis B virus core particles carrying HIV type 1 epitopes, *AIDS Res. Hum. Retroviruses, 13*, 41, 1997.

245. Von Brunn, A. et al., The principal neutralizing determinant (V3) of HIV-1 induces HIV-1-neutralizing antibodies upon expression on HBcAg particles, in *Vaccines 93. Modern Approaches to New Vaccines Including Prevention of AIDS,* Ginsberg, H. S. et al., Eds., Cold Spring Harbor Laboratory Press, Cold Spring Harbor, NY, 1993, pp. 159–165.

246. Von Brunn, A. et al., Principal neutralizing domain of HIV-1 is highly immunogenic when expressed on the surface of hepatitis B core particles, *Vaccine, 11*, 817, 1993.

247. Ulrich, R. et al., Characterization of chimeric core particles of HBV containing foreign epitopes, in *Vaccines 93. Modern Approaches to New Vaccines Including Prevention of AIDS,* Ginsberg, H.S. et al., Eds., Cold Spring Harbor Laboratory Press, Cold Spring Harbor, NY, 1993, pp. 323–328.

248. Berzins, I. R. et al., Sites in the core antigen of HBV allowing insertion of foreign epitopes, in *Vaccines 94,* Norrby, E. et al., Eds., Cold Spring Harbor Laboratory Press, Cold Spring Harbor, New York, 1994, pp. 301–308.

249. Ulrich, R. et al., A new strategy to generate mosaic hbcag particles presenting foreign epitopes, in *Vaccines 97,* Brown, F. et al., Eds., Cold Spring Harbor Laboratory Press, Cold Spring Harbor, NY, 1997, pp. 235–240.

250. Ulrich, R. et al., Chimaeric HBV core particles carrying a defined segment of Puumala hantavirus nucleocapsid protein evoke protective immunity in an animal model, *Vaccine,* 16, 272, 1998.

251. Ulrich, R. et al., New chimaeric hepatitis B virus core particles carrying hantavirus (serotype Puumala) epitopes: immunogenicity and protection against virus challenge, *J. Biotechnol.,* 73, 141, 1999.

252. Touze, A. et al., Baculovirus expression of chimeric hepatitis B virus core particles with hepatitis E virus epitopes and their use in a hepatitis E immunoassay, *J. Clin. Microbiol.,* 37, 438, 1999.

253. Nekrasova, O. V. et al., Bacterial synthesis of immunogenic epitopes of foot-and-mouth disease virus fused either to human necrosis factor or to hepatitis B core antigen (in Russian), *Bioorg. Khim.,* 23, 118, 1997.

254. Yoshikawa, A. et al., Chimeric hepatitis B virus core particles with parts or copies of the hepatitis C virus core protein, *J. Virol.,* 67, 6064, 1993.

255. Claeys, H. et al., Association of hepatitis C virus carrier state with the occurrence of hepatitis C virus core antibodies, *J. Med. Virol.,* 36, 259, 1992.

256. Wu, C. L. et al., Hepatitis C virus core protein fused to hepatitis B virus core antigen for serological diagnosis of both hepatitis C and hepatitis B infections by ELISA, *J. Med. Virol.,* 57, 104, 1999.

257. Claeys, H. et al., Localization and reactivity of an immunodominant domain in the NS3 region of hepatitis C virus, *J. Med. Virol.,* 45, 273, 1995.

258. Dawson, J. A. and Macrina, F. L., Construction and immunologic evaluation of a *Porphyromonas gingivalis* subsequence peptide fused to hepatitis B virus core antigen, *FEMS Microbiol. Lett.,* 175, 119, 1999.

259. Charles, I. G. et al., Identification and characterization of a protective immunodominant B cell epitope of pertactin (P.69) from *Bordetella pertussis, Eur. J. Immunol.,* 21, 1147, 1991.

260. Beterams, G. et al., Packaging of up to 240 subunits of a 17 kDa nuclease into the interior of recombinant hepatitis B virus capsids, *FEBS Lett.,* 481, 169, 2000.

261. Kratz, P. A. et al., Native display of complete foreign protein domains on the surface of hepatitis B virus capsids, *Proc. Natl. Acad. Sci. U.S.A.,* 96, 1915, 1999.

262. Schödel, F. et al., Recombinant HBV core particles carrying immunodominant B-cell epitopes of the HBV preS2-region, in *Vaccines 90,* Brown, F. et al., Eds., Cold Spring Harbor Laboratory Press, Cold Spring Harbor, NY, 1990, pp. 193–198.

263. Schödel, F. et al., Hybrid Hepatitis-B Virus Core/pre-S Particles Expressed in Live Attenuated *Salmonellae* for Oral Immunization, in *Vaccines 91,* Brown, F. et al., Eds., Cold Spring Harbor Laboratory Press, Cold Spring Harbor, NY, 1991, pp. 319–325.

264. Schödel, F. et al., A virulent *Salmonella* expressing hybrid hepatitis B virus core/pre-S genes for oral vaccination, *Vaccine,* 11, 143, 1993.

265. Clarke, B. E. et al., Expression and immunological analysis of hepatitis-B core fusion particles carrying internal heterologous sequences, in *Vaccines 91,* Chanock, R. M. et al., Eds., Cold Spring Harbor Laboratory Press, Cold Spring Harbor, NY, 1991, pp. 313–318.

266. Wizemann, H. et al., Polyhistidine-tagged hepatitis B core particles as carriers of HIV-1/gp120 epitopes of different HIV-1 subtypes, *Biol. Chem.,* 381, 231, 2000.

267. Borisova, G. et al., Spatial structure and insertion capacity of immunodominant region of hepatitis B core antigen, *Intervirology,* 39, 16, 1996.

268. Borisova, G. et al., Hybrid hepatitis B virus nucleocapsid bearing an immunodominant region from hepatitis B virus surface antigen, *J. Virol.,* 67, 3696, 1993.

269. Makeeva, I. V. et al., Heterologous epitopes in the central part of the hepatitis B virus core protein (in Russian), *Mol. Biol. (Moscow),* 29, 211, 1995.

270. Loktev, V. B. et al., Design of immunogens as components of a new generation of molecular vaccines, *J. Biotechnol.,* 44, 129, 1996.

271. Tindle, R. W. et al., Chimeric hepatitis B core antigen particles containing B- and Th-epitopes of human papillomavirus type 16 E7 protein induce specific antibody and T-helper responses in immunised mice, *Virology*, 200, 547, 1994.

272. Londono, L. P. et al., Immunisation of mice using *Salmonella typhimurium* expressing human papillomavirus type 16 E7 epitopes inserted into hepatitis B virus core antigen, *Vaccine*, 14, 545, 1996.

273. Street, M. et al., Differences in the effectiveness of delivery of B- and CTL-epitopes incorporated into the hepatitis B core antigen (HBcAg) c/e1-region, *Arch. Virol.*, 144, 1323, 1999.

274. Schödel, F. et al., Immunity to malaria elicited by hybrid hepatitis B virus core particles carrying circumsporozoite protein epitopes, *J. Exp. Med.*, 180, 1037, 1994.

275. Schödel, F. et al., Hybrid hepatitis B virus core-pre-S proteins synthesized in avirulent *Salmonella typhimurium* and *Salmonella typhi* for oral vaccination, *Infect. Immunol.*, 62, 1669, 1994.

276. Schödel, F. et al., Development of recombinant *Salmonellae* expressing hybrid hepatitis B virus core particles as candidate oral vaccines, *Dev. Biol. Stand.*, 82, 151, 1994.

277. Schödel, F. et al., Hybrid hepatitis B virus core antigen as a vaccine carrier moiety. II. Expression in avirulent *Salmonella* spp. for mucosal immunization, *Adv. Exp. Med. Biol.*, 397, 15, 1996.

278. Schödel, F. et al., Hepatitis B virus core and e antigen: immune recognition and use as a vaccine carrier moiety, *Intervirology*, 39, 104, 1996.

279. Schödel, F. et al., Immunization with hybrid hepatitis B virus core particles carrying circumsporozoite antigen epitopes protects mice against *Plasmodium yoelii* challenge, *Behring. Inst. Mitt.*, 98, 114, 1997.

280. Tacket, C. O. et al., Safety and immunogenicity in humans of an attenuated *Salmonella typhi* vaccine vector strain expressing plasmid-encoded hepatitis B antigens stabilized by the Asd-balanced lethal vector system, *Infect. Immunol.*, 65, 3381, 1997.

281. Boulter, N. R. et al., *Theileria annulata* sporozoite antigen fused to hepatitis B core antigen used in a vaccination trial, *Vaccine*, 13, 1152, 1995.

282. Boulter, N. et al., Evaluation of recombinant sporozoite antigen SPAG-1 as a vaccine candidate against *Theileria annulata* by the use of different delivery systems, *Trop. Med. Int. Health*, 4, A71, 1999.

283. Karpenko, L. I. and Ilyichev, A. A., Chimeric hepatitis B core antigen particles as a presentation system of foreign protein epitopes (in Russian), *Vestn. Ross. Akad. Med. Nauk*, 3, 6, 1998.

284. Karpenko, L. I. et al., Isolation and study of recombinant strains of *Salmonella typhimurium* SL 7207, producing HBcAg and HBcAg-HBs (in Russian), *Vopr. Virusol.*, 45, 10, 2000.

285. Chambers, M. A. et al., Chimeric hepatitis B virus core particles as probes for studying peptide-integrin interactions, *J. Virol.*, 70, 4045, 1996.

286. Borschukova, O. et al., Modified hepatitis B core particles as possible vaccine carriers, in *Vaccines-97*, Brown, F. et al., Eds., Cold Spring Harbor Laboratory Press, Cold Spring Harbor, NY, pp. 33–37.

287. Borisova, G. et al., Behavior of a short preS1 epitope on the surface of hepatitis B core particles, *Biol. Chem.*, 380, 315, 1999.

288. Preikschat, P. et al., Expression, assembly competence and antigenic properties of hepatitis B virus core gene deletion variants from infected liver cells, *J. Gen. Virol.*, 80, 1777, 1999.

289. Koletzki, D. et al., unpublished data, 1998.

290. Sominskaya, I. et al., Determination of the minimal length of preS1 epitope recognized by a monoclonal antibody which inhibits attachment of hepatitis B virus to hepatocytes, *Med. Microbiol. Immunol. (Berlin)*, 181, 215, 1992.

291. Koletzki, D. et al., Mosaic hepatitis B virus core particles allow insertion of extended foreign protein segments, *J. Gen. Virol.*, 78, 2049, 1997.

292. Kazaks, A. et al., unpublished data, 1997.

293. Wizemann, H. and von Brunn, A., Purification of *E. coli*-expressed HIS-tagged hepatitis B core antigen by Ni^{2+}-chelate affinity chromatography, *J. Virol. Methods*, 77, 189, 1999.

294. Kost, T. A. and Condreay, J. P., Recombinant baculoviruses as expression vectors for insect and mammalian cells, *Curr. Opin. Biotechnol.*, 10, 428, 1999.

295. Rohrmann, G. F., Polyhedrin structure, *J. Gen. Virol.*, 67, 1499, 1986.

296. McLinden, J. H. et al., Expression of foreign epitopes on recombinant occlusion bodies of baculoviruses, *Vaccine*, 10, 231, 1992.

297. Eason, J. E. et al., Effects of substituting granulin or a granulin-polyhedrin chimera for polyhedrin on virion occlusion and polyhedral morphology in *Autographa californica* multinucleocapsid nuclear polyhedrosis virus, *J. Virol.*, 72, 6237, 1998.

298. Zhou, J. et al., Expression of vaccinia recombinant HPV 16 L1 and L2 ORF proteins in epithelial cells is sufficient for assembly of HPV virion-like particles, *Virology,* 185, 251, 1991.

299. Zhou, J. et al., Synthesis and assembly of infectious bovine papillomavirus particles *in vitro, J. Gen. Virol.,* 74, 763, 1993.

300. Kirnbauer, R. et al., Papillomavirus L1 major capsid protein self-assembles into virus-like particles that are highly immunogenic, *Proc. Natl. Acad. Sci. U.S.A.,* 89, 12180, 1992.

301. Rose, R. C. et al., Expression of human papillomavirus type 11 L1 protein in insect cells: *in vivo* and *in vitro* assembly of viruslike particles, *J. Virol.,* 67, 1936, 1993.

302. Kirnbauer, R. et al., Efficient self-assembly of human papillomavirus type 16 L1 and L1-L2 into virus-like particles, *J. Virol.,* 67, 6929, 1993.

303. Volpers, C. et al., Assembly of the major and the minor capsid protein of human papillomavirus type 33 into virus-like particles and tubular structures in insect cells, *Virology,* 200, 504, 1994.

304. Rose, R. C. et al., Serological differentiation of human papillomavirus types 11, 16 and 18 using recombinant virus-like particles, *J. Gen. Virol.,* 75, 2445, 1994.

305. Christensen, N. D. et al., Human papillomavirus types 6 and 11 have antigenically distinct strongly immunogenic conformationally dependent neutralizing epitopes, *Virology,* 205, 329, 1994.

306. Sapp, M. et al., Analysis of type-restricted and cross-reactive epitopes on virus-like particles of human papillomavirus type 33 and in infected tissues using monoclonal antibodies to the major capsid protein, *J. Gen. Virol.,* 75, 3375, 1994.

307. Sasagawa, T. et al., Synthesis and assembly of virus-like particles of human papillomaviruses type 6 and type 16 in fission yeast *Schizosaccharomyces pombe, Virology,* 206, 126, 1995.

308. Hofmann, K. J. et al., Sequence determination of human papillomavirus type 6a and assembly of virus-like particles in *Saccharomyces cerevisiae, Virology,* 209, 506, 1995.

309. Jansen, K. U. et al., Vaccination with yeast-expressed cottontail rabbit papillomavirus (CRPV) virus-like particles protects rabbits from CRPV-induced papilloma formation, *Vaccine,* 13, 1509, 1995.

310. Nardelli-Haefliger, D. et al., Human papillomavirus type 16 virus-like particles expressed in attenuated *Salmonella typhimurium* elicit mucosal and systemic neutralizing antibodies in mice, *Infect. Immunol.,* 65, 3328, 1997.

311. Li, M. et al., Expression of the human papillomavirus type 11 L1 capsid protein in *Escherichia coli*: characterization of protein domains involved in DNA binding and capsid assembly, *J. Virol.,* 71, 2988, 1997.

312. Zhang, W. et al., Expression of human papillomavirus type 16 L1 protein in *Escherichia coli*: denaturation, renaturation, and self-assembly of virus-like particles *in vitro, Virology,* 243, 423, 1998.

313. Heino, P. et al., Human papillomavirus type 16 capsid proteins produced from recombinant Semliki Forest virus assemble into virus-like particles, *Virology,* 214, 349, 1995.

314. Roden, R. B. et al., *In vitro* generation and type-specific neutralization of a human papillomavirus type 16 virion pseudotype, *J. Virol.,* 70, 5875, 1996.

315. Unckell, F. et al., Generation and neutralization of pseudovirions of human papillomavirus type 33, *J. Virol.,* 71, 2934, 1997.

316. Touze, A. and Coursaget, P., *In vitro* gene transfer using human papillomavirus-like particles, *Nucleic Acids Res.,* 26, 1317, 1998.

317. Zhao, K. N. et al., DNA packaging by L1 and L2 capsid proteins of bovine papillomavirus type 1, *Virology,* 243, 482, 1998.

318. Hagensee, M. E. et al., Three-dimensional structure of vaccinia virus-produced human papillomavirus type 1 capsids, *J. Virol.,* 68, 4503, 1994.

319. Baker, T. S. et al., Structures of bovine and human papillomaviruses. Analysis by cryoelectron microscopy and three-dimensional image reconstruction, *Biophys. J.,* 60, 1445, 1991.

320. Trus, B. L. et al., Novel structural features of bovine papillomavirus capsid revealed by a three-dimensional reconstruction to 9 Å resolution, *Nat. Struct. Biol.,* 4, 413, 1997.

321. Paintsil, J. et al., Carboxyl terminus of bovine papillomavirus type-1 L1 protein is not required for capsid formation, *Virology,* 223, 238, 1996.

322. Chen, Y. et al., Mutant canine oral papillomavirus L1 capsid proteins which form virus-like particles but lack native conformational epitopes, *J. Gen. Virol.,* 79, 2137, 1998.

323. McCarthy, M. P. et al., Quantitative disassembly and reassembly of human papillomavirus type 11 viruslike particles *in vitro, J. Virol.,* 72, 32, 1998.

324. Breitburd, F. et al., Immunization with viruslike particles from cottontail rabbit papillomavirus (CRPV) can protect against experimental CRPV infection, *J. Virol.,* 69, 3959, 1995.

325. Christensen, N. D. et al., Immunization with viruslike particles induces long-term protection of rabbits against challenge with cottontail rabbit papillomavirus, *J. Virol.,* 70, 960, 1996.

326. Kirnbauer, R. et al., Virus-like particles of bovine papillomavirus type 4 in prophylactic and therapeutic immunization, *Virology,* 219, 37, 1996.

327. Sundaram, P. et al., Intracutaneous vaccination of rabbits with the cottontail rabbit papillomavirus (CRPV) L1 gene protects against virus challenge, *Vaccine,* 15, 664, 1997.

328. Balmelli, C. et al., Nasal immunization of mice with human papillomavirus type 16 virus-like particles elicits neutralizing antibodies in mucosal secretions, *J. Virol.,* 72, 8220, 1998.

329. Dupuy, C. et al., Nasal immunization of mice with human papillomavirus type 16 (HPV-16) virus-like particles or with the HPV-16 L1 gene elicits specific cytotoxic T lymphocytes in vaginal draining lymph nodes, *J. Virol.,* 73, 9063, 1999.

330. Nardelli-Haefliger, D. et al., Mucosal but not parenteral immunization with purified human papillomavirus type 16 virus-like particles induces neutralizing titers of antibodies throughout the estrous cycle of mice, *J. Virol.,* 73, 9609, 1999.

331. Rose, R. C. et al., Oral vaccination of mice with human papillomavirus virus-like particles induces systemic virus-neutralizing antibodies, *Vaccine,* 17, 2129, 1999.

332. Rudolf, M. P. et al., Induction of HPV16 capsid protein-specific human T cell responses by virus-like particles, *Biol. Chem.,* 380, 335, 1999.

333. Zhang, L. F. et al., HPV6b virus like particles are potent immunogens without adjuvant in man, *Vaccine,* 18, 1051, 2000.

334. Evans, T. G. et al., A Phase 1 study of a recombinant viruslike particle vaccine against human papillomavirus type 11 in healthy adult volunteers, *J. Infect. Dis.,* 183, 1485, 2001.

335. Muller, M. et al., Chimeric papillomavirus-like particles, *Virology,* 234, 93, 1997.

336. Schafer, K. et al., Immune response to human papillomavirus 16 L1E7 chimeric virus-like particles: induction of cytotoxic T cells and specific tumor protection, *Int. J. Cancer,* 81, 881, 1999.

337. Jochmus, I. et al., Chimeric virus-like particles of the human papillomavirus type 16 (HPV 16) as a prophylactic and therapeutic vaccine, *Arch. Med. Res.,* 30, 269, 1999.

338. Greenstone, H. L. et al., Chimeric papillomavirus virus-like particles elicit antitumor immunity against the E7 oncoprotein in an HPV16 tumor model, *Proc. Natl. Acad. Sci. U.S.A.,* 95, 1800, 1998.

339. Liu, X. S. et al., Mucosal immunisation with papillomavirus virus-like particles elicits systemic and mucosal immunity in mice, *Virology,* 252, 39, 1998.

340. Peng, S. et al., Papillomavirus virus-like particles can deliver defined CTL epitopes to the MHC class I pathway, *Virology,* 240, 147, 1998.

341. Chackerian, B. et al., Induction of autoantibodies to mouse CCR5 with recombinant papillomavirus particles, *Proc. Natl. Acad. Sci. U.S.A.,* 96, 2373, 1999.

342. Nieland, J. D. et al., Chimeric papillomavirus virus-like particles induce a murine self-antigen-specific protective and therapeutic antitumor immune response, *J. Cell Biochem.,* 73, 145, 1999.

343. Adolph, K. W. et al., Polyoma virion and capsid crystal structures, *Science,* 203, 1117, 1979.

344. Liddington, R. C. et al., Structure of simian virus 40 at 3.8- Å resolution, *Nature,* 354, 278, 1991.

345. Stehle, T. et al., Structure of murine polyomavirus complexed with an oligosaccharide receptor fragment, *Nature,* 369, 160, 1994.

346. Yan, Y. et al., Structure determination of simian virus 40 and murine polyomavirus by a combination of 30-fold and 5-fold electron-density averaging, *Structure,* 4, 157, 1996.

347. Stehle, T. et al., The structure of simian virus 40 refined at 3.1 Å resolution, *Structure,* 4, 165, 1996.

348. Belnap, D. M. et al., Conserved features in papillomavirus and polyomavirus capsids, *J. Mol. Biol.,* 259, 249, 1996.

349. Siray, H. et al., Capsid protein-encoding genes of hamster polyomavirus and properties of the viral capsid, *Virus Genes,* 18, 39, 1999.

350. Leavitt, A. D. et al., Polyoma virus major capsid protein, VP1. Purification after high level expression in *Escherichia coli, J. Biol. Chem.,* 260, 12803, 1985.

351. Salunke, D. M. et al., Self-assembly of purified polyomavirus capsid protein VP1, *Cell,* 46, 895, 1986.

352. Salunke, D. M. et al., Polymorphism in the assembly of polyomavirus capsid protein VP1, *Biophys. J.,* 56, 887, 1989.

353. Garcea, R. L. et al., Site-directed mutation affecting polyomavirus capsid self-assembly *in vitro*, *Nature*, 329, 86, 1987.

354. Moreland, R. B. et al., Characterization of the DNA-binding properties of the polyomavirus capsid protein VP1, *J. Virol.*, 65, 1168, 1991.

355. Cai, X. et al., Expression and purification of recombinant polyomavirus VP2 protein and its interactions with polyomavirus proteins, *J. Virol.*, 68, 7609, 1994.

356. Delos, S. E. et al., Expression of the polyomavirus minor capsid proteins VP2 and VP3 in *Escherichia coli*: *in vitro* interactions with recombinant VP1 capsomeres, *J. Virol.*, 69, 7734, 1995.

357. Chen, X. S. et al., Interaction of polyomavirus internal protein VP2 with the major capsid protein VP1 and implications for participation of VP2 in viral entry, *EMBO J.*, 17, 3233, 1998.

358. Rodgers, R. E. et al., Purification of recombinant budgerigar fledgling disease virus VP1 capsid protein and its ability for *in vitro* capsid assembly, *J. Virol.*, 68, 3386, 1994.

359. Ou, W. C. et al., The major capsid protein, VP1, of human JC virus expressed in *Escherichia coli* is able to self-assemble into a capsid-like particle and deliver exogenous DNA into human kidney cells, *J. Gen. Virol.*, 80, 39, 1999.

360. Voronkova, T. and Ulrich, R., unpublished data, 1998.

361. An, K. et al., Use of the baculovirus system to assemble polyomavirus capsid-like particles with different polyomavirus structural proteins: analysis of the recombinant assembled capsid-like particles, *J. Gen. Virol.*, 80, 1009, 1999.

362. Delos, S. E. et al., Expression of the polyomavirus VP2 and VP3 proteins in insect cells: coexpression with the major capsid protein VP1 alters VP2/VP3 subcellular localization, *Virology*, 194, 393, 1993.

363. Kosukegawa, A. et al., Purification and characterization of virus-like particles and pentamers produced by the expression of SV40 capsid proteins in insect cells, *Biochim. Biophys. Acta*, 1290, 37, 1996.

364. Sandalon, Z. and Oppenheim, A., Self-assembly and protein-protein interactions between the SV40 capsid proteins produced in insect cells, *Virology*, 237, 414, 1997.

365. Pawlita, M. et al., DNA encapsidation by viruslike particles assembled in insect cells from the major capsid protein VP1 of B-lymphotropic papovavirus, *J. Virol.*, 70, 7517, 1996.

366. Chang, D. et al., Self-assembly of the JC virus major capsid protein, VP1, expressed in insect cells, *J. Gen. Virol.*, 78, 1435, 1997.

367. Goldmann, C. et al., Molecular cloning and expression of major structural protein VP1 of the human polyomavirus JC virus: formation of virus-like particles useful for immunological and therapeutic studies, *J. Virol.*, 73, 4465, 1999.

368. An, K. et al., Avian polyomavirus major capsid protein VP1 interacts with the minor capsid proteins and is transported into the cell nucleus but does not assemble into capsid-like particles when expressed in the baculovirus system, *Virus Res.*, 64, 173, 1999.

369. Siray, H. and Ulrich, R., unpublished data, 1998.

370. Forstova, J. et al., Polyoma virus pseudocapsids as efficient carriers of heterologous DNA into mammalian cells, *Hum. Gene Ther.*, 6, 297, 1995.

371. Gillock, E. T. et al., Polyomavirus major capsid protein VP1 is capable of packaging cellular DNA when expressed in the baculovirus system, *J. Virol.*, 71, 2857, 1997.

372. Soeda, E. et al., Enhancement by polylysine of transient, but not stable, expression of genes carried into cells by polyoma VP1 pseudocapsids, *Gene Ther.*, 5, 1410, 1998.

373. Braun, H. et al., Oligonucleotide and plasmid DNA packaging into polyoma VP1 virus-like particles expressed in *Escherichia coli*, *Biotechnol. Appl. Biochem.*, 29, 31, 1999.

374. Sasnauskas, K. et al., Yeast cells allow high-level expression and formation of polyomavirus-like particles, *Biol. Chem.*, 380, 381, 1999.

375. Sasnauskas, K. et al., Generation of recombinant virus-like particles of human and non-human polyomaviruses in yeast *Saccharomyces cerevisiae*, *Intervirology*, in press.

376. Szomolanyi-Tsuda, E. et al., T-Cell-independent immunoglobulin G responses *in vivo* are elicited by live-virus infection but not by immunization with viral proteins or virus-like particles, *J. Virol.*, 72, 6665, 1998.

377. Schmidt, U. et al., Site-specific fluorescence labelling of recombinant polyomavirus-like particles, *Biol. Chem.*, 380, 397, 1999.

378. Gleiter, S. et al., Changing the surface of a virus shell fusion of an enzyme to polyoma VP1, *Protein Sci.*, 8, 2562, 1999.

379. Ulrich, R. et al., Hamster polyomavirus-encoded proteins: gene cloning, heterologous expression and immunoreactivity, *Virus Genes,* 12, 265, 1996.

380. Siray, H. et al., An immunodominant, cross-reactive B-cell epitope region is located at the C-terminal part of the hamster polyomavirus major capsid protein VP1, *Viral Immunol.,* 13, 533, 2000.

381. Gedvilaite, A. et al., Formation of immunogenic virus-like particles by inserting epitopes into surface-exposed regions of hamster polyomavirus major capsid protein, *Virology,* 273, 21, 2000.

382. Lasher, H. N. and Davis, V. S., History of infectious bursal disease in the U.S.A.— the first two decades, *Avian Dis.,* 41, 11, 1997.

383. Böttcher, B. et al., Three-dimensional structure of infectious bursal disease virus determined by electron cryomicroscopy, *J. Virol.,* 71, 325, 1997.

384. Macreadie, I. G. et al., Passive protection against infectious bursal disease virus by viral VP2 expressed in yeast, *Vaccine,* 8, 549, 1990.

385. Azad, A. A. et al., Physicochemical and immunological characterization of recombinant host-protective antigen (VP2) of infectious bursal disease virus, *Vaccine,* 9, 715, 1991.

386. Vakharia, V. N. et al., Infectious bursal disease virus structural proteins expressed in a baculovirus recombinant confer protection in chickens, *J. Gen. Virol.,* 74, 1201, 1993.

387. Fernandez-Arias, A. et al., Expression of ORF A1 of infectious bursal disease virus results in the formation of virus-like particles, *J. Gen. Virol.,* 79, 1047, 1998.

388. Kibenge, F. S. et al., Formation of virus-like particles when the polyprotein gene (segment A) of infectious bursal disease virus is expressed in insect cells, *Can. J. Vet. Res.,* 63, 49, 1999.

389. Hu, Y. C. et al., Chimeric infectious bursal disease virus-like particles expressed in insect cells and purified by immobilized metal affinity chromatography, *Biotechnol. Bioeng.,* 63, 721, 1999.

390. Wang, M. Y. et al., Self-assembly of the infectious bursal disease virus capsid protein, rVP2, expressed in insect cells and purification of immunogenic chimeric rVP2H particles by immobilized metal-ion affinity chromatography, *Biotechnol. Bioeng.,* 67, 104, 2000.

391. Lombardo, E. et al., VP1, the putative RNA-dependent RNA polymerase of infectious bursal disease virus, forms complexes with the capsid protein VP3, leading to efficient encapsidation into virus-like particles, *J. Virol.,* 73, 6973, 1999.

392. Di, X. et al., X-ray powder pattern analysis of cytoplasmic polyhedrosis virus inclusion bodies, *Virology,* 180, 153, 1991.

393. Lavallee, C. et al., Expression in *Escherichia coli* of the cloned polyhedrin gene of Bombyx mori cytoplasmic polyhedrosis virus, *Protein Expr. Purif.,* 4, 570, 1993.

394. Mori, H. et al., Expression of Bombyx mori cytoplasmic polyhedrosis virus polyhedrin in insect cells by using a baculovirus expression vector, and its assembly into polyhedra, *J. Gen. Virol.,* 74, 99, 1993.

395. Ikeda, K. et al., Characterizations of natural and induced polyhedrin gene mutants of *Bombyx mori* cytoplasmic polyhedrosis viruses, *Arch. Virol.,* 143, 241, 1998.

396. Nakazawa, H. et al., Effect of mutations on the intracellular localization of Bombyx mori cytoplasmic polyhedrosis virus polyhedrin, *J. Gen. Virol.,* 77, 147, 1996.

397. Roy, P., Genetically engineered particulate virus-like structures and their use as vaccine delivery systems, *Intervirology,* 39, 62, 1996.

398. Roy, P. et al., Baculovirus multigene expression vectors and their use for understanding the assembly process of architecturally complex virus particles, *Gene,* 190, 119, 1997.

399. Urakawa, T. and Roy, P., Bluetongue virus tubules made in insect cells by recombinant baculoviruses: expression of the NS1 gene of bluetongue virus serotype 10, *J. Virol.,* 62, 3919, 1988.

400. French, T. J. and Roy, P., Synthesis of bluetongue virus (BTV) corelike particles by a recombinant baculovirus expressing the two major structural core proteins of BTV, *J. Virol.,* 64, 1530, 1990.

401. French, T. J. et al., Assembly of double-shelled, viruslike particles of bluetongue virus by the simultaneous expression of four structural proteins, *J. Virol.,* 64, 5695, 1990.

402. Loudon, P. T. and Roy, P., Assembly of five bluetongue virus proteins expressed by recombinant baculoviruses: inclusion of the largest protein VP1 in the core and virus-like proteins, *Virology,* 180, 798, 1991.

403. Le Blois, H. et al., The expressed VP4 protein of bluetongue virus binds GTP and is the candidate guanylyl transferase of the virus, *Virology,* 189, 757, 1992.

404. Liu, H. M. et al., Interactions between bluetongue virus core and capsid proteins translated *in vitro*, *J. Gen. Virol.,* 73, 2577, 1992.

405. Belyaev, A. S. and Roy, P., Development of baculovirus triple and quadruple expression vectors: co-expression of three or four bluetongue virus proteins and the synthesis of bluetongue virus-like particles in insect cells, *Nucleic Acids Res.,* 21, 1219, 1993.

406. Chuma, T. et al., Expression of the major core antigen VP7 of African horsesickness virus by a recombinant baculovirus and its use as a group-specific diagnostic reagent, *J. Gen. Virol.,* 73, 925, 1992.

407. Maree, S. et al., Expression of the major core structural proteins VP3 and VP7 of African horse sickness virus, and production of core-like particles, *Arch. Virol. Suppl.,* 14, 203, 1998.

408. Moss, S. R. and Nuttall, P. A., Subcore- and core-like particles of Broadhaven virus (BRDV), a tick-borne orbivirus, synthesized from baculovirus expressed VP2 and VP7, the major core proteins of BRDV, *Virus Res.,* 32, 401, 1994.

409. Prasad, B. V. et al., Three-dimensional structure of single-shelled bluetongue virus, *J. Virol.,* 66, 2135, 1992.

410. Hewat, E. A. et al., Structure of bluetongue virus particles by cryoelectron microscopy, *J. Struct. Biol.,* 109, 61, 1992.

411. Hewat, E. A. et al., Three-dimensional reconstruction of baculovirus expressed bluetongue virus core-like particles by cryo-electron microscopy, *Virology,* 189, 10, 1992.

412. Hewat, E. A. et al., Structure of correctly self-assembled bluetongue virus-like particles, *J. Struct. Biol.,* 112, 183, 1994.

413. Basak, A. K. et al., Preliminary crystallographic study of bluetongue virus capsid protein, VP7, *J. Mol. Biol.,* 228, 687, 1992.

414. Burroughs, J. N. et al., Crystallization and preliminary X-ray analysis of the core particle of bluetongue virus, *Virology,* 210, 217, 1995.

415. Grimes, J. et al., The crystal structure of bluetongue virus VP7, *Nature,* 373, 167, 1995.

416. Basak, A. K. et al., Crystal structure of the top domain of African horse sickness virus VP7: comparisons with bluetongue virus VP7, *J. Virol.,* 70, 3797, 1996.

417. Basak, A. K. et al., Structures of orbivirus VP7: implications for the role of this protein in the viral life cycle, *Structure,* 5, 871, 1997.

418. Stuart, D. I. et al., Structural studies of orbivirus particles, *Arch. Virol. Suppl.,* 14, 235, 1998.

419. Hewat, E. A. et al., Three-dimensional reconstruction of bluetongue virus tubules using cryoelectron microscopy, *J. Struct. Biol.,* 108, 35, 1992.

420. Pearson, L. D. and Roy, P., Genetically engineered multi-component virus-like particles as veterinary vaccines, *Immunol. Cell Biol.,* 71, 381, 1993.

421. Roy, P. et al., Long-lasting protection of sheep against bluetongue challenge after vaccination with virus-like particles: evidence for homologous and partial heterologous protection, *Vaccine,* 12, 805, 1994.

422. Loudon, P. T. et al., Expression of the outer capsid protein VP5 of two bluetongue viruses, and synthesis of chimeric double-shelled virus-like particles using combinations of recombinant baculoviruses, *Virology,* 182, 793, 1991.

423. Le Blois, H. et al., Synthesis and characterization of chimeric particles between epizootic hemorrhagic disease virus and bluetongue virus: functional domains are conserved on the VP3 protein, *J. Virol.,* 65, 4821, 1991.

424. Tanaka, S. and Roy, P., Identification of domains in bluetongue virus VP3 molecules essential for the assembly of virus cores, *J. Virol.,* 68, 2795, 1994.

425. Belyaev, A. S. and Roy, P., Presentation of hepatitis B virus preS2 epitope on bluetongue virus core-like particles, *Virology,* 190, 840, 1992.

426. Le Blois, H. and Roy, P., A single point mutation in the VP7 major core protein of bluetongue virus prevents the formation of core-like particles, *J. Virol.,* 67, 353, 1993.

427. Wang, L. F. et al., Topography and immunogenicity of bluetongue virus VP7 epitopes, *Arch. Virol.,* 141, 111, 1996.

428. Tanaka, S. et al., Synthesis of bluetongue virus chimeric VP3 molecules and their interactions with VP7 protein to assemble into virus core-like particles, *Virology,* 214, 593, 1995.

429. Monastyrskaya, K. et al., Effects of domain-switching and site-directed mutagenesis on the properties and functions of the VP7 proteins of two orbiviruses, *Virology,* 237, 217, 1997.

430. Adler, S. et al., Induction of T cell response by bluetongue virus core-like particles expressing a T cell epitope of the M1 protein of influenza A virus, *Med. Microbiol. Immunol. (Berlin)*, 187, 91, 1998.

431. Monastyrskaya, K. et al., Mutation of either of two cysteine residues or deletion of the amino or carboxy terminus of nonstructural protein NS1 of bluetongue virus abrogates virus-specified tubule formation in insect cells, *J. Virol.*, 68, 2169, 1994.

432. Monastyrskaya, K. et al., Characterization and modification of the carboxy-terminal sequences of bluetongue virus type 10 NS1 protein in relation to tubule formation and location of an antigenic epitope in the vicinity of the carboxy terminus of the protein, *J. Virol.*, 69, 2831, 1995.

433. Mikhailov, M. et al., A new form of particulate single and multiple immunogen delivery system based on recombinant bluetongue virus-derived tubules, *Virology*, 217, 323, 1996.

434. Lawton, J. A. et al., Three-dimensional structural analysis of recombinant rotavirus-like particles with intact and amino-terminal-deleted VP2: implications for the architecture of the VP2 capsid layer, *J. Virol.*, 71, 7353, 1997.

435. Petitpas, I. et al., Crystallization and preliminary x-ray analysis of rotavirus protein VP6, *J. Virol.*, 72, 7615, 1998.

436. Mathieu, M. et al., Atomic structure of the major capsid protein of rotavirus: implications for the architecture of the virion, *EMBO J.*, 20, 1485, 2001.

437. Labbe, M. et al., Expression of rotavirus VP2 produces empty corelike particles, *J. Virol.*, 65, 2946, 1991.

438. Zeng, C. Q. et al., Characterization of rotavirus VP2 particles, *Virology*, 201, 55, 1994.

439. Tosser, G. et al., Expression of the major capsid protein VP6 of group C rotavirus and synthesis of chimeric single-shelled particles by using recombinant baculoviruses, *J. Virol.*, 66, 5825, 1992.

440. Crawford, S. E. et al., Characterization of virus-like particles produced by the expression of rotavirus capsid proteins in insect cells, *J. Virol.*, 68, 5945, 1994.

441. Redmond, M. J. et al., Assembly of recombinant rotavirus proteins into virus-like particles and assessment of vaccine potential, *Vaccine*, 11, 273, 1993.

442. Conner, M. E. et al., Rotavirus subunit vaccines, *Arch. Virol. Suppl.*, 12, 199, 1996.

443. Fernandez, F. M. et al., Isotype-specific antibody responses to rotavirus and virus proteins in cows inoculated with subunit vaccines composed of recombinant SA11 rotavirus core-like particles (CLP) or virus-like particles (VLP), *Vaccine*, 14, 1303, 1996.

444. Ciarlet, M. et al., Subunit rotavirus vaccine administered parenterally to rabbits induces active protective immunity, *J. Virol.*, 72, 9233, 1998.

445. O'Neal, C. M. et al., Rotavirus virus-like particles administered mucosally induce protective immunity, *J. Virol.*, 71, 8707, 1997.

446. O'Neal, C. M. et al., Rotavirus 2/6 viruslike particles administered intranasally with cholera toxin, *Escherichia coli* heat-labile toxin (LT), and LT-R192G induce protection from rotavirus challenge, *J. Virol.*, 72, 3390, 1998.

447. Fernandez, F. M. et al., Passive immunity to bovine rotavirus in newborn calves fed colostrum supplements from cows immunized with recombinant SA11 rotavirus core-like particle (CLP) or virus-like particle (VLP) vaccines, *Vaccine*, 16, 507, 1998.

448. Jiang, B. et al., Heterotypic protection from rotavirus infection in mice vaccinated with virus-like particles, *Vaccine*, 17, 1005, 1999.

449. Conner, M. E. et al., Virus-like particles as a rotavirus subunit vaccine, *J. Infect. Dis.*, 174 (Suppl. 1), S88, 1996.

450. Madore, H. P. et al., Biochemical and immunologic comparison of virus-like particles for a rotavirus subunit vaccine, *Vaccine*, 17, 2461, 1999.

451. Crawford, S. E. et al., Heterotypic protection and induction of a broad heterotypic neutralization response by rotavirus-like particles, *J. Virol.*, 73, 4813, 1999.

452. Redmond, M. J. et al., Rotavirus particles function as immunological carriers for the delivery of peptides from infectious agents and endogenous proteins, *Mol. Immunol.*, 28, 269, 1991.

453. Frenchick, P. J. et al., Biochemical and immunological characterization of a novel peptide carrier system using rotavirus VP6 particles, *Vaccine*, 10, 783, 1992.

454. Ijaz, M. K. et al., Priming and induction of anti-rotavirus antibody response by synthetic peptides derived from VP7 and VP4, *Vaccine*, 13, 331, 1995.

455. Both, G. W. et al., Relocation of antigens to the cell surface membrane can enhance immune stimulation and protection, *Immunol. Cell Biol.*, 70, 73, 1992.

456. Reddy, D. A. et al., Rotavirus VP6 modified for expression on the plasma membrane forms arrays and exhibits enhanced immunogenicity, *Virology,* 189, 423, 1992.

457. Estes, M. K. et al., Virus-like particle vaccines for mucosal immunization, *Adv. Exp. Med. Biol.,* 412, 387, 1997.

458. Affranchino, J. L. and Gonzalez, S. A., Deletion mapping of functional domains in the rotavirus capsid protein VP6, *J. Gen. Virol.,* 78, 1949, 1997.

459. Zeng, C. Q. et al., The N terminus of rotavirus VP2 is necessary for encapsidation of VP1 and VP3, *J. Virol.,* 72, 201, 1998.

460. Weldon, R. A. J. et al., Incorporation of chimeric Gag protein into retroviral particles, *J. Virol.,* 64, 4169, 1990.

461. Kakker, N. K. et al., Bovine leukemia virus Gag particle assembly in insect cells: formation of chimeric particles by domain-switched leukemia/lentivirus Gag polyprotein, *Virology,* 265, 308, 1999.

462. Takahashi, R. H. et al., Analysis of human lymphotropic T-cell virus type II-like particle production by recombinant baculovirus-infected insect cells, *Virology,* 256, 371, 1999.

463. Thomsen, D. R. et al., Expression of feline leukaemia virus gp85 and Gag proteins and assembly into virus-like particles using the baculovirus expression vector system, *J. Gen. Virol.,* 73, 1819, 1992.

464. Campbell, S. and Vogt, V. M., *In vitro* assembly of virus-like particles with Rous sarcoma virus Gag deletion mutants: identification of the p10 domain as a morphological determinant in the formation of spherical particles, *J. Virol.,* 71, 4425, 1997.

465. Khorasanizadeh, S. et al., Solution structure of the capsid protein from the human T-cell leukemia virus type-I, *J. Mol. Biol.,* 291, 491, 1999.

466. Campos-Olivas, R. et al., Solution structure and dynamics of the Rous sarcoma virus capsid protein and comparison with capsid proteins of other retroviruses, *J. Mol. Biol.,* 296, 633, 2000.

467. Deminie, C. A. and Emerman, M., Incorporation of human immunodeficiency virus type 1 Gag proteins into murine leukemia virus virions, *J. Virol.,* 67, 6499, 1993.

468. Berkowitz, R. D. et al., Retroviral nucleocapsid domains mediate the specific recognition of genomic viral RNAs by chimeric Gag polyproteins during RNA packaging *in vivo, J. Virol.,* 69, 6445, 1995.

469. Zhang, Y. and Barklis, E., Nucleocapsid protein effects on the specificity of retrovirus RNA encapsidation, *J. Virol.,* 69, 5716, 1995.

470. Campbell, S. and Vogt, V. M., Self-assembly *in vitro* of purified CA-NC proteins from Rous sarcoma virus and human immunodeficiency virus type 1, *J. Virol.,* 69, 6487, 1995.

471. Bowzard, J. B. et al., Importance of basic residues in the nucleocapsid sequence for retrovirus Gag assembly and complementation rescue, *J. Virol.,* 72, 9034, 1998.

472. Bennett, R. P. and Wills, J. W., Conditions for copackaging rous sarcoma virus and murine leukemia virus Gag proteins during retroviral budding, *J. Virol.,* 73, 2045, 1999.

473. Freed, E. O., HIV-1 Gag proteins: diverse functions in the virus life cycle, *Virology,* 251, 1, 1998.

474. Delchambre, M. et al., The Gag precursor of simian immunodeficiency virus assembles into virus-like particles, *EMBO J.,* 8, 2653, 1989.

475. Gonzalez, S. A. et al., Assembly of the matrix protein of simian immunodeficiency virus into virus-like particles, *Virology,* 194, 548, 1993.

476. Yamshchikov, G. V. et al., Assembly of SIV virus-like particles containing envelope proteins using a baculovirus expression system, *Virology,* 214, 50, 1995.

477. Gheysen, D. et al., Assembly and release of HIV-1 precursor Pr55gag virus-like particles from recombinant baculovirus-infected insect cells, *Cell,* 59, 103, 1989.

478. Wagner, R. et al., Expression of HIV-1 autologous p55 and p55/gp120-V3 core particles: a new approach in HIV vaccine development, in *Vaccines 91,* Chanock, R. M. et al., Eds., Cold Spring Harbor Laboratory Press, Cold Spring Harbor, NY, 1991, pp. 109–114.

479. Wagner, R. et al., Studies on processing, particle formation, and immunogenicity of the HIV-1 *gag* gene product: a possible component of a HIV vaccine, *Arch. Virol.,* 127, 117, 1992.

480. Hughes, B. P. et al., Morphogenic capabilities of human immunodeficiency virus type 1 Gag and Gag-Pol proteins in insect cells, *Virology,* 193, 242, 1993.

481. Luo, L. et al., Expression of Gag precursor protein and secretion of virus-like Gag particles of HIV-2 from recombinant baculovirus-infected insect cells, *Virology,* 179, 874, 1990.

482. Rasmussen, L. et al., Characterization of virus-like particles produced by a recombinant baculovirus containing the *gag* gene of the bovine immunodeficiency-like virus, *Virology,* 178, 435, 1990.

483. Morikawa, S. et al., Analyses of the requirements for the synthesis of virus-like particles by feline immunodeficiency virus Gag using baculovirus vectors, *Virology,* 183, 288, 1991.

484. Gonzalez, S. A. et al., Identification of domains in the simian immunodeficiency virus matrix protein essential for assembly and envelope glycoprotein incorporation, *J. Virol.,* 70, 6384, 1996.

485. Giddings, A. M. et al., The matrix protein of HIV-1 is not sufficient for assembly and release of virus-like particles, *Virology,* 248, 108, 1998.

486. Tobin, G. J. et al., Immunologic and ultrastructural characterization of HIV pseudovirions containing Gag and Env precursor proteins engineered in insect cells, *Methods,* 10, 208, 1996.

487. Karacostas, V. et al., Human immunodeficiency virus-like particles produced by a vaccinia virus expression vector, *Proc. Natl. Acad. Sci. U.S.A.,* 86, 8964, 1989.

488. Voss, G. et al., Morphogenesis of recombinant HIV-2 Gag core particles, *Virus Res.,* 24, 197, 1992.

489. Smith, A. J. et al., Human immunodeficiency virus type 1 Pr55gag and Pr160gag-pol expressed from a simian virus 40 late replacement vector are efficiently processed and assembled into viruslike particles, *J. Virol.,* 64, 2743, 1990.

490. Smith, A. J. et al., Requirements for incorporation of Pr160gag-pol from human immunodeficiency virus type 1 into virus-like particles, *J. Virol.,* 67, 2266, 1993.

491. Vernon, S. K. et al., Ultrastructural characterization of human immunodeficiency virus type 1 Gag-containing particles assembled in a recombinant adenovirus vector system, *J. Gen. Virol.,* 72, 1243, 1991.

492. Gross, I. et al., *In vitro* assembly properties of purified bacterially expressed capsid proteins of human immunodeficiency virus, *Eur. J. Biochem.,* 249, 592, 1997.

493. Gross, I. et al., N-Terminal extension of human immunodeficiency virus capsid protein converts the *in vitro* assembly phenotype from tubular to spherical particles, *J. Virol.,* 72, 4798, 1998.

494. Morikawa, Y. et al., *In vitro* assembly of human immunodeficiency virus type 1 Gag protein, *J. Biol. Chem.,* 274, 27997, 1999.

495. Morikawa, Y. et al., Roles of matrix, p2, and N-terminal myristoylation in human immunodeficiency virus type 1 Gag assembly, *J. Virol.,* 74, 16, 2000.

496. Nermut, M. V. et al., Further evidence for hexagonal organization of HIV Gag protein in prebudding assemblies and immature virus-like particles, *J. Struct. Biol.,* 123, 143, 1998.

497. Nermut, M. V. et al., Fullerene-like organization of HIV Gag-protein shell in virus-like particles produced by recombinant baculovirus, *Virology,* 198, 288, 1994.

498. Fuller, S. D. et al., Cryo-electron microscopy reveals ordered domains in the immature HIV-1 particle, *Curr. Biol.,* 7, 729, 1997.

499. Massiah, M. A. et al., Three-dimensional structure of the human immunodeficiency virus type 1 matrix protein, *J. Mol. Biol.,* 244, 198, 1994.

500. Momany, C. et al., Crystal structure of dimeric HIV-1 capsid protein, *Nat. Struct. Biol.,* 3, 763, 1996.

501. Gitti, R. K. et al., Structure of the amino-terminal core domain of the HIV-1 capsid protein, *Science,* 273, 231, 1996.

502. Gamble, T. R. et al., Structure of the carboxyl-terminal dimerization domain of the HIV-1 capsid protein, *Science,* 278, 849, 1997.

503. De Guzman, R. N. et al., Structure of the HIV-1 nucleocapsid protein bound to the SL3 psi-RNA recognition element, *Science,* 279, 384, 1998.

504. Summers, M. F. et al., High-resolution structure of an HIV zinc fingerlike domain via a new NMR-based distance geometry approach, *Biochemistry,* 29, 329, 1990.

505. Darlix, J. L. et al., First glimpses at structure-function relationships of the nucleocapsid protein of retroviruses, *J. Mol. Biol.,* 254, 523, 1995.

506. Roques, B. P. et al., Structure, biological functions and inhibition of the HIV-1 proteins Vpr and NCp7, *Biochimie,* 79, 673, 1997.

507. Zhang, W. H. et al., Functional consequences of mutations in HIV-1 Gag p55 selected by CTL pressure, *Virology,* 203, 101, 1994.

508. Wagner, R. et al., Correlates of protection, antigen delivery and molecular epidemiology: basics for designing an HIV vaccine, *Vaccine,* 17, 1706, 1999.

509. Kelleher, A. D. et al., Safety and immunogenicity of a candidate therapeutic vaccine, p24 virus-like particle, combined with zidovudine, in asymptomatic subjects. Community HIV Research Network Investigators, *AIDS,* 12, 175, 1998.

510. Benson, E. M. et al., Therapeutic vaccination with p24-VLP and zidovudine augments HIV-specific cytotoxic T lymphocyte activity in asymptomatic HIV-infected individuals, *AIDS Res. Hum. Retroviruses,* 15, 105, 1999.

511. Von Poblotzki, A. et al., Identification of a region in the Pr55gag-polyprotein essential for HIV-1 particle formation, *Virology,* 193, 981, 1993.

512. Wagner, R. et al., Assembly and extracellular release of chimeric HIV-1 Pr55gag retrovirus-like particles, *Virology,* 200, 162, 1994.

513. Carriere, C. et al., Sequence requirements for encapsidation of deletion mutants and chimeras of human immunodeficiency virus type 1 Gag precursor into retrovirus-like particles, *J. Virol.,* 69, 2366, 1995.

514. Kattenbeck, B. et al., Inhibition of human immunodeficiency virus type 1 particle formation by alterations of defined amino acids within the C terminus of the capsid protein, *J. Gen. Virol.,* 78, 2489, 1997.

515. Luo, L. et al., Mapping of functional domains for HIV-2 Gag assembly into virus-like particles, *Virology,* 205, 496, 1994.

516. Zhang, Y. and Barklis, E., Effects of nucleocapsid mutations on human immunodeficiency virus assembly and RNA encapsidation, *J. Virol.,* 71, 6765, 1997.

517. Zhang, Y. et al., Analysis of the assembly function of the human immunodeficiency virus type 1 Gag protein nucleocapsid domain, *J. Virol.,* 72, 1782, 1998.

518. Garnier, L. et al., Particle size determinants in the human immunodeficiency virus type 1 Gag protein, *J. Virol.,* 72, 4667, 1998.

519. Wang, C. T. et al., Analysis of minimal human immunodeficiency virus type 1 Gag coding sequences capable of virus-like particle assembly and release, *J. Virol.,* 72, 7950, 1998.

520. Luo, L. et al., Chimeric Gag-V3 virus-like particles of human immunodeficiency virus induce virus-neutralizing antibodies, *Proc. Natl. Acad. Sci. U.S.A.,* 89, 10527, 1992.

521. Griffiths, J. C. et al., Hybrid human immunodeficiency virus Gag particles as an antigen carrier system: induction of cytotoxic T-cell and humoral responses by a Gag:V3 fusion, *J. Virol.,* 67, 3191, 1993.

522. Wagner, R. et al., Induction of cytolytic T lymphocytes directed towards the V3 loop of the human immunodeficiency virus type 1 external glycoprotein gp120 by p55gag/V3 chimeric vaccinia viruses, *J. Gen. Virol.,* 74, 1261, 1993.

523. Wagner, R. et al., Induction of a MHC class I-restricted, CD8 positive cytolytic T-cell response by chimeric HIV-1 virus-like particles *in vivo*: implications on HIV vaccine development, *Behring Inst. Mitt.,* 95, 23, 1994.

524. Wagner, R. et al., Construction, expression, and immunogenicity of chimeric HIV-1 virus-like particles, *Virology,* 220, 128, 1996.

525. Schirmbeck, R. et al., Priming of class I-restricted cytotoxic T lymphocytes by vaccination with recombinant protein antigens, *Vaccine,* 13, 857, 1995.

526. Brand, D. et al., A simple procedure to generate chimeric Pr55gag virus-like particles expressing the principal neutralization domain of human immunodeficiency virus type 1, *J. Virol. Methods,* 51, 153, 1995.

527. Truong, C. et al., Assembly and immunogenicity of chimeric Gag-Env proteins derived from the human immunodeficiency virus type 1, *AIDS Res. Hum. Retroviruses,* 12, 291, 1996.

528. Wagner, R. et al., Safety and immunogenicity of recombinant human immunodeficiency virus-like particles in rodents and rhesus macaques, *Intervirology,* 39, 93, 1996.

529. Luo, L. et al., Induction of V3-specific cytotoxic T lymphocyte responses by HIV Gag particles carrying multiple immunodominant V3 epitopes of gp120, *Virology,* 240, 316, 1998.

530. Kang, C. Y. et al., Development of HIV/AIDS vaccine using chimeric Gag-Env virus-like particles, *Biol. Chem.,* 380, 353, 1999.

531. Rovinski, B. et al., Expression and characterization of genetically engineered human immunodeficiency virus-like particles containing modified envelope glycoproteins: implications for development of a cross-protective AIDS vaccine, *J. Virol.,* 66, 4003, 1992.

532. Deml, L. et al., Increased incorporation of chimeric human immunodeficiency virus type 1 gp120 proteins into Pr55gag virus-like particles by an Epstein-Barr virus gp220/350-derived transmembrane domain, *Virology,* 235, 10, 1997.

533. Deml, L. et al., Recombinant human immunodeficiency Pr55gag virus-like particles presenting chimeric envelope glycoproteins induce cytotoxic T-cells and neutralizing antibodies, *Virology,* 235, 26, 1997.

534. Tobin, G. J. et al., Synthesis and assembly of chimeric human immunodeficiency virus Gag pseudo-virions, *Intervirology,* 39, 40, 1996.

535. Tobin, G. J. et al., Chimeric HIV-1 virus-like particles containing gp120 epitopes as a result of a ribosomal frameshift elicit Gag- and SU-specific murine cytotoxic T-lymphocyte activities, *Virology,* 236, 307, 1997.

536. Wagner, R. et al., Cytotoxic T cells and neutralizing antibodies induced in rhesus monkeys by virus-like particle HIV vaccines in the absence of protection from SHIV infection, *Virology,* 245, 65, 1998.

537. Notka, F. et al., Construction and characterization of recombinant VLPs and Semliki-Forest virus live vectors for comparative evaluation in the SHIV monkey model, *Biol. Chem.,* 380, 341, 1999.

538. Garnier, L. et al., Incorporation of pseudorabies virus gD into human immunodeficiency virus type 1 Gag particles produced in baculovirus-infected cells, *J. Virol.,* 69, 4060, 1995.

539. Henriksson, P. et al., Incorporation of wild-type and C-terminally truncated human epidermal growth factor receptor into human immunodeficiency virus-like particles: insight into the processes governing glycoprotein incorporation into retroviral particles, *J. Virol.,* 73, 9294, 1999.

540. Mellor, J. et al., A retrovirus-like strategy for expression of a fusion protein encoded by yeast transposon Ty1, *Nature,* 313, 243, 1985.

541. Mellor, J. et al., Reverse transcriptase activity and Ty RNA are associated with virus-like particles in yeast, *Nature,* 318, 583, 1985.

542. Wilson, W. et al., Expression strategies of the yeast retrotransposon Ty: a short sequence directs ribosomal frameshifting, *Nucleic Acids Res.,* 14, 7001, 1986.

543. Garfinkel, D. J. et al., Ty element transposition: reverse transcriptase and virus-like particles, *Cell,* 42, 507, 1985.

544. Adams, S. E. et al., The functions and relationships of Ty-VLP proteins in yeast reflect those of mammalian retroviral proteins, *Cell,* 49, 111, 1987.

545. Kingsman, A. J. et al., The genetic organization of the yeast Ty element, *J. Cell Sci. Suppl.,* 7:155–67, 155, 1987.

546. Burns, N. R. et al., Symmetry, flexibility and permeability in the structure of yeast retrotransposon virus-like particles, *EMBO J.,* 11, 1155, 1992.

547. Palmer, K. J. et al., Cryo-electron microscopy structure of yeast Ty retrotransposon virus-like particles, *J. Virol.,* 71, 6863, 1997.

548. Al-Khayat, H. A. et al., Yeast Ty retrotransposons assemble into virus-like particles whose T-numbers depend on the C-terminal length of the capsid protein, *J. Mol. Biol.,* 292, 65, 1999.

549. Luschnig, C. et al., The Gag homologue of retrotransposon Ty1 assembles into spherical particles in *Escherichia coli, Eur. J. Biochem.,* 228, 739, 1995.

550. Kingsman, S. M. and Kingsman, A. J., Polyvalent recombinant antigens: a new vaccine strategy, *Vaccine,* 6, 304, 1988.

551. Adams, S. E. et al., Hybrid Ty virus-like particles, *Int. Rev. Immunol.,* 11, 133, 1994.

552. Adams, S. E. et al., Expression vectors for the construction of hybrid Ty-VLPs, *Mol. Biotechnol.,* 1, 125, 1994.

553. Burns, N. R. et al., Production and purification of hybrid Ty-VLPs, *Mol. Biotechnol.,* 1, 137, 1994.

554. Kingsman, A. J. et al., Yeast retrotransposon particles as antigen delivery systems, *Ann. N.Y. Acad. Sci.,* 754, 202, 1995.

555. Brookman, J. L. et al., Analysis of TYA protein regions necessary for formation of the Ty1 virus-like particle structure, *Virology,* 212, 69, 1995.

556. Martin-Rendon, E. et al., Structural determinants within the subunit protein of Ty1 virus-like particles, *Mol. Microbiol.,* 22, 667, 1996.

557. Brachmann, C. B. and Boeke, J. D., Mapping the multimerization domains of the Gag protein of yeast retrotransposon Ty1, *J. Virol.,* 71, 812, 1997.

558. Brookman, J. L. et al., An immunological analysis of Ty1 virus-like particle structure, *Virology,* 207, 59, 1995.

559. Griffiths, J. C. et al., Induction of high-titer neutralizing antibodies, using hybrid human immunode-ficiency virus V3-Ty viruslike particles in a clinically relevant adjuvant, *J. Virol.,* 65, 450, 1991.

560. Layton, G. T. et al., Induction of HIV-specific cytotoxic T lymphocytes *in vivo* with hybrid HIV-1 V3:Ty-virus-like particles, *J. Immunol.,* 151, 1097, 1993.

561. Harris, S. J. et al., The effects of adjuvants on CTL induction by V3:Ty-virus-like particles (V3-VLPs) in mice, *Vaccine,* 14, 971, 1996.

562. Gilmour, J. E. et al., A novel method for the purification of HIV-1 p24 protein from hybrid Ty virus-like particles (Ty-VLPs), *AIDS,* 3, 717, 1989.

563. Gilmour, J. E. et al., Performance characteristics of a novel immunoassay based on hybrid Ty virus-like particles (Ty-VLPs): rapid differentiation between HIV-1 and HIV-2 infection, *AIDS,* 4, 967, 1990.

564. Mills, K. H. et al., HIV p24-specific helper T cell clones from immunized primates recognize highly conserved regions of HIV-1, *J. Immunol.,* 144, 1677, 1990.

565. Martin, S. J. et al., Immunization of human HIV-seronegative volunteers with recombinant p17/p24:Ty virus-like particles elicits HIV-1 p24-specific cellular and humoral immune responses, *AIDS,* 7, 1315, 1993.

566. Weber, J. et al., Immunogenicity of the yeast recombinant p17/p24:Ty virus-like particles (p24-VLP) in healthy volunteers, *Vaccine,* 13, 831, 1995.

567. Veenstra, J. et al., Immunization with recombinant p17/p24:Ty virus-like particles in human immunodeficiency virus-infected persons, *J. Infect. Dis.,* 174, 862, 1996.

568. Klein, M. R. et al., Gag-specific immune responses after immunization with p17/p24:Ty virus-like particles in HIV type 1-seropositive individuals, *AIDS Res. Hum. Retroviruses,* 13, 393, 1997.

569. Peters, B. S. et al., A pilot phase II study of the safety and immunogenicity of HIV p17/p24:VLP (p24-VLP) in asymptomatic HIV seropositive subjects, *J. Infect.,* 35, 231, 1997.

570. Burns, N. R. et al., Purification and secondary structure determination of simian immunodeficiency virus p27, *J. Mol. Biol.,* 216, 207, 1990.

571. Lehner, T. et al., T- and B-cell functions and epitope expression in nonhuman primates immunized with simian immunodeficiency virus antigen by the rectal route, *Proc. Natl. Acad. Sci. U.S.A.,* 90, 8638, 1993.

572. Lehner, T. et al., Mucosal model of genital immunization in male rhesus macaques with a recombinant simian immunodeficiency virus p27 antigen, *J. Virol.,* 68, 1624, 1994.

573. Lehner, T. et al., Targeted lymph node immunization with simian immunodeficiency virus p27 antigen to elicit genital, rectal, and urinary immune responses in nonhuman primates, *J. Immunol.,* 153, 1858, 1994.

574. Lehner, T. et al., Genital-associated lymphoid tissue in female non-human primates, *Adv. Exp. Med. Biol.,* 371A, 357, 1995.

575. Klavinskis, L. S. et al., Mucosal or targeted lymph node immunization of macaques with a particulate SIVp27 protein elicits virus-specific CTL in the genito-rectal mucosa and draining lymph nodes, *J. Immunol.,* 157, 2521, 1996.

576. Layton, G. T. et al., Induction of single and dual cytotoxic T-lymphocyte responses to viral proteins in mice using recombinant hybrid Ty-virus-like particles, *Immunology,* 87, 171, 1996.

577. Reyburn, H. T. et al., Expression of maedi-visna virus major core protein, p25: development of a sensitive p25 antigen detection assay, *J. Virol. Methods,* 37, 305, 1992.

578. Carey, N. et al., Use of recombinant gp135 to study epitope-specific antibody responses to maedi visna virus, *J. Virol. Methods,* 43, 221, 1993.

579. Lindqvist, C. et al., Immunoaffinity purification of baculovirus-expressed rubella virus E1 for diagnostic purposes, *J. Clin. Microbiol.,* 32, 2192, 1994.

580. Fowler, W. J. et al., Identification of immunodominant regions and linear B cell epitopes of the gE envelope protein of varicella-zoster virus, *Virology,* 214, 531, 1995.

581. Garcia-Valcarcel, M. et al., Induction of neutralizing antibody and T-cell responses to varicella-zoster virus (VZV) using Ty-virus-like particles carrying fragments of glycoprotein E (gE), *Vaccine,* 15, 709, 1997.

582. Garcia-Valcarcel, M. et al., Cloning, expression, and immunogenicity of the assembly protein of varicella-zoster virus and detection of the products of open reading frame 33, *J. Med. Virol.,* 53, 332, 1997.

583. Welsh, M. D. et al., Ability of yeast Ty-VLPs (virus-like particles) containing varicella-zoster virus (VZV)gE and assembly protein fragments to induce *in vitro* proliferation of human lymphocytes from VZV immune patients, *J. Med. Virol.,* 59, 78, 1999.

584. Allsopp, C. E. et al., Comparison of numerous delivery systems for the induction of cytotoxic T lymphocytes by immunization, *Eur. J. Immunol.,* 26, 1951, 1996.

585. Gilbert, S. C. et al., A protein particle vaccine containing multiple malaria epitopes, *Nat. Biotechnol.*, 15, 1280, 1997.

586. Plebanski, M. et al., Protection from *Plasmodium berghei* infection by priming and boosting T cells to a single class I-restricted epitope with recombinant carriers suitable for human use, *Eur. J. Immunol.*, 28, 4345, 1998.

587. Gilbert, S. C. et al., Ty virus-like particles, DNA vaccines and Modified Vaccinia Virus Ankara; comparisons and combinations, *Biol. Chem.*, 380, 299, 1999.

588. Harris, S. J. et al., Prediction of murine MHC class I epitopes in a major house dust mite allergen and induction of T1-type CD8+ T cell responses, *Int. Immunol.*, 9, 273, 1997.

589. Hirschberg, S. et al., CD4(+) T cells induced by virus-like particles expressing a major T cell epitope down-regulate IL-5 production in an ongoing immune response to Der p 1 independently of IFN-gamma production, *Int. Immunol.*, 11, 1927, 1999.

590. Malim, M. H. et al., The production of hybrid Ty:IFN virus-like particles in yeast, *Nucleic Acids Res.*, 15, 7571, 1987.

591. Fiskerstrand, C. E. et al., Cloning, expression and characterization of ovine interleukins 1 alpha and beta, *Cytokine,* 4, 418, 1992.

592. Green, I. R. et al., Expression and characterization of bioactive recombinant ovine TNF-alpha: some species specificity in cytotoxic response to TNF, *Cytokine,* 5, 213, 1993.

593. Bujdoso, R. et al., Molecular cloning and expression of DNA encoding ovine interleukin 2, *Cytokine,* 7, 223, 1995.

594. Natsoulis, G. and Boeke, J. D., New antiviral strategy using capsid-nuclease fusion proteins, *Nature,* 352, 632, 1991.

595. Fassbender, S. et al., Reverse transcriptase activity of an intron encoded polypeptide, *EMBO J.,* 13, 2075, 1994.

596. Poggeler, S. et al., Efficient synthesis of a 72-kDa mitochondrial polypeptide using the yeast Ty expression system, *Biochem. Biophys. Res. Commun.,* 219, 890, 1996.

597. Casal, J. I. et al., Parvovirus-like particles as vaccine vectors, *Methods,* 19, 174, 1999.

598. Casal, J. I., Use of parvovirus-like particles for vaccination and induction of multiple immune responses, *Biotechnol. Appl. Biochem.,* 29, 141, 1999.

599. Bueler, H., Adeno-associated viral vectors for gene transfer and gene therapy, *Biol. Chem.,* 380, 613, 1999.

600. Grimm, D. and Kleinschmidt, J. A., Progress in adeno-associated virus type 2 vector production: promises and prospects for clinical use, *Hum. Gene Ther.,* 10, 2445, 1999.

601. Kajigaya, S. et al., A genetically engineered cell line that produces empty capsids of B19 (human) parvovirus, *Proc. Natl. Acad. Sci. U.S.A.,* 86, 7601, 1989.

602. Brown, C. S. et al., Antigenic parvovirus B19 coat proteins VP1 and VP2 produced in large quantities in a baculovirus expression system, *Virus Res.,* 15, 197, 1990.

603. Brown, C. S. et al., Assembly of empty capsids by using baculovirus recombinants expressing human parvovirus B19 structural proteins, *J. Virol.,* 65, 2702, 1991.

604. Kajigaya, S. et al., Self-assembled B19 parvovirus capsids, produced in a baculovirus system, are antigenically and immunogenically similar to native virions, *Proc. Natl. Acad. Sci. U.S.A.,* 88, 4646, 1991.

605. Salimans, M. M. et al., Recombinant parvovirus B19 capsids as a new substrate for detection of B19-specific IgG and IgM antibodies by an enzyme-linked immunosorbent assay, *J. Virol. Methods,* 39, 247, 1992.

606. Martinez, C. et al., Production of porcine parvovirus empty capsids with high immunogenic activity, *Vaccine,* 10, 684, 1992.

607. Saliki, J. T. et al., Canine parvovirus empty capsids produced by expression in a baculovirus vector: use in analysis of viral properties and immunization of dogs, *J. Gen. Virol.,* 73, 369, 1992.

608. Ruffing, M. et al., Assembly of viruslike particles by recombinant structural proteins of adeno-associated virus type 2 in insect cells, *J. Virol.,* 66, 6922, 1992.

609. Christensen, J. et al., Production of mink enteritis parvovirus empty capsids by expression in a baculovirus vector system: a recombinant vaccine for mink enteritis parvovirus in mink, *J. Gen. Virol.,* 75, 149, 1994.

610. Le Gall-Recule, G. et al., Expression of Muscovy duck parvovirus capsid proteins (VP2 and VP3) in a baculovirus expression system and demonstration of immunity induced by the recombinant proteins, *J. Gen. Virol.,* 77, 2159, 1996.

611. Hernando, E. et al., Biochemical and physical characterization of parvovirus minute virus of mice virus-like particles, *Virology,* 267, 299, 2000.

612. Clemens, D. L. et al., Expression of Aleutian mink disease parvovirus capsid proteins by a recombinant vaccinia virus: self-assembly of capsid proteins into particles, *J. Virol.,* 66, 3077, 1992.

613. Choi, J. Y. et al., High-level expression of canine parvovirus VP2 using Bombyx mori nucleopoly-hedrovirus vector, *Arch. Virol.,* 145, 171, 2000.

614. Rayment, F. B. et al., The production of human parvovirus capsid proteins in *Escherichia coli* and their potential as diagnostic antigens, *J. Gen. Virol.,* 71, 2665, 1990.

615. Luo, M. et al., Preliminary x-ray crystallographic analysis of canine parvovirus crystals, *J. Mol. Biol.,* 200, 209, 1988.

616. Tsao, J. et al., The three-dimensional structure of canine parvovirus and its functional implications, *Science,* 251, 1456, 1991.

617. Wu, H. and Rossmann, M. G., The canine parvovirus empty capsid structure, *J. Mol. Biol.,* 233, 231, 1993.

618. Xie, Q. and Chapman, M. S., Canine parvovirus capsid structure, analyzed at 2.9 Å resolution, *J. Mol. Biol.,* 264, 497, 1996.

619. Agbandje, M. et al., Preliminary X-ray crystallographic investigation of human parvovirus B19, *Virology,* 184, 170, 1991.

620. Agbandje, M. et al., Structure determination of feline panleukopenia virus empty particles, *Proteins,* 16, 155, 1993.

621. Agbandje, M. et al., The structure of human parvovirus B19 at 8 Å resolution, *Virology,* 203, 106, 1994.

622. Chipman, P. R. et al., Cryo-electron microscopy studies of empty capsids of human parvovirus B19 complexed with its cellular receptor, *Proc. Natl. Acad. Sci. U.S.A.,* 93, 7502, 1996.

623. Agbandje-McKenna, M. et al., Functional implications of the structure of the murine parvovirus, minute virus of mice, *Structure.,* 6, 1369, 1998.

624. McKenna, R. et al., Three-dimensional structure of Aleutian mink disease parvovirus: implications for disease pathogenicity, *J. Virol.,* 73, 6882, 1999.

625. Simpson, A. A. et al., The structure of an insect parvovirus (Galleria mellonella densovirus) at 3.7 Å resolution, *Structure,* 6, 1355, 1998.

626. Rosenfeld, S. J. et al., Unique region of the minor capsid protein of human parvovirus B19 is exposed on the virion surface, *J. Clin. Invest.,* 89, 2023, 1992.

627. Tullis, G. E. et al., The minor capsid protein VP1 of the autonomous parvovirus minute virus of mice is dispensable for encapsidation of progeny single-stranded DNA but is required for infectivity, *J. Virol.,* 67, 131, 1993.

628. Bansal, G. P. et al., Candidate recombinant vaccine for human B19 parvovirus, *J. Infect. Dis.,* 167, 1034, 1993.

629. Rosenfeld, S. J. et al., Subunit interaction in B19 parvovirus empty capsids, *Arch. Virol.,* 136, 9, 1994.

630. Saikawa, T. et al., Neutralizing linear epitopes of B19 parvovirus cluster in the VP1 unique and VP1-VP2 junction regions, *J. Virol.,* 67, 3004, 1993.

631. Casal, J. I. et al., Peptide vaccine against canine parvovirus: identification of two neutralization subsites in the N terminus of VP2 and optimization of the amino acid sequence, *J. Virol.,* 69, 7274, 1995.

632. Lopez, D. T. J. et al., Recombinant vaccine for canine parvovirus in dogs, *J. Virol.,* 66, 2748, 1992.

633. Wong, S. et al., Formation of empty B19 parvovirus capsids by the truncated minor capsid protein, *J. Virol.,* 68, 4690, 1994.

634. Kawase, M. et al., Most of the VP1 unique region of B19 parvovirus is on the capsid surface, *Virology,* 211, 359, 1995.

635. Kawase, M. et al., Modest truncation of the major capsid protein abrogates B19 parvovirus capsid formation, *J. Virol.,* 69, 6567, 1995.

636. Hurtado, A. et al., Identification of domains in canine parvovirus VP2 essential for the assembly of virus-like particles, *J. Virol.,* 70, 5422, 1996.

637. Brown, C. S. et al., Chimeric parvovirus B19 capsids for the presentation of foreign epitopes, *Virology,* 198, 477, 1994.

638. Miyamura, K. et al., Parvovirus particles as platforms for protein presentation, *Proc. Natl. Acad. Sci. U.S.A.,* 91, 8507, 1994.

639. Sedlik, C. et al., Immunogenicity of poliovirus B and T cell epitopes presented by hybrid porcine parvovirus particles, *J. Gen. Virol.,* 76, 2361, 1995.

640. Sedlik, C. et al., Recombinant parvovirus-like particles as an antigen carrier: a novel nonreplicative exogenous antigen to elicit protective antiviral cytotoxic T cells, *Proc. Natl. Acad. Sci. U.S.A.,* 94, 7503, 1997.

641. Lo-Man, R. et al., A recombinant virus-like particle system derived from parvovirus as an efficient antigen carrier to elicit a polarized Th1 immune response without adjuvant, *Eur. J. Immunol.,* 28, 1401, 1998.

642. Sedlik, C. et al., Intranasal delivery of recombinant parvovirus-like particles elicits cytotoxic T-cell and neutralizing antibody responses, *J. Virol.,* 73, 2739, 1999.

643. Rueda, P. et al., Minor displacements in the insertion site provoke major differences in the induction of antibody responses by chimeric parvovirus-like particles, *Virology,* 263, 89, 1999.

644. Rueda, P. et al., Engineering parvovirus-like particles for the induction of B-cell, CD4(+) and CTL responses, *Vaccine,* 18, 325, 1999.

645. Hoque, M. et al., Chimeric virus-like particle formation of adeno-associated virus, *Biochem. Biophys. Res. Commun.,* 266, 371, 1999.

646. Spehner, D. et al., Assembly of nucleocapsidlike structures in animal cells infected with a vaccinia virus recombinant encoding the measles virus nucleoprotein, *J. Virol.,* 65, 6296, 1991.

647. Fooks, A. R. et al., High-level expression of the measles virus nucleocapsid protein by using a replication-deficient adenovirus vector: induction of an MHC-1-restricted CTL response and protection in a murine model, *Virology,* 210, 456, 1995.

648. Fooks, A. R. et al., Measles virus nucleocapsid protein expressed in insect cells assembles into nucleocapsid-like structures, *J. Gen. Virol.,* 74, 1439, 1993.

649. Warnes, A. et al., Production of measles nucleoprotein in different expression systems and its use as a diagnostic reagent, *J. Virol. Methods,* 49, 257, 1994.

650. Warnes, A. et al., Expression of the measles virus nucleoprotein gene in *Escherichia coli* and assembly of nucleocapsid-like structures, *Gene,* 160, 173, 1995.

651. Sasnauskas, K., personal communication, 2001.

652. Coronel, E. C. et al., Human parainfluenza virus type 1 matrix and nucleoprotein genes transiently expressed in mammalian cells induce the release of virus-like particles containing nucleocapsid-like structures, *J. Virol.,* 73, 7035, 1999.

653. Buchholz, C. J. et al., The conserved N-terminal region of Sendai virus nucleocapsid protein NP is required for nucleocapsid assembly, *J. Virol.,* 67, 5803, 1993.

654. Myers, T. M. et al., A highly conserved region of the Sendai virus nucleocapsid protein contributes to the NP-NP binding domain, *Virology,* 229, 322, 1997.

655. Errington, W. and Emmerson, P. T., Assembly of recombinant Newcastle disease virus nucleocapsid protein into nucleocapsid-like structures is inhibited by the phosphoprotein, *J. Gen. Virol.,* 78, 2335, 1997.

656. Samuel, D. et al., High level expression of recombinant mumps nucleoprotein in *Saccharomyces cerevisiae* and its evaluation in mumps IgM serology, *J. Med. Virol.,* 66, 123, 2002.

657. Meric, C. et al., Respiratory syncytial virus nucleocapsid protein (N) expressed in insect cells forms nucleocapsid-like structures, *Virus Res.,* 31, 187, 1994.

658. Pinto, R. M. et al., Structures associated with the expression of rabies virus structural genes in insect cells, *Virus Res.,* 31, 139, 1994.

659. Iseni, F. et al., Characterization of rabies virus nucleocapsids and recombinant nucleocapsid-like structures, *J. Gen. Virol.,* 79, 2909, 1998.

660. Porta, C. and Lomonossoff, G. P., Scope for using plant viruses to present epitopes from animal pathogens, *Rev. Med. Virol.,* 8, 25, 1998.

661. Zhao, X. et al., *In vitro* assembly of cowpea chlorotic mottle virus from coat protein expressed in *Escherichia coli* and *in vitro*-transcribed viral cDNA, *Virology,* 207, 486, 1995.

662. Yusibov, V. et al., Purification, characterization, assembly and crystallization of assembled alfalfa mosaic virus coat protein expressed in *Escherichia coli, J. Gen. Virol.,* 77, 567, 1996.

663. Kumar, A. et al., The structure of alfalfa mosaic virus capsid protein assembled as a T = 1 icosahedral particle at 4.0-Å resolution, *J. Virol.,* 71, 7911, 1997.
664. Fukuyama, K. et al., Structure of a T = 1 aggregate of alfalfa mosaic virus coat protein seen at 4.5 Å resolution, *J. Mol. Biol.,* 167, 873, 1983.
665. Yusibov, V. and Loesch-Fries, L. S., Functional significance of three basic N-terminal amino acids of alfalfa mosaic virus coat protein, *Virology,* 242, 1, 1998.
666. Thole, V. et al., Amino acids of alfalfa mosaic virus coat protein that direct formation of unusually long virus particles, *J. Gen. Virol.,* 79, 3139, 1998.
667. Yusibov, V. et al., Antigens produced in plants by infection with chimeric plant viruses immunize against rabies virus and HIV-1, *Proc. Natl. Acad. Sci. U.S.A.,* 94, 5784, 1997.
668. Modelska, A. et al., Immunization against rabies with plant-derived antigen, *Proc. Natl. Acad. Sci. U.S.A.,* 95, 2481, 1998.
669. Green, K. Y., The role of human caliciviruses in epidemic gastroenteritis, *Arch. Virol. Suppl.,* 13, 153, 1997.
670. Jiang, X. et al., Expression, self-assembly, and antigenicity of the Norwalk virus capsid protein, *J. Virol.,* 66, 6527, 1992.
671. Jiang, X. et al., Expression, self-assembly, and antigenicity of a snow mountain agent-like calicivirus capsid protein, *J. Clin. Microbiol.,* 33, 1452, 1995.
672. Dingle, K. E. et al., Human enteric *Caliciviridae*: the complete genome sequence and expression of virus-like particles from a genetic group II small round structured virus, *J. Gen. Virol.,* 76, 2349, 1995.
673. Laurent, S. et al., Recombinant rabbit hemorrhagic disease virus capsid protein expressed in baculovirus self-assembles into viruslike particles and induces protection, *J. Virol.,* 68, 6794, 1994.
674. Nagesha, H. S. et al., Self-assembly, antigenicity, and immunogenicity of the rabbit haemorrhagic disease virus (Czechoslovakian strain V-351) capsid protein expressed in baculovirus, *Arch. Virol.,* 140, 1095, 1995.
675. Sibilia, M. et al., Two independent pathways of expression lead to self-assembly of the rabbit hemorrhagic disease virus capsid protein, *J. Virol.,* 69, 5812, 1995.
676. Laurent, S. et al., Structural, antigenic and immunogenic relationships between European brown hare syndrome virus and rabbit haemorrhagic disease virus, *J. Gen. Virol.,* 78, 2803, 1997.
677. Leite, J. P. et al., Characterization of Toronto virus capsid protein expressed in baculovirus, *Arch. Virol.,* 141, 865, 1996.
678. Green, K. Y. et al., Expression and self-assembly of recombinant capsid protein from the antigenically distinct Hawaii human calicivirus, *J. Clin. Microbiol.,* 35, 1909, 1997.
679. Jiang, X. et al., Expression and characterization of Sapporo-like human calicivirus capsid proteins in baculovirus, *J. Virol. Methods,* 78, 81, 1999.
680. Hale, A. D. et al., Expression and self-assembly of Grimsby virus: antigenic distinction from Norwalk and Mexico viruses, *Clin. Diagn. Lab. Immunol.,* 6, 142, 1999.
681. Boga, J. A. et al., A single dose immunization with rabbit haemorrhagic disease virus major capsid protein produced in *Saccharomyces cerevisiae* induces protection, *J. Gen. Virol.,* 78, 2315, 1997.
682. Pletneva, M. A. et al., Characterization of a recombinant human calicivirus capsid protein expressed in mammalian cells, *Virus Res.,* 55, 129, 1998.
683. Geissler, K. et al., Feline calicivirus capsid protein expression and capsid assembly in cultured feline cells, *J. Virol.,* 73, 834, 1999.
684. Mason, H. S. et al., Expression of Norwalk virus capsid protein in transgenic tobacco and potato and its oral immunogenicity in mice, *Proc. Natl. Acad. Sci. U.S.A.,* 93, 5335, 1996.
685. Tacket, C. O. et al., Human immune responses to a novel Norwalk virus vaccine delivered in transgenic potatoes, *J. Infect. Dis.,* 182, 302, 2000.
686. Li, T. C. et al., Expression and self-assembly of empty virus-like particles of hepatitis E virus, *J. Virol.,* 71, 7207, 1997.
687. Xing, L. et al., Recombinant hepatitis E capsid protein self-assembles into a dual-domain T = 1 particle presenting native virus epitopes, *Virology,* 265, 35, 1999.
688. Prasad, B. V. et al., Three-dimensional structure of calicivirus, *J. Mol. Biol.,* 240, 256, 1994.
689. Prasad, B. V. et al., Structure of Norwalk virus, *Arch. Virol. Suppl.,* 12, 237, 1996.
690. Thouvenin, E. et al., Bivalent binding of a neutralising antibody to a calicivirus involves the torsional flexibility of the antibody hinge, *J. Mol. Biol.,* 270, 238, 1997.

691. White, L. J. et al., Biochemical characterization of a smaller form of recombinant Norwalk virus capsids assembled in insect cells, *J. Virol.,* 71, 8066, 1997.

692. Prasad, B. V. et al., X-ray crystallographic structure of the Norwalk virus capsid, *Science,* 286, 287, 1999.

693. Plana-Duran, J. et al., Oral immunization of rabbits with VP60 particles confers protection against rabbit hemorrhagic disease, *Arch. Virol.,* 141, 1423, 1996.

694. Ball, J. M. et al., Recombinant Norwalk virus-like particles as an oral vaccine, *Arch. Virol. Suppl.,* 12, 243, 1996.

695. Ball, J. M. et al., Oral immunization with recombinant Norwalk virus-like particles induces a systemic and mucosal immune response in mice, *J. Virol.,* 72, 1345, 1998.

696. Ball, J. M. et al., Recombinant Norwalk virus-like particles given orally to volunteers: phase I study, *Gastroenterology,* 117, 40, 1999.

697. Nagesha, H. S. et al., Virus-like particles of calicivirus as epitope carriers, *Arch. Virol.,* 144, 2429, 1999.

698. Fiers, W. et al., Complete nucleotide sequence of bacteriophage MS2 RNA: primary and secondary structure of the replicase gene, *Nature,* 260, 500, 1976.

699. Crowther, R. A. et al., Three-dimensional image reconstructions of bacteriophages R17 and f2, *J. Mol. Biol.,* 98, 631, 1975.

700. Van Duin, J., Single-stranded RNA bacteriophages, in *The Bacteriophages,* Vol. 1, Calendar, R., Ed., Plenum Press, New York, 1988, pp. 117–167.

701. Kastelein, R. A. et al., Effect of the sequences upstream from the ribosome-binding site on the yield of protein from the cloned gene for phage MS2 coat protein, *Gene,* 23, 245, 1983.

702. Mastico, R. A. et al., Multiple presentation of foreign peptides on the surface of an RNA-free spherical bacteriophage capsid, *J. Gen. Virol.,* 74, 541, 1993.

703. Kozlovskaya, T. M. et al., Formation of capsid-like structures as a result of expression of the cloned gene of the envelope protein of the RNA-containing bacteriophage fr (in Russian), *Dokl. Akad. Nauk S.S.S.R.,* 287, 452, 1986.

704. Adhin, M. R. et al., Nucleotide sequence from the ssRNA bacteriophage JP34 resolves the discrepancy between serological and biophysical classification, *Virology,* 170, 238, 1989.

705. Ni, C. Z. et al., Crystal structure of the coat protein from the GA bacteriophage: model of the unassembled dimer, *Protein Sci.,* 5, 2485, 1996.

706. Kozlovska, T. M. et al., Recombinant RNA phage Q beta capsid particles synthesized and self-assembled in *Escherichia coli, Gene,* 137, 133, 1993.

707. Priano, C. et al., A complete plasmid-based complementation system for RNA coliphage Q beta: three proteins of bacteriophages Q beta (group III) and SP (group IV) can be interchanged, *J. Mol. Biol.,* 249, 283, 1995.

708. Lim, F. et al., Translational repression and specific RNA binding by the coat protein of the *Pseudomonas* phage PP7, *J. Biol. Chem.,* 276, 22507, 2001.

709. Valegård, K. et al., Purification, crystallization and preliminary X-ray data of the bacteriophage MS2, *J. Mol. Biol.,* 190, 587, 1986.

710. Valegård, K. et al., Structure determination of the bacteriophage MS2, *Acta Crystallogr. B.,* 47, 949, 1991.

711. Golmohammadi, R. et al., The refined structure of bacteriophage MS2 at 2.8 Å resolution, *J. Mol. Biol.,* 234, 620, 1993.

712. Stonehouse, N. J. et al., Crystal structures of MS2 capsids with mutations in the subunit FG loop, *J. Mol. Biol.,* 256, 330, 1996.

713. Stockley, P. G. et al., Probing sequence-specific RNA recognition by the bacteriophage MS2 coat protein, *Nucleic Acids Res.,* 23, 2512, 1995.

714. Valegård, K. et al., Crystal structure of an RNA bacteriophage coat protein-operator complex, *Nature,* 371, 623, 1994.

715. Valegård, K. et al., The three-dimensional structures of two complexes between recombinant MS2 capsids and RNA operator fragments reveal sequence-specific protein-RNA interactions, *J. Mol. Biol.,* 270, 724, 1997.

716. Lago, H. et al., Dissecting the key recognition features of the MS2 bacteriophage translational repression complex, *Nucleic Acids Res.,* 26, 1337, 1998.

717. Peabody, D. S. and Chakerian, A., Asymmetric contributions to RNA binding by the Thr(45) residues of the MS2 coat protein dimer, *J. Biol. Chem.*, 274, 25403, 1999.

718. Van den Worm, S. H. et al., Crystal structures of MS2 coat protein mutants in complex with wild-type RNA operator fragments, *Nucleic Acids Res.*, 26, 1345, 1998.

719. Lim, F. et al., Altering the RNA binding specificity of a translational repressor, *J. Biol. Chem.*, 269, 9006, 1994.

720. Lim, F. and Peabody, D. S., Mutations that increase the affinity of a translational repressor for RNA, *Nucleic Acids Res.*, 22, 3748, 1994.

721. Convery, M. A. et al., Crystal structure of an RNA aptamer-protein complex at 2.8 Å resolution, *Nat. Struct. Biol.*, 5, 133, 1998.

722. Grahn, E. et al., Crystallographic studies of RNA hairpins in complexes with recombinant MS2 capsids: implications for binding requirements, *RNA*, 5, 131, 1999.

723. Rowsell, S. et al., Crystal structures of a series of RNA aptamers complexed to the same protein target, *Nat. Struct. Biol.*, 5, 970, 1998.

724. Bundule, M. et al., Crystallization of bacteriophage fr and its recombinant capsids, *J. Mol. Biol.*, 232, 1005, 1993.

725. Tars, K. et al., Structure determination of bacteriophage PP7 from *Pseudomonas aeruginosa*: from poor data to a good map, *Acta Crystallogr. D. Biol.*, 56, 398, 2000.

726. Ni, C. Z. et al., Crystal structure of the MS2 coat protein dimer: implications for RNA binding and virus assembly, *Structure.*, 3, 255, 1995.

727. Peabody, D. S. and Lim, F., Complementation of RNA binding site mutations in MS2 coat protein heterodimers, *Nucleic Acids Res.*, 24, 2352, 1996.

728. Pushko, P. et al., Analysis of RNA phage fr coat protein assembly by insertion, deletion and substitution mutagenesis, *Protein Eng.*, 6, 883, 1993.

729. Axblom, C. et al., Structure of phage fr capsids with a deletion in the FG loop: implications for viral assembly, *Virology*, 249, 80, 1998.

730. Pushko, P., unpublished data, 1993.

731. Tars, K., unpublished data, 1997.

732. Stonehouse, N. J. and Stockley, P. G., Effects of amino acid substitution on the thermal stability of MS2 capsids lacking genomic RNA, *FEBS Lett.*, 334, 355, 1993.

733. Peabody, D. S., Subunit fusion confers tolerance to peptide insertions in a virus coat protein, *Arch. Biochem. Biophys.*, 347, 85, 1997.

734. Heal, K. G. et al., Expression and immunogenicity of a liver stage malaria epitope presented as a foreign peptide on the surface of RNA-free MS2 bacteriophage capsids, *Vaccine*, 18, 251, 1999.

735. Voronkova, T. and Ulrich, R., unpublished data, 2000.

736. Kozlovska, T. M. et al., RNA phage Q beta coat protein as a carrier for foreign epitopes, *Intervirology*, 39, 9, 1996.

737. Kozlovska, T. M. et al., Display vectors. II. Recombinant capsid of RNA bacteriophage Qbeta as a display moiety, *Proc. Latv. Acad. Sci.*, 51, 8, 1997.

738. Smiley, B. K. and Minion, F. C., Enhanced readthrough of opal (UGA) stop codons and production of *Mycoplasma pneumoniae* P1 epitopes in *Escherichia coli*, *Gene*, 134, 33, 1993.

739. Vasiljeva, I. et al., Mosaic Qbeta coats as a new presentation model, *FEBS Lett.*, 431, 7, 1998.

740. Cielens, I., unpublished data, 1998.

741. Lim, F. et al., The RNA-binding site of bacteriophage Q beta coat protein, *J. Biol. Chem.*, 271, 31839, 1996.

742. Spingola, M. and Peabody, D.S., MS2 coat protein mutants which bind Q beta RNA, *Nucleic Acids Res.*, 25, 2808, 1997.

743. Pickett, G. G. and Peabody, D. S., Encapsidation of heterologous RNAs by bacteriophage MS2 coat protein, *Nucleic Acids Res.*, 21, 4621, 1993.

744. Wu, M. et al., Cell-specific delivery of bacteriophage-encapsidated ricin A chain, *Bioconjugate Chem.*, 6, 587, 1995.

745. Wu, M. et al., Development of a novel drug-delivery system using bacteriophage MS2 capsids, *Biochem. Soc. Trans.*, 24, 413S, 1996.

746. Wu, M. et al., Specific cytotoxicity against cells bearing HIV1 gp120 antigen by bacteriophage-encapsidated ricin A chain: implications for cell specific drug delivery, *Biochem. Soc. Trans.*, 25, 158S, 1997.

747. Cielens, I. et al., Mutilation of RNA phage Qbeta virus-like particles: from icosahedrons to rods, *FEBS Lett.*, 482, 261, 2000.

748. Van Meerten, D. et al., Peptide display on live MS2 phage: restrictions at the RNA genome level, *J. Gen. Virol.*, 82, 1797, 2001.

749. Bahner, I. et al., Expression of the genome of potato leafroll virus: readthrough of the coat protein termination codon *in vivo*, *J. Gen. Virol.*, 71, 2251, 1990.

750. Nurkiyanova, K. M. et al., Tagging potato leafroll virus with the jellyfish green fluorescent protein gene, *J. Gen. Virol.*, 81, 617, 2000.

751. Lamb, J. W. et al., Assembly of virus-like particles in insect cells infected with a baculovirus containing a modified coat protein gene of potato leafroll luteovirus, *J. Gen. Virol.*, 77, 1349, 1996.

752. Schneemann, A. et al., Use of recombinant baculoviruses in synthesis of morphologically distinct viruslike particles of flock house virus, a nodavirus, *J. Virol.*, 67, 2756, 1993.

753. Selling, B. H. et al., Genomic RNA of an insect virus directs synthesis of infectious virions in plants, *Proc. Natl. Acad. Sci. U.S.A.*, 87, 434, 1990.

754. Price, B. D. et al., Complete replication of an animal virus and maintenance of expression vectors derived from it in *Saccharomyces cerevisiae*, *Proc. Natl. Acad. Sci. U.S.A.*, 93, 9465, 1996.

755. Lorenzi, R. and Burrone, O. R., Sequence-specific antibodies against human IgE isoforms induced by an epitope display system, *Immunotechnology*, 4, 267, 1999.

756. Fisher, A. J. et al., Crystallization and preliminary data analysis of Flock House virus, *Acta Crystallogr. B*, 48, 515, 1992.

757. Fisher, A. J. et al., Crystallization of viruslike particles assembled from flock house virus coat protein expressed in a baculovirus system, *J. Virol.*, 67, 2950, 1993.

758. Wery, J. P. et al., The refined three-dimensional structure of an insect virus at 2.8 Å resolution, *J. Mol. Biol.*, 235, 565, 1994.

759. Bothner, B. et al., Evidence of viral capsid dynamics using limited proteolysis and mass spectrometry, *J. Biol. Chem.*, 273, 673, 1998.

760. Dong, X. F. et al., Particle polymorphism caused by deletion of a peptide molecular switch in a quasiequivalent icosahedral virus, *J. Virol.*, 72, 6024, 1998.

761. Scodeller, E. A. et al., A new epitope presenting system displays a HIV-1 V3 loop sequence and induces neutralizing antibodies, *Vaccine*, 13, 1233, 1995.

762. Schiappacassi, M. et al., V3 loop core region serotyping of HIV-1 infected patients using the FHV epitope presenting system, *J. Virol. Methods*, 63, 121, 1997.

763. Buratti, E. et al., Conformational display of two neutralizing epitopes of HIV-1 gp41 on the Flock House virus capsid protein, *J. Immunol. Methods*, 197, 7, 1996.

764. Buratti, E. et al., Improved reactivity of hepatitis C virus core protein epitopes in a conformational antigen-presenting system, *Clin. Diagn. Lab. Immunol.*, 4, 117, 1997.

765. Gough, K. H. and Shukla, D. D., Nucleotide sequence of Johnsongrass mosaic potyvirus genomic RNA, *Intervirology*, 36, 181, 1993.

766. Jagadish, M. N. et al., Expression of potyvirus coat protein in *Escherichia coli* and yeast and its assembly into virus-like particles, *J. Gen. Virol.*, 72, 1543, 1991.

767. Jagadish, M. N. et al., Site-directed mutagenesis of a potyvirus coat protein and its assembly in *Escherichia coli*, *J. Gen. Virol.*, 74, 893, 1993.

768. Edwards, S. J. et al., High level production of potyvirus-like particles in insect cells infected with recombinant baculovirus, *Arch. Virol.*, 136, 375, 1994.

769. Hammond, J. M. et al., Expression of the potyvirus coat protein mediated by recombinant vaccinia virus and assembly of potyvirus-like particles in mammalian cells, *Arch. Virol.*, 143, 1433, 1998.

770. Ravelonandro, M. et al., Construction of a chimeric viral gene expressing plum pox virus coat protein, *Gene*, 120, 167, 1992.

771. Wypijewski, K. J. et al., Expression of the plum pox coat protein gene in transgenic *Nicotiana tabacum* plants, *Acta Biochim. Pol.*, 42, 97, 1995.

772. Jacquet, C. et al., Use of modified plum pox virus coat protein genes developed to limit heteroencapsidation-associated risks in transgenic plants, *J. Gen. Virol.*, 79, 1509, 1998.

773. Jagadish, M. N. et al., High level production of hybrid potyvirus-like particles carrying repetitive copies of foreign antigens in *Escherichia coli*, *Biotechnology (N.Y.)*, 11, 1166, 1993.

774. Jagadish, M. N. et al., Chimeric potyvirus-like particles as vaccine carriers, *Intervirology*, 39, 85, 1996.

775. Jacquet, C. et al., High resistance and control of biological risks in transgenic plants expressing modified plum pox virus coat protein, *Acta Virol.*, 42, 235, 1998.

776. Fernandez-Fernandez, M. R. et al., Development of an antigen presentation system based on plum pox potyvirus, *FEBS Lett.*, 427, 229, 1998.

777. Okada, Y., Historical overview of research on the tobacco mosaic virus genome: genome organization, infectivity and gene manipulation, *Philos. Trans. R. Soc. London B Biol. Sci.*, 354, 569, 1999.

778. Butler, P. J., Self-assembly of tobacco mosaic virus: the role of an intermediate aggregate in generating both specificity and speed, *Philos. Trans. R. Soc. London B Biol. Sci.*, 354, 537, 1999.

779. Klug, A., The tobacco mosaic virus particle: structure and assembly, *Philos. Trans. R. Soc. London B Biol. Sci.*, 354, 531, 1999.

780. Stubbs, G. et al., Structure of RNA and RNA binding site in tobacco mosaic virus from 4-Å map calculated from X-ray fibre diagrams, *Nature*, 267, 216, 1977.

781. Bhyravbhatla, B. et al., Refined atomic model of the four-layer aggregate of the tobacco mosaic virus coat protein at 2.4-Å resolution, *Biophys. J.*, 74, 604, 1998.

782. Jeng, T. W. et al., Visualization of alpha-helices in tobacco mosaic virus by cryo-electron microscopy, *J. Mol. Biol.*, 205, 251, 1989.

783. Shire, S. J. et al., Preparation and properties of recombinant DNA derived tobacco mosaic virus coat protein, *Biochemistry*, 29, 5119, 1990.

784. Hwang, D. J. et al., Expression of tobacco mosaic virus coat protein and assembly of pseudovirus particles in *Escherichia coli*, *Proc. Natl. Acad. Sci. U.S.A.*, 91, 9067, 1994.

785. Hwang, D. J. et al., Assembly of tobacco mosaic virus and TMV-like pseudovirus particles in *Escherichia coli*, *Arch. Virol. Suppl.*, 9, 543, 1994.

786. Donson, J. et al., Systemic expression of a bacterial gene by a tobacco mosaic virus-based vector, *Proc. Natl. Acad. Sci. U.S.A.*, 88, 7204, 1991.

787. Hamamoto, H. et al., A new tobacco mosaic virus vector and its use for the systemic production of angiotensin-I-converting enzyme inhibitor in transgenic tobacco and tomato, *Biotechnology (N.Y.)*, 11, 930, 1993.

788. Fitchen, J. et al., Plant virus expressing hybrid coat protein with added murine epitope elicits autoantibody response, *Vaccine*, 13, 1051, 1995.

789. Turpen, T. H. et al., Malarial epitopes expressed on the surface of recombinant tobacco mosaic virus, *Biotechnology (N.Y.)*, 13, 53, 1995.

790. Sugiyama, Y. et al., Systemic production of foreign peptides on the particle surface of tobacco mosaic virus, *FEBS Lett.*, 359, 247, 1995.

791. Bendahmane, M. et al., Display of epitopes on the surface of tobacco mosaic virus: impact of charge and isoelectric point of the epitope on virus-host interactions, *J. Mol. Biol.*, 290, 9, 1999.

792. Koo, M. et al., Protective immunity against murine hepatitis virus (MHV) induced by intranasal or subcutaneous administration of hybrids of tobacco mosaic virus that carries an MHV epitope, *Proc. Natl. Acad. Sci. U.S.A.*, 96, 7774, 1999.

793. Casper, S. J. and Holt, C. A., Expression of the green fluorescent protein-encoding gene from a tobacco mosaic virus-based vector, *Gene*, 173, 69, 1996.

794. Verch, T. et al., Expression and assembly of a full-length monoclonal antibody in plants using a plant virus vector, *J. Immunol. Methods*, 220, 69, 1998.

795. Wigdorovitz, A. et al., Protection of mice against challenge with foot and mouth disease virus (FMDV) by immunization with foliar extracts from plants infected with recombinant tobacco mosaic virus expressing the FMDV structural protein VP1, *Virology*, 264, 85, 1999.

796. Turpen, T. H., Tobacco mosaic virus and the virescence of biotechnology, *Philos. Trans. R. Soc. London B Biol. Sci.*, 354, 665, 1999.

797. Yusibov, V. et al., Plant viral vectors based on tobamoviruses, *Curr. Top. Microbiol. Immunol.*, 240, 81, 1999.

9 DNA Vaccines

Sang Won Han, Jane Zveiter de Moraes, Célio Lopes Silva,
Roger Chammas, and Maurício Martins Rodrígues

CONTENTS

9.1 INTRODUCTION

Experimental nucleic acid vaccines have been the subject of considerable attention in recent years because of their theoretical potential to achieve effects similar to live attenuated or live recombinant vaccines without the risks associated with the use of infectious agents. Nucleic acid vaccines are an attractive alternative to conventional protein vaccines because of their ability to induce *de novo* production of antigens in a given tissue following DNA delivery. This strategy avoids many problems associated with live vaccines including reversion of virulence, temperature instability, contaminating adventitious agents, and contraindication in immunologically compromised individuals. The technology is relatively simple. A DNA construct encoding the gene of interest is delivered directly to cells of the organism to be immunized where it is taken up and expressed by the host cells. The endogenously expressed immunogen subsequently induces a protective cellular and humoral immune response in the host.

A distinguishing feature of DNA vaccines, as opposed to the classical subunit vaccines, is the production of the immunogen in host cells. This feature permits antigen processing and presentation by both class I and class II MHC molecules. By contrast, killed whole or protein subunit vaccines undergo the presentation by class II MHC molecules only. These differences in presentation are responsible for DNA vaccines raising both cytolytic T cells and antibody, whereas subunit vaccines mostly elicit antibodies. The importance of raising cytolytic T lymphocytes lies in their ability to directly kill pathogen-containing cells. The use of DNA, a nonliving agent, to raise cytolytic T cells represents a milestone in vaccinology. Moreover, DNA vaccines can be used to direct an immune response toward Th1 or Th2 types of T cells. These two types of T cells affect the types of antibody raised and the nature of the inflammatory cells and cytokines mobilized to fight an infection. Also, the ability of DNA vaccines to produce immunizing antigens inside the host cells yields proteins in its native form, which may explain why high levels of antibodies are elicited against these proteins after DNA immunization. Antibodies that fail to recognize the native form of proteins are frequently ineffective in the containment of invading microorganisms.

DNA vaccine strategy offers several important advantages. In preclinical studies, DNA vaccines were found to elicit long-lasting immune responses similar to live vaccines. DNA vaccines could be produced using similar fermentation, purification, and validation techniques. The capacity to use generic production and verification techniques simplifies vaccine development and production. Also, DNA vaccines remain stable at both high and low temperatures. DNA vaccines can be stored either dry or in aqueous solution. The significant stability of DNA vaccines should facilitate distribution and administration of vaccines and can lead to the elimination of the so-called cold chain, a series of refrigerators required to maintain the viability of vaccines during their distribution. Normally, maintaining the cold chain contributes the most to the cost of vaccination.

Advantages of DNA vaccine technology are as follows:

- Poses no risk of infection
- Raises antibodies against the native forms of proteins
- Raises long-lasting immunoresponses
- Elicits cytolytic T-cell responses
- Ability to prime for T-helper 1 or T-helper 2 directed responses
- Facilitates the use of combination vaccines
- Allows rapid screening of multiple sequence variants of proteins for determination of the efficiency of the protective response
- Offers potential for generic production and manufacture
- Demonstrates good stability at low and high temperatures

Synthetic DNA has been used not only for experimental immunization or for early clinical trials, but also as a tool for immunomodulation. Immunostimulatory DNA sequences are known to exist and may be used to drive the immune response toward developing protective immunoreactivity. Cytokine or co-stimulatory molecule gene transfer protocols are truly promising. These strategies, either alone or combined, may be cost-effective for the treatment of a number of prevalent viral, bacterial, and parasitic diseases. Infectious diseases are not the only targets for genetic immunization or gene therapy–mediated immunomodulation. Recently, DNA inoculation protocols have been designed as well for both virus-associated and nonvirus-associated cancers. Gene transfer techniques are being refined at a rapid pace, as discussed elsewhere. Here we discuss the use of DNA for genetic immunization and gene therapy–based immunomodulation addressing basic mechanisms used in current protocols for genetic immunization and ongoing preclinical and clinical trials for infectious diseases and cancers.

9.2 IMMUNOSTIMULATORY NUCLEIC ACIDS

9.2.1 HISTORICAL BACKGROUND

Mycobacteria have been widely studied with regard to cancer immunotherapy.[1] Several molecular compounds of the *Mycobacterium bovis* strain BCG and of other mycobacteria have been reported to have antitumor activity.[2] However, Tokunaga and colleagues[3,4] were the first to demonstrate an immunostimulatory activity of the DNA fraction MY-1 isolated from *M. bovis BCG*. The MY-1 fraction is composed mainly of single-stranded DNA (ssDNA). The incubation of this DNA with human and mouse cell lines did not reveal any direct cytotoxicity suggesting host-mediated anti-tumor activity. These researchers also showed that the adherent splenocytes, macrophages, and NK cells were stimulated by bacterial DNA. This effect could be inhibited by anti-IFN-γ or anti-IFNα/β antibodies.

To determine the minimal and essential sequences responsible for biological activity several single-stranded oligodeoxynucleotides (SS-ODN) were synthesized based on the mycobacterial genome sequences. These SS-ODNs were capable of augmenting NK cell and macrophage activity and of inducing the production of IFNs. First, Tokunaga and his group synthesized 13 kinds of 45-mer SS-ODNs and found that 6 of them had immunostimulatory activity. Based on these 6 compounds, several shortened SS-ODNs were tested, and finally the researchers were able to attribute the immunostimulation to the palindromic hexamers carrying a CpG dinucleotide.[5–8,90,187,203,225]

Based on the observation that the genomes of bacteria and viruses have a high frequency of unmethylated CpG dinucleotides, Krieg and colleagues[9] studied this effect of SS-ODN carrying the CpG motif on the murine-B lymphocyte system. They found that the presence of CpG in the sequence of SS-ODNs is essential for lymphocyte stimulation, although the nucleotides flanking CpG also play an important role. The SS-ODNs with the consensus sequence 5′-R-R-CpG-Y-Y-3′ (where R is purine, Y is pyrimidine) render the best B-cell activation. Additionally, it was found that cytosine in the CpG must be unmethylated, as occurs frequently in bacteria and viruses, and also the oligonucleotides must have at least eight bases in order to maintain the immunostimulatory activity.

Even though the motif of six nucleotides was suggested to be responsible for immunostimulation, it was noted that sequences flanking this motif also influenced the biological outcome.[10] For example, macrophages treated with the SS-ODN identified as1668 produced IL-6 and IL-12, which are key factors that promote the Th1 response, and also stimulated the secretion of toxic levels of TNF-α. However, Lipford and colleagues[11] described a CpG-containing sequence from the mouse IL-12 p40 intron 1 similar to 1668, which had the potential of inducing IL-12 and IL-6 but whose ability to trigger TNF-α was abolished.

The important role of CpG in the induction of the immune response is becoming clear; however, it should be kept in mind that the above-mentioned studies on CpG were based on the use of SS-ODNs. Raz and colleagues[12] showed the influence of immunostimulatory sequences (ISSs) in plasmid DNA on the outcome of intradermal gene immunization. Two similar plasmid vectors constructed to express the bacterial β-galactosidase gene and carrying different antibiotic selection markers — the vector pACB-Z carrying an ampicillin resistance gene (*amp^R*), and the vector pKCB-Z carrying a gene conferring resistance to kanamycin (*kan^R*) — generated remarkably different immune responses. Mice immunized intradermally with pACB-Z showed a strong Th1 response, whereas pKCB-Z produced a very weak response.

When analyzing DNA sequences of amp^R and kan^R genes, these investigators found many repetitions of the motif 5'-R-R-CpG-Y-Y-3'[9] in the amp^R gene. When this ISS was inserted into pKCB-Z, the immune response was restored to the same level as seen with pACB-Z. This response appeared to increase with the increasing number of ISSs. However, the expression level of the β-gal gene was lower in cells transfected with pACB-Z vector than in cells transfected with pKCB-Z vector. When the neutralizing anti-IFN-α antibody was added to the cells transfected with pACB-Z, the β-gal expression level increased almost two times.[12] Induction of IFN-α expression by ISS is important for naive T-cell differentiation toward Th1.[13] At the same time this induction inhibits IgE synthesis, promotes IgG2a synthesis,[14,15] and antagonizes Th2 cells.[13] However, the presence of ISS may interfere with the expression of a therapeutic gene by induction of the synthesis of IFN-α and proinflammatory cytokines, as was seen in the example described above.

These results indicate that plasmid DNA-derived vectors used for immunization have two distinct units: one unit to direct the synthesis of an antigen and the other to induce the production of interferons and interleukines specific to the Th1 response. Therefore, the ISS motif is important as an adjuvant for immunization but may interfere with gene replacement therapy by reducing the expression of the therapeutic gene.

9.2.2 DIVERSITY OF ISS

The dinucleotide CpG is probably the best-known ISS; however, a variety of poly(oligo)nucleotides had demonstrated immunostimulatory activity long before the identification of the CpG motif. In 1967, naturally present double-stranded RNA (dsRNA) isolated from *Penicillium funiculosum,* or reovirus type-3 virions and synthetic polyinosinic:polycytidilic ribonucleic acid (poly I:C), were found to stimulate an antiviral response in rabbits by inducing interferon production.[16,17] Later, a similar effect was observed in primates.[18] The antiviral activity promoted by dsRNA is likely to be mediated by dsRNA, activated protein kinase induced by interferon, which inhibits both host and viral protein synthesis.[19,20] The other synthetic and natural nucleic acids were later found to stimulate the immune system, such as ssDNA and dsDNA genomes from different bacterial strains, and the synthetic polymer poly(dG):poly(dC).[21,22] Chemically modified nucleic acids, partially thiolated dsRNA poly(I)-mercapto poly(C), and phosphorothioate oligodeoxynucleotides were also shown to activate macrophages, NK, and B lymphocytes.[23,24]

Another example is extended runs of dG sequences, which function as a mitogen for B cells stimulating cell proliferation and antibody production.[25] In addition, the dG sequences can be inserted into an oligomer to enhance the effect of ISSs in the production of macrophage cytokine, which do not occur with the dA, dT, and dC sequences.[26]

The strong induction of cytokine production by dG sequences in macrophage appears to be related to the promotion of cell uptake caused by oligomer binding to the macrophage scavenger receptor (MSR). This receptor is known to bind acetylated or oxidized low-density lipoproteins (LDLs) and to interact with and internalize nucleic acids such as poly(I), poly(G), and oligo(dG).[27–29]

The structure of DNAs rich in dG sequences can form unusual quadruplex DNA by the intrastrand and interstrand interaction.[30–32] This peculiar structure of the dG may be responsible for the occurrence of a specific inducer-cell response. However, if the extended stretches of dG sequences are inserted into the plasmid vector, the induction of immunostimulation activities or the efficiency of vaccination is very low.[33] Unlike the CpG motif, which functions in the oligomer or in the plasmid form,[12] the dG sequences in plasmids may have limited activity because of their low concentration or inability to form quadruplexes. The difference in chemical structure formed by the dG sequences in the oligomer, or in double-stranded form in the plasmid, as well as the sequence length and composition and environmental conditions could affect quadruplex DNA formation and, consequently, its immunostimulant activity.

9.2.3 ISS INDUCES A TH1 RESPONSE

Vaccination with an inactivated virus or its subunit induces a T helper 2 (Th2)-type immune response with high titers of neutralizing antibodies but a low level of cell-mediated immunity. However, if the immunization is performed with naked plasmid DNA containing ISS, the response of the immune system tends to involve Th1 cells. The nature of these immunostimulatory signals emitted by ISS is becoming more understood. Transfection of fresh human macrophages with plasmid DNA-containing ISS or with ISS-SS-ODN, but not with ISS-lacking pDNA or mutated (M)-SS-ODN, resulted in increased production of mRNA transcripts for IFN-α, IFN-β, IL-12, and IL-18.[34] These cytokines are known to induce synthesis of IFN-γ and to participate in the process of differentiation of naive Th cells into the Th1 phenotype.[13,35–38] Additionally, incubation of human macrophages with IFN-γ increased mRNA transcripts for IL-12 and IL-18, suggesting an amplification role for IFN-γ in ISS-induced stimulation. Furthermore, human peripheral blood mononuclear cells transfected with ISS-SS-ODN, but not with the mutated ISS, produced IFN and IL with IFN-α, IFN-β and IL-12 being produced by macrophages and IFN-γ being produced by NK cells.[26,34,39] Therefore, ISSs have a role in inducing cells of the innate immune system, independent of the presence of antigen, to produce a cytokine milieu that promotes a Th1-type response.

Based on the response of cytokine production stimulated by ISSs, a possible mechanism for the induction of a Th1 response by DNA vaccination was postulated[40] (Figure 9.1). Two distinct phases appear to exist in the ISS-stimulated immune response. The first phase is the antigen-independent step consisting of the response of the innate immune system, and the second is the antigen-dependent step involving the adaptive immune response. In the first phase, ISS stimulates the innate immune system to produce IFN-γ and its inducers (IFN-α/β, IL-12, IL-18), preparing a cytokine *milieu* capable of directing differentiation of naive Th cells into Th1 cells. In the presence of the antigens, the second phase is initiated. During this phase even a larger amount of IFN-γ than in the first phase is produced and Th1-associated antibodies are elicited.

A similar immune response was observed when ISSs (SS-ODN or pDNA) were used only as adjuvants in immunization with the proteins. IgG2a antibodies and IFN-γ, which are hallmarks of a Th1 response, were produced, and production of theTh2 response-associated IL-5 was decreased.[34,41] However, inoculation of mice with protein/ISS-DNA and protein/ISS-deficient DNA resulted in comparable levels of the IgG1 and IL-4 production[34] associated with Th2. Although this last finding is not consistent with the model postulated above, the authors defended the model arguing that the dose of antigen employed in the experiment of protein/ISS coinjection was much higher (in the microgram range) than the dose used for a single genetic vaccination (in the range of μg/ml to ng/ml).[42]

Within the context of vaccine development, the ability to direct Th1 or Th2 differentiation of antigen-specific immune responses has significant implications for the treatment of various infectious and autoimmune diseases.[43] In the case of murine leishmaniasis, Th1-biased mouse strains (e.g., C57BL/6) present an IFN-γ-dominated response to parasite antigen and are resistant; whereas mice that present a Th2-dominated response involving IL-4 production (e.g., BALB/c) are susceptible to the parasite infestation.[44–46] By antibody blockade of IL-12 the resistant mice can become sensitive, and by administration of IL-12 or blockade of IL-4 the sensitive mice can become resistant.[47–50] The redirection of Th2-driven disease in BALB/c mice to a Th1-driven resistant phenotype can be obtained by immunization of the animal with CpG-SS-ODN. The conversion of the Th1 phenotype to the Th2 phenotype was associated with IL-12 production and maintained expression of IL-12 β2-chains.[51]

Eosinophilia is a good diagnostic indicator for allergic asthma or atopic response[52] and is characteristic for the Th2 immune response. DNA vaccination using hen-egg ovalbumin gene (OVA) was evaluated for the inhibition of pulmonary eosinophilic inflammation.[53] The OVA gene was placed under the control of a CMV promoter. The plasmid containing the OVA gene and the vector without this gene were used for intradermal injection of BALB/c mice. The inoculated animals

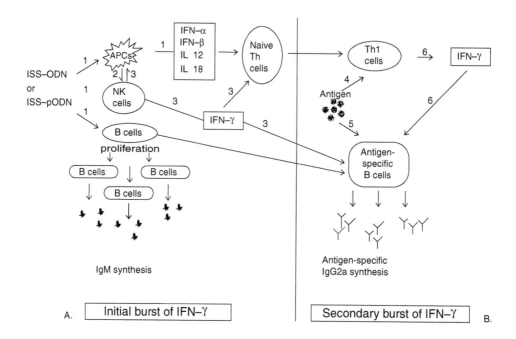

FIGURE 9.1 A mechanism for the induction of a Th1 response by gene vaccination postulated by Tighe et al.[40] (A) During the innate immune response (involving an initial burst of IFN-γ) ISS-ODN or ISS-pDNA triggers the release of IFN-γ by NK cells and inducers of IFN-γ (such as IFN-α, IL-12, and IL-18) by APCs (e.g., macrophages) and acts as mitogens for B cells (Step 1). IFN-α has an amplification role in IL-12 and IL-18 synthesis, and the release of IL-12 from APCs results in further IFN-γ production by NK cells (Step 2). Production of IFN-γ by NK cells acts on both antigen-specific B cells and naive T cells (Step 3). The initial burst of IFN-γ, as measured *in vitro*, is in the pg/ml range. (B) During the adaptive immune response in the presence of an antigen (involving a secondary burst of IFN-γ), the secreted innate cytokines prime naive Th cells to differentiate toward Th1 cells (Step 4). An encounter with the antigen under the influence of secreted innate cytokines induces B cells to switch to IgG2a antibodies (Step 5). Th1 cells produce additional IFN-γ in an antigen-dependent fashion (Th1 response) (Step 6). The secondary burst of IFN-γ, as measured *in vitro*, is in the ng/ml range. Abbreviations: APC, antigen-presenting cell; IFN-γ, interferon γ; IL, interleukin; ISS, immunostimulatory sequence; ISS-ODN, ISS-containing oligonucleotide; ISS-pDNA, ISS-containing plasmid DNA; NK, natural killer; Th1, T helper 1. (From Tighe, H. et al., *Immunol. Today,* 19(2), 89, 1998. With permission from Elsevier Science.)

were subsequently treated with aerosolized OVA. The genetic vaccine was shown to switch from the Th2 immune response to the Th1 immune response. The number of infiltrated eosinophils was decreased in the BAL fluid, lung tissue, and bone marrow after injection with the OVA DNA construct. On the other hand, the injection of saline or vector alone as a control did not elicit this effect. In addition, the production of the Th2 specific IgE and IgG1 was downregulated, while the Th1 characteristic IgG2 response was upregulated. Using a different allergen (Der p5) in a rat model, Hsu et al.[54] showed results similar to immediate pulmonary hypersensitivity. They observed inhibition of histamine secretion and a decrease in IgE production. They also detected a low level of IgE in mice injected intraperitoneally with Der p5 after a prophylactic intramuscular injection of the Der-p5-coding plasmid.[55]

Dendritic cells (DC) are known as "nature's adjuvants" because of their capacity to sensitize naive T cells. Jacob et al.[56] used CpG-SS-ODN and LPS as a control to assess the properties of the Langerhans cell (LC)-like murine fetal-skin-derived DC (FSDDC) *in vitro* and LC *in vivo*. Both SS-ODN and LPS stimulated FSDDC maturation as monitored by decreasing E-cadherin-mediated adhesion, upregulation of MHC class II (MHCII), expression of costimulatory molecules, and acquisition of enhanced accessory cell activity. However, only CpG-SS-ODN-stimulated FSDDC produced large amounts of IL-12. The level of IL-6 and TNF-α was low. Another type of DC, bone-marrow-derived DC (BMDDC), was used by Sparwasser et al.[57] It was shown that BMDDC activated with CpG-SS-ODN produced IL-12 as well as TNF-α and IL-6. This apparent discrepancy was attributed to the different ODN sequences used in these experiments because the CpG-motif-containing ODN — IL-12p40 (synthetic oligomer derived from intron 1 of the mouse IL-12-p40[11] — abolished only the TNF-α release in BMDDC and maintained IL-6 and IL-12 production.[57] This group also showed that the CpG-ODN causes the upregulation of MHCII, CD40, and CD86, but not CD80, and secretion of large amounts of IL-12, IL-6, and TNF-α. This result suggested that the bacterial DNA and CpG-SS-ODN may participate in the activation and maturation of immature DC into professional antigen-presenting cells (APC).

9.3 DNA IMMUNIZATION AGAINST VIRAL INFECTIONS

9.3.1 ANTIBODY-MEDIATED IMMUNE RESPONSES

In many ways DNA immunization mimics viral infection because it delivers the antigen in a similar manner and generates comparable types of immune responses. Not unexpectedly, most of the studies of genetic immunization were performed using viral proteins as model antigens.[58]

Studies using experimental models have proven that the administration of plasmids is a simple way of generating antibodies specific to the most relevant viral antigens.[58–61] The resulting antibodies recognized the native viral proteins and were capable of neutralizing replication of a number of viruses *in vitro*. These observations were very important because they indicated that the antigens expressed *in vivo* can assume a conformation that is similar, if not identical, to the different native viral proteins that contain many conformational epitopes. Another relevant feature of the humoral response generated by genetic immunization is the longevity of antibody production. In mice, serum antibodies can be reproducibly detected for almost the entire life of these animals, and the antibody response can be induced sooner in newborn mice.[62] Whether similar characteristics of the antibody response will be reproduced in nonhuman primates or humans is still under investigation.

In terms of magnitude, the antibody immune response elicited by genetic immunization usually appears to be strong. In many cases, it reproduced the antibody titers obtained during natural viral infections or even during vaccination with licensed vaccines. For example, in the case of influenza virus, injection of as little as 10 μg of plasmids in nonhuman primates generated antibody titers comparable to or higher than those elicited by a full human dose of the licensed vaccines.[63] Similar findings were reported in animals immunized with a plasmid containing the gene encoding the hepatitis B surface antigen when compared to recombinant commercially available vaccines.[64] However, in some cases, antibody titers elicited by genetic immunization were not as high as titers obtained when licensed vaccines or antigens were mixed with new powerful adjuvants.

To enhance the specific antibody response or to reduce the injected amount of DNA-encoded antigens several groups have explored the possibility of coadministering additional plasmids containing genes of cytokines or costimulatory molecules of the immune system. In many cases, coadministration of plasmids encoding for cytokines (IL-2, IL-4, GM-CSF, etc.)[65–68] or co-stimulatory molecules (CD40L, B7.1, B7.2, etc.)[59] significantly improved serum-specific antibody responses in mice. This was particularly relevant when suboptimal doses of DNA-encoded plasmids were used for immunization. Whether this strategy can be of general use for genetic immunization using any viral gene or for immunization of nonhuman primates or humans still requires careful evaluation.

The predominant class of serum antibodies produced after DNA inoculation in mice was IgG. In most cases, a predominance of IgG2a and 2b subclasses could be observed. This type of bias in the antibody response reflects the different types of CD4 helper cells activated during genetic immunization (see below). In some cases, however, IgG1 antibodies were produced in titers similar to those of IgG2 antibodies, perhaps reflecting the activation of other types of CD4 helper cells (see below). In addition to serum antibodies, plasmids delivered by mucosal routes of immunization have been shown to be capable of eliciting mucosal IgA antibodies. So far, this route of immunization has not been thoroughly studied, although it certainly deserves more attention in future studies.

9.3.2 T-Cell-Mediated Immune Responses

Administration of plasmids containing viral genes also elicits potent immune responses mediated by T lymphocytes. Spleen cells isolated from mice immunized with a variety of plasmids proliferated and produced cytokines upon *in vitro* restimulation with the respective antigens or peptides derived from them. The type of immune response observed in CD4 cells varied according to the gene used for vaccination, the route of immunization, and the cytokine-gene contained in plasmids coadministered with DNA-encoded antigen. Cytokines of the Th1 type such as IL-2, IFN-γ, etc. were either the only cytokines produced or the predominant types.[69–73] The accompanying humoral immune response consisted mainly of IgG2 antibodies. However, some plasmids containing other viral genes elicited both types of cytokines, including type-1 and type-2 cytokines such as IL-4 and IL-5. In these cases, both IgG1 and IgG2 antibodies were detected in the sera of immunized mice.[74]

Different routes of immunization can be important in the determination of the pattern of cytokine production and the accompanying humoral immune response. Certain plasmids when delivered by needle injection into the muscle or skin are reported to generate a predominant type I immune response. However, gene gun immunization with the same plasmids leads to a predominant type 2 immune response.[69] Finally, the type of immune response can be modulated by the coadministration of plasmids encoding for cytokines. Th1-inducing cytokines such as IL-12 and IL-18 can direct the immune response toward a Th1 type.[59,68,75–78] In contrast, Th2-inducing cytokines such as IL-4, IL-5, and IL-10 can activate Th2-type cells that normally would be present at a lower rate or be absent.[59,75,68]

Viral infections are classical models for the participation of CD8 cytotoxic T lymphocytes (CTL) in the host-adaptive immune response. The induction of CD8 cells significantly differs from that of CD4 cells. CD8 CTL recognizes peptide-bound MHC class I (MHCI) molecules present on the surface of antigen-presenting cells. MHCI-peptide complexes are more efficiently generated when the antigenic protein is endogenously produced in the host cell cytoplasm. Among the conventional vaccines used in humans only those consisting of attenuated microorganisms can prime the CTL response. Genetic immunization with plasmids can effectively prime these cells by delivering the antigen to the endogenous pathway of antigen presentation through MHCI molecules. So far, most, if not all, plasmids encoding viral antigens generated specific CTL responses in experimental models (mice and nonhuman primates).[60,79–83] These CTL were functional because they could efficiently lyse MHCI-compatible target cells infected with the respective viruses. In addition to lysing the infected target cells, CD8 cells also secreted a variety of inflammatory cytokines including IFN-γ, TNF-α, MIP1-α, MIP1-β, MCP-1, and RANTES, all of which can activate the virus-infected cell or enhance both inflammatory and immune responses against infection.[84] These studies unequivocally demonstrated that DNA immunization is a simple and powerful approach to generating the CD8-mediated immune response and can considerably improve the efficacy of some currently used vaccines.

As observed for the antibody response, coadministration of plasmids containing genes encoding for cytokine genes or costimulatory molecules of the immune system has the potential to improve the magnitude of the CTL response. Several cytokine genes that mediate the immune or inflammatory responses such as GM-CSF, IL-2, IL-18, IFN-γ, TNF-α, MCP-1, and RANTES have been described as capable of enhancing murine CTL-mediated immunity.[59,67,78,85] Among the genes encoding for costimulatory molecules of the immune system, the CD154, CD86, and intracellular adhesion molecule 1 (ICAM-1) demonstrated the potential to improve specific CTL responses.[86–88] As mentioned in connection with antibody response, the enhancing effect was particularly noticeable when a limiting dose of the plasmid-encoded antigen was used during mouse immunization. Again, whether this strategy may be of use in nonhuman primates and humans remains to be tested.

9.3.3 Protective Immunity in Animal Models

The first disease for which protective immunity was demonstrated by genetic immunization was influenza. Intramuscular (i.m.) administration of plasmid encoding for the viral nucleoprotein (NP) elicited specific antibody and CTL and provided significant protective immunity against a heterologous strain of influenza virus.[83] This study had a profound impact on the field of vaccine development against viral, bacterial, and parasitic infections and triggered a number of studies using several genes belonging to a variety of pathogens. The protective immunity elicited by genetic vaccination with the influenza NP gene was dependent mainly on CD4 and CD8 cells.[73,82] Like influenza NP, the protective immunity elicited in mice by vaccination with plasmid encoding for the lymphochoriomeningitis virus (LCMV) NP relied significantly on CD8 CTL.[89,90]

Preclinical studies were also performed with several other plasmids containing genes encoding for antigens that are important targets for antibody-mediated protective immunity during infection with influenza,[91,92] rabies,[93] equine[94] and human herpes,[95–98] cotton-tail rabbit papilloma,[99] respiratory syncytial,[71] Ebola,[100] cytomegalovirus,[101] pseudorabies,[102] rotavirus,[103] Japanese encephalitis,[104] and hepatitis B[64,74,105–107] viruses. In these trials, the protective immunity was predominantly mediated by specific antibodies. Nevertheless, because genetic vaccination is still a new form of generating functional protective immune responses, participation of T lymphocytes cannot be immediately excluded.

Human immunodeficiency virus (HIV) is among the most studied viruses. Successful induction of antibodies as well as CD4 and CD8 CTL specific to several viral proteins has been achieved by genetic immunization of mice.[79] These results stimulated studies using nonhuman primates as models for genetic immunization against HIV infection. Immunization of chimpanzees with a DNA vaccine encoding HIV-1 *env*, *rev*, and *gag/pol* generated antibodies and cell-mediated immune responses specific to these viral antigens. Most importantly, both vaccinated chimpanzees were protected against a challenge with a high dose of the heterologous HIV-1 strain.[108] A potential therapeutic application of genetic vaccination against HIV-1 infection was also examined in experimental models using primates.[109] DNA immunization of HIV-1 infected chimpanzees with plasmid containing the *env/rev* genes boosted the antibody immune response to *env* and significantly reduced the viral load.[110]

Recently, one group of researchers explored the possibility of improving genetic vaccination in nonhuman primates by using different carriers for priming and boosting the immune response. These researchers showed that an intradermal DNA priming of Rhesus macaques followed by a booster injection with recombinant fowl pox virus led to a remarkably strong protection against a series of challenges including a highly virulent primary isolate of the heterologous HIV-1 strain. The observed protective immunity was independent of the neutralizing antibodies to HIV-1 and most likely was mediated by T cells.

Based on these very promising results, a number of clinical trials are under way in several countries including the United States, England, Switzerland, and Sweden. The results of phase I and II trials have recently become available. They demonstrated that DNA administration to HIV-1 seropositive individuals is well tolerated and did not evoke any kind of local or systemic changes in vaccinees. These results also demonstrated that DNA immunization can improve an already existing immune responses (antibody and CTL) to HIV-1 antigens in these individuals.[112,113] The experiments on vaccination of healthy seronegative individuals have not yet been performed.

9.4 DNA VACCINE AGAINST BACTERIAL INFECTIONS

Over the last few years, the results of experiments in which mycobacterial heat shock protein (HSP) antigens were presented to the immune system in a way similar to viral antigens have had significant impact on our understanding of protective immunity against tuberculosis. Tuberculosis kills 3 million people every year.[114,115] The disease is due to respiratory infection with *Mycobacterium tuberculosis* and the World Health Organization has developed strategies to bring tuberculosis under control by using a combination of vaccination with bacillus Calmette-Guérin (BCG) to boost immunity and antibacterial drug treatment to directly kill the bacteria.[116] Despite all these efforts, there are still 10 million new cases in the world every year, mainly in developing countries, and this rate of morbidity and mortality has remained almost unchanged from year to year.[115] Even in the more affluent countries, the rate of this infection is increasing. Furthermore, infection with multidrug-resistant variants of *M. tuberculosis* is increasing at an alarming rate.[115,117] The recent advances in the development of DNA vaccine technology provides new opportunities to control infections with *M. tuberculosis*. Collectively, these new challenges only emphasize the urgency for the development of new drugs to treat tuberculosis and new vaccines to prevent infection.[118]

The BCG vaccine is broadly used to protect against tuberculosis. It is a live vaccine derived from *M. bovis* that has been attenuated by prolonged cultivation in the laboratory. In ten randomized controlled trials of BCG vaccines carried out since 1930 the protective efficacy against tuberculosis has ranged from 0 to 80% in different populations.[117] Thus, BCG is far from being an ideal vaccine against tuberculosis, and in several countries its use has been discontinued not only because of questionable efficacy but also because it prevents the subsequent use of skin sensitivity tests to detect tuberculosis infection.[117] Therefore, a new vaccine composed of only a few key protective antigens needs to be developed.

The approach based on the use of only a few antigens has not been completely successful in the development of a new vaccine against tuberculosis. Over the years, it has been shown that injection of dead BCG or a wide variety of antigenic components, even in large amounts and with an adjuvant, generates only a moderate degree of protective immunity in experimental animals.[119] Therefore, it is not known whether any individual antigen may deliver protection as efficiently as bacterial infection does. Recently, the availability of cloned genes for mycobacterial protein antigens and new expression vectors for mammalian cells has allowed for the development of a new approach; namely, the expression of individual genes directly in antigen-presenting cells (APCs). The rationale for this approach is that BCG, like *M. tuberculosis*, multiplies inside macrophages, one of the main APCs of the immune system, and antigens synthesized in APCs can be presented for immune recognition differently, which evokes a different immune response compared to the way in which extrinsic antigens do.[120] Thus, endogenous antigens from intracellular viruses or bacteria can be presented on the cell surface for recognition by specialized CTL (CD8 T cells), whereas exogenous antigens tend to be presented for recognition by helper T lymphocytes (CD4 T cells). Hence, it was reasoned that the use of genes for those antigens that are likely to be highly expressed by mycobacteria living in macrophages may mimic some of the essential features of the live BCG vaccine.[116,121–]

9.4.1 IMMUNIZATION WITH A RETROVIRAL VECTOR

The HSP65 of *Mycobacterium leprae* was the first cloned mycobacterial antigen gene used in DNA immunization experiments. Fortunately, this protein is highly conserved, antigenically very similar to that of *M. tuberculosis,* and is likely to be produced by mycobacteria under the stress of intracellular residence. A retroviral shuttle vector was used to express the HSP65 antigen in the monocyte-like tumor cell line J774 (Balb/c origin).[122] These cells presented the expressed antigen for specific recognition by T cells of the CD4+, CD8+, and γ/δ TCR types.[116,121–125] When Balb/c mice were injected with transfected cells (J774-hsp65) they acquired a remarkably high degree of protection against a subsequent intraperitoneal or intravenous challenge infection with virulent *M. tuberculosis* H37Rv. The number of bacterial cells declined exponentially in internal organs and was, for example, 100-fold lower in lungs after 5 weeks compared to control animals.[121,123] Injection of this protein with adjuvant was ineffective. The protection was dependent on tumor cell viability and was antigen-specific. For example, there was no protection against challenge with *Listeria monocytogenes*.[121,123,124] The immunizing J774-HSP65 cells were rapidly dying *in vivo* in C57Bl/6 mice (haplotype b), and hence the protective effect was negligible. This observation is consistent with protection dependent on the function of the transfected cells as APCs, rather than a mere antigen source.[125]

The observation of such a high degree of protection with the system described above caused a major conceptual shift that protection can be achieved by vaccinating with a single mycobacterial protein. However, there are several problems associated with direct application of this single-protein concept to vaccine development. In practice, it may be less efficient to use vaccine based on a single microbial protein because not all individuals may respond to any one antigen. For this reason, several other mycobacterial antigens have been tested. The other important point for consideration is that HSP65 may contain cross-reactive epitopes causing autoimmune diseases, and therefore these epitopes must be removed from the vaccine. In addition, these data were obtained using only tumor cells. The process of antigen presentation may significantly differ when APCs other than tumor cells are used.

One way of avoiding the use of tumor cells is to remove normal tissue, introduce a gene *in vitro,* and then replace it into the tissue. This experiment has been performed with bone marrow from 5-fluorouracil-treated BALB/c mice by using the same retroviral construct to transfect stem cells in culture and by transplanting the transfected cells into lethally γ-irradiated recipients.[123] A high proportion of these recipients expressed HSP65 in their peripheral blood cells after 2 weeks, and about half of those had specific delayed-type hypersensitivity (DTH) reactions to the protein (footpad swelling). The challenge infection by intravenous injection of *M. tuberculosis* H37Rv showed that only mice with DTH reactivity were protected. The degree of protection was similar to that observed in the experiments using J774-HSP65: 3 weeks after infection with counts of viable bacteria in liver, spleen, and lungs about 2 log lower in DTH responders than in nonresponders or than in mice reconstituted with normal bone marrow or with marrow transfected with the retroviral vector without the mycobacterial gene.[123] Evidently, features unique to tumor cells were not essential for this protective response.

9.4.2 The Importance of a CTL Response

The salient feature of endogenous antigens is presentation on MHCI for recognition by T cells via CD8-associated receptors. As expected, limiting dilution analysis of T cells in the spleens of J774-HSP65-immunized mice showed that there was an increase in the number of CD8+ T cells that was at least equal to the CD4+ response.[121,123–125] In contrast, aprotein inoculated with an adjuvant selectively increased the frequency of CD4+ cells. Furthermore, 12 CD4+ and 12 CD8+ HSP65-specific T-cell clones were characterized extensively and were shown to have antimycobacterial activity *in vitro.*[125] The inhibition of intracellular tubercle bacilli was observed when supernatants from the clones were used to activate infected macrophages. When these cell clones were placed into cell–cell contact with infected macrophages in the presence of anti-IFN-γ, those cells that had antigen-specific cytotoxicity also inhibited the growth of bacteria. Consequently, two properties were shown to be associated with *in vitro* antimycobacterial activity: IFN-γ production and cytotoxicity. A few representative clones were used in the experiment on adoptive transfer of protection.[125] These CD4+ and CD8+ clones represented a broad spectrum of activities in terms of IFN-γ and IL-4 production and cytotoxicity. The use of some of these clones resulted in bacterial killing. A substantial reduction in the number of bacteria in the internal organs was observed. The most effective clones were CD8+ cytotoxic clones. If a clone was producing IFN-γ, the effect attributable to IFN-γ could be neutralized by giving the animal a monoclonal antibody against IFN-γ. This study demonstrated that CD8+ clones are more effective than CD4+ clones in protection, and both IFN-γ and cytotoxicity contribute to protection, with cytotoxicity perhaps being more important.[125]

Robert Modlin and co-workers[126] have recently discovered that some human cytotoxic CD8+ T lymphocytes can deliver highly microbicidal proteins into infected macrophages and in so doing kill virulent *M. tuberculosis*. These lethal cells can be conventionally MHCI restricted. The essential feature is that they lyse the target macrophages by a granule-mediated perforin mechanism of apoptosis and by codelivering bactericidal proteins. A key mycobactericidal protein delivered by the T cells has been identified as granulysin present in the same granules with perforin.[127] This protein is also found in NK cell granules and has a known potent lethal action against a range of microorganisms and tumor cells.

9.4.3 IMMUNIZATION WITH PLASMID DNA

A disadvantage of the retroviral vector used to generate HSP65 inside APCs is that it depends on its integrating into the nucleus of multiplying cells such as tumor or bone-marrow stem cells. A much simpler and more versatile approach has now become available in the form of direct injection of plasmid DNA. This nucleic acid vaccination approach has dramatically facilitated the global quest for new vaccines against tuberculosis.[6,116,118–120,128–132] DNA vaccination is an alternative means of generating endogenous antigens. The first evidence that this approach could be used to generate protective antituberculosis immunity was reported in 1994.[132]

Plasmids, which were used for nucleic acid vaccination, contained the HSP65 gene under the control of the cytomegalovirus immediate-early gene promoter (pCMV) or the murine hydroxy-methylglutaryl-CoA reductase gene promoter (pHMG). Both plasmids yielded similar results.[132] Most importantly, it was confirmed that immunization with DNA encoding only a single myco-bacterial antigen could confer protection equal to that obtained with live BCG. In a typical test, mice were immunized by injecting 50 μg of DNA-expressing HSP65 into the muscle of each hind leg four times at 2- to 3-week intervals. These mice were challenged by intraperitoneal infection with virulent *M. tuberculosis* 3 weeks after immunization. The results of the infection were assessed in 5 weeks by counting live bacteria in internal organs. Compared to nonimmunized control animals, or those that received plasmid without the mycobacterial gene, mice immunized with HSP65 DNA had about 100-fold fewer bacteria in the lungs. This is very similar to the protective effect obtained in mice given intradermal BCG.[6,116,118–120,128–132]

HSP65 DNA protects to a similar extent in all tested mouse strains such as BALB/c, outbred albino Parkes, and cross-bred CBA/B10. A broad range of protective activity is important for vaccine development and can be explained by the presence of multiple cross-immunoreactive epitopes in the antigen.[128–132] However, a vaccine should probably contain either more than one antigen or more than one gene. Five more mycobacterial candidate genes have been tested separately and in combination as DNA vaccines using pCMV constructs in BALB/c mice. The most efficient can-didates were genes encoding for HSP65, HSP70, or 36 kDa proline-rich antigen. The degree of protection was similar to that obtained with live BCG.[128–132] The combination of plasmids injected simultaneously resulted in greater protection vs. when they were injected separately.[130]

Antibody responses were readily detected by enzyme immunoassay (EIA) 2 weeks after the third dose of plasmid expressing HSP65. A strong lymphoproliferative response to HSP65 antigen was also detected in splenocytes.[129–131] Cells responding after 4-day cultivation with the antigen released IFN-γ. No detectable IL-4 was observed by EIA, indicating a predominantly Th1/Tc1-type response. IL-4 (an indicator of a Th2-type response) became detectable *in vitro* after vaccinating with a plasmid encoding the accessory molecule B7.2 in a 50:50 mixture with HSP65 DNA.[130] A strong Th1/Tc1 bias of the response was further evidenced by RT-PCR analysis of mRNA for cytokines in inguinal lymph nodes draining HSP65 DNA-vaccinated muscle 2 weeks after the fourth injection. No mRNA for IL-4, IL-10, or IL-13 was detectable, whereas the number of IFN-γ and IL-12 mRNA was increased compared to controls receiving just the vector. Splenocytes from HSP65 DNA-vaccinated mice also displayed antigen-specific cytotoxicity against 51Cr-loaded P815 target cells that had been pulsed with synthetic peptides. Two of three peptides representing predicted MHCI restricted T cell epitopes of HSP65 were recognized.[130]

9.4.4 PERSISTENCE OF MEMORY FOR SPECIFIC CYTOTOXICITY AFTER DNA VACCINATION

The endogenous synthesis of the HSP65 antigen (J774-HSP65 or DNA vaccination) or delivery of this antigen into the cytosol (protein-loaded liposomes) generated protection and this effect was associated with a strong CD8+ response in which CD44hi memory-associated cells were prominent.[128] The highest level of protective immunity was found 1 week after immunization with J774-HSP65. However, this level declined substantially within 8 months. Delivery of antigen with liposomes or DNA immunization also generated a substantial early protective immunity similar to BCG. However, the level of immunity elicited by liposomes declined, similar to immunity caused by J774-HSP65, whereas the immune response elicited by hsp65-DNA was sustained at levels similar to the BCG-induced protection. The DNA vaccination also induced an increase in the proportion of CD44hi/CD8+ splenocytes. The count of cytotoxic HSP65-responsive cells among CD44hi/CD8+ splenocytes at different intervals after various vaccination protocols showed that the frequency of these cells was at the highest level (1 of 8) 1 week after immunization. However, this level had substantially declined within 8 months after immunization. Immunization using HSP65-liposomes and HSP65-DNA also resulted in a high frequency of CD44hi cells after 1 week (about 1 of 200). Again, this level subsequently declined over time when HSP65 liposomes were used. The opposite effect was observed after DNA vaccination when the frequency of CD44hi was increased up to 1 of 12 by 8 months. This increased frequency of cytotoxic HSP65-specific CD8+ memory cells was also seen after BCG vaccination. One hypothesis explaining these findings is that only live BCG and DNA vaccination provide for a persistent source of intracellular antigen, which is required to sustain the level of CTL memory and to maintain long-lasting protection.

9.4.5 TYPE 1 CYTOKINES

A striking difference between the immune response to DNA vaccination and the immune response to either BCG or *M. tuberculosis* infection is that DNA induces an almost entirely protective type 1 cytokine response, whereas mycobacterial infections results in the induction of a major component of the noncytotoxic T cells that produce a noncharacteristic type-1 cytokine response.[128] During an infection or after immunization, the numbers of CD4 and CD8 HSP65-reactive T cells were equally increased in the spleen. However, during an infection the majority of these cells were CD44lo, and these cells produce IL-4, whereas after DNA immunization the majority of cells were CD44hi, which produce IFN-γ. After adoptive transfer of protection to naive mice, the total CD8+ cell population purified from spleens of immunized mice was more protective than that from infected mice. When these cells were separated into CD4 and CD8 types, and then into CD44hi and CD44lo types, CD44lo cells were essentially unable to transfer protection.

The most protective CD44hi cells were CD8 cells. The CD8 cells obtained from the immunized mice were much more protective than those obtained from infected mice. Thus, whereas the T cells of CD44lo IL-4-producing phenotype prevailed during infection, the protective T cells of CD8/CD44hi IFN-γ-producing phenotype predominated after immunization. This conclusion was confirmed and extended by analysis of 16 HSP65-reactive T-cell clones from infected mice and 16 from immunized mice. In addition, the most protective clones displayed antigen-specific cytotoxicity.[128]

It can be argued that microbial stress proteins are not good candidates for inclusion in subunit vaccines because of the implication of homologous mammalian proteins participating in autoimmune diseases. Nonetheless, there are potential alternative ways of using these antigens in vaccines: the autoimmune epitopes may be removed from the protein by genetic engineering manipulations,[133] or the immune response may be modified by vaccine formulation or through various delivery procedures to elicit protection without autoimmune reactivity.[134] It has been shown that T cells recognizing mycobacterial HSPs can either protect against infections or potentiate autoimmune disease depending on their phenotype.[134,135] In support of this observation DNA vaccination with mycobacterial HSP65 was recently found to protect the Lewis strain of rats against induction of adjuvant arthritis.[136]

There is an opinion that the immune response to HSP65 does not play an important role in the protective immunity[137] because (a) this antigen cannot be released by bacteria into infected cells early during the infection and is released only by killed bacteria following immune recognition of different bacterial antigen(s), and (b) T cells of the type generated at this stage do not provide protection.[138] The data presented above, however, do not support this opinion. The facts that CD8+ HSP65-reactive T cells can be found in BCG-vaccinated mice, infected macrophages can be lysed by such T-cells *in vitro*, and adoptive transferral facilitates protective immunity *in vivo* all imply that the HSP65 protein is produced in substantial quantities by intracellular mycobacteria. There is a possibility that the cross-immunoreactive endogenous murine HSP60 protein may be induced by the stress of infection and can cause the effects attributable to HSP65. However, this suggestion contradicts the fact that the cytotoxic activity of the CD8+ clone was specific for mycobacterial HSP65.[121,124]

The immunologic properties of HSPs and HSP–peptide complexes have a number of significant implications for the development of DNA vaccination against intracellular infections and for the treatment of preexisting infections with mycobacteria. The observations made with mycobacterial proteins are also relevant to certain parasites.

9.5 DNA VACCINATION AGAINST PARASITIC INFECTIONS

9.5.1 DNA Vaccination in Experimental Models

Since the first report in 1987 it has become well established that CD8 T cells are important factors in host immune protection against several parasitic infections. The obligatory intracellular stages of parasites such as *Plasmodium*, *Trypanosoma cruzi,* and *Leishmania* can deliver antigens to the host cell cytoplasm. Therefore, these antigens can be presented through MHCI molecules to cytotoxic CD8 T cells. The fact that CD8 cells are potent inhibitors of parasitic development prompted investigators to explore naked DNA vaccination as an alternative approach in the development of vaccines against these tropical diseases.

Initial studies on DNA vaccination against a parasitic infection were performed using plasmids containing a gene encoding for the entire circumsporozoite (CS) protein of the rodent malaria parasite, *P. yoelii.* Multiple inoculations with naked DNA resulted in eliciting specific antibodies and CD8 T cells. The most important observation from this experiment was that this immunization provided a significant degree of protective immunity in BALB/c mice against a challenge with *P. yoelii* sporozoites.[139] The protective immunity was genetically restricted and only BALB/c mice (H2^d) were effectively protected. This restriction could be explained, at least in part, by the fact that protective immunity was dependent on the presence of CD8 T cells. Depletion of this subpopulation of lymphocytes completely abrogated the immune protection.[139]

DNA vaccination against *P. yoelii* infection was also successful with a second plasmid carrying the gene encoding for antigen Py17Hep. Immunization with a plasmid containing this gene generated significant protective immunity against a challenge with *P. yoelii* sporozoites in two mouse strains distinct from BALB/c. Similar to immunity induced by the CS gene, protective immunity was dependent on CD8 T cells and could not be achieved in mice treated with anti-CD8 antibodies.[140] In order to circumvent the strong genetic restriction observed after immunization with plasmids containing either CS or PyHep17 genes, several mouse strains were immunized simultaneously with both plasmids. This type of vaccination led to a high level of protective immunity in four of five mouse strains.[140] This experiment highlighted the importance of immunization with multiple malaria antigens to elicit a high degree of protective immunity in individuals with distinct genetic backgrounds. Based on these results, DNA vaccination trials on nonhuman primates and humans were undertaken. These experiments are reviewed later in this section.

The successful use of the naked-DNA approach to vaccinate against experimental malaria infection stimulated other investigators to use a similar approach to generate protective immunity against experimental infection with *Leishmania major* and *T. cruzi*. Immunization of highly susceptible BALB/c mice with a gene encoding for the LACK antigen led to a significant reduction in the size of primary cutaneous lesions caused by *L. major*.[141] In this study the authors compared the naked-DNA vaccination approach with a different type of immunization using a combination of recombinant LACK protein and IL-12. They concluded that DNA vaccination was at least as effective as the recombinant LACK protein and IL-12 combination. Nevertheless, the mechanisms of immune protection in these two systems were completely different. While BALB/c mice developed a strong protective CD4 Th1 response after immunization with recombinant protein, CD8 cells that secreted INF-γ were the main protective cells in naked-DNA-vaccinated mice.[141]

Further studies using this experimental model, where protective immunity generated by recombinant protein in combination with IL-12 was compared to a naked-DNA vaccine, demonstrated that an immune response to DNA immunization persisted longer. The persistent immunity provided by naked-DNA immunization was attributed to its ability to induce IL-12 secretion. Animals immunized with a combination of recombinant protein and a plasmid DNA carrying the IL-12 gene also presented a persistent immunity similar to mice injected with naked-DNA vaccine.[142]

In addition to the LACK protein, another antigen, GP63, expressed by *L. major*, was used in DNA vaccination experiments. Immunization of BALB/c mice with a plasmid encoding for this antigen induced a strong CD4 Th1 immune response and completely blocked the *L. major* infection in one third of the animals.[143]

Recently, DNA vaccination studies were also performed with two distinct genes of *Trypanosoma cruzi*, the causative agent of American trypanosomiasis or Chagas' disease. In both cases, these genes encoded for the major surface antigens of trypomastigote forms of these parasites. Trypomastigotes are exclusively infective forms responsible for the dissemination of infection in the vertebrate hosts. These forms are unable to replicate inside host cells. In one of these studies, a gene encoding for the catalytic domain of the enzyme *trans*-sialidase (TS) was used. Immunization of BALB/c mice with this TS gene elicited specific antibodies and facilitated T-cell activation.[144] The T-cell immune response was later characterized as being mediated mainly by CD4 Th1 and CD8 TC1 cells, which secreted a large amount of INF-γ, but not IL-4 or IL-10.[145] Upon challenge with infective trypomastigotes of *T. cruzi*, the immunized mice had a significant reduction in parasitemia and survived the acute lethal infection.[144]

Similar results were described in BALB/c and C57Bl/6 mice immunized with a plasmid containing the Trypomastigote Surface Antigen-1 gene. The naked DNA vaccination was shown to generate antibodies as well as CD8 CTL specific to parasitic epitopes.[146] After challenge with trypomastigotes of *T. cruzi*, a significant reduction in parasitemia was observed in C57Bl/6 mice, but not in BALB/c mice. Nevertheless, in both cases the DNA-vaccinated mice were more likely to survive acute infection than mice injected with the plasmid vector alone as a control.[146]

9.5.2 Enhancement of DNA-Generated Immune Response to Parasitic Antigens

The earlier studies performed with recombinant viruses expressing a foreign malaria epitope showed that sequential immunization with two distinct live viruses expressing the same CD8 T-cell epitope has the potential to significantly increase the protective immunity mediated by CD8 T cells against malaria.[147,148] Priming of mice with recombinant influenza virus followed by a booster injection with recombinant vaccinia virus, both expressing the same CD8 T-cell epitope, elicited a high degree of protective immunity against malaria infection.[147,148] Similar protective immunity could not be achieved with two doses of only recombinant influenza or vaccinia viruses. This unexpected observation was explained by the fact that mice immunized with the combination of these two different carrier viruses had a frequency of epitope-specific CD8 T cells 10 to 20 times higher than animals immunized twice with one of these carrier viruses.[149] The frequency of epitope-specific

CD8 T cells depends on the sequence of immunization with these two viruses. In mice primed with recombinant vaccinia virus followed by a booster injection with recombinant influenza virus the frequency of CD8 T cells was ten times lower than after priming with influenza virus followed by a vaccinia virus booster.[149] No protective immunity could be observed in this case.[147] These findings were recently extended to other malaria CD8 epitopes expressed in the CS protein of *P. falciparum*, a human malaria parasite.[150] A similar observation has been made when Influenza virus or an HIV CD8 epitope was used, thereby confirming its potential to enhance the specific immune response against multiple pathogens.[149,151]

Based on these observations several authors have explored this unusual approach to increase the CD8 immune response by using two different vectors for immunization. Studies using the CS protein of two distinct rodent malaria parasites, *P. yoelii* and *P. berghei*, were performed in two laboratories independently. In mice that had already been immunized with plasmid containing the CS gene, a boost injection with vaccinia virus containing the same gene provided a much greater specific CD8 response than two injections of any two vectors. The increase in frequency of specific CD8 cells was dependent on the order of immunization because priming with recombinant vaccinia virus followed by a boosting injection of naked DNA did not evoke such a significant increase in the frequency of specific CD8 T cells.[152,153] This protocol of immunization not only led to a significant increase in the CD8 immune response, but it also provided a significant degree of protection against a challenge with *P. berghei* or *P. yoelii*.[152,153]

The immunogenicity of naked-DNA vaccines can be further enhanced by coimmunization with plasmids containing genes of cytokines or costimulatory molecules of the immune system. Using the same rodent malaria model (*P. yoelii*), it was shown that co-injection with a plasmid containing the gene of murine GM-CSF enhanced both the CD8 response and the protective immunity to malaria in mice immunized with the CS gene.[154] Studies using a gene encoding for co-stimulatory molecules of the immune system have not yet been performed in parasitic infections.

9.5.3 DNA IMMUNIZATION OF NONHUMAN PRIMATES AND HUMANS

The successful results of naked-DNA immunization against rodent malaria led to vaccine trials in nonhuman primates and human volunteers. Plasmids containing four distinct genes of *P. falciparum* (PfCSP, PfSSP2, PfLSA-1 and PfExp1) were used to vaccinate rhesus monkeys. The animals were injected with either a single plasmid (PfSSP2) or a mixture of two plasmids (PfCSP/PfSSP2 or PfExp1/PfLSA-1), or a mixture of all four plasmids. Of the 12 monkeys, 9 developed low titers of a specific antibody response to preerythrocytic stages of *P. falciparum*. The *in vitro* CTL responses were detected in at least one of the *P. falciparum* proteins in peripheral blood mononuclear cells (PBMC) of all animals immunized with malaria genes. In contrast, PBMC of animals immunized with a plasmid alone displayed negligible levels of cytotoxicity. The CTL response was mediated by CD8 cells, which also produced INF-γ.[155] These results indicated that, as in the case of rodents, naked-DNA vaccines could elicit a CTL response in nonhuman primates. This study also demonstrated that naked-DNA vaccination could be a feasible approach to the development of a multivalent, multistage vaccine against malaria and perhaps against other parasitic diseases.

A very recent study in humans provided additional support to the observation that naked-DNA vaccines can be successfully used to elicit CTL responses. In this study, groups of five healthy individuals were immunized with different doses of a plasmid containing the PfCSP gene. Individuals received 3 i.m. doses of plasmid, each consisting of 20 to 2500 μg of DNA. Specific CTL responses were detected in two of five individuals immunized with three doses of either 20 or 100 μg of DNA. The higher immunizing doses (500 and 2500 μg of DNA) generated a CTL response in three of five and four of five individuals, respectively. MHC-restricted CD8 T-cells mediated CTL activity. CTL responses were significantly higher in individuals who received either 500 or 2500 μg of DNA. However, no difference was observed in the magnitude of the CTL response after three doses in individuals immunized with these two doses of DNA.[156] In general, the CTL responses were stronger in DNA immunized vaccinees compared to individuals naturally exposed to *P. falciparum* infection. This study strongly supports the idea that naked DNA vaccination can be further developed for human use.

Although studies using naked-DNA immunization are in the initials stages in most parasitic infections, they have already provided evidence that this type of immunization can generate the MHCI-restricted CD8 T cells with strong antiparasitic potential. This approach is especially important because it generates a Th1 CD4 T cell that results in the secretion of IFN-γ, a cytokine that mediates protection against most of the obligatory intracellular parasitic infections.

9.6 DNA VACCINATION AGAINST CANCERS

DNA inoculation represents a novel, promising approach to cancer therapy. Recently, different new strategies based on the latest advances in tumor biology and immunology have been developed for cancer gene therapy.

Carcinogenesis is the pathological process resulting from incurred genetic lesions affecting families of genes that regulate vital cellular functions. These lesions may alter the normal molecular mechanisms regulating cell growth, proliferation, and signal transduction, and through this may lead to malignant transformation. Currently, the notion of cancer as a genetic disease is broadly accepted. Mutated or tumor cells generated at different stages of carcinogenesis, from initiation to tumor progression and the onset of metastasis, are likely to be eradicated by different surveillance mechanisms.

When cancer develops and can be diagnosed, cancer cells usually represent a heterogeneous population of cells that were not removed by homeostatic mechanisms. The transformed cells do not necessarily express genetically altered antigens detectable by the immune system. The virus-associated cancers such as human papillomavirus-associated carcinomas or hepatitis-B-virus-associated hepatocarcinomas, however, are exceptions from this rule in that they always produce immunogenic viral proteins. Nonetheless, cancers are often unable to stimulate an efficient antitumor immune response. For many years, the focus of tumor biology was mainly on the tumor cell itself. Recently, however, it has been found that normal cells such as endothelial cells, stromal fibroblasts, and blood cells also participate in the maintenance and progression of tumors. The modern concept of cancer as a complex cell microenvironment opens new possibilities for the development of new therapeutic strategies targeting cells other than tumor cells.

The range of strategies for cancer therapy based on DNA inoculation is very broad, starting from strategies focusing on tumor cell, to targeting tumor progression, to strategies dealing with tolerance to tumor-specific or tumor-associated antigens, to triggering effector mechanisms for tumor rejection. Some of these strategies can be clearly classified as genetic immunization, while the other strategies are mainly cytoreductive, producing a positive effect in combination with immunization. One of the striking recently developed strategies is the conversion of drug-resistant tumor cells into drug-sensitive cells by genetic transfection. Different variations of this new strategy have been extensively investigated. This section describes and discusses several new approaches that were developed using animal models, some of which are under evaluation at different stages of clinical trials.

9.6.1 TARGETING TUMOR BIOLOGY

Molecular cancer medicine is based on the detection of aberrations in fundamental molecular components that determine normal cellular behavior. Such factors as activation or amplification of oncogenes and progressive inactivation of tumor suppressor genes are usually involved in the malignant progression of tumors.[157] Among gene families that are most clearly implicated in malignant transformation are the tumor suppressor genes, such as *p53*, *p16*, and *rb*, and oncogenes, such as K-*ras*, c-*myc*, and *bcl*-2.

As an illustration consider the p53 tumor-suppressor protein. This protein regulates cell-cycle progression and as a transcription activator is involved in monitoring DNA damage. In case of minor DNA damage in cells with normal level of p53 expression, the cell cycle is arrested in the G1 phase, providing an opportunity for DNA repair. However, if damage is extensive, wild-type p53 triggers apoptosis, the programmed cell death. The p53 mutations are most commonly identified in human cancers and have been extensively studied. It has been demonstrated that the restoration of wild-type p53 function by gene transfer is sufficient to induce either cell-cycle arrest or apoptosis. This effect is not restricted to p53 and was also observed for oncogenes and other tumor suppressor genes. On this basis, the transfer of tumor suppressor genes to cancer cells has been considered one of the most promising approaches for gene therapy.

9.6.1.1 Suppressor Gene Replacement

Using either a liposomal system or adenovirus or retrovirus vectors, the effect of the wild-type *p53* gene on cells has been studied in several models. For example, *in vivo* and *in vitro* experiments using glioma cells demonstrated that the restoration of the p53 wild-type functions resulted in either apoptosis or suppression of tumor growth.[158] Similar results of wild-type gene transfer were obtained for various human cancers such as head and neck tumors, ovarian, prostate, breast, and lung cancers.[159,160] However, despite an accumulated body of information it is still not clear whether the results obtained *in vivo* were due to apoptosis of endothelial cells overexpressing wild-type p53 (antiangiogenic therapy) or to replacement of the normal gene in tumor cells or both. In any case, the results are promising and have encouraged preclinical studies.

Preclinical studies and phase I clinical trials have been conducted for patients with non-small-cell lung cancer. After p53 gene therapy, tumor regression was observed in three out of nine patients.[161,162] Clinical trials aiming at p53 gene replacement have been also carried out on patients with colorectal liver metastasis,[163] hepatocellular carcinoma,[164] and HNSCC (human head and neck squamous cell cancer).[161]

Studies *in vitro* and *in vivo* have revealed a complex interaction between different tumor-suppressor genes. Recent experiments have shown that the progression of glioma may depend on the acquisition of a new phenotype, involving subsequent addition of genetic defects. One of the most frequent additions is the inactivation of tumor-suppressor gene *p16*, which in certain models is associated with tumor invasiveness.[165]

Because genetic lesions in the *p16* tumor-suppressor gene can frequently be found in many lung cancer cell lines, as well as in primary lung cancer tissue,[166] and are the most common genetic alteration in HNSCC,[167] the cytotoxic effect of exogenous *p16* expression on cancer cell proliferation has been evaluated. Different cell lines of human breast, osteosarcoma, cervical, and lung cancer were infected with recombinant adenovirus Ad-p16 containing the wild-type *p16* gene. It was shown that the p16 gene expression induces apoptosis in these cells only when this gene is mutated or deleted. A similar observation has been made for glioma cells retaining wild-type *p53* gene. These cells were less prone to apoptosis.[168] The cancer cell lines containing the mutant or deleted *p16* gene were found to be more sensitive to the cytotoxic effect of Ad-p16. However, this effect was only observed when cells also contained a wild-type retinoblastoma gene (pRB).[169] This conditional relationship between *p16* and pRB was later explained by the finding that both genes function in the same pathway of the cell-cycle control. The expression of rb is negatively regulated by p16. In fact, rb is expressed at higher levels in cell lines lacking functional p16,[170] which is commonly observed in mesothelioma tumors. In these cells, transduction with Ad-p16 led to the accumulation of p16, which resulted in cell-cycle arrest, inhibition of pRB phosphorylation, slower cell growth, and eventual death of the transduced cells.[171]

More recently, another proapoptotic gene, E2F-1, was found to be implicated in the progression of human tumors such as gliomas and medulloblastomas and lung, colon, breast, and ovarian cancer cell lines.[172–174,87,144] E2F-1 is involved in the regulation of the cell cycle. This protein directs the cell cycle to the stage between the G1and S-phases. Nevertheless, overexpression of E2F-1 caused apoptosis, promoting antiglioma activity *in vitro* and *in vivo*.[175] Although not thoroughly understood, it is known that the mechanism of E2F-1-induced apoptosis is not similar to the one observed with p53. E2F-1 is a cellular transcription factor that can be restrained by the rb gene product. It was shown that E2F-1 protein can induce apoptosis in glioma cells independently of the status of *p53*, *p*16, or rb genes, thereby providing an alternative gene therapy approach in those cases where the wild-type p53 is retained.[168,176]

Besides the above-mentioned approaches, the combination of gene therapy with conventional anticancer treatments, such as radiotherapy and chemotherapy, has been investigated. Experiments performed along this line of research demonstrated that adenovirus-mediated p53 gene delivery potentiates the radiation-induced growth inhibition of the experimental brain tumors. Notably, these tumors were reduced in size more than 85% when animals bearing brain tumors were irradiated after intratumoral Ad-*p*53 injections. This combination therapy significantly improved the survival rate among treated animals compared to animals receiving only gene or radiation therapy.[177]

There are several examples of the beneficial combination of gene therapy with conventional chemotherapy. The synergistic effect was demonstrated *in vitro* and *in vivo* for the combination of Ad-*p*53 gene therapy and paclitaxel chemotherapy. The observed significant efficacy of this combination therapy warranted further clinical trials.[160] Another example of successful tumor therapy is the combination of Ad-*p*53 and cisplatin, a DNA-damaging agent. In this case, chemotherapy facilitated Ad-*p*53 induced apoptosis and tumor regression.[178] A phase I clinical trial of this combination therapy demonstrated an improved rate of disease-free survival among patients with advanced non-small-cell lung cancer.[162]

9.6.1.2 Inactivation of Oncogenes

It is known that oncogenes can regulate activity of different cellular factors such as signal transducers, transcription factors, growth factors, or growth-factor receptors. Various mutations or alterations in the expression level of protooncogenes may lead to the induction of cellular transformation. Despite a significant track of research in this area, there are no efficient therapeutic treatments targeting inactivation of oncogenes. One of the most promising technologies, the potential application of which to cancer treatment is being thoroughly explored, is antisense technology using synthetic oligonucleotides and artificial ribozymes.[179–181] Both of these approaches were evaluated in *in vitro* and *in vivo* experiments using c-*myc* gene as target, which is known to be present in human prostate cancer[182] and melanoma cell lines.[183] This technology may be applied to many different oncogenes causing neoplastic transformation of cells. However, the success of this therapy will rest to a significant degree on the resolution of the numerous technical problems such as delivery to the target cells, stability within cells, and high turnover rate.

The antiapoptotic gene *bcl*-2 is the only known oncogene that regulates programmed cell death. The product of this gene is involved in controlling the life span of memory-B cells and plasma cells. It was shown that the overexpression of the *bcl*-2 gene leads to the development of resistance to apoptosis and promotes tumorigenesis. The application of antisense oligonucleotides targeting the *bcl*-2 open reading frame causes a specific downregulation of the *bcl*-2 expression and fosters apoptosis. The success of the *in vitro* and *in vivo* experiments has stimulated the application of this *bcl*-2 antisense therapy to humans. Nine patients who had *bcl*-2-positive relapsed non-Hodgkin lymphoma were subjected to the *bcl*-2 antisense therapy with a noticeable symptomatic improvement. It is important to note that in some of these patients strong evidence of downregulation of the *bcl*-2-gene expression has been found.[184]

9.6.2 AUGMENTING TUMOR-CELL IMMUNOGENICITY

The immune system can be stimulated using various strategies to elicit efficient antibody and cell-mediated immune responses. Although which arm of the immune system is the most effective in the elimination of malignant cells remains unknown, it is clear from several studies that both responses can lead to rejection of tumor cells. It can be foreseen that different strategies may complement each other. However, the development of anticancer vaccines faces several unique problems. One of these problems derives from the fact that cancer cells trigger mechanisms to escape the immune surveillance system, which explains the poor immunogenicity of tumor cells. The tumor microenvironment may also contribute to this diminished immunogenicity. A current concept of the immune system postulates that it discriminates dangerous from nondangerous stimuli.[185,186] In light of this theory, in order to develop anticancer vaccines we need to understand how to train the immune system to recognize cancer cells as dangerous. Below, we discuss a variety of different strategies for gene therapy to stimulate the anticancer immune response.

9.6.2.1 Tumor Antigens

Tumor-associated antigens (TAA) were the very first candidates for the development of cancer vaccines. In fact, many genetically engineered constructs for gene therapy have been developed using genes encoding either a whole TAA or short peptides derived from TAA.[187–189] Carcinoembryonic antigen (CEA),[190,191] prostate cancer surface antigen,[192] and melanoma differentiation-associated antigens[193] are among the first studied vaccine candidates. There are numerous research studies exploring the use of these TAAs for vaccine development. For example, the use of the full-length cDNA encoding for CEA as a DNA vaccine has been examined. This vaccine not only elicited both CEA-specific humoral and cellular immune responses but also provided evident protection against the syngeneic CEA-expressing colon carcinoma in mice.[188,190] The other promising DNA vaccine candidates for cancer therapy are the human papilloma virus 16 E6 and E7 genes, immunization with which can protect against cervical carcinomas.[65,139,194–196,212]

9.6.2.2 Epitope DNA Vaccines

Immunization with the entire antigen may abolish the therapeutic specificity. This seems to be the case for most cancers, since most tumor-associated antigens subtly alter the "self" and thus are present in tumors as well as in normal tissues. Genetic vaccines encoding only the desired epitope have been employed to direct the immune response specifically to the selected region within the TAA. Ciernik et al.[197] showed for the first time that the T-cell epitope-specific cellular immune response can be efficiently induced by the genetic epitope vaccine using the adenovirus E3 leader sequence. This leader sequence facilitated transport of a single epitope into the endoplasmic reticulum, bypassing the need for a TAP transporter. The processed epitope then bound to class I MHC molecules. Currently, this vaccine strategy has evolved into the development of multiepitope DNA vaccines, which include multiple contiguous genes encoding different CD8 epitopes.[198]

9.6.2.3 Idiotypic Determinants

Among a vast number of antigens that have been explored, we should mention the idiotypic determinants of immunoglobulins expressed on the surface of B-cell lymphomas. The idiotypic determinant provides a tumor-specific antigen (TSA) and a target for immunotherapy. Immunization with DNA constructs encoding the idiotype (Id) of a murine B-cell lymphoma induced specific anti-Id antibody responses and protected mice against tumor challenge.[199]

On the other hand, it is known that anti-idiotypic antibodies may mimic antigens of interest and have been used as surrogates for these antigens in immunization procedures.[200–202] More recently, DNA constructs that contain genes encoding for a single chain Fv fragment (scFv), the smallest immunoglobulin fragment that retains a complete antigen-binding site, have been developed.[203] These scFv DNA constructs may be useful mainly in those cases where the mimicked antigen is a very poor immunogen, such as ganglioside-enriched TAA in tumors like melanoma and neuroblastoma. In addition, immunization with scFv genes may stimulate intracellular antibody (intrabody) production, which may prevent the expression of molecules important for tumor progression such as growth factor receptors, e.g., *erb*-2.[204–207]

9.6.2.4 Interleukins

Despite significant efforts invested in the development of DNA constructs for genetic immunization against certain antigens, it was shown that the use of DNA immunization strategy fails to produce the immune response comparable to protein immunization. In these cases, the inclusion of cytokine genes into the DNA constructs noticeably improved the efficacy of immunization. For example, the GM-CSF has been found to improve the efficiency of the immune response against some weak tumor antigens.[199]

In animal models the expression of cytokines by tumor cells after gene transduction confers a greater level of immunogenicity than the constructs lacking the cytokine genes. In fact, tumor cells transduced with cytokine genes (e.g., IL-2, IL-4, IL-6, IL-7, IL-12, TNFα, IFN-γ, or GM-CSF) can cause growth inhibition and tumor rejection, which results in a systemic immunity even against the parental, nontransduced tumor. Interestingly, mouse tumor cells transduced with the IL-2, IL-4, IFN-γ, or GM-CSF gene could induce a delay in growth of transplanted tumors or a rejection of already established metastatic lesions.[75,133,208–211] The cellular and molecular mechanisms by which lymphokines impair tumor growth are not fully understood. In addition to functioning as modulators of immune responses, lymphokines may alter the angiogenic balance within tumors. Lymphokines are among several antiangiogenic substances that are under evaluation for cancer therapy.[212,213]

9.7 IMPROVING ANTIGEN PRESENTATION

Despite the progress in genetic immunization, gene therapy protocols still fail to evoke a strong long-lasting immune response. The main reason for this is the low efficiency of antigen presentation. Usually, DNA vaccine is introduced via an intramuscular or subcutaneous route, which implicates the likely engagement of cells such as myocytes, antigen-presenting cells, and Langerhans cells. Different mechanisms of antigen presentation will be discussed below.

CTL play a crucial role in the immune response to cancer. The molecular basis for CTL-mediated tumor immunity is the recognition by CTL of peptide products of the antigen in the context of class I major histocompatibility complex molecules (MHC I) expressed on the surface of antigen-presenting cells. Many tumors lack MHC I expression on the cell surface. The metastatic cells were suggested to evade the immune responses, mainly through the loss of MHC I. Accordingly, it was shown that the direct transfer of MHC I genes into tumors stimulated immune responses.[214,215] Pioneering studies in this direction have used a gene encoding a foreign MHC I protein, also called transplantation antigen. Interestingly, such foreign antigens induced immune responses to unmodified tumor cells. *In vivo*, this approach has led to a significant reduction in tumor growth and to a complete regression in some cases.[216] An antitumor immune response after gene transfer of a foreign MHC I protein, HLA-B7, was also achieved in HLA-B7-negative patients bearing stage IV melanoma. T-cell analysis showed T-cell migration into treated lesions and enhanced tumor infiltrating lymphocyte reactivity. Local inhibition of tumor growth and a partial remission were also observed.[217] A similar result was reported by Stopeck et al.[218] in patients with metastatic melanoma. Thus, it can be concluded that gene transfer using the DNA-liposome complexes containing an allogeneic MHC protein stimulates local antitumor responses that facilitate the generation of effector cells for cancer immunotherapy.

Besides the MHC I molecules, the cell surface exposes accessory molecules such as B7-1, which are important counterparts in the antigen presentation. Tumor cells expressing B7-1 on their surface are more efficient in stimulating T cells than tumor cells that do not express this antigen.[219] Taking this observation into account, dendritic cells (DC) should be considered as APCs capable of stimulating an antitumor immune response. DC is known to be a crucial professional APC responsible for the induction of a primary cell-mediated immune response. On the other hand, myocytes and tumor cells are ill-equipped for antigen presentation. To overcome this limitation of tumor cells, new strategies, including the *de novo* expression of either MHC I or costimulatory molecules, have been recently developed.

Resting Langerhans cells, highly potent APCs in skin tissue, are inefficient accessory cells for the sensitization phase of the primary T-cell-dependent immune response. However, they may become immunologically mature in the presence of GM-CSF when they acquire most of the features of lymphoid dendritic cells including the increased expression level of MHC I and MHC II molecules. To evaluate the possibility of achieving this transformation by using recombinant DNA technique one group of researchers transfected the Langerhans cells with cDNA encoding for B7-1 or GM-CSF and cDNA encoding for CEA. The observation of augmentation of the CEA-specific lymphoblastic transformation and the specific antibody response supports the feasibility of these approaches and suggests that these approaches may be used in other models.[188]

Currently the clinical application of synthetic tumor peptide-based vaccines are limited to patients producing some specific MHC I molecules. To overcome this limitation DC-expressing tumor peptide–MHC complexes were generated. For this purpose, bone-marrow-derived DCs were transfected with plasmid DNA encoding the TAA, such as viral antigen E7 derived from the human papilloma virus (HPV) 16. A direct insertion of TAA genes into immunostimulatory DC resulted in the endogenous production and processing of the relevant antigenic peptides. When applied as a vaccine, these genetically modified DCs induced antigen-specific CD8+ CTL *in vivo* and promoted the rejection of a subsequent, normally lethal, challenge with an HPV 16-transformed tumor cell line.[220]

A different approach has been explored to develop interleukin-2 gene-transfected fibroblasts that were injected with autologous tumor cells. The functional studies demonstrated that the number of MHC I-restricted CTL directed against the autologous tumor had increased at the immunization site in two of two analyzed melanoma patients. The characterization of T-cell receptors in these lymphocytes infiltrating the vaccination site revealed that the population of lymphocytes consisted predominantly of CTL containing receptors identical to those from the patients' tumor-infiltrating lymphocytes. This finding indicated that the same CTL clone had infiltrated the tumor, circulated in the peripheral blood, and was amplified at the vaccination site.[209]

A number of strategies, including those mentioned above, could elicit the cytotoxic cell response that prevents transplanted tumor growth or progression of experimental metastasis. It is still not clear whether such approaches would prevent tumor growth in models resembling spontaneous carcinogenesis. The tumor microenvironment may restrict the function of CTL.[221,222] Microenvironment remodeling, as has been found in intense postnecrotic inflammatory responses, may induce efficient immune responses.[223] Antiangiogenic strategies as well as suicide-gene-based approaches may be used to induce tumor necrosis and a proinflammatory environment triggering a strong immune response.[224]

9.8 SUICIDE-GENE THERAPY

Delivery of genes that confer sensitivity to cytotoxic drugs (suicide genes) is among the approaches that have earned great attention by virtue of their potential applicability. Enzymes such as Herpes simplex virus thymidine kinase (tk)[225] and *Escherichia coli* cytosine deaminase (cd)[226] are able to convert nontoxic prodrugs into cytotoxic metabolites. Suicide genes are selectively introduced into dividing tumor cells by using a viral vector. Thus, subsequent treatment with drugs leads to the synthesis of a toxic nucleotide by the enzyme expressed in the transfected tumor cell. It was later observed that the killing effect extends to surrounding dividing cells that do not harbor the trans-fected gene, promoting the so-called "bystander effect." This effect seems to be highly selective for growing cells because it blocks DNA synthesis. This approach was originally described for the treatment of brain tumors, where it was showing the most promising results.

In order to improve gene transfer a series of retrovirus and adenovirus vectors has been used. The encouraging results obtained with these vectors[227,228] have allowed the use of suicide-gene/pro-drug systems in the Phase I trials. Several studies on central nervous system malignancies[225] and malignant pleural mesothelioma[229] have been reported. Tumor cytotoxicity and generation of an antitumor immune response was observed in patients who received intravenous ganciclovir therapy after localized administration of the adenovirus vectors carrying the HSV-*tk* gene. The cytotoxic effects were vector dose dependent. However, alterations in size and metabolic activity of tumors were not thoroughly characterized in these studies.

Another relevant observation is that HSV*tk*/ganciclovir-mediated tumor killing can occur via alternative pathways of cell death depending on the cell line. In this way, apoptotic or nonapoptotic mechanisms can be triggered by the cytotoxic effect, which appears to be implicated in the immune response generated after suicide-gene therapy. Recent experimental findings have demonstrated that immune mechanisms involved in these processes of tumor rejection can be of variable efficiency depending on the predominance of nonapoptotic (necrosis) mechanisms.[223]

The success of suicide-gene therapy depends on the complete eradication of tumor cells. According to experimental findings in some models, this effect can be achieved only when both genes, *tk* and *cd*, are coadministered. To explain this finding, it was suggested that a rapid and quantitative antigen release by drug-mediated tumor destruction is necessary for the development of T-cell immunity, which in turn completely destroys tumors.[230]

9.9 PERSPECTIVES FOR THE USE OF SYNTHETIC DNA IN CANCER VACCINATION

Paradigms learned with infectious diseases will allow for new strategies based on genetic vaccina-tion to prevent or treat virus-associated cancers, "infectious and hazardous modified self." Virus-associated cancers represent about 15% of all cancer cases worldwide. Prophylactic vaccination against hepatitis B virus would help prevent some cases of hepatocellular carcinoma; vaccination against HPV would have an impact on cervical cancer incidence. These diseases are prevalent in countries like China and Brazil, where both the social and economic impacts of prophylactic vaccination are evident. Most cancers, however, are not associated with viral infections. The notion that tumor frontiers go beyond the neoplastic cell opens a number of new targets for cancer therapy. We may learn how to reeducate the neoplastic cells, imposing controls on their cell cycle, growth, and cell death. Some of these strategies may even be useful as antiangiogenic approaches as well. The reeducation of the immune system is one of the most promising approaches. Immune cells have all the necessary battery of recognition molecules and cytolytic enzymes to destroy disguised tumor cells. Synthetic DNA strategies may help efficiently direct an immune response toward "noninfectious, but dangerous, modified self." As of June 1999, more than 2000 cancer patients were enrolled in gene-therapy clinical trials (http://www.wiley.com/genetherapy). These clinical trials are based on the strategies discussed above and illustrated in the chart in Figure 9.2. Only 15% of the trials target tumor biology, about 22% are based on suicide-gene therapy, and the remaining 63% are based on various aspects of tumor immunology, from immunomodulation to genetic immunization using known tumor-associated antigens. So far, only 3% of the trials combine strategies (e.g., suicide-gene therapy and cytokine expression in tumor cells). From the experimental research combined strategies hold the greatest promises for effective immunization. Testing the efficiency of these combined approaches in clinical trials are certainly part of the agenda for the next few years.

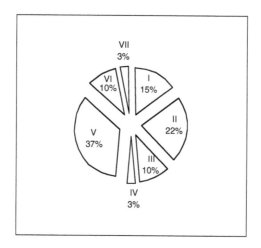

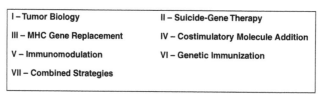

FIGURE 9.2 Current gene therapy clinical trials for cancer treatment. As of June 1999, more than 2000 cancer patients were enrolled in more than 200 approved clinical trials based on gene therapy (http://www.wiley.com/ genetherapy). Strategies were divided into distinct categories. Numbers represent the relative contribution (in percentages) of each strategy. Most tumor-biology-based approaches include trials aiming at replacement of tumor-suppressor genes or blockage of oncogene activities. Suicide-gene therapy is based on the transfer of viral genes rendering tumors sensitive to base analogues as ganciclovir, leading to cell death. About 63% of all current strategies are based on tumor immunology. An increasing number of trials combine strategies in order to test different approaches, which would ultimately improve antitumor immune responses.

REFERENCES

1. Southam, C. M. and Friedman, H., International confernce on immunotherapy of cancer, *Ann. N.Y. Acad. Sci.,* 277, 60, 1976.

2. Ribi, E. et al., Immunotherapy with nonviable microbial components, *Ann. N.Y. Acad. Sci.,* 277, 228, 1976.

3. Shimada, S., Yano, O., and Tokunaga, T., *In vivo* augmentation of natural killer cell activity with a deoxyribonucleic acid fraction of BCG, *Jpn. J. Cancer Res.* 77, 808, 1986.

4. Tokunaga, T. et al., Antitumor activity of deoxyribonucleic acid fraction from *Mycobacterium bovis* BCG. I. Isolation, physicochemical characterization, and antitumor activity, *J. Natl. Cancer Inst.,* 72, 955, 1984.

5. Kataoka, T. et al., Antitumor activity of synthetic oligonucleotides with sequences from cDNA encoding proteins of *Mycobacterium bovis* BCG, *Jpn. J. Cancer Res.,* 83, 244, 1992.

6. Sonehara, K. et al., Hexamer palindromic oligonucleotides with 5'-CG-3' motif(s) induce production of interferon, *J. Interferon Cytokine Res.,* 16, 799, 1996.

7. Tokunaga, T. et al., Synthetic oligonucleotides with particular base sequences from the cDNA encoding proteins of *Mycobacterium bovis* BCG induce interferons and activate natural killer cells, *Microbiol. Immunol.,* 36, 55, 1992.

8. Yamamoto, S. et al., Unique palindromic sequences in synthetic oligonucleotides are required to induce IFN [correction of INF] and augment IFN-mediated [correction of INF] natural killer activity, *J. Immunol.,* 148, 4072, 1992.

9. Krieg, A. M., CpG DNA: a pathogenic factor in systemic lupus erythematosus? *J. Clin. Immunol.,* 15, 284, 1995.

10. Pisetzky, D. S., Immune activation by bacterial DNA: a new genetic code, *Immunology,* 5, 303, 1996.

11. Lipford, G. B., et al., Immunostimulatory DNA: sequence-dependent production of potentially harmful or useful cytokines, *Eur. J. Immunol.,* 27, 3420, 1997.

12. Sato, Y. et al., Immunostimulatory DNA sequences necessary for effective intradermal gene immunization, *Science,* 273, 352, 1996.

13. Brinkmann, V. et al., Interferon alpha increases the frequency of interferon gamma-producing human CD4+ T cells, *J. Exp. Med.,* 178, 1655, 1993.

14. Finkelman, F. D. et al., Regulation by interferon alpha of immunoglobulin isotype selection and lymphokine production in mice, *J. Exp. Med.,* 174, 1179, 1991.

15. Parronchi, P. et al., IL-4 and IFN (alpha and gamma) exert opposite regulatory effects on the development of cytolytic potential by Th1 or Th2 human T cell clones, *J. Immunol.,* 149, 2977, 1992.

16. Field, A. K. et al., Inducers of interferon and host resistance. II. Multistranded synthetic polynucleotide complexes, *Proc. Natl. Acad. Sci. U.S.A.,* 58, 1004, 1967.

17. Tytell, A. A. et al., Inducers of interferon and host resistance. 3. Double-stranded RNA from reovirus type 3 virions (reo 3-RNA), *Proc. Natl. Acad. Sci. U.S.A.,* 58, 1719, 1967.

18. Lampson, G. P. et al., Poly I:C/poly-L-lysine: potent inducer of interferons in primates, *J. Interferon. Res.,* 1, 539, 1981.

19. Meurs, E. et al., Molecular cloning and characterization of the human double-stranded RNA-activated protein kinase induced by interferon, *Cell,* 62, 379, 1990.

20. Zilberstein, A. et al., Isolation of two interferon-induced translational inhibitors: a protein kinase and an oligo-isoadenylate synthetase, *Proc. Natl. Acad. Sci. U.S.A.,* 75, 4734, 1978.

21. Messina, J. P., Gilkeson, G. S., and Pisetsky, D. S., Stimulation of *in vitro* murine lymphocyte proliferation by bacterial DNA, *J. Immunol.,* 147, 1759, 1991.

22. Messina, J. P., Gilkeson, G. S., and Pisetsky, D. S., The influence of DNA structure on the *in vitro* stimulation of murine lymphocytes by natural and synthetic polynucleotide antigens, *Cell Immunol.,* 147, 148, 1993.

23. Cavanaugh, P. F. J., Ho, Y. K., and Bardos, T. J., The activation of murine macrophages and natural killer cells by the partially thiolated double stranded RNA poly(I)-mercapto poly(C), *Res. Commun. Mol. Pathol. Pharmacol.,* 91, 131, 1996.

24. Liang, H. et al., Activation of human B cells by phosphorothioate oligodeoxynucleotides, *J. Clin. Invest.,* 98, 1119, 1996.

25. Pisetsky, D. S. and Reich, C., Stimulation of *in vitro* proliferation of murine lymphocytes by synthetic oligodeoxynucleotides, *Mol. Biol. Rep.,* 18, 217, 1993.

26. Kimura, Y. et al., Binding of oligoguanylate to scavenger receptors is required for oligonucleotides to augment NK cell activity and induce IFN, *J. Biochem. (Tokyo),* 116, 991, 1994.

27. Goldstein, J. L. et al., Binding site on macrophages that mediates uptake and degradation of acetylated low density lipoprotein, producing massive cholesterol deposition, *Proc. Natl. Acad. Sci. U.S.A.,* 76, 333, 1979.

28. Krieger, M. and Herz, J., Structures and functions of multiligand lipoprotein receptors: macrophage scavenger receptors and LDL receptor-related protein (LRP), *Annu. Rev. Biochem.,* 63, 601, 1994.

29. Pearson, A. M., Rich, A., and Krieger, M., Polynucleotide binding to macrophage scavenger receptors depends on the formation of base-quartet-stabilized four-stranded helices, *J. Biol. Chem.,* 268, 3546, 1993.

30. Sen, D. and Gilbert, W., Formation of parallel four-stranded complexes by guanine-rich motifs in DNA and its implications for meiosis, *Nature,* 334, 364, 1988.

31. Sen, D. and Gilbert, W., Novel DNA superstructures formed by telomere-like oligomers, *Biochemistry,* 31, 65, 1992.

32. Williamson, J. R., Raghuraman, M. K., and Cech, T. R., Monovalent cation-induced structure of telomeric DNA: the G-quartet model, *Cell,* 59, 871, 1989.

33. Wloch, M. K. et al., The influence of DNA sequence on the immunostimulatory properties of plasmid DNA vectors, *Hum. Gene Ther.,* 9, 1439, 1998.

34. Roman, M. et al., Immunostimulatory DNA sequences function as T helper-1-promoting adjuvants [see comments], *Nat. Med.,* 3, 849, 1997.

35. Bradley, L. M., Dalton, D. K., and Croft, M., A direct role for IFN-gamma in regulation of Th1 cell development, *J. Immunol.,* 157, 1350, 1996.

36. Okamura, H. et al., Cloning of a new cytokine that induces IFN-gamma production by T cells [see comments], *Nature,* 378, 88, 1995.

37. Trinchieri, G., Interleukin-12: a proinflammatory cytokine with immunoregulatory functions that bridge innate resistance and antigen-specific adaptive immunity, *Annu. Rev. Immunol.,* 13, 251, 1995.

38. Yaegashi, Y. et al., Interferon beta, a cofactor in the interferon gamma production induced by Gram-negative bacteria in mice, *J. Exp. Med.,* 181, 953, 1995.

39. Cowdery, J. S. et al., Bacterial DNA induces NK cells to produce IFN-gamma *in vivo* and increases the toxicity of lipopolysaccharides, *J. Immunol.,* 156, 4570, 1996.

40. Tighe, H. et al., Gene vaccination: plasmid DNA is more than just a blueprint, *Immunol. Today,* 19, 89, 1998.

41. Chu, R. S. et al., CpG oligodeoxynucleotides act as adjuvants that switch on T helper 1 (Th1) immunity, *J. Exp. Med.,* 186, 1623, 1997.

42. Davis, H. L., Michel, M. L., and Whalen, R. G., DNA-based immunization induces continuous secretion of hepatitis B surface antigen and high levels of circulating antibody, *Hum. Mol. Genet.,* 2, 1847, 1993.

43. Finkelman, F. D., Relationships among antigen presentation, cytokines, immune deviation, and autoimmune disease [comment], *J. Exp. Med.,* 182, 279, 1995.

44. Heinzel, F. P. et al., Reciprocal expression of interferon gamma or interleukin 4 during the resolution or progression of murine leishmaniasis. Evidence for expansion of distinct helper T cell subsets, *J. Exp. Med.,* 169, 59, 1989.

45. Reiner, S. L. and Locksley, R. M., The regulation of immunity to *Leishmania major, Annu. Rev. Immunol.,* 13, 151, 1995.

46. Scott, P. et al., Immunoregulation of cutaneous leishmaniasis. T cell lines that transfer protective immunity or exacerbation belong to different T helper subsets and respond to distinct parasite antigens, *J. Exp. Med.,* 168, 1675, 1988.

47. Heinzel, F. P. et al., Endogenous IL-12 is required for control of Th2 cytokine responses capable of exacerbating leishmaniasis in normally resistant mice, *J. Immunol.,* 155, 730, 1995.

48. Heinzel, F. P. et al., Recombinant interleukin 12 cures mice infected with Leishmania major, *J. Exp. Med.,* 177, 1505, 1993.

49. Sadick, M. D. et al., Cure of murine leishmaniasis with anti-interleukin 4 monoclonal antibody. Evidence for a T cell-dependent, interferon gamma-independent mechanism, *J. Exp. Med.,* 171, 115, 1990.

50. Sypek, J. P. et al., Resolution of cutaneous leishmaniasis: interleukin 12 initiates a protective T helper type 1 immune response, *J. Exp. Med.,* 177, 1797, 1993.

51. Zimmermann, S. et al., CpG oligodeoxynucleotides trigger protective and curative Th1 responses in lethal murine leishmaniasis, *J. Immunol.,* 160, 3627, 1998.

52. Weller, P. F., Human eosinophils, *J. Allergy Clin. Immunol.,* 100, 283, 1997.

53. Goodman, J. S. et al., DNA immunotherapeutics: new potential treatment modalities for allergic disease, *Int. Arch. Allergy Immunol.,* 116, 177, 1998.

54. Hsu, C. H. et al., Immunoprophylaxis of allergen-induced immunoglobulin E synthesis and airway hyperresponsiveness *in vivo* by genetic immunization [see comments], *Nat. Med.,* 2, 540, 1996.

55. Hsu, C. H. et al., Inhibition of specific IgE response *in vivo* by allergen-gene transfer, *Int. Immunol.,* 8, 1405, 1996.

56. Jakob, T. et al., Activation of cutaneous dendritic cells by CpG-containing oligodeoxynucleotides: a role for dendritic cells in the augmentation of Th1 responses by immunostimulatory DNA, *J. Immunol.,* 161, 3042, 1998.

57. Sparwasser, T. et al., Bacterial DNA and immunostimulatory CpG oligonucleotides trigger maturation and activation of murine dendritic cells, *Eur. J. Immunol.,* 28, 2045, 1998.

58. Donnelly, J. J., Ulmer, J. B., and Liu, M. A., DNA vaccines, *Life Sci.,* 60, 163, 1997.

59. Cohen, A. D., Boyer, J. D., and Weiner, D. B., Modulating the immune response to genetic immunization, *FASEB J.,* 12, 1611, 1998.

60. Davis, H. L., Plasmid DNA expression systems for the purpose of immunization, *Curr. Opin. Biotechnol.,* 8, 635, 1997.

61. Liu, M. A. et al., DNA vaccines. Mechanisms for generation of immune responses, *Adv. Exp. Med. Biol.,* 452, 187, 1998.

62. Brazolot, M. C. et al., CpG DNA can induce strong Th1 humoral and cell-mediated immune responses against hepatitis B surface antigen in young mice, *Proc. Natl. Acad. Sci. U.S.A.,* 95, 15553, 1998.

63. Donnelly, J. J. et al., Preclinical efficacy of a prototype DNA vaccine: enhanced protection against antigenic drift in influenza virus [see comments], *Nat. Med.,* 1, 583, 1995.

64. Davis, H. L. et al., DNA vaccine for hepatitis B: evidence for immunogenicity in chimpanzees and comparison with other vaccines, *Proc. Natl. Acad. Sci. U.S.A.,* 93, 7213, 1996.

65. Cho, J. H., Lee, S. W., and Sung, Y. C., Enhanced cellular immunity to hepatitis C virus nonstructural proteins by codelivery of granulocyte macrophage-colony stimulating factor gene in intramuscular DNA immunization, *Vaccine,* 17,1136, 1999.

66. Chow, Y. H. et al., Improvement of hepatitis B virus DNA vaccines by plasmids coexpressing hepatitis B surface antigen and interleukin-2, *J. Virol.,* 71, 169, 1997.

67. Geissler, M. et al., Enhancement of cellular and humoral immune responses to hepatitis C virus core protein using DNA-based vaccines augmented with cytokine-expressing plasmids, *J. Immunol.,* 158, 1231, 1997.

68. Sin, J. I. et al., Enhancement of protective humoral (Th2) and cell-mediated (Th1) immune responses against herpes simplex virus-2 through co-delivery of granulocyte-macrophage colony-stimulating factor expression cassettes, *Eur. J. Immunol.,* 28, 3530, 1998.

69. Feltquate, D. M. et al., Different T helper cell types and antibody isotypes generated by saline and gene gun DNA immunization, *J. Immunol.,* 158, 2278, 1997.

70. Lekutis, C. et al., HIV-1 env DNA vaccine administered to rhesus monkeys elicits MHC class II-restricted CD4+ T helper cells that secrete IFN-gamma and TNF-alpha, *J. Immunol.,* 158, 4471, 1997.

71. Li, X. et al., Protection against respiratory syncytial virus infection by DNA immunization, *J. Exp. Med.,* 188, 681, 1998.

72. Martinez, X. et al., DNA immunization circumvents deficient induction of T helper type 1 and cytotoxic T lymphocyte responses in neonates and during early life, *Proc. Natl. Acad. Sci. U.S.A.,* 94, 8726, 1997.

73. Ulmer, J. B. et al., Protective CD4+ and CD8+ T cells against influenza virus induced by vaccination with nucleoprotein DNA, *J. Virol.,* 72, 5648, 1998.

74. Davis, H. L. and Brazolot, M. C., DNA-based immunization against hepatitis B virus, *Springer Semin. Immunopathol.,* 19, 195, 1997.

75. Kim, J. J. et al., Modulation of amplitude and direction of *in vivo* immune responses by co-administration of cytokine gene expression cassettes with DNA immunogens, *Eur. J. Immunol.,* 28, 1089, 1998.

76. Sin, J. I. et al., IL-12 gene as a DNA vaccine adjuvant in a herpes mouse model: IL-12 enhances Th1-type CD4+ T cell-mediated protective immunity against herpes simplex virus-2 challenge, *J. Immunol.,* 162, 2912, 1999.

77. Sin, J. I. et al., *In vivo* modulation of vaccine-induced immune responses toward a Th1 phenotype increases potency and vaccine effectiveness in a herpes simplex virus type 2 mouse model, *J. Virol.,* 73, 501, 1999.

78. Tsuji, T. et al., Enhancement of cell-mediated immunity against HIV-1 induced by coinoculation of plasmid-encoded HIV-1 antigen with plasmid expressing IL-12, *J. Immunol.,* 158, 4008, 1997.

79. Bagarazzi, M. L. et al., Nucleic acid-based vaccines as an approach to immunization against human immunodeficiency virus type-1, *Curr. Top. Microbiol. Immunol.,* 226, 107, 1998.

80. Corr, M. et al., Gene vaccination with naked plasmid DNA: mechanism of CTL priming, *J. Exp. Med.,* 184, 1555, 1996.

81. Donnelly, J. J. et al., DNA vaccines, *Annu. Rev. Immunol.,* 15, 617, 1997.

82. Fu, T. M. et al., Dose dependence of CTL precursor frequency induced by a DNA vaccine and correlation with protective immunity against influenza virus challenge, *J. Immunol.,* 162, 4163, 1999.

83. Ulmer, J. B. et al., Heterologous protection against influenza by injection of DNA encoding a viral protein [see comments], *Science,* 259, 1745, 1993.

84. Kim, J. J. et al., CD8 positive T cells influence antigen-specific immune responses through the expression of chemokines, *J. Clin. Invest.,* 102, 1112, 1998.

85. Kim, J. J. et al., *In vivo* engineering of a cellular immune response by coadministration of IL-12 expression vector with a DNA immunogen, *J. Immunol.,* 158, 816, 1997.

86. Iwasaki, A. et al., Enhanced CTL responses mediated by plasmid DNA immunogens encoding costimulatory molecules and cytokines, *J. Immunol.,* 158, 4591, 1997.

87. Kim, J. J. et al., Engineering DNA vaccines via co-delivery of co-stimulatory molecule genes, *Vaccine,* 16, 1828, 1998.

88. Kim, J. J. et al., Intracellular adhesion molecule-1 modulates beta-chemokines and directly costimulates T cells *in vivo, J. Clin. Invest.,* 103, 869, 1999.

89. Martins, L. P. et al., DNA vaccination against persistent viral infection, *J. Virol.,* 69, 2574, 1995.

90. Yokoyama, M., Zhang, J., and Whitton, J. L., DNA immunization confers protection against lethal lymphocytic choriomeningitis virus infection, *J. Virol.,* 69, 2684, 1995.

91. Kodihalli, S. et al., DNA vaccine encoding hemagglutinin provides protective immunity against H5N1 influenza virus infection in mice, *J. Virol.,* 73, 2094, 1999.

92. Kodihalli, S. et al., Cross-protection among lethal H5N2 influenza viruses induced by DNA vaccine to the hemagglutinin, *J. Virol.,* 71, 3391, 1997.

93. Lodmell, D. L. et al., DNA immunization protects nonhuman primates against rabies virus, *Nat. Med.,* 4, 949, 1998.

94. Ruitenberg, K. M. et al., DNA-mediated immunization with glycoprotein D of equine herpesvirus 1 (EHV-1) in a murine model of EHV-1 respiratory infection, *Vaccine,* 17, 237, 1999.

95. Bourne, N. et al., DNA immunization confers protective immunity on mice challenged intravaginally with herpes simplex virus type 2, *Vaccine,* 14, 1230, 1996.

96. Bourne, N. et al., DNA immunization against experimental genital herpes simplex virus infection, *J. Infect. Dis.,* 173, 800, 1996.

97. McClements, W. L. et al., Immunization with DNA vaccines encoding glycoprotein D or glycoprotein B, alone or in combination, induces protective immunity in animal models of herpes simplex virus-2 disease, *Proc. Natl. Acad. Sci. U.S.A.,* 93, 11414, 1996.

98. Nass, P. H., Elkins, K. L., and Weir, J. P., Antibody response and protective capacity of plasmid vaccines expressing three different herpes simplex virus glycoproteins, *J. Infect. Dis.,* 178, 611, 1998.

99. Donnelly, J. J. et al., Protection against papillomavirus with a polynucleotide vaccine, *J. Infect. Dis.,* 173, 314, 1996.

100. Vanderzanden, L. et al., DNA vaccines expressing either the GP or NP genes of Ebola virus protect mice from lethal challenge, *Virology,* 246,134, 1998.

101. Gonzalez, A. J. et al., DNA immunization confers protection against murine cytomegalovirus infection, *J. Virol.,* 70, 7921, 1996.

102. Haagmans, B. L. et al., Vaccination of pigs against pseudorabies virus with plasmid DNA encoding glycoprotein D, *Vaccine,* 17, 1264, 1999.

103. Herrmann, J. E. et al., Protection against rotavirus infections by DNA vaccination, *J. Infect. Dis.,* 174 (Suppl. 1), S93, 1996.

104. Konishi, E. et al., Induction of protective immunity against Japanese encephalitis in mice by immunization with a plasmid encoding Japanese encephalitis virus premembrane and envelope genes, *J. Virol.,* 72, 4925, 1998.

105. Mancini, M. et al., DNA-mediated immunization in a transgenic mouse model of the hepatitis B surface antigen chronic carrier state, *Proc. Natl. Acad. Sci. U.S.A.,* 93, 12496, 1996.

106. Prange, R. and Werr, M., DNA-mediated immunization to hepatitis B virus envelope proteins: preS antigen secretion enhances the humoral response, *Vaccine,* 17, 617, 1999.

107. Prince, A. M., Whalen, R., and Brotman, B., Successful nucleic acid based immunization of newborn chimpanzees against hepatitis B virus, *Vaccine,* 15, 916, 1997.

108. Boyer, J. D. et al., Protection of chimpanzees from high-dose heterologous HIV-1 challenge by DNA vaccination [see comments], *Nat. Med.,* 3, 526, 1997.

109. Ugen, K. E. et al., Nucleic acid immunization of chimpanzees as a prophylactic/immunotherapeutic vaccination model for HIV-1: prelude to a clinical trial, *Vaccine,* 15, 927, 1997.

110. Boyer, J. D. et al., DNA vaccination as anti-human immunodeficiency virus immunotherapy in infected chimpanzees, *J. Infect. Dis.,* 176, 1501, 1997.

111. Robinson, H. L. et al., Neutralizing antibody-independent containment of immunodeficiency virus challenges by DNA priming and recombinant pox virus booster immunizations, *Nat. Med.,* 5, 526, 1999.

112. Calarota, S. et al., Cellular cytotoxic response induced by DNA vaccination in HIV-1- infected patients, *Lancet,* 351, 1320, 1998.

113. MacGregor, R. R. et al., First human trial of a DNA-based vaccine for treatment of human immuno-deficiency virus type 1 infection: safety and host response, *J. Infect. Dis.,* 178, 92, 1998.

114. Kochi, A., The global tuberculosis situation and the new control strategy of the World Health Organization [editorial], *Tubercle,* 72, 1, 1991.

115. Snider, D. E., Jr., Raviglione, M., and Kochi, A., Global burden of tuberculosis, in *Tuberculosis: Pathogenesis, Protection and Control,* Bloom, B., Ed., ASM Press, Washington, D.C., 1994, pp. 3–11.

116. Silva, C. L., Lukacs, K., and Lowrie, D. B., Major histocompatibility complex non-restricted presentation to CD4+ T lymphocytes of *Mycobacterium leprae* heat-shock protein 65 antigen by macrophages transfected with the mycobacterial gene, *Immunology,* 78, 35, 1993.

117. Fine, P. E., The BCG story, lessons from the past and implications for the future, *Rev. Infect. Dis.,* 11 (Suppl. 2), S353, 1989.

118. Lowrie, D. B., Silva, C. L., and Tascon, R., Progress towards a new tuberculosis vaccine, *Biodrugs,* 10, 201, 1998.

119. Lowrie, D. B., Tascon, R. E., and Silva, C. L., Vaccination against tuberculosis, *Int. Arch. Allergy Immunol.,* 108, 309, 1995.

120. Germain, R. N., MHC-dependent antigen processing and peptide presentation: providing ligands for T lymphocyte activation, *Cell,* 76, 287, 1994.

121. Silva, C. L. and Lowrie, D. B., A single mycobacterial protein (hsp 65) expressed by a transgenic antigen-presenting cell vaccinates mice against tuberculosis, *Immunology,* 82, 244, 1994.

122. Silva, C. L. et al., *Mycobacterium leprae* 65hsp antigen expressed from a retroviral vector in a macrophage cell line is presented to T cells in association with MHC class II in addition to MHC class I, *Microb. Pathog.,* 12, 27, 1992.

123. Silva, C. L. et al., Protection against tuberculosis by bone marrow cells expressing mycobacterial hsp65, *Immunology,* 86, 519, 1995.

124. Silva, C. L. et al., Protection against tuberculosis by passive transfer with T-cell clones recognizing mycobacterial heat-shock protein 65, *Immunology,* 83, 341, 1994.

125. Silva, C. L. et al., Characterization of T cells that confer a high degree of protective immunity against tuberculosis in mice after vaccination with tumor cells expressing mycobacterial hsp65, *Infect. Immun.,* 64, 2400, 1996.

126. Stenger, S. et al., Differential effects of cytolytic T cell subsets on intracellular infection, *Science,* 276, 1684, 1997.

127. Stenger, S. et al., An antimicrobial activity of cytolytic T cells mediated by granulysin, *Science,* 282, 121, 1998.

128. Bonato, V. L. et al., Identification and characterization of protective T cells in hsp65 DNA- vaccinated and *Mycobacterium tuberculosis*-infected mice, *Infect. Immun.,* 66, 169, 1998.

129. Lowrie, D. B. et al., Protection against tuberculosis by a plasmid DNA vaccine, *Vaccine,* 15,834, 1997.

130. Lowrie, D. B., Silva, C. L., and Tascon, R. E., DNA vaccines against tuberculosis, *Immunol. Cell Biol.,* 75, 591, 1997.

131. Lowrie, D. B., Silva, C. L., and Tascon, R. E., Genetic vaccination against tuberculosis, *Springer Semin. Immunopathol.,* 19, 161, 1997.

132. Lowrie, D. B. et al., Towards a DNA vaccine against tuberculosis, *Vaccine,* 12, 1537, 1994.

133. Abdel-Wahab, Z. et al., A phase I clinical trial of immunotherapy with interferon-γ gene modified autologous melanoma cells: monitoring the humoral immune response, *Cancer,* 80, 401, 1997.

134. Anderton, S. M. et al., Differential mycobacterial 65-kDa heat shock protein T cell epitope recognition after adjuvant arthritis-inducing or protective immunization protocols, *J. Immunol.,* 152, 3656, 1994.

135. Cohen, I. R., Autoimmunity to chaperonins in the pathogenesis of arthritis and diabetes, *Annu. Rev. Immunol.,* 9, 567, 1991.

136. Ragno, S. et al., Protection of rats from adjuvant arthritis by immunization with naked DNA encoding for mycobacterial heat shock protein 65, *Arthritis Rheum.,* 40, 277, 1997.

137. Koga, T. et al., T cells against a bacterial heat shock protein recognize stressed macrophages, *Science,* 245, 1112, 1989.

138. Orme, I. M., Andersen, P., and Boom, W. H., T cell response to *Mycobacterium tuberculosis, J. Infect. Dis.,* 167, 1481, 1993.

139. Sedegah, M. et al., Protection against malaria by immunization with plasmid DNA encoding circum-sporozoite protein, *Proc. Natl. Acad. Sci. U.S.A.,* 91, 9866, 1994.

140. Doolan, D. L. et al., Identification and characterization of the protective hepatocyte erythrocyte protein 17 kDa gene of *Plasmodium yoelii,* homolog of *Plasmodium falciparum* exported protein 1, *J. Biol. Chem.,* 271, 17861, 1996.

141. Gurunathan, S. et al., Vaccination with DNA encoding the immunodominant LACK parasite antigen confers protective immunity to mice infected with *Leishmania major, J. Exp. Med.,* 186, 1137, 1997.

142. Gurunathan, S. et al., Vaccine requirements for sustained cellular immunity to an intracellular parasitic infection, *Nat. Med.,* 4,1409, 1998.

143. Walker, P. S. et al., Genetic immunization with glycoprotein 63 cDNA results in a helper T cell type 1 immune response and protection in a murine model of leishmaniasis [see comments], *Hum. Gene Ther.,* 9, 1899, 1998.

144. Costa, F. et al., Immunization with a plasmid DNA containing the gene of trans-sialidase reduces *Trypanosoma cruzi* infection in mice, *Vaccine,* 16, 768, 1998.

145. Rodrigues, M. M. et al., Predominance of CD4 Th1 and CD8 Tc1 cells revealed by characterization of the cellular immune response generated by immunization with a DNA vaccine containing a *Trypanosoma cruzi* gene, *Infect. Immunol.,* 67, 3855, 1999.

146. Wizel, B., Garg, N., and Tarleton, R. L., Vaccination with trypomastigote surface antigen 1-encoding plasmid DNA confers protection against lethal *Trypanosoma cruzi* infection, *Infect. Immunol.,* 66, 5073, 1998.

147. Li, S. et al., Priming with recombinant influenza virus followed by administration of recombinant vaccinia virus induces CD8+ T-cell-mediated protective immunity against malaria, *Proc. Natl. Acad. Sci. U.S.A.,* 90, 5214, 1993.

148. Rodrigues, M. et al., Influenza and vaccinia viruses expressing malaria CD8+ T and B cell epitopes. Comparison of their immunogenicity and capacity to induce protective immunity, *J. Immunol.,* 153, 4636, 1994.

149. Murata, K. et al., Characterization of in vivo primary and secondary CD8+ T cell responses induced by recombinant influenza and vaccinia viruses, *Cell Immunol.,* 173, 96, 1996.

150. Miyahira, Y. et al., Recombinant viruses expressing a human malaria antigen can elicit potentially protective immune CD8+ responses in mice, *Proc. Natl. Acad. Sci. U.S.A.,* 95, 3954, 1998.

151. Gonzalo, R. M. et al., Enhanced CD8+ T cell response to HIV-1 env by combined immunization with influenza and vaccinia virus recombinants, *Vaccine,* 17, 887, 1999.

152. Schneider, J. et al., Enhanced immunogenicity for CD8+ T cell induction and complete protective efficacy of malaria DNA vaccination by boosting with modified vaccinia virus Ankara, *Nat. Med.,* 4, 397, 1998.

153. Sedegah, M. et al., Boosting with recombinant vaccinia increases immunogenicity and protective efficacy of malaria DNA vaccine, *Proc. Natl. Acad. Sci. U.S.A.,* 95, 7648, 1998.

154. Weiss, W. R. et al., A plasmid encoding murine granulocyte-macrophage colony-stimulating factor increases protection conferred by a malaria DNA vaccine, *J. Immunol.,* 161, 2325, 1998.

155. Doolan, D. L. et al., DNA vaccination as an approach to malaria control: current status and strategies, *Curr. Top. Microbiol. Immunol.,* 226, 37, 1998.

156. Wang, R. et al., Induction of antigen-specific cytotoxic T lymphocytes in humans by a malaria DNA vaccine, *Science,* 282, 476, 1998.

157. Stass, A. S. and Mixson, A. J., Oncogenes and tumor suppressor genes: therapeutical implications, *Clin. Cancer Res.,* 3, 2687, 1997.

158. Fueyo, J. and Gomez-Manzano, C., Molecular control of cellular cycle and apoptosis: new treatments for gliomas, *Neurologia,* 13, 349, 1998.

159. Breau, R. L. and Clayman, G. L., Gene therapy for head and neck cancer, *Curr. Opin. Oncol.,* 8, 227, 1996.

160. Nielsen, L. L. et al., Adenovirus-mediated p53 gene therapy and paclitaxel have synergistic efficacy in models of human head and neck, ovarian, prostate, and breast cancer, *Clin. Cancer Res.,* 4, 835, 1998.

161. Roth, J. A. et al., Retrovirus-mediated wild type p53 gene transfer to tumors of patients with lung cancer, *Nat. Med.,* 2, 985, 1996.

162. Roth, J. A. et al., Gene therapy for non-small cell lung cancer: a preliminary report of a phase I trial of adenoviral p53 gene replacement, *Semin. Oncol.,* 25, 33, 1998.

163. Habib, N. A. et al., Contrasting effects of direct p53 DNA injection in primary and secondary liver tumors, *Tumor Targeting*, 1, 295, 1995.
164. Habib, N. A. et al., Preliminary report: the short-term effects of direct p53 DNA injection in primary hepatocellular carcinomas, *Cancer Detect. Prev.*, 20, 103, 1996.
165. Chintala, S. K. et al., Adenovirus-mediated p16/CDKN2 gene transfer suppress glioma invasion *in vitro*, *Oncogene*, 15, 2049, 1997.
166. Lee, J. H. et al., The inhibitory effect of adenovirus-mediated p16INK4A gene transfer on the proliferation of lung cancer cell line, *Anticancer Res.*, 18, 3257, 1998.
167. Rocco, J. W. et al., P16INK4A adenovirus-mediated gene therapy for human head and neck squamous cell cancer, *Clin. Cancer Res.*, 4, 1697, 1998.
168. Gomez-Manzano, C. et al., Gene therapy for gliomas: p53 and E2F-1 proteins and the target of apoptosis, *Int. J. Mol. Med.*, 3, 81, 1999.
169. Craig, C. et al., Effects of adenovirus-mediated p16INK4A expression on cell cycle arrest are determined by endogenous p16 and Rb status in human cancer cells, *Oncogene*, 16, 265, 1997.
170. Fang, X. et al., Expression of p16 induces transcriptional downregulation of RB gene, *Oncogene*, 16, 1, 1998.
171. Frizelle, S. P. et al., Re-expression of p16INK4A in mesothelioma cells results in cell cycle arrest, cell death, tumor suppression and tumor regression, *Oncogene*, 16, 3087, 1998.
172. Hunt, K. K. et al., Adenovirus-mediated overexpression of the transcription factor E2F-1 induces apoptosis in human breast and ovarian carcinoma cell lines and does not require p53, *Cancer Res.*, 57, 4722, 1997.
173. Meng, R. D., Philips, P., and El-Deiry, W. S., P53-independent increase in E2F-1 expression enhances the cytotoxic effects of etoposide and of adriamycin, *Int. J. Oncol.*, 14, 5, 1999.
174. Nip, J. et al., E2F-1 cooperates with topoisomerase II inhibition and DNA damage to selectively augment p53-independent apoptosis, *Mol. Cell Biol.*, 17, 1049, 1997.
175. Fueyo, J. et al., Overexpression of E2F-1 in glioma triggers apoptosis and suppresses tumor growth *in vitro* and *in vivo*, *Nat. Med.*, 4, 685, 1998.
176. Parr, J. et al., Tumor-selective transgene expression *in vivo* mediated by an E2F-responsive adenoviral vector, *Nat. Med.*, 3, 1145, 1997.
177. Badie, B. et al., Adenovirus-mediated p53 gene delivery potentiated the radiation-induced growth inhibition of experimental brain tumors, *J. Neurooncol.*, 37, 217, 1998.
178. Fujiwara, T. et al., Induction of chemosensitivity in human lung cancer cells *in vivo* by adenovirus-mediated transfer of the wild-type p53 gene, *Cancer Res.*, 54, 2287, 1994.
179. Feig, L. A., Strategies for suppressing the function on oncogenic ras protein tumors, *J. Natl. Cancer Inst.*, 85, 1266, 1998.
180. Kita, K. et al., Growth inhibition of human pancreatic cancer cell lines by anti-sense oligonucleotides specific to mutated K-*ras* genes, *Int. J. Cancer*, 80, 553, 1999.
181. Zhang, Y. et al., Retroviral vector-mediated transduction of k-*ras* antisense RNA in human being cancer cells inhibits expression of malignant phenotype, *Hum. Gene Ther.*, 4, 451, 1993.
182. Steiner, M. S. et al., Antisense c-*myc* retroviral vector suppresses established human prostate cancer, *Hum. Gene Ther.*, 9, 747, 1998.
183. Citro, G. et al., C-*myc* antisense oligodeoxynucleotides enhance the efficacy of cisplatin in melanoma chemotherapy *in vitro* and in nude mice, *Cancer Res.*, 58, 283, 1998.
184. Webb, A. et al., BCL-2 antisense therapy in patients with non-Hodgkin lymphoma. *Lancet*, 349, 1137, 1997.
185. Fuchs, E. J. and Matzinger, P., Is cancer dangerous to the immune system? *Sem. Immunol.*, 8, 271, 1996.
186. Matzinger, P., Tolerance, danger, and the extended family, *Annu. Rev. Immunol.*, 12, 991, 1994.
187. Bright, R., Shearer, M., and Kennedy, R., Nucleic acid vaccination against virally induced tumors, *Ann. N.Y. Acad. Sci.*, 772, 241, 1995.
188. Conry, R. M. et al., Selected strategies to augment polynucleotide immunization, *Gene Ther.*, 3, 67, 1996.
189. Wang, B. et al., Immunization by direct DNA inoculation induces rejection of tumor cell challenge, *Hum. Gene Ther.*, 6, 407, 1995.

190. Conry, R. et al., A carcinoembryonic antigen polynucleotide vaccine for human clinical use, *Cancer Gene Ther.*, 2, 33, 1995.

191. McAneny, D. et al., Results of a phase I trial of a recombinant vaccinia virus that expresses carcinoembryonic antigen in patients with advanced colorectal cancer, *Ann. Surg. Oncol.*, 3, 495, 1996.

192. Kim, J. J. et al., Molecular and immunological analysis of genetic prostate specific antigen (PSA) vaccine, *Oncogene,* 17, 3125, 1998.

193. Rosenberg, S. A. et al., Impact of cytokine administration on the generation of antitumor reactivity in patients with metastatic melanoma receiving a peptide vaccine, *J. Immunol.,* 163, 1690, 1999.

194. Gariglio, P. et al., Therapeutic uterine-cervix cancer vaccines in humans, *Arch. Med. Res.,* 29, 279, 1998.

195. Murakami, M. et al., Induction of specific CD8$^+$ T-lymphocyte response using a human papillomavirus-16 E6/E7 fusion protein and autologous dendritic cells, *Cancer Res.,* 59, 1184, 1999.

196. Van Hall, T. et al., Cryptic open reading frames in plasmid vector backbone sequences can provide highly immunogenic cytotoxic T-lymphocyte epitopes, *Cancer Res.,* 58, 3087, 1998.

197. Ciernik, I. F., Berzofsky, J. A., and Carbone, D. P., Induction of cytotoxic T lymphocytes and antitumor immunity with DNA vaccines expressing single T cell epitopes, *J. Immunol.,* 156, 2369, 1996.

198. Suhrbier, A., Multi-epitope DNA-vaccines, *Immunol. Cell Biol.,* 75, 402, 1997.

199. Syrengelas, A. D., Chen, T. T., and Levy, R., DNA immunization induces protective immunity against B-cell lymphoma, *Nat. Med.,* 2, 1038, 1996.

200. Chakraborty, M. et al., Induction of human breast cancer-specific antibody responses in cynomolgus monkeys by a murine monoclonal anti-idiotype antibody, *Cancer Res.,* 55, 525, 1995.

201. Mittelman, A. et al., Kinetics of the immune and regression of metastatic lesions following development of humoral anti-high molecular weight-melanoma associated antigen immunity in three patients with advanced malignant melanoma immunized with mouse anti-idiotypic monoclonal antibody MK2–23, *Cancer Res.,* 54, 415, 1994.

202. Moraes, J. Z. et al., Anti-idiotypic monoclonal antibody AB3, reacting with the primary antigen (CEA), can localize in human colon carcinoma xenografts as efficiently as AB1, *Int. J. Cancer,* 57, 586, 1994.

203. Hayden, M. S., Gililand, L. K., and Ledbetter, J. A., Antibody engineering, *Curr. Opin. Immunol.,* 9, 201, 1997.

204. Beerli, R. R., Wells, W., and Hynes, N. E., Intracellular expression of single chain antibodies reverts Erb B-2 transformation, *J. Biol. Chem.,* 269, 23931, 1994.

205. Cohen, P. A., Mani, J. C., and Lane, D. P., Characterization of a new intrabody directed against the N-terminal region of human p53, *Oncogene,* 17, 2445, 1998.

206. Dachs, G. U. et al., Targeting gene therapy to cancer: a review, *Oncol. Res.,* 9, 313, 1997.

207. Marasco, W. A., Intrabodies: turning the humoral immune system outside in for intracellular immunization, *Gene Ther.,* 4, 11, 1997.

208. Elder, E. M., Lotze, M. T., and Whiteside, T. L., Successful culture and selection of cytokine gene-modified human dermal fibroblasts for the biologic therapy of patients with cancer, *Hum. Gene Ther.,* 7, 479, 1996.

209. Mackensen, A. et al., Induction of tumor-specific cytotoxic T lymphocytes by immunization with autologous tumor cells and interleukin-2 gene transfected fibroblasts, *J. Mol. Med.,* 75, 290, 1997.

210. Stingl, G. et al., Phase I study to the immunotherapy of metastatic malignant melanoma by a cancer vaccine consisting of autologous cancer cells transfected with the IL-2 gene. *Hum. Gene Ther.,* 7, 551, 1996.

211. Sun, Y. et al., Vaccination with IL-12 gene-modified autologous melanoma cells: preclinical results and a first clinical phase I study, *Gene Ther.,* 5, 481, 1998.

212. Dong, Z. et al., Suppression of angiogenesis, tumorigenicity, and metastasis by human prostate cancer cells engineered to produce interferon-beta, *Cancer Res.,* 59, 872, 1999.

213. Klagsbrun, M., Angiogenesis and cancer. AACR Special Conference in Cancer Research, *Cancer Res.,* 59, 487, 1999.

214. Henze, G. et al., Immunotherapy of acute lymphoblastic leukemia by vaccination with autologous leukemic cells transfected with cDNA expression plasmid coding for an allogeneic HLA class I antigen combined with interleukin-2 treatment, *J. Mol. Med.,* 76, 215, 1998.

215. Hui, K. M. et al., Phase I study of immunotherapy of cutaneous metastases of human carcinoma using allogeneic and xenogeneic MHC DNA-liposome complexes, *Gene Ther.,* 4, 790, 1997.

216. Plautz, G. E. et al., Immunotherapy of malignancy by *in vivo* gene transfer into tumors, *Proc. Natl. Acad. Sci. U.S.A.*, 90, 4645, 1993.

217. Nabel, G. L. et al., Immune response in human melanoma after transfer of an allogeneic class I major histocompatibility complex gene with DNA-liposome complexes, *Proc. Natl. Acad. Sci. U.S.A.*, 93, 15388, 1996.

218. Stopeck, A. T. et al., Phase I study of direct gene transfer of an allogeneic histocompatibility antigen, HLA-B7, in patients with metastatic melanoma, *J. Clin. Oncol.*, 15, 341, 1997.

219. Townsend, S. E. et al., Specificity and longevity of antitumor immune responses induced by B7-transfected tumors, *Cancer Res.*, 54, 6477, 1994.

220. Tüting, T. et al., Genetically modified bone marrow-derived dendritic cells expressing tumor-associated viral or "self" antigens induce antitumor immunity *in vivo*, *Eur. J. Immunol.*, 27, 2702, 1997.

221. Ganss, R. and Hanahan, D., Tumor microenvironment can restrict the effectiveness of activated antitumor lymphocytes, *Cancer Res.*, 58, 4673, 1998.

222. Wick, M. et al., Antigenic cancer cells grow progressively in immune hosts without evidence for T cell exhaustion or systemic anergy, *J. Exp. Med.*, 186, 229, 1997.

223. Melcher, A. et al., Tumor immunogenicity is determined by the mechanism of cell death via induction of heat shock protein expression, *Nat. Med.*, 4, 581, 1998.

224. Dellabona, P. et al., Vascular attack and immunotherapy: a "two hits" approach to improve biological treatment of cancer [editorial], *Gene Ther.*, 6, 153, 1999.

225. Eck, S. L. et al., Treatment of advanced CNS malignancies with the recombinant adenovirus H5.010RSVTK: a phase I trial, *Hum. Gene Ther.*, 7, 1465, 1996.

226. Crystal, R. G. et al., Phase I study of direct administration of a replication deficient adenovirus vector containing the *E. coli* cytosine deaminase gene to metastatic colon carcinoma of the liver in association with oral administration of the pro-drug 5-fluorocytosine, *Hum. Gene Ther.*, 8, 985, 1997.

227. Caruso, M. et al., Regression of established macroscopic liver metastases after *in situ* transduction of a suicide gene, *Proc. Natl. Acad. Sci. U.S.A.*, 90, 7024, 1993.

228. Ram, Z. et al., Therapy of malignant brain tumors by intratumoral implantation of retroviral vector-producing cells, *Nat. Med.* 3, 1354, 1997.

229. Sterman, D. H. et al., Adenovirus-mediated herpes simplex virus thymidine kinase/gancyclovir gene therapy in patients with localized malignancy: results of a phase I clinical trial in malignant mesothelioma, *Hum. Gene Ther.*, 9,1083, 1998.

230. Uckert, W. et al., Double suicide gene (cytosine deaminase and herpes simplex virus thymidine kinase) but not single gene transfer allows reliable elimination of tumor cells *in vivo*, *Hum. Gene Ther.*, 9, 855, 1998.

10 Toward Designer Diagnostic Antigens

Joy Chih-Wei Chang

CONTENTS

10.1 INTRODUCTION

In the modern world the detection of infectious agents is often very complex. Only 10 to 15 years ago the major means for detecting viral infections were the cultivation of virus in tissues *in vitro* or the identification of a specific immunoresponse to the virus infection, which typically involved the detection of antibodies by using two simple and popular formats — enzyme immunoassay and radioimmunoassay. Today, however, diagnostics has become one of the most dynamic research areas involving the development of many different approaches such as molecular diagnostics or detection of nucleic acids, chip technology or microarray technology, molecular sensors, rapid assays, and noninvasive technologies for detection of antibodies in several different body fluids other than blood. All of these new developments, however, have not changed the fact that serodiagnostics or antibody detection remains one of most important diagnostic subject areas. The most essential component of serodiagnostic assays is the antigen used as a target responsible for specific antibody binding. For many years the development of diagnostic assays for the detection of disease-specific antibodies depended exclusively on the availability of natural antigens obtained from biological specimens such as body fluid, stool, or sputum, or from *in vitro* propagation from infected human materials.

The advent of recombinant DNA technology has dramatically changed the face of serodiagnostics as a discipline. There has been a trend from using antigens derived from natural or *in vitro* sources toward developing antigens specifically designed as diagnostic reagents using rational approaches to the design and construction of antigens containing specific characteristics relevant for immunodiagnostics. Although the rational design of antigenic targets is still in development, several recent events in this area are reviewed in this chapter.

0-8493-1426-7/03/$0.00+$1.50
© 2003 by CRC Press LLC

10.2 EVOLUTION FROM "FOUND ANTIGENS" TO "IMITATED ANTIGENS"

There are numerous examples in the extant literature describing the development of successful immunoassays that utilize abundant amounts of antigens derived from natural sources. The hepatitis A virus (HAV) serves as a perfect example. This virus can be propagated in tissue culture, and therefore the HAV antigen can be rather simply obtained in amounts sufficient for the development and manufacture of HAV antibody detection assays.[1-4] Another example of using abundant natural antigens in diagnostics is the assay for the detection of adenovirus respiratory tract infections. Since adenovirus can be easily isolated from the pharyngeal area and propagated in a cell culture system, the adenovirus antigens can be readily isolated from cell culture in large quantities required for the development of sensitive, specific, and rapid test for the diagnosis of adenovirus respiratory tract infections.[5,6] Dengue fever can also be diagnosed using antigens isolated from peripheral blood leukocytes of infected patients.[7] These antigens have been successfully used to build different immunoassay formats.[8,9]

Natural antigens can be obtained not only from viruses but also from other infectious agents such as *Chlamydia trachomatis,* which causes female genital infection that may result in infertility. Major outer membranes of different strains of *C. trachomatis* propagated in cell culture have been used to generate monoclonal antibodies that have been used to develop a diagnostic assay for the detection of this infection.[10] Another successful example of using natural antigens in diagnostics is *Giardia lamblia*, the organism that causes diarrhea. The *G. lamblia*-specific antigen, GSA 65, isolated from the stool of infected patients presents a valuable and opportunistic source of antigen for the development of sensitive and specific diagnostic tests for the detection of giardiasis.[11]

The list of natural antigens readily available for diagnostics is impressive. However, the other list of infections and diseases for which natural antigens are not readily available is prodigious. The serological diagnosis of these diseases and infections has been awaiting a different source of diagnostic reagents for assay development. This new source of antigens has been provided in many cases by recombinant DNA technology, which is relatively old news. Nonetheless, potential advances and applications of this technology are far from being exhausted. The more we learn about the organization and flow of genetic information, the more new and different applications for recombinant DNA can be imagined. One new and exciting area is the development of recombinant proteins containing specifically designed properties for use in the development of diagnostic assays.

There are many successful examples of the construction of recombinant antigens for a variety of viral agents,[12-18] bacteria,[19-24] and parasitic pathogens.[25,26] Some of these proteins have been used for diagnostic assay development. Below are several examples of the successful application of recombinant DNA technology to diagnostics.

The hepatitis C virus (HCV) cannot be isolated from the serum of patients in large quantities, and it cannot be propagated in cell culture and, consequently, is not available for the assay development. Therefore, recombinant proteins have been engineered and used to build diagnostic assays. The first cloned HCV genomic fragment designated as region 5-1-1 was isolated from the serum of a patient infected with HCV in 1989.[27] The antigen encoded by this 5-1-1 region was found to be immunoreactive and was used to develop the first generation immunoassay for the detection of anti-HCV activity in serum specimens.[28] As more and more antigenic regions were discovered within the HCV polyprotein, more recombinant antigens were constructed and added to the assay to improve its specificity and sensitivity. Currently, the HCV assay contains recombinant antigens comprising antigenic regions from the core NS3, NS4, and NS5 proteins. The first generation of the HCV enzyme immunoassay used only one antigen, C100-3, derived from the NS4 protein. Two antigens, C22 and C33c, derived from core and NS3 proteins, respectively, were subsequently added in the second-generation assay. This resulted in a significant improvement in assay sensitivity and specificity. Recently, a new recombinant antigen derived from the HCV NS5 protein was added, which resulted in a slight increase of sensitivity of the third-generation HCV assay. [29]

Like HCV, hepatitis B virus (HBV) cannot be propagated in tissue culture *in vitro*.[30] Therefore, the first diagnostic assays were constructed using the hepatitis B surface antigen (HBsAg) derived from the plasma of infected patients, which often is found in very large amounts during HBV infection.[30] However, detection of HBsAg alone is not sufficient for the reliable diagnosis of all the different HBV infections. Another important diagnostic antigen is the hepatitis B core antigen (HBcAg). In the early stage of HBV infection, antibodies of the IgM class against the HBcAg appear first in the plasma of infected patients in virtually all HBV infections and persist after the disappearance of HBsAg and before the appearance of detectable antibody to HBsAg (anti-HBs). Therefore, in the absence of detectable HBsAg and anti-HBs, IgM anti-HBc may be the only serological marker of recent infection and of potentially infectious blood.[31–34] The HBcAg was obtained using bacterial cloning techniques and was first applied for the development of a radio-immunoassay for the detection of anti-HBc.[35] Later, this anti-HBc test was reformulated into an enzyme immunoassay.

The human immunodeficiency virus (HIV) is an example of when natural antigens can be obtained from cell culture, but because of the dangerous nature of this virus using *in vitro* techniques as a source for this antigen for diagnostic assay development is demanding. As a result, recombinant DNA technology is practically the only suitable source for obtaining HIV antigens. Almost simultaneously with the discovery of HIV a cell culture system was developed for the propagation and isolation of this virus from patients with acquired immunodeficiency syndrome (AIDS).[36] Although this cell culture system opened the way for the routine detection of HIV and related cytopathic variants in patients with AIDS and pre-AIDS symptoms, and although this system provided a large amount of this virus for research, it was inevitable for the reasons described above that a bacterial expression system for obtaining HIV antigens as diagnostic reagents for HIV had to be developed. The progress in this area was rapid. In 1985, HIV gene products of structural genes such as core and envelope were thoroughly characterized as antigenic targets for the detection of HIV antibodies in serum specimens from AIDS patients.[37] Soon after this, the HIV recombinant envelope protein[38–41] and the HIV core protein[40,41] were expressed in bacterial cells, purified, characterized, and used to develop recombinant HIV diagnostic assays. Since then, four generations of HIV diagnostic tests have been developed using recombinant antigens. Although synthetic peptides have replaced HIV recombinant proteins in current screening assays, there is an array of HIV recombinant antigens derived from *core, pol,* and *env* genes, which are currently in use in HIV confirmatory assays.

These few examples illustrate a trend in the diagnostic field, namely, a departure from the practice of using antigens that can be readily obtained from infected tissues toward the use of antigens that are prepared using only fragments of viral genomes. The advantages of using "man-made" antigens are clear and were the subject of an extensive discussion several years ago.[42,43] There were, however, some problems associated with using recombinant proteins for diagnostic purposes. The major problem is that the antigenic properties of viral proteins cannot always be modeled with recombinant polypeptides expressed in bacterial cells, the most established and easily maintained expression system available today. Because of very limited knowledge on how protein structure affects antigenic properties and what cellular parameters contribute to the assembly of this structure, it is impossible to grasp real control over this process. Nonetheless, there are some laborious ways, although not very reliable, of obtaining recombinant proteins with the desired antigenic properties.

One of the most successfully used methods is the expression of antigens in systems that are most closely related to human tissues infected with the pathogen. One of the very first examples of using this strategy was the expression of immunologically active HBsAg in heterologous systems. When the HBV genome was cloned with a plasmid vector in bacteria, the gene S encoding for

HBsAg was transferred into an expression bacterial vector and expressed in *E. coli*. However, the expression product of the gene S demonstrated only marginal immunoreactivity, which was subsequently shown to be unsuitable for either diagnostic or vaccine development.[44] It was hypothesized that since HBsAg expressed in bacterial cells is not assembled into virus-like particles (VLP), the form in which HBsAg is usually found in the blood of HBV-infected patients, it could not efficiently model the corresponding antigenic epitopes.

Further experiments showed that when expressed in mammalian cells transfected with recombinant plasmids bearing the HBV gene S, HBsAg forms VLPs, which were immunologically very similar, if not identical, to the plasma-derived HBsAg.[45–48] However, the most important discovery for use in vaccine and diagnostic development was made using yeast expression systems. It was found that HBsAg expressed in yeast cells also could be assembled into VLPs and, most importantly, that this antigen retained immunologic properties suitable for the development of diagnostic assays and vaccine.[49–56] These first experiments significantly impacted the application of recombinant DNA technology to diagnostics and especially to vaccine development.

As pointed out above, in the absence of a comprehensive understanding of the relationship between protein structure and function, the predominant concept for making vaccine candidates or efficient diagnostic reagents using recombinant DNA was to obtain proteins in a form that maximally resembles the native structure. It was anticipated that imitation of this structure would guarantee reproduction of the desired immunologic properties. In addition to HBsAg, this concept governed many projects to obtain recombinant antigens. One of the latest examples is the hepatitis E virus (HEV) antigen derived from the use of the baculovirus expression system.[57–60] The product of expression of the HEV open reading frame 2 (ORF2) encoding for structural proteins assembled in insect cells into VLP, which efficiently modeled the corresponding antigenic epitopes.[58–60] These VLPs were used to develop new diagnostic assays and vaccine candidates.[58–60] Alternatively, a different approach was used to develop ORF2 antigens modeling the HEV-neutralizing antigenic epitope(s) in bacterial cells.[61] These authors trimmed the ORF2-encoded antigen from both the N and C termini and tested antibodies against all of these trimmed proteins using an HEV *in vitro* neutralization assay. One of the fragments of the HEV ORF2 protein was found to be a very potent vaccine candidate[61] and diagnostic reagent.[62] Although often successful, these approaches are extremely labor-intensive and can be economically prohibitive.

To some extent, both of these approaches are not much different from the search for a source of natural antigens from infected tissues. Undoubtedly, the use of recombinant DNA technology accelerated this search while removing the necessity to work with infectious material, which ultimately allowed for more control during the production process. Despite many examples, production of antigens by DNA technology remains a search-and-find mission, rather than one that can be used to rationally design and construct diagnostically relevant antigens.

10.3 FIRST STEPS TOWARD "DESIGNER ANTIGENS": SYNTHETIC GENES

Without a doubt the evolution from natural antigens to genetically engineered antigens dramatically advanced diagnostic research, which created multiple opportunities to develop new diagnostic assays because such antigens when used as diagnostic targets were readily available and reproducible. However, the imitation of antigens using heterologous expression systems was often very challenging. In addition to the aforementioned technical problem of reproducing protein conformation in a manner consistent with modeling of antigenic epitopes in a functionally active form, there remain many other problems associated with both nonspecific immunoreactivity and strain-specific variation in antigenic properties. The solution to these problems is almost impossible without access to a technology capable of unrestrained modifications to the antigen primary structure. This technology has been developed by Khorana and co-workers, who pioneered the basic technology necessary for the chemical synthesis of oligonucleotides, appropriate purification methods, and the joining of single-stranded DNA fragments by DNA ligase in DNA duplex fragments.

In the 1970s, Khorana and his co-workers synthesized the coding region of the structural gene for the precursor of a tyrosine suppresser transfer RNA from *Escherichia coli*.[63–69] Later, they cloned the synthetic tyrosine suppresser tRNA gene in *E. coli* with a plasmid and a bacteriophage as vectors. This synthetic gene demonstrated both functional and biological activity *in vitro* and *in vivo*.[68,70] However, without a way to express proteins from genes, synthetic DNA was of a limited value. In 1977, Itakura and colleagues demonstrated for the first time bacterial expression of a functional mammalian peptide hormone, somatostatin, from a chemically synthesized gene.[71] Since synthetic genes encoding for a functional and biological peptide or protein could be cloned and expressed, research activity in this area dramatically increased· profoundly affecting many scientific disciplines including immunodiagnostics.

In 1987, a German group of researchers made a synthetic gene for potential application in the development of diagnostics for human T-cell leukemia virus type-1 (HTLV1) infection.[72] This synthetic gene encoded for an 88-amino-acid fragment of the envelope protein. To obtain a DNA molecule encoding for a protein containing only 88 aa, 35 short oligonucleotides were designed and chemically synthesized. The first half of the gene was assembled using 17 oligonucleotides by a one-step ligation reaction, and the second half of the gene was assembled from 18 oligonucleotides by a separate one-step ligation reaction. The two gene fragments were then ligated together to form a full-size gene. The sequence of the gene was confirmed by sequencing and the gene was successfully cloned and expressed in bacteria. However, the antigenic properties of the protein were not characterized at that time.

Since the primary structure of oligonucletides can be freely designed, the nucleotide sequence of any synthetic gene can be engineered to contain new restriction sites without changing the amino acid sequence. In 1988, Nassal and colleagues took advantage of this strategy and chemically synthesized a 560-bp long gene for the HBV core protein.[73] This gene was constructed using 11 pairs of chemically synthesized oligonucleotides. The primary structure of this gene was designed in such a way that 27 unique restriction endonuclease recognition sites were incorporated into the body of the gene without changing the amino acid sequence of the encoded protein. The rationale for such a modification of the nucleotide sequence was to simplify mutagenesis of the gene structure by replacing short DNA fragments using restriction endonucleases. HBV core protein was expressed in a soluble form in *E. coli*. It formed particles closely resembling the native HBV core. When this synthetic gene was transferred into the viral genome, transient expression in a hepatoma cell line yielded protein indistinguishable from the native gene products. This synthetic gene provided not only a useful tool for studies on the structure and function of the isolated HBV core protein but also another way to produce large amounts of the HBV core antigen as a diagnostic reagent.

The list of synthetic genes encoding for diagnostically relevant proteins is significantly long. Different research groups used different strategies to assemble polydeoxynucleotide fragments from synthetic oligodeoxynucleotides. However, all of these approaches may be classified into two basic strategies. The first strategy is based on the use of overlapping complementary synthetic oligonucleotides that represent both strands of the entire DNA fragment to be synthesized. After annealing, nicks between oligonucleotides are repaired by DNA ligase. The DNA fragments can then be cloned directly[63,74–77] or after amplification by PCR.[78,79] Although this method is very sensitive to the secondary structure of oliginucleotides and has very low efficiency because of numerous adverse complementary interactions between oligonucleotides, this method has been used in many laboratories. The other strategy uses the property of DNA polymerase to fill in gaps in annealed pairs of oligonucleotides.[80–83] This strategy is less sensitive to the effects of adverse complementary reactions between oligonucleotides and secondary structure of oligonucleotides and appears to be more efficient than the ligase-based strategy. However, the size of the resulting DNA fragments is limited by the use of only two oligonucleotides at a time during synthesis.

For assembly of longer DNA molecules, the synthesized double-stranded fragments either must be digested with restriction endonucleases for further assembly by DNA ligation or should be assembled by PCR. The application of PCR has dramatically improved the efficiency of the polymerase gap-filling strategy for DNA synthesis.[84,85] There are different variations of this strategy. One, which is most frequently used, is when a double-stranded polynucleotide is synthesized from a set of pairs of oligonucleotides, with each pair being designed in such a way that the 3'-ends of oligonucleotides within the pair contain complementary sequences and the 5'-terminal region of one oligonucleotide from the pair is complementary to the 5'-terminal region of one oligonucleotide from the adjacent pair. This design allows for the assembly of a long polynucleotide from short oligonucleotides by PCR.

During the first cycle of PCR, the complementary 3'-ends of each pair are annealed and extended by DNA polymerase to generate a double-stranded DNA. During the following cycles of PCR, the complementary 3'-ends of these extended oligonucleotides from adjacent pairs are annealed and extended. During subsequent PCR cycles, the final DNA product is amplified using PCR primers designed from the terminal regions of the full-size DNA fragment. With minor modifications this approach has been used to construct several synthetic genes.[86,87] PCR has also been used to assist in the DNA ligase assembly of double-stranded polynucleotides.[78,79] In this case, a polynucleotide is assembled by annealing oligonucleotides followed by nick repair by DNA ligase. Two PCR primers flanking the desired sequence are used to amplify the full-size product.

A different strategy from those described above was used to construct a synthetic gene encoding for the hepatitis C virus (HCV) nucleocapsid protein.[88] In this study, the gene was assembled by using a unique sequential cyclic mechanism. The first step of this mechanism is identical to the first step of the PCR-assisted DNA polymerase gap-filling strategy, when two oligonucleotides are annealed by their 3'-ends and extended by DNA polymerase (see above). However, each of the subsequent steps is different. At least one oligonucleotide in this pair is designed to contain a restriction endonuclease recognition site of BstXI[88] or a chemical blocking group at some distance from the very 5'-end responsible for preventing T7 phage gene 6 DNA exonuclease from completely digesting one strand of a double-stranded DNA starting from the 5'-end.[89]

Thus, the digestion of a newly created double-stranded DNA with BstXI or T7 exonuclease, which is present in the mixture with DNA polymerase and oligonucleotides, resulted in the formation of a 3'-terminal single-stranded region. This single-stranded region is designed to be complementary to the 3'-end sequence of the next oligonucleotide, which can be annealed to this single-stranded region. After annealing, the single-stranded oligonucleotide became a template for a DNA polymerase extension reaction primed by the 3'-end of the partially double-stranded DNA synthesized during the previous round of the extension reaction. This new double-stranded product of the DNA extension reaction will become a new substrate for BstXI or DNA exonuclease digestion, which will subsequently expose a new single-stranded region complementary to the next oligonucleotide in the incubation mixture. Thus, the process of DNA synthesis is a cyclic reaction adding one stretch of sequences during each cycle.[88,89] The specificity of oligonucleotide addition for each cycle is encoded in each 3'-terminal single-stranded sequence. By combining an entire set of oligonucleotides with all the required enzymatic activities, such as DNA polymerase and BstXI or exonuclease, in one tube and incubating the mixture at the appropriate temperature, long DNA fragments can be synthesized in one step. A synthetic gene encoding the HCV nucleocapsid protein was constructed in this manner.[88]

The application of synthetic genes to obtaining recombinant proteins using genetic engineering approaches has numerous advantages over the application of natural genes using similar approaches. All of these advantages stem from the unlimited potential of synthetic-DNA technology in making DNA of any primary structure. One example of the application of synthetic genes to resolve an expression problem caused by the sequence of a natural gene is the expression of extracellular superoxide dismutase (SOD) in a heterologous expression system with the goal of obtaining a specific antiserum for the detection of human SOD.[90] SOD is an enzyme that plays an important role in protecting human spermatozoa from peroxidative damage. In some cases, there is a correlation between SOD enzyme concentration and sperm quality.[91,92] The epididymis is the major site of synthesis and secretion of large amounts of SOD. However, SOD isoenzymes are also found in other tissues like leukocytes and macrophages.

To study the distribution of SOD in the male reproductive tract and distinguish it from other related SOD isoenzymes a synthetic gene encoding for the part of human SOD that exhibits maximum disparity with other SOD isoenzymes was constructed. In this particular study, the use of synthetic DNA was especially important because the authentic sequence is extremely GC-rich. The GC-rich regions can cause frequent premature termination of cDNA synthesis since reverse transcriptase is very sensitive to GC content of the copied RNA. Subsequently, GC-rich regions can diminish the level of gene expression in *E. coli*. To avoid these problems and improve expression of the SOD gene in *E. coli*, the primary structure of this gene was redesigned to reduce the GC content by using codons optimal for *E. coli*.[90]

The problem of suboptimal protein expression in heterologous systems can frequently be resolved by changing the primary structure of the gene to optimize codon usage. For example, it was noticed that a major obstacle to optimal expression of *Plasmodium faciparum* genes in transfected cells in the mammalian host is the dramatic difference in codon usage between *P. falciparum* and mammals.[93] The A + T content in the genome of *P. falciparum* is 80% vs. 59% in humans. Except for Met and Trp, each amino acid can be encoded by two to six different synonymous codons. The frequency at which these synonymous codons are used differs among different organisms. It was noticed that this codon bias affected gene expression.[94,95] The preferential use of synonymous codons was linked to the fact that different tRNA species are present within cells in different relative concentrations,[96,97] which vary from one organism to another and even during the development of the same organism.[98]

One estimate of codon bias is called the codon adaptation index (CAI).[99] The CAI measures the resemblance of codon usage in a gene to the most frequently used codons in a set of highly expressed genes. A recent study showed that the CAI for synthetic genes encoding the same T-cell epitope from *P. yoelii* correlated with the level of expression *in vitro* and with T-cell response in mice.[100] A similar observation had been made for the *P. falciparum* synthetic genes encoding for the receptor-binding domain of the 175 kDa erythrocyte-binding protein and for the 42 kDa C-terminal fragment of the *P. falciparum* merozoite surface protein 1. Both synthetic genes were designed to have a high CAI. The results obtained in this study strongly indicated that optimizing codon usage in DNA can improve protein expression for organisms whose codon usage differs substantially from that of the heterologous expression system.[93] Thus, synthetic genes proved to be of significant value whenever the need to modify the primary structure of genes arises. However, the major advantages of using synthetic genes are not limited by these very few examples of improvement in expression levels. It seems that synthetic-DNA technology may be used in any discipline that requires protein engineering to accomplish many different desired outcomes.

10.4 FIRST DESIGNER DIAGNOSTIC ANTIGENS

As mentioned above, producing recombinant antigens, even those obtained using synthetic genes with improved expression properties, still remains an effort of "found antigens" and, therefore, shares both the advantages and disadvantages of using these antigens as antigenic targets in the development of diagnostics tests. It is known that 3 to 4% of antibodies specific to an infectious agent may also recognize some host-specific proteins.[101] Examples of such cross-reactivity can be found in many viral infections. Molecular mimicry was found between a protein of the human immunodeficiency virus and human brain proteins.[102] Similarities between virus-specific antigenic epitopes and host-specific proteins, which resulted in antibody cross-reactivity, were implicated in some cases of autoimmunity.[103]

Some nonspecific antigenic epitopes may immunoreact with a significant number of serum specimens. For example, a region at amino acid position 515–530 of the hepatitis E virus ORF2-encoded protein was found to immunoreact with approximately 10% of sera from normal blood donors residing in the region of the world where HEV is not endemic. These sera did not have anti-HEV activity detectable by immunoblot analysis or by a synthetic peptide EIA.[104] Regions of nonspecific antigenic reactivity can be found within different HCV proteins.[105] One study showed that more than 30% of the 543 synthetic peptides derived from different regions of the HCV polyprotein of different genotypes was recognized by antibodies from healthy blood donors.[106] These observations suggest that nonspecific immunoreactivity of some antigenic epitopes may significantly reduce the specificity of antibody binding by antigens and, as a result, may negatively affect the overall specificity of the diagnostic assay. Taking this view into consideration, it seems reasonable to use only selected short antigenic regions to build a diagnostic assay because this strategy would allow for significant reduction in the number of antigenic epitopes possessing nonspecific immunoreactivity.

The strategy of excluding nonspecific epitopes from diagnostic antigens was first used in the construction of an HEV artificial mosaic antigen as an engineered or designer diagnostic reagent.[104] This strategy involves the use of synthetic peptides to study the linear antigenic epitopes of proteins, selection of short sequences that model broadly and strongly immunoreactive diagnostically relevant antigenic epitopes, and the design and construction of a synthetic gene encoding an artificial polypeptide composed of these short antigenically reactive regions. The mosaic antigen constructed by this strategy imitates not the structure of natural antigens, but rather the immunologic function of the natural antigens without regard to other functions that may be associated with these antigens. This approach produces proteins containing only antigenic epitopes relevant to immunodiagnostics and eliminates epitopes that may be involved in nonspecific reactivity.

One of the advantages of artificial mosaic proteins is their ability to connect, in one polypeptide chain, different antigenic regions from the same protein and different antigenic epitopes from more than one protein, such as those that may belong to different variants of the same pathogen. As a demonstration of this strategy the HEV mosaic protein was constructed from three antigenically active dominant regions from the protein encoded by ORF2, one antigenically active region from protein encoded by ORF3 of the Burmese HEV strain, and one antigenically active region from the protein encoded by ORF3 of the Mexican HEV strain.[104] All of the antigenic regions included in this HEV mosaic protein were separated by short stretches of glycines to limit the effect of one domain on the folding of another. This protein was expressed in *E. coli* and used to develop a diagnostic assay for the detection of anti-HEV activity in serum specimens.[107]

A similar strategy was employed to construct a designer hepatitis-B-virus surface antigen (HBsAg).[108] As in the case described above, the antigenic properties of the HBsAg epitopes were first successfully modeled with short synthetic peptides.[109] The sequences of these peptides were then used to design and construct a multiepitope protein, which, according to the authors' report, retained the immunologic properties of the introduced antigenic epitopes.[108] It appears that successful imitation of antigenic epitopes with synthetic peptides often guarantees proper modeling of antigenic epitopes with artificial proteins. However, proper folding may not always be achieved in artificial antigens composed of various antigenic epitopes derived from different regions and proteins. For example, it was reported that the HEV mosaic protein containing a small deletion of only a few amino acids resulted in a protein demonstrating a drastically diminished ability to bind HEV-specific antibody.[86] This observation suggests that proper modeling of antigenic epitopes with designer proteins may require closer attention to the secondary and tertiary structure and, since the relationship between structure and antigenic properties is not well understood, may require the construction of several variants of artificial antigens before finding one containing the expected properties.

A different strategy was used to design and construct a multicomponent chimeric antigen for diagnosis of canine visceral leishmaniasis.[110] Leishmaniases are a spectrum of diseases caused by different species of the genus *Leishmania*. The parasite antigens of diagnostic relevance were identified using immunoscreening of a *L. infantum* expression library with sera from dogs with active disease. Most of the characterized antigens were found to belong to evolutionarily conserved protein families. However, major diagnostically relevant B-cell epitopes were identified within regions of these proteins that are specific to the parasite. The multicomponent chimeric antigen was not made from synthetic oligonucleotides in this case, but rather it was constructed from cloned PCR fragments encoding small regions of 20 to 106 aa long from the *L. infantum* antigens LiP2a, LiP2b, LiP0, LiH2A, and LiH2B separated by linker regions of different sequences.[110] This engineered protein was purified and tested as a diagnostic reagent. A serological evaluation of this multiple-epitope protein by EIA revealed sensitivity values of 79 to 93% and specificity values of 96 to 100% in the diagnosis of visceral leishmaniasis, indicating that this protein is a valuable antigen for the development of diagnostic assays.

Thus, a new concept of artificial composite antigens composed of antigenic epitopes derived from different proteins or even from different strains of the same pathogen was developed and validated. Despite the simplicity and often undesirable outcomes, the new principle of designer diagnostic antigens proved to be extremely useful for the construction of a new generation of diagnostic reagents that, instead of imitating just the structure of the protein, model only one property relevant to diagnostics; namely, the property to bind antibody elicited to a pathogen during an infection in a sensitive and specific manner. This is a significant methodological departure from the very recent practice of genetic engineering diagnostic antigens based mainly on the ability to model the overall structure of the natural protein. The new concept suggests that the most direct way of obtaining a desired function in an engineered protein is to model this function using only the required structural components in order to faithfully reproduce the desired function. This novel concept of a semirational approach to building diagnostic antigens was found to be significantly beneficial in addressing an emerging new challenge in diagnostics, sequence heterogeneity of pathogens.

10.5 NEW DIAGNOSTIC CHALLENGES: SEQUENCE HETEROGENEITY

The development of nucleic acid sequencing technology in recent years has fostered an explosion of new information on sequence heterogeneity of the genomes of various organisms. The study of sequence heterogeneity has become a very popular and interesting research field in virology and molecular biology. However, besides being an impelling area of research, sequence heterogeneity of different viral or microbial strains or isolates poses significant and different diagnostic challenges. It raises the important question of how sequence heterogeneity affects the antigenic properties of viral and microbial proteins used for the detection of antibodies. Should sequence heterogeneity be factored into the development of diagnostic assays? This question is addressed below using the hepatitis C virus (HCV) as an example.

HCV causes approximately 70% of non-A, non-B hepatitis worldwide.[111,112] The HCV genome is a single-stranded positive sense RNA of ~9500 nucleotides that encodes for a single polyprotein of ~3000 amino acids.[112–115] The N-terminal part of the polyprotein is processed into three structural proteins: nucleocapsid or core protein and two envelope proteins, E1 and E2. The C-terminal part is processed into six nonstructural proteins: NS2, NS3, NS4a, NS4b, NS5a, and NS5b proteins.[113,115–119] HCV represents a group of highly heterogeneous viruses.[120–122] Extensive studies of HCV heterogeneity have resulted in the classification of HCV into six major genotypes.[122–124] The nucleotide sequence homology between genotypes is ~70%, which is similar to that between different serotypes of other flaviviruses.[125] Sequence heterogeneity is not uniformly distributed across the entire HCV genome. The 5′-terminal part of the HCV genome, including the 5′-noncoding regions and core gene, is the most conserved, whereas the E1, E2, NS4, and NS5a regions are very variable.[123,124,126] In particular, the 5′-terminal region of the E2 gene encoding for the hypervariable region (HVR1) is especially heterogeneous.[118,127] The NS2, NS3, and NS5b genes are less variable than the E1, E2, NS4, and NS5a genes.[123]

The affect of sequence heterogeneity on the antigenic properties of different HCV proteins was studied by several groups of researchers.[125,128–137] A large number of diagnostically pertinent antigenic regions have been identified within the HCV core protein,[138–142] E1/E2 proteins,[143–145] NS3 protein,[146,147] NS4 protein,[111,146,147] and NS5 protein.[147–149] The HCV core, NS3, NS4 and NS5 proteins are presently used as diagnostic targets in various commercially available assays for the detection of anti-HCV activity in serum specimens.[122,138,140,146,148–152]

Since the discovery of the very first immunoreactive region, designated 5-1-1, from the HCV polyprotein, this region has been subjected to very extensive analysis of its antigenic properties. Region 5-1-1 is derived from the HCV NS4 protein and is very heterogeneous. Sequence similarity of this region varies from 32.4 to 98.6%.[121] One study, although based on a small subset of sequences, has shown that the NS4a nucleotide sequence is only slightly more conserved than the E1 sequence, with evolutionary distances ranging from 0.46 to 0.80 between major genotypes[134] compared to 0.43 to 0.85 for E1,[126] and much more heterogeneous than core sequences with a range of evolutionary distances from 0.13 to 0.25.[128] As will be shown later, such significant heterogeneity of the primary structure of the HCV NS4 protein has had a substantial effect on its immunoreactivity.

Alarming reports of genotype-specific variation of the antigenic properties started emerging for different regions of the HCV polyprotein.[87,125,128–137,154,155] The NS4 recombinant protein C100–3, comprising region 5-1-1 of genotype 1, was found to be less immunoreactive or nonimmunoreactive with antibodies in serum specimens from patients infected with HCV genotypes 2 and 3 compared to the immunoreactivity with the serum specimens from patients infected with genotype 1.[130] For example, 93% of patients infected with HCV subtype 1a and 79% with HCV subtype 1b had detectable antibodies to C100-3, whereas only 34% of patients infected with HCV genotype 2 had detectable antibodies to this antigen.[132] Further experiments demonstrated that synthetic peptides derived not only from the heterogeneous region 5-1-1 but also from the very much more conserved HCV core protein display genotype-specific immunoreactivity.[125,129,134] Because recombinant proteins and peptides used in many commercial HCV serological diagnostic assays uniformly belong to genotype 1, these observations have raised concerns regarding the sensitivity of current serological assays in detecting antibodies to different HCV genotypes and have inspired more research on the antigenic heterogeneity of HCV proteins. Simultaneously, these results were utilized in the development of serological assays for the discrimination of HCV genotypes, based entirely on the specificity of antibodies to several genotype-specific antigenic epitopes of the HCV core and NS4 and NS5 proteins.[125,134,156]

The most extensive study on antigenic heterogeneity was made for the HCV NS4 protein, which contains two strong antigenic regions. One region, region 5-1-1, comprises the C-terminal part of NS4a and the N-terminal part of NS4b.[157] The other region, designated region 59, is located at the C terminus of the NS4b protein.[147,158] Antigenic epitopes for both regions can be efficiently modeled with synthetic peptides of different sizes.[125,134,147,155] Antigenic analysis of synthetic peptides derived from the consensus sequence of region 5-1-1 of genotype 1, 2, and 3 revealed that this region contains both genotype-independent and genotype-specific antigenic epitopes.[125] These data were confirmed and extended by using overlapping synthetic peptides derived from actual (not consensus) sequences from different HCV genotypes.[155] More than 70 unique sequences of region 5-1-1 representing HCV genotypes 1 to 6 were used to design 377 overlapping synthetic peptides. The sequence similarity between these sequences varied from 32.4 to 98.6%.

The amino acid sequence of region 59 is less heterogeneous than that of region 5-1-1.[159,160] Only seven unique sequences of region 59 (sequence similarity 80.6 to 93.5%) representing genotypes 1 to 3 were used to design 66 overlapping synthetic peptides.[155] All peptides were tested against a panel of serum specimens obtained from patients infected with HCV genotypes 1 to 5. The data demonstrated that immunoreactive peptides fell into two groups. One group, represented by N-terminal peptides, demonstrated genotype-independent immunoreactivity; the other group, from the central part of region 5-1-1, showed strict genotype specificity. Peptides derived from region 59 did not show genotype-specific immunoreactivity.

The most alarming observation made in this study was that peptides derived from precisely the same region of the NS4 protein but from different HCV sequence variants demonstrated a very large range of breadth of immunoreactivity with serum specimens. Peptides derived from some regions showed a tenfold difference in the number of serum specimens detected. Thus, in addition to the identification of epitopes important for the efficient detection of either genotype-independent or genotype-specific HCV antibodies, careful examination of two antigenic regions of the NS4 protein strongly suggested that different sequence variants of diagnostically important antigens may display different sensitivity in the ability to detect antibody, and therefore special attention should be paid to the selection of the most diagnostically efficient variants of antigens.[155]

As the above discussion demonstrates, synthetic peptides have been instrumental in the study of antigenic heterogeneity. Unfortunately, not all diagnostically important antigenic regions can be modeled with synthetic peptides in an immunologically active form. For example, antigenic epitopes of the HCV NS3 protein could be modeled only with recombinant proteins of at least 90 to 100 aa[161] and, conversely, could not be modeled with short synthetic peptides.[147] Consequently, a different strategy was employed to study the immunoreactivity of the HCV NS3 protein. This same strategy was used to evaluate the antigenic hetereogeneity of the NS5a antigenic epitopes, although these epitopes may also be modeled with short synthetic peptides.[147]

The immunodominant antigenic domains of the HCV NS3 and NS5a proteins were localized at position 1357–1459 aa and 2212–2313 aa, respectively.[87,162,163] To determine the effect of sequence heterogeneity on the antigenic properties of these two regions, 46 unique NS3 sequences and 84 unique NS5a sequences were selected from GenBank. Percent homology between different NS3 sequences varied from 73.9 to 99.1% and between NS5a sequences from 49.0 to 97.1%. These sequences were used to design synthetic genes. Two major criteria were used to select sequences for gene synthesis from those sequences retrieved from GenBank: (1) maximum evolutionary distances between sequences and (2) representation of all HCV genotypes.

Based on these criteria 12 sequences from the 46 NS3 sequences and 14 sequences from the 84 NS5a sequences were selected. Synthetic genes were designed and constructed by PCR from synthetic oligonucleotides. All of these genes were expressed in *E. coli* as hybrid proteins with glutathione *S*-transferase. All 26 proteins were purified and tested against panels of anti-HCV positive sera obtained from patients infected with HCV genotypes 1 to 6. Different NS3 proteins immunoreacted with 38 to 79% of anti-c33-positive serum specimens. The most immunoreactive NS3 protein was derived from HCV genotype 6. The least immunoreactive NS3 protein was derived from the HCV genotype 1b. Not one protein was found that demonstrated strict genotype-specific immunoreactivity.[162,163] Different NS5a proteins immunoreacted with 15 to 93% across all anti-NS5-positive serum specimens.

Like the HCV NS3 proteins, the NS5a proteins did not show strict genotype specific immunoreactivity.[87] Nonetheless, it is obvious from the above results that sequence heterogeneity has a profound effect on the antigenic properties of these proteins. One remarkable example of how small amino acid changes may affect the antigenic properties of proteins is the immunoreactivity of two subtype 1b NS5a proteins, which, despite 86% similarity (only 12 aa position difference) demonstrated drastically different immunoreactivity with genotype 5 serum specimens: one protein immunoreacted with all genotype 5 specimens, whereas the other protein did not immunoreact with a single specimen of this genotype.[87]

Thus, the data obtained by several groups of researchers strongly suggest that sequence heterogeneity of HCV proteins affects the antigenic properties of these proteins. Different sequence variants of these proteins may display significantly different rates of specific immunoreactivity with antibodies. This observation is very important for the development of new generations of diagnostic tests for the detection of anti-HCV activity in serum specimens.

10.6 HCV DESIGNER DIAGNOSTIC ANTIGENS

Since different sequence variants of diagnostically relevant HCV proteins display different efficiencies of antibody binding, and since not one variant has been found that is capable of detecting all anti-HCV-positive serum specimens, antigenic heterogeneity presents a considerable challenge to HCV diagnostic development. The HCV diagnostic assay requires more than one antigen as a diagnostic target and may need more than one variant of the same antigen to improve the efficiency of antibody detection. These requirements pose significant technological problems. A potential resolution of these problems would be the development of a designer antigen that binds with equally high efficiency antibodies against all HCV sequence variants. One of the approaches to obtaining such an antigen is the previously discussed strategy of artificial mosaic antigens.

One such antigen, an artificial HCV NS4 antigen composed of 17 small antigenic regions representing HCV genotypes 1 through 5, was recently designed and constructed.[164] Eleven antigenic regions were derived from the 5-1-1 region, and six other regions were derived from region 59. Region 5-1-1 contains two distinct antigenic domains at position 1691–1710 aa and 1712–1733 aa.[147] Five sequences derived from the first domain and six sequences derived from the second antigenic domain were selected to be designed into the mosaic protein. All 17 regions were arranged within the NS4 mosaic protein in an order that satisfied two principles: (1) Regions would be scattered across the artificial protein and (2) the predicted secondary structure for each individual region within this protein would be identical to the predicted secondary structure of the same region within the native HCV polyprotein. The amino acid sequence was reverse-translated into nucleotide sequence using optimal *E. coli* codons.

The gene encoding for this artificial antigen was assembled from synthetic oligonucleotides by a new method designated as restriction enzyme assisted ligation (REAL). This method is based on assembling genes by sequential cloning of DNA segments with a specially designed vector. After insertion into this vector each segment is flanked by recognition sites of the class IIS restriction endonucleases. These recognition sites were arranged in such a way that as restriction enzymes recognize these sites outside of these segments cleavage occurs inside of the segments, generating sticky ends complementary for adjacent segments and thereby allowing two segments to assemble together without any additional or undesirable linker sequences. This approach is especially suitable for assembly of genes composed of repeat sequences or when intermediate products of assembly should be studied as well as a full-size product.

The full-length synthetic gene and all intermediate products of gene assembly were expressed in *E. coli*. Using site-specific antibodies raised against synthetic peptides, it was shown that all regions for which sequence-specific antibodies were obtained were accessible to antibody binding. The diagnostic relevance of this artificial antigen was examined by testing this antigen with HCV seroconversion panels and with a panel of anti-HCV-positive and -negative sera. The artificial NS4 antigen specifically detected anti-NS4 antibodies with almost equal efficiency from serum specimens obtained from patients infected with HCV of different genotypes. The strategy employed to construct this antigen may be applied to the design and construction of other artificial antigens with similar improved diagnostic properties.

A concept similar to that described above was developed by Yagi et al.[165] The artificial protein CepCM designed by this group of researchers included two regions derived from the NS3 protein and antigenic epitopes derived from core and NS4 proteins of HCV genotypes 1 and 2. This protein demonstrated high sensitivity and specificity in detecting HCV antibodies. However, a test to evaluate its efficiency in detecting anti-HCV activity in patients infected with different HCV genotypes was not conducted. Another artificial recombinant protein, designated as multiple-epitope-fusion antigen (MEFA-6), was constructed by Chien at al.[29] To improve the sensitivity of existing HCV recombinant antigens, MEFA-6 was constructed of major immunodominant epitopes from seven functional regions of the HCV genome such as core (region at position 10–53 aa), E1 (position 303–320 aa), E2 (position 405–444 aa), NS3 (position 1192–1457 aa), region 5-1-1 (position 1689–1735 aa), and regions 59 (position 1901–1940 aa) and NS5 (2278–2313). The core sequence was duplicated and the region 5-1-1 sequences were triplicated in the MEFA-6. All sequences, except for two copies of the region 5-1-1, were derived from the HCV subtype 1a. One copy of region 5-1-1 was of subtype 2b and the other of genotype 3.

By using epitope-specific monoclonal and polyclonal antibodies, the authors of this study demonstrated that MEFA-6 exposes all of the major antigenic epitopes built into this protein. This antigen was capable of detecting HCV at a two- to fourfold greater dilutional sensitivity than a previously developed EIA, which uses a chimeric-fusion polyprotein made up of the NS3-NS4-core regions.[146] The MEFA-6 anti-HCV assay developed by Chien et al.[29] exhibits a sensitivity and specificity equivalent to other commercially available and licensed EIAs.

All of the commercial anti-HCV EIAs that have been developed to date have two major challenges. First, sensitivity of assays should be improved to reduce or eliminate the window period (i.e., the time period between the first detection of viremia and the detection of antibody) and such assays are now appearing for the simultaneous detection of HIV core antigen and anti-HIV, antibody detection in immunosuppressed patients, and detection of antibodies to different HCV genotypes. Second, by using several synthetic peptides or recombinant proteins as antigenic targets assay manufacture is more complex and consequently more costly, which ultimately slows the development of new assays.

On one hand, the development of more sensitive assays for the detection of anti-HCV in serum specimens, especially from patients infected with different HCV genotypes, requires the use of more antigenic epitopes, as evidenced by the release of each new generation HCV antibody assay. On the other hand, the inclusion of more and more antigenic targets makes the manufacture of these assays more complicated and may reduce the overall specificity of these assays as well. The artificial proteins described above represent a new strategy that includes (1) careful selection of small diagnostically relevant regions among the pathogen proteins and (2) combining these regions into one diagnostic antigen. It seems that this strategy solves both types of problem and therefore may provide better antigens to improve immunodiagnostics in the future.

10.7 CONCLUSION

Over the last decade, diagnosis of infectious diseases has made tremendous strides in the development of new concepts for the detection of important serological markers of infections. Numerous new formats and detection systems have been devised and developed for the specific identification of antibodies elicited in the response to pathogen invasion. The pathogen-specific antigens are the most essential components of all diagnostic formats. Over the last 20 years, significant progress has been made in genetic engineering of diagnostic reagents. Although very impressive and nonetheless important, the major achievements were, in the area of modeling linear antigenic epitopes and in the development of expression systems that can faithfully reproduce the antigen macrostructure, responsible for conformational antigenic epitopes. However, new challenges have arisen, such as sequence heterogeneity and the closely related heterogeneity of antigenic properties, nonspecific cross-immunoreactivity of antigenic epitopes, and very complex requirements for designing antigens containing properties consistent with new diagnostic formats, especially homogeneous assay formats and molecular sensor technology, all of which places a very heavy demand on the development of new strategies for making antigens with predetermined diagnostically pertinent properties. This chapter has presented the very first strategies for designing and constructing new artificial antigens addressing at least some of the aforementioned challenges.

The fundamental principles that were used to build new artificial antigenic targets with improved diagnostic properties were the careful selection of antigenic regions for strength and breadth of immunoreactivity and assembling all the diagnostically relevant regions into one recombinant protein. Although apparently simple in concept, this approach requires significant effort to identify and select the best antigenic regions. The second principle involving assembly of these selected regions into one artificial antigen in such a way that all antigenic regions built into the final protein remain accessible for antibody binding is trivial in comparison. Today, the design and construction of diagnostic antigens are making only their very first steps toward protein engineering. Tomorrow, diagnostic research will move forward from the simple search for appropriate antigens to the rational design of antigens for the development of new sophisticated approaches to the detection of infectious diseases.

REFERENCES

1. Gauss-Muller, V., Frosner, G. G., and Deinhardt, F., Propagation of hepatitis A virus in human embryo fibroblasts, *J. Med. Virol.*, 7, 233, 1981.
2. Bishop, N. E., Hugo, D. L., Borovec, S. V., and Anderson, D. A., Rapid and efficient purification of hepatitis A virus from cell culture, *J. Virol. Methods*, 47, 203, 1994.
3. Wheeler, C. M., Fields, H. A., Schable, C. A., Meinke, W. J., and Maynard, J. E., Adsorption, purification and growth characteristics of hepatitis A virus strain HAS-15 propagated in fetal rhesus monkey kidney cells, *J. Clin. Microbiol.*, 23, 434, 1986.
4. Binn, L. N., Lemon, S. M., Marchwicki, R. H., Redfield, R. R., Gates, N. L., and Bancroft, W. H., Primary isolation and serial passage of hepatitis A virus strains in primate cell cultures, *J. Clin. Microbiol.*, 20, 28, 1984.
5. Hijazi, Z., Pacsa, A., Eisa, S., El Shazli, A., and Abd el-Salam, R. A., Laboratory diagnosis of acute lower respiratory tract viral infections in children, *J. Trop. Pediatr.*, 42, 276, 1996.
6. Tsutsumi, H., Ouchi, K., Ohsaki, M., Yamanaka, T., Kuniya, Y., Takeuchi, Y., Nakai, C., Meguro, H., and Chiba, S., Immunochromatography test for rapid diagnosis of adenovirus respiratory tract infections: comparison with virus isolation in tissue culture, *J. Clin. Microbiol.*, 37(6), 2007, 1999.
7. Scott, R. M., Nisalak, A., Cheamudon, U., Seridhoranakul, S., and Nimmannitya, S., Isolation of dengue viruses from peripheral blood leukocytes of patients with hemorrhagic fever, *J. Infect. Dis.*, 141, 1, 1980.
8. Cardosa, M. J. and Tio, P. H., Dot enzyme immunoassay: an alternative diagnostic aid for dengue fever and dengue haemorrhagic fever, *Bull World Health Organ,* 69, 741, 1991.
9. Kittigul, L., Meethien, N., Sujirarat, D., Kittigul, C., and Vasanavat, S., Comparison of dengue virus antigens in sera and peripheral blood mononuclear cells from dengue infected patients, *Asian Pac. J. Allergy Immunol.*, 15, 187, 1997.
10. Batteiger, B. E., Newhall, W. J., Terho, P., Wilde, C. E., and Jones, R. B., Antigenic analysis of the major outer membrane protein of *Chlamydia trachomatis* with murine monoclonal antibodies, *Infect. Immunol.*, 53, 530, 1986.
11. Rosoff, J. D. and Stibbs, H. H., Isolation and identification of a *Giardia lamblia* specific stool antigen (GSA 65) useful in coprodiagnosis of giardiasis, *J. Clin. Microbiol.*, 23, 905, 1986.
12. Yelverton, E., Norton, S., Obijeski, J. F., and Goeddel, D. V., Rabies virus glycoprotein analogs: biosynthesis in *Escherichia coli*, *Science*, 219, 614, 1983.
13. Charnay, P., Mandart, E., Hampe, A., Fitoussi, F., Tiollais, P., and Galibert, F., Localization on the viral genome and nucleotide sequence of the gene coding for the two major polypeptides of the hepatitis B surface antigen (HBs Ag), *Nucleic Acids Res.*, 7, 335, 1979.
14. Emtage, J. S., Tacon, W. C., Catlin, G. H., Jenkins, B., Porter, A. G., and Carey, N. H., Influenza antigenic determinants are expressed from haemagglutinin genes cloned in *Escherichia coli*, *Nature*, 283, 171, 1980.
15. Davis, A. R., Nayak, D. P., Ueda, M., Hiti, A. L., Dowbenko, D., and Kleid, D. G., Expression of antigenic determinants of the hemagglutinin gene of a human influenza virus in *Escherichia coli*, *Proc. Natl. Acad. Sci. U.S.A.*, 78, 5376, 1981.
16. Rose, J. K. and Shafferman, A., Conditional expression of the vesicular stomatitis virus glycoprotein gene in *Escherichia coli*, *Proc. Natl. Acad. Sci. U.S.A.*, 78, 6670, 1981.
17. Venkatesan, S., Elango, N., and Chanock, R. M., Construction and characterization of cDNA clones for four respiratory syncytial viral genes, *Proc. Natl. Acad. Sci. U.S.A.*, 80, 1280, 1983.

18. Berman, P. W., Dowbenko, D., Lasky, L. A., and Simonsen, C. C., Detection of antibodies to herpes simplex virus with a continuous cell line expressing cloned glycoprotein D, *Science*, 222, 524, 1983.

19. Stamm, L. V., Folds, J. D., and Bassford, P. J., Jr., Expression of *Treponema pallidum* antigens in *Escherichia coli* K-12, *Infect. Immunol.*, 36, 1238, 1982.

20. Norgard, M. V. and Miller, J. N., Cloning and expression of *Treponema pallidum* (Nichols) antigen genes in *Escherichia coli, Infect. Immunol.*, 42, 435, 1983.

21. Scott, J. R. and Fischetti, V. A., Expression of streptococcal M protein in *Escherichia coli*, *Science*, 22, 758, 1983.

22. Vodkin, M. H. and Leppla, S. H., Cloning of the protective antigen gene of *Bacillus anthracis, Cell*, 34, 693, 1983.

23. Taylor, M. A., McIntosh, M. A., Robbins, J., and Wise, K. S., Cloned genomic DNA sequences from *Mycoplasma hyorhinis* encoding antigens expressed in *Escherichia coli*, *Proc. Natl. Acad. Sci. U.S.A.*, 80, 4154, 1983.

24. Engleberg, N. C., Drutz, D. J., and Eisenstein, B. I., Cloning and expression of *Legionella pneumophila* antigens in *Escherichia coli, Infect. Immunol.*, 44, 222, 1984.

25. Ellis, J., Ozaki, L. S., Gwadz, R. W., Cochrane, A. H., Nussenzweig, V., Nussenzweig, R. S., and Godson, G. N., Cloning and expression in *E. coli* of the malarial sporozoite surface antigen gene from *Plasmodium knowlesi, Nature*, 302, 536, 1983.

26. Kemp, D. J., Coppel, R. L., Cowman, A. F., Saint, R. B., Brown, G. V., and Anders, R. F., Expression of *Plasmodium falciparum* blood-stage antigens in *Escherichia coli*: detection with antibodies from immune humans, *Proc. Natl. Acad. Sci. U.S.A.*, 80, 3787, 1983.

27. Choo, Q., Kuo, G., Weiner, A., Overby, L., Bradley, D., and Houghton, M., Isolation of a cDNA clone derived from a blood-borne non-A, non-B viral hepatitis genome, *Science*, 244, 359, 1989.

28. Choo, Q. L., Weiner, A. J., Overby, L. R., Kuo, G., Houghton, M., and Bradley, D. W., Hepatitis C virus: the major causative agent of viral non-A, non-B hepatitis, *Br. Med. Bull.*, 46, 423, 1990.

29. Chien, D. Y., Arcangel, P., Medina-Selby, A., Coit, D., Baumeister, M., Nguyen, S., George-Nascimento, C., Gyenes, A., Kuo, G., and Valenzuela, P., Use of a novel hepatitis C virus (HCV) major-epitope chimeric polypeptide for diagnosis of HCV infection, *J. Clin. Microbiol.*, 37, 1393, 1999.

30. Burrell, C. J., Mackay, P., Greenaway, P. J., Hofschneider, P. H., and Murray, K., Expression in *Escherichia coli* of hepatitis B virus DNA sequences cloned in plasmid pBR322, *Nature*, 279, 43, 1979.

31. Hoofnagle, J. H., Gerety, R. J., and Barker, L. F., Antibody to hepatitis-B-virus core in man, *Lancet*, 2, 869, 1973.

32. Szmuness, W., Hoofnagle, J. H., Stevens, C. E., and Prince, A. M., Antibody against the hepatitis type B core antigen. A new tool for epidemiologic studies, *Am. J. Epidemiol.*, 104, 256, 1976.

33. Hoofnagle, J. H., Seeff, L. B., Bates, Z. B., Gerety, R. J., and Tabor, E., *Serologic Responses in HB, Viral Hepatitis*, Franklin Institute Press, Philadelphia, 1978, p. 219.

34. Krugman, S., Overby, L. R., Mushahwar, I. K., Ling, C. M., Frosner, G. G., and Deinhardt, F., Viral hepatitis, type B. Studies on natural history and prevention re-examined, *N. Engl. J. Med.*, 300, 101, 1979.

35. Peutherer, J. F., MacKay, P., Ross, R., Stahl, S., and Murray, K., Use of the hepatitis B core antigen produced in *Escherichia coli* in an assay for anti-HBc, *Med. Lab. Sci.*, 38, 355, 1981.

36. Popovic, M., Sarngadharan, M. G., Read, E., and Gallo, R. C., Detection, isolation and continuous production of cytopathic retroviruses (HTLV-III) from patients with AIDS and pre-AIDS, *Science*, 224, 497, 1984.

37. Robey, W. G., Safai, B., Oroszlan, S., Arthur, L. O., Gonda, M. A., Gallo, R. C., and Fischinger, P. J., Characterization of envelope and core structural gene products of HTLV-III with sera from AIDS patients, *Science*, 228, 593, 1985.

38. Burke, D. S., Brandt, B. L., Redfield, R. R., Lee, T. H., Thorn, R. M., Beltz, G. A., and Hung, C. H., Diagnosis of human immunodeficiency virus infection by immunoassay using a molecularly cloned and expressed virus envelope polypeptide. Comparison to Western blot on 2707 consecutive serum samples, *Ann. Intern. Med.*, 106, 671, 1987.

39. Thorn, R. M., Beltz, G. A., Hung, C. H., Fallis, B. F., Winkle, S., Cheng, K. L., and Marciani, D. J., Enzyme immunoassay using a novel recombinant polypeptide to detect human immunodeficiency virus env antibody, *J. Clin. Microbiol.*, 25, 1207, 1987.

40. Dawson, G. J., Heller, J. S., Wood, C. A., Gutierrez, R. A., Webber, J. S., Hunt, J. C., Hojvat, S. A., Senn, D., Devare, S. G., and Decker, R. H., Reliable detection of individuals seropositive for the human immunodeficiency virus (HIV) by competitive immunoassays using *Escherichia coli*-expressed HIV structural proteins, *J. Infect. Dis.*, 157, 149, 1988.

41. Ferroni, P., Tagger, A., Pasquali, M., and Profeta, M. L., HIV antibody screening and confirmatory testing of Italian blood donors, One-year experience of a reference center, *Vox Sang*, 55, 143, 1988.

42. Kirnbauer, R., Papillomavirus-like particles for serology and vaccine development, *Intervirology*, 39, 54, 1996.

43. Jiang, X., Wilton, N., Zhong, W. M., Farkas, T., Huang, P. W., Barrett, E., Guerrero, M., Ruiz-Palacios, G., Green, K. Y., Green, J., Hale, A. D., Estes, M. K., Pickering, L. K., and Matson, D. O., Diagnosis of human caliciviruses by use of enzyme immunoassays, *J. Infect. Dis.*, 181 (Suppl. 2), S349, 2000.

44. MacKay, P., Pasek, M., Mafazin, M., Kovacic, R. T., Allet, B., Stahl, S., Gilbert, W., Schaller, H., Bruce, S., and Murray, K., Production of immunological active surface antigens of hepatitis B virus by *Escherichia coli*, *Proc. Natl. Acad. Sci. U.S.A.*, 78, 4510, 1981.

45. Asselbergs, F. A., Will, H., Wingfield, P., and Hirschi, M., A recombinant Chinese hamster ovary cell line containing a 300-fold amplified tetramer of the hepatitis B genome together with a double selection marker expresses high levels of viral protein, *J. Mol. Biol.*, 189, 401, 1986.

46. Nozaki, C., Miyanohara, A., Fujiyama, A., Hamada, F., Ohtomo, N., and Matsubara, K., Two mammalian cell systems for propagation of the hepatitis B virusgenome in extrachromosomal and chromosomally integrated states: production of the surface and e antigens, *Gene*, 38, 39, 1985.

47. Samanta, H. and Youn, B. W., Expression of hepatitis B virus surface antigen containing the pre-S region in mammalian cell culture system, *Vaccine*, 7, 69, 1989.

48. Diot, C., Gripon, P., Rissel, M., and Guguen-Guillouzo, C., Replication of hepatitis B virus in differentiated adult rat hepatocytes transfected with cloned viral DNA, *J. Med. Virol.*, 36, 93, 1992.

49. Clements, M. L., Miskovsky, E., Davidson, M., Cupps, T., Kumwenda, N., Sandman, L. A., West, D., Hesley, T., Ioli, V., and Miller, W., Effect of age on the immunogenicity of yeast recombinant hepatitis B vaccines containing surface antigen (S) or PreS2 + S antigens, *J. Infect. Dis.*, 170, 510, 1994.

50. El-Reshaid, K., Al-Mufti, S., Johny, K. V., and Sugathan, T. N., Comparison of two immunization schedules with recombinant hepatitis B vaccine and natural immunity acquired by hepatitis B infection in dialysis patients, *Vaccine*, 12, 223, 1994.

51. Miskovsky, E., Gershman, K., Clements, M. L., Cupps, T., Calandra, G., Hesley, T., Ioli, V., Ellis, R., Kniskern, P., and Miller, W., Comparative safety and immunogenicity of yeast recombinant hepatitis B vaccines containing S and pre-S2 + S antigens, *Vaccine*, 9, 346, 1991.

52. Vyas G. N., A brief overview of the new vaccines against hepatitis B virus infection: immunogenic gene products and peptide analogs of antigenic epitopes, *Dev. Biol. Stand.*, 63, 141, 1986.

53. Harford, N., Cabezon, T., Crabeel, M., Simoen, E., Rutgers, A., and De Wilde, M., Expression of hepatitis B surface antigen in yeast, *Dev. Biol. Stand.*, 54, 125, 1983.

54. Shen, L. P., Wang, E. L., Pan, T. C., Dai, P. H., Li, Z. P., and Li, Y. Y., The expression of HBsAg gene in yeast under GAL-10 promoter control, *Sci. Sin. [B]*, 29, 856, 1986.

55. Araki, K., Shiosaki, K., Araki, M., Chisaka, O., and Matsubara, K., The essential region for assembly and particle formation in hepatitis B virus surface antigen produced in yeast cells, *Gene*, 89, 195, 1990.

56. Kuroda, S., Miyazaki, T., Otaka, S., and Fujisawa, Y., *Saccharomyces cerevisiae* can release hepatitis B virus surface antigen (HBsAg) particles into the medium by its secretory apparatus, *Appl. Microbiol. Biotechnol.*, 40, 333, 1993.

57. Tsarev, S. A., Tsareva, T. S., Emerson, S. U., Kapikian, A. Z., Ticehurst, J., London, W., and Purcell, R. H., ELISA for antibody to hepatitis E virus (HEV) based on complete open-reading fram-2 protein expressed in insect cells: identification of HEV infection in primates, *J. Infect. Dis.*, 168, 3369, 1993.

58. Zhang, Y., McAtee, P., Yarbough, P. O., Tam, A. W., and Fuerst, T., Expression, characterization, and immunoreactivities of a soluble hepatitis E virus putative capsid protein species expressed in insect cells, *Clin. Diagn. Lab. Immunol.*, 4, 423, 1997.

59. Li, T. C., Yamakawa, Y., Suzuki, K., Tarsumi, M., Razak, M. A., Uchida, T., Takeda, N., and Miyamura, T., Expression and self-assembly of empty virus-like particles of hepatitis E virus, *J. Virol.*, 71, 7207, 1997.

60. Riddell, M. A., Li, F., and Anderson, D. A., Identification of immunodominant and conformational epitopes in the capsid protein of hepatitis E virus by using monoclonal antibodies, *J. Virol.*, 74, 8011, 2000.

61. Meng, J., Dai, X., Chang, J. C., Lopareva, E., Pillot, J., Fields, H. A., and Khudyakov, Y. E., Identification and characterization of the neutralization epitope(s) of the hepatitis E virus, *Virology*, 288, 203, 2001.

62. Obriadina, A., Meng, J. H., Lopareva, E. N., Spelbring, J., Burkov, A., Krawczynski, K., Khudyakov, Y. E., and Fields, H. A., Antigenic properties of recombinant proteins of hepatitis E virus, in *Proceedings* of *the 10th International Symposium on Viral Hepatitis and Liver Disease*, Margolis, H. S., Alter, M., Conlon, R., Dienstag, J. L., and Liang, T. J., Eds., International Medical Press, London, 2002, 117.

63. Agarwal, K. L., Buchi, H., Caruthers, M. H., Gupta, N., Khorana, H. G., Kleppe, K. K., Kumar, A., Ohtsuka, E., Rajbhandary, U. L., van de Sande, J. H., Sgaramella, V., Weber, H., and Yamada, T., Total synthesis of the gene for an alanine transfer ribonucleic acid from yeast, *Nature*, 227, 27, 1970.

64. Khorana, H. G., Agarwal, K. L., Buchi, H., Caruthers, M. H., Gupta, N. K., Kleppe, K., Kumar, A., Otsuka, E., RajBhandary, U. L., Van de Sande, J. H., Sgaramella, V., Terao, T., Weber, H., and Yamada, T., Studies on polynucleotides, 103. Total synthesis of the structural gene for an alanine transfer ribonucleic acid from yeast, *J. Mol. Biol.*, 72, 209, 1972.

65. Khorana, H. G., Agarwal, K. L., Besmer, P., Buchi, H., Caruthers, M. H., Cashion, P. J., Fridkin, M., Jay, E., Kleppe, K., Kleppe, R., Kumar, A., Loewen, P. C., Miller, R. C., Minamoto, K., Panet, A., RajBhandary, U. L., Ramamoorthy, Sekiya, T., Takeya, T., and van de Sande, J. H., Total synthesis of the structural gene for the precursor of a tyrosine suppressor transfer RNA from *Escherichia coli*. 1. General introduction, *J. Biol. Chem.*, 251, 565, 1976.

66. Sgaramella, V. and Khorana, H. G., CXII. Total synthesis of the structural gene for an alanine transfer RNA from yeast. Enzymic joining of the chemically synthesized polydeoxynucleotides to form the DNA duplex representing nucleotide sequence 1 to 20, *J. Mol. Biol.*, 72, 427, 1972.

67. Belagaje, R., Brown, E. L., Gait, M. J., Khorana, H. G., and Norris, K. E., Total synthesis of a tyrosine suppressor transfer RNA gene. XIII. Synthesis of deoxyribopolynucleotide segments corresponding to the nucleotide sequence −1 to −29 in the promoter region, *J. Biol. Chem.*, 254, 5754, 1979.

68. Sekiya, T., Contreras, R., Takeya, T., and Khorana, H. G., Total synthesis of a tyrosine suppressor transfer RNA gene. XVII. Transcription, *in vitro*, of the synthetic gene and processing of the primary transcript to transfer RNA, *J. Biol. Chem.*, 254, 5802, 1979.

69. Sekiya, T., Takeya, T., Brown, E. L., Belagaje, R., Contreras, R., Fritz, H. J., Gait, M. J., Lees, R. G., Ryan, M. J., Khorana, H. G., and Norris, K. E., Total synthesis of a tyrosine suppressor transfer RNA gene. XVI. Enzymatic joinings to form the total 207-base pair-long DNA, *J. Biol. Chem.*, 254, 5787, 1979.

70. Ryan, M. J., Brown, E. L., Sekiya, T., Kupper, H., and Khorana, H. G., Total synthesis of a tyrosine suppressor tRNA gene. XVIII. Biological activity and transcription, *in vitro*, of the cloned gene, *J. Biol. Chem.*, 254, 5817, 1979.

71. Itakura, K., Hirose, T., Crea, R., Riggs, A. D., Heyneker, H. L., Bolivar, F., and Boyer, H. W., Expression in *Escherichia coli* of a chemically synthesized gene for the hormone somatostatin, *Science*, 198, 1056, 1977.

72. Rosenthal, A., Ulrich, R., Billwitz, H., Frank, R., and Blocker, H., Chemical-enzymatic synthesis, cloning and expression of a synthetic gene coding for an env protein fragment of the human T-cell leukemia virus type 1, *Nucleic Acids Symp. Ser.*, 18, 33, 1987.

73. Nassal, M., Total chemical synthesis of a gene for hepatitis B virus core protein and its functional characterization, *Gene*, 66, 279, 1988.

74. Bell, L. D., Smith, J. C., Derbyshire, R., Finlay, M., Johnson, I., Gilbert, R., Slocombe, P., Cook, E., Richards, H., Clissold, P., Meredith, D., Powell-Jones, C. H., Dawson, K. M., Carter, B. L., and McCullagh, K. G., Chemical synthesis, cloning and expression in mammalian cells of a gene coding for human tissue-type plasminogen activator, *Gene*, 63, 155, 1988.

75. Denefle, P., Kovarik, S., Guitton, J. D., Cartwright, T., and Mayaux, J. F., Chemical synthesis of a gene coding for human angiogenin, its expression in *Escherichia coli* and conversion of the product into its active form, *Gene*, 56, 61, 1987.

76. Edge, M. D., Green, R. A., Heathcliffe, G. R., Meacock, P. A., Schuch, W., Scanlon, D. P., Atkinson, T. C., Newton, R. C., and Markhom, A. F., Total synthesis of a human leukocyte interferon gene, *Nature*, 292, 756, 1981.

77. Engels, J. and Uhlmann, E., Gene synthesis, *Adv. Biochem. Eng. Biotechnol.*, 37, 73, 1988.

78. Jayaraman, K. and Shah, J., PCR mediated gene synthesis, *Nucleic Acid Res.*, 17, 4403, 1987.
79. Jayaraman, K., Finger, S. A., Shah, J., and Fyles, J., Polymerase chain reaction-mediated gene synthesis: synthesis of a gene encoding for isozyme C of horseradish peroxidase, *PNAS*, 88, 4084, 1991.
80. Rossi, J. J., Kierzek, R., Huang, T., Walker, P. A., and Itakura, K., An alternate method for synthesis of double-stranded DNA segments, *J. Biol. Chem.*, 257, 9226, 1982.
81. Scarpulla, R. C., Narang, S. A., and Wu, R., Use of a new retrieving adaptor in the cloning of a synthetic human insulin A-chain gene, *Anal. Biochem.*, 121, 356, 1982.
82. Uhlmann, E., An alternative approach in gene synthesis: use of long self-priming oligonucleotides for the construction of double-stranded DNA, *Gene*, 71, 29, 1988.
83. Weiner, M. P. and Sheraga, H. A., A method for the cloning of unpurified single-stranded oligonucleotides, *Nucleic Acids Res.*, 17, 7113, 1989.
84. Jayaraman, K. and Puccini, C. J., A PCR-mediated gene synthesis strategy involving the assembly of oligonucleotides representing only one of the strands, *Biotechniques*, 12, 392, 1992.
85. Majumder, K., Ligation-free gene synthesis by PCR: synthesis and mutagenesis of multiple loci of a chimeric gene encoding OmpA signal peptide and hirudin, *Gene*, 110, 89, 1992.
86. Khudyakov, Y. E., Favorov, M. O., Khudyakov, N. S., Cong, M., Holloway, B. P., Padhye, N., Lambert, S. B., Jue, D. L., and Fields, H. A., Artificial mosaic protein containing antigentic epitopes of hepatitis E virus, *J. Virol.*, 68, 7067, 1994.
87. Dou, X. G., Talekar, G., Chang, J., Dai, X., Li, L., Bonafonte, M. T., Holloway, B., Fields, H. A., and Khudyakov, Y. E., Antigenic heterogeneity of the hepatitis C virus NS5A protein, *J. Clin. Microbiol.*, 40, 61, 2002.
88. Khudyakov, Y. E., Fields, H. A., Favorov, M. O., Khudyakova, N. S., Bonafonte, M. T., and Holloway, B., Synthetic gene for the hepatitis C virus nucleocapsid protein, *Nucleic Acids Res.*, 21, 2747, 1993.
89. Khudyakov, Y. E., Fields, H. A., Cyclic reactions for the synthesis of artificial DNA, in *Nucleic Acid Amplification Technologies,* Lee, H., Morse, S., and Olsvik, O., Eds., BioTechniques Books, Natick, MA, 1997, 245.
90. Williams, K., Frayne, J., McLaughlin, E. A., and Hall, L., Expression of extracellular superoxide dismutase in the human male reproductive tract, detected using antisera raised against a recombinant protein, *Mol. Hum. Reprod.*, 4, 235, 1998.
91. Alvarez, J. G., Touchstone, J. C., Blasco, L., and Storey, B. T., Spontaneous lipid peroxidation and production of hydrogen peroxide and superoxide in human spermatozoa. Superoxide dismutase as major enzyme protectant against oxygen toxicity, *J. Androl.*, 8, 338, 1987.
92. Storey, B. T., Biochemistry of the induction and prevention of lipoperoxidative damage in human spermatozoa, *Mol. Hum. Reprod.*, 3, 203, 1997.
93. Narum, D. L., Kumar, S., Rogers, W. O., Fuhrmann, S. R., Liang, H., Oakley, M., Tat, A., Sim, B. K. L., and Hoffman, S. L., Codon optimization of gene fragments encoding *Plasmodium falciparum* merzoite proteins enhances DNA vaccine protein expression and immunogenicity in mice, *Infect. Immun.*, 69, 7250, 2001.
94. Gony, M. and Gautier, C., Codon usage in bacteria: correlation with gene expressivity, *Nucleic Acids Res.*, 10, 7055, 1982.
95. Sharp, P. M., Tuohy, T. M. F., and Mosurski, K. R., Codon usage in yeast: cluster analysis clearly differentiates highly and lowly expressed genes, *Nucleic Acids Res.*, 14, 5125, 1986.
96. Ikemura, T., Correlation between the abundance of *Escherichia coli* transfer RNAs and the occurrence of the respective codons in its protein genes, *J. Mol. Biol.*, 146, 1, 1981.
97. Ikemura, T., Correlation between the abundance of yeast transfer RNAs and the occurrence of the respective codons in protein genes. Differences in synonymous codon choice patterns of yeast and *Escherichia coli* with reference to the abundance of isoaccepting transfer RNAs, *J. Mol. Biol.*, 158, 573, 1982.
98. Kurland, C. G., Strategies for efficiency and accuracy in gene expression, *Trends Biochem. Sci.,* 4, 126, 1987.
99. Sharp, P. M. and Li, W. H., The codon adaptation index — a measure of directional synonymous codon usage bias, and its potential applications, *Nucleic Acids Res.*, 15, 1281, 1987.
100. Nagata, T., Uchijima, M., Yoshida, A., Kawashima, M., and Koide, Y., Codon optimization effect on translational efficiency of DNA vaccive in mammalian cells: analysis of plasmid DNA encoding a CTL epitope derived from microotganisms, *Biochem. Biophys. Res. Commun.*, 261, 445, 1999.

101. Srinivasappa, J., Saegusa J., Prabhakar, B. S., Gentry, M. K., Buchmeier, M. J., Wiktor, T. J., Koprowski, H., Oldstone, M. B., and Notkins, A. L., Molecular mimicry: frequency of reactivity of monoclonal antiviral antibodies with normal tissues, *J. Virol.*, 57, 397, 1986.

102. Trujillo, J. R., McLane, M. F., Lee, T. H., and Essex, M., Molecular mimicry between the human immunodeficiency virus type 1 gp120 V3 loop and human brain proteins, *J. Virol.*, 67, 7711, 1993.

103. Fujinami, R. S. and Oldstone, M. B., Molecular mimicry as a mechanism for virus-induced autoimmunity, *Immunol. Res.,* 8, 3, 1989.

104. Khudyakov, Y. E., Favorov, M. O., Jue, D. L., Hine, T. K., and Fields, H. A., Immunodominant antigenic regions in a structural protein of the hepatitis E virus, *Virology*, 198, 390, 1994.

105. Dow, B. C., Buchanan, I., Munro, H., Follett, E. A. C., Davidson, F., Prescott, L. E., Yap, P. L., and Simmonds, P., Relevance of RIBA-3 supplementary test to HCV PCR positive and genotypes for HCV confirmation of blood donors, *J. Med. Virol.*, 49, 132, 1996.

106. Rodriguez-Lopez, M., Riezu-Boj, J. I., Berasain, C., Civeira, M. P., Prieto, J., and Borras-Cuesta, F., Immunogenicity of variable regions of hepatitis C virus proteins: selection and modification of peptide epitopes to assess hepatitis C virus genotypes by ELISA, *J. Gen. Virol.*, 80, 727, 1999.

107. Favorov, M. O., Khudyakov, Y. E., Fields, H. A., Khudyakova, N. S., Padhue, N., Alter, M. J., Mast, E., Polish, L., Yashina, T. L., Yarasheva, D. M., and Margolis, H. S., Enzyme immunoassay for the detection of anti-HEV activity based on synthetic peptides, *J. Virol. Methods*, 46, 237, 1994.

108. Kumar, V., Bansal, V. J., Rao, K. V., and Jameel, S., Hepatitis B virus envelope epitopes: gene assembly and expression in *Escherichia coli* of an immunologically reactive novel multiple-epitope polypeptide 1 (MEP-1), *Gene*, 110, 137, 1992.

109. Manivel, V., Ramesh, R., Panda, S. K., and Rao, K. V. S., A synthetic peptide spontaneously self-assembles to reconstruct a group-specific, conformational determinant of hepatitis B surface antigen, *J. Immunol.,* 148, 4006, 1992.

110. Soto, M., Requena, J. M., Quijada, L., and Alonso, C., Multicomponent chimeric antigen for sero-diagnosis of canine visceral leishmaniasis, *J. Clin. Microbiol.*, 36, 58, 1998.

111. Kuo, G., Choo, Q. L., Alter, H. J., Gitnick, G. L., Redeker, A. G., Purcell, R. H., Miyamura, T., Dienstag, J. L., Alter, M. J., Stevens, C. E. et al., An assay for circulating antibodies to a major etiologic virus of human non-A, non-B hepatitis, *Science*, 244, 362, 1989.

112. Choo, Q. L., Kuo, G., Weiner, A. J., Overby, L. R., Bradley, D. W., and Houghton, M., Isolation of a cDNA clone derived from a blood-borne non-A, non-B viral hepatitis genome, *Science*, 244, 359, 1989.

113. Choo, Q. L., Richman, K. H., Han, J. H., Berger, K., Lee, C., Dong, C., Gallegos, C., Coit, D., Medina-Selby, A., Barr, P., Weiner, A. J., Bradley, D.W., Kuo, G., and Houghton, M., Genetic organization and diversity of the hepatitis C virus, *Proc. Natl. Acad. Sci. U.S.A.*, 88, 2451, 1991.

114. Kato, N., Hijikata, M., Ootsuyama, Y., Nakagawa, M., Ohkoshi, S., Sugimura, T., and Shimotohno, K., Molecular cloning of the human hepatitis C virus genome from Japanese patients with non-A, non-B hepatitis, *Proc. Natl. Acad. Sci. U.S.A.*, 87, 9524, 1990.

115. Takamizawa, A., Mori, C., Fuke, I., Manabe, S., Murakami, S., Fujita, J., Onishi, E., Andoh, T., Yoshida, I., and Okayama, H., Structure and organization of the hepatitis C virus genome isolated from human carriers, *J. Virol.* 65, 1105, 1991.

116. Miller, R. H. and Purcell, R. H., Hepatitis C virus shares amino acid sequence similarity with pestiviruses and flaviviruses as well as members of two plant virus subgroups, *Proc. Natl. Acad. Sci. U.S.A.*, 87, 2059, 1990.

117. Takeuchi, K., Kubo, Y., Boonmar, Y., Waranabe, Y., Katayama, T., Choo, Q. L., Kuo, G., Houghton, M., Saito, E., and Miyamura, T., The putative nucleocapsid and envelope protein genes of hepatitis C virus determined by comparison of the nucleotide sequences of two isolates derived from an experimentally infected chimpanzee and healthy human carriers, *J. Gen. Virol.*, 71, 3027, 1990.

118. Hijikata, M., Kato, N., Ootsuyama, Y., Nakagawa, M., Ohkoshi, S., and Shimotohno, K., Hypervariable regions in the putative glycoprotein of hepatitis C virus, *Biochem. Biophys. Res. Commun.*, 175, 220, 1991.

119. Houghton, M., Weiner, A., and Han, J., Molecular biology of the hepatitis C viruses: implications for diagnosis, development and control of viral disease, *Hepatology*, 14, 381, 1991.

120. Dusheiko, G., Schmilovitz-Weiss, H., Brown, D., McOmish, F., Yap, P. L., Sherlock, S., McIntyre, N., and Simmonds, P., Hepatitis C virus genotypes: an investigation of type-specific differences in geographic origin and disease, *Hepatology*, 19, 13, 1994.

121. Simmonds, P., Variability of hepatitis C virus, *Hepatology*, 21, 570, 1995.

122. Bukh, J., Miller, R. H., and Purcell, R. H., Genetic heterogeneity of hepatitis C virus: quasi-species and genotypes, *Semin. Liver Dis.*, 15, 41, 1995.

123. Okamoto, H., Kurai K., Okada, S., Yamamoto, K., Lizuka, H., and Tanaka, T., Full-length sequence of a hepatitis C virus genome having poor homology to reported isolates: comparative study of four distinct genotypes, *Virology*, 88, 331, 1992.

124. Simmonds, P., Holmes, E. C., Cha, T. A., Chan, S. W., McOmish, F., Irvine, B., and Beall, E., Classification of hepatitis C virus into six major genotypes and a series of subtypes by phylogenetic analysis of the NS5 region, *J. Gen. Virol.*, 74, 2391, 1993.

125. Simmonds, P., Rose, K. A., Graham, S., Chan, S. W., McOmish, F., Dow, B. C., Follett, E. A. C., Yap, P. L., and Marsden, H., Mapping of serotype-specific, immunodominant epitopes in the NS-4 region of hepatitis C virus (HCV): use of type-specific peptides to serologically differentiate infections with HCV types 1, 2 and 3, *J. Clin. Mocrobiol.*, 31, 1493, 1993.

126. Bukh, J., Purcell, R. H., and Miller, R. H., At least 12 genotypes of hepatitis C virus predicted by sequence analysis of the putative E1 gene of isolates collected worldwide, *Proc. Natl. Acad. Sci. U.S.A.*, 90, 8234, 1993.

127. Weiner, A. L., Brauer, M. J., Rosenblatt, J., Tung, J., Crawford, K., and Bonino, F., Variable and hypervariable domain are found in the regions of HCV corresponding to the laivirus envelope and NS1 proteins and the pestivirus envelope glcoproteins, *Virology*, 180, 842, 1991.

128. Chan, S. W., Simmonds, P., Comish, F., Yap, P. L., Mitchell, R., Dow, B., and Follett, E., Serological responses to infection with three different types of hepatitis C virus, *Lancet*, 338, 1391, 1991.

129. Machida, A., Ohnuma, H., Tsuda, F., Munekata, E., Tanaka, T., Akahane, Y., Okamoto, H., and Mishiro, S., Two distinct subtypes of hepatitis C virus defined by antibodies directed to the putative core protein, *Hepatology*, 16, 886, 1992.

130. McOmish, F., Chan, S. W., Dow, B. C., Gillon, J., Frame, W. D., Crawford, R. J., Yap, P. L., Follett, E. A. C., and Simmonds, P., Detection of three types of hepatitis C virus in blood donors: investigation of type-specific differences in serologic reactivity and rate of alanine aminotransferase abnormalities, *Transfusion*, 33, 7, 1993.

131. McOmish, F., Yap, L., Dow, B. C., Follett, E. A. C., Seed, C., Keller, A. J., Cobain, T. J., Krusius, T., Kolho, E., Naukkarinen, R., Lin, C., Lai, C., Leong, S., Medgyesi, G. A., Hejjas, M., Kiyokawa, H., Fukada, K., Cuypers, T., Saeed, A. A., Al-Rasheed, A. M., Lin, M., and Simmonds, P., Geographical distribution of hepatitis C virus genotypes in blood donors: an international collaborative survey, *J. Clin. Microbiol*, 32, 884, 1994.

132. Nagayama, R., Tsuda, F., Okamoto, H., Wang, Y., Mitsui, T., Tanaka, T., Miyakawa, Y., and Mayumi, M., Genotype dependence of hepatitis C virus antibodies detectable by the first-generation enzyme-linked immunobsorbent assay with C 100-3 protein, *J. Clin. Invest.*, 92, 1529, 1993.

133. Alonso, C., Qu, D., Lamelin, J. P., De Sanjose, S., Vitvitski, L., Li, J., Berby, F., Lambert, V., Cortey, M. L., and Trepo, C., Serological responses to different genotypes of hepatitis C virus in France, *J. Clin. Microbiol.*, 32, 211, 1994.

134. Bhattacherjee, V., Prescott, L. E., Pike, I., Rodgers, B., Bell, H., El-Zayadi, A. R., Kew, M. C., Conradie, J., Lin, C. K., Marsden, H., Saeed, A. A., Parker, D., Yap, P. L., and Simmonds, P., Use of NS-4 peptides to identify type-specific antibody to hepatitis C virus genotypes 1, 2, 3, 4, 5 and 6, *J. Gen. Virol.*, 76, 1737, 1995.

135. Zein, N. N., Rakela, J., and Persin, D. H., Genotype-dependent serologic reactivities patients infected with hepatitis C virus in the United States, *Mayo Clin. Proc.*, 70, 449, 1995.

136. Zein, N. N., Germer, J. J., Went, N. K., Schimek, C. M., Thorvilson, J. N., Mitchelland, P. S., and Persing, D. H., Indeterminate results of the second-generation hepatitis C virus (HCV) recombinant immunoblot assay: significance of high-level c22–3 reactivity and influence of HCV genotypes, *J. Clin. Microbiol.*, 35, 311, 1997.

137. Dhaliwal, S. K., Prescott, L. E., Dow, B. C., Davidson, F., Brown, H., Yap, P. L., Follett, E. A. V., and Simmonds, P., Influence of viraemia and genotype upon serological reactivity in screening assays for antibody to hepatitis C virus, *J. Med. Virol.*, 48, 184, 1996.

138. Muraiso, K., Hijikata, M., Ohkoshi, S., Cho, M. J., Kikuchi, M., Kato, N., and Shimotohno, K., A structural protein of hepatitis C virus expressed in *E. coli* facilitates accurate detection of hepatitis C virus, *Biochem. Biophys. Res. Commun.*, 172, 511, 1990.

139. Chiba, J., Ohba, H., Matsuura, Y., Watanabe, Y., Katayama, T., Kikuchi, S., Saito, I., and Miyamura, T., Serodiagnosis of hepatitis C virus (HCV) infection with an HCV core protein molecularly expressed by a recombinant baculovirus, *Proc. Natl. Acad. Sci. U.S.A.*, 88, 4641, 1991.

140. Hosein, B., Fang, C. T., Popovsky, M. A., Ye, J., Zhang, M., and Wang, C. Y., Improved serodiagnosis of hepatitis C virus infection with synthetic peptide antigen from capsid protein, *Proc. Natl. Acad. Sci. U.S.A.*, 88, 3647, 1991.

141. Nasoff, M. S., Zebedee, S. L., Inchauspe, G., and Prince, A. M., Identification of an immunodominant epitope within the capsid protein of hepatitis C virus, *Proc. Natl. Acad. Sci. U.S.A.*, 88, 5462, 1991.

142. Sallberg, M., Ruden, U., Wahren, B., and Magnius, L. O., Immunodominant regions within the hepatitis C virus core and putative matrix proteins, *J. Clin. Microbiol.*, 30, 1989, 1992.

143. Lok, A. S., Chien, D., Choo, Q. L., Chan, T. M., Chiu, E. K., Cheng, I. K., Houghton, M., and Kuo, G., Antibody response to core, envelope and nonstructural hepatitis C virus antigens: comparison of immunocompetent and immunosuppressed patients, *Hepatology*, 18, 497, 1993.

144. Selby, M. J., Choo, Q. L., Berger, K., Kuo, G., Glazer, E., Eckart, M., Lee, C., Chien, D., Kuo, C., and Houghton, M., Expression, identification and subcellular localization of the proteins encoded by the hepatitis C viral genome, *J. Gen. Virol.*, 74, 1103, 1993.

145. Ray, R., Khanna, A., Lagging, L. M., Meyer, K., Choo, Q. L., Ralston, R., Houghton, M., and Becherer, P. R., Peptide immunogen mimicry of putative E1 glycoprotein-specific epitopes in hepatitis C virus, *J. Virol.*, 68, 4420, 1994.

146. Chien, D. Y., Choo, Q. L., Tabrizi, A., Huo, C., McFarland, J., Berger, K., Lee, C., Shuster, J. R., Nguyen, T., Moyer, D. L., Tong, M., Furuta, S., Omata, M., Tegteier, G., Alter, H., Schiff, E., Jeffers, L., Houghton, M., and Kuo, G., Diagnosis of hepatitis C virus (HCV) infection using an immuno-dominant chimeric polyprotein to capture circulating antibodies: reevaluation of the role of HCV in liver disease, *Proc. Natl. Acad. Sci. U.S.A.*, 89, 10011, 1992.

147. Khudyakov, Y. E., Khudyakova, N. S., Jue, D. L., Lambert, S. B., Fang, S., and Fields, H. A., Linear B-cell epitopes of the NS3-NS4-NS5 proteins of the hepatitis C virus as modeled with synthetic peptides, *Virology*, 206, 666, 1995.

148. Mori, S., Ohkoshi, S., Hijikata, M., Kato, N., and Shimotohno, K., Serodiagnostic assay of hepatitis C virus infection using viral protease expressed in *Escherichia coli*, *Jpn. J. Cancer Res.*, 83, 264, 1992.

149. Riezu-Boj, J. I., Parker, D., Civeira, M. P., Phoppard, D., Corbishley, T. P., Camps, J., Castilla, A., and Prieto, J., Detection of hepatitis C virus antibodies with new recombinant antigens: assessment in chronic liver diseases, *J. Hepatol.*, 15, 309, 1992.

150. Van Der Poel, C. L., Cuypers, H. T., Reesink, H. W., Weiner, A. J., Quan, S., Di Nello, R., Van Boven, J. J. P., Winkel, I., Mulder-Folkerts, D., Exel-Oehers, P. J., Schaasberg, W., Leentvaar-Kuypers, A., Polito, A., Houghton, M., and Lelie, P. N., Confirmation of hepatitis C virus infection by new four-antigen recombinant immunoblot assay, *Lancet*, 337, 317, 1991.

151. Inoue, Y., Suzuki, R., Matsuura, Y., Harda, S., Chiba, J., Watanabe, Y., Saito, I., and Miyamura, T., Expression of the amino-terminal half of the NS1 region of the hepatitis C virus genome and detection of an antibody to the expressed protein in patients with liver diseases, *J. Gen. Virol.*, 73, 2151, 1992.

152. Berasain, C., Garcia-Granero, M., Riezu-Boj, J., Civeira, M. P., Prieto, J., and Borras-Cuesta, F., Detection of anti-hepatitis C virus antibodies by ELISA using synthetic peptides, *J. Hepatol.*, 18, 80, 1993.

153. Yuki, N., Hayashi, N., Mita, E., Hagiwara, H., Oshita, M., Ohkawa, K., Katayama, K., Kasahara, A., Fusamoto, H., and Kamada, T., Clinical characteristics and antibody profiles of chronic hepatitis C patients: relation to hepatitis C virus genotypes, *J. Med. Virol.*, 45, 162, 1995.

154. Neville, J. A., Prescott, L. E., Bhattacherjee, V., Adams, N., Pike, I., Rodgers, B., El-Zayadi, A., Hamid, S., Dusheiko, G. M., Saeed, A. A., Haydon, G. H., and Simmonds, P., Antigenic variation of core, NS3 and NS5 proteins among genotypes of hepatitis C virus, *J. Clin. Microbiol.*, 35, 3062, 1997.

155. Chang, J. C., Seidel, C., Ofenloch, B., Jue, D. L., Fields, H. A., and Khudyakov, Y. E., Antigenic heterogeneity of the hepatitis C virus NS4 protein as modeled with synthetic peptides, *Virology*, 257, 177, 1999.

156. Zhang, Z. X., Yun, Z., Chen, M., Sonnerborg, A., and Sallberg, M., Evaluation of a multiple peptide assay for typing of antibodies to the hepatitis C virus: relation to genomic typing by the polymerase chain reaction, *J. Med. Virol.*, 45, 50, 1995.

157. Grakoui, A., Wychowski, C., Lin, C., Feinstone, S. M., and Rice, C. M., Expression and identification of hepatitis C virus polyprotein cleavage products, *J. Virol.*, 67, 1385, 1993.

158. Wienhues, U., Ihlenfeldt, H. G., Seidel, C., Schmitt, U., Kraas, W., and Jung, G., Characterization of a linear epitope in the nonstructural region 4 of hepatitis C virus with reactivity to seroconversion antibodies, *Virology*, 245, 281, 1998.

159. Chamberlain, R. W., Adams, N., Saeed, A. A., Simmonds, P., and Elliott, R. M., Complete nucleotide sequence of a type 4 hepatitis C virus variant, the predominant genotype in the Middle East, *J. Gen. Virol.*, 78, 1341, 1997.

160. Stuyver, L., van Arnhem, W., Wyseur, A., and Maertens, G., Cloning and phylogenetic analysis of the core, E2 and NS3/NS4 regions of the hepatitis C virus type 5a, *Biochem. Biophys. Res. Commun.*, 202, 1308, 1994.

161. Hwang, L., Yang, P., Lai, M., Chiang, B., Kao, J., Wang, J., Lee, S., Chian, H., Chi, W., Chu, Y., Chen, P., and Chen, D., Identification of human antigenic determinants in the hepatitis C virus NS3 protein, *J. Infect. Dis.*, 174, 173, 1996.

162. Chang, J. C. et al., unpublished data, 2002.

163. Khudyakov, Y. E., Dou, X. G., Chang, J. C., Talekar, G., Jue, D., and Fields, H. A., Impact of sequence heterogeneity on antigenic properties of hepatitis C virus (HCV) proteins, *in Proceedings of the 10th International Symposium on Viral Hepatitis and Liver Disease*, Margolis, H. S., Alter, M., Conlon, R., Dienstag, J. L., and Liang, T. J., Eds, International Medical Press, London, 2002, 381.

164. Chang, J. C., Ruedinger, B., Cong, M., Lambert, S., Lopareva, E., Purdy, M., Holloway, B. P., Jue, D. L., Ofenloch B, B., Fields, H. A., and Khudyakov, Y. E., Artificial NS4 mosaic antigen of hepatitis C virus, *J. Med. Virol.*, 59, 437, 1999.

165. Yagi, S., Kashiwakuma, T., Yamaguchi, K., Chiba, Y., Ohtsuka, E., and Hasegawa, A., An epitope chimeric antigen for the hepatitis C virus serological screening test, *Biol. Pharm. Bull.*, 19, 1254, 1996.

11 Recombinant Antibodies

Ramón Montaño and Flor H. Pujol

CONTENTS

11.1 INTRODUCTION

Since their discovery at the end of the 19th century,[1] antibodies (Abs) have become important biological tools because of their multiple and versatile applications in basic research and their enormous potential as reagents for diagnosis, prophylaxis, and therapy. A convenient way of sensing this expectation is to consider the history of the Nobel Prize for Physiology or Medicine. The first Nobel laureate in this category was Emil von Behring "for his work on serum therapy ... by which he has placed in the hands of the physician a victorious weapon against illness and deaths." Since then, 15 more Nobel Prizes have been awarded to individuals who have made significant contributions to immunology. On six occasions (P. Ehrlich in 1908, K. Landsteiner in 1930, R. Porter/G. Edelman in 1972, R. Yalow in 1977, G. Kohler/C. Milstein in 1984, and S. Tonegawa in 1987) those prizes were awarded to individuals whose research field was directly related to the antibody molecule. Polyclonal antibodies ("antisera") were the protagonists during the period of serology and serotherapy. Monoclonal antibodies (mAbs) appeared in the mid-1970s to become the prime tools for powerful analytical techniques such as flow cytometry and immunoenzymatic assays. More recently, the advent of molecular biology and recombinant DNA techniques has made possible the creation of recombinant antibodies (rAbs). These artificial molecules are expected to fulfill hopes and needs in an area still elusive to more traditional Abs: the development of effective prophylactic, therapeutic, and *in vivo* diagnostic reagents for use in humans.

Recombinant Ab technologies are a good example of how curiosity for understanding basic phenomena has yielded information useful for inventing novel molecules of practical interest. For instance, an understanding of the composition, structure, and organization of the genetic complexes coding for Ab molecules in different species[2] has contributed to the development and application of methods for assembling an antigen (Ag) binding moiety of interest as a functional Ab molecule. Ag binding moieties are encoded by variable (V) genetic elements contained within these genetic complexes, and many have been expressed in the context of almost any desired human immunoglobulin (Ig) class or subclass.[3] A number of mammalian expression systems are now available for the expression of a set of cloned V genes.[4] Several eukaryotic cell lines have been shown to support proper expression, synthesis, assembly, and post-translational processing of rAbs. Detailed knowledge of the genomics, architecture, and life cycle of lambda[5,6] and filamentous bacteriophage[7,8] has allowed for the efficient expression of foreign Ab-encoding DNA in bacteria.[9,10] This has made it possible to display highly diverse repertoires of Ab binding sites[11,12] and to mimic the selection strategy used by the humoral immune system. Currently, Ab-phage-display technology offers one efficient way for creating artificial human Abs with reduced immunogenicity and large Ab repertoires for the generation of catalytic Abs. At the same time, the development of methods for creating recombinant DNA molecules, particularly those based on the polymerase chain reaction (PCR), has revolutionized the entire field of molecular biology, including the design and production of rAbs.

RAbs are artificial molecules that constitute a rather heterogeneous group of recombinant proteins. Two major kinds of rAbs can be discriminated. The first includes entire Ab molecules, similar to those found naturally, where the two structural elements allowing Ab function — the Fab and Fc portions — are present. Chimeric and humanized Abs are representative examples of this subgroup. The second type is composed of a more heterogeneous group, where novel Ab-based molecules are designed as single, autonomous, recombinant entities (scFv, Fab, diabodies, triabodies, etc.), or fusion proteins (molecules combining either Fab- or Fc-associated functions with novel properties provided by a toxin, an enzyme, a cell receptor, etc.).

Although the use of recombinant antibody-based molecules in the industrial/medical field is still incipient, many potential applications are currently being evaluated in both academic centers and biotechnology companies, and numerous others appear on the horizon. At present, at least 30% of biological proteins undergoing clinical trials are rAbs.[13] In this chapter, we attempt to summarize the current and major trends of this constantly evolving field where basic science and technological development meet, underscoring those areas that have had more impact in immunology and immunotechnology.

11.2   RECOMBINANT ANTIBODIES: WHY?

Several areas of research and development have driven progress in the field of rAbs. On the basic research side there is an interest in understanding how structural features of an Ab molecule correlate with function and in what aspects the antigen–antibody interaction resembles the reaction between an enzyme and its substrate. On the other hand, there is also great practical interest in developing Abs and Ab-derived molecules as therapeutics for *in vivo* use in humans, in producing Abs against substances with poor immunogenicity or for which immunization is not easily achieved, as well as in developing antibodies with enzyme-like catalytic abilities. As will become evident in the following pages, rAb systems offer excellent working models for all of these situations.

11.3 SOLVING THE PROBLEM OF IMMUNOGENIC MONOCLONAL ANTIBODIES

With the advent of monoclonal antibody/hybridoma technology,[14] many *in vitro* Ab-based techniques and diagnostic tests have been developed and are in wide use. Therefore, it seems reasonable to say that in the context of *in vitro* diagnosis and basic research the utilization of rodent monoclonal antibodies (mAbs) has led to many successful applications. However, *in vivo* applications of rodent mAbs in humans have not been so successful. The main reason for this is the foreign nature of rodent mAbs, which implies a short Ab survival time in humans, which, in turn, hampers the achievement of a desired effect. The most prominent, though not the only, aspect of this foreign character is the rodent mAb immunogenicity, i.e., the appearance of a human antimouse antibody (HAMA) response subsequent to the injection of mouse mAbs into humans.[15,16]

The production and use of human mAbs would be the obvious solution to the problem of rodent mAbs immunogenicity. Several strategies have been attempted to develop human mAbs, including the combination of hybridoma technology with Epstein–Barr virus transformation[17,18] or the application of the CD40 system to generate long-lasting cultures of human B cells.[19,20] Unfortunately, the production of human mAbs has been of limited success, due to a variety of problems.[21-23] Consequently, the development of rAbs has found one of its main propelling forces in the need for reducing rodent mAb immunogenicity, to produce efficient, long-lasting therapeutic Abs for human use. Conveniently, the organization of the structural genes encoding Ab molecules and the modular structural/functional organization of these proteins facilitate enormously its engineering, both at the genetic and protein levels. Chimerization and humanization are the principal approaches followed for reducing rodent mAb immunogenicity.

11.3.1 CHIMERIZATION

A Chimeric Ab (chAb) is an artificial molecule in which the constant portions of the heavy and light chains come from a human Ig and the variable regions V_H and V_L from a rodent mAb (Figure 11.1). The idea behind the construction of a chAb is that constant domains, particularly those included in the so-called Fc region, are the most immunogenic portions of the antibody molecule.[24] When the constant domains of a murine mAb are replaced by the human counterpart, the target specificity is maintained, but the HAMA response is expected to be lower, thereby allowing effector functions to occur more efficiently. Moreover, since the replaced Fc region is of human origin, optimal mediation of effector functions would be expected once the Ab is introduced into humans.

Morrison and colleagues[25] and Boulianne[26] pioneered the technique of antibody chimerization. The procedure implies the cloning of the rodent V_H and V_L genes of interest and the insertion of the cloned genes into mammalian expression vectors containing genes encoding the constant portion of the human H and L chains. These vectors are finally used to stably transfect a selected cell line. Currently, the cloning of V_H and V_L genes is performed by reverse transcriptase-PCR (RT-PCR) using total RNA or mRNA obtained from a hybridoma cell line as starting material, and oligonucleotides complementary to the flanking 3′ and 5′ ends of each gene for priming. Families of oligonucleotides are now available for cloning mouse and human V_H and V_L genes. These families are able to amplify most, if not all, of the functional V genes of both heavy and light chains.[27-39] These oligonucleotides usually contain restriction sites to facilitate the subcloning of PCR-amplified

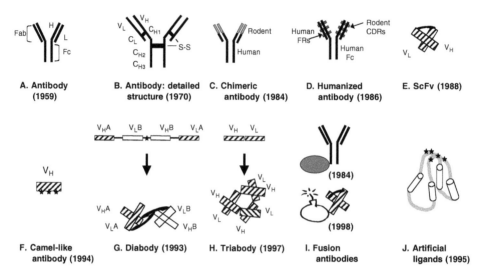

FIGURE 11.1 From natural to artificial antibodies. (A) General structure of an IgG Ab, as described by Porter.[200] Fab (with antigen-binding activity) and Fc (crystalizable) regions are shown. (B) A more detailed view of an Ab, showing the variable (V) and constant (C) domains of the light (L) and heavy (H) chains. Only interchain disulfide bonds (S–S) are shown.[167,200] (C) Chimeric antibody, where the rodent V domains of a desired specificity have been transplanted into a human Ab molecule.[25] (D) Humanized Ab, where rodent complementary-determining regions (CDRs) from V domains have been grafted into a human Ab molecule.[45] Human Ab frameworks (FRs) may need to be mutated or specifically selected to maintain affinity. (E) Single chain fragment V. Nucleotide sequences coding for V_H and V_L are cloned as a single peptide, using a genetic linker coding for neutral amino acids (generally a 15-mer) to facilitate appropriate pairing of both V domains.[61,62] (F) Camel-like Ab. Based on what is found naturally in camel Abs, artificial human V_H fragments have been mutated to reduce the hydrophobicity of the molecule, favoring its solubility and stability as a monomer.[81] The stars indicate amino acid residues mutated. (G) Diabody. DNA sequences coding for two heavy and two light chains, with identical (bivalent) or different (bispecific) specificities are cloned as scFv in a single vector to produce two independent scFv. The linker used to join V_H and V_L sequences codes for only five amino acid residues. The star indicates a stop codon, introduced to generate two separate scFvs, which form functional homo- or heterodimers. (H) Triabody. DNA regions coding for V_H and V_L domains are cloned as a single chain in a vector, but without a linker. After protein synthesis, individual chains are unstable and spontaneously arrange to form a trimer.[71] (I) Fusion Abs. DNA coding for a selected enzyme, toxin, or another protein is fused to the heavy or light chain of an Ab or to an scFv (97,201). (J) Artificial ligands as potential substitutes for Abs. Stars indicate discontinuous residues randomized by patch engineering in different loops of the protein. Those residues are known to be held together in the native protein.[192,195]

V regions into expression vectors. Substantial progress has been made in the development of "cassette" vectors, where V genes can be easily cloned in the context of almost any desired human C_H and C_L[4,40] and then transferred to a variety of available cell lines. Similarly, advances in gene transfer technology have greatly facilitated the efficient insertion of foreign DNA into several different eukaryotic cell lines.

Early works proved that chAbs assemble properly, bind to the specific antigen, and mediate effector functions.[25,26] Furthermore, when compared to their monoclonal equivalents, a reduction in the immunogenicity of the chAb molecules was also documented.[41] Once the feasibility of the approach was tested, chAbs specific to a wide variety of targets have been produced in many laboratories around the world.[32] Successful examples of this technology in the clinical field are ReoPro or abciximab (developed by Centocor, Malvern, PA, marketed by Eli Lilly, Indianapolis, IN), IDEC-C2B8 or rituximab (developed by IDEC Pharmaceuticals and Genentech, South San Francisco, CA), and infliximab from Centocor. ReoPro is a Fab fragment of a mouse-human chAb with specificity for the human platelet GPIIb/IIIa glycoprotein and is used for prevention of acute cardiac ischemia following coronary angioplasty. IDEC-C2B8 is a mouse-human anti-CD20 chAb used for the treatment of non-Hodgkin B-cell lymphoma. Infliximab is a mouse-human chimeric antihuman TNFα Ab that has proved to be effective for the treatment of Crohn's disease.[42]

11.3.2 HUMANIZATION

In chAbs the rodent V regions are left intact. Thus, the specificity and the affinity of the resulting molecule is preserved. Unfortunately, some chimeric molecules have been reported to induce a human anti-chAb antibody (HACA) response when injected into humans. This residual immunogenicity has been attributed to the presence of foreign, immunogenic epitopes in both complementary-determining regions (CDR) and framework regions (FR) of the rodent V Ig domain.[43,44]

A step forward in the efforts to reduce rodent MAbs immunogenicity is the so-called Ab humanization or CDR grafting. The humanized Ab (hAb) technology was pioneered in G. Winter's laboratory.[45] The rationale behind it is based on the belief that the antigen-combining site in the V region of the Ab, i.e., the paratope, is made up of the spatial combination of hypervariable loops formed within the V_H and V_L CDRs, the FRs only playing a secondary role as a scaffold. Consequently, rodent CDRs are grafted at the DNA level into human FRs, conferring the desired specificity to a molecule that is otherwise human, with the advantage that rodent FR-associated epitopes disappear from the resultant recombinant molecule (Figure 11.1). In this approach, it is necessary to know the entire sequence of the rodent V domains and the location of the DNA sequence corresponding to the three rodent CDRs. This information is used to design and create synthetic V genes, where rodent CDR and selected human FR sequences are combined. The synthetic humanized V genes are then inserted into adequate expression vectors, containing the rest of the structural information needed for the synthesis of an entire human light or heavy chain. The rest of the technique is very much the same as for chimeric antibodies.

Unfortunately, CDR grafting is not always accompanied by "affinity grafting," i.e., the described manipulation commonly results in a reduced affinity of the hAb when compared to the original Ab molecule.[24,26] The explanation for this behavior soon became evident: the idea of the CDRs as the only crucial and determinant elements that endow the variable region with its antigen-binding properties (specificity and affinity) is not generally correct. FR amino acid residues in the CDR vicinity usually interact with the hypervariable loops,[47] those interactions being especially relevant for maintenance of the Ab affinity. Furthermore, in some cases a particular backbone is necessary for the proper conformation of a particular loop.[48–50] Also, computer modeling has shown that in some cases FR residues in the vicinity of the CDRs may directly interact with the antigen.[51] Thus, to avoid negatively affecting the Ab affinity during the process of humanization, it was clear that certain features of the original rodent FRs, such as the primary sequence and predicted spatial interactions, must be analyzed and compared with those from human V region candidates.

This comparative analysis should allow for the selection of the most "homologous" human V region to be used as the recipient, and should additionally allow for the identification of potentially deleterious human FR amino acid residues among those making contact with residues of the transplanted CDRs, or that could interact directly with the antigen. The next step of the process would be the replacement of those human residues with their corresponding murine counterparts. The use of computer modeling and analysis of available antibody structures thus led to a more rational and systematic Ab humanization. For example, in a procedure called veneering or resurfacing[52] the problem of V-region immunogenicity is approached in an interesting way. Antibody structure analysis is performed to identify the surface-exposed residues of the rodent framework, and only these residues are replaced by their corresponding human counterparts. It has been reported that two resurfaced Abs, one containing 9 and the other containing 13 amino acid substitutions, do have binding affinities similar to the original murine Abs.[53] However, the immunogenicity of these resurfaced antibodies remains to be determined.

While producing a chAb is a relatively straightforward procedure, the creation of an hAb is comparatively laborious and time-consuming, and maintenance of affinity depends ultimately on trial and error.[54] Furthermore, the main reason for making hAbs, i.e., the ablation of rodent immunogenicity, is not always achieved since some residual immunogenicity has still been reported for "fully" humanized Ab molecules.[55] Apparently, this residual immunogenicity stems from idiotype-related epitopes. Despite this, a substantial number of rodent V regions have been humanized.[24] Among them, it is worth mentioning daclizumab and basiliximab, two hAb specific for the α chain of the human IL-2 receptor that have been successfully tested in preventing acute rejection of renal allografts.[56,57] These hAbs are already in the market under the name of Zenapax (dacliximab from Roche) and Simulect (basiliximab from Novartis Pharmaceuticals Corp.). Another example is trastuzumab (commercially available under the name of Herceptin from Genentech), which is a humanized anti-HER2 Ab effective in the treatment of breast tumors that overexpress the HER2-Neu oncoprotein.[58]

11.4 ANTIBODY-DERIVED, RECOMBINANT MOLECULES

To this point, we have dealt with the production of entire molecules of recombinant antibody. Nevertheless, there are a substantial number of examples where molecular biology and recombinant DNA techniques have been used to engineer novel Ab-based molecules either as single, autonomous, recombinant entities, or as fusion proteins. The next section of this chapter deals with those kinds of molecules, the ways they have been created, and their potential applications.

11.4.1 Ab Fragments

Functional Ab fragments with Ag-binding activity were successfully produced for the first time in 1988 by two different groups, one in Germany[59] and the other in the United States.[60] They simultaneously reported the bacterial expression of fully functional Fv (a V_H–V_L pair) and Fab fragments specific for, respectively, phosphorylcholine and a human-cancer-associated antigen. The key feature of the systems used by these two groups was to direct the transport of the newly synthesized polypeptide chain to the bacterial periplasmic space, where a more favorable environment for folding of disulfide-containing polypeptides is found, thereby avoiding the technical problems associated with intracellular formation of inclusion bodies. They were able to do this by fusing the PelB, ompA, and phoA bacterial signal peptide sequences to the N-terminal region of the expressed Ab fragment polypetide chains.

11.4.1.1 Single Chain Fv (scFv)

In the same year of 1988 two reports appeared in the literature describing the production in *E. coli* of recombinant versions of the Fv fragment where the V_H and V_L domains were physically joined through a small, flexible peptide linker.[61,62] Single-chain Fv (scFv) fragments were reported to retain Ag specificity and to be more stable than Fv (Figure 11.1). They have become one of the preferred formats for expressing recombinant Ag binding specificities,[63] for building Ab libraries,[11] and for preparing Ab fusion proteins.[64] Besides expression in *E. coli*, scFv have also been produced in *Pichia pastoris*,[65] filamentous fungi,[66] insect cells,[67] and mammalian cells.[68]

11.4.1.2 Diabodies, Bispecific Antibodies, and Triabodies

Fab and scFv may be of limited use for some applications because of their monovalency, which leads to a modest intrinsic affinity. Multivalent versions of these Ab fragments should display greater avidity. In addition, rAbs with dual specificity are particularly tempting for some applications such as immunodiagnosis and therapy. Several strategies have been evaluated to produce dimers or multimers of rAbs, many of them involving the chemical cross-linking of two individual rAb fragments after their separate production and purification. However, a more effective genetic strategy has been developed recently. This strategy is based on the use of the scFv format but reducing or even eliminating the linker oligopeptide originally designed to align and favor proper V_H–V_L pairing. For example, by using a 5-mer linker to link V_H with V_L domains from two different Abs A and B, i.e., V_HA–V_LB and V_HB–V_LA, two independent, nonfunctional scFv fragments are synthesized. However, these fragments are able to pair properly, forming a heterodimer where both antigen-binding specificities are restored and found to be physically associated,[69,70] giving rise to a bispecific Ab. Similarly, bivalent Abs can be created using V genes with just one specificity. Moreover, trivalent polymers (triabodies) can be obtained eliminating the linker (Figure 11.1).[13,71]

11.4.1.3 Minibodies

The "minibody" is a curious example of two different structures that have been designated by the same name. In 1994, Tramontano and colleagues[72] described the design of a completely artificial polypeptide, hereinafter called "artificial minibody," based on the structure of the Ab variable region[73] that was able to display ligand-binding properties. Interestingly, this artificial 61-residue beta-protein, consisting of a beta-pleated framework and two hypervariable regions, retained some desirable features of immunoglobulin variable domains, such as tolerance to sequence variability in selected regions of the protein, and consequently the capacity to "evolve." Artificial minibody libraries have been prepared, and particular specificities with high affinity have been selected from them.[74]

A different kind of minibody, also designated as a small immune protein (SIP), was described in 1996. This one is a variant of the scFv fragment where the scFv is joined to a CH3 domain via a cysteine-containing linker. Bivalent homodimers of this recombinant polypeptide are readily produced in myeloma or CHO cells and display excellent properties for tumor imaging and pharmacokinetics.[75,76] Although not tested yet, in theory bispecific heterodimers of these interesting peptides could be prepared by applying "knobs-into-holes" technology.[77]

11.4.1.4 Camel-Like Abs

The discovery that a great diversity of heavy-chain dimers devoid of light chains circulates in camel blood[78,79] motivated a group of researchers to generate single-domain Abs. To mimic the structure of camel Abs, mutations in the human V_H domain were introduced in residues 44 (G44E), 45 (L45R), and 47 (W47G), thereby preventing binding to the V_L domain, and the third hypervariable loop was randomized and displayed on phages (Figure 11.1). Using this approach, soluble Ag-binding V_H domains displaying high affinity were readily expressed in *E. coli*.[80] These results, along with those mentioned in the previous section about the artificial minibody approach, in some way challenge the existing consensus idea considering the V_H–V_L pair as the smallest, functionally efficient, Ag-binding unit. Natural V domains possess a hydrophobic surface that is important for V_H–V_L pairing, and V domains alone are prone to nonspecific interactions because this surface is exposed. The results obtained with "camelized" abs and artificial minibodies show that when V_H domains are mutated to reduce its hydrophobicity, the resultant V segments are able to express Ag-binding functionality.

11.4.1.5 Ab Fusion Proteins

Recombinant DNA technology has made it possible to conceive, design, and elaborate artificial molecules called fusion proteins where functional or structural motifs from two or more natural proteins are combined. Several features make the antibody molecule an ideal candidate for the elaboration of fusion proteins. First, its unique specificity and diversity are desirable for targeting a virtually unlimited number of molecules with marginal risk of cross-reactivity or nonspecific reactions. Second, its biological properties are attractive for certain applications where "customized" recombinant molecules with particular properties related to effector functions, half-life, and size

TABLE 11.1

Comparison between Phage and Ribosome Display Techniques for the Construction of rAb Expression Libraries

Feature	Phage Display	Ribosome Display
Library size	Smaller	Larger
	(Limited by transformation efficiency)	(Limited only by the molecular diversity present in the test tube)
Format	scFv and Fab	scFv (so far)
Possibility of nonadverted gene deletion	Yes	No
	(Due to transformation and cloning)	
Addition of genetic diversity	Requires additional design and cloning	Already incorporated into the procedure by extra rounds of PCR
Library screening and enrichment of relevant specificity/affinity	Requires cycles of scFv elution and *E. coli* reinfection	Next cycle can be started *in situ* without elution steps

are desired. Finally, the domain organization of antibodies greatly facilitates the design and elaboration of recombinant hybrid molecules, both at the genetic and protein levels. Thus, it is not surprising that a substantial number of fusion proteins have been created combining portions of Ab molecules themselves,[81,82] or with toxins,[83,84] interleukins,[85] adhesion molecules,[86] extracellular matrix components,[87] growth factors,[88] hormones,[89] superantigens,[90] CD molecules,[91] cell receptors,[92–94] and even lipids.[95] Some attractive examples with practical potentials are summarized in Table 11.1.

Some important aspects of design should be considered when designing an Ab-fusion protein. For instance, because the Ab molecule is a protein composed of H and L polypeptide chains, one consideration is where to perform the fusion. Obviously, attachment to the heavy chain needs to be done for Ab fusion proteins using the Fc portion of the Ab. This is the case for the so-called immunoadhesins[86] and some interleukin-Fc fusion proteins.[96] When the Fab is the portion to be used, both heavy and light chains may be employed. Nonetheless, even for these cases the heavy chain has been the preferred candidate, and different modalities have been used (after CH1, after the hinge, etc.). More recently, fusions to scFvs have been published.[97]

The relative position or location of the "parent" peptides in the final construct is also a major consideration for some Ab fusion proteins. A few examples have shown that fusion to the C terminus of the Ab heavy chain can render the protein "partner" inactive. This is the case for the nerve growth factor[88] and the immune costimulatory protein B7-1.[92] It is likely that these proteins require N-terminal processing or folding for activity. When this is known or suspected *a priori*, fusion at the N terminus of the Ab heavy chain should be considered. However, fusion to the N-terminal region of the Fab portion may be problematic because it is difficult to anticipate its effect on the Ag–Ab interaction, i.e., Ag recognition and binding could be affected by the presence of an extra structure nearby.

Finally, because of the bifunctional character and complexity of Ab fusion proteins, the selection of an adequate expression system is another important consideration. So far, the expression system preferred has been the mammalian cell,[65,98] although insect cells also have been used successfully.[99] With the advent of Ab phage display technology, bacterial expression of Ab-fusion proteins is also becoming popular. Expression systems will be discussed in more detail in the next section.

11.5 PRODUCTION AND EXPRESSION SYSTEMS

A variety of systems are now available for the production of rAbs. The system most widely used for the generation of Ab fragments is the bacteria *E. coli*. However, mammalian and several other eukaryotic cells are used as expression systems for entire rAb molecules or fusion proteins. Even entire organisms such as plants and transgenic animals are currently tested as systems for production of rAbs. Each system has advantages and disadvantages.

11.5.1 Bacterial Expression Systems

While the production of an entire Ab molecule is best achieved within a eukaryotic host such as a mammalian[100] or a plant[101] cell, bacteria are the preferred host for producing Ab fragments.[102] The absence of an adequate intracellular microenvironment allowing proper post-translational processing of the nascent Ab polypeptide chains, i.e., protein folding, disulfide bond formation, and N- and O-linked glycosylation, is probably the main obstacle impeding the efficient production of whole, functional Ab molecules in prokaryotes.

Escherichia coli was one of the first expression systems available for rAb, and it is still in wide use for the selection and expression of rAb fragments. Initially, the efficiency of recovery of functional rAbs by using bacteria was variable and frequently disappointing.[103] The two main limitations encountered in expressing rAbs in *E. coli* were correct folding of the bivalent Ab and the absence of glycosylation. In the first attempts, heavy and light chains were expressed as inclusion bodies in separate bacterial clones, and a delicate analysis of the reduction/oxidation conditions was needed for each Ab to be expressed.[104,105] More recently, rAbs have been targeted for expression at the bacterial surface by fusion to outer membrane lipoproteins or peptidoglycan.[106–109] Although this expression strategy requires further cleavage of the rAb, it is particularly useful for Ab screening.

Nonetheless, once the advantages of bacterial periplasmic expression were realized,[60,61] a more common strategy for *E. coli* expression of high levels of active rAbs was generally adopted. The periplasmic expression of Ab fragments follows an assembly pathway similar to that of conventional antibodies through the endoplasmic reticulum.[60,105] As mentioned above, several advantages are offered by periplasmic expression of rAbs, including the production of active molecules in a small volume and in a subcellular compartment relatively resistant to proteolytic attack.

The prokaryotic Lac and Tac promoters are the most commonly used promoters for Ab expression in *E. coli*. However, induction and fermentation conditions must be fine-tuned for each rAb expression system to produce adequate levels of active Ab.[105] At present, *E. coli* provides the most affordable and friendly system for producing Ab fragments, with undisputed advantages for genetic manipulation, transformation, expression of Ab repertoires, growth, and fermentation.

11.5.1.1 Lambda Phage

One year after the successful production of functional Ag-binding Ab fragments in bacteria,[60,61] the generation of a cDNA library containing the mouse immunoglobulin Ab repertoire was achieved by using an expression system that combined bacteriophage Lambda and *E. coli*.[9] In this elegant work, separate cDNA libraries for light chains and Fd fragments (V_HC_H1) were prepared by performing RT-PCR amplification of mRNA obtained from the spleen of an immunized mouse. After separate cloning in two λ phage vectors, these libraries were combined to build a combinatorial library of Fab fragments containing around 10^7 clones. The presence of clones expressing Ag-specific Fab fragments was then shown. Several papers followed describing the construction of similar λ phage human Ab libraries from which Ag-binding specific Ab fragments were isolated.[110–113] Lambda bacteriophages offer advantages over conventional plasmids in the preparation of diverse libraries because of their high packaging efficiency and rate of infection. A more convenient way to exploit these and other phage properties was adopted when filamentous instead of λ bacteriophages were employed in the building of Ab libraries.

11.5.1.2 Filamentous Phage Display

The application of phage-display technology[114] for expressing Ab fragments[10] undoubtedly revolutionized the field of rAbs and constitutes a significant addition to the range of rAb technologies.

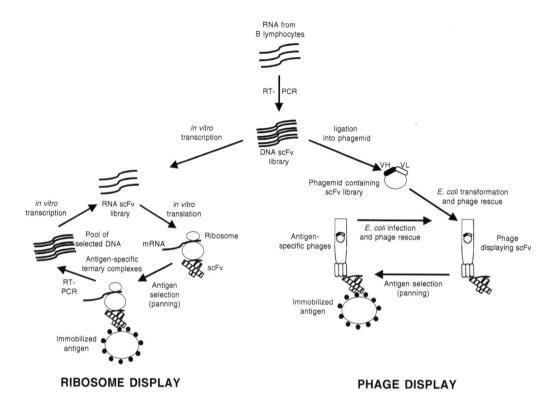

RIBOSOME DISPLAY   **PHAGE DISPLAY**

FIGURE 11.2 Schematic diagram showing major steps for the generation of Ab repertoires through phage or ribosome display technology. Both systems allow for linking of a particular phenotype with its encoding genotype. In ribosome display, this is achieved by formation of a ternary complex between the ribosome, the mRNA, and the newly synthesized scFv (ribosome display). A recombinant phage is the particle used for linking in phage display. Both methods share the initial steps where total RNA or mRNA is extracted from a B-lymphocyte population and then used as a template for reverse transcription and PCR (RT-PCR). In this way, a DNA library is generated containing the Ab repertoire present in the B-cell population. Ribosome display goes through a cell-free cycling sequence. Phage display depends on the repeated incorporation of foreign DNA into bacteria. Many details and variants have been omitted for simplicity.

When a functional Ag-binding Ab domain is displayed on the surface of filamentous phage, the resulting recombinant phage binds to Ag. Rare phages with desired specificity can be selected and isolated after a panning procedure (Figure 11.2). Thus, filamentous phage-based Ab expression libraries are superior to λ ones for screening of large and diverse libraries and for selecting rare specificities from them. Moreover, the display of Ag-binding Ab domains on the surface of a filamentous phage allows for linking of the Ag-binding protein (phenotype) with its encoding genetic information (genotype).

The filamentous coliphages (M13, fd, and f1) are single-stranded, circular DNA phages capable of infecting F′ episome-bearing, male *E. coli*. Two different proteins of the phage coat have been exploited for surface display of Ab fragments, pVIII and pIII. pVIII is the building block of the phage coat (over 2000 copies per phage particle), while pIII is less well represented (3 to 5 copies per virion) but is responsible for phage infectivity.[115] Specific Ab fragments as well as human, mouse, macaque, rabbit, and chicken Ab repertoires have been genetically fused to both pVIII[116–118] and pIII phage proteins.[119] Expression associated with pVIII leads to a phage particle with multivalent binding properties, while pIII-associated expression is usually monomeric, i.e., one scFv-pIII fusion protein per virion. Consequently, the pIII format is usually adopted when Ag binders with high intrinsic affinity are desired.

TABLE 11.2
Examples of Antibody-Fusion Proteins

Ab-Fusion Protein	Genetic Manipulation	Purpose	Ref.
Polymeric IgG	Fusion of the tailpiece of the μ heavy chain to the carboxyl terminus of the γ heavy chain	To enhance IgG effector functions such as complement activation, agglutination, etc.	81
T-bodies: Ab-T-cell-associated molecule	Fusion of an scFv to the cytosolic domain of a ZAP-70 tyr kinase (Syk) through a CD8 transmembrane domain	To redirect the specificity of cytotoxic T cells	93 94
Ab-enzyme (ADEPT)	Fusion of a CD20-specific scFv to the enzyme β-glucuronidase	To direct enzyme prodrug therapy to CD20-bearing B lymphoma cells	198
Ab-enzyme (binary immunotoxin)	Fusion of the two subunits of the human RNase A to two different scfv	To generate a bispecific antibody conjugate, with RNase activity for tumor targeting and killing	199

Filamentous phage-display technology has adopted numerous variants. In the context of producing Ab libraries is worth mentioning the combinatorial infection/*in vivo* recombination approach,[120] which has made it possible to build huge Ab repertoires and identify Ab fragments with nanomolar affinities from them.[121] Another interesting variant of Ab phage display is the so-called selectively infective phage (SIP), where the binding event is directly coupled to the restoration of phage infectivity.[103,122]

11.5.2 Ribosome Display

The intrinsic ability of the Ab phage-display technology to link a phenotype (an scFv or Fab Ab fragment) with its corresponding genotype (V_H–V_L or $V_H C_H 1/V_L C_L$ genes) is one of the key features that make this technique so attractive. Thus, when a particular Ab specificity is selected from the enormous diversity represented in a repertoire library (during the process of panning or selection), the genes encoding this specificity are also selected.

Recently, a novel way for ensuring Ab phenotype–genotype linking called ribosome or polysome display has been reported.[123,124] Ribosome display consists of repetitive cycles of *in vitro* transcription, translation, reverse transcription, and DNA synthesis coupled to an Ag selection event performed after the translation step (Figure 11.2). During the *in vitro* translation, a stabilized ternary complex is formed between an mRNA molecule, a nascent (but already folded and functional) Ab polypeptide, and the ribosome. Thus, the ribosome acts as a linker between the phenotype (the Ab fragment recently translated) and the genotype (represented by the mRNA). This ternary complex is selected in the next step. The relevant Ag binding moieties, usually scFvs, are selected against Ag in a panning procedure, and after the irrelevant specificities are washed out the selected complexes are disrupted, the mRNA is used for a coupled step of reverse transcription and DNA synthesis (an RT-PCR reaction), and the next cycle started over. Recently, it has been shown that ternary complexes can be used directly for the RT-PCR reaction,[124] reaching 10^4- to 10^5-fold enrichments of relevant specificities in a single selection step.

Several advantages may be envisaged for Ab-ribosome display over phage display (Table 11.2). The size of a phage-display library is limited by the bacterial transformation efficiency. Powerful approaches have been developed to increase this size. For example, combinatorial libraries can be generated by coinfection with phage-derived vectors containing separate heavy- and light-chain repertoires.[125] Alternatively, the diversity of the library has been increased by using bacterial recombinases to generate *in vivo* a process of combinatorial infection/recombination, such as those mediated by the *lox*-cre[120] or *att*-int[126] systems. However, even with those approaches the library size obtained is in the order of 10^{11} clones. In theory, larger libraries may be constructed with ribosome display,[127] and this is especially important when building Ab repertoires where diversity is a necessity. In addition, ribosome display avoids transformation and cloning steps that are laborious and may result in loss of diversity. On the contrary, additional genetic diversity through Taq-polymerase-mediated mutagenesis can be readily introduced by ribosome display without the need for cloning. Because no elution step is necessary, relatively (and intrinsically) low affinity scFv-binders may be selected (for example, in the case of some carbohydrate specificities). Despite these attractive features, it is appropriate to say that ribosome display is a relatively new technology.

11.5.3 YEAST AND FUNGI

Saccharomyces cerevisiae and *Aspergillus awamori* can be considered organisms attractive for the industrial production of rAbs. These lower eukaryotes have been used for a long time in several production processes, acquiring the GRAS (generally regarded as safe) status. The yeast *S. cerevisiae,* however, has been tested for the production of rAbs with limited success. An inadequate secretion and improper folding are the putative reasons no efficient production of rAbs has been obtained in *S. cerevisiae.*[66]

Better results have been obtained with the closely related yeast *P. pastoris.* Some salient features of this methyltropic organism as an expression system include the fact that proteins are stably expressed by integration of the recombinant gene into the yeast genome and the expression of secreted proteins can be induced with methanol. A series of expression vectors are commercially available providing homologous recombination of the rAb-encoding gene. Those vectors also include signals for transcription, termination, polyadenilation, and selectable markers. Transformation is performed either by fusion of spheroplasts or electroporation.[128] Up to 200 mg/l of scFv has been secreted in protein-free, minimal ethanol medium.[65] This system is suitable for expression of rAb fragments. Larger proteins have not yet been expressed in *P. pastoris.*[66]

Because they can serve as robust protein secretors, filamentous fungi represent another attractive option. In particular, the fungus *A. awamori* has been successfully used in the expression of several scFv fragments. The yield of rAbs obtained is, however, generally lower when compared to *P. pastoris,* probably due to the great number of proteases produced by molds. Differences in the codon usage in yeast and fungi compared to higher eukaryotic organisms may also affect the production of rAbs.[66]

11.5.4 INSECT CELLS

Insect cell lines derived from *Drosophila* or *Spodoptera* have been successfully used in the production of rAbs.[64] A stable cell line coexpressing the heavy and light chains of a humanized mAb against the F antigen of respiratory syncytial virus has been obtained by transfection of *D. melanogaster* S2 cells. The heavy-chain gene, either alone or in combination with the light chain gene (each cloned in a separate *Drosophila* expression plasmid), has been transfected, and transfectomas have been obtained secreting rAb with the same affinity as the original mAb. Interestingly, complete Abs and heavy-chain dimers were efficiently secreted from *Drosophila* cells via a heat shock cognate protein-mediated pathway.[129] This heat-shock protein is highly homologous to mammalian BiP,[130] one of the chaperon proteins responsible for mediating the secretory pathway of Igs.[131,132] Unlike what is observed in vertebrates, heavy chains alone were successfully secreted by *Drosophila* cells, implying some evolutionary differences between the two secretory pathways,[129] but at the same time offering the opportunity to express heavy-chain dimers correctly folded, secreted, and free of light chains.

The baculovirus expression system is widely used for production of recombinant proteins because of its efficiency in expressing high-molecular-weight proteins and because proteins expressed inside insect cells exhibit folding patterns that somewhat resemble those found within mammalian cells. rAbs have been successfully expressed in baculovirus expression systems.[133] *Spodoptera frugiperda* (Sf9) cells may be infected with two recombinant baculovirus carrying separately the heavy- and light-chain cDNAs, or double-recombinant virus constructions have also been used.[134,135] It has been reported that the baculovirus expression system allows for the production of rAbs or Ab-fusion proteins, in levels three to ten times higher than myeloma cells, the recombinant molecule retaining both affinity and effector functions.[99] These results are encouraging. Nonetheless, O- and N-linked glycosylation of proteins within insect cells results in carbohydrate structures where mannose residues are predominant.[135] This is a glycosylation pattern not commonly found in mammalian cells. Consequently, it would be desirable to evaluate the *in vivo* half-life, biodistribution, and immunogenicity of the baculovirus-derived rAbs since these parameters may be adversely affected by the glycosylation pattern produced in insect cells.

11.5.5 MAMMALIAN CELLS

Mammalian cells were the first system used for the production of rAbs.[136] Monkey kidney cells transformed by the oncogenic SV40 virus (COS cells), Chinese hamster ovary (CHO) cells, and myeloma cells have been successfully transfected with vectors containing Ab cDNA under the control of viral or cellular promoter and enhancer elements. An appropriate selectable marker is included to turn these otherwise transient transfection systems into stable ones. Transfection has been performed with DEAE dextran calcium phosphate, electroporation, or even retroviral transduction. Correct disulfide bonding, chaperone-mediated folding, and glycosylation are expected in mammalian cells. Those aspects can become critical when specific functional features must be preserved. Despite this, expression levels obtained have been quite variable, depending on the vector, DNA constructs, and the nature and specificity of the rAb. Mammalian cells are particularly useful for the expression of bispecific Ab derivatives and Ab fusion proteins because the complex proteins produced are expected to fold more properly than in bacterial systems.[64,100]

Another attractive variant among mammalian expression systems is the production of intracellular Abs for gene therapy.[64,137] The design of an intracellular rAb requires replacing the Ab leader sequence by other specific localization sequence and the use of specific viral vectors for transfection of the target cell type.[137,138] Intracellular rAbs represent a promising approach to interfering with viral life cycles.[139]

11.5.6 TRANSGENIC ANIMALS

Transgenic animals are interesting for the mass production of human antibodies or antibody fragments. Transgenic mice have been produced expressing human antibody genes instead of their own antibody repertoire.[140,141] These animals may be immunized to produce polyclonal or monoclonal Abs. Alternatively, rAbs can be obtained using their B cells as starting material and applying any of the current methods available.[13] Although human Abs were readily produced in transgenic mice, the level of expression remained suboptimal until the endogenous mouse Ab expression system was abolished.[142–144] By using transgenic mice and knockout strains, which do not express endogenous heavy or K chains, human rAbs are produced in the order of hundreds of micrograms per milliliter of mouse serum.[145]

Because the generation of high-affinity Abs depends on somatic mutation, the expectation of producing functional Abs from transgenic mice was based on the assumption that somatic hypermutation, together with successful immunoglobulin gene rearrangements, would be operative when human genes are inserted into the mouse immunoglobulin loci. The successful production of human Abs in transgenic mice has proved that these mechanisms are indeed functional.[142,145,146]

The transference of human immunoglobulin genes into mice has been achieved by building transgenes in plasmid-, YAC- (yeast artificial chromosome), and bacteriophage-derived vectors.

When transgenic miniloci accommodated only up to 100 kilobases (kb) of human immunoglobulin genes, reasonable titers of functional Abs were achieved.[142,145] Even with miniloci displaying a limited number of V genes, large immunoglobulin repertoires were obtained. This was explained by arguing that not only random combination of multiple D and J elements with V genes was occurring, but also the phenomenon of N-region diversity, which yields mutations at the junctional regions and reflects nontemplated nucleotide additions.

The advent of YACs allowed the introduction of larger portions of the human Ig gene loci into transgenic mice and consequently the functional expression of a larger number of human V genes. Up to 1.3 megabases (Mb) of DNA have been successfully introduced into the mouse genome[147,148] using YACs. Thus, a satisfactory display of diversity in this model has been achieved. Since microinjection of YACs is technically difficult (because of the viscosity of such large genome fragments), lipofection and, particularly, spheroplast fusion are the methods of choice for introducing YACs into mouse embryonic stem cell lines.[142]

Recently, a 380-kb region of the human λ light-chain locus has been expressed in transgenic knockout mice. In contrast to what is observed in normal mice, where K chains are used more frequently than λ chain, human Igλ was expressed by most of the spleen B cells. These results suggest that the bias observed in the usage of a specific light chain in the mouse is more dependent on the configuration of the locus than on specific genetic regulations.[144]

An interesting feature of the transgenic mouse model is that it offers the possibility of producing human antibodies against human proteins or antigens that show poor immunogenicity in humans such as tumor-derived antigens.[145,149–151] Another attractive aspect of Ab production in transgenic animals is the possibility of scaling up the production at relatively low cost. For example, by expressing rAbs in goat's milk, up to 25 g of Ab can be obtained per liter of milk. Some differences in the glycosylation pattern of milk-derived recombinant proteins have been documented, but these differences might not affect the effector functions of the Abs expressed in this system.[152]

11.5.7 PLANTIBODIES

Transgenic plants have already proved to be valuable tools for the production of recombinant proteins.[153] The feasibility of producing rAbs in transgenic plants was tested more than a decade ago.[154] Attractive qualities of this expression system are the flexibility for scaling up, the low cost of production, and the stability of the rAb at room temperature. The most common technique used to introduce antibody genes into plant cells is the use of *Agrobacterium tumefasciens* transformed with recombinant, tumor-inducing plasmids that contain the appropriate genes and selectable markers. Each Ig chain is generally expressed separately in different plants and the complete rAb is assembled in a progeny plant by cross-pollination. Ig genes have been generally expressed under the control of the constitutive 35S promoter from cauliflower mosaic virus.[155] Both light and heavy chains have also been introduced simultaneously into a single plant by double transformation or by cloning both genes into a single vector.[156–158]

Leaves have been successfully used for the expression of several rAbs. However, storage organs such as seeds and tubercles have been targeted as the optimal compartments to maintain the long-term stability of the rAbs. In both compartments, vectors targeting retention in the endoplasmic reticulum have led to a high expression level of rAbs. Moreover, rAbs fold in much the same configuration as the original Ab.[159] The majority of the applications of rAbs in transgenic plants have been focused either on the modulation of plant physiology or on evaluating potential protection against plant pathogens, generating an efficient tool for studying plant metabolism.

Most of the Ab configurations (from Fv to multimeric molecules) have been successfully expressed in plants. However, since plant expression represents a cumbersome and time-consuming approach compared to expression in bacteria, plants would not be the system of choice for producing small fragments of rAbs.[101] Plant expression systems are attractive for the expression of multimeric antibodies, since the ability to produce full-length Abs is retained, and the assembly mechanisms are similar to those found in mammalian cells. A recent study showed that mouse IgG expressed in transgenic tobacco plants exhibited N-linked glycosylation broadly similar to what is found in mammalian cells. However, some plant-specific oligosaccharide structures that were detected[160] may generate unwanted immunogenic reactivities. Of particular interest is a fully assembled secretory IgA, obtained by sexual crosses of four transgenic *Nicotiana tabacum* plants,[161] which could be used for mucosal immunotherapy. Additionally, two therapeutic rAbs expressed in plants, one IgG and one IgA, are currently in clinical trials. These antibodies are specific to a *Streptococcus mutans* adhesion protein, and initial results indicate that they are effective in mediating specific protection in humans against oral streptococcal colonization.[162,163]

11.5.8 GLYCOSYLATION

The expression systems described above are promising either for large-scale production or for the generation of specific rAbs. However, many of them still fail to provide a glycosylation completely compatible with human use. This may result in an absence or inadequate performance of effector functions, immune recognition of the foreign rAb, or undesirable pharmacokinetic properties.[164]

The glycosylation produced by baculovirus, yeasts, and even plants seems adequate for generating rAbs displaying effector functions such as phagocytosis, antibody-dependent cell cytotoxicity, and, to a lesser extent, complement activation.[165–167] However, rAb immunogenicity or antigenicity is a more complex and delicate issue for these and other expression systems. One potential problem is the presence of the structure Galα1-3Galβ1-4GlcNac-R, also called galactosyl epitope. It is known that relatively high levels of natural antigalactosyl (anti-Gal) Abs circulate in human blood.[168] The galactosyl epitope is widely expressed in cells and glycoproteins from nonprimate mammals. Moreover, natural anti-Gal Abs efficiently recognize this epitope, as has been shown in models of discordant xenotransplantation.[169] It has been postulated that the presence of the galactosyl epitope is one of the reasons the half-life of nonhuman Abs is intriguingly short in parenteral therapeutic procedures.[170] When galactosyl epitopes are found in a potentially clinical rAb, a possible solution would be to transfer its production to a system lacking α1,3-galactosyltransferase, the enzyme responsible for the generation of this epitope.[170] CHO-derived cell lines seem attractive in this sense because they do not express this transferase.[171] Transgenic mice knockouts for this enzyme have also been produced[172,173] and might be useful for *in vivo* production of therapeutic rAbs. Alternatively, cells or animals that have been manipulated genetically to combine the expression of α-galactosidase and α1,2-fucosyltransferase, which leads to an optimal reduction in the galactosyl epitope,[174] may be used for production of less immunogenic/antigenic rAbs.

It is important to note, however, that the presence of the galactosyl xenoepitope may not be the unique factor affecting pharmacokinetic parameters of rAbs when administered to humans.[175] Insect cells and yeast, for example, produce glycoproteins with high mannose content.[135] These mannose oligosaccharide structures, as well as other glycosylation variants displaying terminal galactose or *N*-acetylglucosamine instead of sialic acid, are more rapidly removed from serum via hepatic receptors.[171] Oligosaccharide structures with repeated residues might also behave as efficient T-cell-independent immunogens. On the other hand, even the transformation event associated with the establishment of a specific human cell line may result in an altered glycosylation pattern in the produced glycoproteins.[171] In conclusion, for immunotherapeutic purposes, even if the protein backbone of a specific rAb has been optimized in terms of minimal immunogenicity and mediation of effector functions, the effectiveness of the reagent may still depend on an appropriate glycosylation of the molecule.

11.6 CATALYTIC ANTIBODIES

The diversity of conformations that can be adopted at an Ab paratope for recognition of foreign molecules or even single epitopes within a molecule is enormous.[176] Thus, it is tempting to speculate about the existence of natural Abs with enzyme-like catalytic abilities. The first catalytic antibodies were identified in 1986.[177,178] These Abs were obtained by standard hybridoma technology and were capable of mediating ester hydrolysis reactions. The results of initial characterization studies were promising.

The general strategy for producing catalytic Abs has been to generate Abs against a transition-state analogue. Designing transition-state analogues is difficult because by definition transition states cannot be synthesized. On the other hand, most of the reactions successfully catalyzed by Abs are single substrate reactions. The strategy for conducting more complex, bimolecular reactions has been to construct combining sites capable of recognizing and activating two substrates.[179] However, most of the issues at present addressed in antibody catalysis involve the chemical modification of a single small substrate, which may be considered a hapten for immunization purposes. Thus, one difference between conventional enzymes and catalytic Abs is that the immunization procedure to produce the latter generally requires the coupling of the hapten to an immunogenic carrier protein. Consequently, the interaction of the catalytic Ab with its putative substrate is limited and not enveloping as generally occurs between a substrate and an enzyme.[180] Additionally, one significant obstacle to the production of catalytic antibodies is the limited number of Abs with catalytic activity obtained through conventional immunization procedures. Even if numerous Abs are produced recognizing the target substrate with high affinity, only a small fraction of them will be capable of mediating the expected catalytic reaction. This limitation imposes the need for creating large repertoires of Abs from which efficient catalytic reagents may be obtained. Thus, the generation of rAbs without previous immunization may improve the production of catalytic Abs more closely related to conventional enzymes at the combining site. Sophisticated ways of selecting catalytic Abs over specific, high-affinity — but not catalytic — Abs are also needed.

11.6.1 RECOMBINANT CATALYTIC ANTIBODIES

Large Ab expression libraries, particularly those displayed on the surface of filamentous phage, appear well suited to the selection of high affinity Abs and for searching enzyme-like catalytic rAbs. Fv fragments obtained from catalytic Abs preserve their catalytic activity.[181,182] Moreover, site-directed mutagenesis of a catalytic Ab has allowed the generation of Fv fragments showing a catalytic rate 45 times higher than the original Fv.[181] More recently, it has been shown that catalytic Abs may facilitate reactions not catalyzed by any known enzyme. An interesting example is an antibody that mediates sequential retro-aldol-retro-Michael reactions.[183] Using this antibody, a new strategy for chemotherapeutic regimes has been explored recently: the antibody-directed enzyme prodrug therapy (ADEPT). The principle of this new therapeutic approach is based on a broadly applicable drug-masking chemistry that operates in conjunction with the broad-scope catalytic Ab, which removes the generic drug-masking groups at the specific target site. Humanization of this unique catalytic Ab should permit its application in immunotherapy or gene therapy.[184]

Combinatorial libraries show promise for the isolation of catalytic Abs, as very large panels of Abs may be obtained from them and are susceptible to careful screening or selection.[185] Current affinity maturation techniques used for the production of rAbs such as chain shuffling[186] and site-directed mutagenesis may be used to increase the affinity of catalytic rAbs. These techniques also offer a means for further understanding catalytic mechanisms at a molecular level.[185,187]

In relation to screening rAbs with catalytic potential, several interesting alternatives are being explored. For example, it is worth mentioning the use of mechanism-based inhibitors instead of substrates. After the combination of inhibitors with phage-displayed Abs, a catalytic reaction occurs, giving rise to a product capable of forming a covalent bond in the combining site. Once covalent bonding captures the "catalytic" phages, noncatalytic phages are washed away and the catalytic ones are released by cleavage of the chemical bond (a disulfide bridge, for example). Thus, clones producing catalytic rAbs are available for subsequent rounds of selection or Ab expression.[188] Another interesting strategy is genetic selection through auxotrophic complementation. This strategy was shown to be feasible in a model of rAbs with orotate decarboxylase activity. Active rAbs, derived from a combinatorial cDNA library, were selected by complementation of an auxotroph strain of *E. coli* in a pyrimidine-free, orotate-supplemented medium. One rAb is sufficient to decarboxylate orotate to uracil, producing a pyrimidine source for growing of the bacteria.[189] In another attempt to link the catalytic activity with the replication process, a technique has been developed in which phage infectivity is restored by the covalent catalytic activity of the rAb.[190] A contrasting strategy consists of expressing recombinant scFv catalytic Abs that confer sensitivity to a toxic compound that is generated from a prodrug in a reaction catalyzed by an rAb.[191]

11.7 CURRENT ISSUES

11.7.1 GENERATING HIGH-AFFINITY rABS

Special emphasis has been placed on applying Ab-phage-display technology for improving Ab affinity, and several approaches have been conceived and assayed. For this purpose, filamentous phage display using the phage protein pIII is the most convenient format at present. Protocols of random and parsimonious mutagenesis at specific sequences of the V_H domain (particularly within the CDR3) have been used with success. The elaboration of combinatorial Fab libraries from separate expression libraries of light and Fd/heavy chains, combined with rounds of Ag selection and recombination using one of the libraries over the population of Ag-selected phages, i.e., chain shuffling, is also a powerful approach for identifying high-affinity binders.[186] SIP and combinatorial infection/*in vivo* recombination (*lox*-cre or *att*-int approaches) have already been mentioned, and their application to the development of artificial or natural (naive or immune) Ab libraries is expected to yield high-affinity Abs. Ribosome display introduces the interesting feature that each round of selection is followed by a PCR reaction, where more diversity may be readily introduced using a nonproofreading DNA polymerase such as Taq polymerase.

11.7.2 PATCH ENGINEERING

In contrast to peptide libraries, where variation is introduced into one continuous sequence of a protein, patch engineering involves the simultaneous randomization of several discontinuous regions of the polypeptide chain that are known to be held together in the folded protein (Figure 11.1).

Patch engineering technology has been applied to generate camel-like rAbs or minibodies (see above) and may be applied to other proteins either to alter a preexisting binding activity or to create an additional binding function.[192] For instance, protein Z, a variant of staphylococcal protein A, has been adapted by patch engineering to bind a variety of proteins such as *Taq* DNA polymerase, human insulin, and apolipoprotein A-1.[193] A randomized library of a cellulose-binding protein, a small biomolecule from the fungus *Trichoderma reesei*, generated recombinant proteins that bound to bovine alkaline phosphatase.[194] Other proteins, such as cytochrome b_{562}, have also been modified through random mutagenesis in two surface loops that do not participate in heme binding, and mutants were selected for binding to a hapten while maintaining the original heme binding capacity.[195] Binding proteins generated by patch engineering may replace Abs in certain applications. Furthermore, small folded proteins may be synthesized chemically and additional diversity provided by using nonnatural amino acids or other modifications.[194]

11.7.3 NEW GENERATION OF ARTIFICIAL LIGANDS: ANTICALINS

Other potential substitutes for antibodies have been described recently.[196] Lipocalins are a family of natural proteins associated with the storage or transport of diverse compounds.[197] These proteins are composed of eight conserved antiparallel β strands joined by four loops. These loops accommodate amino acid substitutions and form a binding pocket that interacts with a specific ligand for each lipocalin. Structural analyses have revealed which amino acid residues are located at the center of the ligand pocket. Using parsimonious or random mutagenesis, lipocalins may be reshaped by combinatorial protein designed to generate appropriate ligands for selected proteins or haptens. In a recent study, 16 critical residues from the ligand pocket were randomly mutated to generate a combinatorial library of "anticalins." The recombinant anticalins displayed ligand affinities similar to mature Abs.[196] This approach may be used to generate anticalins with therapeutic potential by using as a starting protein a lipocalin of human origin. Future combinatorial protein design of lipocalins will determine if anticalins are indeed an alternative to Abs.

REFERENCES

1. Behring, E. von and Kitasato, S., Über das Zustandekommen der Diphtherie-Immunität und der Tetanus-Immunität bei Thieren, *Dtsch. Med. Wochenschr.,* 16, 1113, 1890.
2. Tonegawa, S., Somatic generation of antibody diversity, *Nature,* 302, 575, 1983.
3. Morrison, S. L., *In vitro* antibodies: strategies for production and application, *Annu. Rev. Immunol.,* 10, 239, 1992.
4. Coloma, M. J. et al., Novel vectors for the expression of antibody molecules using variable regions generated by polymerase chain reaction, *J. Immunol. Methods,* 152, 89, 1992.
5. Buchwald, M., Murialdo, H., and Siminovitch, L., The morphogenesis of bacteriophage lambda. II. Identification of the principal structural proteins, *Virology,* 42, 390, 1970.
6. Earnshaw, W. C. and Casjens, S. R., DNA packaging by the double-stranded DNA bacteriophages, *Cell,* 21, 319, 1980.
7. Rasched, I. and Overer, E., Ff coliphages: structural and functional relationships, *Microbiol. Rev.,* 50, 401, 1986.
8. Russel, M., Filamentous phage assembly, *Mol. Microbiol.,* 5, 1607, 1991.
9. Huse, W. D. et al., Generation of a large combinatorial library of the immunoglobulin repertoire in phage lambda, *Science,* 246, 1275, 1989.
10. McCafferty, J. et al., Phage antibodies: filamentous phage displaying antibody variable domains, *Nature,* 348, 552, 1990.
11. Rader, C. and Barbas, C. F., III, Phage display of combinatorial antibody libraries, *Curr. Opin. Biotechnol.,* 8, 503, 1997.
12. Vaughan, T. J. et al., Human antibodies with sub-nanomolar affinities isolated from a large non-immunized phage display library, *Nat. Biotechnol.,* 14, 309, 1996.
13. Hudson, P. J., Recombinant antibody fragments, *Curr. Opin. Biotechnol.,* 9, 395, 1998.
14. Kohler, G. and Milstein, C., Continuous culture of fused cells secreting antibodies of predefined specificity, *Nature,* 256, 495, 1975.
15. Shawler, D. L. et al., Human immune response to multiple injections of murine monoclonal IgG, *J. Immunol.,* 135, 1530, 1985.
16. Schroff, R. et al., Human anti-murine immunoglobulin responses in patients receiving monoclonal antibody therapy, *Cancer Res.,* 45, 879, 1985.
17. Kozbor, D. and Roder, J. C., The production of monoclonal antibodies from human lymphocytes, *Immunol. Today,* 4, 72, 1983.
18. Steinitz, M. et al., EB virus induced B lymphocyte cell lines producing specific antibody, *Nature,* 269, 420, 1977.
19. Kwekkeboom, J. K. et al., CD40 plays an essential role in the activation of human B cells by murine EL4B5 cells, *Immunology,* 79, 439, 1993.
20. Kwekkeboom, J. K. et al., An efficient procedure for the generation of human monoclonal antibodies based on activation of human B lymphocytes by a murine thymoma cell line, *J. Immunol. Methods,* 160, 117, 1993.
21. Glassy, M. C., Production methods for generating human monoclonal antibodies, *Hum. Antibodies Hybridomas,* 4, 154, 1993.
22. James, K. and Bell, G. T., Human monoclonal antibody production, current status and future prospects, *J. Immunol. Methods,* 100, 5, 1987.
23. Thompson, K. M., Human monoclonal antibodies, *Immunol. Today,* 9, 113, 1988.

24. Winter, G. and Harris, W. J., Humanized antibodies, *Immunol. Today,* 14, 243, 1993.

25. Morrison, S. L. et al., Chimeric human antibody molecules: mouse antigen binding domains with human constant region domains, *Proc. Natl. Acad. Sci. U.S.A.,* 81, 6851, 1984.

26. Boulianne, G. L., Hozumi, N., and Schulman, M. J., Production of functional chimeric mouse/human antibody, *Nature,* 312, 643, 1984.

27. Andersen, P. S., Orum, H., and Engberg, J., One-step cloning of murine Fab gene fragments independent of IgH isotype for phage display libraries, *BioTechniques,* 20, 340, 1996.

28. Campbell, M. J. et al., Use of family specific leader region primers for PCR amplification of the human heavy chain variable region gene repertoire, *Mol. Immunol.,* 29, 193, 1992.

29. Coloma, M. J. et al., Primer design for the cloning of immunoglobulin heavy-chain leader-variable regions from mouse hybridoma cells using the PCR, *BioTechniques,* 11, 152, 1991.

30. Gavilondo-Cowley, J. V. et al., Specific amplification of rearranged immunoglobulin variable region genes from mouse hybridoma cells, *Hybridoma,* 9, 407, 1990.

31. Larrick, J. W. et al., Rapid cloning of rearranged immunoglobulin genes from human hybridoma cells using mixed primers and the polymerase chain reaction, *Biochem. Biophys. Res. Commun.,* 160, 1250, 1989.

32. Larrick, J. W. et al., Polymerase chain reaction using mixed primers: cloning of human monoclonal antibody variable region genes from single hybridoma cells, *Bio/Technology,* 7, 934, 1989.

33. Larrick, J. W. et al., Therapeutic human antibodies derived from PCR amplification of B-cell variable regions, *Immunol. Rev.,* 130, 69, 1992.

34. Marks, J. D. et al., Oligonucleotide primers for polymerase chain reaction amplification of human immunoglobulin variable genes and design of family-specific oligonucleotide probes, *Eur. J. Immunol.,* 21, 985, 1991.

35. Orlandi, R. et al., Cloning immunoglobulin variable domains for expression by the polymerase chain reaction, *Proc. Natl. Acad. Sci. U.S.A.,* 86, 3833, 1989.

36. Sastry, L. et al., Cloning of the immunological repertoire in *Escherichia coli* for generation of monoclonal catalytic antibodies: construction of a heavy chain variable region-specific cDNA library, *Proc. Natl. Acad. Sci. U.S.A.,* 86, 5728, 1989.

37. Sblattero, D. and Bradbury, A., A definitive set of oligonucleotide primers for amplifying human V regions, *Immunotechnology,* 3, 271, 1998.

38. Welschof, M. et al., Amino acid sequence based PCR primers for amplification of rearranged human heavy and light chain immunoglobulin variable region genes, *J. Immunol. Methods,* 179, 203, 1995.

39. Montaño, R. A. and Morrison, S. L., unpublished results, 1997.

40. Montaño, R. A. and Morrison, S. L., unpublished results, 1998.

41. Lobuglio, A. F. et al., Mouse/human chimeric monoclonal antibody in man: kinetics and immune response, *Proc. Natl. Acad. Sci. U.S.A.,* 86, 4220, 1989.

42. Mouser, J. F. and Hyams, J. S., Infliximab: a novel chimeric monoclonal antibody for the treatment of Crohn's disease, *Clin. Ther.,* 21, 932, 1999.

43. Adair, J. R., Engineering antibodies for therapy, *Immunol. Rev.,* 130, 5, 1992.

44. Brüggemann, M. et al., The immunogenicity of chimeric antibodies, *J. Exp. Med.,* 170, 2153, 1989.

45. Jones, P. T. et al., Replacing the complementary determining regions in a human antibody with those from a mouse, *Nature,* 321, 522, 1986.

46. Hurle, M. R. and Gross, M., Protein engineering techniques for antibody humanization, *Curr. Opin. Biotechnol.,* 5: 428, 1994.

47. Wu, T. T. and Kabat, E. A., An analysis of the sequences of the variable regions of Bence Jones proteins and myeloma light chains and their implications for antibody complementarity, *J. Exp. Med.,* 132, 211, 1970.

48. Foote, J. and Winter, G., Antibody framework residues affecting the conformation of the hypervariable loops, *J. Mol. Biol.,* 224, 487, 1992.

49. Kettleborough, C. A. et al., Humanization of a mouse monoclonal antibody by CDR-grafting: the importance of framework residues on loop conformation, *Protein Eng.,* 4, 773, 1991.

50. Saul, F. A. and Poljak, R. J., Structural patterns at residue positions 9, 18, 67 and 82 in the VH framework regions of human and murine immunoglobulins, *J. Mol. Biol.,* 230, 15, 1993.

51. Queen, C. et al., A humanized antibody that binds to the interleukin 2 receptor, *Proc. Natl. Acad. Sci. U.S.A.,* 86, 10029, 1989.

52. Padlan, E. A., A possible procedure for reducing the immunogenicity of antibody variable domains while preserving their ligand-binding properties, *Mol. Immunol.*, 28, 489, 1991.

53. Roguska, M. A. et al., Humanization of murine monoclonal antibodies through variable domain resurfacing, *Proc. Natl. Acad. Sci. U.S.A.*, 91, 969, 1994.

54. Presta, L., Antibody engineering, *Curr. Opin. Biotechnol.*, 3, 394, 1992.

55. Hakimi, J. et al., Reduced immunogenicity and improved pharmacokinetics of humanized anti-Tac in cynomolgus monkeys, *J. Immunol.*, 147, 1352, 1991.

56. Ekberg, H. et al., Zenapax (Daclizumab) reduces the incidence of acute rejection episodes and improves patient survival following renal transplantation. On behalf of the No. 14874 and No. 14393 Zenapax Study Groups, *Transp. Proc.*, 31, 267, 1999.

57. Onrust, S. V. and Wiseman, L.R., Basiliximab, *Drugs*, 57, 207, 1999.

58. Shak, S., Overview of the trastuzumab (Herceptin) anti-HER2 monoclonal antibody clinical program in HER2-overexpressing metastatic breast cancer. Herceptin Multinational Investigator Study Group, *Semin. Oncol.*, 26, 71, 1999.

59. Skerra, A. and Plückthun, A., Assembly of a functional immunoglobulin Fv fragment in *Escherichia coli*, *Science*, 240, 1038, 1988.

60. Better, M. et al., *Escherichia coli* secretion of an active chimeric antibody fragment, *Science*, 240, 1041, 1988.

61. Bird, R. E. et al., Single-chain antigen-binding proteins, *Science*, 242, 423, 1988.

62. Huston, J. S. et al., Protein engineering of antibody binding sites: recovery of specific activity in an anti-digoxin single-chain Fv analogue produced in *Escherichia coli*, *Proc. Natl. Acad. Sci. U.S.A.*, 85, 5879, 1988.

63. Wright, A., Shin, S. U., and Morrison, S. L., Genetically engineered antibodies: progress and prospects, *Crit. Rev. Immunol.*, 12, 125, 1992.

64. Hayden, M. S., Gilliland, L. K., and Ledbetter, J. A., Antibody engineering, *Curr. Opin. Immunol.*, 9, 201, 1997.

65. Luo, D. et al., High level secretion of single-chain antibody in *Pichia* expression system, *Biotechnol. Techn.*, 11, 759, 1997.

66. Frenken, L. G. J. et al., Recent advances in the large-scale production of antibody fragments using lower eukaryotic microorganisms, *Res. Immunol.*, 149, 589, 1998.

67. Holvoet, P. et al., Characterization of a chimeric plasminogen activator consisting of a single chain Fv fragment derived from a fibrin fragment D-dimer-specific antibody and a truncated single chain urokinase, *J. Biol. Chem.*, 266, 19717, 1991.

68. Rode, H. J. et al., Cell surface display of a single-chain antibody for attaching polypeptides, *BioTechniques*, 21, 652, 1996.

69. Holliger, P., Prospero, T., and Winter, G., "Diabodies": small bivalent and bispecific antibody fragments, *Proc. Natl. Acad. Sci. U.S.A.*, 90, 6444, 1993.

70. Holliger, P. and Winter, G., Diabodies: small bispecific antibody fragments, *Cancer Immunol. Immunother.*, 45, 128, 1997.

71. Kortt, A. A. et al., Single chain Fv fragments of anti-neuraminidase antibody NC10 containing five and ten residue linkers form dimers and with zero residue linker a trimer, *Protein Eng.*, 10, 423, 1997.

72. Tramontano, A. et al., The making of the minibody: an engineered beta-protein for the display of conformationally constrained peptides, *J. Mol. Recogn.*, 7, 9, 1994.

73. Chotia, C. et al., Structural repertoire of the human VH segments, *J. Mol. Biol.*, 227, 799, 1992.

74. Martin, F. et al., The affinity-selection of a minibody polypeptide inhibitor of human interleukin-6, *EMBO J.*, 13, 5303, 1994.

75. Hu, S. et al., Minibody: a novel engineered anti-carcinoembrionic antigen antibody fragment (single-chain Fv-CH3) which exhibits rapid, high level targeting of xenografts, *Cancer Res.*, 56, 3055, 1996.

76. Li, E. et al., Mammalian cell expression of dimeric small immune proteins, *Protein Eng.*, 10, 731, 1997.

77. Ridgway, J. B. B., Presta, L. G., and Carter, P., "Knobs-into-holes" engineering of antibody CH3 domains for heavy chain heterodimerization, *Protein Eng.*, 9, 617, 1996.

78. Hamers-Casterman, C. et al., Naturally occurring antibodies devoid of light chains, *Nature*, 363, 446, 1993.

79. Nguyen, V. K. et al., Loss of splice consensus signal is responsible for the removal of the entire CH1 domain of the functional camel IGG2A heavy-chain antibodies, *Mol. Immunol.*, 36, 515, 1999.

80. Davies, J. and Riechmann, L., Antibody VH domains as small recognition units, *Bio/Technology*, 13, 475, 1995.

81. Smith, R. I. F. and Morrison, S. L., Recombinant polymeric IgG: an approach to engineering more potent antibodies, *Bio/Technology*, 12, 683, 1994.

82. Zhu, Z., Lewis, G. D., and Carter, P., Engineering high affinity humanized anti-p185HER2/anti-CD3 bispecific F(ab′)2 for efficient lysis of p185HER2 overexpressing tumor cells, *Int. J. Cancer*, 62, 319, 1995.

83. Deonarain, M. P. and Epenetos, A. A., Design, characterization and anti-tumour cytotoxicity of a panel of recombinant, mammalian ribonuclease-based immunotoxins, *Br. J. Cancer*, 77, 537, 1998.

84. Vallera, D. A., Panoskaltsis-Mortari, A., and Blazar, B. R., Renal dysfunction accounts for the dose limiting toxicity of DT390anti-CD3sFv, a potential new recombinant anti-GVHD immunotoxin, *Protein Eng.*, 10, 1071, 1997.

85. Penichet, M. L., Harvill, E. T., and Morrison, S. L., An IgG3-IL-2 fusion protein recognizing a murine B cell lymphoma exhibits effective tumor imaging and antitumor activity, *J. Interferon Cytokine Res.*, 18, 597, 1998.

86. Ashkenazi, A. and Chamow, S. M., Immunoadhesins as research tools and therapeutic agents, *Curr. Opin. Immunol.*, 9, 195, 1997.

87. Oritani, K. et al., Matrix glycoprotein SC1/ECM2 augments B lymphopoiesis, *Blood*, 90, 3404, 1997.

88. McGrath, J. P. et al., Bifunctional fusion between nerve growth factor and a transferrin receptor antibody, *J. Neurosci. Res.*, 47, 123, 1997.

89. Johnson, G. A. et al., Baculovirus-insect cell production of bioactive choriogonadotropin-immunoglobulin G heavy-chain fusion proteins in sheep, *Biol. Reprod.*, 52, 68, 1995.

90. Weiner, L. M., Alpaugh, R. K., and von Mehren, M., Redirected cellular cytotoxicity employing bispecific antibodies and other multifunctional binding proteins, *Cancer Immunol. Immunother.*, 45, 190, 1997.

91. Gregersen, J. et al., A CD4:immunoglobulin fusion protein with antiviral effects against HIV, *Arch. Virol.*, 111, 29, 1990.

92. Challita-Eid, P. M. et al., A B7.1-antibody fusion protein retains antibody specificity and ability to activate via the T cell costimulatory pathway, *J. Immunol.*, 160, 3419, 1998.

93. Eshhar, Z., Tumor-specific T-bodies: towards clinical application, *Cancer Immunol. Immunother.*, 45, 131, 1997.

94. Fitzer-Attas, C. J. et al., Harnessing Syk family tyrosine kinases as signaling domains for chimeric single chain of the variable domain receptors: optimal design for T cell activation, *J. Immunol.*, 160, 145, 1998.

95. Kobatake, E. et al., Fluoroimmunoassay based on immunoliposomes containing genetically engineered lipid-tagged antibody, *Anal. Chem.*, 69, 1295, 1997.

96. Zheng, X. X. et al., IL-2 receptor-targeted cytolytic IL-2/Fc fusion protein treatment blocks diabetogenic autoimmunity in nonobese diabetic mice, *J. Immunol.*, 163, 4041, 1999.

97. Lode, H. N. et al., Immunocytokines: a promising approach to cancer immunotherapy, *Pharmacol. Ther.*, 80, 277, 1998.

98. Trill, J. J., Shatzman, A. R., and Ganguly, S., Production of monoclonal antibodies in COS and CHO cells, *Curr. Opin. Biotechnol.*, 6, 553, 1995.

99. Bei, R., Schlom, J., and Kashmiri, S. V. S., Baculovirus expression of a functional single-chain immunoglobulin and its IL-2 fusion protein, *J. Immunol. Methods*, 186, 245, 1995.

100. Morrison, S. L. and Oi, V. T., Genetically engineered antibody molecules, *Adv. Immunol.*, 44, 65, 1989.

101. Ma, J. K. and Hein, M. B., Antibody production and engineering in plants, *Ann. N.Y. Acad. Sci.*, 792, 72, 1996.

102. Plückthun, A. et al., Producing antibodies in *Escherichia coli*: from PCR to fermentation, in *Antibody Engineering: A Practical Approach*, McCafferty, J., Hoogenboom, H. R., and Chiswell, D. J., Eds., IRL Press, Oxford, 1996, chap. 10.

103. Dall'Acqua, W. and Carter, P., Antibody engineering, *Curr. Opin. Struct. Biol.*, 8, 443, 1998.

104. Buchner, J. and Rudolph, R., Renaturation, purification and characterisation of recombinant Fab fragments produced in *Escherichia coli*, *Bio/Technology*, 9, 157, 1991.

105. Harrison, J. S. and Keshavarz-Moore, E., Production of antibody fragments in *Escherichia coli*, *Ann. N.Y. Acad. Sci.*, 782, 143, 1996.

106. Daugherty, P. S. et al., Development of an optimized expression system for the screening of antibody libraries displayed on the *Escherichia coli* surface, *Protein Eng.*, 12, 613, 1999.

107. Dubel, S. et al., A family of vectors for surface display and production of antibodies, *Gene*, 128, 97, 1993.

108. Francisco, J. A. et al., Production and fluorescence-activated cell sorting of *Escherichia coli* expressing a functional antibody fragment on the external surface, *Proc. Natl. Acad. Sci. U.S.A.*, 90, 10444, 1993.

109. Fuchs, P. et al., Targeting recombinant antibodies to the surface of *Escherichia coli*: fusion to a peptidoglycan associated lipoprotein, *Bio/Technology*, 9, 1369, 1991.

110. Caton, A. J. and Koprowski, H., Influenza virus hemagglutinin-specific antibodies isolated from a combinatorial expression library are closely related to the immune response of the donor, *Proc. Natl. Acad. Sci. U.S.A.*, 87, 6450, 1990.

111. Mullinax, R. L. et al., Identification of human antibody fragment clones specific for tetanus toxoid in a bacteriophage l immunoexpression library, *Proc. Natl. Acad. Sci. U.S.A.*, 87, 8095, 1990.

112. Portolano, S. et al., A human Fab fragment specific for thyroid peroxidase generated by cloning thyroid lymphocyte-derived immunoglobulin genes in a bacteriophage l library, *Biochem. Biophys. Res. Commun.*, 179, 372, 1991.

113. Williamson, R. A., Persson, M. A., and Burton, D. R., Expression of a human monoclonal anti-(rhesus D) Fab fragment in *Escherichia coli* with the use of bacteriophage l vectors, *Biochem. J.*, 277, 561, 1991.

114. Smith, G. P., Filamentous fusion phage: novel expression vectors that display cloned antigens on the surface of the virion, *Science*, 228, 1315, 1985.

115. Model, P. and Russel, M., Filamentous bacteriophage, in *The Bacteriophages*, vol. 2, Valdner, R., Ed., Plenum Press, New York, 1988.

116. Chang, C. N., Landolfi, N. F., and Queen, C., Expression of antibody Fab domains on bacteriophage surfaces. Potential use for antibody selection, *J. Immunol.*, 147, 3610, 1991.

117. Kang, A. S. et al., Linkage of recognition and replication functions by assembling combinatorial antibody Fab libraries along phage surfaces, *Proc. Natl. Acad. Sci. U.S.A.*, 88, 4363, 1991.

118. Wang, L. et al., Cloning of anti-Gal Fabs from combinatorial phage display libraries: structural analysis and comparison of Fab expression in pComb3H and pComb8 phage, *Mol. Immunol.*, 34, 609, 1997.

119. Hoogenboom, H. R. et al., Antibody phage display technology and its applications, *Immunotechnology*, 4, 1, 1998.

120. Waterhouse, P. et al., Combinatorial infection and *in vivo* recombination: a strategy for making large phage antibody repertoires, *Nucleic Acids Res.*, 21, 2265, 1993.

121. Griffiths, A. D. et al., Isolation of high affinity human antibodies directly from large synthetic repertoires, *EMBO J.*, 13, 3245, 1994.

122. Spada, S. and Plückthun, A., Selectively infective phage (SIP) technology: a novel method for *in vivo* selection of interacting protein-ligand pairs, *Nat. Med.*, 3, 694, 1997.

123. Hanes, J. and Plückthun, A., *In vitro* selection and evolution of functional proteins by using ribosome display, *Proc. Natl. Acad. Sci. U.S.A.*, 94, 4937, 1997.

124. He, M. and Taussig, M. J., Antibody-ribosome-mRNA (ARM) complexes as efficient selection particles for *in vitro* display and evolution of antibody combining sites, *Nucleic Acids Res.*, 25, 5132, 1997.

125. Hoogenboom, H. R. et al., Multi-subunit proteins on the surface of filamentous phage: methodologies for displaying antibody (Fab) heavy and light chains, *Nucleic Acids Res.*, 19, 4133, 1991.

126. Geoffroy, F., Sodoyer, R., and Aujame, L., A new phage display system to construct multicombinatorial libraries of very large antibody repertoires, *Gene*, 151, 109, 1994.

127. Jermutus, L., Ryabova, L. A., and Plückthun, A., Recent advances in producing and selecting functional proteins by using cell-free translation, *Curr. Opin. Biotechnol.*, 9, 534, 1998.

128. Pennell, C. A. and Eldin, P., *In vitro* production of recombinant antibody fragments in *Pichia pastoris*, *Res. Immunol.*, 149, 599, 1998.

129. Kirkpatrick, R. B. et al., Heavy chain dimers as well as complete antibodies are efficiently formed and secreted from *Drosophila* via a BiP-mediated pathway, *J. Biol. Chem.*, 270, 19800, 1995.

130. Rubin, D. M. et al., Genomic structure and sequence analysis of *Drosophila melanogaster* HSC70 genes, *Gene*, 128, 155, 1993.

131. Gething, M. J. and Sambrook, J., Protein folding in the cell, *Nature*, 355, 33, 1992.

132. Haas, I. G. and Wabl, M., Immunoglobulin heavy chain binding protein, *Nature*, 306, 387, 1983.

133. Hasemann, C. A. and Capra, J. D., High-level production of a functional immunoglobulin heterodimer in a baculovirus expression system, *Proc. Natl. Acad. Sci. U.S.A.*, 87, 3942, 1990.

134. Zu, P. et al., Antibody production in baculovirus infected insect cells, *Bio/Technology*, 8, 651, 1990.

135. Martin, B. M. et al., Glycosylation and processing of high levels of active human glucocerebrosidase in invertebrate cells using a baculovirus expression vector, *DNA*, 7, 99, 1988.

136. Neuberger, M. S., Expression and regulation of immunoglobulins heavy chain gene transfected into lymphoid cells, *EMBO J.*, 2, 1373, 1983.

137. Cattaneo, A. and Biocca, S., The selection of intracellular antibodies, *Trends Biotechnol.*, 17, 115, 1999.

138. Biocca, S. et al., Intracellular immunization with cytosolic recombinant antibodies, *Bio/Technology*, 12, 396, 1994.

139. Heintges, T., zu Putlitz, J., and Wands, J. R., Characterization and binding of intracellular antibody fragments to the hepatitis C virus core protein, *Biochem. Biophys. Res. Commun.*, 263, 410, 1999.

140. Brüggemann, M. et al., A repertoire of monoclonal antibodies with human heavy chains from transgenic mice, *Proc. Natl. Acad. Sci. U.S.A.*, 86, 6709, 1989.

141. Taylor, L. D. et al., A transgenic mouse that expresses a diversity of human sequence heavy and light chain immunoglobulins, *Nucleic Acids Res.*, 20, 6287, 1992.

142. Brüggemann, M. and Neuberger, M. S., Strategies for expressing human antibody repertoires in transgenic mice, *Immunol. Today*, 8, 391, 1996.

143. Green, L. L. et al., Antigen-specific human monoclonal antibodies from mice engineered with human Ig heavy and light chain YACs, *Nat. Genet.*, 7, 13, 1994.

144. Popov, A. V. et al., A human immunoglobulin lambda locus is similarly well expressed in mice and humans, *J. Exp. Med.*, 189, 1611, 1999.

145. Lonberg, N. et al., Antigen-specific human antibodies from mice comprising four distinct genetic modifications, *Nature*, 368, 856, 1994.

146. Brüggemann, M. and Taussig, M. J., Production of human antibody repertoires in transgenic mice, *Curr. Opin. Biotechnol.*, 8, 455, 1997.

147. Mendez, M. J. et al., Functional transplant of megabase human immunoglobulin loci recapitulates human antibody response in mice, *Nat. Genet.*, 15, 146, 1997.

148. Zou, X. et al., Dominant expression of a 1.3 Mb human IgK locus replacing mouse light chain production, *FASEB J.*, 10, 1227, 1996.

149. Fishwild, D. M. et al., High-avidity human IgK monoclonal antibodies from a novel strain of minilocus transgenic mice, *Nat. Biotechnol.*, 14, 845, 1996.

150. Wagner, S. D. et al., The diversity of antigen-specific monoclonal immunoglobulin gene mini loci, *Eur. J. Immunol.*, 24, 2672, 1994.

151. Yang, X. D. et al., Eradication of established tumors by a fully human monoclonal antibody to the epidermal growth factor receptor without concomitant chemotherapy, *Cancer Res.*, 59, 1236, 1999.

152. Young, M. W. et al., Production of recombinant antibodies in the milk of transgenic animals, *Res. Immunol.*, 149, 609, 1998.

153. Owen, M. R. L. and Pen, J., *Transgenic Plants: A Production System for Industrial and Pharmaceutical Proteins*, John Wiley, Chichester, U.K., 1996.

154. Hiatt, A., Cafferkey, R., and Bowdish, K., Production of antibodies in transgenic plants, *Nature*, 342, 76, 1989.

155. Ma, J. K. and Hein, M. B., Plant antibodies for immunotherapy, *Plant Physiol.*, 109, 341, 1995.

156. De Neve, M. et al., Assembly of an antibody and its derived antibody fragment in *Nicotiana* and *Arabidopsis, Transgenic Res.*, 2, 227, 1993.

157. During, K. et al., Synthesis and self-assembly of a functional antibody in transgenic *Nicotiana tabacum*, *Plant Mol. Biol.*, 15, 281, 1990.

158. Van Engelen, F. A. et al., Coordinate expression of antibody subunit genes yields high levels of functional antibodies in roots of transgenic tobacco, *Plant Mol. Biol.*, 26, 1701, 1994.

159. Conrad, U. and Fiedler, U., Compartment-specific accumulation of recombinant immunoglobulins in plant cells: an essential tool for antibody production and immunomodulation of physiological functions and pathogen activity, *Plant Mol. Biol.*, 38, 101, 1998.

160. Cabanes-Macheteau, M. et al., N-Glycosylation of a mouse IgG expressed in transgenic tobacco plants, *Glycobiology*, 9, 365, 1999.

161. Ma, J. K. et al., Generation and assembly of secretory antibodies in plants, *Science*, 268, 716, 1995.

162. Larrick, J. W. et al., Production of antibodies in transgenic plants, *Res. Immunol.*, 149, 603, 1998.

163. Ma, J. K. et al., Characterization of a recombinant plant monoclonal secretory antibody and preventive immunotherapy in humans, *Nat. Med.*, 4, 601, 1998.

164. Colcher, D. et al., Pharmacokinetics and biodistribution of genetically engineered antibodies, *Q. J. Nucl. Med.*, 42, 225, 1998.

165. Dorai, H. et al., A glycosylated chimeric mouse/human IgG1 antibody retains some effector function, *Hybridoma*, 10, 211, 1991.

166. Jefferis, R., Lund, J., and Goodall, M., Recognition sites on human IgG for Fcgamma receptors, the role of glycosylation, *Immunol. Lett.*, 44, 111, 1995.

167. Wright, A. and Morrison, S. L., Effect of altered CH2-associated carbohydrate structure on the functional properties and *in vivo* fate of chimeric mouse-human immunoglobulin G1, *J. Exp. Med.*, 180, 1087, 1994.

168. Galili, U. et al., One percent of human circulating B lymphocytes are capable of producing the natural anti-Gal antibody, *Blood*, 82, 2485, 1993.

169. Galili, U., Interaction of the natural anti-Gal antibody with alpha-galactosyl epitopes: a major obstacle for xenotransplantation in humans, *Immunol. Today*, 14, 480, 1993.

170. Borrebaeck, C. A. K., Malmborg, A. C., and Ohlin, M., Does endogenous glycosylation prevent the use of mouse monoclonal antibodies as cancer therapeutics? *Immunol. Today*, 14, 477, 1993.

171. Jenkins, N. and Curling, E. M. A., Glycosylation of recombinant proteins: problems and prospects, *Enzyme Microb. Technol.*, 16, 354, 1994.

172. Tearle, R. G. et al., The alpha-1,3-galactosyltransferase knockout mouse. Implications for xenotransplantation, *Transplantation*, 61, 13, 1996.

173. Thall, A. D., Maly, P., and Lowe, J. B., Oocyte Gal alpha 1,3 Gal epitopes implicated in sperm adhesion to the zona pellucida glycoprotein ZP3 are not required for fertilization in the mouse, *J. Biol. Chem.*, 270, 21437, 1995.

174. Osman, N. et al., Combined transgenic expression of alpha-galactosidase and alpha1,2-fucosyltransferase leads to optimal reduction in the major xenoepitope Gal alpha(1,3)Gal, *Proc. Natl. Acad. Sci. U.S.A.*, 94, 14677, 1997.

175. Tanigawara, Y. et al. Pharmacokinetics in chimpanzees of recombinant human tissue-type plasminogen activator produced in mouse C127 and Chinese hamster ovary cells, *Chem. Pharm. Bull.*, 38, 517, 1990.

176. Lerner, R. A., Benkovic, S. J., and Schultz, P. G., At the crossroads of chemistry and immunology: catalytic antibodies, *Science*, 252, 659, 1991.

177. Pollack, S. J., Jacobs, J. W., and Schultz, P. G., Selective chemical catalysis by an antibody, *Science*, 234, 1570, 1986.

178. Tramontano, A., Janda, K. D., and Lerner, R. A., Catalytic antibodies, *Science*, 234, 1566, 1986.

179. Smithrud, D. B. et al., Investigations of an antibody ligase, *J. Am. Chem. Soc.*, 119, 278, 1997.

180. Smithrud, D. B. and Benkovic, S. J., The state of antibody catalysis, *Curr. Opin. Biotechnol.*, 8, 459, 1997.

181. Baldwin, E. and Schultz, P. G., Generation of a catalytic antibody by site-directed mutagenesis, *Science*, 245, 1104, 1989.

182. Gibbs, R. A. et al., Construction and characterization of a single-chain catalytic antibody, *Proc. Natl. Acad. Sci. U.S.A.*, 88, 4001, 1991.

183. Wagner, J., Lerner, R. A., and Barbas, C.F., III, Efficient aldolase catalytic antibodies that use the enamine mechanism of natural enzymes, *Science*, 270, 1797, 1995.

184. Shabat, D. et al., Multiple event activation of a generic prodrug trigger by antibody catalysis, *Proc. Natl. Acad. Sci. U.S.A.*, 96, 6925, 1999.

185. Posner, B. et al., Catalytic antibodies: perusing combinatorial libraries, *Trends Biochem. Sci.*, 19, 145, 1994.

186. Hoogenboom, H. R., Designing and optimizing library selection strategies for generating high-affinity antibodies, *Trends Biotechnol.*, 15, 62, 1997.

187. Kast, P. et al., Exploring the active site of chorismate mutase by combinatorial mutagenesis and selection: the importance of electrostatic catalysis, *Proc. Natl. Acad. Sci. U.S.A.*, 93, 5043, 1996.

188. Janda, K. D. et al., Chemical selection for catalysis in combinatorial antibody libraries, *Science*, 275, 945, 1997.

189. Smiley, J. A. and Benkovic, S. J., Selection of catalytic antibodies for a biosynthetic reaction from a combinatorial cDNA library by complementation of an auxotrophic *Escherichia coli*: antibodies for orotate decarboxylation, *Proc. Natl. Acad. Sci. U.S.A.*, 91, 8319, 1994.

190. Gao, C. et al., Making chemistry selectable by linking it to infectivity, *Proc. Natl. Acad. Sci. U.S.A.*, 94, 11777, 1997.

191. Smiley, J. A. and Benkovic, S. J., Expression of an orotate decarboxylating catalytic antibody confers 5-fluoroorotate sensitivity to a pyrimidine auxotrophic *Escherichia coli*: an example of intracellular prodrug activation, *J. Am. Chem. Soc.*, 117, 3877, 1995.

192. Smith, G. P., Patch engineering: a general approach for creating proteins that have new binding activities, *Trends Biochem. Sci.*, 23, 457, 1998.

193. Nord, K. et al., Binding proteins selected from combinatorial libraries of an alpha-helical bacterial receptor domain, *Nat. Biotechnol.*, 15, 772, 1997.

194. Smith, G. P. et al., Small binding proteins selected from a combinatorial repertoire of knottins displayed on phage, *J. Mol. Biol.*, 277, 317, 1998.

195. Ku, J. and Schultz, P. G., Alternate protein frameworks for molecular recognition, *Proc. Natl. Acad. Sci. U.S.A.*, 92, 6552, 1995.

196. Beste, G. et al., Small antibody-like proteins prescribed ligand specificities derived from the lipocalin fold, *Proc. Natl. Acad. Sci. U.S.A.*, 96, 1898, 1999.

197. Flower, D. R., The lipocalin protein family: structure and function, *Biochem. J.*, 318, 1, 1996.

198. Haisma, H. J. et al., Construction and characterization of a fusion protein of single-chain anti-CD20 antibody and human beta-glucuronidase for antibody-directed enzyme prodrug therapy, *Blood*, 92, 184, 1998.

199. Dubel, S., Reconstitution of human pancreatic RNase from two separate fragments fused to different single chain antibody fragments: on the way to binary immunotoxins, *Tumor Targeting*, 4, 37, 1999.

200. Porter, R. R., Structural studies of immunoglobulins, *Science*, 180, 713, 1973.

201. Neuberger, M. S., Williams, G. T., and Fox, R. O., Recombinant antibodies possessing novel effector functions, *Nature*, 312, 604, 1984.

Index

A

9 780367 395810